Mathematical Methods for Curves and Surfaces II

Innovations in Applied Mathematics

An international series devoted to the latest research in modern areas of mathematics, with significant applications in engineering, medicine, and the sciences.

Series Editor:
Larry L. Schumaker
Stevenson Professor of Mathematics
Vanderbilt University

Previously published titles include

Mathematical Methods for Curves and Surfaces (1995)

Curves and Surfaces with Applications in CAGD (1997)

Surface Fitting and Multiresolution Methods (1997)

Mathematical Methods for Curves and Surfaces II

Lillehammer, 1997

EDITED BY

Morten Dæhlen
SINTEF
Oslo, Norway

Tom Lyche
Institutt for informatikk
University of Oslo
Oslo, Norway

Larry L. Schumaker
Department of Mathematics
Vanderbilt University
Nashville, Tennessee

VANDERBILT UNIVERSITY PRESS
Nashville & London

First Edition 1998
98 99 00 01 02 5 4 3 2 1

This publication is made from high quality paper that meets the minimum requirements of American National Standard for Information Sciences—Permanence of Paper for Printed Library Materials. ∞

Library of Congress Cataloging-in-Publication Data

Mathematical methods for curves and surfaces II : Lillehammer, 1997 /
edited by Morten Dæhlen, Tom Lyche, Larry L. Schumaker. — 1st ed.
p. cm. — (Innovations in applied mathematics)
Papers from an international conference held July 3–8, 1997, in Lillehammer, Norway.
Includes bibliographical references and index.
ISBN 0-8265-1315-8 (alk. paper)
1. Curves on surfaces—Mathematical models—Congresses.
2. Surfaces—Mathematical models—Congresses. I. Dæhlen, Morten, 1959– . II. Lyche, Tom. III. Schumaker, Larry L., 1939– .
IV. Series.
QA565.M372 1998
516.3'52—dc21 98-15040
CIP

CONTENTS

PREFACE

During the week of July 3–8, 1997 an international conference on *Mathematical Methods for Curves and Surfaces* was held at Lillehammer, Norway. The conference was attended by 144 mathematicians from 29 countries. There were 9 one hour survey talks, 32 talks in 8 minisymposia, and 75 contributed research talks. This volume contains papers based on the invited lectures, along with 47 full length refereed research papers. Unfortunately, there was not enough space to include all of the papers presented at the conference.

Topics discussed in this book include Bézier curves and surfaces, bivariate splines, blending canal surfaces, cascade algorithms, constructing patches, convexity preservation, curve design, cyclides, de Casteljau's algorithm, elastic splines, fairing, fractal modelling, generalized B-splines, global illumination, image processing, interpolation, iterative refinement of triangulations, nested finite elements, parametric curves, quasi-interpolants, rational curves, refinable splines, reverse engineering, spherical PH-curves, spherical data fitting, spline interpolation, subdivision, surface reconstruction, variational splines, and wavelets.

As a new innovation, this volume includes an automatically generated *index* which points to the author-emphasized words in the various papers. The user should be aware that no attempt has been made to find all appearances of a given word — a reference is included only if the word was specifically marked by some author.

We would like to thank SINTEF, The Research Council of Norway, and the University of Oslo for their support of the conference. We would like to thank Mariane Rein, Ewald Quak, and Frederik Tyvand for their help in organizing and running the conference. Finally, we are also grateful to Gerda Schumaker for her help with preparing the proceedings.

Nashville, January 9, 1998

PARTICIPANTS

In the following list of participants, the number(s) in brackets indicate the page(s) on which the author's contribution(s) begin.

Gudrun Albrecht [1], *Technische Universität München, Mathematisches Institut, D-80290 München, Germany* [albrecht@mathematik.tu-muenchen.de].

Erlend Arge, *Numerical Objects, P. O. Box 124 Blindern, Forskningsveien 1, N-0314 Oslo, Norway* [sea@nobjects.com].

John Kofi Arthur, *SINTEF Applied Mathematics, P. O. Box 124 Blindern, Forskningsveien 1, N-0314 Oslo, Norway* [john.arthur@math.sintef.no].

Chandrajit Bajaj, *Texas Inst. Comp. Appl. Math., Taylor 2.422, University of Texas, Austin, TX 72712* [bajaj@ticam.utexas.edu].

Christoph Baumgarten [17], *Dresden University of Technology, Fac. of Computer Science, IBDR-LSDB, D-01062 Dresden, Germany* [baumgart@is2201.inf.tu-dresden.de].

Michel Bercovier [351], *Pole Universitaire Léonard de Vinci, 92916 Paris La Défense Cedex, France* [michel.bercovier@devinci.fr].

Przemyslaw Bogacki [25], *Old Dominion University, Dept. of Mathematics and Statistics, Norfolk, VA, 23529, USA* [bogacki@math.odu.edu].

Georges-Pierre Bonneau, *LMC-CNRS, P. B. 53, F-38041 Grenoble, France* [bonneau@imag.fr].

Carl de Boor, *University of Wisconsin, CS 1210, W. Dayton St., Madison WI 53706-1685, USA* [deboor@cs.wisc.edu].

Len Bos, *University of Calgary, Department of Mathematics, Calgary T2N 1N4, Canada* [lpbos@math.ucalgary.ca].

Karl-Heinz Brakhage, *RWTH Aachen, Inst. für Geometrie und Prakt. Math., Templergraben 55, D-52056 Aachen, Germany* [brakhage@igpm.rwth-aachen.de].

Jeff Brown, *UNCW, Math. Department, 601 S. College Rd., Wilmington, North Carolina 28403, USA* [brownj@cms.uncwil.edu].

Guido Brunnett [33], *Universität Kaiserslautern, Postfach 3049, D-67653 Kaiserslautern, Germany* [brunnett@informatik.uni-kl.de].

Daniel Cardenas-Morales, *Universidad de Jaen, Dpto. Matematicas, Paraje La Lagunillas, 23071 Jaen, Spain* [cardenas@ujaen.es].

Jesus M. Carnicer [41], *Universidad de Zaragoza, Edificio de Matematicas, Planta 1a, E-50009 Zaragoza, Spain* [carnicer@posta.unizar.es].

Rob Champion [49], *La Trobe University, P. O. Box 199, 3552 Bendigo, Australia* [r.champion@latrobe.edu.au].

Costanza Conti [55,63], *Università di Firenze, Dipartimento di Engergetica, Via C. Lombroso 6 17, I-50134 Firenze, Italy* [costanza@fonty.de.unifi.it].

Mariantonia Cotronei, *Universita di Messina, Dipartimento di Matematica, Salita Sperone 31, I-98166 Messina, Italy* [marianto@riscme.unime.it].

Oleg Davydov [71], *Universität Dortmund, Fachbereich Mathematik, Lehrstuhl VIII, 44221 Dortmund, Germany* [oleg.davydov@math.uni-dortmund.de].

Stefano De Marchi, *Universita di Udine, Dipartimento de Matematica e Informatica, Via delle Scienze 206, I-33100 Udine, Italy* [demarchi@dimi.uniud.it].

Ulrich Dietz [79,253], *Technische Universität Darmstadt, Schlossgartenstr. 7, D-64289 Darmstadt, Germany* [udietz@mathematik.tu-darmstadt.de].

Tor Dokken, *SINTEF Applied Mathematics, P. O. Box 124 Blindern, Forskningsveien 1, N-0314 Oslo, Norway* [tor.dokken@math.sintef.no].

Morten Dæhlen, *SINTEF Applied Mathematics, P. O. Box 124 Blindern, Forskningsveien 1, N-0314 Oslo, Norway* [morten.daehlen@math.sintef.no].

Matthias Eck, *ICEM Systems GmbH, Kuesterstrasse 8, D-30519 Hannover, Germany* [matthias.eck@icem.de].

Bernhard Elsässer [87], *Technische Universität Darmstadt, Schlossgartenstr. 7, D-64289 Darmstadt, Germany* [elsaesser@mathematik.tu-darmstadt.de].

Rida T. Farouki [95], *University of Michigan, Dept. of Mechanical Engineering and Applied Mechanics, Ann Arbor, MI 48109, USA* [farouki@engin.umich.edu].

Jean-Charles Fiorot [175], *Université de Valenciennes, ENSIMEV, B. P. 311 Le Mont Houy, 59304 Valenciennes Cedex, France* [fiorot@univ-valenciennes.fr].

Michael S. Floater [183], *SINTEF Applied Mathematics, P. O. Box 124 Blindern, N-0314 Oslo, Norway* [michael.floater@math.sintef.no].

Laurent Fuchs, *Université Louis Pasteur, Departement Informatique, 8 rue Rene Descartes, F-67084 Strasbourg Cedex, France* [fuchs@dpt-info.u-strasbg.fr].

Ioan Gansca, *Technical University of Cluj-Napoca, Department of Mathematics, str. C. Daicoviciu nr. 15, Cluj-Napoca 3400, Romania* [leon@cs.ubbcluj.ro].

Olivier Gibaru [175], *CMRAO, 8 Bld Louis XIV, F-59046 Lille, France* [gibaru@lille.ensam.fr].

Ron Goldman, *Rice University, Dept. of Computer Science, Houston, Texas, 77251, USA* [rng@cs.rice.edu].

Ana Gonzalez, *Universidad Autonoma Metropolitana, Av. Michoacan y La Purisima Iztapalapa, Mexico D.F. 09340, Mexico* [ltd@xanum.uam.mx].

Tim N.T. Goodman [191,213], *University of Dundee, Dept. of Mathematical and Computer Sciences, DD1 4HN Dundee, Scotland-UK* [tgoodman@mcs.dundee.ac.uk].

Laura Gori, *Universita La Sapienza - Roma, Dept. Memomat, via A. Scarpa, 16, I-00161 Roma, Italy* [gori@dmmm.uniroma1.it].

Steven Gortler, *Harvard University, Aiken-13, 33 Oxford St., 02138 Cambridge, USA* [sjg@cs.harvard.edu].

Jens Gravesen [221], *Technical University of Denmark, Dept. of Mathematics, Building 303, DK-2800 Lyngby, Denmark* [j.gravesen@mat.dtu.dk].

Günther Greiner, *Universität Erlangen-Nürnberg, IMMD IX Graphische Datenverarbeitung, Am Weichselgarten 9, 91058 Erlangen, Germany* [greiner@informatik.uni-erlangen.de].

Markus Gross, *Swiss Federal Institute of Technology, ETH-Zurich, Computer Science Department, 8092 Zurich, Switzerland* [grossm@inf.ethz.ch].

Andreas Görg, *Technische Universität Darmstadt, Schlossgartenstr. 7, D-64289 Darmstadt, Germany* [goerg@mathematik.tu-darmstadt.de].

Hans Hagen [429], *University of Kaiserslautern, F. B. Informatik, Postfach 3049, D-67653 Kaiserslautern, Germany* [hagen@informatik.uni-kl.de].

Yvon Halbwachs, *SINTEF Applied Mathematics, P. O. Box 124 Blindern, Forskningsveien 1, N-0314 Oslo, Norway* [yvon.halbwachs@math.sintef.no].

Bernd Hamann [229], *University of California at Davis, Dept. of Computer Science, Davis, CA 95616-8562, USA* [hamann@cs.ucdavis.edu].

Doug Hardin, *Vanderbilt University, Dept. of Mathematics, Nashville, TN 37240, USA* [hardin@math.vanderbilt.edu].

Erich Hartmann [237], *Technische Universität Darmstadt, Schlossgartenstr. 7, D-64289 Darmstadt, Germany* [ehartmann@mathematik.tu-darmstadt.de].

Markus Hegland [245], *Australian National University, Comp. Science Lab. RSISE, ACT 0200 Canberra, Australia* [markus.hegland@anu.edu.au].

Masatake Higashi [303], *Toyota Technological Institute, 2-12-1, Hisakata, Tempaku-ku, Nagoya 468, Japan* [higashi@toyota-ti.ac.jp].

Øyvind Hjelle, *SINTEF Applied Mathematics, P. O. Box 124 Blindern, Forskningsveien 1, N-0314 Oslo, Norway* [oyvind.hjelle@math.sintef.no].

Hugues Hoppe, *Microsoft Research, One Microsoft Way, Redmond, WA 98052, USA* [hhoppe@microsoft.com].

Josef Hoschek [253], *Darmstadt University of Technology, Schlossgartenstr. 7, D-64289 Darmstadt, Germany* [hoschek@mathematik.tu-darmstadt.de].

Armin Iske, *CoCreate Software GmbH, Posener Str. 1, D-71065 Sindelfingen, Germany* [armin_iske@hp.com].

Colin G. Johnson, *Napier University, Department of Mathematics, 219 Colinton Road, Edinburgh, EH11 1NW, Scotland-UK* [colinj@maths.napier.ac.uk].

Bert Jüttler [263], *Technical University of Darmstadt, Dept. of Mathematics, Schlossgartenstr. 7, D-64289 Darmstadt, Germany* [juettler@mathematik.tu-darmstadt.de].

Hermann Kellermann, *SINTEF Applied Mathematics, P. O. Box 124 Blindern, Forskningsveien 1, N-0314 Oslo, Norway* [hermann.kellermann@math.sintef.no].

Daniel Keren, *Haifa University, Dept. of Computer Science, 31905 Haifa, Israel* [dkeren@mathcs2.haifa.ac.il].

Su-Sun Kim, *Hanyang Women's Junior College, Dept. of Computer Science, 17 Hangang-Dong, Sungdong-Gu, Seoul 1333-793, South-Korea* [sskim@hydns.hywoman.ac.kr].

Ron Kimmel [271], *University of California, Lawrence Berkeley National Laboratory, Berkeley CA 94720, USA* [ron@math.lbl.gov].

Ha-Jine Kimn, *Ajou University, School of Information and Computer Engineering, 5 Woncheon-Dong Paldal-Gu, Suwon 442-749, South-Korea* [hjkimn@garam.kreonet.re.kr].

Leif Kobbelt [279], *University of Erlangen, Computer Graphics Group, Am Weichselgarten 9, 91058 Erlangen, Germany* [kobbelt@informatik.uni-erlangen.de].

Ljublisă M. Kocić [287], *University of Niš, Dept. of Mathematics, P. O. Box 73, 18000 Niš, Yugoslavia* [kole@laplace.elfak.ni.ac.yu].

Krasimir Kolarov, *Interval Research Corp., 1801 Page Mill Road, Bldg. C, Palo Alto, California 94304, USA* [kolarov@interval.com].

Jernej Kozak [167], *University of Ljubljana, Dept. of Mathematics, Jadranska 19, 1000 Ljubljana, Slovenija* [jernej.kozak@fmf.uni-lj.si].

Rimvydas Krasauskas [359], *Dept. of Mathematics, Vilnius University, Naugarduko 24, 2006 Vilnius, Lithuania* [rimvydas.krasauskas@maf.vu.lt].

Ulf Jacob Krystad, *SINTEF Applied Mathematics, P. O. Box 124 Blindern, Forskningsveien 1, N-0314 Oslo, Norway* [ulf-jacob.krystad@math.sintef.no].

Frans Kuijt, *University of Twente, Fac. of Applied Mathematics, P. O. Box 217, Enschede, NL-7500 AE, The Netherlands* [f.kuijt@math.utwente.nl].

Mitsuru Kuroda [303], *Toyota Technological Institute, 2-12-1, Hisakata, Tempaku-ku, Nagoya 468-8511, Japan* [kuroda@toyota-ti.ac.jp].

Johannes Kåsa, *SINTEF Applied Mathematics, P. O. Box 124 Blindern, Forskningsveien 1, N-0314 Oslo, Norway* [johannes.kasa@math.sintef.no].

Pierre-Jean Laurent, *Université Joseph Fourier, LMC-IMAG, B. P. 53, 38041 Grenoble Cedex 09, France* [pjl@imag.fr].

Alain Le Méhauté [311], *Université de Nantes, Dept. of Mathematics, 2 rue de la Houssiniére, F-44322 Nantes Cedex 3, France* [alm@math.univ-nantes.fr].

Seng Luan Lee [191,213], *Dept. of Mathematics, National University of Singapore, Singapore 119260* [matleesl@math.nus.sg].

Hervé Legrand, *Matra Datavision, 53 Avenue de l'Europe, F-13082 Aix en Provence, France* [m-chiabudini@aix.matra-dtv.fr].

Stefan Leopoldseder, *Technische Universität Wien, Wiedner Hauptstrasse 8-10/113, A1040 Wien, Austria* [stefan@geometrie.tuwien.ac.at].

Will Light, *University of Leicester, University Road, Leicester LE1 7RH, UK* [pwl@mcs.le.ac.uk].

Suresh Lodha, *University of California, Santa Cruz, 202 Dickens Way, CA 95064, USA* [lodha@cse.ucsc.edu].

Gabor Lukács [319], *Hungarian Academy of Sciences, Computer and Automation Research Institute, Kende U. 13-17, P. O. Box 63, H-1518 Budapest, Hungary* [lukacs@sztaki.hu].

Tom Lyche, *University of Oslo, Dept. of Informatics, P. O. Box 1080 Blindern, N-0316 Oslo, Norway* [tom@ifi.uio.no].

William C. Lynch, *Interval Research Corp., 1801 Page Mill Road, Bldg. C, Palo Alto, California 94304, USA* [lynch@interval.com].

Esmeralda Mainar [327], *Universidad de Zaragoza, Edificio de matematicas Planta 1a, E-50009 Zaragoza, Spain* [esme@posta.unizar.es].

Leonor Malva, *Dept. de Matematica da Fac. de Ciencias e Tecnologia da Universidade de, Coimbra, Largo de Dinis, 300 Coimbra, Portugal* [mcorreia@mail.telepac.pt].

Petter Jon Mandt, *University of Oslo, Dept. of Informatics, P. O. Box 1080 Blindern, N-0316 Oslo, Norway* [pettema@ifi.uio.no].

Stephen Mann [335], *University of Waterloo, Computer Science Dept., 200 University Ave W., Waterloo N2L 2R7, Canada* [smann@uwaterloo.ca].

Miljenko Marusic, *University of Zagreb, Dept. of Mathematics, Bijenicka 30, 10000 Zagreb, Croatia* [miljenko.marusic@math.hr].

John C. Mason [9], *University of Huddersfield, School of Computing and Mathematics, Queensgate, Huddersfield HD1 3DH, UK* [j.c.mason@hud.ac.uk].

Tanya Matskewich [351], *The Hebrew University of Jerusalem, Inst. of Computer Science, Givat Ram, Jerusalem 91904, Israel* [fisa@cs.huji.ac.il].

Christoph Mäurer [359], *Technical University of Darmstadt, Schlossgartenstr. 7, D-64289 Darmstadt, Germany* [cmaeurer@mathematik.tu-darmstadt.de].

Marie-Laurence Mazure, *Université Joseph Fourier, LMC-IMAG, B. P. 53, 38041 Grenoble Cedex 09, France* [mazure@imag.fr].

Even Mehlum, *SINTEF Applied Mathematics, P. O. Box 124 Blindern, Forskningsveien 1, N-0314 Oslo, Norway* [even.mehlum@math.sintef.no].

Ingrid Melinder, *KTH, Inst. Numerisk analys och datalogi, NADA; KTH, S-144 00 Stockholm, Sweden* [melinder@nada.kth.se].

Antonio Jesus Lopez Moreno, *Universidad de Jaen, E.U.P. Linares, c Alfonso X el Sabio 28, 23700 Linares, Spain* [ajlopez@piturda.ujaen.es].

Francisco-Javier Munos-Delgado, *Universidad de Jaen, Dpto. Matematicas, Avda. Madrid 35, E-23071 Jaen, Spain* [fdelgado@ujaen.es].

Knut Mørken, *University of Oslo, Dept. of Informatics, P. O. Box 1080 Blindern, N-0316 Oslo, Norway* [knutm@ifi.uio.no].

Jens Olav Nygaard, *University of Oslo, Dept. og Mathematics, P. O. Box 1053 Blindern, N-0316 Oslo, Norway* [jnygaard@math.uio.no].

Marco Paluszny [367], *Universidad Central de Venezuela, Centro CGGA, Apartado 47809, Los Chaguaramos, Caracas 1041, Venezuela* [marco@ciens.ucv.ve].

Miguel Fernandez Pasadas [295], *Universidad de Granada, Departamento de Matemática Aplicada, Severo Ochoa s/n, 18071 Granada, Spain* [mpasadas@goliat.ugr.es].

Jörg Peters, *Purdue University, Dept. of Computer Science, West-Lafayette, IN 47907-1398, USA* [jorg@cs.purdue.edu].

Francesca Pitolli, *Universita La Sapienza - Roma, Dept. Memomat, via A. Scarpa, 16, I-00161 Roma, Italy* [pitolli@dmmm.uniroma1.it].

Gerlind Plonka [375], *Fachbereich Mathematik, Universität Duisburg, 47048 Duisburg, Germany* [plonka@math.uni-duisburg.de].

Janet F. Poliakoff [401], *The Nottingham Trent University, Dept. of Computing, Nottingham, NG1 4BU, UK* [jfp@doc.ntu.ac.uk].

Daniel Potts, *Institut für Mathematik, Medizinische Universität Lübeck, Wallstr. 40, D-64289, Germany* [potts@informatik.mu-luebeck.de].

Luigia Puccio, *University of Messina, Dept. of Mathematics, Salita Sperone, 31, C. da Papardo, I-98166 Messina, Italy* [gina@imeuniv.unime.it].

Ewald Quak, *SINTEF Applied Mathematics, P. O. Box 124 Blindern, Forskningsveien 1, N-0314 Oslo, Norway* [ewald.quak@math.sintef.no].

Christophe Rabut [55,63], *Institut National de Sciences Appliquees, Dept. de Génie Mathématique, Complexe Scientifique de Rangueil, F-31077 Toulouse Cedex 4, France* [rabut@gmm.insa-tlse.fr].

Thomas Randrup, *Odense Steel Shipyard Ltd., P. O. Box 176, DK-55100 Odense C, Denmark* [thr@ma-oss.dk].

Ulrich Reif, *University of Stuttgart, Pfaffenwaldring 57, D-70569 Stuttgart, Germany* [reif@mathematik.uni-stuttgart.de].

Shmuel Rippa, *Orbothech Ltd., P. O. Box 215, 81102 Yavne, Israel* [shmulik@orbothech.co.il.].

Mladen Rogina, *University of Zagreb, Dept. of Mathematics, Bijenicka 30, 10000 Zagreb, Croatia* [rogina@math.hr].

Milvia Rossini, *Universita di Milano, Dip. di Matematica, via Saldini 50, I-20133 Milano, Italy* [rossini@vmimat.mat.unimi.it].

John A. Roulier, *University of Connecticut, Dept. of Computer Science and Engineering U-155, Storrs, CT 06269-3155, USA* [jrou@brc.uconn.edu].

Malcolm Sabin [409], *Cambridge University, DAMTP, Silver Str., Cambridge CB5 9EP, UK* [mas33@damtp.cam.ac.uk].

Tsuyoshi Saitoh [303], *Tokyo Denki University, 2-2 Kanda-Nisiki-Cho, Chiyoda, 101-8547 Tokyo, Japan* [saitoh@saitohlab2.c.dendai.ac.jp].

Kestutis Salkauskas, *The University of Calgary, 2500 University Drive N.W., Calgary, Alberta T2N 1N4, Canada* [ksalkaus@math.ucalgary.ca].

Maria Lucia Sampoli [343], *Università di Firenze, Dip. di Energetica "S. Stecco", Via C. Lombroso 6/17, 50134 Firenze, Italy* [sampoli@fonty.de.unifi.it].

Ramon F. Sarraga, *General Motors Research, P. O. Box 9055, Warren Michgan 48090-955, USA* [sarraga@gmr.com].

Robert Schaback [417], *Georg–August–Universität Göttingen, Institut für Numerische und Angew. Mathematik, Lotzestr. 16-18, D-37083 Göttingen, Germany* [schaback@math.uni-goettingen.de].

Gerik Scheuermann [429], *University of Kaiserslautern, F. B. Informatik, Postfach 3049, D-67653 Kaiserslautern, Germany* [scheuer@informatik.uni-kl.de].

Mike Schneider [437], *Technical University of Darmstadt, Schlossgartenstr. 7, D-64289 Darmstadt, Germany* [mschneider@mathematik.tu-darmstadt.de].

Peter Schröder, *California Institute of Technology, 1200 E. Californa Bvd., Pasadena, CA 91125, MS-256-80, USA* [ps@cs.caltech.edu].

Larry L. Schumaker [117], *Vanderbilt University, Dept. of Mathematics, Nashville, TN 37240, USA* [s@mars.cas.vanderbilt.edu].

Hans-Peter Seidel [445], *University of Erlangen, Computer Graphics, Am Weichselgarten 9, D-91058 Erlangen, Germany* [seidel@informatik.uni-erlangen.de].

Alba C. Simoncelli [287], *Università degli Studi di Napoli Federico II, Dipartimento di Matematica ed Applicazioni, Via Cinthia, Monte S. Angelo, 80126 Napoli, Italy* [simoncel@matna2.dma.unina.it].

Vibeke Skytt [461], *SINTEF Applied Mathematics, P. O. Box 124 Blindern, Forskningsveien 1, N-0314 Oslo, Norway* [vibeke.skytt@math.sintef.no].

Nir Sochen [469], *Faculty of Electrical Engineering, Technion – The Technology Institute of Israel, Technion City, Haifa, 32000, Israel* [sochen@ee.technion.ac.il].

Vasily Strela [375], *Imperial College, Dept. of Mathematics, 180 Queen's Gate, London SW7 2BZ, UK* [strela@ic.ac.uk].

Kyrre Strøm, *SINTEF Applied Mathematics, P. O. Box 124 Blindern, Forskningsveien 1, N-0314 Oslo, Norway* [kyrre.strom@math.sintef.no].

Christian Tarrou, *University of Oslo, Dept. of Informatics, P. O. Box 1080 Blindern, N-0316 Oslo, Norway* [christian.tarrou@math.sintef.no].

Demetri Terzopoulos, *University of Toronto, Dept. of Computer Science, 6 King's College Road, Pratt Building, Toronto, Ontario M5S 3H5, Canada* [dt@cs.toronto.edu].

Sylvain Thery, *Université Louis Pasteur, Departement Informatique, 7 rue Rene Descartes, F-67084 Strasbourg Cedex, France* [thery@dpt-info.u-strasbg.fr].

Juan José Torrens [295,477], *Universidad Publica de Navarra, Departamento de Matematica e Informatica, Campus de Arrosadia s/n, E-31006 Pamplona, Spain* [jjtorrens@upna.es].

Francisco Tovar, *Universidad Central de Venezuela, Centro CGGA Apartado 47809, Los Chaguaramos, Caracas 1041, Venezuela* [ftovar@jade.ciens.ucv.ve].

Leonardo Traversoni, *Universidad Autonoma Metropolitana, Av. Michoacan y La Purisima Iztapalapa, D.F. 09340, Mexico* [ltd@xanum.uam.mx].

Kenji Ueda [485], *Ricoh Company Ltd., 1-1-17 Koishikawa, Bunkyo-ku, Tokyo 112-0002, Japan* [ueda@src.ricoh.co.jp].

Hassan Ugail [493], *University of Leeds, Dept. of Appl. Mathematical Studies, Woodhouse Lane, Leeds LS29JT, UK* [hassan@amsta.leeds.ac.uk].

Sergey B. Vakarchuk, *Institute of Geotechnical Mechanics, Simferopolskaja St. 2A, 320600 Dniepropetrovsk, Ukraine* [nanu@igtm.dp.ua].

Támas Várady [501], *Hungarian Academy of Sciences, Computer and Automation Research Institute, Kende U. 13-17, P. O. Box 63, H-1111 Budapest, Hungary* [varady@sztaki.hu].

Ravi C. Venkatesan, *Systems Research Corporation, 25 Goodwill Society, Baner Road, Aundh, 411007 Pune, India* [rcv_dac@giaspn01.vsnl.net.in].

Oleg Volpin [351], *The Hebrew University of Jerusalem, Inst. of Computer Science, Givat Ram, Jerusalem 91904, Israel* [oleg@cs.huji.ac.il].

Marshall Walker [529], *York University, Dept. of Computer Science and Mathematics, Atkinson College, Toronto, M3J 1P3, Canada* [marshall.walker@mathstat.yorku.ca].

Johannes Wallner [537], *Technische Universität Wien, Wiedner Hauptstrasse 8-10/113, A1040 Wien, Austria* [hannes@geometrie.tuwien.ac.at].

Yumiko Watanabe [303], *Tokyo Denki University, 2-2 Kanda-Nisiki-Cho, Chiyoda, 101-8547 Tokyo, Japan* [watanabe@saitohlab2.c.dendai.ac.jp].

Geir Westgaard [545], *Technische Universität Berlin, ISM Sekr. SG 10, Salzufer 17-19, D-10587 Berlin, Germany* [westgaard@ism.tu-berlin.de].

Peter Wingren, *Umeå University, Dept. of Mathematics, S-901 87 Umeå, Sweden* [peterw@abel.math.umu.se].

Fujiichi Yoshimoto, *Wakayama University, Department of Computer and Communication Sciences, Sakaedani 930, Wakayama 610, Japan* [fuji@sys.wakayama-u.ac.jp].

Rainer Zeifang, *CoCreate Software GmbH, Posener Str. 1, D-71065 Sindelfingen, Germany* [rainer_zeifang@hp.com].

Denis Zorin, *California Institute of Technology, 1200 E. California Blvd., Pasadena, California 91125, USA* [dzorin@gg.caltech.edu].

A Practical Classification Method for Rational Quadratic Bézier Triangles with Respect to Quadrics

Gudrun Albrecht

Abstract. An algorithm for deciding whether a given rational triangular Bézier patch of degree 2 lies on a quadric surface is analysed from the numerical point of view. Despite the numerical treatment of the problem, *explicit* conditions for an *exact* decision are given.

§1. Introduction

Quadric surfaces or quadrics, i.e. algebraic surfaces of second order, are very popular surfaces in applications such as mechanical engineering and architecture. Because of their appealing and practical shapes as well as their simple implicit and parametric representations, they are widely used, see e.g. [5,6], and they are a substantial part of CAD and solid modeling systems. With the introduction of rational Bézier and NURBS surfaces, it became necessary to relate quadrics to this new form of surface representation. Therefore, in the last years the representation of quadrics in rational Bézier form has been an important issue in the field of CAGD.

Unlike conic sections where every rational quadratic Bézier segment lies on a conic section, not every rational biquadratic rectangular or quadratic triangular Bézier patch lies on a quadric surface, see e.g. [5].

Based on a theoretical result obtained by Degen [4], there exists an easy algorithm [1] for *(i)* testing whether a given rational triangular Bézier patch of degree 2 lies on a quadric or not, and *(ii)* for even finding the quadric's affine type.

The purpose of this paper is to numerically analyse part *(i)* of this algorithm, which consists in the solution of a system of 12 polynomial equations and the test whether four of the unknowns are zero or not. Although this problem is solved numerically, it is possible to *exactly* decide whether the unknowns in question are zero or not by using a floating point representation

Mathematical Methods for Curves and Surfaces II
Morten Dæhlen, Tom Lyche, and Larry L. Schumaker (eds.), pp. 1–8.

ISBN 0-8265-1315-8.

with a sufficiently big exponent range and a sufficiently large mantissa. Explicit values for the required bounds for the exponent and the mantissa are given. Sect. 2 provides the necessary tools from numerical analysis, namely the floating point representation and J. F. Canny's GAP Theorem. After formulating the problem, Sect. 3 thus explains the procedure for obtaining the exact decision and contains the detailed derivation of the required numerical bounds.

§2. Preliminaries

The purpose of this section is to provide the necessary tools and notations for a numerical treatment of the question whether a given rational triangular Bézier patch of degree 2 lies on a quadric surface or not. Since we calculate with floating point numbers, we need (see e.g. [7])

Definition 1. *The set $\mathcal{M}$ of normalized floating point numbers in the number system with basis β is given by*

$$\mathcal{M} := \{M\beta^{E-t} |\, M = 0 \text{ or } \beta^{t-1} \leq |M| < \beta^t\,;\, |E| \leq e\,;\, E \in \mathbb{Z}\,;\, t, e \in \mathbb{N}\}, \quad (1)$$

where M is the so–called mantissa *of length t and E the* exponent.

Thus, the quantities e and t describe the order of magnitude and the accuracy of a floating point number, respectively, as specified in

Lemma 2.

a) *If σ is an upper bound for the smallest positive floating point number and Σ a lower bound for the biggest positive floating point number representable in $\mathcal{M}$, then*

$$e > \max\{log_\beta(\sigma^{-1}) - 1, log_\beta(\Sigma)\}.$$

b) *If $rd(x)$ is the floating point approximation of the exact number x (obtained by correct rounding), then*

$$\frac{|rd(x) - x|}{|x|} \leq \frac{\beta\beta^{-t}}{2} =: eps$$

and

$$|rd(x) - x| \leq \frac{\beta^{E-t}}{2},$$

where eps is the so–called relative machine accuracy.

We also need J. F. Canny's GAP–Theorem [3], p.70 in order to exactly decide whether numerically calculated solutions of a polynomial system of equations are zero or not.

Theorem 3. *Let $\mathcal{P}(d,c)$ be the class of integral polynomials of degree d and maximum coefficient magnitude c. Let $f_i(x_1,...,x_n) \in \mathcal{P}(d,c)$, $i = 1,...,n$ be a collection of n polynomials in n variables which has only finitely many solutions when homogenized. If $(\alpha_1,...,\alpha_n)$ is a solution of the system, then for any j either $\alpha_j = 0$ or $|\alpha_j| > (3dc)^{-nd^n} =: \delta^{-1}$.*

According to Canny's proof [3], even $\delta^{-1} < |\alpha_j| < \delta$. For the floating point representation, according to Lemma 2a) this implies a sufficiently large exponent, namely

$$e > log_\beta(\delta). \tag{2}$$

§3. Quadric or not — Conditions for an Exact Numerical Decision

We are given a nonplanar rational triangular Bézier patch of degree 2

$$\mathbf{B} : \mathbf{x}(u,v,w) = \frac{\sum\limits_{i+j+k=2} w_{ijk}\mathbf{b}_{ijk}B^2_{ijk}(u,v,w)}{\sum\limits_{i+j+k=2} w_{ijk}B^2_{ijk}(u,v,w)} \in \mathbb{R}^3, \tag{3}$$

where (u,v,w) are barycentric coordinates $(u+v+w=1)$, and $B^2_{ijk}(u,v,w) = 2!/(i!j!k!)u^iv^jw^k$ are Bernstein polynomials. $\mathbf{b}_{ijk} = (b^1_{ijk}, b^2_{ijk}, b^3_{ijk})^T \in \mathbb{R}^3$ are the control points and $w_{ijk} \in \mathbb{R}^+ \setminus \{0\}$ the weights.

Based on the practical criterium presented in [1] for deciding whether the patch (3) lies on a quadric or not the purpose of this section is to specify conditions on the floating point representation to be used in the numerical calculations, in order to permit an *exact* decision. As in [1] we assume the control points $\mathbf{b}_{002}$, $\mathbf{b}_{020}$, $\mathbf{b}_{200}$, and $\mathbf{b}_{011}$ to be not coplanar without loss of generality, and for $i,j,k \in \{0,1,2\}$, $i+j+k=2$ we introduce

$$\tilde{\mathbf{b}}_{ijk} := \frac{2!}{i!j!k!}(w_{ijk}, w_{ijk}\mathbf{b}_{ijk})^T, \tag{4}$$

as well as

$$B := (\tilde{\mathbf{b}}_{200}, \tilde{\mathbf{b}}_{020}, \tilde{\mathbf{b}}_{002}, \tilde{\mathbf{b}}_{011}) \in \mathbb{R}^{4,4}. \tag{5}$$

We then consider the equation system

$$\begin{aligned}
\mathbf{F}_{101}(\mathbf{y}_{101}) &:= B\mathbf{y}_{101} - \tilde{\mathbf{b}}_{101} = 0,\\
\mathbf{F}_{110}(\mathbf{y}_{110}) &:= B\mathbf{y}_{110} - \tilde{\mathbf{b}}_{110} = 0,\\
g_0(\alpha_0, \mathbf{y}_{101}) &:= \alpha_0 - f^0(\mathbf{y}_{101}) = 0,\\
g_1(\alpha_1, \mathbf{y}_{101}, \mathbf{y}_{110}) &:= \alpha_1 - f^1(\mathbf{y}_{101}, \mathbf{y}_{110}) = 0,\\
g_2(\alpha_2, \mathbf{y}_{101}, \mathbf{y}_{110}) &:= \alpha_2 - f^2(\mathbf{y}_{101}, \mathbf{y}_{110}) = 0,\\
g_3(\alpha_3, \mathbf{y}_{110}) &:= \alpha_3 - f^3(\mathbf{y}_{110}) = 0,
\end{aligned} \tag{6}$$

consisting of 12 equations for 12 unknowns, where

$$\mathbf{y}_{101} = (y^0_{101}, y^1_{101}, y^2_{101}, y^3_{101})^T, \quad \mathbf{y}_{110} = (y^0_{110}, y^1_{110}, y^2_{110}, y^3_{110})^T,$$

and

$$\begin{aligned} f^0(\mathbf{y}_{101}) &:= -y^0_{101}y^1_{101}y^2_{101} + y^0_{101}(y^3_{101})^2 + y^1_{101}, \\ f^1(\mathbf{y}_{101}, \mathbf{y}_{110}) &:= -y^0_{101}y^1_{101}y^2_{110} - (y^0_{101}y^1_{110} + y^0_{110}y^1_{101})y^2_{101} \\ &\quad + 2y^0_{101}y^3_{101}y^3_{110} + y^0_{110}(y^3_{101})^2 - 2y^3_{101} + y^1_{110}, \\ f^2(\mathbf{y}_{101}, \mathbf{y}_{110}) &:= -y^0_{110}y^1_{110}y^2_{101} - (y^0_{101}y^1_{110} + y^0_{110}y^1_{101})y^2_{110} \\ &\quad + 2y^0_{110}y^3_{101}y^3_{110} + y^0_{101}(y^3_{110})^2 - 2y^3_{110} + y^2_{101}, \\ f^3(\mathbf{y}_{110}) &:= -y^0_{110}y^1_{110}y^2_{110} + y^0_{110}(y^3_{110})^2 + y^2_{110}. \end{aligned} \tag{7}$$

According to [1] we then have

Theorem 4. *The rational triangular quadratic Bézier patch (3) lies on a quadric surface if and only if*

$$\alpha_0 = \alpha_1 = \alpha_2 = \alpha_3 = 0, \tag{8}$$

where α_i $(i = 0, ..., 3)$ are as in (6).

The problem thus consists in testing whether certain components of a solution of a polynomial equation system are zero or not, which is exactly the question treated in Canny's GAP Theorem (see Theorem 3).

In order to be able to apply Theorem 3 to the problem in Theorem 4, we need to transform the polynomial equation system (6) to a system with integral coefficients. This can be achieved by converting the floating point coefficients of (6) to rational numbers and then converting these to integers. In the following we thus assume without loss of generality that the 12 equations of (6) have integral coefficients. Naively applying Theorem 3 to (6) yields the very large δ–value

$$\delta = (9c)^{12 \cdot 3^{12}}, \tag{9}$$

where c is the maximum coefficient magnitude of (6). According to (2) this would imply a very big exponent range for the floating point representation to be used. But due to the special form of (6), we can split it into the following 4 systems

$$\mathbf{F}_{101}(\mathbf{y}_{101}) = 0 \, , \; g_0(\alpha_0, \mathbf{y}_{101}) = 0, \tag{10}$$

$$\mathbf{F}_{101}(\mathbf{y}_{101}) = 0 \, , \; \mathbf{F}_{110}(\mathbf{y}_{110}) = 0 \, , \; g_1(\alpha_1, \mathbf{y}_{101}, \mathbf{y}_{110}) = 0, \tag{11}$$

$$\mathbf{F}_{101}(\mathbf{y}_{101}) = 0 \, , \; \mathbf{F}_{110}(\mathbf{y}_{110}) = 0 \, , \; g_2(\alpha_2, \mathbf{y}_{101}, \mathbf{y}_{110}) = 0, \tag{12}$$

$$\mathbf{F}_{110}(\mathbf{y}_{110}) = 0 \, , \; g_3(\alpha_3, \mathbf{y}_{110}) = 0, \tag{13}$$

where

$$\delta_0 := (9c_0)^{5 \cdot 3^5}, \quad \delta_1 := (9c_1)^{9 \cdot 3^9}, \quad \delta_2 := (9c_2)^{9 \cdot 3^9}, \quad \delta_3 := (9c_3)^{5 \cdot 3^5} \tag{14}$$

are the respective δ–values and c_i the respective maximum coefficient magnitudes. Since $\delta_i \ll \delta$ $(i = 0, 1, 2, 3)$ this splitting yields a much smaller upper exponent bound

$$e > \max_{0 \le i \le 3} \log_\beta(\delta_i) \tag{15}$$

for the floating point representation.

Corollary 5. *A solution* $(\mathbf{y}^{*T}, \alpha_i^*)$, $i \in \{0, ..., 3\}$ *of the corresponding system (10)–(13) satifies*

$$\alpha_i^* = 0 \text{ or } |\alpha_i^*| > \delta_i^{-1}, \tag{16}$$

where δ_i *according to (14) and* $\mathbf{y}^* \in \mathbb{R}^4$ *for* $i = 0, 3$ *and* $\mathbf{y}^* \in \mathbb{R}^8$ *for* $i = 1, 2$.

As a consequence, in order to check (16) for a numerically calculated approximation $(\mathbf{y}^T, \alpha_i = f^i(\mathbf{y}))$ of the exact solution $(\mathbf{y}^{*T}, \alpha_i^* = f^i(\mathbf{y}^*))$, we can use the following criterion:

Test Criterion ([2]): If

$$|f^i(\mathbf{y}) - f^i(\mathbf{y}^*)| < \frac{\delta_i}{2}, \tag{17}$$

then $f^i(\mathbf{y}^*) = 0$ if and only if $|f^i(\mathbf{y})| < \frac{\delta_i}{2}$.

To be able to apply this Test Criterion the accuracy in the calculation of $\mathbf{y}$ is to be chosen such that (17) is satisfied. The determination of the required accuracy is based on

Lemma 6. *Suppose the solution vectors* $\mathbf{y}_{101}^*, \mathbf{y}_{110}^*$ *of the 2 linear equation systems in (6) are bounded by* $|y_{101}^{*j}| \le s^j$, $|y_{110}^{*j}| \le s^{4+j}$ $(j = 0, \ldots, 3)$ *with* $s^k \in \mathbb{R}_+$ $(k = 0, \ldots, 7)$. *Then*

$$|f^i(\mathbf{y}) - f^i(\mathbf{y}^*)| \le \Gamma_i \, \|\mathbf{y} - \mathbf{y}^*\|_2, \qquad i = 0, \ldots, 3,$$

where $\mathbf{y}, \mathbf{y}^* \in \mathbb{R}^4$ *for* $i = 0, 3$ *and* $\mathbf{y}, \mathbf{y}^* \in \mathbb{R}^8$ *for* $i = 1, 2$. Γ_i *depends on* $s^0, \ldots, s^7$ *and on the coefficients of the polynomials* f^i.

Proof: By applying the multivariate mean value theorem to f^i, we obtain

$$f^i(\mathbf{y}) = f^i(\mathbf{y}^*) + \sum_{l=0}^{r} (y^l - y^{*l}) \, f_l^i(\mathbf{y}^* + \theta(\mathbf{y} - \mathbf{y}^*)),$$

where $\theta \in [0, 1]$, $\mathbf{y} = (y^0, \ldots, y^r)^T$, $\mathbf{y}^* = (y^{*0}, \ldots, y^{*r})^T$ with $r = 3$ for $i = 0, 3$ and $r = 7$ for $i = 1, 2$, and $f_l^i = \frac{\partial f^i}{\partial y^l}$. It thus follows that

$$|f^i(\mathbf{y}) - f^i(\mathbf{y}^*)| = \sqrt{\Big(\sum_{l=0}^{r} (y^l - y^{*l}) \, f_l^i(\mathbf{y}^* + \theta(\mathbf{y} - \mathbf{y}^*))\Big)^2}.$$

By applying the multinomial theorem and considering

$$(y^i - y^{*i})^2 \le \sum_{l=0}^{r} ((y^l - y^{*l})^2,$$

$$2(y^i - y^{*i})(y^j - y^{*j}) \le \sum_{l=0}^{r} ((y^l - y^{*l})^2,$$

we obtain

$$|f^i(\mathbf{y}) - f^i(\mathbf{y}^*)| \le G^i(\mathbf{y}^* + \theta(\mathbf{y}-\mathbf{y}^*))\ \|\mathbf{y}-\mathbf{y}^*\|_2, \tag{18}$$

where $G^i(\mathbf{y}^* + \theta(\mathbf{y}-\mathbf{y}^*)) := \sqrt{\sum_{k_0+\ldots+k_r=2}\prod_{l=0}^r f_l^i(\mathbf{y}^* + \theta(\mathbf{y}-\mathbf{y}^*))^{k_l}}$. According to (7) the f_l^i are polynomials of degree 2 and may be written as

$$f_l^i(\mathbf{x}) = \sum_{j=0}^r \sum_{k=0}^r \gamma_{jk}^{il} x^i x^k,$$

where $\mathbf{x} = (x^0, \ldots, x^r)^T$ and the coefficients γ_{jk}^{il} are easily determined from (7). We then multiply out

$$\sum_{k_0+\ldots+k_r=2} \prod_{l=0}^r f_l^i(\mathbf{x})^{k_l} = \sum_{j=0}^r \sum_{k=0}^r \sum_{p=0}^r \sum_{q=0}^r \Gamma_{jkpq}^i x^j x^k x^p x^q,$$

calculate the values Γ_{jkpq}^i accordingly, and consider $|y^{*l} + \theta(y^l - y^{*l})| \le s^l$. We thus obtain

$$G^i(\mathbf{y}^* + \theta(\mathbf{y}-\mathbf{y}^*)) \le \sqrt{\sum_{j=0}^r \sum_{k=0}^r \sum_{p=0}^r \sum_{q=0}^r |\Gamma_{jkpq}^i| s^j s^k s^p s^q} =: \Gamma_i. \ \square \tag{19}$$

Remark: The bounds s^j in Lemma 6 may e.g. be obtained by presolving the linear equation system in (6) with a reasonable but arbitrary machine accuracy *eps* yielding the approximations $\bar{\mathbf{y}}_{101}$, $\bar{\mathbf{y}}_{110}$, for the exact solutions $\mathbf{y}_{101}^*$, $\mathbf{y}_{110}^*$:

$$\bar{\mathbf{y}}_{101} = \mathbf{y}_{101}^*(1+\epsilon)\ ,\ \bar{\mathbf{y}}_{110} = \mathbf{y}_{110}^*(1+\epsilon)\ ,\ |\epsilon| \le M \in \mathbb{R}_+. \tag{20}$$

By choosing

$$\sigma \ge \max\{M\,\|\bar{\mathbf{y}}_{101}\|_2, M\,\|\bar{\mathbf{y}}_{110}\|_2\} \tag{21}$$

we then define for $j = 0, 1, 2, 3$

$$s^j := \max\{|\bar{y}_{101}^j - \sigma|, |\bar{y}_{101}^j + \sigma|\}\ ,\ s^{4+j} := \max\{|\bar{y}_{110}^j - \sigma|, |\bar{y}_{110}^j + \sigma|\}.$$

Due to (20), $\|\bar{\mathbf{y}}_{101}\|_2 \ge \bar{y}_{101}^j$, $\|\bar{\mathbf{y}}_{110}\|_2 \ge \bar{y}_{110}^j\ (j = 0, \ldots, 3)$ and $(1+M)^{-1} \doteq 1 - M$ this guarantees

$$\bar{y}_{101}^j - \sigma \le y_{101}^{*j} \le \bar{y}_{101}^j + \sigma\ ,\ \bar{y}_{110}^j - \sigma \le y_{110}^{*j} \le \bar{y}_{110}^j + \sigma\,.$$

For a reasonably well conditioned problem and a stable algorithm,

$$M = \mathcal{O}(eps).$$

By using Gauß–elimination for the solution of the 2 linear equation systems in (6) and by choosing as coefficient matrix B — according to [1] there are at most 15 possibilities — the one with the lowest condition number we obtain a reasonable value for σ from (21).

Theorem 7. *The question of whether the rational triangular quadratic Bézier patch (3) lies on a quadric surface or not can be answered exactly if the values e and t of the floating point representation (1) used for the calculations satisfy*

$$e > \max_{0 \le i \le 3} log_\beta(\delta_i),$$
$$t > \max_{0 \le i \le 3} log_\beta(\Gamma_i),$$

where δ_i according to (14) and Γ_i according to (19).

Proof: The bound for the length of the mantissa t remains to be explained. Combining the Test Criterion and Lemma 6 yields the condition

$$\|\mathbf{y} - \mathbf{y}^*\|_2 < \frac{\delta_i}{2\Gamma_i}, \qquad i = 0, \ldots, 3.$$

On the other hand, according to Lemma 2b),

$$\|\mathbf{y} - \mathbf{y}^*\|_2 \le \frac{\delta_i \beta^{-t_i}}{2},$$

and we thus obtain the condition

$$t_i > log_\beta(\Gamma_i). \quad \square$$

Acknowledgments. I would like to thank Ch. L. Bajaj for bringing J. F. Canny's GAP Theorem to my attention.

References

1. Albrecht, G., Determination and classification of triangular quadric patches, Comput. Aided Geom. Design, to appear.
2. Bajaj, Ch. L., and G. L. Xu, Piecewise rational approximations of real algebraic curves, J. Comput. Math. **15**, No. 1 (1997), 55–71.
3. Canny, J. F., *The Complexity of Robot Motion Planning*, MIT Press, Cambridge, MA, 1988.
4. Degen, W. L. F., The types of triangular Bézier surfaces, in *The Mathematics of Surfaces VI*, G. Mullineux (ed.), The IMA Conference Series No. 58, Clarendon Press Oxford, 1996, 153–170.
5. Farin, G., *Curves and Surfaces for Computer Aided Geometric Design*, Academic Press Inc., Boston, 1990.
6. Giering, O., and H. Seybold, *Konstruktive Ingenieurgeometrie*, Carl Hanser Verlag, München, 1987.
7. Stoer, J., Bulirsch, R., *Introduction to Numerical Analysis*, Springer, New York, 1993.

Gudrun Albrecht
Zentrum Mathematik
Technische Universität München
D–80290 München, GERMANY
`albrecht@mathematik.tu-muenchen.de`
`http://www-geo.mathematik.tu-muenchen.de/~albrecht/`

An Efficient and Robust Algorithm for Solving the Foot Point Problem

I. J. Anderson, M. G. Cox, A. B. Forbes,
J. C. Mason and D. A. Turner

Abstract. The foot point problem (FPP) is a key element of orthogonal regression and data matching procedures for curves and surfaces, both of which have significant applications in the field of metrology, *i.e.*, the science of measurement. A foot point is an orthogonal projection from a given datum onto a curve or surface, although, in general, we are only interested in the foot point yielding the position on a curve or surface closest to a given datum. The foot point problem involves calculating these closest foot points for a set of data and a specified curve or surface. In this paper, we present a new, efficient and robust algorithm for addressing the FPP for arbitrary data sets and parametric curves. The emphasis is on finding accurate initial parameter estimates as starting values for standard locally convergent non-linear iterative techniques.

§1. Introduction

Given a twice differentiable parametric curve $\boldsymbol{f} = \boldsymbol{f}(u)$, where $u \in [0,1]$, and a data point $\boldsymbol{x}$, we define the (squared) distance between the datum and the curve to be

$$F(u) = (\boldsymbol{f}(u) - \boldsymbol{x})^{\mathrm{T}}(\boldsymbol{f}(u) - \boldsymbol{x}). \tag{1}$$

The foot point problem involves determining the value of the parameter corresponding to the point on the curve closest to the datum, namely that given by

$$\min_u F(u). \tag{2}$$

This corresponds to solving

$$F'(u) = 0, \tag{3}$$

which is a necessary (but not sufficient) condition for a minimum of $F(u)$. The solution of (3) yields the parameter value u^*, and the corresponding foot

Mathematical Methods for Curves and Surfaces II
Morten Dæhlen, Tom Lyche, and Larry L. Schumaker (eds.), pp. 9–16.

ISBN 0-8265-1315-8.

point $f(u^*)$ which represents the point of intersection between the curve and the orthogonal projection from the datum to the curve. Equation (3) is a far-better-conditioned approach than minimizing $F(u)$ [7]. Given that the relative machine precision of the computer arithmetic is η, the limiting accuracy in the solution value of u in the latter is $\sqrt{2\eta|F(u^*)/F''(u^*)|}$, as can be seen by a Taylor expansion of $F(u)$ about the solution u^*, whereas that in the former is close to ηu^*. Thus, for a reasonably scaled problem, only half the precision available is attainable by using function values as the criterion.

The FPP is important in metrology, especially in two key problems — the design problem and the matching problem. In the design problem [6], we gather coordinate data of a prototype part (or some other physical model) and represent the data by a suitable geometric curve which agrees as closely as possible (in some norm) with the data. In the matching problem [3,4], we gather coordinate data of a manufactured part and compare this data with the geometric curve from the design problem, under appropriate changes of variable and orientation. Once this has been carried out, a decision can be made as to whether the manufactured part is fit for purpose. In both cases, the FPP must be solved many times and for a variety of representative data points. These added requirements have an effect on the choice of an efficient algorithm. We note that the FPP, although important, is given little attention in the literature, and it is therefore necessary to discuss algorithms at a fundamental level.

Given an initial estimate of the value of the parameter that minimizes (1), we may use a quadratically convergent method, such as Newton-Raphson [5], in order to compute a solution. Unfortunately, $F(u)$ will in general have a number of local maxima and minima. Hence, given a crude estimate of u^*, Newton-Raphson may well converge to the wrong foot point. Therefore, the ability to compute acceptable solutions depends upon the availability of accurate parameter estimates. For ordered data nominally following the shape of a curve, reasonable approximations to the parametric coordinates of the foot point can usually be obtained by using the normalized cumulative chord length of the data [1]. For unordered data, this approach is not applicable.

In this paper we present a new algorithm for finding fast estimates for the FPP for unordered data, from which Newton-Raphson will typically converge rapidly to an acceptable solution, provided that we can identify a search interval containing the parametric solution. Indications are made of some approaches for finding the correct interval. Occasionally divergence will still occur (or, more frequently, convergence to the wrong foot point) for certain data. In such cases, we describe a refinement of the algorithm that will give an arbitrarily accurate estimate.

§2. Piecewise Linear Interpolation Algorithm

We begin by considering a suitable search interval containing the parametric solution for the foot point u^*. By defining $u = u^* + \epsilon$, where ϵ is sufficiently small, we have

$$F'(u^* + \epsilon) = F'(u^*) + \frac{\epsilon^{2p-1}}{(2p-1)!} F^{(2p)}(u^*) + \mathcal{O}(\epsilon^{2p}),$$

for some integer p (where typically $p = 1$). Since u^* is a minimizer of $F(u)$, we have $F'(u^*) = 0$ and $F^{(2p)}(u^*) > 0$. Therefore, $F'(u) < 0$ for $\epsilon < 0$ and $F'(u) > 0$ for $\epsilon > 0$. Thus, the existence of a local minimum of $F(u)$ is characterized by the sign of $F'(u)$ changing from negative to positive in a small neighbourhood of the solution.

We require a foot point for a single datum $\boldsymbol{x}$. We consider replacing the parameter curve by a piecewise linear interpolant, from which an estimate of the foot point parameter can be generated.

1.1) Select n distinct parameter values $\{u_j\}_{j=1}^n$ with $0 \le u_1 < u_2 < \cdots < u_n \le 1$.

1.2) Evaluate the curve at these parameter values to create reference points $\{\boldsymbol{r}_j = \boldsymbol{f}(u_j)\}_{j=1}^n$, and join these points by chords to form a piecewise linear curve. Define the jth chord to be the line segment connecting $\boldsymbol{r}_j$ and $\boldsymbol{r}_{j+1}$, with $\boldsymbol{r}_{n+1} \equiv \boldsymbol{r}_1$.

1.3) Define the point $\boldsymbol{c}_j$ to be the point on the jth chord closest to the datum. This will be the foot of the orthogonal projection from $\boldsymbol{x}$ onto the chord provided this point lies between $\boldsymbol{r}_j$ and $\boldsymbol{r}_{j+1}$. Otherwise, the closest point will be one of the reference points at the end of the chord segment.

1.4) Choose the initial search interval to be $[u_k, u_{k+1}]$, where $\boldsymbol{c}_k$ is the closest to the datum, $\boldsymbol{x}$, amongst the $\boldsymbol{c}_j$ for each chord.

1.5) Check that $F'(u_k) \le 0$ and $F'(u_{k+1}) \ge 0$, *i.e.*, a solution exists in the initial search interval. See Section 4 for consideration of cases where this condition fails to hold.

1.6) Evaluate the corresponding parameter value of $\boldsymbol{c}_k$ on the line segment. Transform this parameter value into an estimate of u^*.

1.7) Apply the Newton-Raphson method to equation (3) starting from the estimate obtained in step 1.6). See Section 3 for consideration of cases in which the Newton-Raphson method fails to converge.

The points $\boldsymbol{c}_j$ in step 1.4) are calculated using orthogonal projection:

$$\boldsymbol{c}_j = \boldsymbol{r}_j + \lambda_j(\boldsymbol{r}_{j+1} - \boldsymbol{r}_j),$$

where

$$\lambda_j = \frac{(\boldsymbol{r}_{j+1} - \boldsymbol{r}_j)^T(\boldsymbol{x} - \boldsymbol{r}_{j+1})}{\|\boldsymbol{r}_{j+1} - \boldsymbol{r}_j\|^2},$$

provided $0 \le \lambda_j \le 1$. For values of $\lambda_j < 0$ we assign $\lambda_j = 0$, and similarly, for values of $\lambda_j > 1$ we assign $\lambda_j = 1$. The initial estimate for the foot point parameter is chosen to be

$$\hat{u} = u_k + \lambda_k(u_{k+1} - u_k).$$

For a set of m data the algorithm is repeated from step 1.3) for each new datum, since steps 1.1) and 1.2) are independent of $\boldsymbol{x}$. The algorithm requires $\mathcal{O}(mn)$ floating point operations to obtain foot point estimates for all data points, and this is a reasonably fast procedure for small n.

§3. Refined Piecewise Linear Approximation Algorithm

In this section we describe a refined algorithm for solving the FPP when Newton-Raphson fails to converge to u^* from the starting value $\hat{u}$. We assume that $F'(u_k) \leq 0$ and $F'(u_{k+1}) \geq 0$. We then apply the following procedure:

2.1) Evaluate the parametric mid-point of the interval, $\tilde{u} = u_{k+1} + 1/2(u_k - u_{k+1})$ and calculate $F'(\tilde{u})$. If $F'(\tilde{u}) \leq 0$, let the refined search interval be $[\tilde{u}, u_{k+1}]$. Similarly, if $F'(\tilde{u}) > 0$, let the search interval be $[u_k, \tilde{u}]$.

2.2) Draw a chord connecting $\boldsymbol{f}(u_k)$ (or $\boldsymbol{f}(u_{k+1})$) and $\boldsymbol{f}(\tilde{u})$. Obtain an estimate of u^* by applying steps 1.3) and 1.6).

2.3) Apply the Newton-Raphson method to equation (3) starting from the estimate of u^* obtained in step 2.2).

2.4) If divergence is indicated, repeat from step 2.1).

Using this refinement process, the error bound defined by $|u^* - \hat{u}|$ is halved at each iteration, allowing us to attain an arbitrarily accurate estimate, although in practice we switch to Newton-Raphson as soon as possible. Further, by demanding that the sign of $F'(u)$ changes from negative to positive in the interval, we ensure that the above algorithm converges to a minimum of $F(u)$, although, in general, we cannot guarantee that it is the global minimum.

§4. Failure to Find a Satisfactory Search Interval

Although the approach described above works reasonably well, it is easy to construct examples where the appropriate sign change for $F'(u)$ does not occur. In this section, we discuss such instances.

4.1. Multiple Foot Points in the Search Interval

Suppose that we have chosen the search interval $[u_k, u_{k+1}]$. It is possible that this interval contains two foot points, one corresponding to a local minimum of $F(u)$ and the other a local maximum of $F(u)$. Define the parameter value of the local minimum to be $u_{\min}$ and the parameter value of the local maximum to be $u_{\max}$. If $u_{\min} < u_{\max}$, both $F'(u_k)$ and $F'(u_{k+1})$ will be negative. Similarly, if $u_{\min} > u_{\max}$, both $F'(u_k)$ and $F'(u_{k+1})$ will be positive. In either case the condition in step 1.5) of the algorithm will fail. Such cases are most likely to occur when a datum lies close to a portion of the profile with high curvature. Fig. 1 shows a profile that contains three sharp apices. In practical experiments, problems were encountered at each apex.

The sign change condition has failed because the quality of the piecewise linear interpolant is poor and does not reflect accurately the underlying shape of the profile. Consequently, the initial search interval is not satisfactory. It

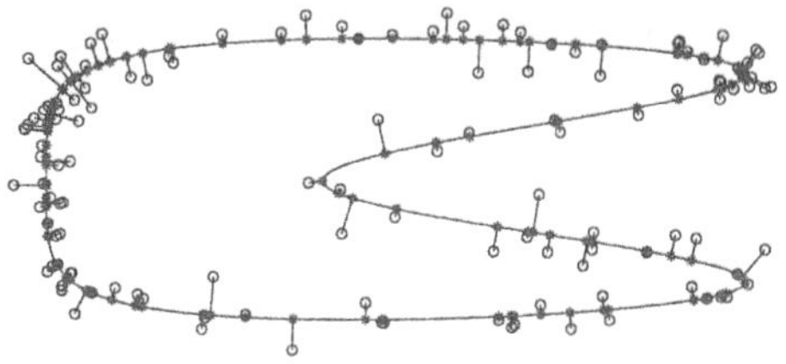

Fig. 1. A synthesised data set obtained by sampling with Gaussian error from the illustrated parametric curve, and the foot points corresponding to these points.

is clear that increasing the number of reference points produces an improved interpolant and hence a better search interval. Since we are only interested in refining the quality of the interpolant in the neighbourhood of the solution only a small number of additional reference points are required. Although we are not guaranteed that the solution lies in the interval $[u_k, u_{k+1}]$, we have good reason to believe that it is close to this region. Thus, we choose to add three new reference points corresponding to the parametric mid-points of the intervals $[u_{k-1}, u_k]$, $[u_k, u_{k+1}]$ and $[u_{k+1}, u_{k+2}]$.

Using this augmented set of parameter values, we generate a new estimate of u^* using the algorithm described in Section 2. We note that since there are only three new reference points, most of the computations involved in producing an estimate for u^* have already been done and we can exploit this to produce an improved parameter estimate in an efficient manner.

4.2. Ambiguous Search Interval

A different problem arises for a datum which lies outside the curve and whose foot point lies close to one of the reference points, $\boldsymbol{r}_k$, say. The closest point to the datum on each of the chords $\boldsymbol{r}_{k-1}\boldsymbol{r}_k$ and $\boldsymbol{r}_k\boldsymbol{r}_{k+1}$ is $\boldsymbol{r}_k$ and so the choice of initial search interval is ambiguous. Extra care needs to be taken to ensure that the interval chosen has the appropriate sign change for $F'(u)$.

An additional problem arises if this ambiguity of interval is combined with multiple foot points in at least one of the intervals. Here, both $F'(u_{k-1})$ and $F'(u_{k+1})$ will have the same sign and it is possible that neither interval will satisfy the condition in step 1.5) of the algorithm. For this case we add two new reference points corresponding to the parametric mid-points of the intervals $[u_{k-1}, u_k]$ and $[u_k, u_{k+1}]$, and we repeat the algorithm as above.

§5. Numerical Examples

In this section, we illustrate the accuracy, efficiency and reliability of the refined piecewise linear interpolation algorithm.

Fig. 1 shows a synthesised data set consisting of 128 points approximating the nominal shape of a geometric curve. The curve has been deliberately chosen for the difficulty it poses for the FPP, since a typical datum possesses several orthogonal projections on the curve. In the figure, the data are represented by small circles and the solid lines drawn from the data show the orthogonal projections onto the closest foot points. The residual distances from the data to the foot points are normally distributed.

$\hat{u}$	Converge to u^*	Converge to min	Converge to max	Diverge	NR steps
0	1	60	67	0	2.5
1/4	68	38	21	1	6.4
1/2	61	12	52	3	10.3
3/4	62	42	23	1	6.8

Tab. 1. Performance of Newton-Raphson method for various choices of the starting value $\hat{u}$.

n	Converge to u^*	Converge to min	Refine	NR steps
8	124	4	40	2.7
16	128	0	12	2.2
32	128	0	4	1.8
64	128	0	3	1.7

Tab. 2. Performance of refined piecewise linear interpolation method.

Tab. 1 highlights the importance of good estimates for the FPP. We consider the success of Newton-Raphson when the same starting value $\hat{u}$ is used for every datum, for various choices of $\hat{u}$. Successive columns denote the numbers of data for which Newton-Raphson, respectively, converges to the closest foot point, converges to a local (but not the global) minimum of $F(u)$, converges to a local maximum of $F(u)$, and diverges. We also give the mean number of Newton-Raphson iterations per datum that are required for convergence (to an absolute accuracy of 5×10^{-4}).

The lack of consistency in the behaviour of the Newton-Raphson method for the various starting values deserves comment. The differing results are a consequence of the local behaviour of the curve near the particular parameter value. For instance, at $u = 0$, for which the corresponding position on the curve is the sharp apex at the centre of Fig. 1, we can see that a large number of data yield a nearby local maximum or minimum of $F(u)$. This accounts for the disastrous performance of $\hat{u} = 0$ as a starting estimate for the closest foot point of a datum.

Tab. 2 represents the performance of the refined piecewise linear interpolation algorithm. We compare the behaviour of the algorithm for $n = 8, 16, 32, 64$ uniformly spaced values of u. We present similar results to those of Tab. 1. However, since there is neither divergence nor convergence to local maxima, as expected, we omit these columns. Tab. 2 also shows the number of data for which we needed to make a refinement of the chord (giving an initial estimate), either to ensure the satisfaction of a FP solution or to follow up the divergence of the Newton-Raphson method from the original estimate. We also give the mean number of Newton-Raphson iterations per datum that are required for convergence.

§5.1. Choice of n and $\{u_j\}$

Two conclusions are apparent from these results. Firstly, the value of n can be kept small, while still achieving reliable estimates for u^*. Indeed, increasing n significantly has only a limited effect of reducing the number of refinements that are necessary. Secondly, the majority of the cases where refinement was necessary correspond to foot points located in regions of high curvature. For these cases, the quality of the piecewise linear interpolant was unsatisfactory, and it is clear that we require a high density of reference points in regions where the curvature is high.

For the special case of parametric spline curves, the curvature is characterized entirely by the control polygon and the knot set. In order to produce a region of high curvature, several vertices of the polygon, and hence several knots, are required. Consequently, a very good indicator of the number and location of suitable reference points can be obtained by defining the initial values of $\{u_j\}_{j=1}^n$ to be the knot set of the parametric spline curve.

§6. Concluding Remarks and Future Work

This paper has formulated the foot point problem (FPP), as well as emphasizing its importance in metrology. While well-known locally convergent methods are always available, the provision of accurate starting values has been overlooked, especially for unordered data sets.

We have provided a two-tiered algorithm. In the first stage, we generate initial estimates of the parameter values in an efficient manner, from which the Newton-Raphson method, for example, will generally converge to the required solution. In the second stage, we present a refined algorithm which overcomes cases such as the Newton-Raphson method diverging.

It is possible to construct examples in which the parametric estimate converges to a local minimum, rather than the global minimum. However, numerical experiments indicate that, given an appropriate choice of n and parameter values $\{u_j\}_{j=1}^n$, the algorithm yields reliable estimates for a reasonable parametric curve and data set.

Future work by the authors will address some of the following issues:

- More efficient and reliable determination of the search interval in the case of large n.
- Automatic parametric partitioning of the curve accounting for local geometry (curvature, etc.).
- A study of the complexity of the overall algorithm, with emphasis on the choice of n to obtain a near-minimal operation count for the complete computation.

- The solution of the FPP within the iterative context of the design and matching problems.
- Application to problems in metrology.

References

1. Anthony, G. T., and M. G. Cox, The National Physical Laboratory's Data Approximation Subroutine Library, in *Algorithms for Approximation*, J. C. Mason, and M. G. Cox (eds.), Clarendon Press, Oxford, 1987, 669–687.
2. Butler, B. P., M. G. Cox, and A. B. Forbes, The reconstruction of workpiece surfaces from probe coordinate data, in *Design and Application of Curves and Surfaces*, R. B. Fisher (ed.), Oxford University Press, 1994, 99–116.
3. Butler, B. P., A. B. Forbes, and P. M. Harris, Algorithms for geometric tolerance assessment, Technical report DITC 228/94, National Physical Laboratory, Teddington, 1994.
4. Butler, B. P., A. B. Forbes, and P. M. Harris, Geometric tolerance assessment problems, in *Advanced Mathematical Tools in Metrology*, P. Carlini, M. G. Cox, R. Monaco, and F. Pavese (eds.), World Scientific, Singapore, 1994, 95–104.
5. Conte, S. D., and C. de Boor, *Elementary Numerical Analysis*, McGraw-Hill, New York, 1981.
6. Forbes, A. B., A separation-of-variables approach to generalized distance regression, in preparation.
7. Gill, P. E., W. Murray, and M. H. Wright, *Practical Optimization*, Academic Press, London, 1981.

I. J. Anderson, J. C. Mason and D. A. Turner
School of Computing and Mathematics
University of Huddersfield
HD1 3DH, UK
`i.j.anderson@hud.ac.uk`
`j.c.mason@hud.ac.uk`
`d.a.turner@hud.ac.uk`

M. G. Cox and A. B. Forbes
CISE
NPL
Teddington
TW11 0LW, UK
`mgc@npl.co.uk`
`abf@npl.co.uk`

Smooth Rational Cubic Spline Interpolation

Christoph Baumgarten and Gerald Farin

Abstract. An interpolation scheme for planar curves is described, obtained by joining parametric rational cubics approximating logarithmic spiral segments. The resulting curves are G^2 continuous and their curvature radius plot is close to piecewise linear. The presented method is globally convexity preserving and provides (almost) logarithmic spiral precision as well as circle precision. A number of examples illustrates the interpolation method.

§1. Introduction

Rational cubic spline curves in Bézier form play an important role in the field of Computer Aided Geometric Design. They can be viewed as a generalization of cubic spline curves since they provide more degrees of freedom in the form of weights. At first glance, this seems to be an advantage. But a designer working with a system based on this class of curves is sometimes overburdened with these degrees of freedom, for the weights are not easy to handle: the relationship between a change of a weight and the resulting change of the curve is not always very intuitive.

The purpose of this work is therefore the development of a rational cubic spline interpolation method for the planar case which automatically produces well-shaped curves – *well-shaped* in the sense of having a continuous curvature or curvature radius plot (a plot of the curvature or radius of curvature *vs* arc length), which contains as few local extrema as possible (compare [5]).

The presented approach consists of several steps: We consider the set S of planar rational cubic spline curves which approximate logarithmic spiral segments within a reasonable precision. As the curvature radius plot of a logarithmic spiral is linear and hence without local extrema, these curves have excellent shape properties. In a first step, curves from S are joined G^1-continuously, where an arbitrarily chosen sequence of data points is matched (as a restriction, three consecutive points must not be collinear). Then the tangents at the data points are rotated until the gaps in the curvature radius plot vanish, i.e., until the new spline curve is G^2. Additional data points are

Mathematical Methods for Curves and Surfaces II
Morten Dæhlen, Tom Lyche, and Larry L. Schumaker (eds.), pp. 17–24.

ISBN 0-8265-1315-8.

inserted where necessary. Special end conditions for the first and last tangents complete the interpolation process.

As the resulting spline curve consists of approximated spiral arcs, the curvature radius plot of the spline curve is piecewise linear (or almost piecewise linear, as we are talking about an approximation). This leads to a marked reduction of local extrema. Furthermore, taking four or more points from a logarithmic spiral as input, an approximation of a segment of the same logarithmic spiral is obtained. This property is called *logarithmic spiral precision* (or almost spiral precision, as we only obtain an approximation of the spiral segment). Since the circle can be considered as a degenerate logarithmic spiral, the method has *circlular precision* as well.

Our interpolation method is a *global method*, since changing the position of one data point affects the whole curve. It is *stable* in the sense that a small change of the data points will produce a small change in the shape of the curve.

It should also be mentioned that the generated spline curves never contain inflection points or straight lines, i.e., the interpolation method is *globally convexity preserving*, the resulting curves are always left turning (even if the input data is right turning, see Fig. 6; right turning interpolating curves however can be obtained in a way analogous to the one described in this paper). If, during interpolation, inflection points or straight lines are desired, the method would have to be extended for these special cases, which is not part of this work (see [5]).

By building interpolating G^2 spline curves from curves taken from the rather specific set $\mathcal{S}$, no additional degrees of freedom have to be specified.

§2. Approximating a Logarithmic Spiral Segment

We consider a *segment* $\boldsymbol{s}$ *of a logarithmic spiral* $\boldsymbol{S} : r = r_0\, e^{k\lambda}, r_0 \in \mathbb{R}^+,\ k, \lambda \in \mathbb{R}$, rotating around a pole $\boldsymbol{o}$ with $\boldsymbol{s}_0$ as start point (rotation angle $\lambda = 0$) and $\boldsymbol{s}_1$ as end point ($\lambda = \phi$, where ϕ is the segment angle with $0 < \phi < 2\pi$). For $k = 0$, $\boldsymbol{s}$ degenerates to a circle segment. In order to approximate $\boldsymbol{s}$ by a planar rational cubic

$$\boldsymbol{b}(t) := \frac{\sum_{j=0}^{3} w_j \boldsymbol{b}_j B_j^3(t)}{\sum_{j=0}^{3} w_j B_j^3(t)}, \qquad t \in [0,1]$$

(determined by four Bézier points $\boldsymbol{b}_j$ and corresponding weights w_j; B_j^3 are Bernstein polynomials of degree 3), we match the zero, first and second order derivatives of $\boldsymbol{b}$ at $\boldsymbol{b}_0$ and $\boldsymbol{b}_3$ with $\boldsymbol{s}$: We set $\boldsymbol{b}_0 := \boldsymbol{s}_0$, $\boldsymbol{b}_3 := \boldsymbol{s}_1$, and the directions of $\boldsymbol{b}_1 - \boldsymbol{b}_0$ and $\boldsymbol{b}_3 - \boldsymbol{b}_2$ to the directions of the tangent vectors of the spiral segment at $\boldsymbol{s}_0$ and $\boldsymbol{s}_1$. Since two weights of a rational cubic may be set to an arbitrary value, we select $w_1 = w_2 := \frac{2}{3}$ [2]. The weights w_0 and w_3 are chosen such that the radius of curvature of $\boldsymbol{b}$ at $\boldsymbol{b}_0$ and $\boldsymbol{b}_3$ is equivalent to the spiral segment's radius of curvature at $\boldsymbol{s}_0$ and $\boldsymbol{s}_1$ (compare the method of osculatory interpolation described in [3]). In order to determine the only

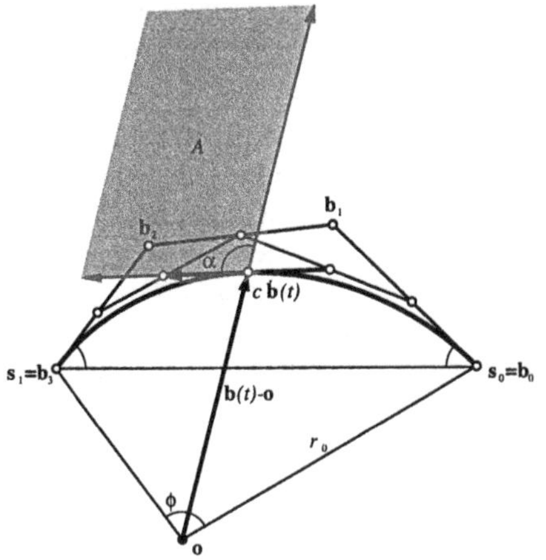

Fig. 1. The condition for $\boldsymbol{b}$: The area A spanned by $\boldsymbol{b}(t)-\boldsymbol{o}$ and $c\,\dot{\boldsymbol{b}}(t)$, $c \neq 0$, and the scalar product of these vectors have to be proportional. (Note that the geometrical meaning of $|\det(\boldsymbol{v}, \boldsymbol{w})|$ is the area spanned by the vectors $\boldsymbol{v}$ and $\boldsymbol{w}$).

remaining unknowns $\|\boldsymbol{b}_1 - \boldsymbol{b}_0\|$ and $\|\boldsymbol{b}_3 - \boldsymbol{b}_2\|$, we exploit the fact that a ray starting in pole $\boldsymbol{o}$ intersects the logarithmic spiral segment $\boldsymbol{s}$ always at the same angle $0 < \alpha = \operatorname{arcot}(k) < \pi$. If we assume our spiral segment to be in parametric form, we may rewrite this property for $\boldsymbol{s}$ as $\forall \lambda \in [0, \phi] : \angle\,(\boldsymbol{s}(\lambda) - \boldsymbol{o}, \dot{\boldsymbol{s}}(\lambda)) = \alpha$. Since the determinant $\det(\boldsymbol{v}, \boldsymbol{w})$ and the scalar product $\boldsymbol{v}\boldsymbol{w}$ of two vectors $\boldsymbol{v}, \boldsymbol{w}$ satisfy the identities $|\det(\boldsymbol{v}, \boldsymbol{w})| = \|\boldsymbol{v}\|\,\|\boldsymbol{w}\| \sin(\angle(\boldsymbol{v}, \boldsymbol{w}))$, and $\boldsymbol{v}\boldsymbol{w} = \|\boldsymbol{v}\|\,\|\boldsymbol{w}\| \cos(\angle(\boldsymbol{v}, \boldsymbol{w}))$, respectively, the property implies $\forall \lambda \in [0, \phi] : \det(\boldsymbol{s}(\lambda) - \boldsymbol{o}, \dot{\boldsymbol{s}}(\lambda)) = \|\boldsymbol{s}(\lambda) - \boldsymbol{o}\|\,\|\dot{\boldsymbol{s}}(\lambda)\| \sin(\alpha)$ as well as $\forall \lambda \in [0, \phi] : (\boldsymbol{s}(\lambda) - \boldsymbol{o})\,\dot{\boldsymbol{s}} = \|\boldsymbol{s}(\lambda) - \boldsymbol{o}\|\,\|\dot{\boldsymbol{s}}(\lambda)\| \cos(\alpha)$. We combine both equations and obtain

$$\forall \lambda \in [0, \phi] : \frac{\det(\boldsymbol{s}(\lambda) - \boldsymbol{o},\, \dot{\boldsymbol{s}}(\lambda))}{(\boldsymbol{s}(\lambda) - \boldsymbol{o})\,\dot{\boldsymbol{s}}(\lambda)} = \frac{\sin(\alpha)}{\cos(\alpha)} = \tan(\alpha) \tag{1}$$

as a property of $\boldsymbol{s}$ (if $\alpha \neq \frac{\pi}{2}$; $\alpha = \frac{\pi}{2}$ has to be handled separately [1]), which will be used as a *condition* for our approximating rational cubic $\boldsymbol{b}$ (see Fig. 1): If

$$\frac{\det(\boldsymbol{b}(t) - \boldsymbol{o},\, \dot{\boldsymbol{b}}(t))}{(\boldsymbol{b}(t) - \boldsymbol{o})\,\dot{\boldsymbol{b}}(t)} = \frac{\sin(\alpha)}{\cos(\alpha)} \tag{2}$$

holds for any $t \in \mathbb{R}$, it holds for $t \in [0, 1]$ and guarantees the spiral property (1) for $\boldsymbol{b}$. But (1) still implies a logarithmic spiral, hence, $\boldsymbol{b}$ will only *approximate* the condition (2). This condition can be transformed into an equation of the form

$$\forall t \in \mathbb{R} :\ \cos(\alpha) \sum_{j=1}^{8} M_j B_{j-1}^7(t) = \sin(\alpha) \sum_{j=1}^{8} N_j B_{j-1}^7(t)\,.$$

Comparing the coefficients yields a system $\cos(\alpha) M_j = \sin(\alpha) N_j$ of 8 equations, where the third equation pair $j = 3, 6$ can be solved for the desired vector lengths. The details of this lengthy computation are presented in detail in [1]. The approximation of $\boldsymbol{s}$ might be of insufficient quality (the curvature radius plot of $\boldsymbol{b}$ is not necessarily monotone). In this case, $\boldsymbol{s}$ is partitioned into subsegments and the approximation is repeated for each subsegment. This way, for any possible spiral segment, an approximating rational cubic spline curves of any desired quality can be generated [1]. We denote the set of planar rational cubic spline curves approximating logarithmic spiral segments with an approximation error (determined using an appropriate error measurement) below a given limit by $\mathcal{S}$.

§3. Smooth Interpolation

Rational cubic spline curves provide, in comparison to their nonrational counterparts, additional weights and therefore more degrees of freedom. In order to facilitate a designer's work, spline methods are of interest which compute as many of these degrees of freedom as possible, while producing excellently shaped curves. In the following, we will present such an approach for the planar case, based on interpolation.

Given is a sequence of data points $\boldsymbol{d}_i$, $i = 0, ..., L$, $L \geq 3$, where three consecutive points are never collinear. In order to compute a well-shaped rational cubic spline curve $\boldsymbol{d}$ interpolating the $\boldsymbol{d}_i$, we proceed stepwise:

(a) We specify tangent vectors $\boldsymbol{t}_i$ corresponding to $\boldsymbol{d}_i$ by using a simple method as for example the *FMILL method* [3]: $\boldsymbol{t}_i := \boldsymbol{d}_{i+1} - \boldsymbol{d}_{i-1}$. The start and the end tangent vectors $\boldsymbol{t}_0$ and $\boldsymbol{t}_L$ may be chosen as tangent vectors related to the two circles that match $\boldsymbol{d}_0$, $\boldsymbol{d}_1$, $\boldsymbol{t}_1$ and $\boldsymbol{t}_{L-1}$, $\boldsymbol{d}_{L-1}$, $\boldsymbol{d}_L$, respectively.

(b) Two neighboring data points $\boldsymbol{d}_i$, $\boldsymbol{d}_{i+1}$ together with their tangent vectors $\boldsymbol{t}_i$, $\boldsymbol{t}_{i+1}$ *uniquely* define a connecting logarithmic spiral segment $\boldsymbol{s}_i$ with $0 < \phi_i < 2\pi$. In order to be able to replace this spiral segment by a spline curve $\tilde{\boldsymbol{s}}_i \in \mathcal{S}$ according to Section 2, the spiral segment's pole has to be computed from $\boldsymbol{d}_i$, $\boldsymbol{d}_{i+1}$, $\boldsymbol{t}_i$ and $\boldsymbol{t}_{i+1}$. This is explained in detail in [1]. Having computed the $\tilde{\boldsymbol{s}}_i$ for all $i = 0, ..., L-1$, we obtain a G^1 continuous rational cubic spline curve $\boldsymbol{d}^1$ that interpolates the given data points. Note that in general, it does not seem to be possible to select the tangent vectors in (a) such that G^2 continuity is guaranteed right away. Several approaches in that direction ended in complex nonlinear systems of equations, which were not computable in closed form.

(c) In order to convert our G^1 continuous spline curve $\boldsymbol{d}^1$ into an "almost G^2 continuous" spline curve $\boldsymbol{d}^2$, we apply the following *rotation algorithm*, denoted in pseudo code, where $\tau > 0$ is a given tolerance which should typically be selected below 10^{-3}. Traversing the spline curve from "left to right" and "right to left" at the same time provides more symmetry in the results for symmetric input data.

```
repeat
  right := 0;  left := L;  precision := sufficient;
  while (right ≤ L) do
    if (rotate(tleft, τ) or rotate(tright, τ))
      precision := not sufficient;
    fi;
    right := right + 1;  left := left - 1;
  od;
until (precision = sufficient);
```

The function rotate(t_i, τ) is the core of the rotation algorithm: The tangent vector $\boldsymbol{t}_i$ is rotated around the corresponding data point $\boldsymbol{d}_i$; $\tilde{\boldsymbol{s}}_{i-1}$ and $\tilde{\boldsymbol{s}}_i$ are recomputed according to the new tangent vector. (The special case of $i = 0$ and $i = L$ will be handled later.) The aim of the rotation is to minimize the difference $|\rho_{1,i-1} - \rho_{0,i}|$, such that $|\rho_{1,i-1} - \rho_{0,i}| < \tau$, where $\rho_{1,i-1}$ and $\rho_{0,i}$ denote the radii of curvature of $\tilde{\boldsymbol{s}}_{i-1}$ and $\tilde{\boldsymbol{s}}_i$ at their common junction point $\boldsymbol{d}_i$. We may rotate $\boldsymbol{t}_i$ in two directions (Fig. 2): If we rotate $\boldsymbol{t}_i$ counterclockwise, the corresponding (positive) rotation angle ν_i has to remain below $\min\{\nu_{0,i}^+ := \angle(\boldsymbol{t}_i, \boldsymbol{d}_{i+1} - \boldsymbol{d}_i), \nu_{1,i}^+ := \angle(\boldsymbol{t}_i, \boldsymbol{t}_{i-1})\}$; if we rotate $\boldsymbol{t}_i$ clockwise, (the negative) ν_i must not reach $\max\{\nu_{0,i}^- := -\angle(\boldsymbol{d}_i - \boldsymbol{d}_{i-1}, \boldsymbol{t}_i), \nu_{1,i}^- := -\angle(\boldsymbol{t}_{i+1}, \boldsymbol{t}_i)\}$. Otherwise, combinations of tangent vectors $\boldsymbol{t}_{i-1}, \boldsymbol{t}_i$, and $\boldsymbol{t}_i, \boldsymbol{t}_{i+1}$ could occur, where $\phi_{i-1}, \phi_i \notin (0; 2\pi)$, i.e., one of the connecting spline curves $\tilde{\boldsymbol{s}}_{i-1}$ or $\tilde{\boldsymbol{s}}_i$ might not exist. This is further explained in [2].

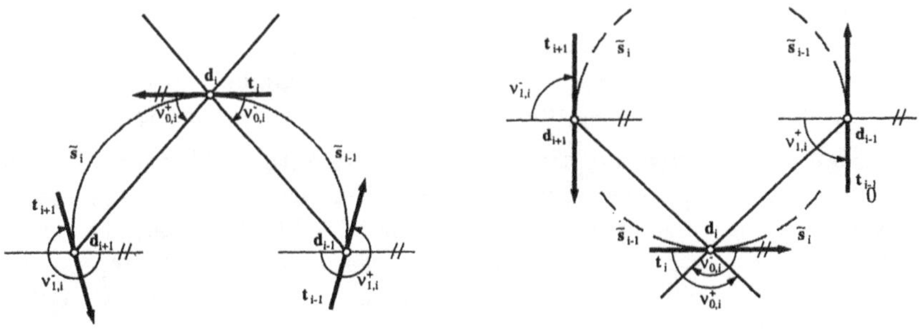

Fig. 2. Rotation of the tangent vector $\boldsymbol{t}_i$.

It is possible that the rotation yields a minimum of the difference $|\rho_{1,i-1} - \rho_{0,i}|$ above the given tolerance τ. In this case, the function rotate inserts, as a side effect, an additional data point ($L := L + 1$): If the segment angle of $\boldsymbol{s}_{i-1}$ exceeds the segment angle of $\boldsymbol{s}_i$, a point from $\tilde{\boldsymbol{s}}_{i-1}$ is taken, otherwise a point from $\tilde{\boldsymbol{s}}_i$. We may, for example, use a junction point generated by the subdivision algorithm mentioned in Section 2 (if $\tilde{\boldsymbol{s}}_{i-1}$ or $\tilde{\boldsymbol{s}}_i$, respectively, was not subdivided, subdivision has to be enforced). It can be shown that the repeated insertion of additional data points leads to $|\rho_{1,i-1} - \rho_{0,i}| < \tau$ for every $i \in \{1, \ldots, L-1\}$ [2]. The insertion of additional data points by `rotate` may be avoided if the data points are specified "densely enough" by the user. In other words, an automatically inserted data point always indicates that not enough data points were specified in the neighborhood of the added data point. The user may, of course, replace the inserted data points by more suitably located data points.

In order to find new first and last tangent vectors $\boldsymbol{t}_0$ and $\boldsymbol{t}_L$, we proceed in much the same way as we did for the other tangent vectors: We simply assume $\rho_{1,-1}$ and $\rho_{0,L}$ to be specified. For $\boldsymbol{t}_0$, the rotation angle ν_0 then has to remain in the interval $(-\angle(\boldsymbol{t}_1, \boldsymbol{t}_0), \angle(\boldsymbol{t}_0, \boldsymbol{d}_1 - \boldsymbol{d}_0))$; for $\boldsymbol{t}_L$, the corresponding ν_L is located within $(-\angle(\boldsymbol{d}_L - \boldsymbol{d}_{L-1}, \boldsymbol{t}_L), \angle(\boldsymbol{t}_L, \boldsymbol{t}_{L-1}))$. If the rotation yields $|\rho_{1,i-1} - \rho_{0,i}| > \tau$ for $i = 0$ or $i = L$, one of $\rho_{1,-1}$ or $\rho_{0,L}$ is too small and we have to add a new data point taken from $\tilde{\boldsymbol{s}}_0$ or $\tilde{\boldsymbol{s}}_{L-1}$. In order to specify $\rho_{1,-1}$, we use $\rho_{0,1}$ and $\rho_{1,1}$ as well as the chord lengths $\tilde{S}_0$ and $\tilde{S}_1$ of the spline curves $\tilde{\boldsymbol{s}}_0$ and $\tilde{\boldsymbol{s}}_1$ computed in a previous iteration step of the rotation algorithm. We select

$$\rho_{1,-1} = \begin{cases} \rho_{0,1} + (\rho_{0,1} - \rho_{1,1}) \frac{\tilde{S}_0}{\tilde{S}_1} =: \rho & \text{if } \rho \geq \delta, \\ \delta & \text{otherwise,} \end{cases}$$

where δ denotes a small positive value. This way, $\tilde{\boldsymbol{s}}_0$ and $\tilde{\boldsymbol{s}}_1$ are related to the same logarithmic spiral if possible. The specification of $\rho_{0,L}$ proceeds in an analogous manner.

The return value of `rotate` is `true` if a rotation has been performed and `false` otherwise.

(d) The "almost G^2" spline curve $\boldsymbol{d}^2$ generated by (c) is transformed into a G^2 spline curve $\boldsymbol{d}$: Since there is a linear dependency between the weights w_3 and w_0 of the rational cubics neighboring the data point $\boldsymbol{d}_i$ and the radii of curvature $\rho_{1,i-1}$ and $\rho_{0,i}$, respectively, the remaining gap between $\rho_{1,i-1}$ and $\rho_{0,i}$ can be closed by appropriately adjusting the weights w_3 and w_0 [2].

Figures 3 – 7 illustrate the interpolation algorithm. All figures include a corresponding curvature or curvature radius plot. Data points are represented by solid squares (or solid circles, if they were added by the interpolation algorithm), Bézier points are shown as hollow circles, and gray circles represent the so-called weight points $\boldsymbol{q}_j := (w_j \boldsymbol{b}_j + w_{j+1} \boldsymbol{b}_{j+1})/(w_j + w_{j+1}), j \in \{0, 1, 2\}$, of a rational cubic.

§4. Concluding Remarks

In this work we have described a globally convexity preserving interpolation scheme that is based on the approximation of logarithmic spiral segments by rational cubics. The scheme therefore can be easily integrated into existing NURBS-based systems [4]. The resulting curves are guaranteed to have G^2 continuity and an almost piecewise linear curvature radius plot. Furthermore, (almost) logarithmic spiral as well as circle precision can be provided. Since the interpolation method does not leave any degrees of freedom while generating excellently shaped curves, it simplifies the work of a designer.

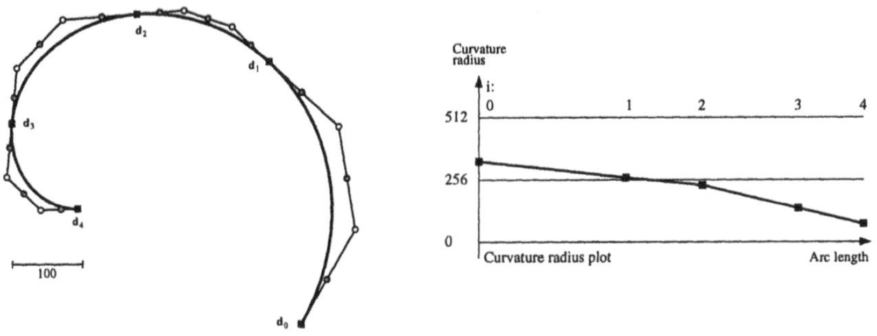

Fig. 3. An example illustrating the interpolation algorithm.

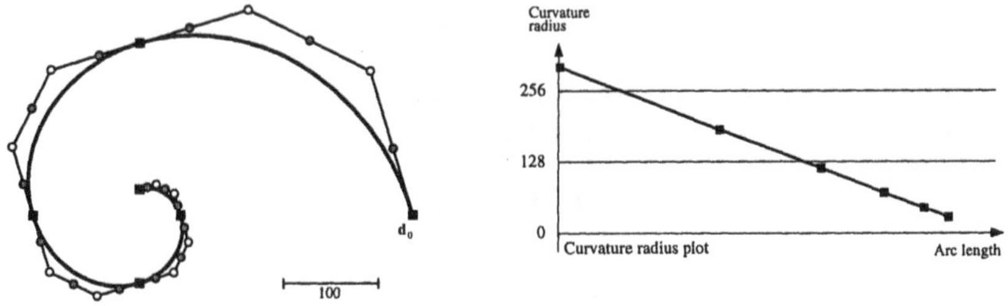

Fig. 4. The (almost) logarithmic spiral precision.

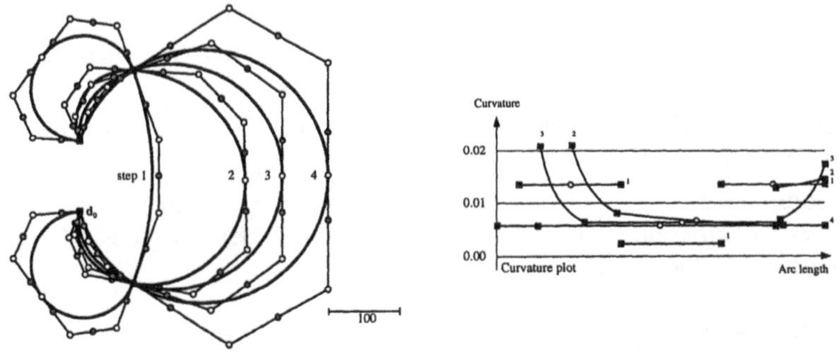

Fig. 5. The circle precision. Several steps of the interpolation algorithm are shown.

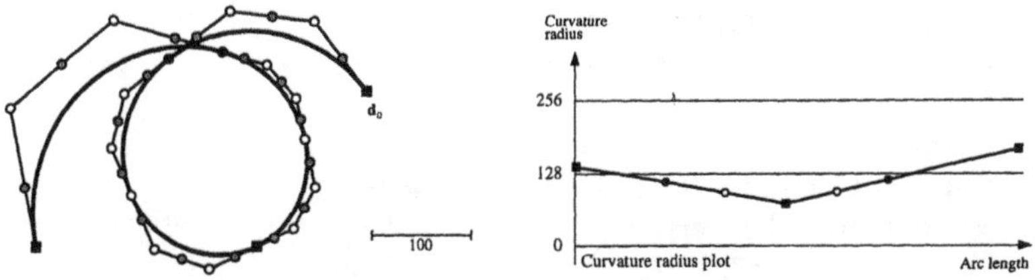

Fig. 6. An example for automatically inserted data points.

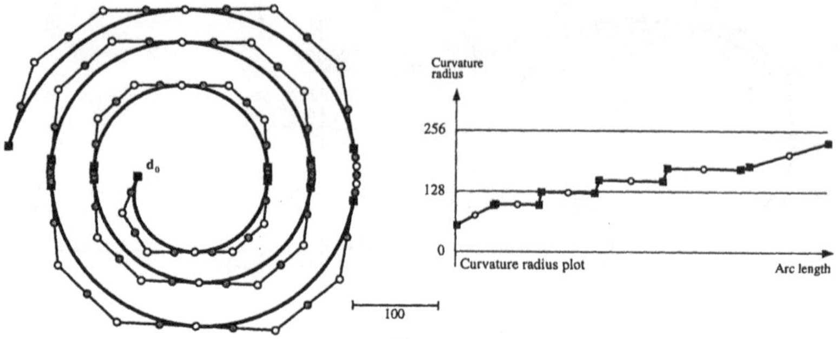

Fig. 7. The distances between consecutive data points differ extremely.

Further research will be done with respect to the number of local extrema in the curvature radius plot. Although this number is significantly reduced by our approach, it is not necessarily the smallest possible one.

Acknowledgments. This research was supported in part by NSF grant DM9123527 to Arizona State University.

References

1. Baumgarten, C., and G. Farin, Approximation of logarithmic spirals, Comput. Aided Geom. Design **14** (1997), 515-532.
2. Baumgarten, C., and G. Farin, Shape Optimization of Piecewise Rational Cubics, Technical Report no. TR-95-009, Arizona State University, 1995.
3. Farin, G., *Curves and Surfaces for Computer Aided Geometric Design*, Fourth Edition, Academic Press, Boston, 1996.
4. Farin, G., *NURB Curves and Surfaces: From Projective Geometry to Practical Use*, AK Peters, Ltd., Wellesley, 1994.
5. Goodman, T., Shape Preserving Interpolation by Planar Curves, in *Advanced Course on Fairshape*, J. Hoschek, and P. Kaklis (eds.), Teubner, Stuttgart, 1996, 29-42.

Christoph Baumgarten
TU Dresden, Informatik
IBDR – LSDB
01062 Dresden, GERMANY
`baumgart@is2201.inf.tu-dresden.de`

Gerald Farin
Dept. of Computer Science and Engineering
Arizona State University
Tempe, AZ 85287, USA
`farin@asu.edu`

Generalized Fréchet Distance Between Curves

Przemyslaw Bogacki and Stanley E. Weinstein

Abstract. The notion of distance between curves is central to geometric design. A class of functionals generalizing the Fréchet distance formula is discussed. Members of this class are shown to possess a significant property related to subdivision.

§1. Introduction

Frequently, in many areas of mathematics, science and engineering the need arises to measure the "distance" between two oriented curves. In particular, this notion is central to computer-aided geometric design (CAGD) or geometric modeling, since parametric curves are the basic building blocks from which most geometric objects are *constructed.* A suitable measure of distance is needed to determine how well one curve or surface, approximates another in applications such as degree reduction or knot removal ([1,2,4,7,10,11]).

This issue was the subject of the recent paper [3]. An intuitive model introduced therein, can be summarized as follows. Two one-way roads (oriented curves), P and Q, are traversed by two drivers, Pamala and Quincy, respectively, who are communicating with one another by cellular phones. Their phone company charges them proportionally to the instantaneous pointwise distance between them. Properly restricting this model (e.g., by requiring that the sum of the speeds be constant), we see that by minimizing the aggregate connection charges we obtain a measure of the closeness of the two oriented curves, P and Q. In addition to this informal model, [3] proposed some important properties of a distance formula, and introduced a family of distance functionals that satisfy these properties.

Section 2 of this paper contains some preliminary material. Desirable properties of a definition of distance are presented in Section 3, along with a brief summary of some of the most popular measures of distance. In Section 4 we investigate L_p analogues of the Fréchet distance, and prove that these functionals possess a property related to the subdivision of curves.

Mathematical Methods for Curves and Surfaces II
Morten Dæhlen, Tom Lyche, and Larry L. Schumaker (eds.), pp. 25–32.

ISBN 0-8265-1315-8.

§2. Preliminaries

We are concerned with defining a distance between two parametric curves. Let the set of curves we are considering be

$$\Omega := \left\{ P \,\middle|\, P : [0,1] \to E^n, \begin{array}{r} \text{either } P \text{ is piecewise smooth and regular} \\ \text{or } P \text{ is a single point} \end{array} \right\},$$

and define the set of parameterizations

$$H = \left\{ \varphi \,\middle|\, \begin{array}{l} \varphi : [0,1] \to [0,1]\,,\, \varphi(0) = 0,\, \varphi(1) = 1, \\ \varphi \text{ is continuous, and strictly increasing} \end{array} \right\}.$$

We shall say that P and Q in Ω represent the same oriented curve, denoted by $P \sim Q$, if and only if there exists a $\varphi \in H$ such that $P(t) = Q(\varphi(t))$ for all $t \in [0,1]$.

Throughout this paper, $\|\cdot\|$ denotes the Euclidean norm in E^n. For each $P \in \Omega$ and each $t \in [0,1]$ let $s_P(t) = \int_0^t \|P'(\tau)\| \, d\tau$ denote the arc length of P restricted to the interval $[0,t]$. Let $\ell_P = s_P(1)$ denote the length of the curve P. If $\ell_P > 0$, then because P is assumed to be regular, $s_P(t)$ is strictly increasing, and thus invertible. Therefore for $\ell_P > 0$, let $\hat{P} := P \circ s_P^{-1}$ denote the representation of P *in terms of arc length*. Thus, $\hat{P} : [0, \ell_P] \to E^n$. For $\ell_P = 0$, (when P is a single point) let $\hat{P} := P$ on the degenerate parameter interval $[0, \ell_P] = [0,0]$.

If $\ell_P \neq 1$ then $\hat{P} \notin \Omega$. This difficulty can be overcome through the use of the reduced length parameterization. Let $\bar{P} \in \Omega$ denote the representation of P *in terms of reduced length* as in [6, p. 142], where $\bar{P}(t) := \hat{P}\,(t\,\ell_P)$ for all $t \in [0,1]$.

§3. Some Popular Definitions of Distance and Their Properties

Several recent papers ([3,6,8,9,10,11,12,13]) have been concerned with various definitions of the distance between two curves. We shall present several of these definitions along with a summary of some of their properties.

We begin with a collection of desirable properties for a distance functional which were presented in [3]. These properties are required to hold for all P, Q and R in Ω. The first two properties are two of the properties of a metric.

M1) (Positive Definiteness): $d(P,Q) \geq 0$, and $d(P,Q) = 0 \Leftrightarrow P \sim Q$.

M2) (Symmetry): $d(P,Q) = d(Q,P)$.

The next four properties make the distance well suited to geometric modeling.

G1) (Parameter Independence): $P \sim R \Rightarrow d(P,Q) = d(R,Q)$.

G2) (Homogeneity): $d(\lambda P, \lambda Q) = |\lambda| \, d(P,Q)$ for all $\lambda \in \mathbb{R}$.

G3) (Euclidean Map Invariance): If $T : E^n \to E^n$ is a Euclidean map then

$$d(T \circ P, T \circ Q) = d(P,Q).$$

G4) (Euclidean Distance Preservation):

$$\min_{p\in P, q\in Q} \|p-q\| \leq d(P,Q) \leq \max_{p\in P, q\in Q} \|p-q\|,$$

where the notation $p \in P$ and $q \in Q$ signifies that p and q are points on the curves P and Q, respectively.

The final two properties relate to the subdivision of curves. Let the notation $P \sim P_1 \vec{\cup} P_2$ signify either that the curve P has been subdivided into two curves P_1 and P_2, such that $P_1(1) = P_2(0)$.

S1) (Weak Subdivision Property): If $P \sim P_1 \vec{\cup} P_2$ and $Q \sim Q_1 \vec{\cup} Q_2$, then

$$d(P,Q) \leq \max\{d(P_1,Q_1), d(P_2,Q_2)\}.$$

S2) (Strong Subdivision Property): Property S1 holds, and if $d(P_1,Q_1) \neq d(P_2,Q_2)$, then

$$d(P,Q) < \max\{d(P_1,Q_1), d(P_2,Q_2)\}$$

unless $\ell_{P_1} + \ell_{Q_1} = 0$ or $\ell_{P_2} + \ell_{Q_2} = 0$.

The positive definite property G1 is relative to the equivalence relation $\sim$. Thus, $d(P,Q) = 0$ implies that P and Q represent the same curve ($P \sim Q$). It does *not* imply that $P(t) = Q(t)$ for all $t \in [0,1]$.

Property G1 implies that the distance between two curves is independent of their representations. If a distance d satisfies properties M2 and G1, and if $P_1 \sim P_2$ and $Q_1 \sim Q_2$ then $d(P_1,Q_1) = d(P_2,Q_2)$. Property G2 implies that if two curves are scaled, then the distance between them changes proportionally. Property G3 implies that the distance between two curves is preserved under any Euclidean map. Euclidean maps preserve the Euclidean distances between points. For this reason, Euclidean maps are sometimes referred to as "rigid body motions". They include translations and rotations as special cases. Property G4 is the only one of these properties which relates to magnitude of the distance. If a distance d satisfies property G4 then for each P and Q in Ω there exist points $p^* \in P$ and $q^* \in Q$ such that $d(P,Q) = \|p^* - q^*\|$.

These subdivision properties are important, if a *distance* is to be used in an adaptive process of curve approximation. They also have impact on the distance between offsets. By mathematical induction, both of the subdivision properties have straightforward extensions to more than two subdivisions of each curve.

Next we examine some of the popular definitions of distance:

- Distances Based on the Given Parameterizations of the Curves

$$d_p(P,Q) = \left(\int_0^1 \|P(t) - Q(t)\|^p \, dt\right)^{1/p} \quad \text{for } 1 \leq p < \infty, \text{ and}$$

$$d_\infty(P,Q) = \sup_{0\leq t\leq 1} \|P(t) - Q(t)\|.$$

These definitions are poorly suited for use in geometric modeling, because they depend on the particular parametric representations of the curves.

- Distances Based on the Reduced Length Parameterizations

$$\bar{d}_p(P,Q) = d_p(\bar{P},\bar{Q}) = \left(\int_0^1 \left\|\bar{P}(t) - \bar{Q}(t)\right\|^p dt\right)^{1/p} \quad \text{for } 1 \le p < \infty, \text{ and}$$
$$\bar{d}_\infty(P,Q) = d_\infty(\bar{P},\bar{Q}) = \sup_{0\le t\le 1} \left\|\bar{P}(t) - \bar{Q}(t)\right\|.$$

Some of the shortcomings of these functionals are presented in [3]. In particular they do not satisfy either subdivision property and thus are not well suited to approximation procedures that are based on subdivision. In particular if any of these functionals is used to measure the distance between curves, given any subdivision $P \sim \vec{\cup}_{j=1,2,\ldots,m} P_j$, no a priori tolerance for the approximation of each segment P_j by Q_j, could guarantee that $Q \sim \vec{\cup}_{j=1,2,\ldots,m} Q_j$ would approximate P to within a prescribed tolerance. Furthermore, examples exist where Q is an offset of P by k units, yet any of these definitions produce a distance $> k$.

- The Hausdorff Distance [9]

$$d_H(P,Q) = \max\left\{\sup_{p\in P}\inf_{q\in Q}\|p-q\|, \sup_{q\in Q}\inf_{p\in P}\|p-q\|\right\}$$

ignores the orientation of the curves.

- The Fréchet Distance [6, pp. 142–146], [12]

$$d_F(P,Q) = \inf_{\varphi,\psi\in H}\sup_{0\le t\le 1}\left\|\bar{P}(\varphi(t)) - \bar{Q}(\psi(t))\right\|.$$

Of all the distance defined above only the Fréchet distance satisfies properties M1, M2, G1-G4 and S1. It does not satisfy property S2. The Fréchet distance is related to the L_∞ norm. In the next section we present L_p analogues of the Fréchet distance.

In the following table, the properties satisfied by the selected distance functionals are indicated by a + sign:

	M1	M2	G1	G2	G3	G4	S1	S2
Distances Based on the Given Parameterizations		+		+		+		
Distances Based on the Reduced Length Param.	+	+	+	+	+	+		
The Hausdorff Distance		+	+	+	+	+	+	
The Fréchet Distance	+	+	+	+	+	+	+	
The L_p Analogues of the Fréchet Distance	+	+	+	+	+	+	+	+

§4. L_p Analogues of the Fréchet Distance

The authors in [3] introduced the following family of distance functionals:

Definition. *Let P and $Q \in \Omega$. For each $1 \leq p < \infty$ define*

$$d_{F,p}(P,Q) := \inf_{\varphi,\psi \in H} \left(\int_0^1 \|\bar{P}(\varphi(t)) - \bar{Q}(\psi(t))\|^p \, \frac{\ell_P \varphi'(t) + \ell_Q \psi'(t)}{\ell_P + \ell_Q} \, dt \right)^{1/p}$$

if $\ell_P + \ell_Q > 0$. Otherwise, i.e. if both P and Q are points, define

$$d_{F,p}(P,Q) := \|P(0) - Q(0)\| .$$

In [3] it was proved that for any P and Q in Ω, $d_{F,p}(P,Q) \uparrow d_F(P,Q)$ as $p \to \infty$. It was also proved that $d_{F,p}$ satisfies all eight of the desirable properties of a distance listed in Section 3, including both subdivision properties. The following theorem presents another property of $d_{F,p}$:

Theorem 1. *Given any P and $Q \in \Omega$ such that $\ell_P + \ell_Q > 0$, and any $\theta \in (0,1)$*

$$d^p_{F,p}(P,Q) = \min \left(\theta d^p_{F,p}(P_1,Q_1) + (1-\theta) d^p_{F,p}(P_2,Q_2) \right),$$

where the minimum is taken over all subdivisions $P \sim P_1 \vec{\cup} P_2$ and $Q \sim Q_1 \vec{\cup} Q_2$ such that $\theta(\ell_P + \ell_Q) = \ell_{P_1} + \ell_{Q_1}$.

Proof:
PART I. We first prove that

$$d^p_{F,p}(P,Q) \leq \inf \left(\theta d^p_{F,p}(P_1,Q_1) + (1-\theta) d^p_{F,p}(P_2,Q_2) \right),$$

for any subdivisions $P_1 \vec{\cup} P_2 \sim P$ and $Q_1 \vec{\cup} Q_2 \sim Q$ such that $\theta(\ell_P + \ell_Q) = \ell_{P_1} + \ell_{Q_1}$. Let $\varphi_1, \varphi_2, \psi_1$ and ψ_2 be arbitrary functions in H, and let T be an arbitrary constant such that $0 < T < 1$. Then define

$$\varphi(t) = \begin{cases} \frac{\ell_{P_1} \varphi_1\left(\frac{t}{T}\right)}{\ell_P}, & t \leq T \\ \frac{\ell_{P_1} + \ell_{P_2} \varphi_2\left(\frac{t-T}{1-T}\right)}{\ell_P}, & t > T \end{cases} \qquad \psi(t) = \begin{cases} \frac{\ell_{Q_1} \psi_1\left(\frac{t}{T}\right)}{\ell_Q}, & t \leq T \\ \frac{\ell_{Q_1} + \ell_{Q_2} \psi_2\left(\frac{t-T}{1-T}\right)}{\ell_Q}, & t > T. \end{cases}$$

Note that if ℓ_P or ℓ_Q are 0, then φ or ψ, respectively, are arbitrary functions from H, instead of being defined as above. We then have,

$$\bar{P}(\varphi(t)) = \begin{cases} \bar{P_1}\left(\varphi_1\left(\frac{t}{T}\right)\right), & t \leq T \\ \bar{P_2}\left(\varphi_2\left(\frac{t-T}{1-T}\right)\right), & t > T, \end{cases} \qquad \bar{Q}(\psi(t)) = \begin{cases} \bar{Q_1}\left(\psi_1\left(\frac{t}{T}\right)\right), & t \leq T \\ \bar{Q_2}\left(\psi_2\left(\frac{t-T}{1-T}\right)\right), & t > T. \end{cases}$$

Consequently,

$$d^p_{F,p}(P,Q) \le \int_0^1 \|\bar{P}(\varphi(t)) - \bar{Q}(\psi(t))\|^p \frac{\ell_P\varphi'(t) + \ell_Q\psi'(t)}{\ell_P + \ell_Q}\,dt$$
$$= \int_0^T \left\|\bar{P}_1\left(\varphi_1\left(\tfrac{t}{T}\right)\right) - \bar{Q}_1\left(\psi_1\left(\tfrac{t}{T}\right)\right)\right\|^p \tfrac{\ell_{P_1}\varphi_1'\left(\frac{t}{T}\right)+\ell_{Q_1}\psi_1'\left(\frac{t}{T}\right)}{\ell_P+\ell_Q}\left(\tfrac{1}{T}\right)dt$$
$$+ \int_T^1 \left\|\bar{P}_2\left(\varphi_2\left(\tfrac{t-T}{1-T}\right)\right) - \bar{Q}_2\left(\psi_2\left(\tfrac{t-T}{1-T}\right)\right)\right\|^p \tfrac{\ell_{P_2}\varphi_2'\left(\frac{t-T}{1-T}\right)+\ell_{Q_2}\psi_2'\left(\frac{t-T}{1-T}\right)}{\ell_P+\ell_Q}\left(\tfrac{1}{1-T}\right)dt$$
$$= \theta\int_0^1 \left\|\bar{P}_1(\varphi_1(t) - \bar{Q}_1(\psi_1(t))\right\|^p \tfrac{\ell_{P_1}\varphi_1'(t)+\ell_{Q_1}\psi_1'(t)}{\ell_{P_1}+\ell_{Q_1}}\,dt$$
$$+(1-\theta)\int_0^1 \left\|\bar{P}_2(\varphi_2(t)) - \bar{Q}_2(\psi_2(t))\right\|^p \tfrac{\ell_{P_2}\varphi_2'(t)+\ell_{Q_2}\psi_2'(t)}{\ell_{P_2}+\ell_{Q_2}}\,dt.$$

Since $\varphi_1, \varphi_2, \psi_1$ and ψ_2 were arbitrary, we have

$$d^p_{F,p}(P,Q) \le \left(\theta d^p_{F,p}(P_1,Q_1) + (1-\theta)d^p_{F,p}(P_2,Q_2)\right).$$

Since this is valid for all subdivisions $P_1\vec{\cup}P_2 \sim P$ and $Q_1\vec{\cup}Q_2 \sim Q$ such that $\theta(\ell_P + \ell_Q) = \ell_{P_1} + \ell_{Q_1}$, part I of the proof is complete.

PART II. We next show that

$$d^p_{F,p}(P,Q) \ge \inf\left(\theta d^p_{F,p}(P_1,Q_1) + (1-\theta)d^p_{F,p}(P_2,Q_2)\right).$$

For any $\varepsilon > 0$, there exist functions $\varphi_\varepsilon, \psi_\varepsilon \in H$ such that

$$d^p_{F,p}(P,Q) \ge \int_0^1 \|\bar{P}(\varphi_\varepsilon(t)) - \bar{Q}(\psi_\varepsilon(t))\|^p \frac{\ell_P\varphi_\varepsilon'(t) + \ell_Q\psi_\varepsilon'(t)}{\ell_P + \ell_Q}dt - \varepsilon.$$

Consider the function $g_\varepsilon(t) := \frac{s_P(\varphi_\varepsilon(t))+s_Q(\psi_\varepsilon(t))}{\ell_P+\ell_Q}$. Since g_ε is in H, it is also invertible. Let $T = g_\varepsilon^{-1}(\theta)$. We determine the subdivisions of P into $P_{\varepsilon,1}, P_{\varepsilon,2}$ and of Q into $Q_{\varepsilon,1}, Q_{\varepsilon,2}$ by setting $\ell_{P_{\varepsilon,1}} = s_P(\varphi_\varepsilon(T))$ and $\ell_{Q_{\varepsilon,1}} = s_Q(\psi_\varepsilon(T))$. Let us define

$$\varphi_{\varepsilon,1}(t) = \tfrac{\ell_P\varphi_\varepsilon(Tt)}{\ell_{P_{\varepsilon,1}}}, \quad \varphi_{\varepsilon,2}(t) = \tfrac{\ell_P\varphi_\varepsilon(T+t(1-T))-\ell_{P_{\varepsilon,1}}}{\ell_{P_{\varepsilon,2}}},$$
$$\psi_{\varepsilon,1}(t) = \tfrac{\ell_Q\psi_\varepsilon(Tt)}{\ell_{Q_{\varepsilon,1}}}, \quad \psi_{\varepsilon,2}(t) = \tfrac{\ell_Q\psi_\varepsilon(T+t(1-T))-\ell_{Q_{\varepsilon,1}}}{\ell_{Q_{\varepsilon,2}}},$$

with the exceptions that for each of these equations whenever the denominator of the right hand side becomes zero, the corresponding left hand side function is chosen arbitrarily from H. By a calculation similar to Part I we obtain

$$d^p_{F,p}(P,Q) \ge \theta d^p_{F,p}(P_{\varepsilon,1},Q_{\varepsilon,1}) + (1-\theta)d^p_{F,p}(P_{\varepsilon,2},Q_{\varepsilon,2}) - \varepsilon.$$

Thus, since this holds for arbitrary $\varepsilon > 0$, we have

$$d^p_{F,p}(P,Q) \ge \inf\left(\theta d^p_{F,p}(P_1,Q_1) + (1-\theta)d^p_{F,p}(P_2,Q_2)\right).$$

PART III. Parts I and II imply

$$d^p_{F,p}(P,Q) = \inf\left(\theta d^p_{F,p}(P_1,Q_1) + (1-\theta) d^p_{F,p}(P_2,Q_2)\right).$$

Finally, we show that the expression under the infimum attains its minimum value. For any $(t^*, s^*) \in [0,1] \times [0,1]$ define the real valued continuous functions

$$\delta_p(t^*, s^*) := d_{F,p}(P_1^*, Q_1^*), \text{ and } \gamma_p(t^*, s^*) := d_{F,p}(P_2^*, Q_2^*),$$

where P_1^* denotes the restriction of P to the interval $[0, t^*]$, P_2^* denotes the restriction of P to the interval $[t^*, 1]$, Q_1^* denotes the restriction of Q to the interval $[0, s^*]$, and Q_2^* denotes the restriction of Q to the interval $[s^*, 1]$. The theorem follows from the fact that the infimums taken above are equivalent to

$$\inf\left\{\theta \left\{\delta_p(t^*, s^*)\right\}^p + (1-\theta)\left\{\gamma_p(t^*, s^*)\right\}^p\right\},$$

where these infimums are taken over the compact set

$$\left\{(t^*, s^*) \in [0,1] \times [0,1] : \frac{\ell_{P_1^*} + \ell_{Q_1^*}}{\ell_P + \ell_Q} = \theta\right\}. \quad \square$$

By induction, Theorem 1 has the following generalization:

Theorem 2. *Given any P and Q in Ω such that $\ell_P + \ell_Q > 0$, and any positive numbers $\theta_1, ..., \theta_m$ such that $\sum_{i=1}^{m} \theta_i = 1$*

$$\{d_{F,p}(P,Q)\}^p = \min\left(\sum_{i=1}^{m} \theta_i \left\{d_{F,p}(P_i, Q_i)\right\}^p\right),$$

where the minimum is taken over all subdivisions $P \sim \vec{\cup}_{j=1,2,...,m} P_j$ and $Q \sim \vec{\cup}_{j=1,2,...,m} Q_j$ such that $\frac{\ell_{P_i} + \ell_{Q_i}}{\ell_P + \ell_Q} = \theta_i$.

§5. Conclusions

Theorems 1 and 2 may offer computational advantages by establishing relationship of the distance between the original (possibly quite complex) curves and the distances between corresponding subdivisions of these curves. The authors are currently exploring algorithms to use the generalizations of the Fréchet distance for both determining the distance between two curves and for approximation of one curve by another (subject to constraints). While these distance formulas tend to be computationally more expensive that some of the other mentioned in Section 3, Theorems 1 and 2 may lead to reductions of this cost.

References

1. Bogacki, P., and S. E. Weinstein, Geometric degree reduction, The Fourth SIAM Conference on Geometric Design, Nashville, TN.
2. Bogacki, P., S. E. Weinstein, and Y. Xu, Degree reduction of Bézier curves by uniform approximation with endpoint interpolation, Computer-Aided Design **27** (1995), 651–661.
3. Bogacki, P., S. E. Weinstein, and Y. Xu, Distances between oriented curves in geometric modeling, Advances in Comp. Math. **7** (1997), 593–621.
4. Eck, M., Least squares degree reduction of Bézier curves, Comput. Aided Geom. Design **27** (1995), 845–851.
5. Emery, J. D., The definition and computation of a metric on plane curves, Computer-Aided Design **18** (1986), 25–28.
6. Ewing, G. M., *Calculus of Variations with Applications*, Dover, New York, 1985.
7. Farin, G., *Curves and Surfaces for Computer Aided Geometric Design. A Practical Guide*, 3rd ed., Academic Press, New York, 1994.
8. Fritsch, F.N., and G. M. Nielson, On the problem of determining the distance between parametric curves, in *Curve and Surface Design*, H. Hagen (ed.), SIAM, Philadelphia, 1992, 123–141.
9. Hausdorff, F., *Set Theory*, Chelsea, New York, 1957.
10. Hoschek, J., Approximate conversion of spline curves, Comput. Aided Geom. Design **4** (1987), 59–66.
11. Hoschek, J., and F. J. Schneider, Approximate spline conversion for integral and rational Bézier and B-spline surfaces, in *Geometry Processing for Design and Manufacturing*, R. E. Barnhill (ed.), SIAM, Philadelphia, 1992, 45–86.
12. Lyche, T. and K. Mørken, A metric for parametric approximation, in *Curves and Surfaces in Geometric Design*, P. J. Laurent, A. Le Méhauté, and L. L. Schumaker (eds.), A.K. Peters, Wellesley, MA, 1994, 311–318.
13. Meek, D.S., and D. J. Walton, Alignment of planar curves, Image and Vision Computing **12** (1994), 305–311.

Przemyslaw Bogacki and Stanley E. Weinstein
Department of Mathematics and Statistics
Old Dominion University
Norfolk, Virginia 23529, USA
`bogacki@math.odu.edu`
`wein@math.odu.edu`

Elastic Splines with Tension Control

Guido Brunnett and Jörg Wendt

Abstract. A precise derivation of the characteristic equations of elastic splines in tension is given. A new fast algorithm for interpolation is presented that allows interactive design with these curves. An example is provided that illustrates the influence of the tension parameters on the curve shape.

§1. Introduction

An elastic curve is the idealized mathematical model of an elastic material that extends mainly in one dimension as, e.g., a thin elastic beam. Based on D. Bernoulli's work, plane elastic curves are defined as the extremals (critical points) of the functional

$$\int_0^L \left(\kappa^2(s) + \sigma\right) ds, \tag{1}$$

where $\kappa = \kappa(s)$ denotes the curvature of the curve as a function of the arc length parameters s and where σ is constant. Applying the calculus of variations to this problem, differential equations describing elastic curves have been derived by Euler and others (see [1,6]).

In the context of CAGD this variational problem has gained widespread attention as a smoothing strategy in variational design and as the origin of the 'true' or nonlinear spline, see e.g. [4,7,8].

In this paper we extend the notion of elastica to *elastic splines with tension control;* i.e. a curve $\boldsymbol{x}$ of variable length L passing through a sequence of points $\boldsymbol{x}(l_i) = P_i, \quad i = 1, \dots, n$, in the plane and minimizing the functional (1), where σ is now a piecewise constant function on the partition $\Delta_n : 0 = l_1 < l_2 < \cdots < l_n = L$.

Section 2 contains a precise derivation of the necessary conditions for this variational problem including the Euler-Lagrange equation, Erdmann's corner conditions and the boundary condition for variable length.

Mathematical Methods for Curves and Surfaces II
Morten Dæhlen, Tom Lyche, and Larry L. Schumaker (eds.), pp. 33–40.

ISBN 0-8265-1315-8.

In Section 3 an algorithm is presented that computes the interpolating elastic spline for a given sequence of points $P_1, \dots, P_n$ and tension parameters $\sigma_1, \dots, \sigma_{n-1}$ for the spline segments. This algorithm is based on the univariate method for elastic curves developed in [4].

This new curve scheme has been integrated into a graphical modelling system that allows interactive design of elastic splines with tension control. The last section provides several an example of the influence of the tension parameters on the curve shape.

§2. Elastic Splines

We consider the following variational problem: Given n points $P_1, P_2, \dots, P_n$ and two unit vectors U, V in the Euclidean plane E^2. Let M denote the set of all arc length parametrized C^1 curves $\boldsymbol{y} : [0, L] \to E^2$ of variable length L satisfying the conditions

i) there is a partition $\Delta_n : 0 = l_1 < l_2 < \dots < l_{n-1} < l_n = L$ of $[0, L]$ with $\boldsymbol{y}_i = \boldsymbol{y}|_{[l_i, l_{i+1}]} \in C^\infty[l_i, l_{i+1}] \quad (i = 1, \dots, n-1)$,

ii) $\boldsymbol{y}(l_i) = P_i, \qquad i = 1, \dots, n,$

iii) $\boldsymbol{y}'(0) = U, \quad \boldsymbol{y}'(L) = V.$

We seek an element of M that provides a local minimum for the functional

$$\sum_{i=1}^{n-1} \int_{l_i}^{l_{i+1}} \left(\kappa_i^2(s) + \sigma_i\right) \, ds,$$

where for $i = 1, \dots, n$, κ_i is the curvature function of $\boldsymbol{y}_i$ and $\sigma_i \in \mathbb{R}^+$ are tension parameters.

In order to apply the calculus of variations, we represent any element $\boldsymbol{y}$ of M by a segmented function $\Psi \in C^0[0, L]$ for the turning angle of the tangent vector of $\boldsymbol{y}$. The segments $\Psi_i := \Psi|_{[l_i, l_{i+1}]}$ of Ψ are related to the curvature of $\boldsymbol{y}$ via $\Psi_i'(s) = \kappa_i(s)$.

The variational problem considered is to minimize the functional

$$\sum_{i=1}^{n-1} \int_{l_i}^{l_{i+1}} \left([\Psi_i'(s)]^2 + \sigma_i\right) \, ds$$

subject to the $2(n-1)$ isoperimetric constraints

$$\int_{l_i}^{l_{i+1}} \cos \Psi(s) \, ds = P_{i+1}^1 - P_i^1, \qquad \int_{l_i}^{l_{i+1}} \sin \Psi(s) \, ds = P_{i+1}^2 - P_i^2,$$

$i = 1, \dots, n-1$.

Let Θ be a solution to the stated problem. Then we embed Θ in a one-parametric family $\Psi(s, \varepsilon)$ of functions based on the partitions

$$\Delta_n(\varepsilon) : 0 = l_1(\varepsilon) < l_2(\varepsilon) < \cdots < l_{n-1}(\varepsilon) < l_n(\varepsilon) = L(\varepsilon)$$

by requiring that $\Psi(s,0) = \Theta(s)$. Furthermore, we require that for all $\varepsilon \in [-\gamma,\gamma]$ $(\gamma \in \mathbb{R}^+)$

$$\Psi(0,\varepsilon) = \Theta(0), \qquad \Psi(L(\varepsilon),\varepsilon) = \Theta(L(0)) \quad \text{and}$$

$$\Psi(s,.) \in C^0([0,L(.)]),\ \Psi_i(s,.) \in C^\infty([l_i(.),l_{i+1}(.)]),\ \Psi_i(.,\varepsilon) \in C^1([-\gamma,\gamma]),$$

$i = 1,\ldots,n-1$.

In the following we denote the knots of the partition $\Delta_n(0)$ of Θ with capital letters without argument: $L_i := l_i(0)$ and $L := L(0)$.

Considering those elements of $\Psi(s,\varepsilon)$ that differ from Θ only on the i-th segment, one obtains that the i-th segment Θ_i of Θ has to be an extremal of the functional

$$I_i = \int_{l_i}^{l_{i+1}} \left([\Psi_i'(s)]^2 + \sigma_i + \lambda_i \cos\Psi_i + \mu_i \sin\Psi_i\right) ds,$$

where the Lagrange multipliers λ_i, μ_i are constants because of the isoperimetric nature of the problem. Thus, Θ_i has to satisfy the Euler-Lagrange equation

$$-\lambda_i \sin\Theta_i + \mu\cos\Theta_i - 2\Theta_i'' = 0. \tag{2}$$

Introducing the constants

$$\lambda_i = 2a_i\cos\Phi_i, \qquad \mu_i = 2a_i \sin\Phi_i, \qquad i = 1,\ldots,n-1, \tag{3}$$

we rewrite the Euler equation (2) in the form

$$\Theta_i''(s) = \kappa_i'(s) = -a_i\sin(\Theta_i(s) - \Phi_i).$$

Multiplying this equation by $2\Theta_i'$ and integrating yields

$$\kappa_i^2(s) = 2a_i\cos(\Theta_i - \Phi_i) + C_i, \tag{4}$$

where C_i denotes an integration constant. It has been shown in [2] that (4) is equivalent to

$$\kappa_i'' + \frac{1}{2}\kappa_i^3 - \frac{1}{2}C_i\kappa_i = 0.$$

Now, we express the considered functional as a function of ε:

$$I(\varepsilon) = \sum_{i=1}^{n-1} \int_{l_i(\varepsilon)}^{l_{i+1}(\varepsilon)} F_i(s,\Psi_i(s,\varepsilon),\Psi_i'(s,\varepsilon))ds$$

with

$$F_i(s,\Psi,\Psi') := [\Psi_i'(s)]^2 + \sigma_i + \lambda_i\cos\Psi_i + \mu_i\sin\Psi_i.$$

Using the notation

$$P_i^- = (L_i, \Theta_i(L_i), \Theta_i'(L_i)), \qquad P_i^+ = (L_{i+1}, \Theta_i(L_{i+1}), \Theta_i'(L_{i+1}))$$

and the validity of the Euler-Lagrange equation for each segment, one obtains

$$I'(0) = \sum_{i=1}^{n-1} \left(\frac{\partial F_i}{\partial \Psi_i'}(P_i^+) \cdot \frac{\partial \Psi_i}{\partial \varepsilon}(L_{i+1}, 0) - \frac{\partial F_i}{\partial \Psi_i'}(P_i^-) \cdot \frac{\partial \Psi_i}{\partial \varepsilon}(L_i, 0) \right. \tag{5}$$
$$\left. + F_i(P_i^+) l_{i+1}'(0) - F_i(P_i^-) l_i'(0) \right).$$

The functions

$$\Gamma_i^-(\varepsilon) := \Psi_i(l_i(\varepsilon), \varepsilon), \qquad \Gamma_i^+(\varepsilon) := \Psi_i(l_{i+1}(\varepsilon), \varepsilon) \qquad (i = 1, \ldots, n-1)$$

express the angles Ψ_i at the parameter values $l_i(\varepsilon)$ and $l_{i+1}(\varepsilon)$ as functions of ε. Since Ψ is continuous one obtains $\Gamma_{i+1}^-(\varepsilon) = \Gamma_i^+(\varepsilon)$ for all ε. Differentiating this identity yields

$$\Gamma_{i+1}^{-'}(\varepsilon) = \Gamma_i^{+'}(\varepsilon). \tag{6}$$

Substituting the formulas

$$\Gamma_i^{-'}(0) = \Theta_i'(L_i) \cdot l_i'(0) + \frac{\partial \Psi_i}{\partial \varepsilon}(L_i, 0),$$
$$\Gamma_i^{+'}(0) = \Theta_i'(L_{i+1}) \cdot l_{i+1}'(0) + \frac{\partial \Psi_i}{\partial \varepsilon}(L_{i+1}, 0)$$

into (5), the necessary condition $I'(0) = 0$ together with (6) and $l_1'(0) = \Gamma_1^{-'}(0) = \Gamma_{n-1}^{+}(0) = 0$, yields the following sets of equations:

$$\frac{\partial F_i}{\partial \Psi_i'}(P_i^+) = \frac{\partial F_{i+1}}{\partial \Psi_{i+1}'}(P_{i+1}^-) \qquad i = 1, \ldots, n-2, \tag{7}$$

$$F_i(P_i^+) - \frac{\partial F_i}{\partial \Psi_i'}(P_i^+) \cdot \Theta_i'(L_{i+1}) \tag{8}$$
$$= F_{i+1}(P_{i+1}^-) - \frac{\partial F_{i+1}}{\partial \Psi_{i+1}'}(P_{i+1}^-)\Theta_{i+1}'(L_{i+1}) \quad i = 1, \ldots, n-2,$$

$$F_{n-1}(P_{n-1}^+) - \frac{\partial F_{n-1}}{\partial \Psi_{n-1}'}(P_{n-1}^+)\Theta_{n-1}'(L_n) = 0. \tag{9}$$

From the definition of F_i one obtains that (7) is equivalent to

$$\kappa_i(L_{i+1}) = \kappa_{i+1}(L_{i+1}) \qquad (i = 1, \ldots, n-2), \tag{10}$$

(8) is equivalent to

$$\sigma_i + \lambda_i \cos \Theta_i(l_{i+1}) + \mu_i \sin \Theta_i(l_{i+1}) = \qquad (11)$$
$$\sigma_{i+1} + \lambda_{i+1} \cos \Theta_{i+1}(l_{i+1}) + \mu_{i+1} \sin \Theta_{i+1}(l_{i+1}) \quad (i = 1, \ldots, n-2)$$

and (9) is equivalent to

$$\kappa_{n-1}^2(l_n) = \lambda_{n-1} \cos \Theta_{n-1}(l_n) + \mu_{n-1} \sin \Theta_{n-1}(l_n) + \sigma_{n-1}. \qquad (12)$$

Using (3) the equations (11) and (12) take the form

$$\sigma_i + 2a_i \cos(\Theta_i(l_{i+1}) - \Phi_i) \qquad (13)$$
$$= \sigma_{i+1} + 2a_{i+1} \cos(\Theta_{i+1}(l_{i+1} - \Phi_{i+1})$$

$(i = 1, \ldots, n-2)$ and

$$\kappa_{n-1}^2(l_n) = \sigma_{n-1} + 2a_{n-1} \cos(\Theta_{n-1}(l_n) - \Phi_{n-1}). \qquad (14)$$

Evaluating the relation (4) equation (13) yields together with (10)

$$\sigma_i - C_i = \sigma_{i+1} - C_{i+1} \qquad i = 1, \ldots, n-2. \qquad (15)$$

In the same way (14) yields $C_{n-1} = \sigma_{n-1}$. Thus, (15) gives

$$C_i = \sigma_i, \qquad i = 1, \ldots, n-2.$$

We summarize the results of this section in the following theorem.

Theorem 1. *Let $\boldsymbol{x}$ denote a solution of the variational problem described. Then $\boldsymbol{x}$ has a continuous curvature function $\kappa : [0, L] \to \mathbb{R}$. The segments $\kappa_i := \kappa|_{[l_i, l_{i+1}]}$ of κ satisfy the differential equations*

$$\kappa_i'' + \frac{1}{2}\kappa_i^3 - \frac{1}{2}\sigma_i \kappa_i = 0, \qquad (16)$$

for $i = 1, \ldots, n-1$.

Any such $\boldsymbol{x}$ according to Theorem 1 is called an elastic spline in tension.

§3. Interpolation with Elastic Splines

In this section we present an algorithm for interpolation of a sequence of n points $P_1, \ldots, P_n$ in the plane with an elastic spline $\boldsymbol{x}$ with tension parameters σ_i $(i = 1, \ldots, n-1)$ and variable length L. Besides the interpolation points we prescribe the tangent directions U, V in the start point rsp. endpoint of the spline.

It follows from the results of Section 2 that the univariate method presented in [4] for interpolating Hermite-type data with a single elastic curve

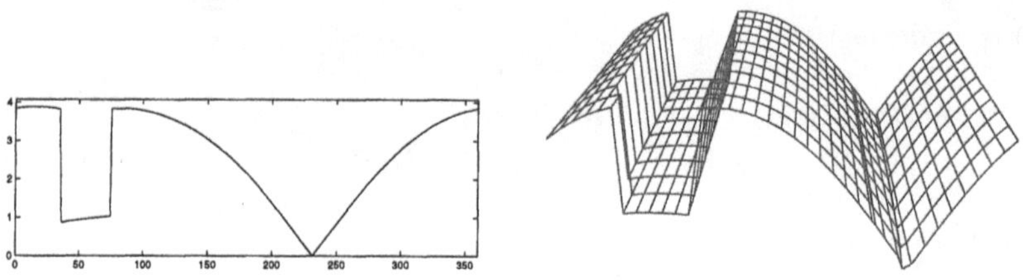

Fig. 1. left: Graph of the function F_j right: F_j evolving over σ-space.

is applicable to compute the segments of the elastic spline provided that the tangents directions $\boldsymbol{U}_i$ of the spline in the knots are known. The univariate method has several advantages compared to alternative strategies based on the solution of a nonlinear system of equations (see e.g. [5]). These include speed and the availability of initial values to compute particular elements of the set of solutions. Therefore it is intended to build the interpolation algorithm based on the univariate method.

The algorithm begins with the estimation of the tangent directions $\boldsymbol{U}_i$ at the interior points P_i $(i = 2, \ldots, n-2)$ which is based on the simple formula

$$\boldsymbol{U}_i(P_i) = P_{i+1} - P_{i-1}. \tag{17}$$

We will explain later on why (17) is a good choice. Note, that the length of $\boldsymbol{U}_i$ has no meaning in the algorithm.

In the second step we use the univariate method to compute initial spline segments $\boldsymbol{x}_i$ interpolating the data

$$\boldsymbol{x}_i(l_i) = P_i \qquad \boldsymbol{x}_i(l_{i+1}) = P_{i+1} \qquad \boldsymbol{x}_i'(l_i) = \boldsymbol{U}_i \qquad \boldsymbol{x}_i'(l_{i+1}) = \boldsymbol{U}_{i+1}, \tag{18}$$

where the tension parameter of $\boldsymbol{x}_i$ is σ_i. As pointed out in [4] for each i several solutions exist. Our criterion to single a particular solution is to use the curve with the smallest value of the energy functional (1).

In the third step we perform a nonlinear minimization that is based on the fact that an elastic spline has to be curvature continuous. If κ_i^0, κ_i^1 denote the curvature value of the i-th spline segment at the startpoint rsp. endpoint the objective function has the form

$$F(\Theta_2, \Theta_3, \ldots, \Theta_{n-1}) = \sum_{i=1}^{n-1} |\kappa_i^1 - \kappa_{i+1}^0|, \tag{19}$$

where Θ_i are the unknown angles of the tangents at the knots.

The left part of Figure 1 shows the graph of the function

$$F_j(\Theta_j) = F(\Theta_{2,\ldots}^0, \Theta_{j-1}^0, \Theta_j, \Theta_{j+1,\ldots}^0, \Theta_{n-1}^0),$$

where all tangents besides at the j-th knot are kept fixed while Θ_j varies between 0° and 360°. This figure illustrates the following characteristic features of F_j: F_j is discontinuous and consists of two pieces of smooth shape, the major piece has exactly one zero $\overline{\Theta}$, and F_j is nearly linear in its neighborhood. This zero corresponds to the wanted solution. Due to the nice shape, the zero is efficiently computed with a Newton method provided that one value of the major piece is available as a start point. The initial value for Θ based on (17) is a good choice because it lies not only on the wanted piece but in fact quite close to the root. The right part of figure 1 shows how F_j changes over σ-space. The ridge on the right side of the surface corresponds to the zeros of F_j while the straight line that appears as a net irregularity to left of the ridge corresponds to the initial values for Θ based on (17).

The minor piece of F_j appears in a vicinity of $\overline{\Theta} - 180°$, i.e. for tangent directions far off the direction indicated by the point sequence. The discontinuity between the two pieces is due the fact that they involve qualitatively different solutions of the interpolation problem.

The nice behaviour of F_j suggests to perform the minimization of (19) by changing only one angle $\Theta_i \quad (i = 2, \ldots, n-1)$ at a time. In each minimization step the following operations are performed:

- Compute the curvature discontinuities $\triangle_i = \kappa_i^1 - \kappa_{i+1}^0$ for $i = 1, \ldots, n-2$. Let $\triangle_j$ denote the largest one.
- Compute the zero $\overline{\Theta}_j$ of the linear approximation of F_j at the current value of Θ_j.
- Fill in new spline segments s_{j-1} and s_j into the sequence $s_1, \ldots, s_{n-1}$ of elastic curves according to the tangent direction $U_j = \left(\cos \overline{\Theta}_j, \sin \overline{\Theta}_j\right)$.

To speed up the algorithm, we perform most of the iteration steps based on a table of precomputed solutions. This table provides for a data set P_i, Θ_i $(i = 1, \ldots, n)$ and σ_i $(i = 1, \ldots, n-1)$ the corresponding values κ_i $(i = 1, \ldots, n)$, thus, it allows to evaluate F directly.

In the first pass of the minimization we evaluate F only based on table entries. Once the sum of curvature discontinuities is smaller than a prescribed value, the second pass of the minimization is started. It computes the spline segments with the univariate method. Start points for the univariate root finder are constructed from the parameter values of the resulting spline of the first pass.

The final step of the minimization is performed until a zero of F has been obtained or another stopping criterion is fullfilled. Due to the local setup of the minimization procedure, each iteration step is performed fast but a high number of iterations (ca. $5 \times$ numbers of knots) have to be executed before convergence occurs.

Note, that elastic splines only exist for a choice of the tension parameters within certain bounds that depend on the data. If tension parameters are provided by the user that do not allow interpolation with an elastic spline, the system suggests tension parameters for which interpolation is possible.

On an SGI O2 with R10000 processor, our algorithm computes an elastic spline interpolating 100 data points in 1 second.

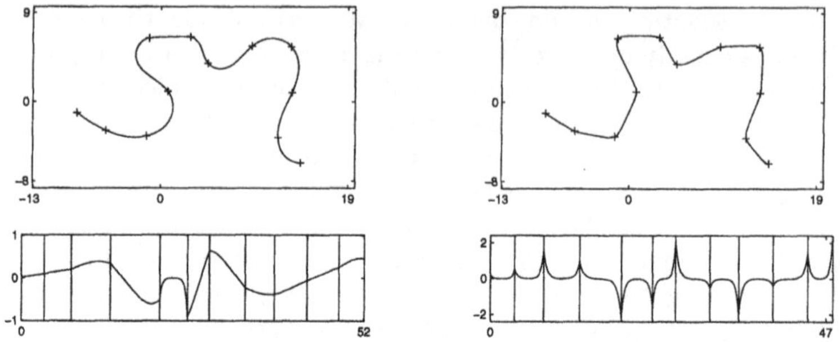

Fig. 2. Elastic splines in tension and their curvature plot.

The following example illustrates the influence of the tension parameters on the shape of an elastic spline. The left curve in figure 2 is an elastic spline with vanishing tension parameters for all segments besides of the fifth segment. The right curve shows the result for high tension values on each segment of the spline.

References

1. Birkhoff, G., and C. de Boor, Piecewise polynomial interpolation and approximation, *in Approximation of Functions*, H. L. Garabedian (ed.) Elsevier, Amsterdam, 1965, 164–190.
2. Brunnett, G., The curvature of plane elastic curves, Technical Report NPS-MA-93-013, Naval Postgraduate School.
3. Brunnett, G., and J. Kiefer, Interpolation with minimal energy splines, Computer-Aided Design **26** (1994), 137–144.
4. Brunnett, G., and J. Wendt, A univariate method for plane elastic curves, Comput. Aided Geom. Design **14** (1997), 273–292.
5. Edwards, J. A., Exact equations of the nonlinear spline, ACM Trans. Math. Software **18** (1992), 174–192.
6. Lee, E. H., and G. E. Forsythe, Variational study of nonlinear spline curves, SIAM Review, **15** (1975), 120–133.
7. Mehlum, E., Nonlinear splines, in *Computer Aided Geometric Design*, R. E. Barnhill and R. F. Riesenfeld (eds.), Academic Press, New York, 1974, 173–207.
8. Mehlum, E., Appell and the Apple (nonlinear splines in space), in *Mathematical Methods for Curves and Surfaces*, M. Dæhlen, T. Lyche, and L. L. Schumaker (eds.), Vanderbilt University Press, Nashville, 1995, 365–384.

Guido Brunnett
Dept. of Computer Science, University of Kaiserslautern
Postfach 30 49, 67653 Kaiserslautern, GERMANY
`brunnett@informatik.uni-kl.de`

Sign Consistency and Shape Properties

J. M. Carnicer

Abstract. The most common shape preserving properties of curves in CAGD are analyzed using a technique based on the sign consistency properties of associated collocation matrices. The interrelation among some simple shape properties is studied. Some results on monotonicity and convexity preservation are reviewed.

§1. Introduction, Preliminary Definitions and Notations

As usual in CAGD we shall deal with a set of curves represented in the form

$$\mathbf{z}(t) = \sum_{i=0}^{n} u_i(t)\mathbf{p}_i, \tag{1}$$

where $\mathbf{p}_i \in \mathbb{R}^s$, $i = 0, \ldots, n$. The polygon $\mathbf{p}_0 \cdots \mathbf{p}_n$ will be called the control polygon and the points $\mathbf{p}_i$, $i = 0, \ldots, n$, the control points. It is a natural requirement that the shape of the curve $\mathbf{z}$ is related to the shape of the corresponding control polygon. The shape preserving properties of each representation (1) depends on the system of functions $\mathbf{u} = (u_0, \ldots, u_n)$, $u_i: [a, b] \to \mathbb{R}$, $i = 0, \ldots, n$.

For instance, a simple property which is demanded for curve control is the convex hull property, that is the points of the curve always lie inside the convex hull of the control polygon. It is well known that a representation (1) has the convex hull property if and only if all functions u_i are nonnegative and add up to one. These kinds of systems are usually called blending systems.

Let us now introduce some matrix notation. For notational convenience we shall assume that all vectors, which we shall write in boldface type, are row vectors unless stated the contrary. Given a sequence of vectors $\mathbf{p}_0, \ldots, \mathbf{p}_n$, we shall denote by

$$M\begin{pmatrix} \mathbf{p} \\ 0, \ldots, n \end{pmatrix}$$

Mathematical Methods for Curves and Surfaces II
Morten Dæhlen, Tom Lyche, and Larry L. Schumaker (eds.), pp. 41–48.

ISBN 0-8265-1315-8.

the matrix whose rows are precisely $\mathbf{p}_0, \dots, \mathbf{p}_n$. This allows us to write (1) using matrix notation as

$$\mathbf{z}(t) = \mathbf{u}(t)P, \tag{2}$$

where $P := M \begin{pmatrix} \mathbf{p} \\ 0,\dots,n \end{pmatrix}$ is the matrix of control points.

If a geometric property only involves one point, then (2) is sufficient to describe the situation. However, we shall deal with shape properties which depend on the relative position of several points of the curve (or vertices of the polygon). For this purpose we introduce the collocation matrices

$$M \begin{pmatrix} \mathbf{u} \\ t_1, \dots, t_k \end{pmatrix} = M \begin{pmatrix} u_0, \dots, u_n \\ t_1, \dots, t_k \end{pmatrix} = \begin{pmatrix} u_0(t_1) & \cdots & u_n(t_1) \\ \vdots & & \vdots \\ u_0(t_k) & \cdots & u_n(t_k) \end{pmatrix}, \quad t_1 < \cdots < t_k. \tag{3}$$

Any s-dimensional curve $\mathbf{z}$ can be considered as a system of s functions and has corresponding collocation matrices. So, we can derive from (2) the matrix equation

$$M \begin{pmatrix} \mathbf{z} \\ t_1, \dots, t_k \end{pmatrix} = M \begin{pmatrix} \mathbf{u} \\ t_1, \dots, t_k \end{pmatrix} M \begin{pmatrix} \mathbf{p} \\ 0, \dots, n \end{pmatrix} \tag{4}$$

which relates a collocation matrix of the system of functions formed by the components of the curve $\mathbf{z}$, the collocation matrix of the system $\mathbf{u}$ at the same points and the matrix of control points.

Usually the geometric properties of a curve can be expressed in terms of determinants of collocation matrices. The Cauchy-Binet formula allows us to write the determinant of the product of two matrices in terms of the determinants of the factor matrices. Applying the Cauchy-Binet formula to (4) when $k = s$ yields

$$\det M \begin{pmatrix} \mathbf{z} \\ t_1, \dots, t_s \end{pmatrix} = \sum_{j_1 < \cdots < j_s} \det M \begin{pmatrix} u_{j_1}, \dots, u_{j_s} \\ t_1, \dots, t_s \end{pmatrix} \det M \begin{pmatrix} \mathbf{p} \\ j_1, \dots, j_s \end{pmatrix}. \tag{5}$$

Formula (5) allows us to relate geometric properties of $\mathbf{z}$ with properties of the minors of the matrix $M \begin{pmatrix} \mathbf{u} \\ t_1,\dots,t_s \end{pmatrix}$. This technique has been extensively used in [5,4,1,2] to derive properties of the curves in terms of the control polygon assuming that certain hypotheses on the collocation matrices (3) hold.

Most blending systems used in CAGD for curve generation, such as the Bernstein basis or the B-spline basis, are totally positive systems. Let us recall that a totally positive matrix is a matrix such that all of its minors are nonnegative. A system of functions $\mathbf{u}$ is called a totally positive system, if all their collocation matrices are totally positive. Totally positive systems enjoy the variation diminishing property: for a given hyperplane the curve does not cross the hyperplane more often than the control polygon does.

The variation diminishing property is perhaps the simplest and most important shape property because it leads to the idea of shape preserving representations (see [6]). In fact, the variation diminishing property is stronger than many other shape properties: monotonicity or convexity preservation. From the variation diminishing property it follows that if a (planar) control polygon crosses any line at most twice, then the corresponding curve (1) shares the same property. This is a standard argument (cf. [6]) showing that the variation diminishing property implies that the curves generated by convex polygons are also convex.

A relevant property of matrices which is weaker than total positivity is sign consistency. A sign consistent matrix is a matrix such that all minors of a given order have the same nonstrict sign. This property is analyzed in connection with the variation diminishing property in Chapter 5 of [8].

This paper uses equation (5) to derive shape preserving properties of sign consistent systems of blending functions. In Section 2, positively oriented polygons and curves are defined. The connection between sign consistency and the preservation of the orientation is discussed. In Section 3 the most common shape preserving properties are discussed in order to describe weaker properties than total positivity leading to shape preservation.

§2. Positively Oriented Polygons

A logical requirement for the representations (1) is that they have to be invariant under translations or even under affine transformations.

Definition 1. *A system of functions* $\mathbf{u}$ *defined on* $[a, b]$ *is* affine invariant *if for any affine map* $h : \mathbb{R}^s \to \mathbb{R}^k$, *the curve* $\mathbf{z}$ *defined by (1) satisfies*

$$h(\mathbf{z}(t)) = \sum_{i=0}^{n} u_i(t) h(\mathbf{p}_i), \tag{6}$$

that is, the curve obtained by any affine transformation h *has a representation of the form (1), where the control polygon for* $h(\mathbf{z}(t))$ *is the transformed control polygon,* $h(\mathbf{p}_0) \cdots h(\mathbf{p}_n)$.

It can be easily shown that a system of functions $\mathbf{u}$ is affine invariant if and only if the functions u_i add up to one, that is,

$$\sum_{i=0}^{n} u_i(t) = 1, \quad \forall t \in [a, b]. \tag{7}$$

Definitions 2. *A matrix is* sign-consistent (SC^k) *of order* k *if all minors of order* k *have the same nonstrict sign. We shall denote by* SC^k_+ *the subclass of* SC^k *matrices with all minors of order* k *being nonnegative. A system of functions is* SC^k_+ *if all collocation matrices are* SC^k_+. *A polygon* $\mathbf{p}_0 \cdots \mathbf{p}_n$, $\mathbf{p}_i \in \mathbb{R}^s$ *is* positively oriented *if the matrix of control points* $M\begin{pmatrix} \mathbf{p} \\ 0,1,\ldots,n \end{pmatrix}$ *is* SC^s_+. *A curve* $\mathbf{z}$ *in* $\mathbb{R}^s$ *is* positively oriented *if all collocation matrices of* $\mathbf{z}$ *are* SC^s_+.

Let us recall that each s-dimensional curve can be seen as a system of functions. So, a curve is positively oriented curve if and only if it is a SC^s_+ system.

Theorem 3. *Let* $\mathbf{u}$ *be a system of functions. Then the following statements are equivalent:*

(i) $\mathbf{u}$ *is* SC_+^s.

(ii) *If* $\mathbf{p}_0 \cdots \mathbf{p}_n$ *is a positively oriented polygon in* $\mathbb{R}^s$ *then* $\mathbf{z}$ *given by (1) is a positively oriented curve.*

Proof: Let us first see that (i) implies (ii). Since (5) holds, we see that the determinant of the matrix (4) (with $k = s$) is a sum of products of nonnegative numbers and hence it is nonnegative for all $t_1 < \cdots < t_s$.

Conversely, choose the following positively oriented polygon $\mathbf{p}_0 \cdots \mathbf{p}_n$ in $\mathbb{R}^s$:

$$\mathbf{p}_{j_1} = (1, 0, \ldots, 0), \ \ldots, \ \mathbf{p}_{j_s} = (0, \ldots, 0, 1), \ \text{and}$$
$$\mathbf{p}_j = (0, \ldots, 0) \ \text{for all } j \in \{0, \ldots, n\} \setminus \{j_1, \ldots, j_s\}.$$

Then we have

$$M \begin{pmatrix} \mathbf{z} \\ t_1, \ldots, t_s \end{pmatrix} = M \begin{pmatrix} u_{j_1}, \ldots, u_{j_s} \\ t_1, \ldots, t_s \end{pmatrix},$$

and since $\mathbf{z}$ has to be a positively oriented curve, we deduce that

$$\det M \begin{pmatrix} u_{j_1}, \ldots, u_{j_s} \\ t_1, \ldots, t_s \end{pmatrix} \geq 0,$$

for all $j_1 < \cdots < j_s$, $t_1 < \cdots < t_s$. □

Corollary 4. *Let* $\mathbf{u}$ *be a* SC_+^k*-system of functions satisfying (7),* $\mathbf{p}_0 \cdots \mathbf{p}_n$ *a polygon in* $\mathbb{R}^s$ *and* $h : \mathbb{R}^s \to \mathbb{R}^k$ *any affine map. If* $h(\mathbf{p}_0) \cdots h(\mathbf{p}_n)$ *is a positively oriented polygon, then* $h(\mathbf{z}(t))$ *is a positively oriented curve.*

§3. Monotonicity and Convexity Properties

Let us first analyze a simple shape property: monotonicity. In [4], monotonicity preserving systems were introduced.

Definition 5. *A system of functions* $(u_0, \ldots, u_n)$ *is* monotonicity preserving *if* $\sum_{i=0}^n c_i u_i$ *is a increasing function, for any* $c_0 \leq \cdots \leq c_n$.

Monotonicity preserving systems $\mathbf{u}$ satisfying (7) have the following geometrical interpretation. Let $\mathbf{p}_0 \cdots \mathbf{p}_n$ be a control polygon in $\mathbb{R}^s$, and let $\mathbf{z}$ be the curve defined by (1). Let $h : \mathbb{R}^s \to \mathbb{R}$ any surjective affine map, which can be interpreted as a projection of the space onto some line. Then if the projection of the control polygon onto this line is increasing, that is, $h(\mathbf{p}_0) \leq \cdots \leq h(\mathbf{p}_n)$, then applying the monotonicity preserving property to (6) we deduce that that $h(\mathbf{z}(t))$ is increasing, that is the projection of the curve onto the same line is increasing.

Defining now $h : \mathbb{R} \to \mathbb{R}^2$ by $h(c) = (1, c)$ we see that Definition 5 is equivalent to saying that if $h(c_0) \cdots h(c_n)$ is a positively oriented polygon then $h(\sum_{i=0}^n c_i u_i)$ is a positively oriented curve. A simple application of Corollary 4 gives the following result.

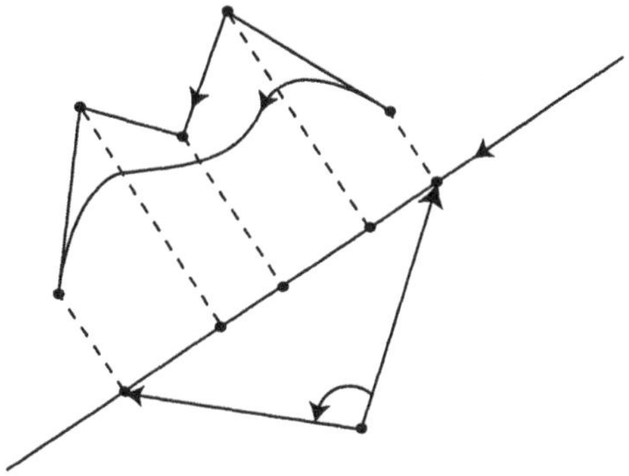

Fig. 1. Preservation of the sense of rotation and of monotonicity.

Proposition 6. *If* **u** *satisfies (7) and is a SC_+^2 system, then it is monotonicity preserving.*

Let us consider another shape property, the preservation of the sense of a rotation.

Definitions 7. *A curve* $\mathbf{z}: [a, b] \to \mathbb{R}^2$ turns counterclockwise *around a point* $\mathbf{c}$ *if* $\mathbf{z}(t) - \mathbf{c}$ *is positively oriented. A polygon* $\mathbf{p}_0 \cdots \mathbf{p}_n$ *turns counterclockwise around a point* $\mathbf{c}$ *if* $(\mathbf{p}_0 - \mathbf{c}) \cdots (\mathbf{p}_n - \mathbf{c})$ *is a positively oriented polygon. A system of functions* preserves the sense of rotation *if it transforms polygons turning counterclockwise around a point into curves turning counterclockwise around the same point.*

As another simple consequence of Theorem 3 we derive the following

Proposition 8. *Let* **u** *be a system of functions satisfying (7). Then* **u** *preserves the sense of rotation if and only if it is* SC_+^2.

Proof: The fact that **u** preserves the sense of rotation around $\mathbf{c} = \mathbf{0}$ is equivalent to saying that it transforms positively oriented polygons into positively oriented curves, and by Theorem 3 this in turn is equivalent to SC_+^2. On the other hand, preservation of the sense of rotation around the origin is, due to the affine invariance, equivalent to the preservation of the sense of rotation around any other center $\mathbf{c} \in \mathbb{R}^2$. □

From the previous proposition, we see that the preservation of the sense of the rotation implies the preservation of the monotonicity. Figure 1 illustrates the preservation of the monotonicity and the preservation of the sense of rotation.

Probably, the most useful shape-preserving property of affine invariant representations is the preservation of convexity. Let us analyze now the preservation of global convexity. In [6] a convex planar curve (resp., polygon) is described as a curve (resp., polygon) such that any line of the plane crosses the curve at most twice. In that paper, it was shown that if **u** enjoys the variation diminishing property, then the curves generated by convex polygons have to be convex. Global convexity can be regarded as a reformulation in terms of orientation of polygons of the above definition of convexity.

Definition 9. *Let $h: \mathbb{R}^2 \to \mathbb{R}^3$ be given by $h(\mathbf{p}) = (1, \mathbf{p})$. A curve $\mathbf{z} : [a, b] \to \mathbb{R}^2$ not contained in a line is* globally convex *if $h(\mathbf{z})$ is positively oriented. A polygon $\mathbf{p}_0 \cdots \mathbf{p}_n$ not contained in a line is globally convex if $h(\mathbf{p}_0) \cdots h(\mathbf{p}_n)$ is a positively oriented polygon. A system is global convexity preserving if it transforms globally convex polygons into globally convex curves.*

Using Corollary 4 again, we obtain a sufficient condition for a system to be global convexity preserving. Linear independence ensures that if the control polygon is not contained in a line, then the curve is not contained in that line.

Proposition 10. *If $\mathbf{u}$ is a SC_+^3 system of linearly independent functions satisfying (7), then it is global convexity preserving.*

A locally convex curve $\mathbf{z}(t)$ can be defined as a curve such that for each t there exists a neighbourhood in which the restriction of the curve to that neighbourhood is globally convex and not contained in a line. For smooth curves, this property is closely related with the nonnegativity of the curvature. However, simple examples show that local convexity is not preserved by the most common curves, unless some limitations on the number of turns of the polygon are imposed (see [7]).

However, if the projection of a curve onto a line is strictly increasing, local convexity and global convexity are equivalent.

Proposition 11. *Let $\mathbf{z} : [a, b] \to \mathbb{R}^2$ be a curve. Assume that there exists a surjective affine map $h : \mathbb{R}^2 \to \mathbb{R}$ such that $h(\mathbf{z}(t))$ is a strictly increasing function. Then the following properties are equivalent:*
(i) *$\mathbf{z}$ is locally convex,*
(ii) *$\mathbf{z}$ is globally convex.*

Proof: Clearly (ii) implies (i). Conversely, assume that $\mathbf{z}(t) = (x(t), y(t))$ is locally convex. The affine map h is of the form $h(x, y) = \alpha x + \beta y + \gamma$ for some $(\alpha, \beta) \neq 0$. Let $\varphi(t) := \alpha x(t) + \beta y(t) + \gamma$ and $\psi(t) := -\beta x(t) + \alpha y(t)$. Then we have

$$(1, \varphi(t), \psi(t)) = (1, x(t), y(t)) \begin{pmatrix} 1 & \gamma & 0 \\ 0 & \alpha & -\beta \\ 0 & \beta & \alpha \end{pmatrix},$$

and then

$$\det M \begin{pmatrix} 1, \varphi(t), \psi(t) \\ t_0, t_1, t_2 \end{pmatrix} = (\alpha^2 + \beta^2) \det M \begin{pmatrix} 1, x(t), y(t) \\ t_0, t_1, t_2 \end{pmatrix}. \tag{8}$$

Let us define

$$\tau := \sup\{t \in [a, b] \mid \mathbf{z} \text{ is globally convex on } [a, t]\}. \tag{9}$$

Since there exists a neighbourhood of a such that $\mathbf{z}$ is globally convex on it, we know that the set on the right hand side of (9) is nonempty and that $\tau > a$.

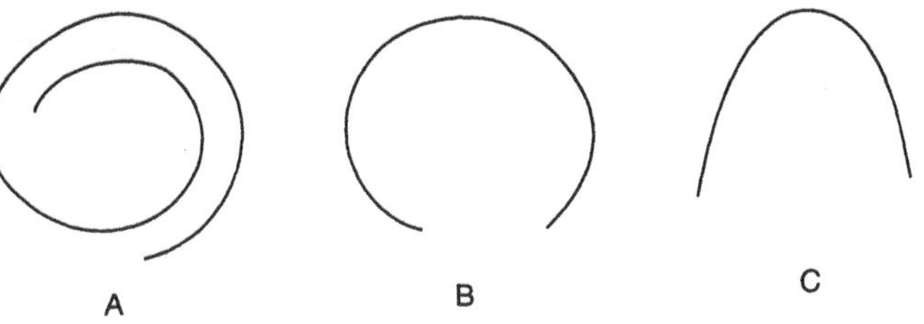

Fig. 2. Convex curves.

By the local convexity of $\mathbf{z}$, there exists $\varepsilon > 0$ such that $\mathbf{z}$ is globally convex on $[\tau-\varepsilon, \tau+\varepsilon] \cap [a, b]$. Now let us see that $(1, \mathbf{z}(t))$ is positively oriented on $[a, \sigma]$, where $\sigma := \min(b, \tau + \varepsilon)$ and then it will be globally convex on that interval. By (8) it is sufficient to show that any collocation matrix $M\binom{1,\varphi,\psi}{t_0,t_1,t_2}$ has nonnegative determinant. If all the points t_0, t_1, t_2 are in $[a, \tau)$ or all of them are in $[\tau - \varepsilon, \sigma]$, there is nothing to prove. Otherwise, some of the points will be in $[a, \tau)$ and others in $(\tau - \varepsilon, \sigma]$. We may add two points in $(\tau - \varepsilon, \tau)$ and obtain a sequence $s_0 < s_1 < s_2 < s_3 < s_4$ in $[a, \sigma]$ containing t_0, t_1, t_2 such that s_i, s_{i+1}, s_{i+2} belong either to $[a, \tau)$ or to $[\tau - \varepsilon, \sigma]$, $i = 0, 1, 2$. Now all the 2×2 minors involving the first and second columns of collocation matrix $A := M\binom{1,\varphi,\psi}{s_0,s_1,s_2,s_3,s_4}$ are positive because φ is strictly increasing, whereas the 3×3 minors using consecutive rows are nonnegative. From Theorem 3.2 of Chapter 2 of [8], all 3×3 minors of A have to be nonnegative. Since $\det M\binom{1,\varphi,\psi}{t_0,t_1,t_2}$ is one of those minors, we have deduced that $(1, \varphi, \psi)$ is a positively oriented curve on $[a, \sigma]$. Then $\mathbf{z}$ is globally convex on $[a, \sigma]$. From (9) $\tau = \sigma = b$ and $\mathbf{z}$ is globally convex on $[a, b]$. □

Figure 2 shows different types of convex curves. Curve A is a locally convex curve but not globally convex, curves B and C are globally convex. The projection of curve C onto some line is a strictly increasing function. However, for the curve B such a projection does not exist, because it turns an angle greater than π.

The previous result motivates another shape property: geometric convexity, a stronger property than global convexity. Roughly speaking, a curve is geometrically convex if it is globally convex (or globally concave) and it turns an angle less than π. In this case, there exists a line in the plane, such that the projection of the curve onto that line is an increasing function. This shape property has been suggested by Goodman, who introduced the concept of regularity in connection with the problem of counting inflections of curves [6,7], and was further studied by Carnicer, García-Esnaola and Peña. For more details about the preservation of the geometric convexity, see [2,3].

Finally, let us briefly discuss why the most common examples of curves used in CAGD enjoy the shape properties discussed in this paper. It is well-known (cf. [5]) that the Bernstein basis, the B-spline basis and the bases used for generating rational Bézier and rational B-spline (NURBS) curves as well as β-splines (see [6]) are totally positive and in particular SC_+^k for all $k \leq n$. On the other hand, all these bases satisfy (7). Therefore, all these systems are monotonicity and global convexity preserving, that is all of them preserve the most common shape properties.

Acknowledgments. This research was partially supported by DGICYT PB96–0730.

References

1. Carnicer, J. M., M. García-Esnaola, and J. M. Peña, Convex curves from convex control polygons, in *Mathematical Methods for Curves and Surfaces*, M. Dæhlen, T. Lyche, and L. L. Schumaker (eds.), Vanderbilt University Press, Nashville, 1995, 63–72.
2. Carnicer, J. M., M. García-Esnaola, and J. M. Peña, Convexity of rational curves and total positivity, J. Comput. Appl. Math. **71** (1996) 365–382.
3. Carnicer, J. M., M. García-Esnaola, and J. M. Peña, Bases with convexity preserving properties, in *Advanced Topics in Multivariate Approximation*, F. Fontanella, K. Jetter and P.-J. Laurent (eds.), World Scientific Publishing Co., 1996, 17–31.
4. Carnicer, J. M. and J. M. Peña, Monotonicity preserving representations, in *Curves and Surfaces in Geometric Design*, P.-J. Laurent, A. Le Méhauté, and L. L. Schumaker (eds.), A. K. Peters, Wellesley MA, 1994, 83–90.
5. Carnicer, J. M. and J. M. Peña, Total positivity and optimal bases, in *Total Positivity and its Applications*, M. Gasca and C. A. Micchelli (eds.), Kluwer Academic Publishers, 1996, 133–155.
6. Goodman, T. N. T., Shape preserving representations, in *Mathematical Methods in Computer Aided Geometric Design*, T. Lyche and L. L. Schumaker (eds.), Academic Press, New York, 1989, 333–357.
7. Goodman, T. N. T., Inflections on curves in two and three dimensions, Comput. Aided Geom. Design **8** (1991), 37–50.
8. Karlin, S., *Total Positivity,* Vol. I, Stanford University Press, Stanford, 1968.

J. M. Carnicer
Departamento de Matemática Aplicada
Universidad de Zaragoza
Edificio de Matemáticas, planta 1a.
50009 Zaragoza, SPAIN
carnicer@posta.unizar.es

Actuarial Science and Variational Splines

R. Champion, C. T. Lenard, and T. M. Mills

Abstract. The inspiration for some of the important ideas in the development of univariate spline theory, particularly variational spline methods, originated in the field of actuarial science. The purpose of this paper is to sketch briefly the historical development of smoothing techniques that played a key role in the evolution of splines.

§1. Introduction

Actuarial scientists have a long history over nearly two centuries of developing and applying nonparametric smoothing methods, in particular to graduate observed rates of mortality

$$m_t = \frac{A_t}{E_t}, \tag{1}$$

where A_t is the number of deaths of people aged t, and E_t is the effective population aged t. Figure 1 depicts recent Australian mortality data smoothed with a cubic variational spline.

When I. J. Schoenberg [17] sought to overcome numerical instability in missile trajectory computations in the 1940s by fitting a smooth analytic function to tabulated data, he defined cardinal spline functions to solve his problem. In doing so Schoenberg was inspired by smoothing algorithms used in the actuarial community. He cites Whittaker and Robinson [22] which includes a chapter devoted to moving average graduation, and a paper in which T. N. E. Greville [9] systematized osculatory polynomial formulae applied by actuaries to produce smooth interpolating and graduating curves. These articles lead the reader to over a century of literature, almost entirely within the actuarial discipline, devoted to smoothing.

Mathematical Methods for Curves and Surfaces II
Morten Dæhlen, Tom Lyche, and Larry L. Schumaker (eds.), pp. 49–54.

ISBN 0-8265-1315-8.

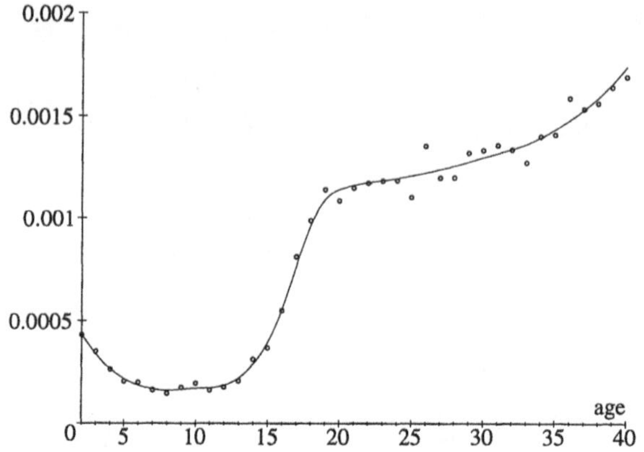

Fig. 1. Australian male rate of mortality 1992-95.

§2. Discrete Methods

A plethora of moving average formulae of general form

$$\hat{z}_t = \sum_{j=-m}^{m} c_j z_{t-j} \tag{2}$$

were investigated, particularly by British actuaries, during the 19th and early 20th century. J. Finlaison [8], who used a 9-point formula in a report to the House of Commons in 1829, is identified as the pioneer of moving average graduation. Such formulae were developed by various means including averages of interpolating polynomial values, local polynomial least squares, truncated finite difference expansions, repeated averaging and optimisation. Formulae with band-widths as high as 27 were employed! Aspects of the theory and history of these methods with references to original papers may be found in [2,5,11,13,14,19,22,23]. Seal [19] discusses methods that can be related (sometimes tenuously) to piecewise cubic functions, Cleveland and Loader [5] focus on the relationship of these methods to modern local polynomial smoothers, whereas Macauley [13] introduced actuarial research into time series analysis in the 1930s.

During the 19th century there appear to be only two recognised contributions to the field of nonparametric smoothing from outside actuarial science. These were by G. V. Schiaparelli [16] (canals of Mars fame), probably the first to use local least squares polynomial methods, and G. H. Darwin [4] (son of Charles) who was perhaps the first to investigate bivariate formulae. Both sought to smooth astronomical and meteorological data.

Initially, smoothing performance was measured by the sum of squares of differences of the graduated sequence. Third differences, $\sum_{t=1}^{n-3}(\Delta^3\hat{z}_t)^2$, were the popular choice. Towards the end of the century attention was focused on characteristics of the formulae themselves, primarily the smoothness of the *coefficient curve* measured by $\sum_{j=-m}^{m-3}(\Delta^3 c_j)^2$, for example. The 21-point formula due to J. Spencer in 1904 [20] was shown, among formulae exact for all sequences of cubic polynomial values, to be close to optimal in this sense.

§3. Variational Methods

Optimisation criteria were applied to develop many classes of discrete smoothers. Local least squares algorithms were obtained initially by minimising the expected error variance, apparently independently, by Schiaparelli [16], E. L. De Forest [6], J. P. Gram who indicated earlier unpublished work of L. Oppermann (see [11]), and others [19]. The optimisation problem is to find formulae exact for polynomials of given degree for which $\sum_{j=-m}^{m} c_j^2$ is minimal. De Forest, who was the first to emphasise the smoothness of the coefficient curve, modified this condition to minimise $\sum_{j=-m}^{m-3}(\Delta^4 c_j)^2$.

In 1919 E. T. Whittaker [22], building on the foundations laid by actuarial scientists, proposed a method of graduation that is now recognised as the precursor of variational smoothing splines. To balance the desires for smoothness and faithfulness to the data, Whittaker used maximum likelihood arguments that led to minimising

$$G = \sum_{t=1}^{n}(z_t - \hat{z}_t)^2 + \lambda \sum_{t=1}^{n-3}(\Delta^3 \hat{z}_t)^2 \tag{3}$$

to determine the most probable values of mortality.

§4. Continuous Methods

Smooth curves drawn by hand to reduce irregularities in observed data were discussed qualitatively in the scientific literature during the 19th century. Generally however, parametric least squares fitting prevailed in practice. It was often argued in the actuarial community that a sophisticated approach to hand drawn graduating curves, known as the graphic method, was more appropriate than arithmetic methods using moving averages. In 1886 T. B. Sprague [21] set out a semi-quantitative procedure and showed that his graduation performed better than arithmetic methods of the day according to the accepted smoothness criteria. The literature during the period 1870–1890 reveals a vigorous debate between graphic and arithmetic graduation schools.

A second use of smooth curves by actuaries was interpolation by osculating polynomial arcs. Sprague (1880) adopted a method of C. Hermite and used quintic arcs with 2 continuous derivatives. Early this century cubic arcs, usually with one continuous derivative, were used to interpolate and smooth data, particularly in the USA, Britain and Scandinavia. In fact W. A. Jenkins [12] used graduating cubic polynomials with two continuous derivatives, in other words a smoothing spline. In a sense, osculatory polynomial curves were a compromise between the arithmetic and continuous methods. Greville [9], in his review in 1944, developed a general theory and classification of various formulae proposed by actuaries.

§5. Splines

In his original constructive characterisation of spline functions as piecewise degree k polynomial functions with at least $k-1$ continuous derivatives, Schoenberg [17] combined the computational simplicity and smoothness characteristics of osculatory polynomial functions with the moving average expression (2). He replaced the smooth coefficient sequence, c_j, with a basis spline M, and to apply Fourier methods he augmented the data set using appropriate conditions beyond the real data. The moving average thus became the cardinal spline

$$F(t) = \sum_{i=-\infty}^{\infty} z_i M(t-i). \tag{4}$$

Polynomial splines burst into prominence in the 1960s as a simple and efficient tool for representing curves and surfaces, and approximating functions generally. Since then an industry has grown around the theory of splines and their application.

Variational splines have been an integral part of this development, but are undoubtedly less well known. Variational splines are defined as solutions of optimisation problems, rather than as smooth functions constructed in a piecewise manner. The natural, univariate, interpolating spline functions were shown by de Boor [1] to be the functions σ_m that, among all interpolating functions s in $H^{(m)}(I)$, minimize

$$G_m(s) = \int_I s^{(m)}(t)^2 dt. \tag{5}$$

Holladay [10], unaware of the spline literature at the time, had obtained the result for cubic splines some 6 years earlier. In 1964 Schoenberg, armed with de Boor's result and Whittaker's method of graduating mortality, generalised the interpolation problem to a smoothing one. He replaced the discrete smoothness measure in Whittaker's expression (3) with the energy integral in (5) to define smoothing variational splines as the functions $\sigma_{m,\lambda}$ that realize the minimum in $H^{(m)}(I)$ of

$$G_{m,\lambda}(s) = \sum_{i=1}^{n} (z_i - s(t_i))^2 + \lambda \int_I s^{(m)}(t)^2 dt. \tag{6}$$

Reinsch [15] obtained the result for cubic splines ($m = 3$) independently of Schoenberg, and pioneered algorithms for computing the splines. The univariate variational splines discussed here are the same polynomial splines that arise from a constructive approach. However generalisations of the optimisation problem often yield simple piecewise functions with useful approximation properties that are not polynomial functions (See [3] for a brief survey.) Thus the constructive and variational approaches to splines follow different directions from a common foundation.

It is clear that actuarial science has contributed greatly to the fundamental ideas on the path to the spline industry, particularly in the case of variational splines. It is also evident that spline methods and other nonparametric smoothing methods share characteristics, having developed in different directions from common roots in the problem of graduation of actuarial data.

References

1. de Boor, C., Best approximation properties of spline functions of odd degree, J. Math. Mech. **12** (1963), 747–749.
2. Borgan, Ø., On the theory of moving averages, Scand. Actuar. J. (1979), 83–105.
3. Champion, R., C. T. Lenard, and T. M. Mills, An introduction to abstract splines, The Mathematical Scientist **21** (1996), 6–26.
4. Darwin, G. H., On fallible measures of variable quantities, and on the treatment of meeorological observations, Phil. Magazine (Fifth Series) **4** (1877), 1–14.
5. Cleveland, W. S. and C. Loader, Smoothing by local regression: principles and methods, in *Statistical Theory and Computational Aspects of Smoothing*, W. Härdle and M. G. Schimek (eds.), Physica-Verlag, Heidelberg, 1996, 10–49.
6. De Forest, E. L., On some methods of interpolation applicable to the graduation of irregular series, such as mortality tables, &c., &c., in the *Annual Report of the Board of Regents of the Smithsonian Institute for the year 1871*, Government Printing Office, Washington, 1873, 275–339.
7. De Forest, E. L., Additions to a memoir on methods of interpolation applicable to the graduation of irregular series, in the *Annual Report of the Board of Regents of the Smithsonian Institute for the year 1874*, Government Printing Office, Washington, 1873, 319–353.
8. Finlaison, J., Report on the evidence and elementary facts on which the tables of life annuities are founded, House of Commons, London, 1829.
9. Greville, T. N. E., The general theory of osculatory interpolation, Trans. Actuar. Soc. Amer. **45** (1944), 202–265.
10. Holladay, J. C., A smoothest curve approximation, Math. Table Aids Comput. **11** (1957), 233–243.
11. Hoem, J. M., The reticent trio: some little known early discoveries in life insurance mathematics by L. H. F. Oppermann, T. N. Thiele and J. P. Gram, Internat. Statist. Rev. **51** (1983), 213–221.
12. Jenkins, W. A., Graduation based on a modification of osculatory interpolation, Trans. Actuar. Soc. Amer. **28** (1927), 198–215.
13. Macauley, F. R., *The Smoothing of Time Series*, National Bureau of Economic Research, New York, 1931.

14. Miller, M. D., *Elements of Graduation*, Actuarial Society of America, New York, 1946.
15. Reinsch, C. H., Smoothing by spline functions, Numer. Math. **10** (1967), 177–183.
16. Schiaparelli, G. V., Sul modo di ricavare la vera espressione delle leggi della natura dalle curve empiriche, Effemeride Astronomiche di Milano per l'Anno (1867), 3-56.
17. Schoenberg, I. J., Contributions to the problem of approximation of equidistant data by analytic functions, Quart. Appl. Math. **4** (1946), 45–99.
18. Schoenberg, I. J., Spline functions and the problem of graduation, Proc. Nat. Acad. Sci. USA **52** (1964), 947–950.
19. Seal, H. L., Graduation by piecewise cubic polynomials: a historic review, Blätter der DGVM **15** (1981), 89–114.
20. Spencer, J., On the graduation of rates of sickness and mortality presented by the experience of the Manchester Unity of Oldfellows during the period 1893–97, J. Inst. Actuar. **38**, 334–343.
21. Sprague, T. B., Explanation of a new formula for interpolation, J. Inst. Actuar. **22** (1880), 270-285.
22. Whittaker, E. T., and G. Robinson, *The Calculus of Observations*, Blackie & Son Ltd., London, 4th edition, 1944.
23. Wolfenden, H. H., On the development of formulae for graduation by linear compounding, with special reference to the work of Erastus L. De Forest, Trans. Actuar. Soc. Amer. **26** (1925), 81–121.

Rob Champion, Christopher T. Lenard, and Terry Mills
La Trobe University
PO Box 199,
Bendigo, 3552, AUSTRALIA
`r.champion@latrobe.edu.au`
`c.lenard@latrobe.edu.au`
`t.mills@latrobe.edu.au`

Shape-Preserving Quasi-Interpolating Univariate Cubic Splines

C. Conti, R. Morandi and C. Rabut

Abstract. In this paper we present a strategy to construct a shape-preserving quasi-interpolant function expressed as a linear combination of cubic splines where the coefficients of the combination are given by the data. The quasi-interpolant is shown to be linear-reproducing and monotone and/or convex conforming to the data.

§1. Introduction

In this paper a strategy to construct a quasi-interpolant shape preserving spline function fitting a set of values $\{(x_i, f(x_i))\}_{i=1}^n$, $x_i \in \mathbb{R}$, $f(x_i) \in \mathbb{R}$ is presented. The quasi-interpolant is expressed as $\sigma = \sum_{i=1}^n f(x_i)C_i + p_1$, where $\{C_i\}_{i=1}^n$ are natural cubic splines, $p_1 \in \mathbb{P}_1$ and σ satisfies the $\mathbb{P}_1$-reproduction property in $[x_1, x_n]$, that is for any $p \in \mathbb{P}_1$, $\sum_{i=1}^n p(x_i)C_i(x) + p_1(x) = p(x)$, $\forall x \in [x_1, x_n]$. In addition, σ satisfies the shape-preserving properties, that is if the data are monotone and/or convex, then σ is monotone and/or convex too.

The problem of constructing a quasi-interpolant shape-preserving function of this form was solved by Beatson and Powell in [2] by using multiquadric functions, i.e. by linear combinations of $\phi_j(x) = [c^2 + (x - x_j)^2]^{1/2}$. However, this quasi-interpolant requires the derivatives of f at the endpoints x_1, x_n, and so a new shape-preserving quasi interpolant multiquadric function was proposed by Wu and Schaback in [9].

In this paper we assume that the data $\{f(x_i)\}_{i=1}^n$ come from a function belonging to the Hilbert space $D^{-2}L^2(\mathbb{R})$ which is the space of all functions with 2nd derivatives in $L^2(\mathbb{R})$. The Hilbert space $D^{-2}L^2(\mathbb{R})$ is endowed with the scalar product

$$(f, g)_2 = \int_{\mathbb{R}} f''(x)g''(x)dx + f(a_1)g(a_1) + f(a_2)g(a_2),$$

Mathematical Methods for Curves and Surfaces II
Morten Dæhlen, Tom Lyche, and Larry L. Schumaker (eds.), pp. 55–62.

ISBN 0-8265-1315-8.

where a_1, a_2 are two distinct points of $\mathbb{R}$. Following [3,4] the quasi-interpolant is obtained by approximating the right-hand side of the identity $f(x) = (f, H(x, \bullet))_2$, where H is the reproducing kernel of the Hilbert space. The obtained quasi-interpolant has the form $\sigma = \sum_{i=1}^{n} f(x_i)C_i + p_1$, with $p_1 \in \mathbb{P}_1$, and is $\mathbb{P}_1$-reproducing and shape-preserving. Concerning the error bound, in [2] it is proved that if $f \in C^2$ and $c = \mathcal{O}(h)$ with $h = \max_{i=1,\ldots,n-1}\{h_i = x_{i+1} - x_i\}$, the error of the multiquadric quasi-interpolant is $\mathcal{O}(h^2 log(h))$. In this paper, with the help of the $\mathbb{P}_1$-reproduction property, we conclude that for any $f \in C^2$ the error of the quasi-interpolant spline function σ is $\mathcal{O}(h^2)$.

The paper is divided into the following sections. In Section 2 some known properties of natural cubic splines and of the reproducing kernel of the Hilbert space $D^{-2}L^2(\mathbb{R})$ are presented. In Section 3 the strategy to construct the shape preserving quasi-interpolant spline function is described, and some graphs are shown to illustrate the behaviour of the basis functions. In Section 4 the $\mathbb{P}_1$-reproduction of the quasi-interpolant is proved, and the shape-preserving properties are investigated. Finally, in Section 5 some applications of the method to classical data sets are shown to illustrate the behaviour of the proposed method.

§2. Variational Splines: Space Setting

Let $D^{-2}L^2(\mathbb{R})$ be the Hilbert space of functions with second derivatives in $L^2(\mathbb{R})$. This space is endowed with the scalar product

$$(f, g)_2 = \int_{\mathbf{R}} f''(y)g''(y)dy + f(a_1)g(a_1) + f(a_2)g(a_2), \tag{1}$$

where $\{a_1,\ a_2\}$ are two distinct points in $\mathbb{R}$. This scalar product induces the norm $||f||_2 = ((f, f)_2)^{\frac{1}{2}}$. We will also use the semi-norm $|f|_2 = (\int_{\mathbf{R}}(f'')^2)^{\frac{1}{2}}$. Let $A \subset \mathbb{R}$ be a countable set with two distinct points. If $\{z_a\}_{a\in A}$ are real numbers, a *natural cubic spline* σ may be defined by the following conditions [6]:

$$\begin{cases} \sigma \in D^{-2}L^2(\mathbb{R}), \\ \forall a \in A \ \ \sigma(a) = z_a, \\ \forall f \in D^{-2}L^2(\mathbb{R}) \ \ (\forall a \in A \ f(a) = z_a) \rightarrow |\sigma|_2 \le |f|_2. \end{cases}$$

The reproducing kernel of the Hilbert space $D^{-2}L^2(\mathbb{R})$ is the map H from $\mathbb{R} \times \mathbb{R}$ to $\mathbb{R}$ satisfying

$$\begin{aligned} &i)\ \forall x \in \mathbb{R}, H(x, \bullet) \in D^{-2}L^2(\mathbb{R}),\ \forall y \in \mathbb{R}, H(\bullet, y) \in D^{-2}L^2(\mathbb{R}), \\ &ii)\ \forall x \in \mathbb{R}, \forall f \in D^{-2}L^2(\mathbb{R}),\ f(x) = (f, H(x, \bullet))_2. \end{aligned} \tag{2}$$

It can be proved that if $A_0 = \{a_1,\ a_2\}$, $H(x, y)$ is defined as

$$\begin{aligned} H(x,y) =& v(x-y) - \sum_{a\in A_0} p_a(x)v(a-y) - \sum_{a\in A_0} p_a(y)v(x-a) \\ &+ \sum_{a\in A_0}\sum_{b\in A_0} p_a(x)p_b(y)v(a-b) + \sum_{a\in A_0} p_a(x)p_a(y) \end{aligned} \tag{3}$$

where v can be, for example, the function $v(x) = \frac{|x|^3}{12}$, or the function $v(x) = \frac{(x)_+^3}{6}$ and p_a, $\forall a \in A_0$, are the linear polynomials $p_{a_1} = \frac{a_2 - \bullet}{a_2 - a_1}$, $p_{a_2} = \frac{\bullet - a_1}{a_2 - a_1}$ satisfying the conditions $p_{a_1}(a_1) = 1$, $p_{a_1}(a_2) = 0$, $p_{a_2}(a_2) = 1$, $p_{a_2}(a_1) = 0$.

In [6] it is proved that every natural cubic spline can be written as $\sigma = \sum_{a \in A} \mu_a H(\bullet, a) + q_1$, where q_1 is a polynomial of degree at most 1 and $\{\mu_a\}_{a \in A}$ are real coefficients satisfying the condition $\sum_{a \in A} \mu_a p(a) = 0$, for any p in $\mathbb{P}_1$. On the other hand, from (3) we also get that every cubic spline can be expressed as $\sigma = \sum_{a \in A} \mu_a v(\bullet - a) + p_1$, where $p_1 \in \mathbb{P}_1$ with $\{\mu_a\}_{a \in A}$ satisfying $\sum_{a \in A} \mu_a p(a) = 0$. Furthermore, any function of the previous form is a cubic spline interpolating the points $\{(b, \sigma(b))\}_{b \in A}$. The points of A are called *knots* of the spline σ.

Definition 1. *Let $X = D^{-2}L^2(\mathbb{R})$ and let E be an approximation operator so that $E : X \to X$. Let A be a countable subset of $\mathbb{R}$. E is called $\mathbb{P}_k$-*reproducing *if $\forall p \in \mathbb{P}_k$ $E(p) = p$.*

§3. How to build a Quasi-Interpolant Spline Function

In this section a method is discussed to obtain a cubic spline function quasi-interpolating the values $\{(x_i, f(x_i))\}_{i=1,\ldots,n}$. The approximation is done in two steps. Firstly we derive a simple interpolant of the data, namely f^*, which is a piecewise polynomial belonging to $D^{-2}L^2(\mathbb{R})$; secondly we note that from (2) the function f^* can be expressed as

$$f^* = \int_{\mathbb{R}} (f^*)''(y) \frac{\partial^2 H(\bullet, y)}{\partial y^2} dy + f^*(a_1) H(\bullet, a_1) + f^*(a_2) H(\bullet, a_2), \tag{4}$$

so that f can be approximated by discretization of the right-hand side of (4). Let f be known at the points x_i, $i = 1, \ldots, n$. The function $f^* \in D^{-2}L^2(\mathbb{R})$ can be defined as

$$f^*(x) = \begin{cases} f(x_1) + \frac{f(x_2) - f(x_1)}{x_2 - x_1}(x - x_1), & x < x_1 \\ p_3^{f,i}(x), \ i = 1, \ldots, n-1, & x \in [x_i, x_{i+1}] \\ f(x_n) + \frac{f(x_n) - f(x_{n-1})}{x_n - x_{n-1}}(x - x_n), & x > x_n \end{cases} \tag{5}$$

where, in each interval $[x_i, x_{i+1}]$, $p_3^{f,i}$ is a polynomial of degree 3 interpolating f at x_i, x_{i+1} and with first derivatives at x_i, x_{i+1} given by $(p_3^{f,i})^{(1)}(x_i) = \frac{f(x_i) - f(x_{i-1})}{x_i - x_{i-1}}$, $(p_3^{f,i+1})^{(1)}(x_{i+1}) = \frac{f(x_{i+1}) - f(x_i)}{x_{i+1} - x_i}$. Furthermore, we construct two sets of equidistant points $x_{i-1} = x_1 - i \cdot h_1$, $\forall i < 1$, and $x_{n+i} = x_n + i \cdot h_{n-1}$, $\forall i > 1$, where $h_i = x_{i+1} - x_i$, $i = 1, \ldots, n-1$. Then from (3) and (4) the function f^* can be written in the form

$$f^* = \int_{\mathbb{R}} (f^*)''(y) \frac{\partial^2 H(\bullet, y)}{\partial y^2} dy + f^*(a_1) p_{a_1} + f^*(a_2) p_{a_2}$$

with $p_{a_1} = \frac{a_2-\bullet}{a_2-a_1}$, $p_{a_2} = \frac{\bullet-a_1}{a_2-a_1}$. Choosing $a_1 = x_1$, $a_2 = x_n$ and considering the linearity of f^* outside $[x_1, x_n]$, we can write

$$f^* = \sum_{i\in\mathbb{Z}} \int_{x_i}^{x_{i+1}} f^{*\prime\prime}(y) \frac{\partial^2 H(\bullet, y)}{\partial y^2} dy + p_1^f,$$

where $p_1^f = f(x_1)p_{x_1} + f(x_n)p_{x_n}$ is the linear polynomial interpolating f at x_1 and at x_n. From (5) we get

$$f^* = \sum_{i=1}^{n-1} \int_{x_i}^{x_{i+1}} (p_3^{f,i})''(y) \frac{\partial^2 H(\bullet, y)}{\partial y^2} dy + p_1^f.$$

Now, in each interval $[x_i, x_{i+1}], i = 1, \ldots, n-1$, we substitute $H(x, \bullet)$ with the polynomial of degree two interpolating $H(x, x_{i-1})$, $H(x, x_i)$, $H(x, x_{i+1})$ with constant second derivative equal to $D^2_{2,i}H(x, x_i) = \gamma^i_{-1}H(x, x_{i-1}) + \gamma^i_0 H(x, x_{i1}) + \gamma^i_1 H(x, x_{i+1})$, where $\gamma^i_{-1} = \frac{2}{h_{i-1}(h_{i-1}+h_i)}$, $\gamma^i_0 = \frac{-2}{h_{i-1}h_i}$ and $\gamma^i_1 = \frac{2}{h_i(h_{i-1}+h_i)}$. Thus we obtain the following approximation of f,

$$\tilde{f} = \sum_{i=1}^{n-1} D^2_{2,i}H(\bullet, x_i) \int_{x_i}^{x_{i+1}} (f^*)''(y) dy + p_1^f.$$

After integration, we get

$$\tilde{f} = \sum_{i=1}^{n-1} D^2_{2,i}H(\bullet, x_i)((p_3^{f,i+1})^{(1)}(x_{i+1}) - (p_3^{f,i})^{(1)}(x_i)) + p_1^f,$$

and therefore

$$\tilde{f} = \sum_{i=1}^{n-1} D^2_{2,i}H(\bullet, x_i)(\frac{1}{h_{i-1}} f(x_{i-1}) - \frac{(h_i + h_{i-1})}{h_i h_{i-1}} f(x_i) + \frac{1}{h_i} f(x_{i+1})) + p_1^f,$$

where $D^2_{2,i}H(\bullet, x_i) = D^2_{2,i}v(\bullet - x_i) - p_{x_1}D^2_{2,i}v(x_1 - x_i) - p_{x_n}D^2_{2,i}v(x_n - x_i)$. Considering $v(x) = \frac{|x|^3}{12}$, we get

$$D^2_{2,i}v(x - x_i) = \begin{cases} -\frac{x-x_i}{2} - \frac{h_{i-1}-h_i}{6} & x \le x_{i-1}, \\ \gamma^i_{-1}\frac{|x-x_{i-1}|^3}{6} + \gamma^i_0\frac{|x-x_i|^3}{6} + \gamma^i_1\frac{|x-x_{i+1}|^3}{6} & x_{i-1} < x < x_{i+1}, \\ +\frac{x-x_i}{2} + \frac{h_{i-1}-h_i}{6} & x \ge x_{i+1}. \end{cases}$$

Because of the uniformity of the added knots outside the interval $[x_1, x_n]$, we can conclude that in $[x_1, x_n]$ the functions $D^2_{2,i}H(x, x_i)$ are equal to zero if $i \le 0$ and if $i \ge n+1$. This implies that if we define the functions

$$\begin{aligned} K_i =& \frac{1}{h_i} D^2_{2,i+1}H(\bullet, x_{i+1}) - \frac{(h_i + h_{i-1})}{h_i h_{i-1}} D^2_{2,i}H(\bullet, x_i) + \\ & \frac{1}{h_{i-1}} D^2_{2,i-1}H(\bullet, x_{i-1}), \end{aligned} \tag{6}$$

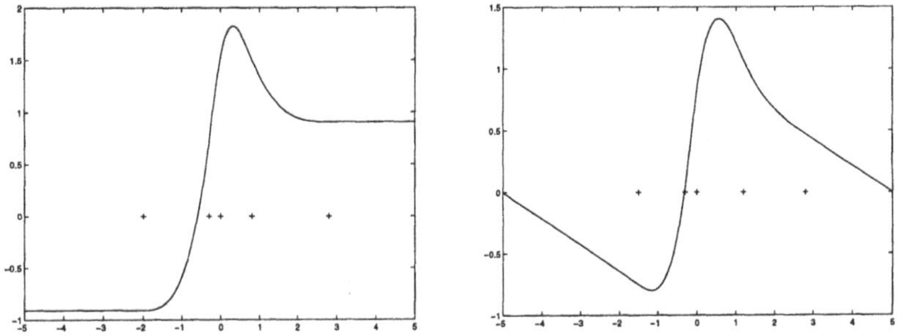

Fig. 1. Basis function C_i and K_i, respectively.

also these functions are equal to zero in $[x_1, x_n]$ if $i \leq -1$, $i \geq n+2$. As from (5) we get $f^*(x_0) = 2f(x_1) - f(x_2)$ and $f^*(x_{n+1}) = 2f(x_n) - f(x_{n-1})$, for sake of simplicity, we define f equal to f^* at x_0 and at x_n. Now it is easy to check that in $[x_1, x_n]$ the approximation $\tilde{f}$ can be written as

$$\tilde{f} = \sum_{i=0}^{n+1} f(x_i) K_i + p_1^f. \tag{7}$$

Let C_i be the function

$$C_i = \frac{1}{h_i} D^2_{2,i+1} v(\bullet - x_{i+1}) - \frac{(h_i + h_{i-1})}{h_i h_{i-1}} D^2_{2,i} v(\bullet - x_i) + \frac{1}{h_{i-1}} D^2_{2,i-1} v(\bullet - x_{i-1}).$$

The following theorem can be easily proved.

Theorem 1. *The functions* $\{C_i\}_{i \in \mathbb{Z}}$ *and* $\{K_i\}_{i \in \mathbb{Z}}$ *previously defined are natural cubic splines with knots* $\{x_{i-2},\ x_{i-1},\ x_i,\ x_{i+1},\ x_{i+2}\}$.

Figure 1 shows the graph of C_i and K_i.

Remark 1. If the knots are uniformly distributed, the functions $\{C_i\}_{i\in\mathbb{Z}}$ reduce to the usual cubic *B-splines*. In fact in this case, after a little algebra we get

$$C_i = \frac{1}{h^3}(v(\bullet - x_{i+2}) - 4v(\bullet - x_{i+1}) + 6v(\bullet - x_i) - 4v(\bullet - x_{i-1}) + v(\bullet - x_{i-2})).$$

From the definition of C_i, the function $\tilde{f}$ can be expressed as

$$\tilde{f} = \sum_{i=0}^{n+1} f(x_i) C_i - p_{x_1} \sum_{i=0}^{n+1} f(x_i) C_i(x_1) - p_{x_n} \sum_{i=0}^{n+1} f(x_i) C_i(x_n) + p_1^f,$$

or in a more concise form as $\tilde{f} = f_C - p_1^{f_C} + p_1^f$, where f_C is the quasi-interpolant function

$$f_C = \sum_{i=0}^{n+1} f(x_i) C_i. \tag{8}$$

§4. Properties of the Approximations $\tilde{f}$ and f_C

In this the section properties of the approximants $\tilde{f}$ and f_C are discussed.

Theorem 2. *Let $\tilde{f}$ and f_C be the functions defined by (7) and (8) respectively. If $f \in \mathbb{P}_1$ then $\tilde{f} = f_C = f$ in $[x_1, x_n]$.*

Theorem 3. *If f is a monotone function, then the quasi-interpolant spline f_C defined by (8) is monotone in $[x_1, x_n]$.*

Proof: Let $f_C = \sum_{i=0}^{n} f(x_i)C_i$. This function can also be written in the form

$$\begin{aligned} f_C = & \sum_{i=1}^{n+1} \frac{f(x_i) - f(x_{i-1})}{h_{i-1}} (D^2_{2,i-1}v(\bullet - x_{i-1}) - D^2_{2,i}v(\bullet - x_i)) \\ & + \frac{1}{h_{n+1}} f(x_{n+1})(D^2_{2,n+2}v(\bullet - x_{n+2}) - D^2_{2,n+1}v(\bullet - x_{n+1})) \\ & - \frac{1}{h_{-1}} f(x_0)(D^2_{2,0}v(\bullet - x_0) - D^2_{2,-1}v(\bullet - x_{-1})). \end{aligned}$$

On the other hand, it is easy to check that in $[x_1, x_n]$,

$$D^2_{2,k}v(x - x_k) = \frac{x - x_k}{2}, \; k = -1, 0, \; D^2_{2,k}v(x - x_k) = \frac{x_k - x}{2}, \; k = n+1, n+2,$$

so that f_C can be written as

$$f_C = \sum_{i=1}^{n+1} \frac{f(x_i) - f(x_{i-1})}{h_{i-1}} (D^2_{2,i-1}v(\bullet - x_{i-1}) - D^2_{2,i}v(\bullet - x_i)) + \frac{f(x_{n+1}) + f(x_0)}{2}.$$

After differentiation we obtain

$$f_C' = \sum_{i=1}^{n+1} \frac{f(x_i) - f(x_{i-1})}{h_{i-1}} (D^2_{2,i-1}v'(\bullet - x_{i-1}) - D^2_{2,i}v'(\bullet - x_i)).$$

It is easy to prove that the function $B_{2,i} = D^2_{2,i-1}v'(\bullet - x_{i-1}) - D^2_{2,i}v'(\bullet - x_i)$ is a positive function with support on $[x_{i-2}, x_{i+1}]$. □

Theorem 4. *If f is a convex (concave) function then the quasi-interpolant splines $\tilde{f}$ and f_C, defined by (7) and (8) respectively, are convex (concave).*

Proof: Firstly, we note that $\tilde{f}'' = f_C''$. Let be $f_C(x) = \sum_{i=0}^{n} f(x_i)C_i(x)$. This function can also be written as

$$\begin{aligned} f_C = & f(x_0)\alpha_0 + f(x_1)\alpha_1 + f(x_{n-1})\alpha_{n-1} + f(x_n)\alpha_n \\ & + \sum_{i=1}^{n} \Big(\frac{1}{h_{i-1}} f(x_{i-1}) - \frac{h_{i-1} + h_i}{h_i h_{i-1}} f(x_i) + \frac{1}{h_i} f(x_{i+1})\Big) D^2_{2,i}v(\bullet - x_i), \end{aligned} \tag{9}$$

where $\alpha_i, i = 0, 1, n-1, n$, are linear boundary functions because of the uniformity of the added knots. Differentiating (9) twice, we obtain

$$f_C'' = \sum_{i=1}^{n} \Big(\frac{1}{h_{i-1}} f(x_{i-1}) - \frac{h_{i-1} + h_i}{h_i h_{i-1}} f(x_i) + \frac{1}{h_i} f(x_{i+1})\Big) D^2_{2,i}v''(\bullet - x_i).$$

The functions $B_{1,i} = D^2_{2,i}v''(\bullet - x_i)$ are positive with support on $[x_{i-1}, x_{i+1}]$.
□

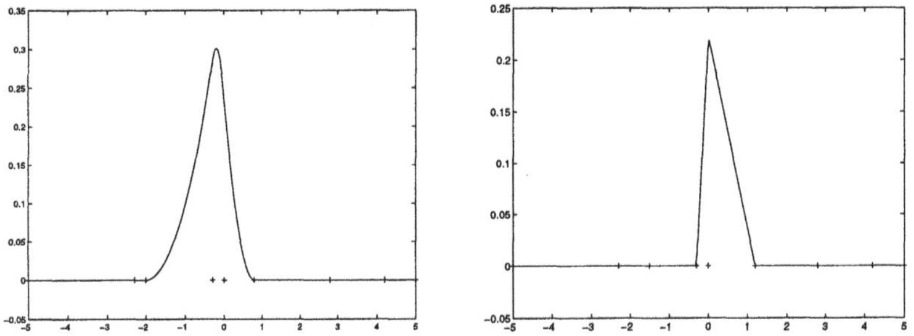

Fig. 2. Basis functions $B_{2,i}$ and $B_{1,i}$.

§5. Examples

In this section some graphs are shown to illustrate the behaviour of the shape-preserving spline functions f_C and $\tilde{f}$ for classical data sets ([5,7,8]). In all pictures the symbol ' + ' denotes the data.

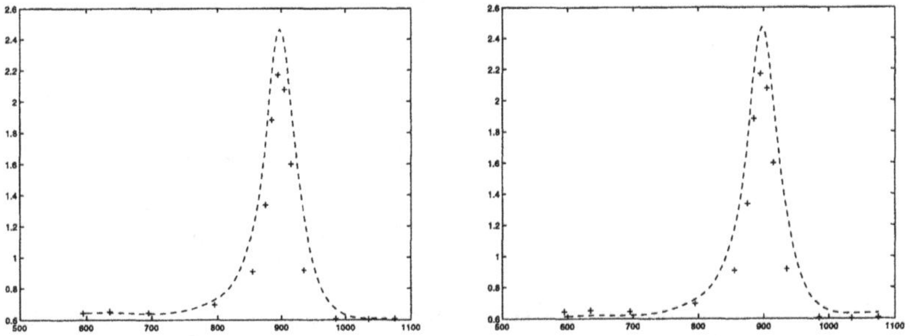

Fig. 3. $\tilde{f}$ and f_C approximants.

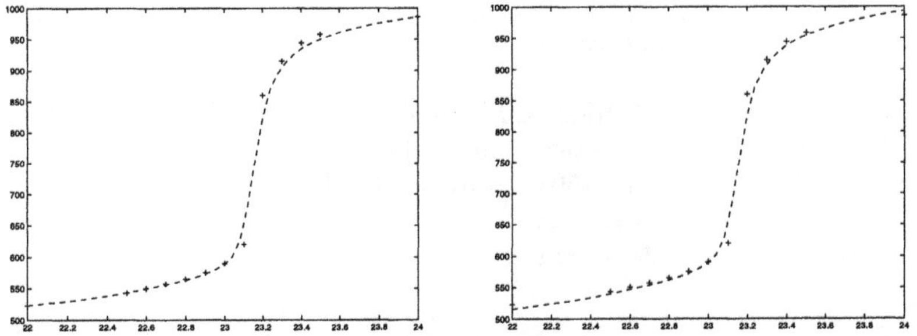

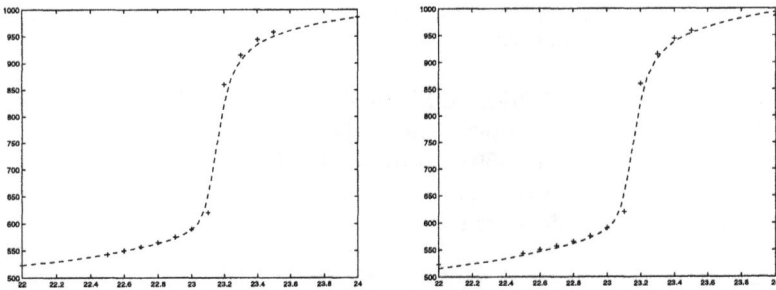

Fig. 4. $\tilde{f}$ and f_C approximants.

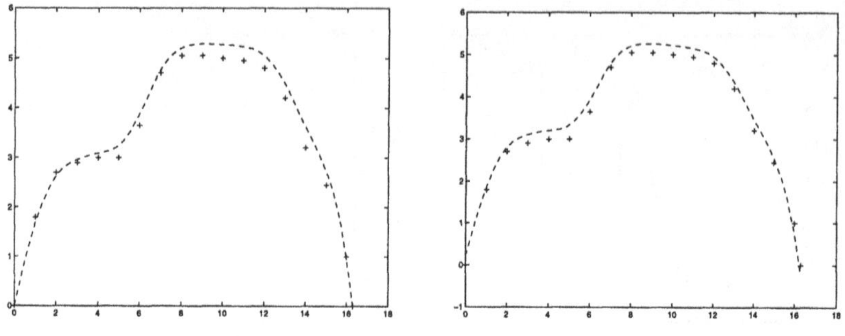

Fig. 5. $\tilde{f}$ and f_C approximants.

References

1. Aronszajn, N., Theory of reproducing kernels, Trans. Amer. Math. Soc. **68** (1950), 337–404.
2. Beatson R. K. and M. J. D. Powell, Univariate multiquadric approximation, Construc. Approx. **8** (1992), 275–288.
3. Benbourhim, M. N., Spline approximant in the Hilbert subspace of $C(\Omega)$, Numer. Math. **49** (1986), 291–303.
4. Conti, C., Funzioni Spline Quasi-Interpolanti: Approssimazione di Curve e Superfici., *Ph. D. thesis*, 1997
5. de Boor, C., *A Practical Guide to Splines*, Springer 1978.
6. Duchon, J., Interpolation des fonctions de deux variables suivant le principe de la flexion plaques minces, Rev. Française Automat. Informat. Rech. Opér., Anal. Numer. **10**, 12 (1976), 5–12.
7. Pruess S., Alternatives to the exponential spline in tension, Math. Comp. **33** (1979), 1273–1281.
8. Roulier J. A., Constrained interpolation, SIAM J. Sci. Stat. Comp. **1** (1980), 333–344.
9. Wu Z. and R. Schaback, Shape-preserving properties and convergence of univariate multiquadric quasi-interpolation, Acta Mathematicae Sinica, Study Bulletin **10** (1994), 441–446.

Costanza Conti and Rossana Morandi
Dipartimento di Energetica, Università di Firenze,
via C. Lombroso 6/17, -50134 Firenze, ITALY
`conti@fonty.de.unifi.it`
`morandi@ingfil.de.unifi.it`

Christophe Rabut
Département de Génie Mathémathique I.N.S.A., Complexe
Scientifique de Rangueil, 31077 Toulouse-Cédex 4, FRANCE
`rabut@gmm.insa-tlse.fr`

A Technique for the Construction of Generalized B-splines

Costanza Conti and Christophe Rabut

Abstract. For any integer values of n and k (usually $n > k$) and any real function f, we give explicit formulae to derive a discrete approximation of $f^{(k)}(b)$ as a linear combination of pointwise values of the function. This approximation is $\mathbb{P}_{k+n}$-exact in the sense that if f is any polynomial of degree at most n, the obtained approximation is exactly $f^{(k)}(b)$. This approximation is a generalization of the usual divided difference. We then use this approximation in order to derive some generalization of polynomial B-splines and we use these functions in order to improve the linear system obtained in interpolation in quite usual cases.

§1. Introduction

In order to extend some of the nice properties of the usual B-splines, we use the conjunction of the three following observations. First, these functions are obtained divided difference of order k applied to a function φ_t defined for any x in $\mathbb{R}$ by $\varphi_t(x) = \frac{1}{(k-1)!}(x-t)_+^{k-1}$. Second, the divided difference of order k is an approximation of the kth derivative, and third, the observation that the kth derivative of $\varphi_t(x)$ is exactly the Dirac distribution in t. Combining these three considerations, we see that the polynomial B-splines are obtained as some discretization of the kth derivative applied to a function whose k^{th} derivative is exactly the Dirac distribution.

Our extension will be done in three directions. First we use a "better" approximation of the kth derivative; "better" means here that the approximation is exact when applied to a polynomial of degree less than or equal to n, with $n > k$, while the divided difference is exact when applied to a polynomial of degree less than or equal to k. Second we use the discretization of a differential polynomial $q(\mathrm{D})$, instead of the kth derivative, applied to a fundamental solution of $q(\mathrm{D})$, *i.e.* a function φ such that $(q(\mathrm{D}))\varphi$ is the Dirac

Mathematical Methods for Curves and Surfaces II
Morten Dæhlen, Tom Lyche, and Larry L. Schumaker (eds.), pp. 63–70.

ISBN 0-8265-1315-8.

distribution. Last we apply this to functions φ_t^γ which are approximations of the functions mentioned before φ_t in the sense that $\lim_{\gamma\to 0}\varphi_t^\gamma = \varphi_t$. Thus, we obtain "Bell shaped functions" in the sense that they are mainly centered around the point of the approximation and are decaying at infinity. Furthermore, in the case of degree $2m-1$ polynomial splines with equidistant knots, the so-obtained functions may be used for approximation of equidistant data without solving a linear system by setting $\sigma = \sum_{i\in\mathbb{Z}^d} y_i\, B_h(\bullet - ih)$ where the data are the points $(ih, y_i)_{i\in\mathbb{Z}}$, and B_h is obtained with a step h discretization; this approximation is $\mathbb{P}_n$-exact whenever the discretization is $\mathbb{P}_{2m+n}$-exact, where $n \le 2m-1$.

The first part of this paper is devoted to a $\mathbb{P}_{k+n}$-exact generalization of the kth divided difference, and to the generalization to a discretization of a differential polynomial. In the second part we give the definition of the generalized B-splines, and properties of these functions. In the third part we give some examples, and last in the fourth part we examine the extension to the multidimensional case, which includes some of the most commonly used radial basis functions.

Notation.
A is subset of $\mathbb{R}$. A generic point of A will be denoted by a, or c, while b is some fixed point in $\mathbb{R}$. $[A,b]$ is the smallest closed interval in $\mathbb{R}$ containing $A\cup\{b\}$. We denote by $\mathrm{d}(A,b)$ the quantity $\sup_{x\in[A,b]}|b-x|$, and k and n are integers; D^k is the kth derivative: $\mathrm{D}^k f = f^{(k)}$. For ℓ in $\mathbb{N}$, $\mathbb{P}_\ell$ is the set of real polynomials of degree at most ℓ. $\mathcal{F}$ denotes the set of all functions from $\mathbb{R}$ to $\mathbb{R}$.

§2. Generalized Divided Difference

Theorem 1. *Suppose card $(A) = k+n+1$, and let $(\lambda_a^b)_{a\in A}$ be $k+n+1$ real numbers. Suppose that the $(\lambda_a^b)_{a\in A}$ are such that for any p in $\mathbb{P}_{k+n}$,*

$$\sum_{a\in A}\lambda_a^b\, p(a) \;=\; p^{(k)}(b). \tag{1}$$

Then the $(\lambda_a^b)_{a\in A}$ are given uniquely by

$$\lambda_a^b \;=\; \begin{cases} \dfrac{k!}{\prod_{c\in A-\{a\}}(a-c)} & \text{if } n=0 \\[2ex] \dfrac{k!\sum_{\substack{A_{ab}\subset A-\{a,b\}\\ card\,(A_{ab})=n}}\prod_{c\in A_{ab}}(b-c)}{\prod_{c\in A-\{a\}}(a-c)} & \text{if } n>0. \end{cases} \tag{2}$$

Proof: Let $c\in A$, and let p_c be the degree $k+n$ Lagrangean polynomial on A and c such that $p_c(c)=1$ and $p_c(a)=0$ for all $a\in A-\{c\}$. By applying (1) to p_c, we get $\lambda_c = p_c^{(k)}(b)$, which is (2) by expliciting $p_c^{(k)}(b)$.

Definition. *Let $(\lambda_a^b)_{a\in A}$ be defined by (2). Then, for any f in $\mathcal{F}$, the quantity*

$$\mathrm{D}_{n,A}^{k,b} f = \sum_{a\in A} \lambda_a^b f(a) \tag{3}$$

*is called the A-*divided difference of order k and level $k+n$ in b of f.

Theorem 2. The A-divided difference of order k and level $k+n$ in b of f defined in (3) satisfies the following properties:

i) *Suppose $n=0$, then $\mathrm{D}_{n,A}^{k,b} f$ does not depend on b, and is precisely the kth divided difference of f based on A.*

ii) *Let $C=\left(\sum_{a\in A} |\lambda_a^b|\right)(\mathrm{d}(A,b))^{n+1}/(n+1)!$ and suppose $f\in C^{k+n+1}([A,b])$. Then $\left|\mathrm{D}_{n,A}^{k,b} f - f^{(k)}(b)\right| \le C \sup_{x\in[A,b]} |f^{(k+n+1)}(x)|$.*

iii) *Let $A_h=\{b+(a-b)h \mid a\in A\}$ and $[A_h,b]$ be the smallest interval in $\mathbb{R}$ containing $A_h\cup\{b\}$. Then $\mathrm{D}_{n,A_h}^{k,b} f = \sum_{a\in A} h^{-k}\lambda_a^b f(b-(a-b)h)$. Furthermore,*

$$\left|\mathrm{D}_{n,A_h}^{k,b} f - f^{(k)}(b)\right| \le C\, h^{n+1} \sup_{x\in[A_h,b]} |f^{(k+n+1)}(x)|.$$

Proof: i) When $n=0$, the values of $(\lambda_a^b)_{a\in A}$ as given in (2) do not depend on b and are precisely the well-known coefficients of the usual k^{th} divided difference.

ii) is obtained from Taylor's formula of order $k+n$ applied to each point of A, by using (1), and by bounding the remainder.

iii) Using (2), the values $\lambda_{a_h}^h$ such that $\mathrm{D}_{n,A_h}^{k,b} f = \sum_{a_h\in A_h} \lambda_{a_h}^h f(a_h)$ satisfies $\lambda_{a_h}^h = h^{-k}\lambda_a$ if $a_h = b+(a-b)h$, so we get the result by using ii).

Remark. This theorem shows that the A-divided difference of order k and level $k+n$ in b when applied to a smooth function can be considered as an approximation of $f^{(k)}(b)$ using only point evaluations of f. The rate of convergence is a $\mathcal{O}(h^{n+1})$, and so the approximation is better when n is larger. It is a generalization of the divided difference which is obtained for $n=0$; this generalization depends on b when $n\neq 0$.

We now extend this to a differential polynomial:

Definition. *Suppose* card $(A)=k+n+1$. *Let q be a polynomial of degree k and let $c=(c_j)_{j=0,\dots k}$ be such that $q(x)=\sum_{j=0}^{k} c_j x^j$. Then the quantity $\left(q(\mathrm{D})\right)_{n,A}^{b} f = \sum_{j=0}^{k} c_j\, \mathrm{D}_{n,A}^{j,b} f$ is called the A-*discretization of $q(\mathrm{D})f$ of order $k+n$ in b.

§3. Generalized B-splines

We now want to derive some functions which generalize, in some sense, the usual B-spline. Let q be a real valued polynomial of degree k such that $q(0)=0$, and let $q(\mathrm{D})$ be the associated differential operator. Let the function $\varphi \in \mathcal{F}$ be such that $q(\mathrm{D})\varphi=\delta$, where δ is the Dirac distribution. For some real numbers γ, let $\varphi_\gamma \in \mathcal{F}$ be such that $\lim_{\gamma\to 0}\varphi_\gamma = \varphi$ and $\lim_{|x|\to\infty} \frac{\varphi_\gamma(x)}{\varphi(x)} = 1$ (possibly $\varphi_\gamma=\varphi$).

For some fixed $b \in \mathbb{R}$, we define the function $B_b(x)=(q(\mathrm{D}))^b_{n,A}\varphi_\gamma(\bullet - x)$. We should obtain some approximation, say, of δ_b, the Dirac distribution in b, since $(q(\mathrm{D}))^b_{n,A}\varphi_\gamma$ can be seen as an approximation of $\big(q(\mathrm{D})\varphi_\gamma\big)(b)$, φ_γ as some approximation of φ and $q(\mathrm{D})\varphi=\delta$. Besides, since $q(0)=0$, $q(\mathrm{D})$ has a regularization effect, while $(q(\mathrm{D}))^b_{n,A}$ "tries to cancel" the effect of $q(\mathrm{D})$ without altering the regularity of φ. So B_b can be considered as some regularized approximation of the Dirac distribution, or equivalently, we can say that B_b "tries to imitate" the Dirac distribution, while being a linear combination of translates of φ_γ, and thus keeping the regularity of φ_γ. So B_b should be decaying at infinity, mainly "centered" around b, and have some other properties of the usual B-splines. We can also expect that the higher n is, the better is the "approximation of the Dirac distribution", and so the more "centered" B_b should be around b. Actually, we will see that the rate of the decay of B_b is higher when n is bigger. This idea was first used by Nira Dyn and David Levin in [4], where in two dimensions they applied a discretization of the biharmonic operator to the function $\varphi(x)=\|x\|^2 \ln \|x\|^2$ (they used divided differences, *i.e.* in the present notation, with $n=0$ and $\gamma=0$).

Definition. *Suppose* card $(A)=k+n+1$, *and let* $B_b(x)=(q(\mathrm{D}))^b_{n,A}\varphi_\gamma(x-\bullet)$. *Then* B_b *is called the* generalized B-spline of order n in b associated with A and the function φ_γ.

Theorem 3. *For the function* B_b *being defined as above, we have:*

i) *the function* B_b *is decaying at infinity; more precisely:*

$$B_b(x) = o\left(|x|^{-n-1}\right) \qquad \text{as } |x| \to \infty. \tag{4}$$

ii) *Suppose* $\gamma=0$, *and let* j *be an integer. Then,*

$$\int_{\mathbb{R}} x^j\, B_b(x)\, dx \;=\; \begin{cases} 1, & \text{if } j=0, \\ 0, & \text{if } 0<j\le n. \end{cases} \tag{5}$$

Proof: i) The proof is done in [2, p.24] when $\gamma=0$; it consists in proving that the Fourier transform $\widehat{B_b}$ of B_b is in $C^{n+1}(\mathbb{R})$. For $\gamma \neq 0$, we use the relation $\lim_{|x|\to\infty} \frac{\varphi_\gamma(x)}{\varphi(x)} = 1$.

ii) To prove (5), we first prove the relation $\widehat{B_b}(0) = 1$ if $j=0$, and $\mathrm{D}^\alpha \widehat{B_b}(0) = 0$ if $0<j\le n$; this is obtained by using (1) and (3).

Remark. The relations (4) and (5) show that B_b is a "bell shaped function", *i.e.* a decaying function whose integral over $\mathbb{R}$ is finite. Furthermore, when

$\gamma=0$, the larger n is, the more moments of B_b are equal to 0. In this sense, the larger n is, the better B_b is an approximation of the Dirac distribution. Since φ_γ is a kind of approximation of φ, we have the same qualitative conclusion for φ_γ, in a weaker sense since the moments of B_b are usually not equal to 0 when $\gamma\neq 0$.

§ 4. Applications

We now use A such that card $(A) > k+n+1$ and, for any a in A, $A_a \subset A$ is a subset of A which depends on a and satisfies card $(A_a)=k+n+1$ (usually the points of A_a are close to a). $B_a = \left(q(\mathrm{D})\right)^a_{n,A_a}\varphi_\gamma(\bullet - x)$ is the generalized B-spline of order n associated with A_a and the function φ_γ.

We show in this section how these "generalized B-splines" can be used to improve the quality of the linear system to be solved in order to determine an interpolating function of the form $\sum\limits_{a\in A} \mu_a\varphi_\gamma(\bullet - a)$. Then we show that in the polynomial spline case, if the data are equidistant, the approximation of a function f by $\sigma=\sum\limits_{a\in A} f(a)\,B_a$ is better when using large n, up to a certain level. Finally, we present some B_b functions for various φ_γ.

4.1 Improving a Linear System for Interpolation

Let A be a discrete set containing a $\mathbb{P}_{k+n}$-unisolvent set. We now want to determine a function σ of the form $\sigma=\sum_{a\in A}\mu_a\varphi_\gamma(\bullet - a)$ such that for any c in A, $\sigma(c)=y_c$. To do so, we can of course solve the linear system $\forall c\in A,\ \sum_{a\in A}\mu_a\varphi_\gamma(c-a) = y_c$. However, this system is usually badly conditioned.

To improve this system, let us use B_a as defined in the introduction of this section, and let us remark that if $\sigma = \sum_{a\in A}\mu'_a B_a$, then σ is also of the form $\sigma = \sum_{a\in A}\mu_a\varphi_\gamma(\bullet - a)$. So, in order to determine the function σ that interpolates the points $(c, y_c)_{c\in A})$, we can either solve the linear system $\forall c\in A,\ \sum_{a\in A}\mu_a\varphi_\gamma(c-a) = y_c$, or solve the linear system $\forall a\in A,\ \sum_{a\in A}\mu'_a B_a(c) = y_c$. Due to the general shape of the functions B_a (particularly to the decay of B_a at infinity), the matrix of the second linear system is most likely better centered around its diagonal than the one of the first system. In this sense, the system in $(\mu'_a)_{a\in A}$ is generally numerically "better" (more stable for some rounding errors) than the one in $(\mu_a)_{a\in A}$. In fact, this was the motivation of Nira Dyn and David Levin in [4] (with $\gamma=n=0$ in the particular case of thin plate splines). The efficiency is all the better when n is larger and γ is smaller. An example for $\gamma\neq 0$ is given below (§4.3.2) with $\varphi_\gamma(x)=\sqrt{x^2+\gamma^2}$.

4.2 Odd Degree Polynomial B-splines of Order n on Cardinal Data

We present here the particular case of polynomial splines with cardinal data of step h.

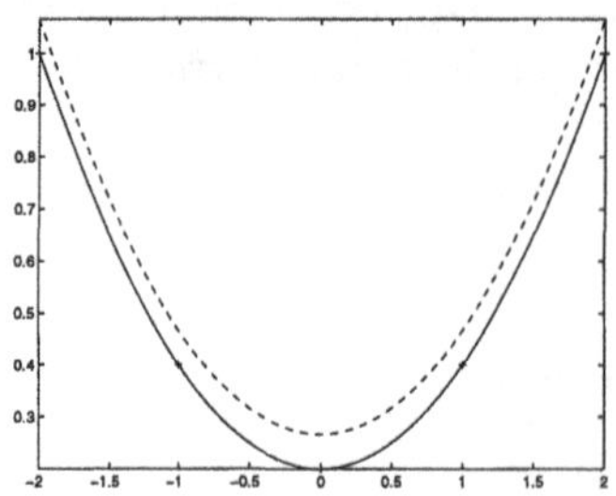

Fig. 1. Approximation of a parabola

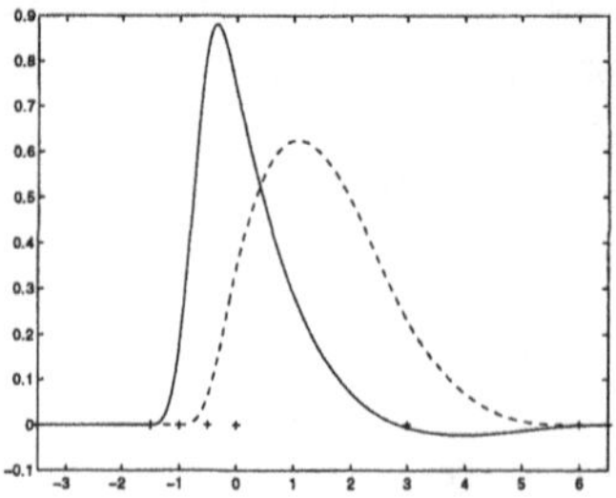

Fig. 2. M_i (– –) and N_i . (—) .

Definition. *Let $m \in \mathbb{N}$ be such that $m \neq 0$, and let $n \in \mathbb{N}$ be such that $n < m$. Let φ be such that for any x in $\mathbb{R}$, $\varphi(x) = \frac{1}{2\cdot(2m-1)!}|x|^{2m-1}$. φ satisfies the relation $\varphi^{(2m)} = \delta$. Let $A_h^n = \{jh \mid j \in \mathbb{Z}^d \; |j| \le m+n\}$, and let B_h^n be defined by $B_h^n = h\,\mathrm{D}_{2n,A_h^n}^{2m,0}\,\varphi$ B_h^n is called the* polynomial B-spline of degree $2m-1$ of order n and step h.

B_h^n is a generalization of the usual odd degree polynomial B-splines, and share many of their properties, including the following (see [5] for proofs):

Theorem 4.

i) *The B-spline approximation using B_h^n is $\mathbb{P}_{2n+1}$-reproducing, i.e. for any $p \in \mathbb{P}_{2n+1}$, $\sum_{j\in\mathbb{Z}} p(jh)\, B_h^n(\bullet - jh) = p$.*

ii) *Let $\ell \in \mathbb{N}_*$ and $f \in C^\ell(\mathbb{R})$ be such that all derivatives of f up to the order ℓ are bounded over $\mathbb{R}$. Then the following relation holds:*

$$\left\| \textstyle\sum_{j\in\mathbb{Z}} f(jh)\, B_h^n(\bullet - jh) - f \right\|_\infty = \begin{cases} \mathrm{O}(h^\ell) & \text{if } \ell \le 2n+1 \\ \mathrm{O}(h^{2n+2}|\ln h| & \text{if } l \ge 2n+2 \end{cases}$$

as h goes to 0.

Fig. 1 shows the approximation of a parabola using B_h (resp. B_h^1), where the points $(ih, p(ih))_{i\in\mathbb{Z}}$ are denoted by $\times$, the dashed line is for $\sigma_{h,0}^p$, and the solid line is for $\sigma_{h,0}^p$.

4.3 Examples with Non-regular Data

4.3.1 Cubic Splines. Let $N \in \mathbb{N}$ and $A = \{a_{-6}, a_{-5}, \ldots, a_{N+6}\}$. Moreover, for $-3 \le i \le N+3$ and $-1 \le k \le 1$ let $A_i^k = \{a_{i-2-k}, a_{i-1-k}, \ldots, a_{i+2+k}\}$. Let $\varphi(x) = |x|^3/12$ (so we have $\varphi^{(4)} = \delta$). Let $B_i = \mathrm{D}_{0,A_i^0}^{4,a_i}\varphi$ and $C_i(x) = \mathrm{D}_{2,A_i^1}^{4,a_i}\varphi$. As we can see in Fig. 2 , B_i and C_i are "bell shaped" functions. Actually the function $M_i(x) = \frac{a_{i+2}-a_{i-2}}{4} B_i$ is known as the usual cubic spline with knots A_i^0 (see for example [1], or [6]). In the same way, for equidistant knots, $N_i = h\,C_i$, where $h = a_{i+1} - a_i$, can be used as a B-spline. Because of the local support of B_h, the results obtained in 4.2.iiib are easily extendable for finitely many equidistant data.

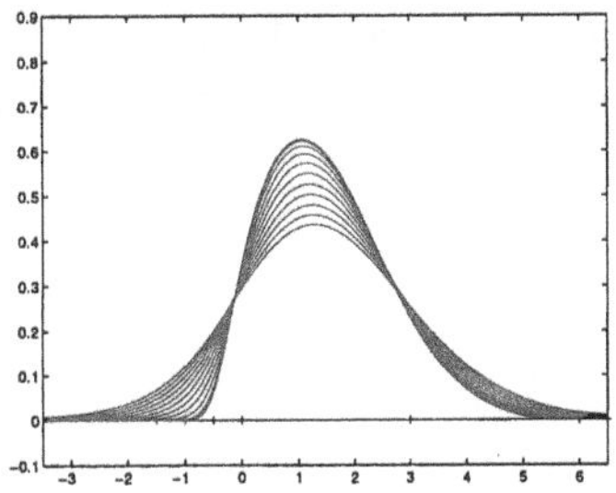

Fig. 3. B_i^γ for $\gamma=0.,.2,\ldots,2$

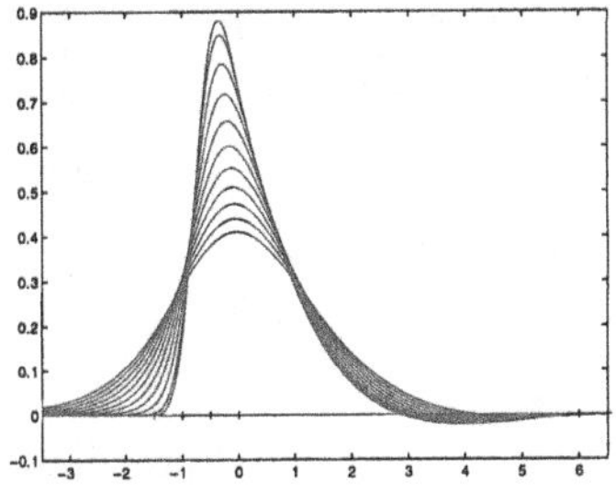

Fig. 4. C_i^γ for $\gamma=0.,.2,\ldots,2.$

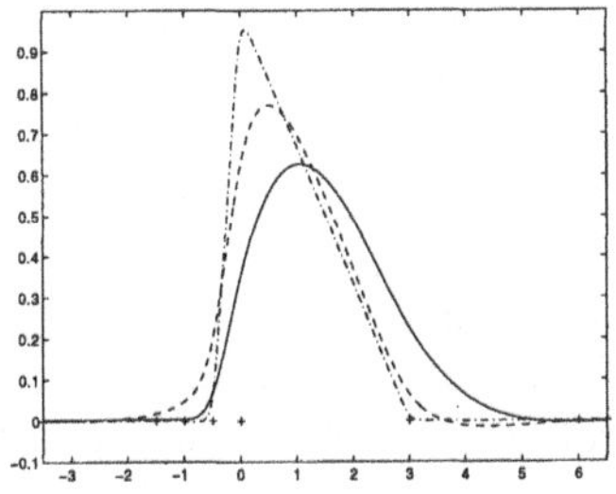

Fig. 5. Splines under tension ($\rho=0,.2,\ldots,2$).

4.3.2 Univariate "Shifted Pseudocubic Splines". We use the same A and A_i^k as above. For some real number γ, let $\varphi_\gamma(x) = \left(\sqrt{x^2+\gamma^2}\right)^3/12$ (φ_γ satisfies the requirements in the introduction). Let $B_i^\gamma = \mathrm{D}^{4,a_i}_{0,A_i^0}\varphi_\gamma$ and $C_i^\gamma = \mathrm{D}^{4,a_i}_{2,A_i^2}\varphi_\gamma$. The curves of B_i^γ (resp.C_i^γ) for $\gamma = 0.2, 0.4, \ldots, 2$ are shown in Fig. 3 (resp. Fig. 4). The smaller γ is, the closer to B_i (resp.C_i) B_i^γ (resp. C_i^γ) is, and the bigger $B_i^\gamma(0)$ (resp. $C_i^\gamma(0)$) is.

4.3.3 Splines under Tension. We use same A and A_i^k as above. Let $\rho \in \mathbb{R}$, and let $\varphi(x) = \frac{-1}{2\rho^3}(\mathrm{e}^{-\rho|x|} + \mathrm{e}^{\rho|x|})$. Then $(\mathrm{D}^4 + \rho\mathrm{D}^2)\varphi = \delta$. The function $B_i = \frac{a_{i+2}-a_{i-2}}{4}\mathrm{D}^{4,a_i}_{0,A_i^0} + \rho\frac{a_{i+1}-a_{i-1}}{2}\mathrm{D}^{2,a_i}_{0,A_i^{-1}})\varphi$ is shown in Fig. 5 for $\rho = 0.1$ (continuous line), $\rho=2$ (line $--$), and $\rho=15$ (line $-\cdot-\cdot$). B_i is a spline under tension, as defined for example in [7], and is mainly concentrated around the knot 0. The smaller (resp. the larger) ρ is, the closer the cubic (resp. linear) B-spline B_i is.

§5. Extension to Many Dimensions and Scattered Data

The divided difference of order n and the discretization of $q(\mathrm{D})f$ of order n can be extended to many dimensions. We give some details of the extension. Let $\alpha = (\alpha_1, \ldots, \alpha_d)$ be a multi-index. If $A = A_1 \times A_2 \times \ldots \times A_d$ with card $(A_i) = \alpha_i + n + 1$ for $i = 1, \ldots, d$, we can define $\mathrm{D}^{\alpha,b}_{n,f}$ the αth level n divided difference in b of f by $D^{\alpha,b}_{n,A} f = D^{\alpha_1,b}_{n,A_1} f \, D^{\alpha_2,b}_{n,A_2} f \ldots D^{\alpha_d,b}_{n,A_d} f$ where, for any i in $[1,d]$, f in $D^{\alpha_i,b}_{n,A_i} f$ is considered as a function of the unique i^{th} variable. We then get explicit values of the coefficients $(\lambda^b_a)_{a\in A}$ such that $D^{\alpha,b}_{n,A} f = \sum\limits_{a\in A} \lambda^b_a f(a)$. If the points of A are scattered, we can also define the order α and level n divided difference in b of f in the form $D^{\alpha,b}_{n,A} f = \sum\limits_{a\in A} \lambda^b_a f(a)$, but in this case the coefficients $(\lambda^b_a)_{a\in A}$ are not explicitly known and are obtained by solving a linear system. Proceeding as in Section 3, we can improve the linear system to be solved in order to determine the interpolating function of the form $\sum\limits_{a\in A} \mu_a f(\bullet - a)$.

All this includes many commonly used radial basis functions, such as the m-harmonic splines ($\varphi = a_{m,d} \, \|\bullet\|^{2m-d}$ if d is odd, $\varphi = a_{m,d} \|\bullet\|^{2m-d} \ln \|\bullet\|$ if d is even where $a_{m,d}$ is some known constant, see [3]), the shifted polyharmonic splines ($\varphi_\gamma = \varphi(\sqrt{\|\bullet\|^2 + \gamma^2})$ where φ is the above defined function, but not all of them, since for example the multiquadrics $\varphi_\gamma = \sqrt{\|\bullet\|^2 + \gamma^2}$ when d is even are not included in E (but they are included when d is odd since then they are the shifted polyharmonic splines). The results in theorem 4.2.i and ii. are similar for cardinal data polyharmonic splines in any dimension.

References

1. de Boor, C., *A Practical Guide to Splines*, Springer Verlag, New York, 1978.
2. Conti, C., Funzioni Spline Quasi-interpolanti: Approssimazione di Curve e Superfici, Thesis, Firenze, 1997.
3. Duchon, J., Interpolation des fonctions de 2 variables suivant le principe de la flexion des plaques minces, Rev. Française Automat. Informat. Rech. Opér., Anal. Numer. **10** (1976), 5–12.
4. Dyn, N., and D. Levin, Iterative solution of system originating from integral equations and surface interpolation, SIAM J. Numer. Anal. **20** (1983), 377–390.
5. Rabut, C., B-splines polyharmoniques cardinales : Interpolation, quasi-interpolation, filtrage, thesis, Toulouse, 1990.
6. Schumaker, L. L., *Spline Functions: Basic Theory*, Wiley, New York, 1981.
7. Schweikert, D.G., An interpolation curve using a spline in tension, J. Math. Phys. **45** (1966), 312–317.

Costanza Conti
Dipartimento di Energetica
Università di Firenze
via Lomboso 6/17
I-50134 Firenze, ITALY
costanza@fonty.de.unifi.it

Christophe Rabut
Dépt de Génie Mathématique
INSA, Toulouse
Complexe Scientifique de Rangueil
31077 Toulouse Cédex, FRANCE
rabut@gmm.insa-tlse.fr

Locally Linearly Independent Basis for C^1 Bivariate Splines of Degree $q \geq 5$

Oleg Davydov

Abstract. We construct a locally linearly independent basis for the space $S_q^1(\Delta)$ ($q \geq 5$). Bases with this property were available only for some subspaces of smooth bivariate splines.

§1. Introduction

Let $\Omega \subset \mathbb{R}^2$ be a simply connected polygonal domain, and let Δ denote a triangulation of Ω consisting of N triangles, V vertices and E edges. Given $0 \leq r < q$, consider the linear space of bivariate polynomial splines of degree q and smoothness r,

$$S_q^r(\Delta) := \{s \in C^r(\Omega) : \; s_{|_T} \in \Pi_q \; \text{ for all triangles } T \in \Delta\},$$

where

$$\Pi_q := \operatorname{span}\{x^i y^j : \; i \geq 0, \; j \geq 0, \; i + j \leq q\}$$

is the space of bivariate polynomials of total degree q. To simplify notation, we set

$$d_q := \dim \Pi_q = \frac{(q+1)(q+2)}{2}, \qquad q = 0, 1, \ldots .$$

The question of identifying the dimension of $S_q^r(\Delta)$ was first considered by Strang [14]. Morgan & Scott [11] showed that

$$\dim S_q^1(\Delta) = d_q N - (d_q - d_{q-2})E_I + d_1 V_I + \sigma, \quad q \geq 5, \tag{1}$$

where V_I and E_I denote the number of interior vertices and interior edges respectively, and σ is the number of singular vertices of Δ, *i.e.*, those interior vertices for which the adjacent edges of each attached edge are collinear, so

Mathematical Methods for Curves and Surfaces II
Morten Dæhlen, Tom Lyche, and Larry L. Schumaker (eds.), pp. 71–78.

ISBN 0-8265-1315-8.

that exactly four triangles share a singular vertex and their union is a quadrilateral with the diagonals drawn in.

Moreover, in [11] a nodal basis for $S_q^1(\Delta)$ ($q \geq 5$) was constructed. This means that the functions in $S_q^1(\Delta)$ were determined by their values and derivatives at points in Ω (nodal values). An important feature of the Morgan-Scott basis is that each basis function is supported at most in the star of a vertex, *i.e.*, the union of all triangles sharing the vertex.

For many further results on the dimension and bases for the spaces $S_q^r(\Delta)$ and their superspline subspaces, see [1–5,9,10,12,13,15] and references therein.

As it has been shown recently [7], the property of *local linear independence* of a system of functions plays an important role in the problems of multivariate spline interpolation. Because of this, the question of existence of locally linearly independent systems of spline-functions was considered in [8]. Particularly, it was proved in [8, Section 3.5] that for any $\rho \geq 2r$ and $q \geq 2\rho+1$, the space of supersplines

$$S_q^{r,\rho}(\Delta) := \{s \in S_q^r(\Delta) : \; s \in C^\rho(v) \; \text{ for all vertices } v \in \Delta\}$$

admits a locally linearly independent basis. However, the Morgan-Scott basis for $S_q^1(\Delta)$ is easily seen not to be locally linearly independent (see Remark 11).

The aim of this paper is to provide a locally linearly independent basis for $S_q^1(\Delta)$, $q \geq 5$ (see Theorem 8). We note that our construction in fact differs from that of [11] only in the choice of second order derivatives at the vertices.

§2. Locally Linearly Independent Systems

Let K be a topological space, and let $F(K)$ denote the linear space of all real functions on K. We set

$$\operatorname{supp} f := \overline{\{t \in K : \; f(t) \neq 0\}}, \quad f \in F(K).$$

Definition 1. A system $\{u_1, \ldots, u_n\} \subset F(K)\backslash\{0\}$ is said to be locally linearly independent if for any $t \in K$ and any neighborhood $B(t)$ of t there exists an open set B' such that $t \in B' \subset B(t)$ and the subsystem

$$\{u_i : \; B' \cap \operatorname{supp} u_i \neq \emptyset\}$$

is linearly independent on B'.

For some other possible definitions of locally linearly independent systems of functions as well as their examples, see [7,8]. This notion turned out to be particularly important for the theory of almost interpolation.

Definition 2. Let $U \subset F(K)$ be a finite-dimensional linear space, $\dim U = n$. A set $T = \{t_1, \ldots, t_n\} \subset K$, is called an almost interpolation set with respect to U if for any system of neighborhoods B_i of t_i, $i = 1, \ldots, n$, there exist points $t_i' \in B_i$ such that $T' = \{t_1', \ldots, t_n'\}$ is admissible for Lagrange interpolation from U, *i.e.*, for any given data $\{y_1, \ldots, y_n\}$ there exists a unique function $u \in U$ satisfying

$$u(t_i') = y_i, \quad i = 1, \ldots, n.$$

Theorem 3. [7] *Let* $\{u_1, \ldots, u_n\} \subset F(K)$ *be a locally linearly independent system and* $U = \operatorname{span}\{u_1, \ldots, u_n\}$. *A set* $T = \{t_1, \ldots, t_n\} \subset K$, *is an almost interpolation set w.r.t.* U *if and only if there exists some permutation* σ *of* $\{1, \ldots, n\}$ *such that*

$$t_i \in \operatorname{supp} u_{\sigma(i)}, \quad i = 1, \ldots, n.$$

Another useful feature of a locally linearly independent system is that it forms a least supported basis for its span.

Theorem 4. [6] *A system* $\{u_1, \ldots, u_n\} \subset F(K) \setminus \{0\}$ *is locally linearly independent if and only if* $\{u_1, \ldots, u_n\}$ *is a least supported basis for* $U = \operatorname{span}\{u_1, \ldots, u_n\}$, *i.e., for every basis* $\{v_1, \ldots, v_n\}$ *of* U *there exists a permutation* σ *of* $\{1, \ldots, n\}$ *such that*

$$\operatorname{supp} u_i \subset \operatorname{supp} v_{\sigma(i)}, \quad i = 1, \ldots, n.$$

The following characterization of local linear independence of a system of splines in $S_q^r(\Delta)$ is an immediate consequence of [7, Theorem 3.5].

Theorem 5. *Let* $\{s_1, \ldots, s_n\} \subset S_q^r(\Delta)\setminus\{0\}$ *and* $\Pi_q \subset S := \operatorname{span}\{s_1, \ldots, s_n\}$. *Then* $\{s_1, \ldots, s_n\}$ *is locally linearly independent if and only if*

$$\operatorname{card}\{i : T \subset \operatorname{supp} s^{[i]}\} = d_q, \quad \textit{for every triangle } T \in \Delta. \tag{2}$$

Proof: It follows from [7, Theorems 3.5 and 3.4, 2)] that a basis for S is locally linearly independent if and only if the local dimension of S at any point t in the interior of an arbitrary triangle $T \in \Delta$ equals the number of basis functions which are supported on a neighborhood of t. Since $\Pi_q \subset S$, it is easy to see that the local dimension at such a point is always d_q and that every spline $s \in S$ which is supported on a neighborhood of t, is also supported on the whole triangle T, so that $\operatorname{card}\{i : T \subset \operatorname{supp} s^{[i]}\}$ equals the number of basis functions supported on a neighborhood of t. □

§3. A Basis for $S_q^1(\Delta)$ ($q \geq 5$)

Following the notation introduced in [11], we consider "edge derivatives" of splines $s \in S_q^1(\Delta)$. Let v be a vertex in Δ, let e_1, e_2 be two consecutive edges attached to v, and let T be the triangle with vertex v and edges e_1, e_2. By the first and, respectively, second e_i-derivative of s at v we mean

$$s_{e_i}(v) := \frac{\partial s}{\partial r_i}(v) \quad \text{and} \quad s_{e_i^2}(v) := \frac{\partial^2 (s_{|T})}{\partial r_i^2}(v), \quad i = 1, 2,$$

where r_i is the unit vector in the e_i direction away from v. Furthermore, by the cross (e_1, e_2)-derivative of s at v we mean

$$s_{e_1e_2}(v) := \frac{\partial^2(s_{|T})}{\partial r_1 \partial r_2}(v).$$

For every edge $e \in \Delta$ we choose one (of two possible) unit vectors orthogonal to e and denote it by $r^{\perp}$. Then the edge normal derivative of s at any point $z \in e$ is defined by

$$s_{e^{\perp}}(z) := \frac{\partial s}{\partial r^{\perp}}(z).$$

Let e', e, e'' be three consecutive edges attached to a vertex v. Denote by θ' and θ'' the angles between e and e' and between e and e'' respectively. Then the second e-derivative and the cross derivatives of s at v stay in the following relation

$$\sin(\theta' + \theta'')s_{e^2}(v) = \sin\theta'' s_{ee'}(v) + \sin\theta' s_{ee''}(v). \tag{3}$$

(See equation (III) in [11].) Particularly, if e is degenerate at v, *i.e.*, $\theta' + \theta'' = \pi$, then we have

$$\sin\theta'' s_{ee'}(v) = -\sin\theta' s_{ee''}(v). \tag{4}$$

For every vertex $v \in \Delta$, let $T_v^1, \ldots, T_v^{n_v}$ be all triangles attached to v and numbered in counterclockwise order (starting from a boundary triangle if v is a boundary vertex). Denote by e_i the common edge of T_v^{i-1} and T_v^i, $i = 2, \ldots, n_v$. If v is an interior vertex, $e_1 = e_{n_v+1}$ denote the common edge of T_v^1 and $T_v^{n_v}$. Otherwise, e_1 and e_{n_v+1} are the boundary edges (attached to v) of T_v^1 and $T_v^{n_v}$ respectively. We now consider the following set of nodal values:

1) for each vertex $v \in \Delta$, the nodal values $s(v)$, $s_x(v)$, and $s_y(v)$,
2) for each edge $e \in \Delta$, the nodal values $s(z_e^{0,1}), \ldots, s(z_e^{0,q-5})$, where $\{z_e^{0,1}, \ldots, z_e^{0,q-5}\}$ is a set of distinct points in the interior of e,
3) for each edge $e \in \Delta$, the nodal values $s_{e^{\perp}}(z_e^{1,1}), \ldots, s_{e^{\perp}}(z_e^{1,q-4})$, where $\{z_e^{1,1}, \ldots, z_e^{1,q-4}\}$ is a set of distinct points in the interior of e,
4) for each triangle $T \in \Delta$, the nodal values $s(z_T^1), \ldots, s(z_T^{d_{q-6}})$, where $\{z_T^1, \ldots, z_T^{d_{q-6}}\} \subset \operatorname{int} T$ is a set of points admissible for Lagrange interpolation from Π_{q-6},
5) for each vertex $v \in \Delta$, the nodal values $s_{e_ie_{i+1}}(v)$ for all $i \in \{1, \ldots, n_v\}$ such that e_i is nondegenerate at v,
6) for each vertex $v \in \Delta$, the nodal values $s_{e_i^2}(v)$ for all i such that e_i is degenerate at v,
7) for each boundary vertex $v \in \Delta$, the nodal values $s_{e_1^2}(v)$ and $s_{e_{n_v+1}^2}(v)$,
8) for each singular vertex $v \in \Delta$, the nodal value $s_{e_1e_2}(v)$.

While our sets of nodal values of type 1–4 are the same as those of the paper [11], the remaining nodal values are chosen differently. However, it is easy to see that the number of nodal values in 5)–8) is n_v if v is a nonsingular interior vertex, $n_v + 1$ if v is a singular vertex, and $n_v + 2$ if v is a boundary vertex, which equals the number of nodal values of type 5 and 6 in [11]. Therefore, a calculation in [11] shows that the total number of nodal values listed in 1)–8) is equal to the dimension of $S_q^1(\Delta)$.

Lemma 6. *Suppose that* $s \in S_q^1(\Delta)$ *has all nodal values in 1)–8) equal to zero. Then* $s \equiv 0$.

Proof: Following the argumentation in [11], we only need to show that for each vertex $v \in \Delta$,

$$s_{e_i^2}(v) = 0, \quad i = 1, \dots, n_v + 1, \quad \text{and} \quad s_{e_i e_{i+1}}(v) = 0, \quad i = 1, \dots, n_v.$$

Let us first consider cross derivatives. Since the nodal values of type 5 are zero, we have $s_{e_i e_{i+1}}(v) = 0$ for all i such that e_i is nondegenerate at v. If e_i is degenerate at v, then it follows from (4) that

$$\frac{1}{\sin\theta_i} s_{e_i e_{i+1}}(v) = -\frac{1}{\sin\theta_{i-1}} s_{e_{i-1} e_i}(v),$$

where θ_i denotes the angle between e_i and e_{i+1}. If now e_{i-1} is nondegenerate at v, then $s_{e_{i-1} e_i}(v) = 0$ and hence $s_{e_i e_{i+1}} = 0$. If both e_i and e_{i-1} are degenerate at v, but e_{i-2} is nondegenerate, then

$$\frac{1}{\sin\theta_i} s_{e_i e_{i+1}}(v) = -\frac{1}{\sin\theta_{i-1}} s_{e_{i-1} e_i}(v) = \frac{1}{\sin\theta_{i-2}} s_{e_{i-2} e_{i-1}}(v) = 0,$$

hence $s_{e_i e_{i+1}} = 0$. Finally, if at least three edges are degenerate at v, then v is necessarily a singular vertex, $n_v = 4$ and all four edges e_1, e_2, e_3 and e_4 are degenerate at v. Then

$$\frac{1}{\sin\theta_1} s_{e_1 e_2}(v) = -\frac{1}{\sin\theta_2} s_{e_2 e_3}(v) = \frac{1}{\sin\theta_3} s_{e_3 e_4}(v) = -\frac{1}{\sin\theta_4} s_{e_4 e_1}(v).$$

Since the nodal value of type 8 is zero, we deduce from the last equation that all four cross derivatives at v are zero.

It remains to show that all second e_i-derivatives are also zero at v. Since the nodal values of type 6 and 7 are zero, we only consider those i for which e_i lies in the interior of Ω and is nondegenerate at v. Then $\sin(\theta_i + \theta_{i-1}) \neq 0$, so that by (3),

$$s_{e_i^2}(v) = \frac{\sin\theta_i}{\sin(\theta_i + \theta_{i-1})} s_{e_i e_{i+1}}(v) + \frac{\sin\theta_{i-1}}{\sin(\theta_i + \theta_{i-1})} s_{e_i e_{i-1}}(v).$$

We have already proved that $s_{e_i e_{i+1}}(v) = s_{e_i e_{i-1}}(v) = 0$. Therefore, $s_{e_i^2}(v) = 0$. □

Let us number the nodal values $1, \ldots, D$, where

$$D := \dim S_q^1(\Delta) = d_q N - (d_q - d_{q-2})E_I + d_1 V_I + \sigma.$$

Then, in view of Lemma 6, it follows from basic linear algebra that for each $j \in \{1, \ldots, D\}$ there exists a unique spline $s^{[j]} \in S_q^1(\Delta)$ which has the j-th nodal value equal to 1 and all the other nodal values zero. Moreover, it is clear that $\{s^{[1]}, \ldots, s^{[D]}\}$ is a basis for $S_q^1(\Delta)$. We say that a basis function $s^{[j]}$ is of type 1–7 or 8 if the j-th nodal value is of the corresponding type.

We now describe the supports of the basis functions $s^{[j]}$.

Lemma 7.

1) *The support of a basis function $s^{[j]}$ of type 1 is the star of the vertex v.*
2) *The support of a basis function $s^{[j]}$ of type 2 is the union of the two triangles sharing the edge e.*
3) *The support of a basis function $s^{[j]}$ of type 3 is the union of the two triangles sharing the edge e.*
4) *The support of a basis function $s^{[j]}$ of type 4 is the triangle T.*
5) *The support of a basis function $s^{[j]}$ of type 5 is either the union of the triangles T_v^{i-1}, T_v^i and T_v^{i+1} if the edge e_{i+1} is nondegenerate at v, or the union of the triangles T_v^{i-1}, T_v^i, T_v^{i+1} and T_v^{i+2} if the edge e_{i+1} is degenerate at v, but e_{i+2} is nondegenerate at v, or the union of the triangles T_v^{i-1}, T_v^i, T_v^{i+1}, T_v^{i+2} and T_v^{i+3} if the edges e_{i+1} and e_{i+2} both are degenerate at v, but e_{i+3} is nondegenerate at v. (If v is a boundary vertex, then the triangles with a superscript not in $\{1, \ldots, n_v\}$ should be omitted.)*
6) *The support of a basis function $s^{[j]}$ of type 6 is the union of the two triangles sharing the edge e_i.*
7) *The support of a basis function $s^{[j]}$ of type 7 is one of the triangles T_v^1 or $T_v^{n_v}$.*
8) *The support of a basis function $s^{[j]}$ of type 8 is the star of the vertex v.*

Proof: The statements 1)–4) are obvious. In order to prove 5)–8), we argue in the same way as in the proof of Lemma 6, except that one of the nodal values of $s \in S_q^1(\Delta)$ is 1 while the others are zero. Suppose, for example, that $s^{[j]}$ is of type 5 and the corresponding nodal value is $s_{e_\mu e_{\mu+1}}(v)$, with both e_μ and $e_{\mu+1}$ nondegenerate at v. Then $s^{[j]}_{e_\mu e_{\mu+1}}(v) = 1$ and all the other nodal values of $s^{[j]}$ are zero. The same calculation as in the proof of Lemma 6 shows that

$$s^{[j]}_{e_i^2}(v) = 0, \quad i = 1, \ldots, n_v + 1, \quad i \notin \{\mu, \mu + 1\},$$

$$s^{[j]}_{e_i e_{i+1}}(v) = 0, \quad i = 1, \ldots, n_v, \quad i \neq \mu,$$

and

$$s^{[j]}_{e_\mu^2}(v) = \frac{\sin\theta_\mu}{\sin(\theta_\mu + \theta_{\mu-1})} \neq 0, \quad s^{[j]}_{e_{\mu+1}^2}(v) = \frac{\sin\theta_\mu}{\sin(\theta_\mu + \theta_{\mu+1})} \neq 0,$$

from which it follows that $\operatorname{supp} s^{[j]} = T_v^{\mu-1} \cup T_v^{\mu} \cup T_v^{\mu+1}$. □

As soon as Lemma 7 is established, it is easy to check that the basis $\{s^{[1]}, \ldots, s^{[D]}\}$ satisfies (2), so that the following theorem is valid.

Theorem 8. *The above-constructed basis* $\{s^{[1]}, \ldots, s^{[D]}\}$ *for* $S_q^1(\Delta)$ $(q \geq 5)$ *is locally linearly independent.*

In view of Theorems 3 and 4, the next two statements immediately follow from Theorem 8.

Corollary 9. *Let* $\{t^{[1]}, \ldots, t^{[D]}\} \subset \Omega$ *be a set of distinct points such that* $t^{[j]} \in \operatorname{supp} s^{[j]}$, $j = 1, \ldots, D$, *where* $\operatorname{supp} s^{[j]}$ *are described in Lemma 7. Then* $\{t^{[1]}, \ldots, t^{[D]}\}$ *is an almost interpolation set with respect to* $S_q^1(\Delta)$.

Corollary 10. $\{s^{[1]}, \ldots, s^{[D]}\}$ *is a least supported basis for* $S_q^1(\Delta)$.

Remark 11. The original scheme by Morgan and Scott makes use of one cross derivative and all except one edge derivatives at each nonsingular interior vertex v. Let us denote the exceptional edge by e. It is not difficult to see that all basis functions corresponding to second order derivatives at v are supported on both triangles T' and T'' sharing e. Hence, there are at least $d_q + n_v - 3$ basis functions supported on T', where n_v is the number of triangles attached to v. By Theorem 5 it then follows that the Morgan-Scott basis fails to be locally linearly independent as soon as there exists a nonsingular interior vertex v with $n_v > 3$.

Acknowledgments. The research was supported in part by a Research Fellowship from the Alexander von Humboldt Foundation.

References

1. Alfeld, P., On the dimension of multivariate piecewise polynomials, in *Numerical Analysis 1985*, D. F. Griffiths and G. A. Watson (eds.), Longman Scientific and Technical (Essex), 1986, 1–23.
2. Alfeld, P., B. Piper, and L. L. Schumaker, Minimally supported bases for spaces of bivariate piecewise polynomials of smoothness r and degree $d \geq 4r + 1$, Comput. Aided Geom. Design **4** (1987), 105–123.
3. Alfeld, P., B. Piper, and L. L. Schumaker, An explicit basis for C^1 quartic bivariate splines, SIAM J. Numer. Anal. **24** (1987), 891–911.
4. Alfeld, P., and L. L. Schumaker, On the dimension of bivariate spline spaces of smoothness r and degree $d = 3r + 1$, Numer. Math. **57** (1990), 651–661.
5. Billera, L. J., Homology of smooth splines: Generic triangulations and a conjecture of Strang, Trans. Amer. Math. Soc. **310** (1988), 325–340.
6. Carnicer, J. M., and J. M. Peña, Least supported bases and local linear independence, Numer. Math. **67** (1994), 289–301.

7. Davydov, O., M. Sommer, and H. Strauss, On almost interpolation and locally linearly independent bases, preprint.
8. Davydov, O., M. Sommer, and H. Strauss, Locally linearly independent systems and almost interpolation, in *Multivariate Approximation and Splines*, G. Nürnberger, J. W. Schmidt, and G. Walz (eds.), ISNM, Birkhäuser, 1997, 59–72.
9. Hong, D., Spaces of bivariate spline functions over triangulation, Approx. Theory Appl. **7** (1991), 56–75.
10. Ibrahim, A., and L. L. Schumaker, Super spline spaces of smoothness r and degree $d \geq 3r + 2$, Constr. Approx. **7** (1991), 401–423.
11. Morgan, J., and R. Scott, A nodal basis for C^1 piecewise polynomials of degree $n \geq 5$, Math. Comp. **29** (1975), 736–740.
12. Schumaker, L. L., On spaces of piecewise polynomials in two variables, in *Approximation Theory and Spline Functions*, S. P. Singh, J. H. W. Burry, and B. Watson (eds.), Reidel (Dordrecht), 1984, 151–197.
13. Schumaker, L. L., On super splines and finite elements, SIAM J. Numer. Anal. **26** (1989), 997–1005.
14. Strang, G., Piecewise polynomials and the finite element method, Bull. Amer. Math. Soc. **79** (1973), 1128–1137.
15. Whiteley, W., The combinatorics of bivariate splines, in *Applied Geometry and Discrete Mathematics, Victor Klee Festschrift*, P. Gritzmann and B. Sturmfels (eds.), DIMACS series, AMS (Providence), 1991, 587–608.

Oleg Davydov
Universität Dortmund
Fachbereich Mathematik
Lehrstuhl VIII
44221 Dortmund, GERMANY
oleg.davydov@math.uni-dortmund.de

Fair Surface Reconstruction from Point Clouds

Ulrich Dietz

Abstract. This paper describes a method for reconstructing fair B-spline surfaces from scattered data points. The points are assumed to have arbitrary boundaries and varying densities; additionally normal vector information may be given. The basic idea of the presented algorithm is to obtain a smooth approximation surface by simulating the deep-drawing process of a metal sheet. We only prescribe a distance tolerance and an angle tolerance, all other quantities are computed automatically during the iterative algorithm. This includes the parametrization of the points, the appropriate degree of smoothing as well as the necessary number of patches. The well-known thin plate energy is utilized and supplied with a local support to obtain better results. The whole reconstruction algorithm is largely linear and stable.

§1. Introduction

The two important parts of Reverse Engineering are the segmentation of the measured points and the computation of fair surfaces to each subset of points. A good overview concerning this topic can be found in [15]. The segmentation process should take into consideration the underlying geometry. This means the point set is split at feature lines or styling lines, and regions are created normally occurring in the engineering process of constructing a mathematical model with help of several surfaces. Hermann [7] called this kind of segmentation functional decomposition (see also [15]). In general the individual parts are arbitrary shaped (not tri-/rect-angular) with holes inside. A suitable approximation scheme has to deal with these arbitrary boundaries, holes in the interior (inner boundaries), measurement errors, and also with rapidly varying point densities, coming e.g. from different views.

The presented method interprets the surface to be calculated as a metal sheet and simulates the industrial process of deep-drawing. The iterative surface approximation algorithm starts with a least squares fitting plane as a surface of lowest bending energy. Then successively a sequence of surfaces is computed (see Fig. 1) minimizing thin plate energy while deforming towards the points which serve as a mould. The iteration stops when prescribed distance and angle tolerances are satisfied. However, the iteration will fail to

Mathematical Methods for Curves and Surfaces II
Morten Dæhlen, Tom Lyche, and Larry L. Schumaker (eds.), pp. 79–86.

ISBN 0-8265-1315-8.

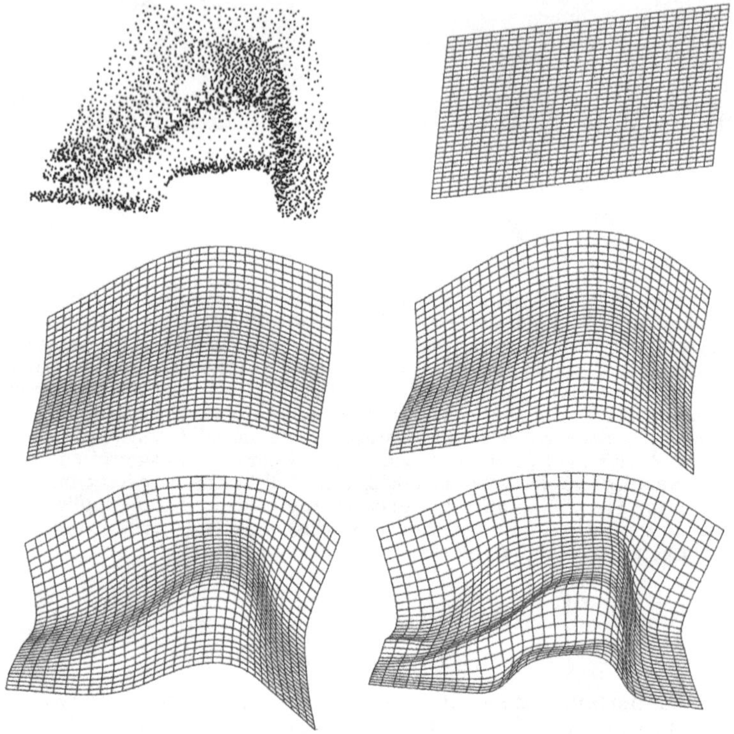

Fig. 1. 3234 data points with surfaces of iteration steps 1, 7, 10, 14, 18.

converge to a reasonable surface if the geometry of the points is complicated. As the surface representation we use tensor product B-splines (see [2,9])

$$X(u,v) = \sum_{i=0}^{n} \sum_{j=0}^{m} \boldsymbol{d}_{ij} N_{ir}(u) N_{js}(v) \tag{1}$$

because of their benefits in parametric surface fitting and because of the wide usage in current CAD systems. The final approximation surface is then trimmed according to the boundaries of the point set as described in [10]. Together with the trimmed surfaces for the other point subsets, we have an up-to-a-tolerance G^1 description of the measured object.

Methods similar to our approach can be found in [5,7]. Different approaches but dealing with the same problem are described in [6,13,14]. Our approach is basically build on a previous paper [8]. However, now we are able to satisfy normal constraints and to locally control the surface fairness.

The objective functional we are minimizing is a linear combination of the sum of squared distance errors, the sum of squared angle deviations, the well-known thin plate energy, and a penalty term controlling the local influence of the energy term. To avoid nonlinear optimization, we minimize the objective functional with respect to each group of unknowns (control points, point parameter values, etc.) in turns. So every iteration step consists of three separate parts:

(1) linear surface approximation (positions and normals),
(2) Newton-like point parameter corrections,
(3) linear weight function correction.

In the following sections each part of the algorithm is described in detail.

§2. Parametrization

The crucial step in parametric surface fitting is the assignment of *good* parameter values (u_i, v_i) to the given points $\boldsymbol{P}_i$, $i = 1, \ldots, M$. The quality of the final approximation surface strongly depends on the selection of an appropriate parametrization. Furthermore, the topology on the point set is only defined by the locations of these point parameter values in the parametric domain of the surface (see [5]). In the literature many proposals and heuristics can be found for how to compute appropriate initial parametrizations for the purpose of parametric surface fitting [3,5,9,11]. Most of these methods guarantee a valid parametrization (but not optimal in the sense of orthogonal distance vectors). However, they rely on a special topological structure, which we do not assume. On the other side, our method may fail to provide a reasonable parametrization if the geometry of the points is too complicated.

In the present method we start our iterations with a least squares fitting plane, and so an initial (and for the plane optimal) parametrization can simply be obtained by orthogonal projection of the points onto this plane. In the subsequent iteration steps the plane is successively deformed to the final shape. The parametrization is adapted after each iteration step according to the actual surface with the help of the so-called *parameter correction* [9]. After each deformation of the approximation surface an optimal parametrization (orthogonal distance vectors) is restored by minimizing the total error sum with respect to the parameters (u_i, v_i). The necessary conditions yield the 2-dimensional nonlinear systems

$$\begin{aligned}(\boldsymbol{P}_i - \boldsymbol{X}(u_i, v_i))\boldsymbol{X}_u(u_i, v_i) = 0,\\ (\boldsymbol{P}_i - \boldsymbol{X}(u_i, v_i))\boldsymbol{X}_v(u_i, v_i) = 0\end{aligned} \tag{2}$$

for each point $\boldsymbol{P}_i$. They can be solved efficiently with a damped Gauss-Newton method with correction terms of type

$$\begin{pmatrix}\Delta u\\ \Delta v\end{pmatrix} = \begin{pmatrix}\boldsymbol{X}_u^2 & \boldsymbol{X}_u\boldsymbol{X}_v\\ \boldsymbol{X}_u\boldsymbol{X}_v & \boldsymbol{X}_v^2\end{pmatrix}^{-1}\begin{pmatrix}(\boldsymbol{P}-\boldsymbol{X})\boldsymbol{X}_u\\ (\boldsymbol{P}-\boldsymbol{X})\boldsymbol{X}_v\end{pmatrix}. \tag{3}$$

A known problem of parameter correction is that convergence is only local. But since we adapt the parametrization after each small deformation of the surface, we get a much more global character than for traditional parameter correction. An additional benefit of this process is that it produces nearly isometric parametrizations. Unlike the most available approximation methods we have here no need for a sophisticated initial parametrization; suitable parameter values are automatically computed during the iterative algorithm.

§3. Approximation

Once we have determined suitable parameter values (u_i, v_i), $i = 1, \dots, M$ for the data points $\boldsymbol{P}_i$, we can attack the problem to compute the control points $\boldsymbol{d}_{ij}$ of a smooth approximation surface $\boldsymbol{X}(u, v)$. The positional deviation of the surface to the points is measured as usual by the total error sum

$$E_{\text{dist}} = \sum_{i=1}^{M} \|\boldsymbol{P}_i - \boldsymbol{X}(u_i, v_i)\|^2. \tag{4}$$

To describe the smoothness of the surface, we use fairness functionals which is quite common in the CAGD community. We have the choice between parametrization independent, but highly nonlinear functionals (e.g. [12]) describing some physical based energies and simplified quadratic functionals, which are dependent on surface parametrization (e.g. [1]). The benefit of the latter functionals is that the minimization leads to a linear system of equations which can be solved very efficiently. The energy functionals [4] proposed by Greiner have a somewhat intermediate status since they are parametrization independent and quadratic; the computational costs are also somewhere in-between. For the sake of computational efficiency, we use the well-known thin plate energy

$$E_{\text{fair}} = \iint \lambda(u, v)(\boldsymbol{X}_{uu}^2 + 2\boldsymbol{X}_{uv}^2 + \boldsymbol{X}_{vv}^2)\, du\, dv \tag{5}$$

weighted by the real-valued B-spline function

$$\lambda(u, v) = \sum_{i=0}^{n} \sum_{j=0}^{m} w_{ij} N_{ir}(u) N_{js}(v) \geq 0. \tag{6}$$

This functional surface λ allows to control locally the influence of the energy term. In flat regions λ can be chosen large with a strong smoothing effect; near edges λ has to be small to allow an appropriate approximation result. For fixed $\lambda(u, v)$, E_{fair} is obviously quadratic in the control points $\boldsymbol{d}_{ij}$.

The angle deviations between the prescribed unit normal vectors $\boldsymbol{N}_i$, $i = 1, \dots, M$ and the surface normal vectors can be approximately measured with the quadratic functional [16]

$$E_{\text{norm}} = \sum_{i=1}^{M} (\boldsymbol{N}_i \cdot \boldsymbol{X}_{u_i})^2 + (\boldsymbol{N}_i \cdot \boldsymbol{X}_{v_i})^2. \tag{7}$$

An improved functional describing the angle deviations which is still quadratic but independent on the surface parametrization is developed in §4.

Now we combine all these functionals and set up the objective functional

$$E = E_{\text{dist}} + \mu E_{\text{norm}} + E_{\text{fair}} + \gamma E_{\text{cdev}} \tag{8}$$

of our approximation step. The last expression

$$E_{\mathrm{cdev}} = \iint (\lambda_0 - \lambda(u,v))^2 \, du \, dv \tag{9}$$

in (8) is necessary to avoid the trivial solution $\lambda(u,v) \equiv 0$, which means no fairing is applied at all. The variable λ_0 is a positive constant; μ, γ are positive scalar penalty factors controlling the influence of the corresponding terms. Each of the introduced functionals is quadratic in the unknown control points $\boldsymbol{d}_{ij}$.

The necessary conditions for a minimum of E with respect to the $\boldsymbol{d}_{ij}$'s yield the linear equation system

$$(\boldsymbol{A}_{\mathrm{dist}} + \mu \boldsymbol{A}_{\mathrm{norm}} + \boldsymbol{A}_{\mathrm{fair}})\, \boldsymbol{d} = \boldsymbol{B}, \tag{10}$$

where the $3(n+1)(m+1) \times 1$-vector $\boldsymbol{d}$ contains all control points $\boldsymbol{d}_{ij}$. The $3(n+1)(m+1) \times 3(n+1)(m+1)$-matrices $\boldsymbol{A}_{\mathrm{dist}}$, $\boldsymbol{A}_{\mathrm{norm}}$ and $\boldsymbol{A}_{\mathrm{fair}}$ are symmetric, positive semi-definite, and due to the local support of the B-spline basis functions sparse and banded. At least three non-collinear points and $\lambda(u,v) > 0$ guarantee the system (10) has a unique solution. Because of the occurring dot products in E_{norm}, the system (10) has triple size and bandwidth as for a usual least squares fit. Beside size and arrangement of the matrix/vector elements $\boldsymbol{A}_{\mathrm{dist}}\boldsymbol{d} = \boldsymbol{B}$ is a usual normal equation system [9]. $\boldsymbol{A}_{\mathrm{fair}}$ is a so-called energy matrix with matrix elements

$$a_{gh} = \sum_{c=0}^{n} \sum_{d=0}^{m} w_{cd} \left(U_{ikc}^{22} V_{jld}^{00} + 2U_{ikc}^{11} V_{jld}^{11} + U_{ikc}^{00} V_{jld}^{22} \right) \tag{11}$$

with $g = 3(i+k(n+1))+e$, $h = 3(j+l(n+1))+e$, $e = 0, \ldots, 2$ and abbreviations

$$\begin{aligned} U_{ikc}^{\alpha\beta} &= \int N_{ir}^{(\alpha)}(u) N_{kr}^{(\beta)}(u) N_{cr}(u) \, du, \\ V_{jld}^{\alpha\beta} &= \int N_{js}^{(\alpha)}(v) N_{ls}^{(\beta)}(v) N_{ds}(v) \, dv. \end{aligned} \tag{12}$$

All remaining matrix elements of $\boldsymbol{A}_{\mathrm{fair}}$ are zero. Finally matrix $\boldsymbol{A}_{\mathrm{norm}}$ contains elements of type

$$\begin{aligned} \sum_{t=1}^{M} N_{tc} N_{td} (& N_{ir}'(u_t) N_{js}(v_t) N_{kr}'(u_t) N_{ls}(v_t) + \\ & N_{ir}(u_t) N_{js}'(v_t) N_{kr}(u_t) N_{ls}'(v_t)), \end{aligned} \tag{13}$$

where N_{tc}/N_{td} denote the c-th/d-th coordinate of the unit normal vector $\boldsymbol{N}_t$. So far we have a linear surface approximation step for the control points $\boldsymbol{d}_{ij}$ for fixed μ, γ, λ_0 and $\lambda(u,v)$.

Now we compute the optimal function $\lambda(u,v)$ by minimizing E with respect to the scalar control points w_{ij} while fixing the control points $\boldsymbol{d}_{ij}$ and the scalars μ and λ_0. We call this step weight correction. It is a constrained optimization problem due to $\lambda(u,v) > 0$, but it can simply be solved by considering the unconstrained problem and then choosing the penalty factor γ in such a way that $\min_{u,v} \lambda(u,v) = \sigma > 0$. The necessary conditions of the unconstrained problem yield a linear system of equations with solution

$$\boldsymbol{w} = \boldsymbol{\lambda}_0 - \frac{1}{2\gamma} \boldsymbol{B}^{-1}(\boldsymbol{d}^T \boldsymbol{D} \boldsymbol{d}). \tag{14}$$

Here $\boldsymbol{w}$, $\boldsymbol{d}$, $\boldsymbol{\lambda}_0$ are the vectors containing all w_{ij}, $\boldsymbol{d}_{ij}$, λ_0. $\boldsymbol{B}$ is an energy matrix and $\boldsymbol{D}$ an energy tensor; both can be computed exactly.

§4. Normal Vector Fields

In automotive industries the quality of a surface is checked with the help of reflection lines. They are closely related to the normal vectors of the surface. So it is desirable not only to be able to approximate point clouds but additionally normal vector fields $\boldsymbol{N}_i$, $(|\boldsymbol{N}_i| = 1)$, $i = 1, \ldots, M$ [7,16]. A probably more important reason for the approximation of normal vectors is to hold at least numerically G^1 continuity at trimmed surface boundaries where the several surfaces of our reconstruction meet. The direct minimization of the angle deviations $\sum \alpha_i^2 := \sum \angle^2(\boldsymbol{N}_i, \boldsymbol{X}_{N_i})$ between prescribed normals and surface normals $\boldsymbol{X}_{N_i}$ is highly nonlinear.

The commonly used functional (7) is quadratic and leads to a linear system of equations [16]. Unfortunately, this expression is not invariant to surface parametrization. The paper [7] proposes improved functionals not depending on the lengths of the partial derivatives $\boldsymbol{X}_{u_i}, \boldsymbol{X}_{v_i}$, but still on the angles between them. The following functional also takes these angles into account; it is invariant to surface parametrization and quadratic in the control points:

$$E_{\text{Norm}} = \sum_{i=1}^{M} (\boldsymbol{N}_i \cdot \boldsymbol{X}_{r_i})^2 + (\boldsymbol{N}_i \cdot \boldsymbol{X}_{s_i})^2 \tag{15}$$

with

$$\boldsymbol{X}_{r_i} | \boldsymbol{X}_{s_i} = \left(\frac{\boldsymbol{X}_{u_i}}{|\hat{\boldsymbol{X}}_{u_i}|} \pm \frac{\boldsymbol{X}_{v_i}}{|\hat{\boldsymbol{X}}_{v_i}|} \right) \Big/ \left| \frac{\hat{\boldsymbol{X}}_{u_i}}{|\hat{\boldsymbol{X}}_{u_i}|} \pm \frac{\hat{\boldsymbol{X}}_{v_i}}{|\hat{\boldsymbol{X}}_{v_i}|} \right|, \tag{16}$$

where $\hat{\boldsymbol{X}}_{u_i}, \hat{\boldsymbol{X}}_{v_i}$ denote quantities of the surface from a previous iteration step. It holds that

$$|\boldsymbol{X}_{r_i}| \approx |\boldsymbol{X}_{s_i}| \approx 1, \;\; \boldsymbol{X}_{r_i} \overset{\perp}{\approx} \boldsymbol{X}_{s_i}, \;\; \boldsymbol{X}_{N_i} \boldsymbol{X}_{r_i} = \boldsymbol{X}_{N_i} \boldsymbol{X}_{s_i} = 0. \tag{17}$$

The functional E_{Norm} is a good approximation of the exact functional $\sum \alpha_i^2$ for small angles α_i, since $\alpha_i^2 \approx \sin^2 \alpha_i$ and sum and difference of two unit vectors are orthogonal. The minimization of (15) leads to a linear equation system with matrix elements similar to (13). The system matrix is also symmetric, positive semi-definite, banded and has triple size and bandwidth of the usual least squares system.

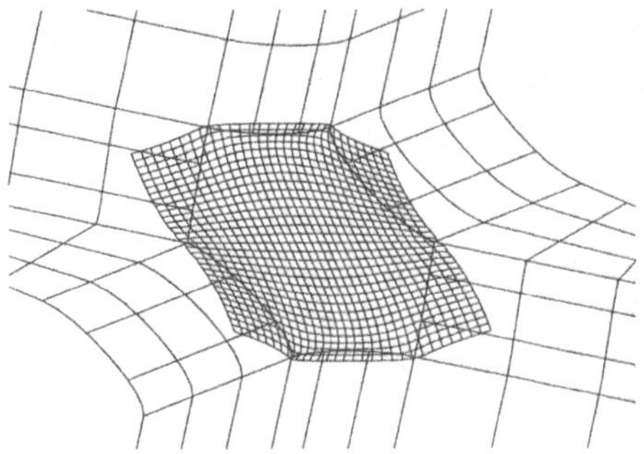

Fig. 2. Untrimmed approximation surface filling a polygonal hole; points and normal vectors only given on the boundaries of the hole.

§5. Algorithm

The deep-drawing process is now described in a more algorithmic way with help of the notation of the previous sections.

- *Input* are the points $\boldsymbol{P}_i$ and the normal vectors $\boldsymbol{N}_i$. The user specifies distance tolerance ϵ, angle tolerance δ and polynomial degrees r, s.
- The variables $\lambda_0 \gg 1$ and $w_{ij} := \lambda_0$ are *initialized* to force the the first approximation surface to be nearly planar. The number of patches is set to a maximal value; μ is set to zero (no normal vector approximation).
- The points $\boldsymbol{P}_i$ are *parametrized* with values (u_i, v_i) by orthogonal projection to a least squares fitting plane.
- The *iteration loop* starts. Linear approximation steps for $\boldsymbol{d}_{ij}$ alternate with parameter corrections for (u_i, v_i) and linear weight corrections for w_{ij}. If ϵ is satisfied, the normal vector approximation is included by setting $\mu := \epsilon/\delta$; if furthermore δ is satisfied the iteration stops. Otherwise the influence of the energy functional is reduced by setting $\lambda_0 := \lambda_0/2$.
- In a *post processing* step the necessary number of patches to meet the tolerances is determined with a binary search. Finally the approximation surface is trimmed at the boundaries of the point set.

Acknowledgments. This work was supported by the German *Bundesministerium für Bildung, Wissenschaft, Forschung und Technologie.* The author thanks Prof. Hoschek, Prof. Várady and Dr. Greiner for helpful discussions.

References

1. Bloor, M. I. G., M. J. Wilson, and H. Hagen, The smoothing properties of variational schemes for surface design, Comput. Aided Geom. Design **12** (1995), 381–394.
2. Farin, G., *Curves and Surfaces for CAGD: A Practical Guide*, Academic Press, 4th edition, 1997.

3. Floater, M. S., Parametrization and smooth approximation of surface triangulations, Comput. Aided Geom. Design **14** (1997), 231–250.
4. Greiner, G., Curvature approximation with application to surface modeling, in *Advanced Course on FAIRSHAPE*, J. Hoschek, and P. Kaklis (eds.), Teubner, Stuttgart, 1996, 241–251.
5. Greiner, G., and K. Hormann, Interpolating and approximating scattered 3d data with hierarchical tensor product B-splines, in *Surface Fitting and Multiresolution Methods*, A. Le Méhauté, C. Rabut, and L. L. Schumaker (eds.), Vanderbilt Univ. Press, Nashville, 1997, 163–172.
6. Hagen, H., S. Heinz, and A. Nawotki, Variational design with boundary conditions and parameter optimized surface fitting, in *Geometric Modeling: Theory and Practice*, W. Straßer, R. Klein, and R. Rau (eds.), Springer, 1997.
7. Hermann, T., Z. Kovács, and T. Várady, Special applications in surface fitting, in *Geometric Modeling: Theory and Practice*, W. Straßer, R. Klein, and R. Rau (eds.), Springer, 1997.
8. Hoschek, J., and U. Dietz, Smooth B-spline surface approximation to scattered data, in *Reverse Engineering*, J. Hoschek, and W. Dankwort (eds.), Teubner, 1996, 143–152.
9. Hoschek, J., and D. Lasser, *Fundamentals of Computer Aided Geometric Design*, AK Peters, Wellesley, 1994.
10. Hoschek, J., and F.-J. Schneider, Approximate spline conversion for integral and rational Bézier and B-spline surfaces, in *Geometry Processing for Design and Manufacturing*, R. E. Barnhill (ed.), SIAM, 1992, 45–86.
11. Ma, W., and J. P. Kruth, Parametrization of randomly measured points for least squares fitting of B-spline curves and surfaces, Computer-Aided Design **27** (1995), 663–675.
12. Moreton, H. P., and C. H. Séquin, Functional optimization for fair surface design, Computer Graphics **26** (1992), 167–176.
13. Sinha, S. S., and B. G. Schunck, A two-stage algorithm for discontinuity-preserving surface reconstruction, IEEE Trans. Pattern Anal. and Machine Intelligence **14** (1992), 36–55.
14. Terzopoulos, D., The computation of visible-surface representations, IEEE Trans. Pattern Anal. and Machine Intelligence **10** (1988), 417–438.
15. Várady, T., R. R. Martin, and J. Cox, Reverse engineering of geometric models – an introduction, Computer-Aided Design **29** (1997), 255–268.
16. Ye, X., T. R. Jackson, and N. M. Patrikalakis, Geometric design of functional surfaces, Computer-Aided Design **28** (1996), 741–752.

Ulrich Dietz
Technische Universität Darmstadt
Schloßgartenstraße 7
64289 Darmstadt, GERMANY
`udietz@mathematik.tu-darmstadt.de`

Approximation with Rational B-spline Curves and Surfaces

Bernhard Elsässer

Abstract. A new approach for approximating a point set with a rational B-Spline curve or surface is presented. With the help of an error function similar to the Euclidian square distance, positive weights and the control points are optimized by solving only one linear system.

§1. Introduction

Many of today's computer aided design systems for free-form curve and surface modelling use rational parametric representations because of their larger degree of freedom and their ability to represent conics exactly. In the field of reverse engineering an important problem is the approximation of digitized points by parametric curves and surfaces. The approximation with integral polynomial curves and surfaces (e.g. integral B-spline curves and surfaces) minimizing the Euclidian square distances between the points with given parameter values and the curve (surface) leads to a linear system. In the rational case, however, the minimization of the Euclidian square distances (error function) yields a nonlinear system [2,3,4].

In our new approach for rational approximation, we solve only a linear system by choosing a new error function which is similar to the Euclidian square distances. In addition we use parameter correction to reduce the approximation error even more.

§2. Approximation with Rational B-spline Curves

We will consider the following problem. Suppose $m+1$ points $\mathbf{P}_i = (p_{ix}, p_{iy}, p_{iz})^\top \in \mathbb{R}^3$ with parameter values t_i are given. A rational B-spline curve

$$\mathbf{X}(t) = \frac{\sum_{j=0}^{n} w_j \mathbf{d}_j N_{jk}(t)}{\sum_{j=0}^{n} w_j N_{jk}(t)} \quad , \quad \mathbf{d}_j = \begin{pmatrix} d_{jx} \\ d_{jy} \\ d_{jz} \end{pmatrix}$$

Mathematical Methods for Curves and Surfaces II
Morten Dæhlen, Tom Lyche, and Larry L. Schumaker (eds.), pp. 87–94.

ISBN 0-8265-1315-8.

is required which approximates the points $\mathbf{P}_i$. Unknowns for the approximation are the control points $\mathbf{d}_j$ and the weights w_j. The B-spline functions N_{jk} of order k are defined over the knot sequence

$$\mathbf{T} = (\tau_0 = \cdots = \tau_{k-1}, \tau_k, \ldots, \tau_n, \tau_{n+1} = \cdots = \tau_{n+k}) \;,\; n \geq k-1.$$

The parameter values t_i satisfy the *Schoenberg-Whitney* conditions. To maintain the well-known properties for B-splines the weights should satisfy $w_j > 0$ which is the aim of this approach.

The approximation procedure depends on the choice of the error function. To get a linear system as a necessary condition for the minimization of the error function $\bar{F}$, the unknowns for the optimization must be linear or quadratic in $\bar{F}$. Therefore, we introduce as non-Euclidian error measurement

$$\bar{F} := F + \lambda G \quad (\lambda > 0),$$

where

$$F := \sum_{i=0}^{m} Q(\bar{\mathbf{X}}(t_i), \bar{\mathbf{P}}_i)$$

and

$$Q(\bar{\mathbf{X}}, \bar{\mathbf{Y}}) := \sum_{i,j,k,l=0}^{3} a_{ij}^{kl}\, x_i x_j y_k y_l. \tag{1}$$

Here $\bar{\mathbf{X}}$, $\bar{\mathbf{Y}}$ are the homogeneous coordinates of points $\mathbf{X}, \mathbf{Y} \in \mathbb{R}^3$, and

$$\bar{\mathbf{X}}(t) = \begin{pmatrix} \sum_{j=0}^{n} w_j\, N_{jk}(t) \\ \sum_{j=0}^{n} \bar{d}_{jx}\, N_{jk}(t) \\ \sum_{j=0}^{n} \bar{d}_{jy}\, N_{jk}(t) \\ \sum_{j=0}^{n} \bar{d}_{jz}\, N_{jk}(t) \end{pmatrix} \quad , \quad \bar{\mathbf{P}}_i = c_i \begin{pmatrix} 1 \\ p_{ix} \\ p_{iy} \\ p_{iz} \end{pmatrix} \tag{2}$$

are the homogeneous coordinates of $\mathbf{X}(t)$, $\mathbf{P}_i$. Q should be a distance function ($Q \geq 0$), the function G should control the weights w_j which always have to be positive in order to maintain the convex hull property for B-spline curves. λ is a penalty factor. $\bar{d}_{jx} := w_j d_{jx}$, $\bar{d}_{jy} := w_j d_{jy}$, $\bar{d}_{jz} := w_j d_{jz}$, w_j are unknowns of $\bar{F}$. The function G and the factors c_i will be defined later.

The distance function Q should satisfy the following conditions:

(i) Q is a quadratic form in $\bar{\mathbf{X}}$ resp. in $\bar{\mathbf{Y}}$

(ii) $Q(\bar{\mathbf{X}}, \bar{\mathbf{Y}}) = Q(\bar{\mathbf{Y}}, \bar{\mathbf{X}})$ (symmetry)

(iii) $Q(\bar{\mathbf{X}}, \alpha\bar{\mathbf{X}}) = 0 \quad (\alpha \in \mathbb{R})$ (interpolation property)

(iv) $Q(\bar{\mathbf{X}}, \bar{\mathbf{Y}}) > 0$, if $\bar{\mathbf{Y}} \neq \alpha\bar{\mathbf{X}} \quad (\alpha \neq 0,\ \bar{\mathbf{X}}, \bar{\mathbf{Y}} \neq \mathbf{0})$

The last two conditions are essential properties for a distance function.

It is shown in [1] that all functions Q defined as in (1) satisfying the conditions (i)-(iv) have the representation

$$Q(\bar{\mathbf{X}}, \bar{\mathbf{Y}}) = \bar{\mathbf{z}}^\top \mathbf{C} \bar{\mathbf{z}} \quad \text{with} \quad \bar{\mathbf{z}} = \begin{pmatrix} x_0 y_1 - x_1 y_0 \\ x_0 y_2 - x_2 y_0 \\ x_0 y_3 - x_3 y_0 \\ x_1 y_2 - x_2 y_1 \\ x_1 y_3 - x_3 y_1 \\ x_2 y_3 - x_3 y_2 \end{pmatrix} \tag{3}$$

and $\mathbf{C} \in \mathbb{R}^{6,6}$ is a positive definite matrix. It is sufficient to choose $\mathbf{C}$ as the identity matrix, the simplest positive definite matrix. Additionally, we assume that

$$\bar{\mathbf{P}}_i = \frac{1}{\sqrt{1 + p_{ix}^2 + p_{iy}^2 + p_{iz}^2}} \begin{pmatrix} 1 \\ p_{ix} \\ p_{iy} \\ p_{iz} \end{pmatrix}. \tag{4}$$

Using the identity matrix for $\mathbf{C}$ and inserting (4) in (3), it follows that

$$F = \sum_{i=0}^{m} \bar{\mathbf{X}}^2(t_i) - (\bar{\mathbf{X}}(t_i) \cdot \bar{\mathbf{P}}_i)^2. \tag{5}$$

F has the following important properties:

Theorem 1.

(i) *F is invariant with respect to rotations of $\mathbb{R}^3$,*

(ii) *F is approximately the Euclidian square distance for $w_j = 1$ and $|\mathbf{P}_i| < \epsilon$:*

$$F \approx \sum_{i=0}^{m} (\mathbf{X}(t_i) - \mathbf{P}_i)^2. \tag{6}$$

Proof: Let $x_0(t)$ denote the first and $\bar{\mathbf{X}}'(t)$ the last 3 components of $\bar{\mathbf{X}}(t)$. (i): Rotating the points $\mathbf{P}_i$ and the control points $\mathbf{d}_j$ with a orthogonal rotation matrix $\mathbf{D}$ yields

$$\begin{aligned} F &= \sum_{i=0}^{m} x_0^2(t_i) + \bar{\mathbf{X}}'^\top(t_i)\mathbf{D}^\top\mathbf{D}\bar{\mathbf{X}}'(t_i) \\ &\quad - \frac{1}{1 + \mathbf{P}_i^\top\mathbf{D}^\top\mathbf{D}\mathbf{P}_i}(x_0(t_i) + \bar{\mathbf{X}}'^\top(t_i)\mathbf{D}^\top\mathbf{D}\mathbf{P}_i)^2 \\ &= \sum_{i=0}^{m} x_0^2(t_i) + \bar{\mathbf{X}}'^2(t_i) - \frac{1}{1 + \mathbf{P}_i^2}(x_0(t_i) + \bar{\mathbf{X}}'(t_i) \cdot \mathbf{P}_i)^2. \end{aligned}$$

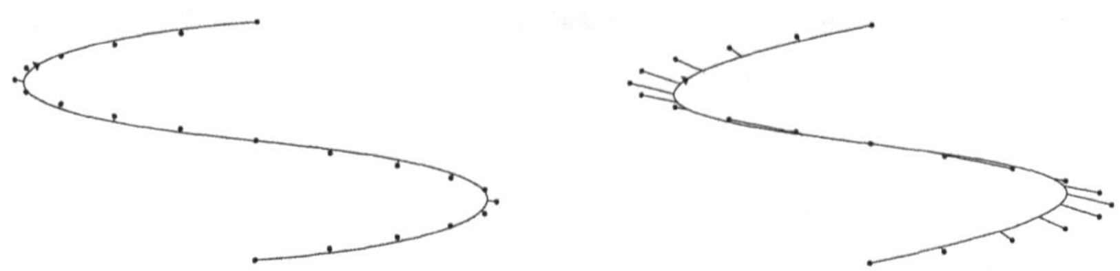

Fig. 1. Approximation with scaled points and rescaling afterwards (left hand side). Approximation with unscaled points (right hand side). Points span about 160 × 80 mm.

This completes the proof for (i).
(ii): Let $\times$ denote the vector product in $\mathbb{R}^3$. Then we obtain

$$
\begin{aligned}
F &= \sum_{i=0}^{m} x_0^2(t_i) + \bar{\mathbf{X}}'^2(t_i) - \frac{1}{1+\mathbf{P}_i^2}\left(x_0(t_i) + \bar{\mathbf{X}}'(t_i)\cdot\mathbf{P}_i\right)^2 \\
&= \sum_{i=0}^{m} \left(\bar{\mathbf{X}}'(t_i) \times \frac{\mathbf{P}_i}{|\mathbf{P}_i|}\right)^2 + \frac{1}{1+\mathbf{P}_i^2}\left((\bar{\mathbf{X}}'(t_i) - x_0(t_i)\mathbf{P}_i)\cdot\frac{\mathbf{P}_i}{|\mathbf{P}_i|}\right)^2 \\
&= \sum_{i=0}^{m} (\bar{\mathbf{X}}'(t_i) - x_0(t_i)\mathbf{P}_i)^2 - \frac{\mathbf{P}_i^2}{1+\mathbf{P}_i^2}\left((\bar{\mathbf{X}}'(t_i) - x_0(t_i)\mathbf{P}_i)\cdot\frac{\mathbf{P}_i}{|\mathbf{P}_i|}\right)^2 .
\end{aligned}
$$

For $|\mathbf{P}_i| < \epsilon$ and $w_j = 1$ the last line of the calculation implies (ii). □
It suffices to choose $\epsilon = 1/2$ (the approximation order of (6) is $O(\max|\mathbf{P}_i|^2))$ in order to satisfy (6). Property (6) shows the similarity of the non-Euclidian function (5) to the Euclidian square distances.

From the geometric point of view it is very important that the approximation curve is affine invariant with respect to the points $\mathbf{P}_i$. Therefore, the following coordinate transformation steps are required:

(i) origin of the new coordinate system is the center of gravity of the $\mathbf{P}_i$,

(ii) $|\mathbf{P}_i| \leq 1/2 \quad (i = 0, \ldots, m)$.

In addition, the scaling (ii) guarantees property (6) as well as the minimization of error vectors which are (approximately) perpendicular to the approximating curve. To see the significant differences of the approximations with or without scaling, see Figure 1. The left-hand side of Figure 1 represents the approximation curve and a point set which was scaled before minimizing $\bar{F}$ (after the minimization the resulting curve and the points were rescaled). The right-hand side of Figure 1 shows the approximation without scaling. In both parts of Figure 1 error vectors are drawn which connect the points $\mathbf{P}_i$ with the corresponding curve points $\mathbf{X}(t_i)$ used at the end of the minimization of $\bar{F}$ (together with parameter correction, see below). On the left hand the error vectors are approximately perpendicular to the curve, whereas on the right hand the much longer vectors in general are not perpendicular. In both cases,

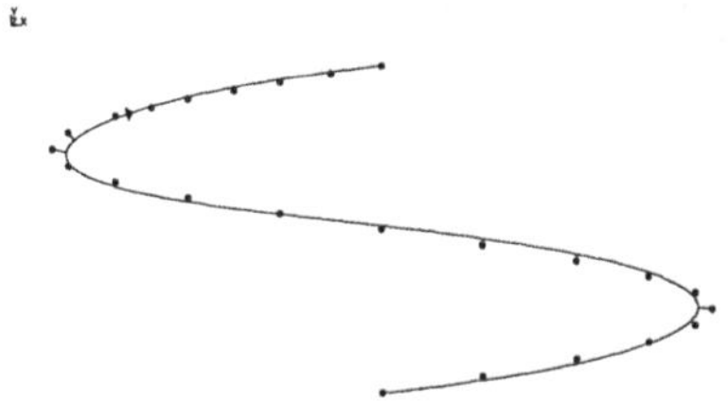

Fig. 2. Approximation with additional inserted points.

however, the maximal error values of the non-Euclidian error function $\bar{F}$ are nearly the same.

Figure 2 shows, if additional points are inserted between adjacent data points in the second half of the left curve in Figure 1, that the approximation curve (with coordinate transformation (i), (ii) before minimizing $\bar{F}$ and retransformation afterwards) is still similar to the left curve in Figure 1. In this sense the method is stable against changes of the center of gravity of the points.

Now we introduce the function G which should control the weights w_j:

$$G := \sum_{i=0}^{m}(\sum_{j=0}^{n} w_j\, N_{jk}(t_i) - 1)^2.$$

As part of the necessary conditions for a minimum of $\bar{F}$, we obtain

$$\frac{\partial \bar{F}}{\partial w_j} = \frac{\partial F}{\partial w_j} + \lambda \frac{\partial G}{\partial w_j} = 0. \tag{7}$$

Determining the limit of (7) for $\lambda \to \infty$, one gets the system

$$\frac{\partial G}{\partial w_j} = 0,$$

which has the solutions $w_j = 1$. So if $\lambda > 0$ is chosen large enough, we can achieve positive weights when $\bar{F}$ is minimized.

The necessary conditions

$$\frac{\partial \bar{F}}{\partial \bar{d}_{jx}} = 0, \; \dots \; , \frac{\partial \bar{F}}{\partial w_j} = 0$$

for a minimum of $\bar{F}$ lead to the system

$$(\mathbf{A}_1 + \lambda \mathbf{A}_2)\mathbf{x} = \lambda \mathbf{b}\ ,\ \mathbf{A}_1, \mathbf{A}_2 \in \mathbb{R}^{4n+4,4n+4} \tag{8}$$

$$\mathbf{x} = (\bar{\mathbf{d}}_0, \dots, \bar{\mathbf{d}}_n, w_0, \dots, w_n)^\top\ ,\ \bar{\mathbf{d}}_j = (\bar{d}_{jx}, \bar{d}_{jy}, \bar{d}_{jz})^\top.$$

In [1] we proved that for each fixed $\lambda > 0$ system (8) is linear and has a unique solution $\mathbf{x} \neq \mathbf{0}$ which is a global minimum of $\bar{F}$. During the solution of (8) a suitable λ can be estimated to achieve positive weights $w_j \in [w_{min}, w_{max}]$, $0 < w_{min} < 1 < w_{max}$ (see [1]).

To reduce the approximation error parameter correction is also used: Corrected parameter values t'_i with

$$\bar{F}(t'_0, \dots, t'_m) = \sum_{i=0}^{m} D(\bar{\mathbf{X}}(t'_i), \bar{\mathbf{P}}_i) \longrightarrow \min$$

are required. The control points and weights are fixed. The necessary conditions imply the nonlinear equations

$$\frac{\partial \bar{F}}{\partial t'_i} = \frac{\partial D(\bar{\mathbf{X}}(t'_i), \bar{\mathbf{P}}_i)}{\partial t'_i} = 0 \quad (i = 0, \dots, m). \tag{9}$$

The whole approximation process consists of the two steps: solving the linear system (8), and correcting the parameter values (solve nonlinear equations (9)), which are repeated iteratively.

§3. Approximation with Rational B-spline Surfaces

Now we extend the approximation problem to surfaces. Suppose $m+1$ points $\mathbf{P}_i = (p_{ix}, p_{iy}, p_{iz})^\top \in \mathbb{R}^3$ with parameter values (u_i, v_i) are given. Due to the reasons mentioned in the curve case, we assume that $\mathbf{P}_i$ are scaled such that $|\mathbf{P}_i| \leq 1/2$ and the center of gravity of the points is the origin of the coordinate system. A rational B-spline surface

$$\mathbf{X}(u,v) = \frac{\sum_{j=0}^{n}\sum_{l=0}^{r} w_{jl}\mathbf{d}_{jl}\, N_{jk}(u) N_{lp}(v)}{\sum_{j=0}^{n}\sum_{l=0}^{r} w_{jl}\, N_{jk}(u) N_{lp}(v)}, \quad \mathbf{d}_{jl} = \begin{pmatrix} d_{jlx} \\ d_{jly} \\ d_{jlz} \end{pmatrix}$$

approximating the points is required. N_{jk}, N_{lp} are B-spline functions of order k resp. p defined over the knot sequences

$$\mathbf{U} = (\mu_0 = \dots = \mu_{k-1}, \mu_k, \dots, \mu_n, \mu_{n+1} = \dots = \mu_{n+k}),$$
$$\mathbf{V} = (\nu_0 = \dots = \nu_{k-1}, \nu_k, \dots, \nu_r, \nu_{r+1} = \dots = \mu_{r+p}),$$
$$n \geq k-1, \quad r \geq p-1.$$

The parameter values u_i and v_i satisfy *Schoenberg-Whitney* conditions with respect to $\mathbf{U}$ and $\mathbf{V}$.

Now we transfer the results of the curve approximation to the surface case: In (5) $\bar{\mathbf{X}}(t_i)$ has to be replaced by

$$\bar{\mathbf{X}}(u,v) = \begin{pmatrix} \sum_{j=0}^{n}\sum_{l=0}^{r} w_{jl}\, N_{jk}(u)N_{lp}(v) \\ \sum_{j=0}^{n}\sum_{l=0}^{r} \bar{d}_{jlx}\, N_{jk}(u)N_{lp}(v) \\ \sum_{j=0}^{n}\sum_{l=0}^{r} \bar{d}_{jly}\, N_{jk}(u)N_{lp}(v) \\ \sum_{j=0}^{n}\sum_{l=0}^{r} \bar{d}_{jlz}\, N_{jk}(u)N_{lp}(v) \end{pmatrix}.$$

In G the denominator of $\mathbf{X}(t)$ has to be replaced by the denominator of $\mathbf{X}(u,v)$, and we obtain the error function $\bar{F}$

$$\bar{F} := \sum_{i=0}^{m} \bar{\mathbf{X}}^2(u_i,v_i) - (\bar{\mathbf{X}}(u_i,v_i)\cdot\bar{\mathbf{P}}_i)^2 + \lambda G,$$
$$G := \sum_{i=0}^{m} \Big(\sum_{j=0}^{n}\sum_{l=0}^{r} w_{jl} N_{jk}(u_i)N_{lp}(v_i) - 1\Big)^2. \tag{10}$$

The $\bar{\mathbf{P}}_i$ are defined as in (2). The unknowns for the minimization of $\bar{F}$ are $\bar{d}_{jlx} := w_{jl}d_{jlx}$, $\bar{d}_{jly} := w_{jl}d_{jly}$, $\bar{d}_{jlz} := w_{jl}d_{jlz}$, w_{jl}. Analogously to the curve case, we use also parameter correction. Again the approximation process has two iterative steps: minimize (10) (solve a linear system) and correct the parameter values. At the end of the approximation the points and the resulting surface are rescaled.

§4. Test Examples

To test the quality of the new approach, we compare it with the method of Schneider [2]. He minimized the Euclidian square distances between the points and the rational B-spline surface (error function), a nonlinear optimization problem. With the help of the pseudo inverse he linearized the error function. This method has certain disadvantages: for a larger number of points ($m > 100$) it is very slow, it uses a lot of computer memory and the degrees of the B-spline surface are restricted by 5. For comparison two test examples were used:

1) the surface BRODE, a 9×9 Bézier patch which spans about 14×10 mm.
2) the surface SEITE1, a bicubic set of 9×7 Bézier patches which span about 2200×500 mm.

Both examples were benchmarks at the Conference on Curves and Surfaces in Chamonix, 1993 [2]. The approximation points were sampled on these surfaces. The tables below show the approximation results. The maximal (Euclidian) error in the tables is the error after rescaling.

To summarize the results listed in the tables one can say that compared with the method of Schneider,

- the new approach has similar approximation errors.
- the new approach is much faster (up to a factor of 15).

Moreover, the degrees of the surfaces of new approach are in principle not restricted, and in the new approach only a small amount of computer memory is used.

BRODE: 180 sampled points				
new approach				
degree	segm.	max. error	iter.	CPU: HP 9000/735
3×3	2×1	0.035 mm	60	10.3 sec
3×3	4×2	0.012 mm	25	15.2 sec
5×5	1×1	0.021 mm	10	11.9 sec
method Schneider				
3×3	2×1	0.044 mm	20	166.4 sec
3×3	3×2	0.011 mm	20	275.5 sec
5×5	1×1	0.0095 mm	1	25.9 sec

SEITE1: 764 sampled points				
new approach				
degree	segm.	max. error	iter.	CPU: HP 9000/735
3×3	4×18	0.17 mm	6	479.8 sec
5×5	1×11	0.16 mm	9	206.7 sec
method Schneider				
3×3	4×18	0.22 mm	5	> 2000 sec
5×5	1×11	0.33 mm	5	> 2000 sec

References

1. Elsässer, B., Approximation mit rationalen B-Spline Kurven und Flächen, Dissertation, TU Darmstadt, 1998.
2. Hoschek, J., and Schneider, F.-J., Approximate conversion and data compression of integral and rational B-spline surfaces, in *Curves and Surfaces in Geometric Design*, P.-J. Laurent, A. Le Méhauté, and L. L. Schumaker (eds.), A. K. Peters, Wellesley MA, 1994, 241–250.
3. Ma, W., and Kruth, J.-P., Mathematical modelling of free-form curves and surfaces, in *Curves and Surfaces in Geometric Design*, P.-J. Laurent, A. Le Méhauté, and L. L. Schumaker (eds.), A. K. Peters, Wellesley MA, 1994, 319–326.
4. Pratt, M. J., Goult, R. J., and Ye, L., On rational parametric curve approximation, Comput. Aided Geom. Design **10** (1993), 363–377.

Bernhard Elsässer
TU Darmstadt
FB Mathematik AG 3
Schloßgartenstraße 7
D-64289 Darmstadt, GERMANY
`elsaesser@mathematik.tu-darmstadt.de`

Cycles upon Cycles: An Anecdotal History of Higher Curves in Science and Engineering

Rida T. Farouki and Joanne Rampersad

Abstract. Although they are not amenable to *exact* representations in modern CAD systems, certain transcendental curves — generated by the rolling motions of lines and circles — have played key historical roles in a remarkable variety of scientific and technological applications. We offer an eclectic survey of the origins and applications of such curves, encompassing the use of epicycloids and circle involutes in defining gear tooth profiles; cycloids as the geometrical and mechanical basis of Huygens' isochronous pendulum clock, and as the solution of the brachistochrone problem; and the use of epitrochoids in describing planetary orbits and the combustion chamber of the Wankel rotary engine. These case studies provide sobering evidence that, compared to modern endeavors, whatever "computer-aided geometric design research" of the pre-computer era lacked in labor-saving computational gadgetry was compensated for by insight and ingenuity.

§1. Introduction

The computer revolution has sparked a resurgent interest in the analytic or "computational" approach to geometry that was inaugurated in 1637, when René Descartes introduced the notion of coordinate systems in the appendix *La géométrie* to his philosophical treatise *Discours de la méthode pour bien conduire sa raison et chercher la vérité dans les sciences.* This phenomenon is most vigorously expressed in the field of computer-aided geometric design (CAGD), which is concerned with representing and manipulating geometrical information for practical scientific and engineering purposes.

Perhaps the most basic issue that computer-aided design systems must address is the choice of functional forms that define their internal geometrical representations. The chosen forms must accommodate the common "simple" geometries *exactly*, and also offer sufficient flexibility for "free-form" designs. Moreover, they must permit the reduction of key geometrical procedures to tractable algorithms. The *rational* curves and surfaces (i.e., those that can be parameterized by rational functions) emerge as clear winners in this respect.

Mathematical Methods for Curves and Surfaces II
Morten Dæhlen, Tom Lyche, and Larry L. Schumaker (eds.), pp. 95–116.

ISBN 0-8265-1315-8.

Digital computers are, after all, only capable of finite sequences of arithmetic operations — which amount merely to rational function evaluations.

The predilection for curves of "simple" functional form is an echo from the earliest days of analytic geometry. Descartes, for example, considered loci defined by finite (parametric or implicit) algebraic equations to be *geometrical* curves. Loci that do not admit such definitions, but have simple kinematical or other intuitive descriptions, he considered to be *mechanical* curves — with the implication that they lie beyond the realm of geometry proper; we now call them *transcendental* curves. Indeed, Cartesian coordinates were criticized by contemporaries like Gilles Personne de Roberval, and later by Isaac Newton, on the grounds that they do not furnish algebraic descriptions for curves that were then well studied, such as the cycloid (see §5 and §6 below).

The emphasis in CAGD on rational curves has encouraged a widespread ignorance of (or, at least, lack of appreciation for) the historical and practical significance of non–rational algebraic and transcendental loci in science and technology. Beyond lines and circles, for example, the most important curve in engineering is arguably the circle involute (see §4) — a locus that cannot be exactly described in CAD systems. In this brief survey we aim, by a series of historical vignettes, to restore such curves to a semblance of their former glory. The ingenuity of earlier generations of "computational geometers" (despite — or even, perhaps, *because of* — their lack of computers) in utilizing these loci to explicate the workings of nature and serve the economic needs of society offers us some valuable lessons in humility.

§2. Rolling Motions of Lines and Circles

The curves that interest us are most easily conceived through kinematical descriptions, involving the notion of smooth bodies that *roll without slipping* against each other. If the bodies have plane boundary curves $\mathcal{C}$ and $\mathcal{C}'$, initially in contact at a point that we label P on $\mathcal{C}$ and P' on $\mathcal{C}'$, the no–slip condition means that, throughout their relative motion, the contact point must have the same instantaneous velocity along the common tangent line, when considered independently as a point of $\mathcal{C}$ and of $\mathcal{C}'$. Consequently, if Q on $\mathcal{C}$ and Q' on $\mathcal{C}'$ identify the contact point after a finite time interval elapses, the length of the curvilinear arc PQ on $\mathcal{C}$ must equal that of $P'Q'$ on $\mathcal{C}'$ (Figure 1).

We take $\mathcal{C}$ fixed and $\mathcal{C}'$ rolling against it, and consider the trajectory of a point of $\mathcal{C}'$ — or, more generally, a point of a "moving plane" carried with $\mathcal{C}'$ during its motion (the curve $\mathcal{C}$ is known as the *centrode* [15] for this motion: at each instant, we may consider $\mathcal{C}'$ to be rotating about a center on it). Loci generated in this manner are called roulettes [20,21] — invoking just lines and circles, for example, they encompass the following instances:

A cycloid is the locus traced by a fixed point on a circle C that rolls without slipping on a fixed base line L.

A circle involute is the locus traced by a fixed point on a line L that rolls without slipping on a fixed base circle C.

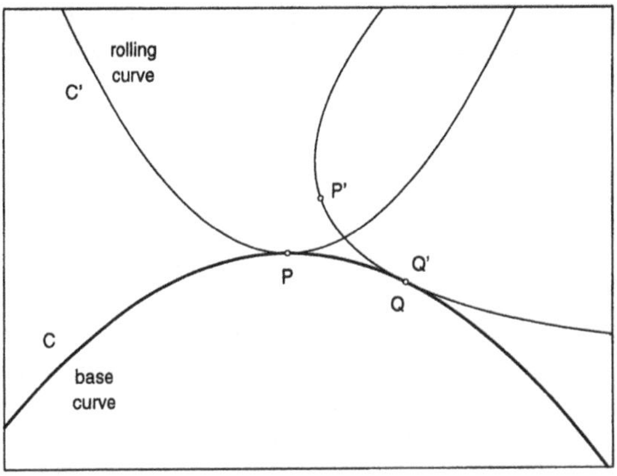

Fig. 1. The arcs PQ and $P'Q'$ are equal when C' rolls without slipping on C.

An epicycloid/hypocycloid is the locus traced by a fixed point on a circle A that rolls without slipping on the exterior/interior of a fixed base circle B.

An epitrochoid/hypotrochoid is the locus traced by a fixed point on a radial line extending from the center of a circle A that rolls without slipping on the exterior/interior of a fixed base circle B.

The cycloid and circle involute are "dual" to each other, in the following sense. Consider a line L and a circle C in contact at a point P — interpreting P as a fixed point of C, which rolls on L, it generates a cycloid; conversely, regarding P as a fixed point of L, which rolls on C, we obtain a circle involute (the latter is often defined as the locus traced by the end of a taut string that is "unwrapped" from C; this is exactly equivalent to our definition).

Cycloids and circle involutes may, in fact, be regarded as special instances of epicycloids/hypocycloids. If, in the definition of the latter, we take a finite–radius circle C for A and an infinite–radius circle — i.e., a line L — for B, we recover the definition of a cycloid. On the other hand, identifying A with L and B with C yields the definition of a circle involute. Note that, if A or B is a circle of infinite radius — i.e., is actually a line — the distinction between rolling on the "exterior" and the "interior" becomes immaterial (the prefixes *epi* and *hypo* are from the Greek for "on" and "in" — or above and below).

Finally, we also note that epicycloids/hypocycloids are special instances of epitrochoids/hypotrochoids, in which the distance of the fixed point P from the center of the circle A is equal to its radius. Hence, the above definitions may be viewed as forming a hierarchy — wherein each level identifies a more general family of loci, subsuming the loci described in preceding levels. One can also define a "plain" (neither epi– nor hypo–) trochoid through an obvious generalization of the cycloid, but this curve does not concern us here.

The kinematical description of these curves is, in fact, the source of their scientific and technological utility in contexts ranging from gears and pendula to celestial mechanics. Our intent is to visit each of them at greater length, sojourning long enough to impart an idea of how they arose in basic problems of physics and engineering. As we shall see, applications in the latter spheres often spurred new developments in mathematics, rather than vice–versa.

§3. Epicycloids — Gear Tooth Profiles

We take a base circle B of radius b with center $(0, b)$ and imagine a second circle A of radius a initially in contact with B at the origin and subsequently rolling in an anticlockwise sense about it. Now the coordinates of the center of A change from $(0, -a)$ to $((b + a)\sin\theta, b - (b + a)\cos\theta)$ as it traverses an angle θ about the center of B. Moreover, to prevent slip during this motion, A must rotate about its center by an anticlockwise angle ϕ, such that $a\phi = b\theta$, relative to the line joining the centers of A and B. Adding the displacement $(-a\sin(\theta + \phi), a\cos(\theta + \phi))$ incurred by this rotation to the center location of A, we see that the fixed point of A initially in contact with B traces the locus

$$x(\theta) = (b + a)\sin\theta - a\sin k\theta, \quad y(\theta) = b - (b + a)\cos\theta + a\cos k\theta,$$

where $k = 1 + b/a$, as A rolls without slipping on B. The epicycloid is periodic (and rational) if k is a rational number — otherwise it is transcendental. An example is shown in Figure 2.

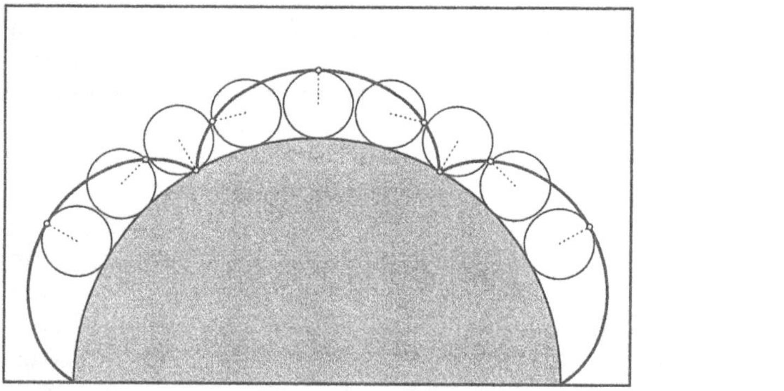

Fig. 2. Epicycloid generated by a fixed point on a circle rolling on a base circle.

Gears are perhaps the most fundamental components of all machinery — they provide great flexibility in power transmission, and control over speed and torque, with minimal dissipation. To accomplish this, however, demands considerable subtlety in the geometry and fabrication of gear teeth.

The epicycloid was historically the first gear–tooth profile known to yield conjugate action of meshing gears — i.e., a strictly constant angular–velocity ratio of the gears is maintained, at each instant during the engagement and disengagement of successive teeth. Although the circle involute (see §4 below) is now recognized to be a superior tooth profile, epicycloidal gears persisted for many years — they were used, for example, in the Model T Ford.

The inability of primitive gears — simply "pegged wheels" — to smoothly and reliably transmit large torques lead to recognition of the need for curved tooth profiles that satisfy the kinematical condition of conjugate action. This problem attracted the attention of famous Renaissance scientists — Leonardo da Vinci (1452–1519) sketched gears with curved teeth; and Girolamo Cardano (1501–1576), who discovered the solution of the cubic equation, also published what was perhaps the earliest "scientific" study of gears [7].

Although epicycloids had been studied by Albrecht Dürer [10] in 1525, the French mathematician Girard Desargues (1591–1661) and Danish astronomer Ole Rømer (1644–1710) are usually credited with their first use in describing gear–tooth profiles [31]. Rømer is famous for having demonstrated the finite speed of light by timing Jupiter's moons; we know of his work on gears only indirectly through the comments of Leibniz and others. Desargues, a proficient architect and engineer, is now recognized as a founder of projective geometry, though his ideas were considered obscure by contemporaries — he spoke freely of "points at infinity" without a rigorous basis for their treatment [14].

Desargues constructed a water–raising device at the château of Beaulieu, near Paris, that incorporated the first documented use of epicycloidal gears. However, he left no written record of it: the device was sketched and described by Huygens in a letter dated October 29, 1671 (see §5 for more on Huygens). It was also described [18] by Philippe de La Hire (1640–1718), an admirer of Desargues, who was subsequently charged with repairing the device.

de La Hire is credited [31] with securing firm mathematical foundations for epicycloidal gears, through the publication of his *Traité des Épicycloides* [18] in 1694. He also recognized that the circle involute is a form of epicycloid — corresponding to a generating circle of infinite radius — and claimed it to be the "best" gear tooth profile (though more than a century would elapse before this idea was adopted as common practice).

Epicycloidal gear teeth were further refined [6] by Charles Étienne Louis Camus (1699–1768). Some of his illustrations of epicycloidal gears are shown in Figure 3 — note that only the upper portions of the teeth are epicycloids; the lower portions are simply linear radial extensions. Camus enunciated the need for conjugate action in rather convoluted language (according, at least, to his English translator John Isaac Hawkins):

> A machine which does not go uniformly, and whose parts make variable efforts on each other, when a force constantly equal is applied to it, requires, in order to go and to overcome a given resistance, that there should be applied to it a power, the absolute force of which may decrease or increase, according to the situations, more or less favourable, of the pieces of which it is composed; and if it be required, that the force applied to this machine should be constant, the power ought to be capable of moving it in the most disadvantageous situation of its parts. Hence, the force which would be sufficient to move a machine in a mean situation, between the one most favourable and the one least favourable to its parts, would not be sufficient to make

it go in all situations possible. On the other hand, a machine, the parts of which, in regard to each other, are continually in situations equally advantageous, will always go, when there is applied to it a mean moving power, which would be incapable of keeping the former in motion in all situations which might be given to its different parts.

We may therefore consider as the best form that can be given to the teeth of wheels of any machine, that which will cause the teeth to be always, in regard to each other, in situations equally favourable; and which, consequently, will give the machine the property of being moved uniformly by a power constantly equal.

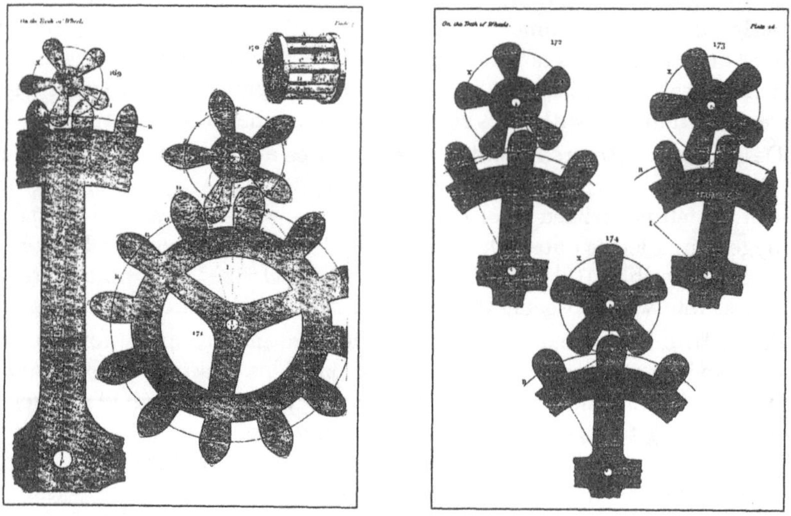

Fig. 3. Examples of epicycloidal gear designs by C. É. L. Camus (1733).

§4. Circle Involutes — Improved Gear Teeth

Take a base circle C with radius r and center at $(0, r)$. We imagine a line L, initially in contact with C at the origin and subsequently rolling about it in an anticlockwise sense. In the absence of slip, the fixed point on L initially in contact with C must be at distance $r\theta$ along L from the new contact point when L rolls through angle θ. From the contact point, which moves from $(0,0)$ to $(r\sin\theta, r - r\cos\theta)$, we measure a length $r\theta$ backward along the tangent direction $(\cos\theta, \sin\theta)$ and thereby obtain the parameterization

$$x(\theta) = r(\sin\theta - \theta\cos\theta), \quad y(\theta) = r(1 - \cos\theta - \theta\sin\theta)$$

for a circle involute (see Figure 4). One can easily verify that the above agrees with the epicycloid equation in the limit $a \to \infty$ (with $b = r$). The presence of

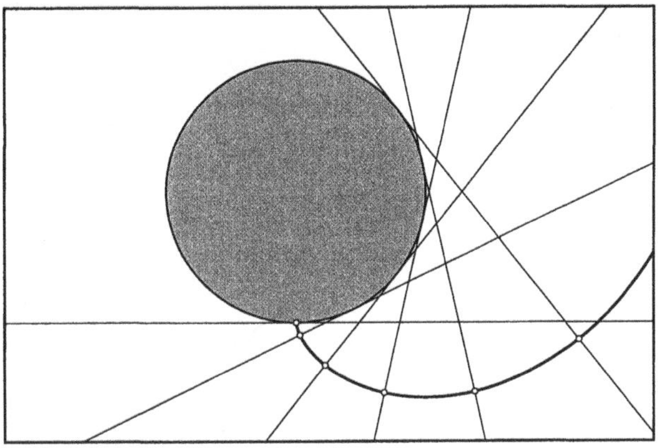

Fig. 4. Circle involute generated by a fixed point on a line rolling on a base circle.

θ alone *and* as an argument of trigonometric functions means that this curve is *transcendental* — it cannot be exactly described by algebraic equations.

While the epicycloid was the first gear–tooth form known to give precise conjugate action, primitive straight–toothed gears can be found in astrolabes, sundials, and clockwork from the Greek, Byzantine, and Islamic civilizations [12,25]. A "defect" of epicycloidal gears, however, is that they require an exact spacing between the gear centers to achieve true conjugate action. One respect in which involute teeth are superior (as noted by de La Hire) to epicycloidal teeth is that they relax this constraint: conjugate action is guaranteed for *any* distance between the gears — provided, of course, that they mesh.

The great Swiss mathematician Leonhard Euler (1707–1783) was the first [11] to give a precise mathematical enunciation of the principles of gear design (conjugate action and minimization of sliding motion between teeth to reduce friction and wear), and to fully develop the geometry of involute teeth. There remained, however, a large gap between Euler's mathematical ideal and what engineers were capable of fabricating, that was bridged a century later through efforts of the Englishmen John Hawkins and Robert Willis (as late as 1812, it was argued in James White's *Mémoire* that gear teeth need not be accurately formed, since they would "wear in" to the proper shape).

Figure 5 illustrates the basic geometry of involute gear teeth. Involutes to the *base circles*, B_1 and B_2, of two meshing gears are shown at successive instants. When gear 1 rotates clockwise, its involute "pushes" that of gear 2, causing the latter to rotate anticlockwise. It can be shown that the angular velocities of the gears satisfy $\omega_2/\omega_1 = -\rho_1/\rho_2$, where ρ_1, ρ_2 are the base–circle radii. Moreover, the contact point of the involutes traverses the *pressure line* — the common tangent of the two base circles — at fixed speed. Also shown (dashed) are the *pitch circles*, whose radii divide the distance between the gear centers in the ratio $r_1 : r_2 = \rho_1 : \rho_2$. The action of the two involutes against each other involves pure rolling at the *pitch point*, where the pressure line cuts the line of centers, and a combination of rolling and sliding elsewhere. The

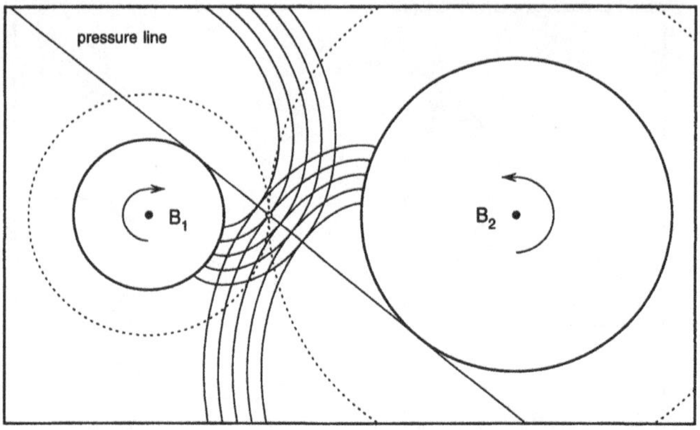

Fig. 5. Contact point of involutes to gear base circles moves along pressure line.

conjugate–action relation $\omega_2/\omega_1 = -r_1/r_2$ does not depend on the spacing of the gear centers (although the pressure angle does).

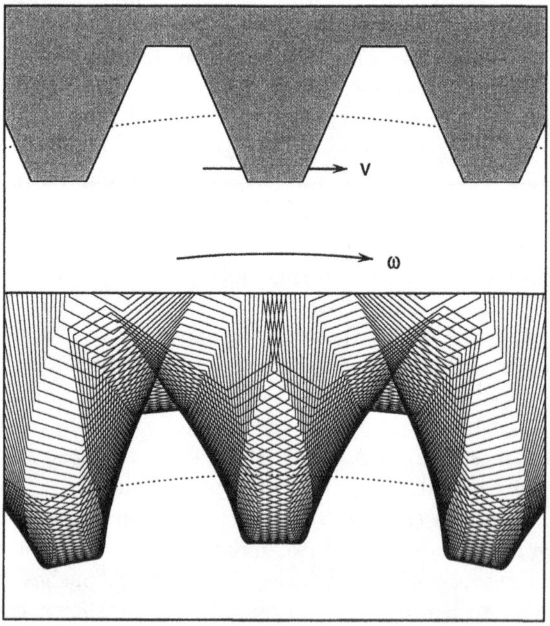

Fig. 6. Generation of involute gear teeth (the dotted curve is the pitch circle).

It might seem, at first, that involute teeth would be rather difficult to cut into a gear "blank" at consistently high precision. Just as involute teeth yield a simple kinematical relation (conjugate action) between gears, however, it is possible to fabricate them by appropriate motions of a simple cutter. Figure 6 shows a *rack cutter* engaging a gear blank. The cutter has straight–sided teeth — the correct involute form of an infinite–radius gear. It reciprocates in and out of the plane, and is gradually fed radially toward the center of the blank. Meanwhile, the blank turns at angular speed ω and the cutter is fed forward at speed v, with $v/\omega = r$ (the pitch–circle radius). The desired involute teeth are then seen, in a reference frame rotating with the blank, to emerge as the *envelope* of successive instantaneous cutter positions (Figure 6).

§5. Cycloids I — Huygens' *Horologium Oscillatorium*

Take the x–axis for the base line L and consider a circle C of radius r, initially in contact with L at the origin and subsequently rolling in a clockwise sense along it. To avoid slip, C must travel a horizontal distance $r\theta$ on rolling through angle θ — i.e., its center must move from $(0, r)$ to $(r\theta, r)$. During this motion, the coordinates of the fixed point on C initially in contact with L, relative to the center of C, change from $(0, -r)$ to $(-r\sin\theta, -r\cos\theta)$. Hence, the locus traced by this point — a cycloid — has the parameterization

$$x(\theta) = r(\theta - \sin\theta), \quad y(\theta) = r(1 - \cos\theta).$$

Like the circle involute, the cycloid (Figure 7) is a transcendental curve.

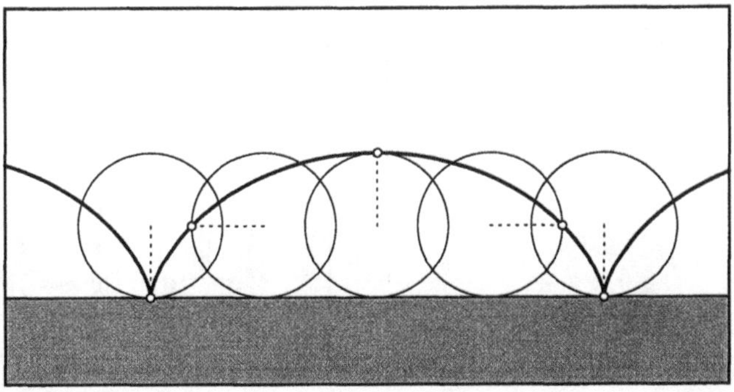

Fig. 7. Cycloid generated by a fixed point on a circle rolling on a base line.

Although cycloids were known to the ancient Greeks, the first substantive study was probably by Nicholas of Cusa (1401–1464). Galileo (1564–1642) and his student Torricelli (1608–1647) were interested in finding the area beneath a cycloidal arc ($3\pi r^2$). Among other eminent (pre–calculus) cycloid aficionados, we mention Descartes, Fermat, Roberval, and Wren. The extent of interest in "cycloid research" during the 17th century is suggested by a competition, sponsored by Blaise Pascal (1623–1662), to prove certain of its properties — none of the entries were deemed prizeworthy. Pascal also wrote a *Histoire de la Roulette* that provoked acrimonious nationalistic arguments.

(Pascal had, in fact, abandoned science in 1654 in order to devote himself to theology — his interest re–emerged in 1658 when, during a sleepless night incurred by toothache, he occupied himself with cycloid investigations. Apart from discovering several new properties, this activity appears [9] to have cured his ailment — the dental profession, however, has subsequently been loath to prescribe mathematical exertions in lieu of fillings and extractions.)

The 1673 study *Horologium Oscillatorium* [16] by the Dutch physicist and mathematician Christiaan Huygens (1629–1695) is undoubtedly the crowning achievement (Figure 8) of cycloid investigations. Though not as widely known as his *Traité de la Lumière* of 1678, which introduced the wave theory of light, this earlier treatise is perhaps Huygens' greatest work — it not only identified an ingenious use of cycloids in timekeeping, but in doing so founded a novel branch of differential geometry: the theory of *evolutes* and *involutes*. Huygens expresses this view, in dedicating his work to King Louis XIV, as follows:

> ... I have given preference to the investigation of those things which can be found useful either to making life more comfortable or to the understanding of nature ... I have pursued this double goal with greater success in the invention of my clock than anything else.

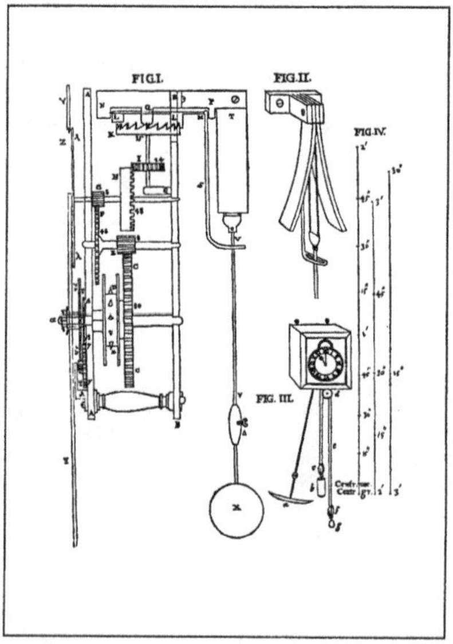

Fig. 8. Title page and drawing from Huygens' *Horologium Oscillatorium* (1673).

The design of chronometers of sufficient accuracy for marine navigation was an urgent technological problem in Huygens' time [28]. Counting swings of a simple pendulum was a well-known principle of timekeeping — but, since the period of these swings depends on their amplitude, the simple pendulum clock is susceptible to errors due to decay or perturbation of its oscillations. Huygens sought, first, to identify the trajectory of a mass that will oscillate in a gravitational field with a period *precisely* independent of amplitude and, second, a physical means to realize such oscillations. Cycloids were the key to *both* these problems in designing an "isochronous" pendulum clock.

In the usual treatment of the simple pendulum (a mass m suspended on a string of length ℓ, as in Figure 9) one invokes the "small–angle approximation" — assuming that $\theta \ll 1$, we set $\sin\theta \approx \theta$ in the well–known equation of motion

$$\frac{\mathrm{d}^2\theta}{\mathrm{d}t^2} + \frac{g}{\ell}\sin\theta = 0, \tag{1}$$

where g is the acceleration due to gravity. This yields simple harmonic motion, $\theta(t) = \alpha\cos(2\pi t/T_0)$, with a period $T_0 = 2\pi\sqrt{\ell/g}$ that is *independent of the amplitude* α. In this model, we need only know g and the pendulum length ℓ accurately to calibrate a pendulum clock — the mass m of the bob, and the magnitude of the "kick" that sets it in motion, are immaterial.

The *exact* solution of (1), however, involves the *Jacobian elliptic functions* [19] and has the amplitude–dependent period

$$T(\alpha) = 4\sqrt{\ell/g}\int_0^{\pi/2}\frac{\mathrm{d}\phi}{\sqrt{1-\sin^2\frac{1}{2}\alpha\,\sin^2\phi}}$$

(which can be verified by simple energy arguments). This is a *complete elliptic integral of the first kind* that cannot be resolved into elementary functions of α — its value must be looked up in mathematical tables, or it may be computed to any desired accuracy from the infinite series

$$\frac{T(\alpha)}{T_0} = 1 + \left(\frac{1}{2}\right)^2\sin^2\tfrac{1}{2}\alpha + \left(\frac{1\cdot 3}{2\cdot 4}\right)^2\sin^4\tfrac{1}{2}\alpha + \left(\frac{1\cdot 3\cdot 5}{2\cdot 4\cdot 6}\right)^2\sin^6\tfrac{1}{2}\alpha + \cdots$$

(which converges for all α). If $\alpha = 10°$, for example, calibrating a pendulum clock using T_0 incurs a discrepancy of about 3 minutes/day, and this increases to nearly 30 minutes/day for $\alpha = 30°$. Even if $T(\alpha)$ for a nominal amplitude α were used instead of T_0 for calibration, any change in α due to friction or perturbations will induce systematic errors in the indicated time.

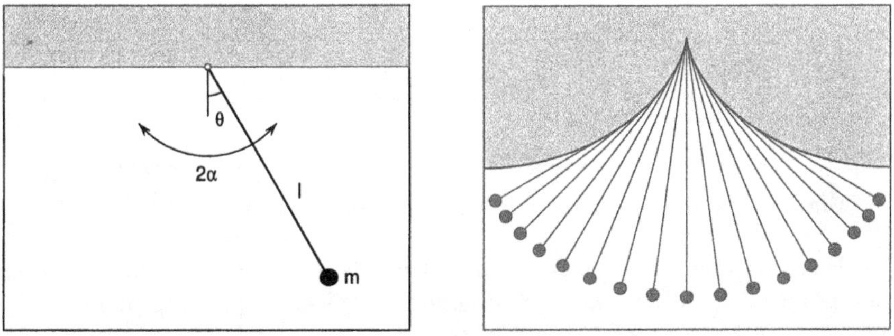

Fig. 9. The simple pendulum (left) and isochronous cycloidal pendulum (right).

Huygens sought to design an "isochronous" pendulum clock — one that has a period *exactly* independent of its amplitude (not just in the limiting case of small oscillations). He was aware that, ignoring friction, the oscillation of a pendulum of fixed length ℓ is equivalent to the motion of a small particle that rolls on the inside of a spherical bowl of radius ℓ — both motions are governed simply by the exchange of potential and kinetic energy along a *circular* path. Recognizing that this principle generalizes to other paths, Huygens sought a bowl shape that would cause particles to exhibit isochronous rolling motions — i.e., the time taken for the particle to reach the bottom of the bowl would be independent of the point from which it was released.

By ingenious geometrical arguments (the calculus was not yet developed), Huygens discovered that a particle rolling in an "inverted" cycloid exhibits the desired isochronous property. The question was thus: how can one construct a "variable–length" pendulum, whose bob will execute a cycloidal trajectory? The solution to this problem that Huygens found is truly amazing — one uses another cycloid to "continuously adjust" the pendulum length! Consider the two cycloids defined for $0 \le \phi \le 2\pi$ by

$$C_1 : \quad x_1(\phi) = r(\phi - \sin\phi), \quad y_1(\phi) = r(\cos\phi - 1),$$

$$C_2 : \quad x_2(\phi) = r(\phi + \sin\phi), \quad y_2(\phi) = r(1 - \cos\phi). \tag{2}$$

Actually, C_1 and C_2 are just appropriately–positioned segments of a *unique* (infinitely extended) cycloid. Figure 10 shows their generation by two circles of radius r rolling from left to right in contact with and below the lines $y = 0$ and $y = 2r$, respectively. The initial common point $(0, 0)$ of the circles, considered as a fixed point of each, traces out the cycloids C_1 and C_2.

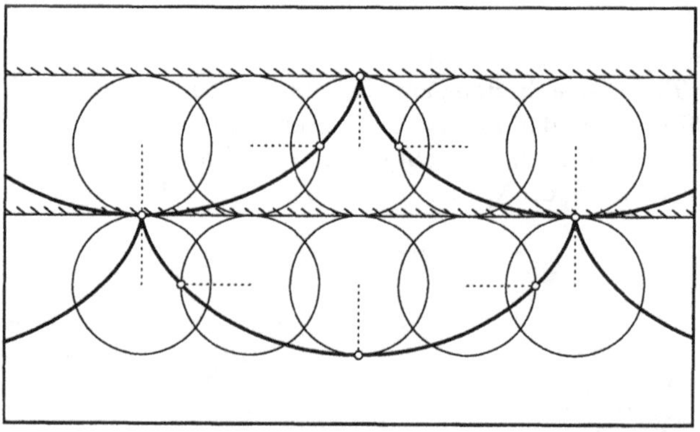

Fig. 10. An involute–evolute pair of cycloids C_1 (lower) and C_2 (upper).

Huygens found that the cycloids (2) satisfy an involute–evolute relation: each point of C_2 is the *center of curvature* for the corresponding point on C_1; conversely, each point of C_1 may be regarded as the *end of a taut string that is "unwrapped" from* C_2 up to the corresponding point on the latter. His idea, then, was to use C_2 as a "cycloidal jaw" with the pendulum string attached at its cusp (at $\phi = \pi$). As the pendulum oscillates, the string alternately wraps and unwraps about the left and right cheeks of this jaw (Figure 9) and hence the trajectory of the bob is the involute of the jaw profile, i.e., it is the cycloid C_1. By the isochronicity of cycloidal motion, this ensures an oscillation period *precisely* independent of the amplitude. Huygens expressed this [16] as:

> **PROPOSITION XXV** *On a cycloid whose axis is erected on the perpendicular and whose vertex is located at the bottom, the times of descent, in which a body arrives at the lowest point at the vertex after having departed from any point on the cycloid, are equal to each other; and these times are related to the time of a perpendicular fall through the whole axis of the cycloid with the same ratio by which the semicircumference of a circle is related to its diameter.*

Considering the cycloidal path C_1 of the bob as the envelope of its circles of curvature, one may regard the bob as following a sequence of instantaneous circular motions, with radii and centers determined by the point along C_2 to which the string has unwrapped. Note that the amplitude–independent period of the cycloidal pendulum, $T = 4\pi\sqrt{r/g}$, is determined by the generating circle radius r, rather than the string length ℓ. This agrees with the small–amplitude period $T_0 = 2\pi\sqrt{\ell/g}$ for a simple pendulum of length $\ell = 4r$ (which corresponds to half the length of one cycloidal arch).

Figure 8 offers convincing evidence that Huygens was as concerned with the engineering details of accurate clocks as with establishing the theory of evolutes and involutes. Unfortunately, however, the tools at his disposal were no match for the precision of his mathematics — the "cycloidal clocks" that he constructed did not realize their full potential.

§6. Cycloids II — The Brachistochrone

While Huygens relied entirely on geometrical arguments, the cycloid soon arose in the archetypal problem of an important branch of the calculus — now known as the *calculus of variations.* This is concerned with methods for finding the function $y(x)$ on $x \in [\,x_0, x_1\,]$ that minimizes an integral of the form

$$\int_{x_0}^{x_1} F(x, y, y')\,\mathrm{d}x\,,$$

where $y' = \mathrm{d}y/\mathrm{d}x$ and F is some known (differentiable) function of x, y, y'. The modern approach to such a "variational problem" is to reduce it to the second–order (non–linear) ordinary differential equation

$$\frac{\partial F}{\partial y} - \frac{\mathrm{d}}{\mathrm{d}x}\frac{\partial F}{\partial y'} = 0\,,$$

— known as the *Euler–Lagrange equation* [3] — for $y(x)$, subject to boundary conditions $y(x_0) = y_0$ and $y(x_1) = y_1$.

In 1696 the Swiss mathematician Johann Bernoulli (1667–1748) proposed a problem of this nature as an international competition. Consider a particle rolling on the inside a smooth bowl: the particle is released (Figure 11) from a point $P = (-\ell, h)$ on the bowl at height h above its lowest point $O = (0,0)$. Bernoulli's question was: what cross–sectional shape $y(x)$ minimizes the time

$$T = \int_{-\ell}^{0} \sqrt{\frac{1 + y'^2(x)}{2g\,[\,h - y(x)\,]}}\,\mathrm{d}x$$

taken for the particle to roll on the bowl from P to O?

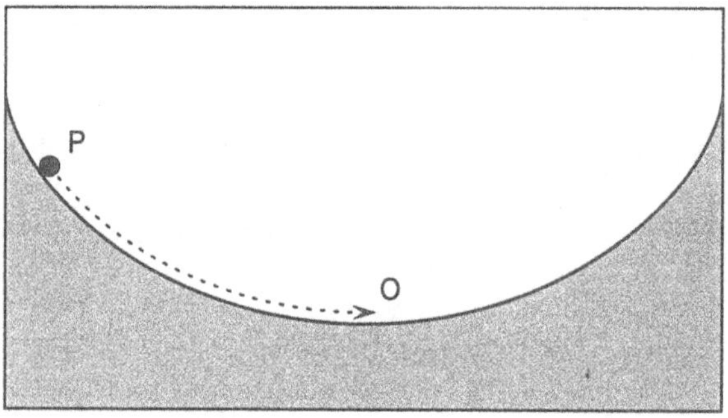

Fig. 11. The problem of the brachistochrone.

Huygens had shown that the isochrone or tautochrone (curve of *constant* or *equal* time) is the cycloid — the locus that Bernoulli sought was known as the brachistochrone, from the Greek for *least* time. Bernoulli approached this problem through a rather ingenious physical analogy: noting that the speed v of the particle at the point (x, y) depends only on the potential energy released in reaching it — i.e., on the local y coordinate, but *not* the actual shape of its path up to that point — he realized [24] that light propagation in a stratified medium (e.g., the atmosphere) is governed by a similar principle.

By reasoning that the variation of the speed of a light ray with the density (more properly, the refractive index) of a medium causes it to follow a curved path, Bernoulli was able to set up and solve the differential equation describing the shape of the brachistochrone; the result was — a *cycloid*!! He published his solution under the title "Curvatura radii in diaphanis non uniformibus ..." (The curvature of a ray in a non–uniform medium ...) in the *Acta Eruditorum* of May 1697, and was evidently feeling very pleased with himself:

> ... We have a just admiration for Huygens, because he was the first to discover that a heavy point on an ordinary cycloid falls in the same time, whatever the position from which the motion begins. But the reader will be greatly amazed, when I say that exactly this cycloid, or *tautochrone* of Huygens, is our required *brachistochrone* ... I discovered a wondrous agreement between the curved path of a light ray in a continuously varying medium and our *brachistochrone*.

Bernoulli's friend Gottfried Wilhelm Leibniz (1646–1716) also solved the problem, and the two colluded to embarrass their antagonist, Isaac Newton (1642–1727), with whom a feud was raging over precedence in inventing the calculus. So the challenge was sent to Newton, then Master of the Royal Mint in London. Though he had largely abandoned scientific matters, Newton felt compelled to defend his honor: by staying up all night on returning from his official duties, he managed to solve the problem. Bernoulli and Leibniz were chastened on promptly receiving an anonymous solution in English!

The May 1697 issue of *Acta Eruditorum* also contained the contribution "Solutio problematum fraternorum ..." (Solution of my brother's problem) by Jakob Bernoulli, and correct solutions by Leibniz, Newton, and Tschirnhaus — also an erroneous one by the Marquis de l'Hôpital. Galileo, much earlier, had also studied cycloids, though in his celebrated *Dialogue* [13] championing a heliocentric theory of the solar system, he mistakenly identifies the *circle* as the trajectory that yields isochronous oscillations of minimum period.

§7. Epitrochoids I — Orbital Paths of the Planets

The epitrochoid is a generalization of the epicycloid in which, instead of taking a fixed circumferential point of a circle A rolling on a base circle B, we consider the locus traced by a point at arbitrary distance h along a fixed radial line emanating from the center of A. Let A and B have radii a and b, the latter with fixed center $(0, b+a-h)$ and the former with center initially at $(0, -h)$. Consider a radial spoke of length h, initially vertical, that A carries with it in rolling on B — by arguments similar to those of §3 one finds that the tip of this spoke, initially at $(0,0)$, traces the epitrochoid curve

$$x(\theta) = (b+a)\sin\theta - h\sin k\theta\,, \quad y(\theta) = b+a-h - (b+a)\cos\theta + h\cos k\theta$$

where, again, $k = 1+b/a$. The epitrochoid is periodic (and rational) when k is rational. Choosing $h = a$ evidently yields the epicycloid as a special instance (Figure 12), while the case $a = b$ corresponds to a *limaçon of Pascal* [20,21].

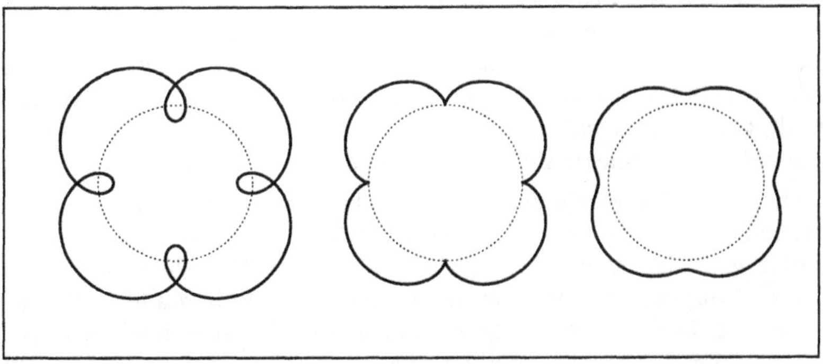

Fig. 12. Epitrochoids with $b = 4a$ and $h > a$ (left), $h = a$ (middle), $h < a$ (right).

The ancient Greeks developed a quite sophisticated geocentric cosmology, based on compounded circular motion, that originated with the Pythagoreans, was further developed by Aristotle, and culminated in Alexandria around 140 A.D. with the *Almagest* [26] of Claudius Ptolemy. Ptolemy's treatise remained the most authoritative source on astronomy until the Renaissance — its title is a corruption of the original Greek, through Arabic translations by the 9th–century Islamic scholars Ishaq ibn Hunayn and Thabit ibn Qurra.

The goal of Greek cosmology was to explain the observed motions of the five known planets or "wandering stars" (Mercury, Venus, Mars, Jupiter, and Saturn) relative to the background of fixed stars. The Pythagoreans imagined that the planets were carried along their orbits by crystalline spheres. Having already discovered an intimate connection between music and mathematics — namely, the correspondence of "pleasing" harmonies with integral divisions of a vibrating string — they envisaged such regularities also governing celestial phenomena. Indeed, their mystical philosophy embraced the idea of an audible *music of the spheres* emanating from the planetary motions.

The planets do not always exhibit orderly eastward motion; occasionally they reverse and travel *retrograde* (westward) for a while. We now recognize this as a parallax effect, incurred by the Earth's orbital motion, but the Greeks sought an explanation compatible with their geocentric perspective.

Being predisposed to "perfect" circular motions, Ptolemy discovered that planetary paths could be accurately described by a model in which planets move on small circles, or *epicycles*, whose centers traverse larger circles called *deferents* (the "perfection" of this system was compromised by the fact that, to secure best agreement with observations, Ptolemy had to displace the Earth from the deferent center, and have the epicycle centers moving with constant angular velocity relative to another eccentric point, called the *equant*).

The final demise of geocentric theories of the universe was triggered by the publication of *De Revolutionibus Orbium Coelestium* by Nicolaus Copernicus [8] in 1543 (the year of his death). However, Copernicus retained deferents and epicycles in his heliocentric theory: his contribution was more of a change in philosophical outlook than a quantitative advance in explaining planetary motions. The correct characterization of orbits, as ellipses traced at variable speed, had to await the empirical laws of Johannes Kepler (1571–1630) for a geometrical and kinematical description, and Newtonian mechanics [22] for a dynamical explanation (contrary to popular opinion, Newton's demonstration that elliptical orbits result from inverse–square gravitational forces was based on *geometrical* arguments rather than calculus — see also [2]).

The deferent/epicycle description of planetary orbits (motion on a circle whose center traces another circle) seems rather different from the epitrochoid definition given in §2, but it is not difficult to verify that the two are actually equivalent. Suppose the deferent and epicycle have radii d and e. We choose the center of the former at $(0, d-e)$, and that of the latter initially at $(0, -e)$ with the planet at the 12 o'clock position on it. Then, if the motions on both deferent and epicycle are anticlockwise, with k full revolutions on the latter for each revolution on the former, the orbit is described by

$$x(\theta) = d\sin\theta - e\sin k\theta, \quad y(\theta) = d - e - d\cos\theta + e\cos k\theta. \qquad (3)$$

One can easily see that this is the epitrochoid defined by parameters $a = d/k$, $b = d(k-1)/k$, and $h = e$ (conversely, $d = a + b$, $e = h$, and $k = 1 + b/a$).

The deferent/epicycle description (3) does, in fact, encompass the proper form of planetary orbits (namely, ellipses) as a special case. Taking $k = -1$ in (3) we obtain $x(\theta) = (d+e)\sin\theta$, $y(\theta) = (d-e)(1-\cos\theta)$, which corresponds to an ellipse with center $(0, d-e)$ and semi-axes $d+e$ and $|d-e|$. But, since $k = 1 + b/a$, one is inclined to ask: what does a negative k value signify?

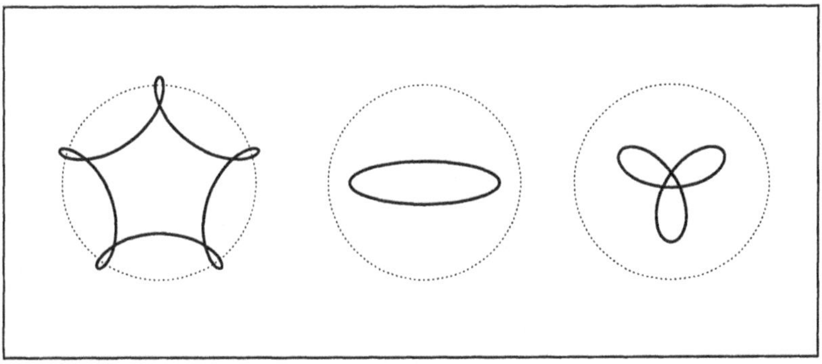

Fig. 13. Hypotrochoids with $b = 5a$ (left), $2a$ (middle), $3a/2$ (right) and $h = 0.28b$.

If we follow through the epitrochoid derivation, but consider A rolling on the interior rather than the exterior of B, we obtain the *hypotrochoid*

$$x(\theta) = (b-a)\sin\theta - h\sin k\theta, \quad y(\theta) = b - a - h - (b-a)\cos\theta + h\cos k\theta$$

where $k = 1 - b/a$. This is, in fact, precisely what one expects from formally substituting $-a$ in lieu of a in the epitrochoid equation. The difference in the parameter $k = 1 \pm b/a$ between epitrochoids and hypotrochoids arises because, when A rolls in an anticlockwise sense about B, to prevent slip it must rotate *clockwise* or *anticlockwise* about its own center, according to whether it rolls on the *interior* or *exterior* of B. Thus, an epitrochoid with $k < 1$ is actually a hypotrochoid — and, in particular, hypotrochoids with $b = 2a$ are all ellipses with semi-axes $a + h$ and $|a - h|$; see Figure 13.

Traces of Pythagorean mysticism persisted in Kepler's early works. In his *Mysterium Cosmographicum* [17] of 1596, he correlates the five planetary orbits with the inscribed/circumscribed spheres of the "perfect" or Platonic solids — the tetrahedron, cube, octahedron, dodecahedron, and icosahedron.

§8. Epitrochoids II — The Wankel Rotary Engine

Returning to Earth, we find the epitrochoid arising as a key ingredient of an ingenious internal-combustion engine design, championed during the first half of the 20th century by the German engineer Felix Wankel. The aim of an IC engine is to convert thermal combustion energy into mechanical energy of motion. A gun, for example, is a primitive IC engine, in which heat released in burning gunpowder is (partly) transformed into kinetic energy of the bullet. Usually, however, one requires a continuous and controllable production of mechanical power — typically in the form of a rotating shaft.

A feasibility study for an IC engine was performed in 1680 by our versatile friend, Christiaan Huygens, who conceived the idea [29] of having a gunpowder explosion drive a piston mounted in a cylinder. Although less dangerous fuels are preferred, this *linear* piston–in–cylinder motion is now ubiquitous in IC engine design. A basic drawback of this approach is that, in order to provide a *rotary* output motion, several pistons must be coupled to a rather complex and finely–balanced mechanism (connecting rod, crank, and crankshaft).

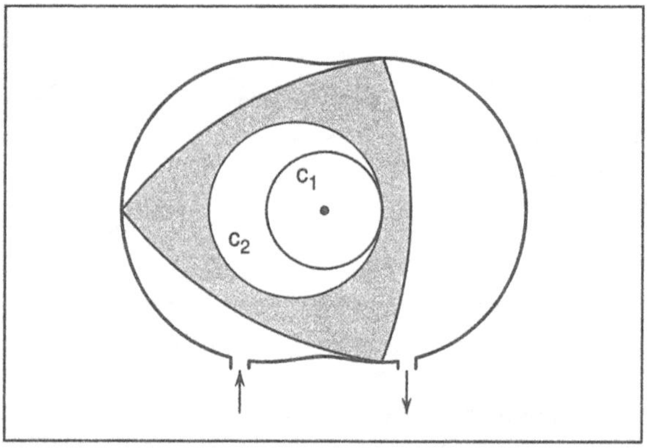

Fig. 14. The epitrochoidal chamber and rotor (shaded) of the Wankel engine.

Wankel's aim was to produce rotary motion *directly* from the energy of a burning gas [1,23]. The usual pistons and cylinders are replaced by a rotor mounted in a chamber, both specially shaped such that at each instant the space between them is divided into three distinct volumes. As the rotor turns, the successive phases of (a) induction and compression of the air/fuel mixture; (b) ignition of the compressed gas and forcing of the rotor by its pressure; and (c) exhaust of the spent fuel occur in these volumes. Sophisticated geometries of the rotor and chamber (see Figure 14) are required to accomplish this.

The rotor is essentially a "curvilinear triangle" that rotates eccentrically within the chamber, its vertices constantly in touch with the perimeter. Power is transmitted by an annular (internal–toothed) gear mounted inside the rotor that meshes with a spur gear on the drive shaft. The pitch circles of these gears are labelled c_2 and c_1, respectively, in Figure 14 — their radii must have a 3:2 ratio for a two–lobed chamber, as shown here. Also shown in Figure 14 are inlet and outlet ports for the fuel and exhaust gases.

The geometry of the Wankel engine may be understood as follows. The pitch circles of the rotor and drive–shaft gears roll without slip against each

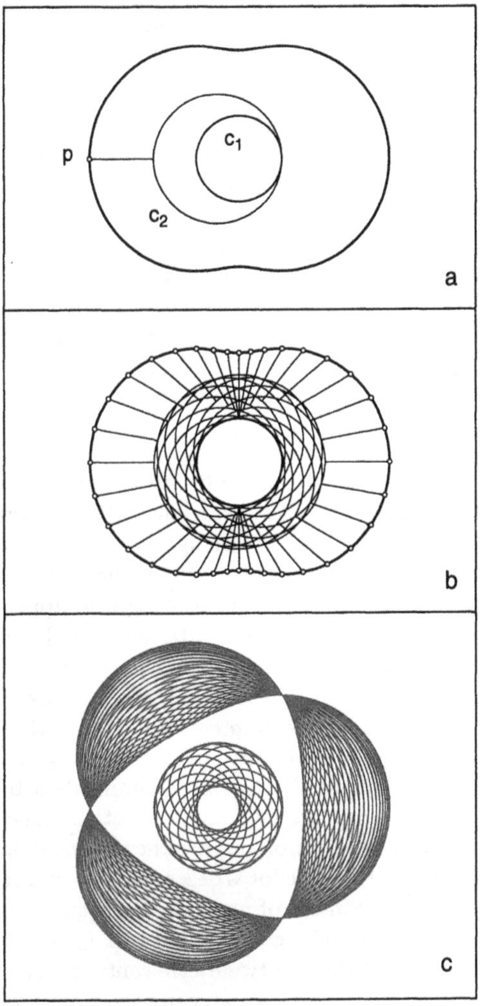

Fig. 15. Geometry of the Wankel rotary engine.

other. Hence, a vertex of the rotor describes (Figure 15a) the locus of a point **p** carried by a circle c_2 that rolls without slip on another circle c_1 — i.e., the epitrochoid shown in Figure 15b. This defines the shape of the Wankel engine chamber (note that the rolling circle *contains* the fixed circle; one can verify that this is equivalent to the epitrochoid definition in §2 above).

Consider now the rotor shape. We imagine ourselves sitting in a reference frame attached to the rotor during its motion — from this perspective, the epitrochoidal chamber appears to revolve eccentrically about us. In order for the rotor to fit within the chamber at all times, the *envelope* of instantaneous relative positions of the chamber must not encroach on it (the envelope of a family of curves may be interpreted [4,5] as the locus of intersection points of "neighboring" members, or the locus that is tangent to each member).

In Figure 15c, which illustrates a sequence of positions of the epitrochoid relative to the rotor, the "curvilinear triangle" shape is immediately apparent as (a subset of) their envelope. Hence, the precise mathematical description of the rotor is rather involved — it is (part of) the envelope of a one-parameter family of epitrochoids. While this complicated geometry obviously makes the rotary engine more difficult to manufacture, it offers distinct advantages over ordinary piston engines — fewer parts; smoother operation; and reduced bulk for a given horsepower. The geometry in Figures 14 and 15 is actually just one instance among many epitrochoidal or hypotrochoidal chamber and associated rotor shapes [30] that can be used to design engines, pumps, and compressors.

§9. Closure

There is a natural tendency, in the evolution of science and technology, to regard the accomplishments of past generations with a (perhaps subconscious) air of smugness and condescension. Their ignorance of knowledge revealed in the interim, and their deprivation of the technological gadgetry — computers, especially — now deemed indispensible to progress, appear to us as pitiful and even intolerable circumstances. This attitude is cultivated, above all, by a total neglect of the cultural and historical aspects of mathematics, science, and technology in the education of successive generations.

Yet a careful and detached assessment indicates that such chauvinism is utterly groundless. Within the context of the prevailing scientific milieu, the geometrical ingenuity of our forebears was undiminished by a lack of means to "crunch numbers" — and it is questionable whether the emergence of CAD systems has had a beneficial effect on our own geometrical/analytical abilities. Moreover, as suggested by the anecdotes described above, these past endeavors are characterized by the treatment of mathematics, science, and technology as an organic and mutually nourishing whole, in stark contrast to the increasing specialization and fragmentation of twentieth–century knowledge.

Although it now has a perhaps greater urgency, our iconoclastic message is not new — we cannot match the eloquence of its expression by the English poet Alexander Pope (1688–1744) in his *Essay on Man* [27]:

Trace science then, with modesty thy guide;
First strip off all her equipage of pride,
Deduct what is but vanity, or dress,
Or learning's luxury, or idleness;
Or tricks to show the stretch of human brain,
Mere curious pleasure, or ingenious pain:
Expunge the whole, or lop th'excrescent parts
Of all, our vices have created arts:
Then see how little the remaining sum,
Which served the past, and must the times to come!

References

1. Ansdale, R. F. and D. J. Lockley, *The Wankel RC Engine*, Barnes & Co., New York, 1968.
2. Arnol'd, V. I., *Huygens and Barrow, Newton and Hooke* (translated by E. J. F. Primrose), Birkhäuser, Basel, 1990.
3. Bliss, G. A., *Lectures on the Calculus of Variations*, Univ. of Chicago Press, 1946.
4. Boltyanskii, V. G., *Envelopes*, MacMillan, New York, 1964.
5. Bruce, J. W. and P. J. Giblin, *Curves and Singularities*, Cambridge Univ. Press, 1984.
6. Camus, C. É. L., Sur la figure des dents des roues et des ailes des pignons, in *Histoires et Mémoires de l'Academie des Sciences*, Paris, 1733; translated into English by J. I. Hawkins as *A Treatise on the Teeth of Wheels: Demonstrating the Best Forms Which Can Be Given to Them for the Purposes of Machinery, Such as Mill–Work and Clock–Work*, E. & F. N. Spon, London, 1806.
7. Cardano, G., *De Rerum Varietate*, Basel, 1557.
8. Copernicus, N., *De Revolutionibus Orbium Coelestium*, 1543; translated into English by C. G. Wallis as *On the Revolutions of Heavenly Spheres*, Prometheus Books, Amherst, NY, 1995.
9. Dunham, W., *Journey Through Genius: The Great Theorems of Mathematics*, Wiley, New York, 1990.
10. Dürer, A., *Unterweisung der Messung mit dem Zirkel und Rechtscheit*, Nürnberg, 1525.
11. Euler, L., De aptissima figura rotarum dentibus tribuenda, in *Academiæ Scientiarum Imperiales Petropolitanæ, Novi Commentarii* t. V (1754–1755), 299–316; De figura dentium rotarum, *idem.* t. XI (1765), 207–231.
12. Field, J. V. and M. T. Wright, Gears from the Byzantines: A portable sundial with calendrical gearing, *Annals of Science* **42** (1985), 87–138.
13. Galilei, G. (translated by S. Drake), *Dialogue Concerning the Two Chief World Systems*, 2nd edition, Univ. of California Press, 1967.
14. Gillespie, C. C. (ed.), *Dictionary of Scientific Biography* Vol. 4, Scribner, New York, 1970, 46–51.
15. Hilbert, D. and S. Cohn–Vossen (translated by P. Nemenyi), *Geometry and the Imagination*, Chelsea, New York, 1952.
16. Huygens, C., *Horologium Oscillatorium sive De Motu Pendulorum ad Horologia Aptato Demonstrationes Geometricæ*, Paris, 1673; translated into English by R. J. Blackwell as *The Pendulum Clock, or Geometrical Demonstrations Concerning the Motion of Pendula as Applied to Clocks*, Iowa State Univ. Press, Ames, IA, 1986.

17. Kepler, J., *Mysterium Cosmographicum*, Tübingen, 1596; translated into English by A. M. Duncan as *The Secret of the Universe*, Abaris Books, New York, 1981.
18. La Hire, P. de, *Traité des Épicycloides et Leur Usages Dans la Mécanique*, Paris, 1694; and *Traité de Mécanique*, Paris, 1695.
19. Lawden, D. F., *Elliptic Functions and Applications*, Springer, New York, 1989.
20. Lawrence, J. D., *A Catalog of Special Plane Curves*, Dover, New York, 1972.
21. Lockwood, E. H., *A Book of Curves*, Cambridge Univ. Press, 1967.
22. Newton, I., *Philosophiæ Naturalis Principia Mathematica*, 1687; see also S. Chandrasekhar, *Newton's Principia for the Common Reader*, Oxford Univ. Press, 1995.
23. Norbye, J. P., *The Wankel Engine: Design, Development, Applications*, Chilton Book Co., Philadelphia, 1971.
24. Polya, G., *Mathematics and Plausible Reasoning, Volume 1: Induction and Analogy in Mathematics*, Princeton Univ. Press, 1954.
25. Price, D. J. de S., Gears from the Greeks: The Antikythera mechanism — a calendar computer from ca. 80 BC, *Trans. Amer. Philosoph. Soc.* (new series) **64** no. 7 (1974), 1–70.
26. Ptolemy, C., *The Almagest* (translated and annotated by G. J. Toomer), Springer, New York, 1984.
27. Rogers, P. (ed.), *Alexander Pope: A Critical Edition of the Major Works*, Oxford Univ. Press, 1993.
28. Sobel, D., *Longitude: The True Story of a Lone Genius Who Solved the Greatest Scientific Problem of His Time*, Walker, New York, 1995.
29. Usher, A. P., *A History of Mechanical Inventions*, Dover, New York, 1988.
30. Wankel, F., *Rotary Piston Machines*, Iliffe Books Ltd., London, 1965.
31. Woodbury, R. S., *History of the Gear-Cutting Machine — A Historical Study in Geometry and Machines*, MIT Press, Cambridge, MA, 1958.

Rida T. Farouki and Joanne Rampersad
Department of Mechanical Engineering and Applied Mechanics,
University of Michigan, Ann Arbor, MI 48109, USA.
`farouki@engin.umich.edu`
`jvramper@engin.umich.edu`

Scattered Data Fitting on the Sphere

Gregory E. Fasshauer and Larry L. Schumaker

Abstract. We discuss several approaches to the problem of interpolating or approximating data given at scattered points lying on the surface of the sphere. These include methods based on spherical harmonics, tensor-product spaces on a rectangular map of the sphere, functions defined over spherical triangulations, spherical splines, spherical radial basis functions, and some associated multi-resolution methods. In addition, we briefly discuss sphere-like surfaces, visualization, and methods for more general surfaces. The paper includes a total of 206 references.

§1. Introduction

Let S be the unit sphere in $\mathbb{R}^3$, and suppose that $\{v_i\}_{i=1}^n$ is a set of scattered points lying on S. In this paper we are interested in the following problem:

Problem 1. *Given real numbers* $\{r_i\}_{i=1}^n$, *find a (smooth) function* s *defined on* S *which interpolates the data in the sense that*

$$s(v_i) = r_i, \qquad i = 1, \ldots, n, \tag{1}$$

or approximates it in the sense that

$$s(v_i) \approx r_i, \qquad i = 1, \ldots, n. \tag{2}$$

Data fitting problems where the underlying domain is the sphere arise in many areas, including e.g. geophysics and meteorology where the sphere is taken as a model of the earth. The question of whether interpolation or approximation should be carried out depends on the setting, although in practice measured data are almost always noisy, in which case approximation is probably more appropriate.

In most applications, we will want s to be at least continuous. In some cases we may want it to be C^1 so that the associated *surface* $\mathcal{F} := \{s(v)v : v \in S\}$ is tangent plane continuous.

Mathematical Methods for Curves and Surfaces II

Morten Dæhlen, Tom Lyche, and Larry L. Schumaker (eds.), pp. 117–166.

ISBN 0-8265-1315-8.

The aim of this paper is to survey the spectrum of methods which have been developed (mostly in the past ten years) for solving Problem 1. The paper is divided into sections as follows:

1) Introduction,
2) Spherical harmonics,
3) Methods based on mapping the sphere to a rectangle,
4) Methods based on dividing the sphere into subsets consisting of spherical triangles,
5) Spherical splines (piecewise spherical harmonics),
6) Methods based on linear combinations of radial basis functions,
7) Multiresolution methods,
8) Additional topics, including sphere-like surfaces, general surfaces, visualizing surfaces on surfaces, and numerical quadrature on the sphere.

We conclude the paper with a bibliography containing 206 references.

§2. Spherical Harmonics

Many classical interpolation and approximation methods are based on polynomials. The appropriate analog of polynomials on the sphere are called *spherical harmonics*. They can be defined in several different (equivalent ways) which we now discuss. For details, see e.g. [24,80,105,118,165].

Let $\mathcal{P}_d$ be the space of trivariate polynomials of total degree at most d, and let $\mathcal{H}_d := \mathcal{P}_d|_S$ be its restriction to the sphere. A trivariate polynomial p is called homogeneous of degree d provided $p(\lambda x, \lambda y, \lambda z) = \lambda^d p(x,y,z)$ for all $\lambda \in \mathbb{R}$. It is called harmonic provided $\Delta p \equiv 0$, where Δ is the Laplace operator defined by $\Delta f := (D_x^2 + D_y^2 + D_z^2) f$.

Definition 2. *The linear space*

$$H_d := \{p|_S \;:\; p \in \mathcal{P}_d \text{ and } p \text{ is homogeneous of degree } d \text{ and harmonic}\}$$

is called the space of spherical harmonics of exact degree d.

It is well known (see e.g. [24], page 314) that the dimension of H_d is $2d+1$, and that it is the eigenspace corresponding to the eigenvalue $\lambda_d = -d(d+1)$ of the Laplace-Beltrami operator Δ^* on S given by

$$\Delta^* f = \frac{1}{\sin^2\theta} D_\phi^2 f + \frac{1}{\sin\theta} D_\theta\,(\sin\theta\, D_\theta f), \tag{3}$$

where $\theta \in [0,\pi]$ and $\phi \in [0, 2\pi]$ are the spherical coordinates of a point on S. It is also known that H_d is the orthogonal complement of $\mathcal{H}_{d-1}$ in the space $\mathcal{H}_d$ with respect to the L_2-inner product on S. Using this fact repeatedly, it follows that

$$\mathcal{H}_d = H_d \oplus H_{d-1} \oplus \cdots \oplus H_0, \tag{4}$$

and thus $\dim(\mathcal{H}_d) = (d+1)^2$.

For applications, it is important to have an explicit basis for H_d. The classical construction (cf. [24,80,105,118,165]) is based on the spherical coordinates θ and ϕ. Let P_d be the Legendre polynomial of degree d normalized such that $P_d(1) = 1$, and let P_d^ℓ be the associated Legendre function of degree d and order ℓ defined by

$$P_d^\ell(x) = (1-x^2)^{\ell/2} D_x^\ell P_d(x), \qquad -1 \le x \le 1.$$

Then the functions

$$\begin{aligned} Y_{d,2\ell+1}(\theta,\phi) &:= \cos(\ell\phi)\, P_d^\ell(\cos\theta), \qquad \ell = 0,\dots,d, \\ Y_{d,2\ell}(\theta,\phi) &:= \sin(\ell\phi)\, P_d^\ell(\cos\theta), \qquad \ell = 1,\dots,d, \end{aligned}$$

form an orthogonal (but not orthonormal) basis for H_d. Each of the $Y_{d,\ell}$ can be expanded in terms of sine and cosine functions. The formulae are simple for $d = 0,1$. Indeed, $Y_{0,1}(\theta,\phi) = 1$ and

$$\begin{aligned} Y_{1,1}(\theta,\phi) &= \cos(\theta), \\ Y_{1,2}(\theta,\phi) &= \sin(\phi)\sin(\theta), \\ Y_{1,3}(\theta,\phi) &= \cos(\phi)\sin(\theta). \end{aligned}$$

The formulae become increasingly complicated for larger values of d.

As shown in [70], it is also possible to construct a basis for H_d directly in terms of Cartesian coordinates. Every trivariate polynomial p in $\mathcal{P}_d$ which is homogeneous of degree d has the form

$$g(x,y,z) = \sum_{i+j+k=d} a_{ijk} x^i y^j z^k. \tag{5}$$

Now g will be harmonic if and only if the coefficients of all powers of x, y, z which remain in the equation $\Delta g = 0$ are zero. Then a basis for H_d can be constructed by finding linearly independent vectors a satisfying the corresponding homogeneous system of equations $Ca = 0$, where a is the coefficient vector in (5). Applying this process for $d = 0, 1, 2$, one finds the bases

$$\begin{aligned} H_0 &= \mathrm{span}\{1\}, \\ H_1 &= \mathrm{span}\{x,y,z\}, \\ H_2 &= \mathrm{span}\{xy, xz, yz, x^2-y^2, x^2-z^2\}. \end{aligned}$$

Both of the above basis constructions are somewhat cumbersome. Some extremely convenient basis functions (which depend on certain rotation invariant spherical barycentric coordinates) will be constructed in Sect. 5.2 for the spaces

$$\mathcal{B}_d := \begin{cases} H_0 \oplus H_2 \oplus \cdots \oplus H_{2k}, & d = 2k, \\ H_1 \oplus H_3 \oplus \cdots \oplus H_{2k+1}, & d = 2k+1. \end{cases} \tag{6}$$

One of the things that make spherical harmonics interesting is the fact that smooth functions on S can be approximated well by combinations of spherical harmonics (of sufficiently high degree). One way to construct such approximations is via the spherical harmonic expansion (also called the Laplace expansion)

$$f_d = \sum_{k=0}^{d} \Big[\sum_{\ell=0}^{k} a_{k,\ell} Y_{k,2\ell+1} + \sum_{\ell=1}^{k} b_{k,\ell} Y_{k,2\ell}\Big], \tag{7}$$

where

$$\begin{aligned} a_{k,\ell} &= \frac{(2k+1)(k-\ell)!}{2\pi(k+\ell)!} \int_0^{2\pi} \int_0^{\pi} f(\theta,\phi) Y_{k,2\ell+1}(\theta,\phi) \sin\theta \, d\theta \, d\phi, \\ b_{k,\ell} &= \frac{(2k+1)(k-\ell)!}{2\pi(k+\ell)!} \int_0^{2\pi} \int_0^{\pi} f(\theta,\phi) Y_{k,2\ell}(\theta,\phi) \sin\theta \, d\theta \, d\phi. \end{aligned} \tag{8}$$

This is the analog of the usual Fourier series expansion on the unit circle. It can be shown that for any function $f \in L_2(S)$, $\|f - f_d\|_2 \to 0$ as $d \to \infty$, while for functions $f \in C^2(S)$, $\|f - f_d\|_\infty \to 0$, see [24], page 513, and also [145,194–196].

There is an extensive literature on the general question of how well smooth functions defined on the sphere can be approximated by spherical harmonics, including both direct and inverse theorems [5,13,44,50,51,71,83,94,98–101,118,125,128–132,144,145,148–152,157–162,178,186–188,190–192,202,203]. We have space here for just one result which can be regarded as an analog of the classical *Jackson's theorem* for polynomial approximation. The statement concerns functions lying in a spherical analog of the classical Sobolev space. Following [100], let $W_p^r(S)$ be the space of all functions $f \in L_p(S)$ such that

$$\|\delta_\gamma^\kappa (\Delta^*)^\rho f\|_{L_p(S)} \le M\gamma^{r-2\rho}, \qquad 0 < \gamma < \pi,$$

where δ_γ^κ is a recursively defined spherical difference (computed as the difference between f at some point v and its average on some circle of geodesic radius γ around v), Δ^* is the Laplace-Beltrami operator (3), M is some constant independent of γ, and the integers ρ and κ satisfy $2\kappa > r - 2\rho > 0$. Note that the definition of $W_p^r(S)$ depends on the choice of κ and γ.

Theorem 3. *There exists a constant $C > 0$ such that for every $f \in W_p^r(S)$, there is a function $g \in \mathcal{H}_d$ such that*

$$\|f - g\|_{L_p(S)} \le C d^{-r} \|f\|_{W_p^r(S)}.$$

Theorem 3 is a *direct theorem*. For some related *inverse* theorems, see [100,118]. For other results involving moduli of smoothness on the sphere, see [50,51,132,144].

Formula (7) is not designed for fitting data, but it can be used if there is sufficient data available to create numerical approximations to the integrals in (8). For results based on gridded data, see [7,29].

Since the dimension of $\mathcal{H}_d$ is $n := (d+1)^2$, it is natural to use it to interpolate data at a set of n scattered points on S. Writing f in the form (7), this leads to the linear system

$$\sum_{k=0}^{d}\Big[\sum_{\ell=0}^{k} a_{k,\ell} Y_{k,2\ell+1}(\theta_i,\phi_i) + \sum_{\ell=1}^{k} b_{k,\ell} Y_{k,2\ell}(\theta_i,\phi_i)\Big] = r_i, \qquad i = 1,\ldots,n, \quad (9)$$

for the $(d+1)^2$ coefficients. It is now of interest to identify those point sets for which this system is nonsingular.

Definition 4. *A set of points $\{(\theta_i,\phi_i)\}_{i=1}^n$ for which the system in (9) is nonsingular is called $\mathcal{H}_d$-*unisolvent.

While it is known [118] that unisolvent sets exist, we have not been able to find a general characterization of them in the literature. A special result for points on parallel circles on the sphere can be found in [73]. The papers [84,135,149,150,152] also deal with interpolation.

In contrast, the analogous question of *$\mathcal{B}_d$-unisolvency* for the spaces $\mathcal{B}_d$ appearing in (6) has a very satisfactory answer.

Theorem 5. *Given d, let $m := \binom{d+2}{2}$. Let $v_1,\ldots,v_m$ be a set of points on the sphere. Suppose that for each $1 \le i \le m$, there exists a set of d distinct great circle arcs such that*

1) *v_i does not lie on any of the arcs,*
2) *all of the other v_j lie on at least one of the arcs.*

Then the set $\{v_i\}_{i=1}^m$ is a $\mathcal{B}_d$-unisolvent set.

This result is an analog of a classical result of Chung and Yao [23] (see also [22]) for interpolation by multivariate polynomials. For a proof and related results based on a study of homogeneous polynomials, see [89–91].

Spherical harmonics have been heavily used for fitting data, particularly in geophysics and meteorology. We make no attempt to list application papers here, but for one interesting application, see [134].

We conclude this section with one final remark. Since spherical harmonics are the direct analog of polynomials, one can expect that they suffer from the same problem as univariate or bivariate polynomials — a tendency to oscillate due to their *global* nature and lack of *flexibility*. One way to alleviate this would be to work with *piecewise spherical harmonics*, which is exactly what is done with the spaces of spherical splines discussed in Sect. 5.

§3. Methods Based on Mapping S to a Rectangle

The idea of fitting scattered data on the sphere by converting the problem to one defined on a rectangle has been exploited in several papers, see e.g. [28–30,72,177,193]. Early methods suffered from problems at the poles, but this can be overcome with the use of trigonometric splines as discussed in Sect. 3.3 below.

3.1. Mapping S to a Rectangle

We denote the *north pole* of S by $v_{\mathcal{N}}$, and the *south pole* by $v_{\mathcal{S}}$. The inverse of the mapping

$$\Phi(\theta,\phi) := \begin{pmatrix} \sin(\theta)\cos(\phi) \\ \sin(\theta)\sin(\phi) \\ \cos(\pi-\theta) \end{pmatrix} \tag{10}$$

maps points on $S \setminus \{v_{\mathcal{N}}, v_{\mathcal{S}}\}$ onto

$$R_I := \{(\theta,\phi) \;:\; 0 < \theta < \pi \;\text{ and }\; 0 \le \phi \le 2\pi\}.$$

A function s defined on the rectangle

$$R := \{(\theta,\phi) \;:\; 0 \le \theta \le \pi \;\text{ and }\; 0 \le \phi \le 2\pi\} \tag{11}$$

is well-defined on the entire sphere S if and only if it is 2π-periodic in ϕ and has constant values at both the south and north poles, *i.e.*,

$$s(0,\phi) = r_{\mathcal{S}} \quad\text{and}\quad s(\pi,\phi) = r_{\mathcal{N}}, \qquad 0 \le \phi \le 2\pi, \tag{12}$$

for some $r_{\mathcal{S}}$ and $r_{\mathcal{N}}$.

This identification of the sphere S with the rectangle R makes it possible to recast data fitting problems on S as data fitting problems on R, whereby points $v_i \in S$ are associated with points $(\theta_i, \phi_i) \in R$. For example, to solve the interpolation problem (1), it suffices to construct a function defined on R satisfying (12) and

$$s(\theta_i,\phi_i) = r_i, \qquad i = 1,\dots,n. \tag{13}$$

Suppose s is such a function. In order to make the corresponding surface $\mathcal{F} := \{s(\theta,\phi)\Phi(\theta,\phi) : \; (\theta,\phi) \in R\}$ be smooth, it is necessary to impose some smoothness on s. For example, to get a continuous surface, it is enough to require that $s \in C(R)$ and that it be 2π periodic in ϕ. To get a continuously varying tangent plane (everywhere except at the poles), $s \in C^1(R_I)$ and $D_\theta s$ must be 2π periodic in ϕ.

3.2. Tensor-Product Polynomial Splines

Since the fitting problem has been transformed to a rectangle, it is natural to solve it using *tensor product functions* of the form

$$s(\theta,\phi) := \sum_{i=1}^{M}\sum_{j=1}^{N} c_{i,j} B_i(\theta) T_j(\phi), \tag{14}$$

where $B_1(\theta)$, ..., $B_M(\theta)$ are functions defined on $[0,\pi]$, and $T_1(\phi),\dots,T_N(\phi)$ are 2π-periodic functions defined on $[0, 2\pi]$.

The obvious choice for both sets of functions would be polynomial B-splines. For the B_i one can take B-splines of order m with knots

$$0 = x_1 = \cdots x_m < x_{m+1} < \cdots < x_M < x_{M+1} = \cdots = x_{M+m}. \tag{15}$$

For the T_j one can take periodic B-splines of order n, where the knots are chosen periodically (cf. [175]):

$$\begin{aligned} &0 = y_n < \cdots < y_{N+n} = 2\pi, \\ &y_j = y_{j+N} - 2\pi, \qquad j = 1,\dots,n-1. \end{aligned} \tag{16}$$

3.3. Trigonometric Splines and Tangent Plane Continuity

In many applications one wants the surface $\mathcal{F}$ associated with a function s to be tangent-plane continuous everywhere, including the poles. It was shown in [28] (see also [72]) that this holds if and only if $s \in C^1(R_I)$, s, $D_\theta s$ and $D_\phi s$ are 2π periodic in ϕ, and

$$D_\theta s(\pi,\phi) = A_{\mathcal{N}}\cos(\phi) + B_{\mathcal{N}}\sin(\phi), \quad 0 \le \phi \le 2\pi, \tag{17}$$
$$D_\theta s(0,\phi) = A_{\mathcal{S}}\cos(\phi) + B_{\mathcal{S}}\sin(\phi), \quad 0 \le \phi \le 2\pi, \tag{18}$$

at the north and south poles, respectively, where $A_{\mathcal{S}}$, $B_{\mathcal{S}}$, $A_{\mathcal{N}}$, and $B_{\mathcal{N}}$ are constants.

These conditions cannot be satisfied using tensor-product polynomial splines, and so the methods in [28,72] are based on satisfying them only approximately. However, it was shown in [177] that they can be satisfied exactly if the T_j are chosen to be *periodic trigonometric B-splines* (cf. [175]) of *odd* order m defined on the periodic knot sequence in $[0, 2\pi]$ described above. The reason for the restriction to odd order is that (12) can only be satisfied if the trigonometric spline space contains constants, which is the case only when m is odd.

Theorem 6. [177] *Let s be defined as in (14) where the B_i are polynomial splines of order m with knots (15) and the T_j are periodic trigonometric splines of order 3 with knots (16). Then the resulting surface $\mathcal{F}$ is continuous and tangent plane continuous at all points of S if and only if*

$$\begin{aligned}
c_{1,j} &= r_{\mathcal{S}}\cos\left((y_{j+2} - y_{j+1})/2\right),\\
c_{M,j} &= r_{\mathcal{N}}\cos\left((y_{j+2} - y_{j+1})/2\right),\\
c_{2,j} &= c_{1,j} + (x_{m+1} - x_m)(A_{\mathcal{S}}\alpha_j + B_{\mathcal{S}}\beta_j)/(m-1),\\
c_{M-1,j} &= c_{M,j} - (x_{M+1} - x_M)(A_{\mathcal{N}}\alpha_j + B_{\mathcal{N}}\beta_j)/(m-1),
\end{aligned}$$

where $\alpha_j := \cos((y_{j+1} + y_{j+2})/2)$ and $\beta_j := \sin((y_{j+1} + y_{j+2})/2)$, for $j = 1, \ldots, N$.

3.4. Two-Stage Processes

Tensor-product splines are ideal for fitting gridded data, but not so well-suited to fitting scattered data. One way to handle scattered data is to perform a two-stage process, where the first stage is to use any convenient (local) method to compute values at points on a rectangular grid. This is most conveniently done with a local trigonometric spline quasi-interpolant. An appropriate such quasi-interpolant was given in [177] for $m = 3$. For a general treatment of trigonometric spline quasi-interpolants, see [104].

The fitting methods discussed in this section lend themselves to use in a *multiresolution analysis*. We discuss this in detail in Sect. 7.2.

3.5. Map and Blend Methods

Another way to avoid the problems at the poles is to work with two different rectangles, one of which leaves out regions surrounding the north and south pole, and the other of which leaves out regions surrounding the east pole and west pole (which correspond to the points $(0,1,0)$ and $(0,-1,0)$ in Cartesian coordinates). Then the two fits can be blended together, and if both interpolate (assuming there is no data at the poles), it is even possible to arrange for the blended function to also interpolate. For a discussion of some methods based on this idea, see [54].

3.6. Tensor Trigonometric Splines

It is also possible to choose both of the B_i and T_j in (14) to be trigonometric splines. Since using trigonometric splines on circular arcs is equivalent to using the circular Bernstein-Bézier (CBB) polynomials discussed in [1], this amounts to working with tensor-product CBB-polynomials.

§4. Methods Based on Spherical Triangulations

In this section we discuss several interpolation methods based on spherical triangulations.

4.1. Spherical Triangulations

Any two points $v, w \in S$ determine a unique great circle which splits into two pieces. If v and w are not antipodal, then one of these pieces is shorter – we call it the great circle arc connecting v and w, and denote it by $\langle v, w\rangle$. The length of $\langle v, w\rangle$ is called the geodesic distance from v to w.

Definition 7. *Given three unit vectors v_1, v_2, v_3 which span $\mathbb{R}^3$, the associated* spherical triangle *$T = \langle v_1, v_2, v_3\rangle$ is defined to be the set*

$$T = \{v \in S \; : \; v = b_1 v_1 + b_2 v_2 + b_3 v_3, \quad b_i \geq 0\}. \tag{19}$$

The boundary *of T consists of the three great circle arcs $\langle v_1, v_2\rangle$, $\langle v_2, v_3\rangle$, $\langle v_3, v_1\rangle$. A set of triangles $\Delta := \{T_i\}_1^N$ lying on a sphere S is called a* spherical triangulation *provided that $S = \cup T_i$, and any two triangles intersect only at a common vertex or along a common edge.*

As in the planar case, in general there are many different triangulations associated with a given set of vertices $V := \{v_i\}_{i=1}^n$. However, the well-known Euler formulae $N = 2n-4$ and $E = 3n-6$ hold for any spherical triangulation, where N is the number of triangles and E is the number of edges of Δ.

By choosing appropriate criteria for comparing triangulations, one can define various types of optimal spherical triangulations, including an analog of the classical Delaunay (Thiessen) triangulation [88,127,153]. Fortran code for constructing the Delaunay triangulation on a sphere is discussed in [154] and is available from *netlib*.

4.2. The Hermite Interpolation Problem

Several of the methods to be discussed below actually solve a kind of Hermite interpolation problem where in addition to the data (1), derivative information at each of the data points is also given. To describe this in more detail, we need to discuss the derivative of a function defined on the sphere.

Given a point $v \in S$, let g be a unit vector in $\mathbb{R}^3$ which is tangent to S at v. Then g together with the center of the sphere defines a plane which cuts the sphere along a great circle C passing through v. Now given a function f defined on S, the directional derivative in the direction g at v can be defined as

$$D_g f(v) := \lim_{\substack{w \to v \\ w \in C}} \frac{f(w) - f(v)}{d(v,w)},$$

where $d(v,w)$ is the geodesic distance from v to w. For an alternative definition involving the gradient of a homogeneous extension of f, see Sect. 5.3.

Definition 8. *Let $v_1, \dots, v_n$ be scattered points on the sphere. Suppose that for each $i = 1, \dots, n$, we are given two linearly independent tangent directions g_i, h_i associated with v_i. Then the* Hermite interpolation problem *is to find a function s defined on the sphere so that*

$$s(v_i) = r_i, \quad D_{g_i} s(v_i) = r_i^g, \quad D_{h_i} s(v_i) = r_i^h, \qquad i = 1, \dots, n, \tag{20}$$

where $\{r_i, r_i^g, r_i^h\}_{i=1}^n$ are given real numbers.

In many data fitting situations we will not be given derivative information. However, Hermite interpolation methods can still be applied if we can estimate the derivative information in some reasonable way, see Sect. 5.5.5.

4.3. Transfinite Interpolation Methods

In [87,88] the Hermite interpolation problem (20) is solved by adapting two planar transfinite interpolation methods to the sphere. In particular, a version of the BBG method (see [8]) and of the side vertex method (see [126]) are given. These methods work on one triangle at a time, and are based on first interpolating between its vertices to create a function on the edges of T, and then extending this function to the interior of T by interpolating between a vertex and the opposite side. A variant of these methods which is based on mapping spherical triangles to flat ones instead of working with geodesic distances can be found in [153,154].

4.4. Minimum Norm Networks

The interpolation methods described in the above subsection are based on creating a curve network by simple interpolation. As an alternative, one can build the curve network by minimizing the energy functional (see *e.g.* [127,147])

$$\sigma(f) := \sum_{e_{ij}} \frac{1}{\|v_i - v_j\|} \int [D_{ij} f(t)]^2 dt,$$

where $e_{ij} := \langle v_i, v_j \rangle$ is the great circle arc connecting v_i and v_j, t denotes arc length, and D_{ij} is the derivative in the direction described by the arc e_{ij}.

In [137] a similar energy functional based on chord length instead of geodesic distance was used to construct minimum norm networks in a more general setting.

§5. Spherical Splines

In this section we discuss certain linear spaces of splines defined on spherical triangulations which are natural analogs of the classical polynomial splines on planar triangulations. Our treatment is based on [2,3,4], and follows the notation introduced there. We begin by discussing how to define spherical barycentric coordinates on a spherical triangle and how to use them to define spherical functions which are the analogs of Bernstein-Bézier polynomials. Spaces of spherical splines associated with a spherical triangulation are described in Sect. 5.4.

5.1. Spherical Barycentric Coordinates

For many years people in the CAGD community believed it to be impossible to define barycentric coordinates on a spherical triangle. And indeed, it is impossible (cf. the discussion in [16]) if one insists that they sum to 1. However, it was recognized in [2] that a nice theory can be developed without this condition, and that in fact there is a very natural way to define barycentric coordinates with respect to spherical triangles. (It was later discovered that the same coordinates had been introduced and studied more than 100 years ago by Möbius [116]).

Given a nondegenerate spherical triangle $T := \langle v_1, v_2, v_3 \rangle$, every point $v \in S$ has a unique representation of the form $v = b_1 v_1 + b_2 v_2 + b_3 v_3$. The $b_1(v), b_2(v), b_3(v)$ are called the spherical barycentric coordinates of v relative to T. They are infinitely differentiable functions of v, are nonnegative for all $v \in T$, and satisfy

$$b_i(v_j) = \delta_{ij}, \quad i, j = 1, 2, 3.$$

However, in contrast to the usual barycentric coordinates associated with a planar triangle, they do not add up to 1. Instead,

$$b_1(v) + b_2(v) + b_3(v) > 0, \qquad \text{all } v \in T.$$

As shown in [2], spherical barycentric coordinates are rotation invariant. Several equivalent formulae for computing them in terms of angles and arc lengths can also be found there.

5.2. Spherical Bernstein-Bézier Polynomials

Given a spherical triangle T and an integer d, the associated spherical Bernstein basis functions of degree d are defined to be the functions

$$B^d_{ijk} := \frac{d!}{i!j!k!} b_1^i b_2^j b_3^k, \qquad i + j + k = d.$$

These $\binom{d+2}{2}$ functions are linearly independent [2]. We write $\mathcal{B}_d$ for their span.

It was observed in [2] that the B^d_{ijk} are actually linear combinations of spherical harmonics. The following proposition shows the precise connection with the spherical harmonic spaces H_m introduced in Sect. 2.

Proposition 9. *$\mathcal{B}_0 = \mathcal{H}_0$. Moreover, $\mathcal{H}_d = \mathcal{B}_d \oplus \mathcal{B}_{d-1}$ and*

$$\mathcal{B}_d = \begin{cases} H_0 \oplus H_2 \oplus \cdots \oplus H_{2k}, & d = 2k, \\ H_1 \oplus H_3 \oplus \cdots \oplus H_{2k+1}, & d = 2k+1, \end{cases} \tag{21}$$

for all $d \geq 1$.

Proof: The assertion for $d = 0$ is obvious since both $\mathcal{B}_0$ and $\mathcal{H}_0$ consist only of the constant functions. Since B^d_{ijk} are restrictions of trivariate polynomials of degree at most d to the sphere, it is clear that all of the functions B^d_{ijk} and B^{d-1}_{ijk} lie in $\mathcal{H}_d$. We claim they are linearly independent. Suppose d is *even* and that

$$\sum_{i+j+k=d} a_{ijk} B^d_{ijk}(v) + \sum_{i+j+k=d-1} b_{ijk} B^{d-1}_{ijk}(v) \equiv 0$$

for all $v \in S$. Then examining this expression at $-v$, and using the fact that $B^d_{ijk}(-v) = B^d_{ijk}(v)$ while $B^{d-1}_{ijk}(-v) = -B^{d-1}_{ijk}(v)$, we get

$$\sum_{i+j+k=d} a_{ijk} B^d_{ijk}(v) - \sum_{i+j+k=d-1} b_{ijk} B^{d-1}_{ijk}(v) \equiv 0.$$

This implies that each sum is separately identically zero, and the desired linear independence follows. The case *d odd* is similar. Now since $\dim \mathcal{B}_d + \dim \mathcal{B}_{d-1} = \binom{d+2}{2} + \binom{d+1}{2} = (d+1)^2$, we conclude that $\mathcal{H}_d$ splits into the two subspaces $\mathcal{B}_d$ and $\mathcal{B}_{d-1}$. Then (21) follows from (4) and the fact that the functions in $\mathcal{B}_d$ are homogeneous of even (odd) degree when d is even (odd). □

Combining this result with (4), it follows that

$$\mathcal{H}_d = \mathcal{B}_d \oplus \mathcal{B}_{d-1} \tag{22}$$

and thus

$$\mathcal{B}_{d-1} \not\subset \mathcal{B}_d \qquad \text{but} \qquad \mathcal{B}_{d-2} \subset \mathcal{B}_d. \tag{23}$$

In [2] an expression of the form

$$p = \sum_{i+j+k=d} c_{ijk} B^d_{ijk} \tag{24}$$

is called a spherical Bernstein-Bézier (SBB) polynomial. In view of their definition, it is no surprise that SBB-polynomials can be evaluated efficiently and stably with the usual *de Casteljau algorithm.* The same algorithm can also be used to perform *subdivision*, i.e., to find the coefficients of the three SBB-polynomials which represent p on the three subtriangles formed by splitting T about some point $w \in T$ — see [2].

As in the planar case, it is also possible to perform degree raising on SBB polynomials, except that now the degree has to be raised by two rather than just one. In particular, if p is as in (24), then

$$p = \sum_{i+j+k=d+2} \bar{c}_{ijk} B_{ijk}^{d+2},$$

where

$$\bar{c}_{ijk} = \frac{1}{(d+1)(d+2)} \Big[i(i-1)c_{i-2,j,k} + \beta_{110} ij c_{i-1,j-1,k} + j(j-1)c_{i,j-2,k} + \beta_{101} ik c_{i-1,j,k-1} + k(k-1)c_{i,j,k-2} + \beta_{011} jk c_{i,j-1,k-1} \Big].$$

Here

$$\beta_{011} = \frac{\sin^2 t_1}{\sin^2 \frac{t_1}{2}} - 2, \quad \beta_{101} = \frac{\sin^2 t_2}{\sin^2 \frac{t_2}{2}} - 2, \quad \text{and} \quad \beta_{110} = \frac{\sin^2 t_3}{\sin^2 \frac{t_3}{2}} - 2,$$

where t_i is the arc length of the edge opposite vertex v_i, $i = 1, 2, 3$.

The restriction of an SBB polynomial to an edge e of a spherical triangle results in a univariate function defined on the circular arc e. Such functions are called circular Bernstein-Bézier (CBB) polynomials [1], and in fact are trigonometric polynomials in arc length. They are also of interest for CAGD purposes, see e.g. [76,85,163].

5.3. Joining SBB Polynomials Smoothly

As a first step towards defining a space of splines on a spherical triangulation, one needs to describe how to make two SBB-polynomials defined on adjoining spherical triangles join together smoothly. Following [2], we now describe how this can be done.

In order to talk about smooth joins of SBB polynomials, we need to work with derivatives. First order directional derivatives of spherical functions were introduced above in Sect. 4.2. To generalize this to higher order (mixed) directional derivatives, it is more convenient to define directional derivatives in terms of the Cartesian coordinate system for $\mathbb{R}^3$. It was shown in [3] that if g is a unit tangent vector as in Sect. 4.2, then

$$D_g f(v) = g^T \nabla F(v),$$

where F is a *homogeneous* extension of f to $\mathbb{R}^3$ and ∇F is its gradient. While polynomials have a natural homogeneous extension, a general function f has many homogeneous extensions. However, as shown in [2], if a *tangent direction* g is chosen, then the value of $D_g f(v)$ does not depend on the extension used.

For general functions f, *higher order* (order 2 or greater) directional derivatives are in general dependent on the way in which f is homogeneously extended, see [2]. However, for an SBB-polynomial, there is a convenient formula for its (unique) derivative. Given vectors $g_1, \ldots, g_m$ in $\mathbb{R}^3$, it was shown in [2] that the associated m-th order directional derivative is given by

$$D_{g_1,\ldots,g_m}p := D_{g_1} \cdots D_{g_m}p = \frac{d!}{(d-m)!} \sum_{i+j+k=d-m} c_{ijk}^m B_{ijk}^{d-m}, \tag{25}$$

where c_{ijk}^m are the coefficients obtained by applying the de Casteljau algorithm m times, starting with c_{ijk} and using the spherical barycentric coordinates $b_1(g_\nu), b_2(g_\nu), b_3(g_\nu)$ at the ν-th step.

Theorem 10. [2] *Let $T = \langle v_1, v_2, v_3 \rangle$ and $\widetilde{T} = \langle v_4, v_2, v_3 \rangle$ be two spherical triangles sharing the edge $\langle v_2, v_3 \rangle$, and let $\{B_{ijk}^d\}$ and $\{\tilde{B}_{ijk}^d\}$ be the associated Bernstein-Bézier basis functions. Suppose p and $\tilde{p}$ are SBB-polynomials on T and $\widetilde{T}$ with coefficients $\{c_{ijk}\}$ and $\{\tilde{c}_{ijk}\}$, respectively. Then p and $\tilde{p}$ and all of their directional derivatives up to order m agree on the edge shared by T and $\widetilde{T}$ if and only if*

$$\tilde{c}_{ijk} = \sum_{r+s+t=i} c_{r,j+s,k+t} B_{rst}^i(v_4) \tag{26}$$

for all $i = 0, \ldots, m$ and all j, k such that $i + j + k = d$.

For later use, we note the following formulae [3] for first and second derivatives at the first vertex of T:

$$D_g p(v_1) = d[b_1(g)c_{d,0,0} + b_2(g)c_{d-1,1,0}],$$
$$D_g^2 p(v_1) = d(d-1)[b_1^2(g)c_{d,0,0} + 2b_1(g)b_2(g)c_{d-1,1,0} + b_2^2(g)c_{d-2,2,0}].$$

It is also important to note that the restrictions of derivatives of SBB polynomials to edges of T are CBB polynomials (see the end of Sect. 5.2). For example, if p is cubic and h is a direction which does not lie in the plane through the center of the sphere which contains the edge $e := \langle v_1, v_2 \rangle$, then for any point v on the edge e, the cross-boundary derivative is the quadratic CBB-polynomial

$$D_h p(v) = 3[c_{200}^1 b_1^2(v) + 2c_{110}^1 b_1(v) b_2(v) + c_{020}^1 b_2(v)^2],$$

where the c_{ijk}^1 are computed using one step of the de Casteljau algorithm based on the spherical barycentric coordinates $b_1(h), b_2(h), b_3(h)$ of h relative to T.

5.4. Spherical Splines

Given nonnegative integers r and d, the linear space

$$\mathcal{S}_d^r(\Delta) := \{s \in C^r(S) \;:\; s|_{T_i} \in \mathcal{B}_d,\ i = 1, \ldots, N\}$$

is called the space of spherical splines of smoothness r and degree d. Note that this space is defined in terms of $\mathcal{B}_d$ and not $\mathcal{H}_d$. Thus, if $s \in \mathcal{S}_d^r(\Delta)$, then its pieces are SBB-polynomials whose exact degree is even if d is even, and odd if d is odd, cf. (22)–(23).

The space $\mathcal{S}_d^r(\Delta)$ is the analog of the space of polynomial splines defined on a planar triangulation. It seems particularly appropriate for use on the sphere since in view of Proposition 9, it consists of functions whose pieces are spherical harmonics joined together with global smoothness C^r, and thus has both the smoothness and high degree of flexibility which makes splines the powerful tools they are.

As in the planar case, it is possible to identify the dimension of $\mathcal{S}_d^r(\Delta)$ and construct locally supported bases for them for all values of $d \geq 3r+2$, see [4]. We do not have space to review this theory here. However, we do discuss several scattered data interpolation and approximation methods based on spherical splines in the following sections.

It should be noted that the term *spherical spline* has several different meanings in the literature. It was used in [64,66,71] for certain radial basis type functions which are spherical analogs of the classical *thin-plate splines* – see also [197–200]. Unfortunately, recently the term has also been used for *curves* defined on the surface of the sphere [81].

5.5. Local Spline Interpolation Methods

As shown in [3], the basic interpolation problem (1) can now be solved as follows:

1) construct a triangulation Δ with vertices at the given data points,
2) for each triangle T in Δ, use the data at the vertices (along with additional derivative information if necessary) to define a function s_T defined on T which is a single SBB polynomial on T or a collection of SBB polynomial pieces on some partition of T. Choose s_T to interpolate the data at the vertices of T. With some care one can also make the function s whose restrictions are the s_T have some degree of global smoothness.

The fact that the interpolant s is constructed one triangle at a time insures that the method is *local* in the sense that the restriction of s to a triangle T depends only on the data in that triangle. Methods of this type are called macro-element methods. They have been widely applied in bivariate data fitting, and as observed in [3], every bivariate macro-element method has a natural spherical analog. Here we discuss just three examples:

1) quintic C^1 macro-elements,
2) cubic C^1 elements on the Clough-Tocher split,
3) quadratic C^1 elements on the 6-triangle Powell-Sabin split.

Each of the methods discussed will solve a version of the Hermite interpolation problem (20) associated with a set of vertices $\{v_i\}_{i=1}^n$.

5.5.1. Quintic C^1 Elements

Theorem 11. *Suppose values for*

$$s(v_i),\ D_{g_i}s(v_i),\ D_{h_i}s(v_i),\ D^2_{g_i}s(v_i),\ D_{g_i}D_{h_i}s(v_i),\ D^2_{h_i}s(v_i)$$

are given for $i = 1, \ldots, n$, where g_i and h_i are two linearly independent unit vectors which are tangent to S at v_i. In addition, suppose a value for a cross-boundary derivative at the center of each edge of Δ is given. Then there exists a unique spherical spline $s \in \mathcal{S}^1_5(\Delta)$ which interpolates these data.

This interpolant can be constructed one triangle at a time. Explicit formulae for the Bernstein-Bézier coefficients of each piece of s can be found in [3]. By construction, s satisfies C^2 continuity conditions at each of the vertices. In this sense it is a spherical superspline, see [3]. It is the spherical analog of the classical Argyris element.

5.5.2. Clough-Tocher Element

For each triangle $T = \langle v_1, v_2, v_3 \rangle$ in the triangulation Δ, let

$$\bar{v} := \frac{v_1 + v_2 + v_3}{\|v_1 + v_2 + v_3\|}$$

be its center. If $\bar{v}$ is connected to each of the vertices of T with great-circle arcs, T is split into three spherical subtriangles. This is called the Clough-Tocher split of the triangle. Given a triangulation Δ, let Δ_{CT} be the triangulation obtained by applying the Clough-Tocher split to each triangle of Δ.

Theorem 12. *Suppose function and first derivative information are given at the vertices v_i of the triangulation Δ as in Theorem 11, along with a value for a cross-boundary derivative at the center of each edge of Δ. Then there exists a unique spherical spline $s \in \mathcal{S}^1_3(\Delta_{CT})$ which interpolates these data.*

This interpolant can also be constructed one triangle at a time. Explicit formulae for the Bernstein-Bézier coefficients of each piece of s can be found in [3].

5.5.3. Powell-Sabin Element

Given a triangulation Δ, for each $j = 1, \ldots, N$, let $\bar{v}_j$ be the incenter of the j-th triangle obtained by radially projecting onto S the incenter of the planar triangle with the same vertices. Suppose the incenters of adjacent triangles are connected with great-circle arcs, and that the incenter of each triangle is also connected to each of its three vertices with great-circle arcs. This splits each triangle of Δ into six spherical subtriangles. The resulting triangulation Δ_{PS} is called the Powell-Sabin refinement of Δ.

Theorem 13. *Suppose function and first derivative values are given at each of the vertices of a spherical triangulation Δ as in Theorem 11. Then there exists a unique spline $s \in \mathcal{S}^1_2(\Delta_{PS})$ which interpolates these data.*

Explicit formulae for the Bernstein-Bézier coefficients of each piece of s can be found in [3].

5.5.4. A Hybrid Rational Element

Let $T = \langle v_1, v_2, v_3 \rangle$ be a spherical triangle. Then a hybrid cubic SBB-element was defined in [97] to be a function of the form

$$r(v) = \sum_{i+j+k=3} c_{ijk}(v) B^3_{ijk}(v), \tag{27}$$

where

$$c_{111}(v) = \sum_{\ell=1}^{3} \alpha_\ell A_\ell(v), \tag{28}$$

and c_{ijk} are constants for the other choices of i, j, k. Here A_1, A_2, A_3 are appropriate blending functions. For example, assuming the point v on S has spherical barycentric coordinates b_1, b_2, b_3, one can set

$$A_1(v) := \begin{cases} 0, & v = v_1, v_2, v_3, \\ \dfrac{b_2^m b_3^m}{b_1^m b_2^m + b_2^m b_3^m + b_1^m b_3^m}, & \text{otherwise}, \end{cases} \tag{29}$$

with $A_2(v)$ and $A_3(v)$ defined analogously. Using these elements, the Hermite interpolation problem (20) can be solved as follows:

1) for each triangle T, use the data at the vertices to determine all coefficients c_{ijk} with $(i, j, k) \neq (1, 1, 1)$.
2) compute α_1 so that C^1 continuity between adjoining pieces is guaranteed and $\sum_{i=1}^4 L_i^2$ is minimized, where $L_1, \ldots, L_4$ describe the C^2 continuity conditions across the edge associated with α_1. Repeat to compute α_2 and α_3.

As shown in [97], Step 2 involves solving 3×3 linear systems. The method is exact for cubic SBB-polynomials, *i.e.*, if f is such a function, then its piecewise hybrid cubic interpolant s is identically equal to f. Two other methods for computing the α_ℓ are also discussed in [97]. The same idea can be used to create a C^2 quintic hybrid element [18].

5.5.5. Estimating Derivatives on the Sphere

In order to apply the Hermite interpolation methods described above when only the basic data of Problem 1 is given, it is necessary to first estimate the derivatives at each data site v_i. The natural approach is to use a numerical differentiation rule based on data at points "near" v_i.

One approach is to use trivariate polynomials of low degree, see e.g. [88,153,154]. As a natural alternative, in [97] low degree SBB-polynomials are used instead. This involves choosing a spherical triangle surrounding the point of interest. The effect of varying the size and orientation of this triangle is explored there.

A method for estimating derivatives of functions defined on general surfaces (based on the use of the so-called *exponential map*) can be found in [133].

5.6. Minimal Energy Interpolants

The idea of creating interpolants which minimize some form of energy has been heavily studied (it has been applied to standard univariate splines, tensor-product splines, thin-plate splines, and a variety of other situations). The approach can also be carried out on the sphere. As usual, it has the advantage of producing interpolants which are often smoother than those obtained by other methods, but at a higher computational cost due to the global nature of the process.

Following [3], the starting point is the space of spherical splines $\mathcal{S}_d^0(\Delta)$. Each spline in this space is uniquely defined by the set of coefficients of its SBB-pieces. By continuity, common coefficients along edges of the spherical triangulation can be identified. This leads to a single coefficient vector $\mathbf{c}$ whose length is the dimension of $\mathcal{S}_d^0(\Delta)$. Assuming that the first n coefficients $c_1, \ldots, c_n$ correspond to the values of s at the vertices $v_1, \ldots, v_n$, it is clear that s will satisfy the interpolation conditions (1) provided $c_i = r_i$ for $i = 1, \ldots, n$.

Let Q be a symmetric positive definite matrix, and let $\mathcal{E}(\mathbf{c}) := \mathbf{c}^T Q \mathbf{c}$. Given r, suppose that

$$G\mathbf{c} = 0 \tag{30}$$

describe the smoothness conditions required for $s \in \mathcal{S}_d^0(\Delta)$ to belong to C^r. Then a minimal energy interpolating spline s is defined to be the function in $\mathcal{S}_d^r(\Delta)$ which minimizes $\mathcal{E}(\mathbf{c})$ subject to (30).

The coefficients $\mathbf{c}$ of the minimal energy interpolating spline can be computed by solving the linear system

$$\begin{pmatrix} Q & I & G^T \\ I & 0 & 0 \\ G & 0 & 0 \end{pmatrix} \begin{pmatrix} \mathbf{c} \\ \boldsymbol{\lambda} \\ \boldsymbol{\gamma} \end{pmatrix} = \begin{pmatrix} 0 \\ \mathbf{r} \\ 0 \end{pmatrix}, \tag{31}$$

where I is the $n \times n$ identity matrix and $\mathbf{r} := (r_1, \ldots, r_n)^T$. Here $\boldsymbol{\gamma}, \boldsymbol{\lambda}$ are vectors of Lagrange multipliers.

One way to define the energy functional is to take

$$\mathcal{E}(f) := \int_S (Of)^2 ds, \tag{32}$$

where O is an appropriate differential operator on the sphere, such as

$$O := (\Delta^*)^{m/2}, \tag{33}$$

where m is an even integer and where Δ^* is the Laplace-Beltrami operator (3). The definition of O for the case where m is odd is more complicated and can be found in [198]. The functionals (32) are the same functionals which are minimized in defining *spherical thin plate splines*, see [198–201].

5.7. Discrete Least Squares

In practice, the data fitting problem usually involves noisy measurements $r_i = f(v_i) + \varepsilon_i$ of an unknown function f at n points v_i on the sphere. In this case it is better to approximate rather than interpolate. One approach is to create a least squares fit:

1) choose a triangulation Δ with m vertices $(m < n)$,
2) choose a spline space $\mathcal{S}_d^r(\Delta)$,
3) find the spline $s \in \mathcal{S}_d^r(\Delta)$ which minimizes $L(s) := \sum_{i=1}^n [s(v_i) - r_i]^2$.

There does not seem to be a simple automatic way to choose the vertices for the triangulation. In practice it is tempting to choose $\mathcal{S}_3^1(\Delta)$ for the spline space, despite the fact that it is not known whether interpolation is even possible with this space, see [3] for a discussion. As usual, the coefficients of s can be computed by solving a linear system.

5.8. Penalized Least Squares

In some fitting problems, particularly when the data are especially noisy, it may be useful to replace the standard discrete least squares problem by a penalized least squares problem. The idea is to minimize a combination

$$K(\mathbf{c}) := L(\mathbf{c}) + \lambda \mathcal{E}(\mathbf{c}),$$

where $\mathcal{E}(\mathbf{c})$ is a measure of energy as discussed in Sect. 5.6, and $L(\mathbf{c})$ is the sum of squares of the errors as in the previous section. The parameter λ controls the trade-off between these two quantities, and is typically chosen to be a small positive number, see [74].

5.9. Remarks

We do not have space here to give examples of the various spherical spline fits discussed in Sects. 5.5 – 5.8 above. However, a number of numerical examples are discussed in [3] which contains both figures and tables giving a basic comparison of the methods as relates to storage, exactness, and computational time. The effect of thin triangles and the condition numbers of the interpolation matrices are also discussed there, along with appropriate scaling strategies. The effect of near-singular vertices (a vertex is singular when it is formed by two intersecting circular arcs) is also treated there.

In general, the choice of a method will depend on a variety of factors including 1) is the data noisy, 2) how much data is available, 3) what smoothness is required, and 4) what degree of accuracy is needed? Concerning accuracy, we mention that the experiments in [3,97] suggest that the quintic macro element method has accuracy $\mathcal{O}(h^6)$, the Clough-Tocher and hybrid rational methods have accuracy $\mathcal{O}(h^4)$, and the Powell-Sabin method has accuracy $\mathcal{O}(h^3)$, where h is the maximal diameter of the triangles in Δ.

5.10. Other Methods

Virtually any method which works for polynomial splines based on planar triangulations has an analog for the sphere using spherical splines. To see what has been done in the planar case, see the surveys [55,59,63,174]. Here we mention several additional interesting ideas:

1) *Data Dependent Triangulations.* The essential observation here is that given a *fixed set of vertices*, there are many possible associated triangulations. Thus, in fitting a set of data using splines based on such triangulations, it may be possible to get much better approximations with certain triangulations rather than others. In particular, the classical Delaunay triangulations may be far from best, and in fact long thin triangles might be much more appropriate. In the classical polynomial spline case on planar triangulations, this idea has been explored in detail in [37–39,142,143]. Algorithmically, the procedure works as follows. One starts with some reasonable initial triangulation, for example the Delaunay triangulation. Then to construct a "best" triangulation, edges are *swapped* recursively guided by some measure of goodness of fit or smoothness. Since this procedure is essentially trying to solve a very large nonlinear optimization problem, in practice one cannot expect it to always yield a global best approximation. One interesting approach to avoiding getting stuck at local minima is to employ simulated annealing, see [176]. The entire process carries over immediately to spherical splines.

2) *Knot Insertion and Deletion.* The idea here is to take advantage of the local nature of splines. In particular, if a spline is not doing a good job of fitting in some part of the domain, then one can insert additional knots in that area. Conversely, in regions where the fit is already very good, one can remove knots. Algorithmically, one starts with some initial fit and performs both knot insertion and deletion recursively. For details in the case of classical polynomial splines on planar triangulations, see [92]. Again, the method carries over immediately to spherical splines.

3) *Spherical Simplex Splines.* In the eighties there was considerable interest in certain spaces of *simplex splines.* They arose from the process of trying to create a multivariate analog of the univariate polynomial B-splines. For an extensive survey of the theory up to 1983, see [27]. Recently, analogous spherical simplex splines have been introduced and studied in [124,136].

4) *Natural Neighbor Methods.* Natural neighbor coordinates were introduced by Sibson (see *e.g.* [179]). In [15] analogous coordinates were defined for S, and used to create a locally supported interpolant to scattered data on S. The resulting surface is C^0 at the data sites, and C^1 everywhere else.

§6. Methods Based on Radial Basis Functions

In recent years there has been a great deal of interest in radial basis functions (RBF's) as a tool for interpolation and data fitting. What is usually referred to as a radial basis function is the composition of a univariate function with some sort of distance function. In Euclidean spaces, an RBF with center $v_i \in \mathbb{R}^d$ is a function of the form

$$\hat{\Phi}_i(v) := \hat{\varphi}(\|v - v_i\|), \qquad v \in \mathbb{R}^d, \tag{34}$$

where $\hat{\varphi} : [0, \infty) \to \mathbb{R}$ and $\|\cdot\|$ denotes Euclidean distance.

In order to interpolate or approximate given data, it is natural to take linear combinations of RBF's of the form

$$s(v) = \sum_{i=1}^{N} c_i \, \hat{\varphi}(\|v - v_i\|), \qquad v \in \mathbb{R}^d. \tag{35}$$

The coefficients c_i of this radial basis function expansion are then determined by solving a system of linear equations. The centers are usually chosen to coincide with (some of) the data. This assumption considerably simplifies the analysis of the linear system (see Sect. 6.2 for more details in the spherical case).

An early example of a radial basis function interpolant is provided by Shepard's method (see [174] for a discussion). Other early uses of RBF's were in geological modeling [77-79] (as (reciprocal) multiquadrics), and in variational problems, where – in a special case – they are called thin plate splines, see [32,58,60,106,111] and references therein.

Further interest in RBF's was inspired by the observation [59] that in the bivariate setting, they are among the most effective methods for interpolation and approximation of scattered data. Their use was also encouraged by the fact that radial basis methods are extremely easy to program in any number of variables. We do not attempt to cite the entire classical RBF literature here, but for some typical methods and further references, see [17,33,34,36,40,43,52,115,141,166,180,204].

In analogy with the Euclidean case, cf. (34), it is natural to call

$$\Phi_i(v) := \varphi(d(v, v_i)), \qquad v \in S,$$

a spherical radial basis function (SRBF) provided $\varphi : [0, \pi] \to \mathbb{R}$ and $d(v, v_i)$ is the length of the (shorter) great circle arc connecting v and v_i. This is the geodesic distance between v and v_i. In the classical literature the function $\Phi_i(v) = \varphi(d(v, v_i))$ is called a zonal function with pole v_i.

Given any Euclidean RBF as in (34), there is a natural way to associate a spherical RBF with it. Indeed, since

$$\|v - w\| = \sqrt{2 - 2\, v \cdot w} = 2 \sin \frac{d(v, w)}{2},$$

for any $v, w \in S$, it follows that

$$\hat{\Phi}_i(v) = \hat{\varphi}(\|v - v_i\|) = \Phi_i(v) = \varphi(d(v, v_i)),$$

with

$$\varphi(t) = \hat{\varphi}(2\sin(t/2)), \qquad 0 \le t \le \pi.$$

See [42,96] for some related discussions.

Recently, basis functions of the form

$$\tilde{\Phi}_i(v) = \tilde{\varphi}(v \cdot v_i)$$

have also been considered, see e.g. [73]. Since $v \cdot w = \cos(d(v,w))$, these can also be interpreted as SRBF's associated with the function $\varphi(t) = \tilde{\varphi}(\cos(t))$.

For some recent papers which survey certain aspects of spherical radial basis functions and point to open problems, see [19–21,71].

6.1. Using RBF's in $\mathbb{R}^3$ for the Sphere

One way to use radial basis functions to fit data given at points $v_1, \ldots, v_n$ on the sphere is to consider these data points to lie in $\mathbb{R}^3$. Then if s is any RBF fit to the data based on Euclidean distance between points, we can simply restrict its evaluation to S.

For example, to interpolate values r_i at the points v_i, we can solve the system

$$\sum_{j=1}^{n} c_j \hat{\varphi}(\|v_i - v_j\|) = r_i, \qquad i = 1, \ldots, n.$$

There is no complete classification of those functions $\hat{\varphi}$ for which this system is nonsingular for all choices of $\{v_i\}_{i=1}^n$. However, wide classes of functions $\hat{\varphi}$ have been identified which guarantee that the associated matrix is *positive definite*, which in turn assures its nonsingularity, see e.g. [115]. In this case, $\hat{\varphi}$ is called strictly positive definite.

There are many examples of strictly positive definite RBF's in the literature, and any of these can be used to solve the interpolation problem (1) on the sphere. However, it is highly advantageous to work with locally supported functions since they lead to *sparse* linear systems. Wendland [204] found a class of radial basis functions which are smooth, locally supported, and strictly positive definite on $\mathbb{R}^d$ for some d. They consist of a product of a truncated power function and a low degree polynomial. For example, one can take

$$\hat{\varphi}_h(t) = \left(\frac{h-t}{h}\right)_+^2,$$

$$\hat{\varphi}_h(t) = \left(\frac{h-t}{h}\right)_+^4 \left(\frac{4t+h}{h}\right),$$

$$\hat{\varphi}_h(t) = \left(\frac{h-t}{h}\right)_+^6 \left(\frac{35t^2 + 18rh + 3h^2}{h^2}\right),$$

where h is a (small) positive number. These functions are nonnegative for $t \in [0,h]$, are zero for $t > h$, and belong to C^0, C^2, and C^4, respectively.

Moreover, they are strictly positive definite in $\mathbb{R}^3$. Similar functions with higher-order smoothness, or which are strictly positive definite on $\mathbb{R}^d$ for $d > 3$ can also be constructed. By the remarks at the end of the previous subsection, these functions can be transformed to work directly with geodesic distance on the sphere. Thus, for example, the second function above becomes

$$\varphi_h(t) = \left(1 - \frac{2}{h}\sin\frac{t}{2}\right)_+^4 \left(\frac{8}{h}\sin\frac{t}{2} + 1\right),$$

where t now measures geodesic distance. The support of this function is $[0, \arcsin(h/2)]$. In [205] some other locally supported RBF's in Euclidean spaces are discussed.

6.2. Interpolation with Spherical RBF's

As observed above, one way to get SRBF's is to transform Euclidean RBF's to work with geodesic distance on the sphere. Alternatively, one can study functions φ which work with geodesic distance directly. This approach has spawned a large body of literature which we now briefly review. We begin with interpolation.

For the sphere, the analog of the radial basis function expansion (35) is

$$s(v) = \sum_{i=1}^{N} c_i\, \varphi(d(v, v_i)), \qquad v \in S. \tag{36}$$

To solve the interpolation problem (1) with linear combinations of spherical RBF's, it is necessary to find coefficients $c_1, \ldots, c_n$ such that

$$\sum_{j=1}^{n} c_j\, \varphi(d(v_i, v_j)) = r_i, \qquad i = 1, \ldots, n.$$

As in the Euclidean case, there is no complete characterization of those functions φ for which this $n \times n$ linear system is nonsingular. As before, a convenient sufficient condition for nonsingularity is that the corresponding matrix $A := (\varphi(d(v_i, v_j)))$ be positive definite.

Definition 14. *A continuous function $\varphi : [0, \pi] \to \mathbb{R}$ is called* positive definite of order n on S *if*

$$\sum_{i=1}^{n}\sum_{j=1}^{n} \varphi(d(v_i, v_j))\, c_i c_j \ge 0 \tag{37}$$

for any n points $v_1, \ldots, v_n \in S$, and any $c = [c_1, \ldots, c_n]^T \in \mathbb{R}^n$. If this inequality holds strictly for any nontrivial c, φ is called strictly positive definite of order n. *If φ is (strictly) positive definite of order n for any n, then it is called* (strictly) positive definite.

The problem of identifying positive definite functions on the sphere was addressed already by Schoenberg [170]. In particular, he showed that any φ

which is positive definite on the sphere has an expansion in terms of Legendre polynomials P_k (normalized such that $P_k(1) = 1$) of the form

$$\varphi(t) = \sum_{k=0}^{\infty} a_k P_k(\cos t), \tag{38}$$

with $a_k \geq 0$, and a_k such that $\sum a_k$ converges.

In view of its connection to interpolation on the sphere, the question of which functions φ are *strictly* positive definite (with respect to geodesic distance on S) has received considerable attention recently. It was shown in [206] that a sufficient condition is that all of the a_k in (38) be positive. This condition was recently improved in [171] to allow finitely many zero coefficients; note the comments made by Askey in Math Reviews about a (removable) error in the proof.

As a complement to these sufficient conditions, it has been shown recently that for strict positive definiteness on the circle (and therefore also on the sphere) it is necessary that infinitely many coefficients with even subscripts as well as infinitely many with odd subscripts be positive (see [114] or the survey [20]).

The problem of completely characterizing functions which are strictly positive definite on the sphere remains open. However, C^∞-kernels on $S \times S$ which are strictly positive definite in a distributional sense were completely characterized in [119]. Also, strictly positive definite functions on the infinite-dimensional sphere S^∞ were completely characterized in [113]. The latter results are of interest since any function which is strictly positive definite on S^∞ has the same property on S.

For the purpose of interpolation, strict positive definiteness *of order* n on S suffices. It was shown by [155] and also [146] that a sufficient condition for this is that at least the first $(n+1)/2$ coefficients in (38) are positive. To solve an interpolation problem with a (globally supported) SRBF involves evaluating it n^2 times in order to build the matrix in (37). This means that expansions with $(n+1)/2$ nonzero coefficients are not very practical for larger values of n, and it would be better to work with a φ which has a simple closed form representation. We conclude this section with such an example.

The function

$$\varphi(t) = \frac{1}{\sqrt{1+\gamma^2 - 2\gamma\cos t}}, \qquad 0 < \gamma < 1, \tag{39}$$

is called the spherical reciprocal multiquadric. This infinitely differentiable function generates the Legendre polynomials, and therefore has the series representation (see *e.g.* [189])

$$\varphi(t) = \sum_{k=0}^{\infty} \gamma^k P_k(\cos t),$$

which is known to converge for $|\gamma| < 1$. By the above, it follows that this φ is strictly positive definite on S.

The use of (39) for interpolation on S was suggested by Hardy and Göpfert [78]. Their motivation came from geophysics: this particular function, which they called the multiquadric biharmonic, arises naturally in the calculation of disturbing potentials of the earth.

6.3. Interpolation with Spherical Harmonic Reproduction

In order to insure that SRBF interpolants of the form (36) do a good job of fitting arbitrary smooth functions on the sphere S, it is important that when used to interpolate a spherical harmonic, they give an exact fit. However, in general, this is not the case, and even constants are not fit exactly. This problem can be addressed by working with functions of the form

$$s(v) = \sum_{i=1}^{n} c_i \varphi(d(v, v_i)) + \sum_{k=0}^{m-1} \sum_{\ell=1}^{2k+1} d_{k,\ell} Y_{k,\ell}(v), \qquad v \in S, \tag{40}$$

where $\{Y_{k,\ell}\}$ are the spherical harmonic basis functions for H_k described in Sect. 2.

By viewing SRBF interpolation in appropriate reproducing kernel Hilbert spaces (see Sect. 6.4 below), it can be shown that in constructing s as in (40) to satisfy the interpolation conditions (1), it is natural to enforce the additional conditions

$$\sum_{j=1}^{n} c_j Y_{k,\ell}(v_j) = 0, \qquad k = 0, \ldots, m-1, \ \ell = 1, \ldots, 2k+1. \tag{41}$$

We now discuss conditions under which this system is nonsingular.

Definition 15. *A continuous function $\varphi : [0, \pi] \to \mathbb{R}$ is called* conditionally positive definite of order m on S *if*

$$\sum_{i=1}^{n} \sum_{j=1}^{n} \varphi(d(v_i, v_j))\, c_i c_j \geq 0 \tag{42}$$

for any n points $v_1, \ldots, v_n \in S$ and any $c = [c_1, \ldots, c_n]^T \in \mathbb{R}^n$ satisfying (41). If the points $v_1, \ldots, v_n$ are distinct and (42) is a strict inequality for nonzero c, then φ is called strictly conditionally positive definite of order m on S.

Note that, unfortunately, the term *order* has a different meaning here than it did in Definition 14. The terminology here seems to be well-established, while its other use for positive definiteness stems from [155].

Theorem 16. *Suppose the function φ is strictly conditionally positive definite of order m on S and that the points $v_1, \ldots, v_n$ contain an $\mathcal{H}_{m-1}$-unisolvent subset. Then there is a unique function of the form (40) satisfying (1) and (41).*

Proof: The result follows from the same argument as in the Euclidean case given in *e.g.* [20]. □

As pointed out in Sect. 2, there does not seem to be a satisfactory theory of $\mathcal{H}_m$-unisolvent sets, but a special result can be found in [73]. A sufficient condition for strict conditional positive definiteness of order m is

Theorem 17. [93] *Suppose that*

$$\varphi(t) = \sum_{k=m}^{\infty} a_k P_k(\cos t),$$

where P_k is the Legendre polynomial of degree k, and $a_k = \frac{2k+1}{b_k^2 4\pi}$ for some real numbers $b_k > 0$ with $\sum_{k=m}^{\infty} \frac{2k+1}{b_k^2} < \infty$. Then φ is strictly conditionally positive definite of order m on S.

We note that in the statement of this result in [93] there is a restriction that $b_k \geq 1$ which, according to one of the authors [95], is not needed. A similar result concerning strict conditional positive definiteness in the distributional sense is given in [41]. A complete characterization of strict conditional positive definiteness does not yet exist. Partial results containing various (independent) necessary and sufficient conditions for the case $m = 1$ can be found in [112–114].

We note that, as in the Euclidean case, strictly conditionally positive definite functions of order one can also be used for interpolation without constant precision (see [20]).

The best-known example of a strictly conditionally positive definite function is the spherical multiquadric (see *e.g.* [53,77,138])

$$\varphi(t) = \sqrt{1 + \gamma^2 - 2\gamma \cos t}, \qquad \gamma > 0,$$

where t measures geodesic distance on the sphere. It was shown in [48] that this function is strictly conditionally positive definite of order one. For $\gamma = 1$ it can be seen (using the transformation discussed at the beginning of Sect. 6) that the spherical multiquadric is the restriction to the sphere of the $\mathbb{R}^3$-multiquadric (with multiquadric parameter $c = 0$). Note that the resulting function is not differentiable at its center in this case. For other values of γ this connection with Euclidean RBF's no longer holds.

6.4. Variational Interpretation and Spherical Thin Plate Splines

In $\mathbb{R}^2$ it was observed by Duchon [32] that the function which minimizes a certain energy functional over all smooth functions satisfying the interpolation conditions (1) is a radial basis function which is the kernel of a corresponding reproducing kernel Hilbert space. Such functions are called thin plate splines, and have been extensively studied — see *e.g.* [35,58–60,111,117,197] and references therein. Analogous spaces of thin plate splines have also been developed on the sphere, see *e.g.* [64,66,198,199]. These splines were also called *spherical splines* in the first two papers, but we shall not use this terminology in order to avoid confusion with the completely different spherical splines discussed in Sect. 5.

A complete variational theory for classical RBF's was developed by Madych and Nelson [108,109]. Several authors have recently extended this work to the sphere, see [41,71,73,82,93]. The first and fourth of these papers are based on the fact that spherical harmonics are fundamental solutions of the Laplace-Beltrami operator, whereas the other three rely on a certain sequence $\{b_k\}_{k=m}^{\infty}$ to define appropriate Hilbert spaces in which to formulate the variational problems. The discussion in [71] and [73,93] is quite similar, the main differences being that in [93] (and its follow-up paper [73]) sharper error bounds are obtained. In [73,93] the case of higher-dimensional spheres is also treated. On the other hand, in [71] the properties of the sequence $\{b_k\}_{k=m}^{\infty}$ are given more attention. We will mention some of the ideas of these papers in the next section. The discussion in [41] is the most general since it addresses general Riemannian manifolds.

The main result is that any (conditionally) positive definite function on S can be viewed as the reproducing kernel of an associated Hilbert space, and it can be shown that the interpolant with minimal Hilbert space norm is a spherical radial basis function expansion (with added spherical harmonic terms). Furthermore, this interpolant is unique, and using the variational framework it is possible to give error bounds (see the next section).

6.5. Error Bounds and Stability

Error bounds for interpolation of smooth functions by spherical radial basis functions have recently been announced in [73,82]. While the bounds are very similar, the techniques are quite different. We present the main result of [73] since it is phrased in notation which is similar to what we have used above. It gives error bounds covering conditionally positive definite functions of all orders, and is based on the general theory of Golomb and Weinberger [75].

First we need some additional notation. One can define the associated Hilbert space mentioned in the previous section as the space

$$X_m = \{f \in C(S) \; : \; f = \sum_{k=0}^{\infty} \sum_{\ell=1}^{2k+1} c_{k\ell} Y_{k,\ell} \text{ and } |f|_m < \infty\},$$

where the semi-norm is given by

$$|f|_m = \Big(\sum_{k=m}^{\infty} b_k^2 \sum_{\ell=1}^{2k+1} c_{k\ell}^2 \Big)^{1/2},$$

with $b_k > 0$ such that $\sum_{k=m}^{\infty} \frac{2k+1}{b_k^2} < \infty$.

Theorem 18. *Suppose $\mathcal{V}_h = \{v_1^h, \ldots, v_{n(h)}^h\} \subset S$ contains an $\mathcal{H}_{m-1}$ unisolvent subset and that*

$$\max_{v \in S} \min_{v_i^h \in \mathcal{V}_h} d(v, v_i^h) \leq h.$$

Given $f \in X_m$, let s_h be the unique interpolant to f at the points in $\mathcal{V}_h$ satisfying (41). This interpolant has the form

$$s_h(v) = \sum_{i=1}^{n(h)} c_i \varphi(d(v, v_i^h)) + \sum_{k=0}^{m-1} \sum_{\ell=1}^{2k+1} d_{k\ell} Y_{k,\ell}(v), \qquad v \in S,$$

with φ as in Theorem 17. If $\varphi \in C^\lambda[0, \rho]$ for some $0 < \rho < \pi/2$, then there is a constant $C > 0$ such that

$$\|f - s_h\|_\infty \leq C h^\lambda |f|_m,$$

for h sufficiently small and all $f \in X_m$.

For strictly positive definite SRBF's, analogous error bounds were given in [82]. As mentioned above, the techniques used there are different from those in [73], and have the advantage that they lead to an explicit constant in the error bound. We also mention that [41] contains error bounds on general compact Riemannian manifolds, as well as results for points lying on an equiangular grid on the sphere.

In using SRBF's for interpolating data on the sphere, another issue which has to be addressed is the *stability* of the method. This amounts to investigating the condition number of the interpolation matrix, which in turn reduces to estimating the norm of its inverse. This has been done for a certain type of strictly positive definite functions φ in [121].

Theorem 19. *Assume φ is strictly positive definite with $a_k > 0$ in (38). Let $q \in (0, \pi]$ be the minimal data separation of the data sites $v_1, \ldots, v_n$, and let ν be the smallest integer such that*

$$\nu \geq \frac{24}{1 - \pi^2/80} \frac{\log(1/t(q))}{q^2},$$

where

$$t(q) = \frac{2q}{\pi^3 [\max\{8(\pi^2/q), 25\pi/2\}]}.$$

Then

$$\|A^{-1}\|_2 \leq \frac{C}{\min\{a_0, \min_{1 \leq k \leq \lceil \nu/2 \rceil - 1} \frac{a_k}{\sqrt{k}}\}},$$

where C is a constant. In particular, if $a_k \sim 1/(1 + k^\alpha)$, $\alpha > 1$, then

$$\|A^{-1}\|_2 = O\left[\left(\frac{\log(1/q^2)}{q^2}\right)^{\alpha + 1/2}\right], \qquad q \to 0.$$

The results in this section show that the error becomes smaller as the data sites move closer together, but at the same time the condition number becomes poorer. This is the "trade-off principle" well known from Euclidean radial basis function theory (see [167]).

Fundamental sets of continuous functions on the sphere were discussed in [183,184].

6.6. Locally Supported SRBF's

In this section we briefly discuss some locally supported SRBF's which were constructed directly for the sphere (see [14,71,172]) in contrast to those of Wendland (mentioned in Sect. 6.1) which were originally built for Euclidean space. Given $h > 0$, let

$$\tilde{\varphi}_h^{(k)}(t) = \begin{cases} \left(\dfrac{t-h}{1-h}\right)^k, & \text{for } t \geq h, \\ 0, & \text{otherwise,} \end{cases} \tag{43}$$

with $t = v \cdot w$. Although these functions can be transformed to Wendland's C^0 function by $1 - t \to t$ and $1 - h \to h$, it is not clear whether they are strictly positive definite on S. It is more instructive (cf. [172]) to consider

$$\varphi_h^{(k)}(t) = \begin{cases} \left(\dfrac{\cos t - h}{1-h}\right)^k, & \text{for } t \geq \arccos h, \\ 0, & \text{otherwise,} \end{cases}$$

where now $t = d(v, w) = \arccos(v \cdot w)$ is the geodesic distance. With h replaced by $\cos h$, the cases $k = 1, 2$ were already used in [185].

In [172] it was shown that the functions $(\varphi_h^{(k)})^{(2)}$ obtained by spherical convolution of $\varphi_h^{(k)}$ with itself are strictly positive definite on S. In [14] these functions were used to construct approximations via convolutions to an unknown function f assumed to be Lipschitz continuous on S. Since convolution has a smoothing effect, it is clear that in order to obtain a good approximation, the support of the basis functions should not be chosen too large. The following error estimate (see [14]) reflects this observation:

$$\|f - (\tilde{\varphi}_h^{(k)})^{(2)} *' f\|_\infty \leq C\sqrt{1-h},$$

where $*'$ is used to denote a discrete approximation to the spherical convolution, and $(\tilde{\varphi}_h^{(k)})^{(2)}$ is as in (43). The local support of the basis functions is used to devise a hierarchical approximation method similar to the one described in Sect. 7.5. There seems to have been no attempt made to use these locally supported radial basis functions for interpolation purposes directly (although the possibility of doing so is mentioned in [172]). A major disadvantage of using the functions $(\varphi_h^{(k)})^{(2)}$ is that they have to be computed numerically.

Approximation of functions on the sphere via singular integrals has also been studied, see *e.g.* [13].

Before locally supported basis functions were discovered, local interpolation schemes were constructed by interpolating only to subsets of the data and then blending these partial interpolants together. This idea was outlined in both [117] and [138] for radial basis functions on the sphere.

6.7. Generalized Hermite Interpolation

It is also possible to solve more general interpolation problems with spherical radial basis functions. Suppose we are given data sites $v_1, \ldots, v_n$, as before, but that instead of simple function values at these points, we have data which are generated by a linearly independent set of linear functionals $\{L_i\}_{i=1}^n$. Thus, the problem is to find an interpolant s such that

$$L_i s = z_i, \qquad i = 1, \ldots, n,$$

where $z_i = L_i f$ for some unknown function f defined on S. If the functionals are point evaluation functionals, we have the standard interpolation problem (1). If we consider point evaluations of derivatives, then we have an Hermite interpolation problem. Other linear functionals such as local averages are also possible.

To solve this problem, the interpolant is assumed to be of the form

$$s(v) = \sum_{i=1}^{n} c_i L_i^{(2)} \varphi(d(v, v_i)), \tag{44}$$

where the superscript (2) denotes that L_i is applied to $\varphi \circ d$ as a function of the second variable.

The first to study this problem on the sphere was Narcowich [119]. He showed that if all $a_k > 0$ in (38), then (for a general Riemannian manifold) the generalized Hermite interpolation problem has a unique solution. This class of functions includes *e.g.* the spherical reciprocal multiquadrics, and functions of the form $\varphi(t) = e^{\gamma \cos t}$, $\gamma > 0$. In [46] other classes of basis functions were shown to lead to a unique solution for this problem, namely compositions of conditionally positive definite functions of order one with completely monotone functions, or with such functions whose derivative is completely monotone. These results ensure that the spherical multiquadrics can also be used for generalized Hermite interpolation. An expansion such as (44) seems to have been used first on the sphere by Freeden, see [65,67].

6.8. Discrete Least Squares Approximation

For RBF's in the Euclidean case, discrete least squares approximation was studied in [180]. Essentially, the authors of those papers show that the associated Gramian matrix is nonsingular for some of the most popular radial basis functions (norm, multiquadrics, reciprocal multiquadrics and Gaussians) as long as the data sites are sufficiently well distributed and the centers for the radial basis functions are fairly evenly clustered about the data sites with the diameter of the clusters being relatively small compared to the separation distance of the data. Although there is no literature on least squares approximation with spherical RBF's, similar results can be expected.

In [47] two algorithmic approaches to adaptive least squares fitting on S were compared. The first method is knot insertion, where starting with a coarse approximation (few centers), additional centers are added iteratively at the data location with the largest error component. This process is repeated until a certain tolerance is satisfied. If the initial knots are also chosen at data locations, then there are no problems with the collocation matrix becoming singular.

The second method is usually referred to as knot removal and proceeds in the reverse direction. It is most valuable if a sparse representation of the data is needed, *e.g.*, for many subsequent evaluations. The algorithm starts with a very accurate initial fit and then tries to delete those centers (or basis functions) whose removal results in the smallest error. The procedure ends when a certain tolerance is reached.

The above algorithms can be made more efficient by dealing with several points at a time, and by optimizing the center locations with the help of nonlinear optimization methods. For the Euclidean setting, adaptive least squares fitting was discussed in [61,62].

§7. Multiresolution Methods

In recent years there has been a great deal of interest in *multiresolution methods* for compressing and approximating images, signals, and general functions and data. While an extensive theory exists for univariate and bivariate functions, there are only a few results for functions defined on the sphere S.

7.1. The Basic Multilevel Approach

Let

$$\mathcal{F}_0 \subset \cdots \subset \mathcal{F}_k \subset \mathcal{F} \tag{45}$$

be a nested set of linear spaces, and suppose that for each i, P_i is a linear projection mapping $\mathcal{F}$ to $\mathcal{F}_i$. In general, $P_i s$ should provide an approximation in $\mathcal{F}_i$ to s. Typical choices include L_2 best approximation and interpolation at appropriate points.

Given an s_k at the finest level, *i.e.*, $s_k \in \mathcal{F}_k$, it can be rewritten as $s_k = s_{k-1} + g_{k-1}$, where $s_{k-1} := P_{k-1} s_k \in \mathcal{F}_{k-1}$ and $g_{k-1} := s_k - s_{k-1} \in \mathcal{F}_k$. Repeating this process recursively leads to the multilevel expansion

$$s_k = s_0 + g_0 + \cdots + g_{k-1}, \tag{46}$$

where

$$s_{i-1} := P_{i-1} s_i, \qquad g_{i-1} := s_i - s_{i-1},$$

for $i = 1, \ldots, k$. The term s_0 in the expansion lies in the coarsest space $\mathcal{F}_0$, while the terms $g_{i-1} \in \mathcal{F}_i$ can be regarded as providing finer and finer levels of detail. Assuming that each of these terms is expanded in terms of appropriate basis functions, compression can be achieved by removing the associated coefficients which fall below some prescribed threshold.

Classical wavelet theory is based on choosing P_i to be the L_2 approximation operator which produces a best approximation in $\mathcal{F}_i$. The above decomposition process becomes especially simple if both the spaces $\mathcal{F}_i$ and their orthogonal complements $\mathcal{G}_i$ in $\mathcal{F}_{i+1}$ have simple bases. Often these are spanned by translates and dilates of a small number of functions. Examples of spaces with simple bases will be discussed in Sects. 7.2 and 7.3 below.

Another very useful way to choose the P_i is to take them to be *interpolation operators*, where $P_i s$ is defined by the condition

$$P_i s(v_j^i) = s(v_j^i), \qquad j = 1, \ldots, n_i, \tag{47}$$

where $\mathcal{V}_i := \{v_j^i\}_{j=1}^{n_i}$ are appropriate points in the domain of the functions in $\mathcal{F}_i$, and where

$$\mathcal{V}_0 \subset \cdots \subset \mathcal{V}_k \tag{48}$$

is a prescribed nested sequence of sets of points. A method of this type defined on the sphere will be discussed in Sect. 7.4 below.

It should be pointed out that the above decomposition algorithm is closely related to a classical iterative procedure for creating increasingly accurate approximations to a given function s. Given a nested sequence of spaces as in (45), suppose that Q_i are linear projection operators with $Q_j Q_i = Q_i$ for all $j \geq i$. Suppose that Q_i maps a space $\mathcal{F}$ containing $\mathcal{F}_k$ (and thus all of the spaces in (45)) onto $\mathcal{F}_i$. Then, given $s \in \mathcal{F}$, $Q_i s$ selects an approximation to s from the space $\mathcal{F}_i$. Now, starting with a given function $s \in \mathcal{F}$, first compute $\tilde{s}_0 := Q_0 s \in \mathcal{F}_0$, and then set $\tilde{g}_0 := Q_1(s - \tilde{s}_0)$. This process can be repeated by setting

$$\tilde{g}_{i-1} := Q_i(s - \tilde{s}_{i-1}), \qquad \tilde{s}_i := \tilde{s}_{i-1} + \tilde{g}_{i-1},$$

for $i = 1, \ldots, k$, leading to the multilevel approximation

$$s \approx \tilde{s}_k = \tilde{s}_0 + \tilde{g}_0 + \cdots + \tilde{g}_{k-1}, \tag{49}$$

where $\tilde{s}_0 \in \mathcal{F}_0$ and $\tilde{g}_{i-1} \in \mathcal{F}_i$. As before, $\tilde{s}_0$ can be regarded as a coarse approximation, and the $\tilde{g}_{i-1}$ as providing finer and finer levels of detail. The final $\tilde{s}_k$ is in general not the best approximation of s from $\mathcal{F}_k$, but has the advantage of the convenient multilevel expansion (49). If one starts with $s \in \mathcal{F}_k$, then (49) becomes an exact multilevel expansion of s.

In general, the two algorithms described above are not equivalent. However, there is an important setting where they are.

Theorem 20. *Suppose that a nested sequence of point sets as in (48) is given, and that $P_i = Q_i$ is the operator which maps a function in $\mathcal{F}$ to the unique function in $\mathcal{F}_i$ which interpolates it at the points of $\mathcal{V}_i$ as in (47). Then the above two algorithms are equivalent in the sense that for any $s_k \in \mathcal{F}_k$, the expansions (46) and (49) are exactly the same with $s_0 = \tilde{s}_0$ and $g_i = \tilde{g}_i$ for all $i = 0, \ldots, k-1$.*

Proof: Given $s_k \in \mathcal{F}_k$, it is easy to see by induction that s_i is the unique function in $\mathcal{F}_i$ which interpolates to s_k on the point set $\mathcal{V}_i$. Indeed, assuming this for $i+1$, then

$$s_i(v_j^i) = P_i s_{i+1}(v_j^i) = s_{i+1}(v_j^i) = \cdots = s_k(v_j^i)$$

for all $j = 1, \ldots, n_i$. We can prove the same thing for $\tilde{s}_i$ by induction in the other direction. Indeed,

$$\begin{aligned} \tilde{s}_i(v_j^i) &= \tilde{s}_{i-1}(v_j^i) + Q_i(s_k - \tilde{s}_{i-1})(v_j^i) \\ &= \begin{cases} \tilde{s}_{i-1}(v_j^i), & v_j^i \in \mathcal{V}_{i-1} \\ \tilde{s}_{i-1}(v_j^i) + s_k(v_j^i) - \tilde{s}_{i-1}(v_j^i), & v_j^i \in \mathcal{V}_i \setminus \mathcal{V}_{i-1} \end{cases} = s_k(v_j^i). \end{aligned}$$

The assertion follows. □

The basic multilevel idea has been used in a number of recent papers (see *e.g.* [14,25,49,52,120,173]). We discuss some of these in the following sections.

7.2. A Multiresolution Method Based on Tensor Splines

In this section we discuss a multilevel scheme for the sphere S based on L_2 approximation using the tensor-product polynomial–trigonometric splines defined on the rectangle R defined in Sect. 3. Our discussion follows [103].

The process begins with a tensor polynomial-trigonometric spline s satisfying the conditions of Theorem 6. This insures tangent plane continuity of the associated surface defined on S. To simplify matters, equally spaced knots are used.

For each $i = 0, \ldots, k$, let $\mathcal{F}_i$ be the space spanned by the normalized polynomial B-splines $N_1^3, \ldots, N_{m_i}^3$ of order 3 associated with the knots $\{x_j^i\}_{j=1}^{m_i+3}$ where $x_1^i = x_2^i = x_3^i = 0$, $x_{m_i+1}^i = x_{m_i+2}^i = x_{m_i+3}^i = \pi$, and

$$x_{j+3}^i = jh_i, \qquad j = 1, \ldots, m_i - 2, \tag{50}$$

with $h_i = \pi/(m_i - 2)$ and $m_i = 3 \cdot 2^i + 2$. Similarly, let $\widetilde{\mathcal{F}}_i$ be the space of periodic trigonometric splines of order 3 associated with the knots

$$\tilde{x}_j^i = (j-1)\tilde{h}_i, \qquad j = 1, \ldots, \widetilde{m}_i, \tag{51}$$

where $\tilde{h}_i = 2\pi/\widetilde{m}_i$ and $\widetilde{m}_i = 3 \cdot 2^i$. Here the integer i is a nesting parameter, with the knots at level i being obtained from those at level $i-1$ by inserting new knots at midpoints of knot intervals. These spline spaces satisfy $\mathcal{F}_0 \subset \cdots \subset \mathcal{F}_k$ and $\widetilde{\mathcal{F}}_0 \subset \cdots \subset \widetilde{\mathcal{F}}_k$. Let

$$\mathcal{F}_i = \mathcal{F}_{i-1} \oplus \mathcal{G}_{i-1}, \qquad \widetilde{\mathcal{F}}_i = \widetilde{\mathcal{F}}_{i-1} \oplus \widetilde{\mathcal{G}}_{i-1}.$$

Now the idea is to decompose s into a coarse spline and several terms containing details. Starting with $s_k := s$, the main result of [103] is an algorithm for rewriting s_k as the orthogonal decomposition

$$s_k = s_{k-1} + g_{k-1}^{(1)} + g_{k-1}^{(2)} + g_{k-1}^{(3)},$$

where

$$g_{k-1}^{(1)} \in \mathcal{F}_{k-1} \times \widetilde{\mathcal{G}}_{k-1}, \quad g_{k-1}^{(2)} \in \mathcal{G}_{k-1} \times \widetilde{\mathcal{F}}_{k-1}, \quad g_{k-1}^{(3)} \in \mathcal{G}_{k-1} \times \widetilde{\mathcal{G}}_{k-1}. \quad (52)$$

The key to creating this decomposition is the fact that each of the spaces in this decomposition has a very simple basis in terms of polynomial and trigonometric B-splines and their associated spline wavelets. For the construction of trigonometric spline wavelets (and more general L-spline wavelets), see [102]. Then the *decomposition algorithm* (the process of finding the coefficients in the expansions of each of the functions in (52)) is a simple matter of matrix algebra.

The decomposition step (52) can be repeated until one ends up with a coarsest approximation s_0. Then standard thresholding methods can be applied to eliminate small coefficients in all expansions, thus achieving compression. Some care must be exercised in the thresholding step in order to maintain smoothness at the poles. In particular, at each level, coefficients of splines or wavelets which contribute to the values of $s_k(\theta, \phi)$ or $D_\theta s_k(\theta, \phi)$ for $\theta = 0, \pi$ must be retained.

An associated *reconstruction algorithm* (which is again nothing but matrix multiplication) is also presented, along with various test examples. A similar method (based on exponential splines) was discussed in [26], see also [168].

Another way to avoid problems at the poles is to first split s into two parts $s = f + g$, where f is a homogeneous part satisfying $f(\theta, \phi) = D_\phi f(\theta, \phi) = 0$ for $\theta = 0, \pi$ and all $0 \le \phi \le 2\pi$, see [102].

7.3. A Multiresolution Method Based on SRBF's

In [122,123] wavelets were constructed using spherical radial basis functions. One starts with an infinite sequence of points $v_1, v_2, \ldots$ which is dense in S. Assuming that $\mathcal{V}_i$ consists of the first n_i of these points, it is clear that the $\mathcal{V}_i$ are nested. Given a strictly positive definite function φ on S as in Theorem 19, the associated n_i dimensional spaces

$$\mathcal{F}_i := \text{span}\{\varphi(d(v, v_j))\}_{j=1}^{n_i}$$

are called the sampling spaces. These spaces are nested and their union is dense in $L_2(S)$ (see [184]). In particular, any $f \in \mathcal{F}_{i-1}$ can be represented in $\mathcal{F}_i$ by simply padding its last $n_i - n_{i-1}$ coordinates (with respect to the basis functions listed above) with zeros.

As in the classical wavelet theory, the wavelet spaces $\mathcal{G}_{i-1}$ are defined as the orthogonal complement of $\mathcal{F}_{i-1}$ in $\mathcal{F}_i$, i.e.

$$\mathcal{F}_i := \mathcal{F}_{i-1} \oplus \mathcal{G}_{i-1}.$$

It turns out that there is a convenient way to represent functions in $\mathcal{G}_{i-1}$. Let $\varphi \star \varphi$ be the SRBF defined via spherical convolution by

$$\varphi \star \varphi(d(v, w)) = \int_S \varphi(d(v, u))\varphi(d(u, w))d\omega(u),$$

where $d\omega$ is the surface element for S. It is shown in [123] that $\varphi \star \varphi$ is strictly positive definite on S. Let $A_\star = (\varphi \star \varphi(d(v_j, v_k))_{j,k=1}^{n_i}$ Then a basis for $\mathcal{G}_{i-1}$ is given by the last $n_i - n_{i-1}$ columns of $A_\star^{-1}$, *i.e.*, by those columns corresponding to the centers $v_{n_{i-1}+1}, \ldots, v_{n_i}$.

Decomposition and reconstruction formulae can also be found in [123]. They involve the solution of the interpolation problem involving $A_\star$. Thus, it is desirable that the sampling spaces be spanned by locally supported functions. In [123] one can find several examples for φ along with explicit formulae for $\varphi \star \varphi$. The locally supported functions of [172] (see also Sect. 6.6) and their iterated versions can also be used.

7.4. A Multiresolution Method Based on Spherical Splines $\mathcal{S}_1^0(\Delta)$

In this section and the next we describe multilevel methods for the sphere in which the projection operators P_i of Sect. 7.1 are chosen to be interpolation operators. The method to be described in this section makes use of C^0 spherical splines of degree one.

Suppose $\Delta_0 \subset \Delta_1 \subset \cdots$ is a nested sequence of spherical triangulations which are obtained from a basic triangulation Δ_0 by the following process: to get Δ_i, each triangle in Δ_{i-1} is split into four subtriangles by connecting the midpoints of its adjacent edges. Clearly, the associated spherical spline spaces $\mathcal{S}_1^0(\Delta_i)$ are nested, *i.e.*, $\mathcal{S}_1^0(\Delta_0) \subset \mathcal{S}_1^0(\Delta_1) \subset \cdots$. Let $\mathcal{V}_i$ be the set of vertices of Δ_i, and suppose P_i denotes the interpolating projector onto $\mathcal{S}_1^0(\Delta_i)$.

The decomposition process begins with a spline $s_k \in \mathcal{S}_1^0(\Delta_k)$ at the finest level k. Then s_{k-1} is the spline in $\mathcal{S}_1^0(\Delta_{k-1})$ which *interpolates* to s_k at the vertices of Δ_{k-1}. This process is almost trivial to implement since on each triangle (cf. Sect. 5), any spline s in $\mathcal{S}_1^0(\Delta_{k-1})$ can be expanded in terms of the first degree SBB-polynomials $B_{100}^1, B_{010}^1, B_{001}^1$ associated with that triangle, and the coefficients are just the values of s at those vertices. Thus, the coefficients of s_{k-1} are obtained from those of s_k by simply retaining those associated with vertices of Δ_{k-1} and dropping those associated with midpoints of the edges of triangles in Δ_k.

Since g_{k-1} lies in $\mathcal{S}_1^0(\Delta_k)$, its SBB-coefficients are also easily obtained. Indeed, by linear interpolation, if w is the midpoint of the great circle arc joining two vertices v_i and v_j of Δ_{k-1}, then the value of $g_{k-1}(w)$ is simply

$$g_{k-1}(w) = s_k(w) - s_{k-1}(w) = s_k(w) - \frac{[s_k(v_i) + s_k(v_j)]}{2}.$$

This is the key *decomposition* formula. The associated *reconstruction* formula is

$$s_k(w) = g_{k-1}(w) + \frac{[s_k(v_i) + s_k(v_j)]}{2}.$$

As above, the decomposition process can be repeated until the coarsest level is reached. This gives the decomposition (46). At this point a thresholding process can be applied to remove all small coefficients, thus achieving compression.

This scheme was first presented in [173], along with several other related schemes with different choices of P_i (including certain *lifting schemes* which have additional desirable moment properties). Despite its simplicity, this method is quite effective in compressing data, e.g. see [173] for an example involving compression of a very large set (called ETOPO5) of elevation/bathymetric data taken over the surface of the earth by satellite. As pointed out in [25] (see also [169]) this idea was used already by Faber [45] in 1909. It can be extended to C^1 splines on appropriate Powell-Sabin triangulations, see [25,169].

7.5. A Multilevel Interpolation Method Based on SRBF's

In the context of radial basis functions in Euclidean spaces, the basic multilevel interpolation approach was introduced in [52]. It is also the subject of the recent papers [49,120]. Its application to the sphere is straightforward, and our discussion below focusses on this case.

Let $v_1, \ldots, v_k$ be a sequence of points on S, and let $\mathcal{V}_i$ consist of the first n_i (reasonably uniformly spaced) of these points. Then clearly the $\mathcal{V}_i$ are nested. Given a function s defined on the sphere, applying the method of Sect. 7.1 (cf. [52]) one starts at the coarsest level with the interpolant $s_0 = P_0 s$ of the form

$$s_0(v) = \sum_{j=1}^{n_0} c_j^0 \varphi_{h_0}(\|v - v_j^0\|).$$

At each step an additional amount of detail is computed via $g_{i-1} = P_i(s - s_{i-1})$, where

$$g_{i-1}(v) = \sum_{j=1}^{n_i} c_j^i \varphi_{h_i}(\|v - v_j^i\|).$$

Here h_i is a parameter which is used to scale the support size of the basis functions in accordance with the density of the data sites. The scaling should be done in such a way as to leave the bandwidth of the interpolation matrix (or the number of data sites inside the support of the basis functions) roughly constant throughout the process. This insures that the process will be well-conditioned.

Note that in this implementation the points $v_1, \ldots, v_k$ can be arbitrarily spaced. This is in contrast to the previous section where they were obtained as vertices of the triangles arising in successive refinement of a basic spherical triangulation. Numerically, one can observe that the algorithm has a rate of convergence which is at least linear in the "meshsize", where meshsize is understood as in Theorem 18.

A similar algorithm for *approximation* by singular integrals was proposed in [14]. The authors of that paper obtained their level i approximation by a convolution of the data with an iterate of a locally supported SRBF as described in Sect. 6.6.

The motivation for [120] was to give rates of convergence for the algorithm suggested in [52]. However, the authors of [120] were able to give estimates only for a modified problem which employs convolutions of increasing multiplicity of the basis functions at the various levels, with the most smoothness required at the coarsest level. It is pointed out in [120] that for large problems, one arrives at smaller errors using this version of the multilevel algorithm as compared to solving the problem directly at the finest level with basis functions of relatively little smoothness (in case this is even computationally feasible).

7.6. An Approximate Newton Method

In [49] the decomposition step $s_i = s_{i-1} + g_{i-1}$ in the basic multilevel method was interpreted as an instance of an approximate Newton method, *i.e.*, $s_i - s_{i-1} = g_{i-1}$, where g_{i-1} is interpreted as an approximation to the inverse of the derivative of the mapping which computes the residual at level i. It was shown there (for Euclidean spaces) that an additional smoothing step at each level of the iteration can improve the convergence rate from linear to superlinear. With smoothing, the approximation spaces are no longer nested, and (49) is no longer an identity, although the right-hand side still provides an approximation to s. This interpretation of the algorithm is especially suited for the solution of differential equations.

§8. Additional Topics

In this final section we mention several other topics related to fitting scattered data on the sphere which we do not have space to discuss here in detail.

8.1. Sphere-like Surfaces

All of the above discussion has focused on the unit sphere S. As observed in [2,3], much of it is also valid for the following more general surfaces:

Definition 21. *Given a smooth positive function ρ defined on the unit sphere S, the set $\mathcal{S} := \{u \in \mathbb{R}^3 : u = \rho(v)v,\ v \in S\}$ is called* a sphere-like surface.

When $\rho \equiv 1$, $\mathcal{S}$ reduces to the unit sphere. The earth is an example of a sphere-like surface, although the determination of ρ is no simple matter (and can be regarded as a scattered data fitting problem on the sphere), see [31].

8.2. General Surfaces

The sphere is only one example of an interesting surface for which the problem of fitting scattered data is of importance. The surface could as well be a torus, the wing of an aircraft, or any other physical 3D object. Some work has been done in the more general setting – see e.g. [9–12,56,119,137] and the references therein.

We point out that the simplest approach to fitting scattered data on a general surface — restricting the evaluation of a trivariate method to the surface — is not recommended, since problems are likely to arise as soon as this surface is "thin", such as the wing of an aircraft (cf. [11]).

The problem of creating a Voronoi diagram associated with a set of points on a general surface (which then leads to an associated Delaunay triangulation) has been addressed in [86].

8.3. Visualizing Surfaces Defined on S

As observed above, the standard way (cf. [2,6,56,57,63,127]) to associate a *surface* with a *function* s defined on the sphere is via $\mathcal{F} := \{s(v)v \ : \ v \in S\}$. Thus, $s(v)$ can be identified with the *distance* from the center of the sphere to a point on the associated surface $\mathcal{F}$ if one moves out along the normal vector at v. Alternatively, $s(v)$ can be taken as the distance above S, i.e. $\mathcal{F} := \{(s(v)+1)v \ : \ v \in S\}$.

One way to *visualize* $\mathcal{F}$ is to replace it by a surface consisting of *planar facets*. This can be accomplished by creating a spherical triangulation on S, computing the points $s(v_i)v_i \in \mathbb{R}^3$ associated with the vertices v_i of this triangulation, and then associating a planar facet with each spherical triangle as follows: given T with vertices v_i, v_j, v_k, let F_T be the triangle in $\mathbb{R}^3$ with vertices at $s(v_i)v_i, s(v_j)v_j, s(v_k)v_k$. The resulting faceted surface can then be displayed with standard graphical software. The capability to rotate the surface in 3D is very helpful. This method is easily generalized to general (convex) surfaces.

Another simple way to visualize function values defined on a general surface in $\mathbb{R}^3$ is to triangulate the domain surface as above, and then assign a color mapped from some color scale to each vertex of the triangulation.

There are several other approaches to visualizing surfaces on surfaces including (color coded) contour regions, isophotes, and projections of 4D-graphs. For details, see [57,110,139,140,156,182] and references therein.

8.4. Numerical Quadrature on the Sphere

The question of computing approximate values for integrals of functions defined on S is a classical subject in numerical analysis. The standard approach to creating spherical quadrature formulae can be described as follows. Suppose λ is a linear functional defined on $C(S)$, and suppose P is an approximation process mapping $C(S)$ to an approximating space $\mathcal{A}$. Then given any $f \in C(S)$, $\lambda f \approx \lambda P f$. Such formulae are of particular interest when Pf depends only on values of f at some (scattered) points v_i on S, and when $\lambda P f$ can be explicitly computed in terms of the values $r_i = f(v_i)$.

One approach to developing such quadrature formulae is to use *spherical harmonics* for the approximating functions. If this is done locally on some spherical triangulation, it amounts to using the spherical splines defined in Sect. 5. Useful quadrature formulae have been derived in this way, although in contrast to the case of classical polynomials in Euclidean space, there are no exact formulae for integrals of spherical harmonics. Here we list only a few references [69,107,181].

References

1. Alfeld, P., M. Neamtu, and L. L. Schumaker, Circular Bernstein-Bézier polynomials, in *Mathematical Methods for Curves and Surfaces,* Morten Dæhlen, Tom Lyche, and Larry L. Schumaker (eds), Vanderbilt University Press, Nashville & London, 1995, 11–20.
2. Alfeld, P., M. Neamtu, and L. L. Schumaker, Bernstein–Bézier polynomials on spheres and sphere–like surfaces, Comput. Aided Geom. Design **13** (1996), 333–349.
3. Alfeld, P., M. Neamtu, and L. L. Schumaker, Fitting scattered data on sphere-like surfaces using spherical splines, J. Comput. Appl. Math. **73** (1996), 5–43.
4. Alfeld, P., M. Neamtu, and L. L. Schumaker, Dimension and local bases of homogeneous spline spaces, SIAM J. Math. Anal. **27** (1996), 1482–1501.
5. Babar, C. S. and G. S. Pandey, On approximation to a function by Cesàro-Nörlund means of its ultraspherical series on a sphere, Vikram Math. J. **11** (1991), 29–38.
6. Bajaj, C. and G. Xu, Modeling and visualization of scattered function data on curved surfaces, in *Fundamentals of Computer Graphics*, J. Chen, N. Thalmann, Z. Tang, and D. Thalmann (eds.), World Scientific Publishing Co., 1994, 19–29.
7. Balmino, G., K. Lambeck, and W. M. Kaula, A spherical harmonic analysis of the earth's topography, J. Geophys. Res. **78** (1973), 478–481.
8. Barnhill, R. E., G. Birkhoff, and W. J. Gordon, Smooth interpolation in triangles, J. Approx. Theory **8** (1973), 114–128.
9. Barnhill, R. E. and T. A. Foley, Methods for constructing surfaces on surfaces, in *Geometric Modeling, Methods and Applications,* H. Hagen and D. Roller (eds), Springer Verlag, Berlin, 1991, 1–15.
10. Barnhill, R. E., K. Opitz, and H. Pottmann, Fat surfaces: a trivariate approach to triangle-based interpolation on surfaces, Comput. Aided Geom. Design **9** (1992), 365–378.
11. Barnhill, R. E. and H. S. Ou, Surfaces defined on surfaces, Comput. Aided Geom. Design **7** (1990), 323–336.
12. Barnhill, R. E., B. R. Piper, and K. L. Rescorla, Interpolation to arbitrary data on a surface, in *Geometric Modeling: Algorithms and New Trends,* G. E. Farin (ed), SIAM Publications, Philadelphia, 1987, 281–290.
13. Berens, H., P. L. Butzer, and S. Pawelke, Limitierungsverfahren von Reihen mehrdimensionaler Kugelfunktionen und deren Saturationsverhalten, Publ. Res. Inst. Math. Sci. (Kyoto), Ser. A. **4** (1968), 201–268.

14. Brand, R., W. Freeden, and J. Fröhlich, An adaptive hierarchical approximation method on the sphere using axisymmetric locally supported basis functions, Computing **57** (1996), 187–212.

15. Brown, J. L., Natural neighbor interpolation on the sphere, in *Wavelets, Images, and Surface Fitting,* P.-J. Laurent, A. Le Méhauté, and L. L. Schumaker (eds), A. K. Peters, Wellesley MA, 1994, 67–74.

16. Brown, J. L. and A. J. Worsey, Problems with defining barycentric coordinates for the sphere, Math. Modelling Numer. Anal. **26** (1992), 37–49.

17. Buhmann, M., New developments in the theory of radial basis function interpolation, in *Multivariate Approximation: From CAGD to Wavelets,* Kurt Jetter and Florencio Utreras (eds), World Scientific Publishing, Singapore, 1993, 35–75.

18. Chang, L. H. T., Scattered Data Interpolation Schemes for Curves and Surfaces, dissertation, Univ. Sains Malaysia, 1997.

19. Cheney, E. W., Approximation and interpolation on spheres, in *Approximation Theory, Wavelets and Applications,* S. P. Singh (ed), Kluwer, Dordrecht, Netherlands, 1995, 47–53.

20. Cheney, E. W., Approximation using positive definite functions, in *Approximation Theory VIII, Vol. 1: Approximation and Interpolation,* Charles K. Chui and Larry L. Schumaker (eds), World Scientific Publishing Co., Inc., Singapore, 1995, 145–168.

21. Cheney, E. W. and X. Sun, Interpolation on spheres by positive definite functions, to appear in a volume dedicated to Professor Varma and edited by N. K. Govil, 1997.

22. Chui, C. K. and Hang-Chin Lai, Vandermonde determinants and Lagrange interpolation in $\mathbb{R}^s$, in *Nonlinear and Convex Analysis,* B. L. Lin and S. Simons (eds), Marcel Dekker, New York, 1987, 23–32.

23. Chung, K. C. and T. H. Yao, On lattices admitting unique Lagrange interpolations, SIAM J. Numer. Anal. **14** (1977), 735–743.

24. Courant, D. and D. Hilbert, *Methods of Mathematical Physics, Vol. 1,* Interscience, New York, 1953.

25. Dæhlen, M., T. Lyche, K. Mørken, R. Schneider, and H.-P. Seidel, Multiresolution analysis based on quadratic Hermite interpolation – Part 2: piecewise polynomial surfaces, preprint, 1997.

26. Dahlke, S., W. Dahmen, I. Weinreich, and E. Schmitt, Multiresolution analysis and wavelets on S^2 and S^3, Numer. Func. Anal. Optim. **16** (1995), 19–41.

27. Dahmen, W. and C. A. Micchelli, Recent progress in multivariate splines, in *Approximation Theory IV,* C. Chui, L. Schumaker, and J. Ward (eds), Academic Press, New York, 1983, 27–121.

28. Dierckx, P., Algorithms for smoothing data on the sphere with tensor product splines, Computing **32** (1984), 319–342.

29. Dierckx, P., The spectral approximation of bicubic splines on the sphere, SIAM J. Sci. Statist. Comput. **7** (1986), 611–623.

30. Dierckx, P., Fast algorithms for smoothing data over a disk or a sphere using tensor product splines, in *Algorithms for the Approximation of Functions and Data,* J. C. Mason and M. G. Cox (eds), Oxford Univ. Press, Oxford, 1987, 51–65.

31. Dragomir, V. C. et al., *Theory of the Earths Shape,* Elsevier Sc. Publ., New York, 1982.

32. Duchon, J., Interpolation des fonctions de deux variables suivant le principe de la flexion des plaques minces, RAIRO Anal. Numer. **10** (1976), 5–12.

33. Dyn, N., Interpolation of scattered data by radial functions, in *Topics in Multivariate Approximation,* C. K. Chui, L. L. Schumaker, and F. Utreras (eds), Academic Press, New York, 1987, 47–61.

34. Dyn, N., Interpolation and approximation by radial and related functions, in *Approximation Theory VI,* C. Chui, L. Schumaker, and J. Ward (eds), Academic Press, New York, 1989, 211–234.

35. Dyn, N., D. Levin, and S. Rippa, Surface interpolation and smoothing by "thin plate" splines, in *Approximation Theory IV,* C. Chui, L. Schumaker, and J. Ward (eds), Academic Press, New York, 1983, 445–449.

36. Dyn, N., D. Levin, and S. Rippa, Numerical procedures for surface fitting of scattered data by radial functions, SIAM J. Sci. Statist. Comput. **7** (1986), 639–659.

37. Dyn, N., D. Levin, and S. Rippa, Algorithms for the construction of data dependent triangulations, in *Algorithms for Approximation II,* J. C. Mason and M. G. Cox (eds), Chapman & Hall, London, 1990, 185–192.

38. Dyn, N., D. Levin, and S. Rippa, Data dependent triangulations for piecewise linear interpolation, IMA J. Numer. Anal. **10** (1990), 137–154.

39. Dyn, N., D. Levin, and S. Rippa, Boundary corrections for data dependent triangulations, J. Comput. Appl. Math. **39** (1992), 179–192.

40. Dyn, N. and C. A. Micchelli, Interpolation by sums of radial functions, Numer. Math. **58** (1990), 1–9.

41. Dyn, N., F. Narcowich, and J. D. Ward, Variational principles and Sobolev-type estimates for generalized interpolation on a Riemannian manifold, Constr. Approx., to appear.

42. Dyn, N., F. Narcowich, and J. D. Ward, A framework for interpolation and approximation on Riemannian manifolds, preprint, 1997.

43. Dyn, N. and A. Ron, Radial basis function approximation: from gridded centers to scattered centers, Proc. London Math. Soc. **71(3)** (1995), 76–108.

44. Džafarov, A. S., The inverse problem of the theory of best approximation of functions on a sphere and on a segment (Russian), Akad. Nauk Azerbaĭdžan. SSR Dokl. **26** (1970), 3–6.

45. Faber, G., Über stetige Funktionen, Math. Ann. **66** (1909), 81–94.

46. Fasshauer, G. E., Hermite interpolation with radial basis functions on spheres, preprint, 1995.

47. Fasshauer, G. E., Adaptive least squares fitting with radial basis functions on the sphere, in *Mathematical Methods for Curves and Surfaces,* Morten Dæhlen, Tom Lyche, and Larry L. Schumaker (eds), Vanderbilt University Press, Nashville & London, 1995, 141–150.

48. Fasshauer, G. E., Radial Basis Functions on Spheres, dissertation, Vanderbilt University, 1995.

49. Fasshauer, G. E. and J. W. Jerome, Multistep approximation algorithms: Improved convergence rates through postconditioning with smoothing kernels, preprint, 1997.

50. Fedorov, V. M., Approximation of functions on a sphere, Moscow Univ. Math. Bull. **45** (1990), 17–23.

51. Fedorov, V. M., Approximation of functions on a sphere. II, Moscow Univ. Math. Bull. **46** (1991), 30–36.

52. Floater, M. S. and A. Iske, Multistep scattered data interpolation using compactly supported radial basis functions, J. Comput. Applied Math. **73** (1996), 65–78.

53. Foley, T. A., Interpolation to scattered data on a spherical domain, in *Algorithms for Approximation II,* J. C. Mason and M. G. Cox (eds), Chapman & Hall, London, 1990, 303–310.

54. Foley, T. A., The map and blend scattered data interpolant on the sphere, Comput. Math. Appl. **24** (1992), 49–60.

55. Foley, T. A. and H. Hagen, Advances in scattered data interpolation, Surveys Math. Indust. **4** (1994), 71–84.

56. Foley, T. A., D. A. Lane, G. M. Nielson, R. Franke, and H. Hagen, Interpolation of scattered data on closed surfaces, Comput. Aided Geom. Design **7** (1990), 303–312.

57. Foley, T. A., D. A. Lane, G. M. Nielson, and R. Ramaraj, Visualizing functions over a sphere, IEEE Comp. Graphics & Appl. **10** (1990), 32–40.

58. Franke, R., Smooth interpolation of scattered data by local thin plate splines, Comp. Maths. Appls. **8** (1982), 273–281.

59. Franke, R., Scattered data interpolation: tests of some methods, Math. Comp. **38** (1982), 181–200.

60. Franke, R., Scattered data interpolation using thin plate splines with tension, Comput. Aided Geom. Design **2** (1985), 87–95.

61. Franke, R., H. Hagen, and G. M. Nielson, Least squares surface approximation to scattered data using multiquadric functions, Advances in Comp. Math. **2** (1994), 81–99.

62. Franke, R., H. Hagen, and G. M. Nielson, Repeated knots in least squares multiquadric functions, in *Geometric Modelling – Dagstuhl 1993*, H. Hagen, G. Farin and H. Noltemeier (eds), Springer Verlag, Berlin, 1995, 177–187.

63. Franke, R. and G. M. Nielson, Scattered data interpolation and applications: A tutorial and survey, in *Geometric Modeling, Methods and Applications*, H. Hagen and D. Roller (eds), Springer Verlag, Berlin, 1991, 131–160.

64. Freeden, W., On spherical spline interpolation and approximation, Math. Meth. Appl. Sci. **3** (1981), 551–575.

65. Freeden, W., Spline methods in geodetic approximation problems, Math. Meth. Appl. Sci. **4** (1982), 382–396.

66. Freeden, W., Spherical spline interpolation – basic theory and computational aspects, J. Comput. Appl. Math. **11** (1984), 367–375.

67. Freeden, W., A spline interpolation method for solving boundary value problems of potential theory from discretely given data, Num. Meth. Part. Diff. Eq. **3** (1987), 375–398.

68. Freeden, W. and J. C. Mason, Uniform piecewise approximation on the sphere, in *Algorithms for Approximation II*, J. C. Mason and M. G. Cox (eds), Chapman & Hall, London, 1990, 320–333.

69. Freeden, W. and R. Reuter, Remainder terms in numerical integration formulas of the sphere, in *Multivariate Approximation Theory II*, W. Schempp and K. Zeller (eds), Birkhäuser, Basel, 1982, 151–170.

70. Freeden, W. and R. Reuter, Exact computation of spherical harmonics, Computing **32** (1984), 365–378.

71. Freeden, W., M. Schreiner, and R. Franke, A survey of spherical spline approximation, Surveys Math. Indust. **7** (1997), 29–85.

72. Gmelig-Meyling, R. H. J. and P. Pfluger, B-spline approximation of a closed surface, IMA J. Numer. Anal. **7** (1987), 73–96.

73. Golitschek, M. von and W. A. Light, Interpolation by polynomials and radial basis functions on spheres, preprint, 1997.

74. Golitschek, M. von and L. L. Schumaker, Data fitting by penalized least squares, in *Algorithms for Approximation II*, J. C. Mason and M. G. Cox (eds), Chapman & Hall, London, 1990, 210–227.

75. Golomb, M. and H. F. Weinberger, Optimal approximation and error bounds, in *On Numerical Approximation*, R. E. Langer (ed), University of Wisconsin Press, Madison, 1959, 117-190.

76. Goodman, T. N. T. and S. L. Lee, Interpolatory and variation-diminishing properties of generalized B-splines, Proc. Roy. Soc. Edinburgh Sect. A **96A** (1984), 249–259.

77. Hardy, R. L., Theory and applications of the multiquadric-biharmonic method, Comput. Math. Appl. **19** (1990), 163–208.

78. Hardy, R. L. and W. M. Göpfert, Least squares prediction of gravity anomalies, geodial undulations, and deflections of the vertical with multiquadric harmonic functions, Geophys. Res. Letters **2** (1975), 423–426.

79. Hardy, R. L. and S. A. Nelson, A multiquadric-biharmonic representation and approximation of disturbing potential, Geophys. Res. Letters **13** (1986), 18–21.

80. Hobson, E. W., *The Theory of Spherical and Ellipsoidal Harmonics,* Chelsea, New York, 1955.

81. Hoschek, J. and G. Seemann, Spherical splines, Rev. Francaise Automat. Informat. Rech. Opér., Anal. Numer. **26** (1992), 1–22.

82. Jetter, K., J. Stöckler, and J. D. Ward, Error estimates for scattered data interpolation on spheres, preprint, 1997.

83. Kamzolov, A. I., Approximation of functions on the sphere S^n (Russian), Serdica **10** (1984), 3–10.

84. Kel'zon, A. A, Interpolation on the sphere (Russian), Izv. Vysš. Učebn. Zaved. Matematika **1975** (11), 41–46.

85. Koch, P. E., T. Lyche, M. Neamtu, and L. L. Schumaker, Control curves and knot insertion for trigonometric splines, Advances in Comp. Math. **3** (1995), 11–20.

86. Kunze, R., F.-E. Wolter, and T. Rausch, Geodesic Voronoi diagrams on parametric surfaces, Rpt. 2, Informatik, Univ. Hannover, 1997.

87. Lawson, C. L., Subroutines for C^1 surface interpolation to data defined over the surface of a sphere, Rpt. 487, JPL, Cal. Tech, 1982.

88. Lawson, C. L., C^1 surface interpolation for scattered data on a sphere, Rocky Mountain J. Math. **14** (1984), 177–202.

89. Lee, S. L., The use of homogeneous coordinates in spline functions and polynomial interpolation, preprint.

90. Lee, S. L. and G. M. Phillips, Interpolation on the simplex by homogeneous polynomials, in *Numerical Mathematics,* J. Wilson, R. P. Argarwal, and Y. M. Chow (eds), ISNM Vol. 86, Birkhäuser Verlag, Basel, 1988, 295–305.

91. Lee, S. L. and G. M. Phillips, Construction of lattices for Lagrange interpolation in projective space, Constr. Approx. **7** (1991), 283–297.

92. Le Méhauté, A. and Y. Lafranche, A knot removal strategy for scattered data in $\mathbb{R}^2$, in *Mathematical Methods in Computer Aided Geometric Design,* T. Lyche and L. Schumaker (eds), Academic Press, New York, 1989, 419–426.

93. Levesley, J., W. Light, D. Ragozin, and X. Sun, Variational theory for interpolation on spheres, preprint, 1997.

94. Li, L. Q. and H. Berens, The Peetre K-moduli and best approximation on the sphere (Chinese), Acta Math. Sinica **38** (1995), 589–599.

95. Light, W. A., Private communication, 1998.

96. Light, W. A. and E. W. Cheney, Interpolation by periodic radial basis functions, J. Math. Anal. Appl. **168** (1992), 111–130.
97. Liu, X.-Y. and L. L. Schumaker, Hybrid Bézier patches on sphere-like surfaces, J. Comput. Appl. Math. **73** (1996), 157–172.
98. Lizorkin, P. I., On the approximation of functions on the sphere σ. On the spaces $B^{\alpha}_{p,q}(\sigma)$. (Russian), Dokl. Akad. Nauk **331** (1993), 555–558.
99. Lizorkin, P. I. and S. M. Nikol'skiĭ, Approximation on the sphere in the metric of continuous functions (Russian), Dokl. Akad. Nauk SSSR **272** (1983), 524–528.
100. Lizorkin, P. I. and S. M. Nikol'skiĭ, A theorem concerning approximation on the sphere, Anal. Math. **9** (1983), 207–221.
101. Lizorkin, P. I. and Kh. P. Rustamov, Nikol'skiĭ-Besov spaces on a sphere that are associated with approximation theory, Proc. Steklov Inst. Math. **204** (1994), 149–172.
102. Lyche, T. and L. L. Schumaker, L–spline wavelets, in *Wavelets: Theory, Algorithms, and Applications,* C. Chui, L. Montefusco, and L. Puccio (eds), Academic Press, New York, 1994, 197–212.
103. Lyche, T. and L. L. Schumaker, A multiresolution tensor spline method for fitting functions on the sphere, preprint, 1997.
104. Lyche, T., L. L. Schumaker, and S. Stanley, Quasi-interpolants based on trigonometric splines, J. Approx. Theory, to appear.
105. MacRobert, T. M., *Spherical Harmonics,* Pergamon Press, Oxford, 1967.
106. Madych, W. R. and S. A. Nelson, Multivariate interpolation: a variational theory, manuscript, 1983.
107. Madych, W. R. and S. A. Nelson, Spherical quadrature and inversion of radon transforms, Proc. Amer. Math. Soc. **95** (1985), 453–457.
108. Madych, W. R. and S. A. Nelson, Multivariate interpolation and conditionally positive definite functions, Approx. Theory Appl. **4** (1988), 77–89.
109. Madych, W. R. and S. A. Nelson, Multivariate interpolation and conditionally positive definite functions, II, Math. Comp. **54** (1990), 211–230.
110. Max, N. L. and E. D. Getzoff, Spherical harmonic surfaces, IEEE Comp. Graph. Appl. **8** (1988), 42–50.
111. McMahon, John R. and Richard Franke, Knot selection for least squares thin plate splines, SIAM J. Sci. Statist. Comput. **13(2)** (1992), 484–498.
112. Menegatto, V. A., Interpolation on spherical domains, Analysis **14** (1994), 415–424.
113. Menegatto, V. A., Strictly positive definite kernels on the Hilbert sphere, Appl. Anal. **55** (1994), 91–101.
114. Menegatto, V. A., Strictly positive definite kernels on the circle, Rocky Mountain J. Math. **25** (1995), 1149–1163.

115. Micchelli, C. A., Interpolation of scattered data: distance matrices and conditionally positive definite functions, Constr. Approx. **2** (1986), 11–22.

116. Möbius, A. F., Ueber eine neue Behandlungsweise der analytischen Sphärik, in *Abhandlungen bei Begründung der Königl. Sächs. Gesellschaft der Wissenschaften*, Jablonowski Gesellschaft, Leipzig, 45-86. (See also *A. F. Möbius, Gesammelte Werke*, F. Klein (ed.), vol. **2**, Leipzig, 1886, 1–54.

117. Montès, P., Local kriging interpolation: Application to scattered data on the sphere, in *Curves and Surfaces,* P.-J. Laurent, A. Le Méhauté, and L. L. Schumaker (eds), Academic Press, New York, 1991, 325–329.

118. Müller, C., *Spherical Harmonics,* Springer Lecture Notes in Mathematics, Vol. 17, 1966.

119. Narcowich, F. J., Generalized Hermite interpolation and positive definite kernels on a Riemannian manifold, J. Math. Anal. Appl. **190** (1995), 165–193.

120. Narcowich, F. J., R. Schaback, and J. D. Ward, Multilevel interpolation and approximation, preprint, 1997.

121. Narcowich, F. J., N. Sivakumar, and J. D. Ward, Stability results for scattered data interpolation on Euclidean spheres, Advances in Comp. Math., to appear.

122. Narcowich, F. J. and J. D. Ward, Nonstationary spherical wavelets for scattered data, in *Approximation Theory VIII, Vol. 2: Wavelets and Multilevel Approximation,* C. Chui and L. Schumaker (eds), World Scientific Publishing, Singapore, 1995, 301–308.

123. Narcowich, F. J. and J. D. Ward, Nonstationary wavelets on the m-sphere for scattered data, Appl. Comput. Harmonic Anal. **3** (1996), 324–336.

124. Neamtu, M., Homogeneous simplex splines, J. Comput. Appl. Math. **73** (1996), 173–189.

125. Newman, D. J. and H. S. Shapiro, Jackson's theorem in higher dimensions, in *Über Approximationstheorie,* P. L. Butzer, and J. Korevaar (ed), Birkhäuser, Basel, 1964, 208–219.

126. Nielson, G. M., The side-vertex method for interpolation in triangles, J. Approx. Theory **25** (1979), 318–336.

127. Nielson, G. M. and R. Ramaraj, Interpolation over a sphere based upon a minimum norm network, Comput. Aided Geom. Design **4** (1987), 41–58.

128. Nikol'skiĭ, S. M. and P. I. Lizorkin, Approximation theory on the sphere, Proc. Steklov Inst. Math. **172** (1985), 295–302.

129. Nikol'skiĭ, S. M. and P. I. Lizorkin, On the theory of approximations on a sphere (Russian), Trudy Mat. Inst. Steklov **172** (1985), 272–279, 355.

130. Nikol'skiĭ, S. M. and P. I. Lizorkin, Function spaces on a sphere that are connected with approximation theory (Russian), Mat. Zametki **41** (1987), 509–516, 620.

131. Nikol'skiĭ, S. M. and P. I. Lizorkin, Approximation of functions on the sphere, Math. USSR-Izv. **30** (1988), 599–614.

132. Nikolskiĭ, S. M. and P. I. Lizorkin, Approximation on the sphere – a survey, Banach Center Publ. **22** (1989), 281–292.

133. Opitz, K. and H. Pottmann, Computing shortest paths on polyhedra: applications in geometric modeling and scientific visualization, Int. J. Comput. Geom. Appl. **4** (1994), 165–178.

134. Perrin, F., J. Pernier, O. Bertrand, and J. F. Echallier, Spherical splines for scalp potential and current density mapping, Electroencephalography and clinical neurophsiology **72** (1989), 184–187.

135. Pevnyĭ, A. B., Spherical splines and interpolation on a sphere, Comput. Math. Math. Phys. **35** (1995), 109–112.

136. Pfeifle, R. and H.-P. Seidel, Spherical triangular B-splines with application to data fitting, Comp. Graphics Forum (Proc. Eurographics '95) **14** (1995), C89–C96.

137. Pottmann, H., Interpolation on surfaces using minimum norm networks, Comput. Aided Geom. Design **9** (1992), 51–67.

138. Pottmann, H. and M. Eck, Modified multiquadric methods for scattered data interpolation over a sphere, Comput. Aided Geom. Design **7** (1990), 313–321.

139. Pottmann, H., H. Hagen, and A. Divivier, Visualizing functions on a surface, J. Visual. Comput. Anim. **2** (1991), 52–58.

140. Pottmann, H. and K. Opitz, Curvature analysis and visualization for functions defined on Euclidean spaces or surfaces, Comput. Aided Geom. Design **11** (1994), 655–674.

141. Powell, M. J. D., The theory of radial basis functions in 1990, in *Advances in Numerical Analysis II: Wavelets, Subdivision, and Radial Basis Functions,* W. Light (ed), Oxford University Press, Oxford, 1992, 105–210.

142. Quak, E. and L. L. Schumaker, Cubic spline fitting using data dependent triangulations, Comput. Aided Geom. Design **7** (1990), 293–301.

143. Quak, E. and L. L. Schumaker, Least squares fitting by linear splines on data dependent triangulations, in *Curves and Surfaces,* P.-J. Laurent, A. Le Méhauté, and L. L. Schumaker (eds), Academic Press, New York, 1991, 387–390.

144. Ragozin, D. L., Constructive polynomial approximation on spheres and projective spaces, Trans. Amer. Math. Soc. **162** (1971), 157–170.

145. Ragozin, D. L., Uniform convergence of spherical harmonic expansions, Math. Ann. **195** (1972), 87–94.

146. Ragozin, D. L. and J. Levesley, Zonal kernels, approximations and positive definiteness on spheres and compact homogeneous spaces, in *Curves and Surfaces in Geometric Design,* A. Le Méhauté, C. Rabut, and L. L. Schumaker (eds), Vanderbilt University Press, Nashville TN, 1997, 371–378.

147. Ramaraj, R., Interpolation and display of scattered data over a sphere, M.S. Thesis, Arizona State University, 1986.

148. Reimer, M., Best approximations to polynomials in the mean and norms of coefficient-functionals, in *Multivariate Approximation Theory, ISNM 51*, W. Schempp and K. Zeller (eds), Birkhäuser Verlag, Basel, 1979, 289–304.

149. Reimer, M., Interpolation on the sphere and bounds for the Lagrangian square sum, Resultate der Math. **11** (1987), 144–164.

150. Reimer, M., Interpolation mit sphärischen harmonischen Funktionen, in *Numerical Methods in Approximation Theory Vol. 8, ISNM 81*, L. Collatz, G. Meinardus and G. Nürnberger (eds), Birkhäuser, Basel, 1987, 184–187.

151. Reimer, M. and B. Sündermann, A Remez-type algorithm for the calculation of extremal fundamental systems for polynomial spaces over the sphere, Computing **37** (1986), 43–58.

152. Reimer, M. and B. Sündermann, Günstige Knoten für die Interpolation mit homogenen harmonischen Polynomen, Resultate der Math. **11** (1987), 254–266.

153. Renka, R. J., Interpolation of data on the surface of a sphere, ACM Trans. Math. Software **10** (1984), 417–436.

154. Renka, R. J., Algorithm 623: Interpolation on the surface of a sphere, ACM Trans. Math. Software **10** (1984), 437–439.

155. Ron, A. and X. Sun, Strictly positive definite functions on spheres in Euclidean spaces, Math. Comp. **65** (1996), 1513–1530.

156. Ruhlmann, G. M. and J. C. McKeeman, Local search: a new hidden line elimination algorithm to display spherical coordinate equations, Comput. & Graphics **15** (1991), 535–544.

157. Rustamov, Kh. P., Direct theorems on best L_p-approximation on a sphere S^{n-1} (Russian), Izv. Akad. Nauk Azerbaĭdzhan. SSR Ser. Fiz.-Tekhn. Mat. Nauk **5(5)** (1984), 3–8.

158. Rustamov, Kh. P., Inverse theorems of best L_p-approximation on the sphere S^{n-1} (Russian), Izv. Akad. Nauk Azerbaĭdzhan. SSR Ser. Fiz.-Tekhn. Mat. Nauk **5(6)** (1984), 6–12.

159. Rustamov, Kh. P., On direct and inverse theorems for the best L_p-approximation on a sphere (Russian), Dokl. Akad. Nauk SSSR **294** (1987), 788–791.

160. Rustamov, Kh. P., On the best approximation of functions on the sphere in the metric of $L_p(S^n)$, $1 < p < \infty$, Anal. Math. **17** (1991), 333–348.

161. Rustamov, Kh. P., Approximation of functions on a sphere (Russian), Dokl. Akad. Nauk SSSR **320** (1991), 1319–1325.

162. Rustamov, Kh. P., On the approximation of functions on a sphere (Russian), Izv. Ross. Akad. Nauk Ser. Mat. **57** (1993), 127–148.

163. Sánchez-Reyes, J., Single valued curves in polar coordinates, Comput. Aided Geom. Design **22** (1990), 19–26.

164. Sánchez-Reyes, J., Single-valued surfaces in spherical coordinates, Comput. Aided Geom. Design **11** (1994), 491–517.

165. Sansone, G., *Orthogonal Functions,* Interscience, New York (reprinted 1991 by Dover), 1959.

166. Schaback, R., Creating surfaces from scattered data using radial basis functions, in *Mathematical Methods for Curves and Surfaces,* Morten Dæhlen, Tom Lyche, and Larry L. Schumaker (eds), Vanderbilt University Press, Nashville & London, 1995, 477–496.

167. Schaback, R., Error estimates and condition numbers for radial basis function interpolation, Advances in Comp. Math. **3** (1995), 251–264.

168. Schmitt, E., Wavelets and multiresolution analysis on sphere-like surfaces, SPIE Wavelet Applications in Signal and Image Processing, 1995, 92–101.

169. Schneider, R., Multiresolution analysis basierend auf beschränkten Projektionen mit Anwendungen in der Computergraphik, Diplomarbeit, Univ. Erlangen, 1997.

170. Schoenberg, I. J., Positive definite functions on spheres, Duke Math. J. **9** (1942), 96–108.

171. Schreiner, M., On a new condition for strictly positive definite functions on spheres, Proc. Amer. Math. Soc. **125** (1997), 531–539.

172. Schreiner, M., Locally supported kernels for spherical spline interpolation, J. Approx. Theory **89** (1997), 172–194.

173. Schroeder, P. and W. Sweldens, Spherical wavelets: Efficiently representing functions on the sphere, Computer Graphics Proceedings (SIGGRAPH 95), 161–172.

174. Schumaker, L. L., Fitting surfaces to scattered data, in *Approximation Theory, II,* G. G. Lorentz, C. K. Chui, and L. L. Schumaker (eds), Academic Press, New York, 1976, 203–268.

175. Schumaker, L. L., *Spline Functions: Basic Theory,* Wiley, New York, 1981.

176. Schumaker, L. L., Computing optimal triangulations using simulated annealing, Comput. Aided Geom. Design **10** (1993), 329–345.

177. Schumaker, L. L and C. Traas, Fitting scattered data on spherelike surfaces using tensor products of trigonometric and polynomial splines, Numer. Math. **60** (1991), 133–144.

178. Shalaev, V. V., Sharp estimates for the approximation of functions, continuous on a sphere, by linear operators of convolution type, Ukrainian Math. J. **43** (1991), 523–525.

179. Sibson, R., A brief description of natural neighbor interpolation, in *Interpreting Multivariate Data,* D. V. Barnett (ed), Wiley, New York, 1981, 21–36.

180. Sivakumar, N. and J. D. Ward, On the least squares fit by radial functions to multidimensional scattered data, Numer. Math. **65** (1993), 219–243.
181. Sobolev, S. L., Cubature formulas on the sphere invariant under finite groups of rotations, Soviet Math. **3** (1962), 1307–1310.
182. Suffern, K. G., Perspective views of polar coordinate functions, Computer-Aided Design **24** (1992), 307–315.
183. Sun, X., The fundamentality of translates of a continuous function on spheres, Numerical Algorithms **8** (1994), 131–134.
184. Sun, X. and E. W. Cheney, Fundamental sets of continuous functions on spheres, Constr. Approx. **13** (1997), 245–250.
185. Svensson, S. L., Finite elements on the sphere, J. Approx. Theory **40** (1984), 246–260.
186. Swarztrauber, P. N., On the approximation of discrete scalar and vector functions on the sphere, SIAM J. Numer. Anal. **7** (1970), 934–949.
187. Swarztrauber, P. N., On the spectral approximation of discrete scalar and vector functions on the sphere, SIAM J. Numer. Anal. **16** (1979), 934–949.
188. Swarztrauber, P. N., The approximation of vector functions and their derivatives on the sphere, SIAM J. Numer. Anal. **18** (1981), 191–210.
189. Szegő, G., *Orthogonal Polynomials,* Amer. Math. Soc. Coll. Publ. Vol. XXIII, Providence, 1959.
190. Terekhin, A. P., Approximation in L_p by spherical polynomials, and classes of differentiable functions on a sphere (Russian), Trudy Mat. Inst. Steklov **187** (1989), 216–226.
191. Tissier, G., Approximation d'une fonction définie sur une sphère dans une base de fonctions sphèriques de Legendre (French), Rev. Francaise Automat. Informat. Rech. Opér., Anal. Numer. **5** (1971), 9–28.
192. Tissier, G., Interpolation à plusieurs variables sur la sphère (French), Numer. Math. **19** (1972), 136–145.
193. Traas, C. R., Smooth approximation of data on the sphere with splines, Computing **38** (1987), 177–184.
194. Ugulava, D. K., Approximation of functions on an m-dimensional sphere by the Cesàro means of the Fourier-Laplace series (Russian), Mat. Zametki **9** (1971), 343–353.
195. Ugulava, D. K., On the theory of the approximation of functions on a multidimensional sphere (Russian), Sakharth. SSR Mecn. Akad. Gamothvl. Centr. Šrom. **11** (1972), 139–156.
196. Ugulava, D. K., Approximation of continuous functions on a multidimensional sphere (Russian), Sakharth. SSR Mecn. Akad. Gamothvl. Centr. Šrom. **12** (1973), 36–52.
197. Wahba, G., Convergence rate of thin plate smoothing splines when the data are noisy, in *Smoothing Techniques for Curve Estimation,* T. G. M. Rosenblatt (ed), Springer Verlag, Berlin, 1979, 232–246.

198. Wahba, G., Spline Interpolation and smoothing on the sphere, SIAM J. Sci. Statist. Comput. **2** (1981), 5–16.

199. Wahba, G., Erratum: Spline interpolation and smoothing on the sphere, SIAM J. Sci. Statist. Comput. **3** (1982), 385–386.

200. Wahba, G., Vector splines on the sphere, with application to the estimation of vorticity and divergence from discrete, noisy data, in *Multivariate Approximation Theory II,* W. Schempp and K. Zeller (eds), Birkhäuser, Basel, 1982, 407–429.

201. Wahba, G., Surface fitting with scattered noisy data on Euclidean d–space and on the sphere, Rocky Mountain J. Math. **14** (1984), 281–299.

202. Wang, K. Y. and P. Zhang, Strong uniform approximation on sphere, Beijing Shifan Daxue Xuebao **30** (1994), 321–328.

203. Wehrens, M., Best approximation on the unit sphere in $\mathbf{R}^k$, in *Functional Analysis and Approximation*, Birkhäuser, Basel, 1981, 233–245.

204. Wendland, H., Piecewise polynomial, positive definite and compactly supported radial functions of minimal degree, Advances in Comp. Math. **4** (1995), 389–396.

205. Wu, Z., Multivariate compactly supported positive definite radial functions, Advances in Comp. Math. **4** (1995), 283–292.

206. Xu, Y. and E. W. Cheney, Strictly positive definite functions on spheres, Proc. Amer. Math. Soc. **116** (1992), 977–981.

Acknowledgments. We thank Will Light, Fran Narcowich, Mike Neamtu, and Joe Ward for their helpful comments. The second author was partially supported by NSF Grant 9500643.

Greg Fasshauer
CSAM Department
Illinois Institute of Technology
Chicago, IL 60616, USA
`fass@amadeus.csam.iit.edu`

Larry L. Schumaker
Dept. of Mathematics
Vanderbilt University
Nashville, TN 37240, USA
`s@mars.cas.vanderbilt.edu`

On Spline Interpolation of Space Data

Yu Yu Feng and Jernej Kozak

Abstract. In this paper the interpolation by G^2 continuous Bézier spline curves $\mathbf{B}_n$ in $\mathbb{R}^d$ is outlined. Each segment of the spline curve interpolates r interior and two boundary points. The general approach is followed in detail for the case $d = 3, n = 3, r = 1$. For the single component case, the optimal approximation order is proved, and asymptotic existence established. The interpolation scheme is demonstrated by practical experiments.

§1. Introduction

Let $\mathbb{R}^d$ denote Euclidean space, with the dot product $\cdot$, and the implied norm $\|\cdot\| := \sqrt{\cdot}$. Let $\mathbf{T}_0, \mathbf{T}_1, \ldots, \mathbf{T}_N \in \mathbb{R}^d$ be a sequence of given points that is to be interpolated by a spline curve of degree n. Recall the fact deduced from [6] for the particular case $n = 3$: a cubic planar curve can interpolate six, but a cubic space curve only five points. Thus, in order to make the interpolation problem correct, let us choose d in advance, and let us assume that the points $\mathbf{T}_i$ lie on a smooth regular parametric curve $\mathbf{f} : I = [\alpha, \beta] \subset \mathbb{R} \to \mathbb{R}^d$, with $\mathbf{f}(\xi_i) = \mathbf{T}_i$ for some $\alpha = \xi_0 < \xi_1 < \cdots < \xi_N = \beta$, where $\mathbf{f}$ is a convex "true" $\mathbb{R}^d$ curve, i.e., its $d - 1$ curvatures nowhere vanish. This implies that if the points will be sampled close enough they will be "true" $\mathbb{R}^d$ data. In order to make the spline interpolant parametrisation independent, we shall assume that the polynomial pieces will be joined together by geometric continuity conditions, and we shall restrict ourselves to G^2 spline curves. Let us consider one spline segment. The reparametrisation $t \mapsto at + b$ preserves the degree of the polynomial curve, and one is free to choose two parameter values at which two particular data points will be interpolated. However, the parameter values at which the other data points will be interpolated will be unknown. So it is natural to prescribe that the spline curve interpolates two boundary, and $r \geq 0$ interior points per each spline section. This implies $N = m(r+1)$, where m denotes the number of spline segments. At the boundary, two additional interpolating conditions have to be added. Perhaps it is best to require that

Mathematical Methods for Curves and Surfaces II
Morten Dæhlen, Tom Lyche, and Larry L. Schumaker (eds.), pp. 167–174.

ISBN 0-8265-1315-8.

also tangent directions $\mathbf{d}_0, \mathbf{d}_N$ of $\mathbf{f}$ at α, β are interpolated too. We shall express the interpolating spline curve in the Bézier form. Let $\mathbf{B}_n$ denote the Bézier spline curve with breakpoints $\zeta_\ell := \xi_{\ell(r+1)}, \ell = 0, 1, \ldots, m$, and

$$\mathbf{B}_n^\ell\left(\frac{t-\zeta_{\ell-1}}{\Delta\zeta_{\ell-1}}\right) = \mathbf{B}_n(t), \quad t \in [\zeta_{\ell-1}, \zeta_\ell], \qquad \ell = 1, 2, \ldots, m$$

its polynomial components. Then $\mathbf{B}_n^\ell(t) = \sum_{i=0}^n \mathbf{b}_{n(\ell-1)+i} B_i^n(t),\ t \in [0,1]$, with Bernstein polynomials given as $B_i^n(t) := \binom{n}{i} t^i (1-t)^{n-i}$. The segment $\mathbf{B}_n^\ell$ should interpolate the data points $\mathbf{T}_{(r+1)(\ell-1)}, \mathbf{T}_{(r+1)(\ell-1)+1}, \ldots, \mathbf{T}_{(r+1)\ell}$, i.e., for some $t_{\ell,0} := 0 < t_{\ell,1} < t_{\ell,2} < \cdots < t_{\ell,r} < t_{\ell,r+1} := 1$ one has

$$\mathbf{B}_n^\ell(t_{\ell,i}) = \mathbf{T}_{(r+1)(\ell-1)+i}, \qquad i = 1, \ldots, r, \tag{1}$$

and $\mathbf{b}_{n(\ell-1)} = \mathbf{T}_{(r+1)(\ell-1)}$, $\mathbf{b}_{n\ell} = \mathbf{T}_{(r+1)\ell}$ at the boundary. The geometric continuity will be formulated in the form that does not require additional constants. Let

$$\wedge : \mathbb{R}^d \times \mathbb{R}^d \to \mathbb{R}^{\binom{d}{2}} : \left((x_i)_{i=1}^d, (y_i)_{i=1}^d\right) \mapsto x \wedge y := (x_i y_j - x_j y_i)_{i<j} \tag{2}$$

denote the particular wedge product. A continuous parametric curve $\mathbf{g} = \mathbf{g}(t) \in \mathbb{R}^d$ will be G^1 continuous (up to sign) if

$$\dot{\mathbf{g}}_- \wedge \dot{\mathbf{g}}_+ = 0, \tag{3}$$

and also G^2 continuous if additionally

$$\frac{1}{\|\dot{\mathbf{g}}_-\|^3} \dot{\mathbf{g}}_- \wedge \ddot{\mathbf{g}}_- = \frac{1}{\|\dot{\mathbf{g}}_+\|^3} \dot{\mathbf{g}}_+ \wedge \ddot{\mathbf{g}}_+. \tag{4}$$

Here, $\cdot_\pm$ denotes the left and right limit at the given parameter value. Note that there are only $d-1$ independent equations in (3) and (4). However, it is not possible to determine in advance which $d-1$ equations imply the other $\binom{d-1}{2}$ ones. For example, if $x_i = 0, x_j = 0$, the equation $x_i y_j - x_j y_i = 0$ doesn't say anything about y_j, y_i.

For the particular curve $\mathbf{g} = \mathbf{B}_n$ one has to apply (3) and (4) only at the breakpoints ζ_ℓ, i.e.,

$$\begin{aligned} \mathbf{d}_0 \wedge \Delta\mathbf{b}_0 &= 0, \\ \Delta\mathbf{b}_{n\ell-1} \wedge \Delta\mathbf{b}_{n\ell} &= 0, \qquad \ell = 1, 2, \ldots, m-1, \\ \Delta\mathbf{b}_{nm-1} \wedge \mathbf{d}_N &= 0, \end{aligned} \tag{5}$$

and

$$\frac{\Delta\mathbf{b}_{n\ell-1} \wedge \Delta\mathbf{b}_{n\ell-2}}{\|\Delta\mathbf{b}_{n\ell-1}\|^3} + \frac{\Delta\mathbf{b}_{n\ell} \wedge \Delta\mathbf{b}_{n\ell+1}}{\|\Delta\mathbf{b}_{n\ell}\|^3} = 0, \qquad \ell = 1, 2, \ldots, m-1. \tag{6}$$

Another look at the conditions (1), (5), and (6) reveals that the total number of independent equations would be equal to the number of unknowns if

$$dn - (d-1)r = 3d - 2. \tag{7}$$

Since the numbers d and $d-1$ are relatively prime, the diophantine equation (7) has, for fixed d, always infinite number of positive solutions $r, n \in \mathbb{N}$. Tab. 1 shows some of them. Note that the Hermite G^2 interpolation can be viewed as a limit case for even r.

d	Pairs r,n				
2	0,2	2,3	4,4	6,5	8,6
3	1,3	4,5	7,7	10,9	13,11
4	2,4	6,7	10,10	14,13	18,16

Tab. 1. Admissible pairs (r, n)

The case $d = 2, r = 0, n = 2$ was studied in [8], its modified form in [4], and its rational counterpart in [9]. The limit case of $d = 2, r = 2, n = 3$, i.e., the Hermite interpolation, goes back to [2], and the general case can be found in [5]. The case $d = 3, r = 1, n = 3$ will be studied in this paper.

Let us restrict to the condition $r \le n-1$, i.e., to the lower degree curves. But then one can use (1) to express r interior control points on each curve segment as a function of $(t_{\ell,i})_{i=1}^{r}$ as well as $n-1-r$ remaining free control points, and insert them in (5) and (6). The nonlinear system to solve is now simplified to (5) and (6), where on the ℓ-th segment the vectors involved, $\Delta \mathbf{b}_{n(\ell-1)}$, $\Delta \mathbf{b}_{n\ell-1}$, are functions of $r + (n-1-r)d = 2(d-1)$ unknowns.

There are three main questions that have to be worked out for each interpolation scheme: the existence (and the possible uniqueness) of the solution, the approximation order, and the numerical construction. A few remarks can be made in general for the asymptotic analysis. It is based upon local expansion of the curve $\mathbf{f}$, and the asymptotic behaviour of the solution of the equations (5) and (6). Perhaps the easiest way to produce the local expansion of $\mathbf{f}$ is to use Frenet frame as the local coordinate system, and Frenet-Serret formulas to obtain the expansion. Let $\mathbf{f}$ be parametrised by the arclength s, and let $(\mathbf{e}_i(s))_{i=1}^{d}$ denote the Frenet frame. Then the Frenet-Serret formulas read as

$$\begin{aligned}
\mathbf{e}_1'(s) &= \kappa_1(s)\mathbf{e}_2(s), \\
\mathbf{e}_i'(s) &= -\kappa_{i-1}(s)\mathbf{e}_{i-1}(s) + \kappa_i(s)\mathbf{e}_{i+1}(s), \qquad i = 2, 3, \ldots, d-1, \\
\mathbf{e}_d'(s) &= -\kappa_{d-1}(s)\mathbf{e}_{d-1}(s),
\end{aligned} \tag{8}$$

and the fact that $\mathbf{f}$ is a "true" convex $\mathbb{R}^d$ curve is introduced naturally by expanding the curvatures κ_i, say at 0, as

$$\kappa_i(s) = \kappa_{i0} + \frac{1}{1!}\kappa_{i1}s + \frac{1}{2!}\kappa_{i1}s^2 + \cdots, \quad |\kappa_{i0}| \ge \text{const} > 0. \tag{9}$$

With the help of (8) one computes

$$\begin{aligned} \mathbf{f}''(s) &= \mathbf{e}_1'(s) = \kappa_1(s)\mathbf{e}_2(s), \\ \mathbf{f}'''(s) &= (\kappa_1(s)\mathbf{e}_2(s))' = -\kappa_1^2(s)\mathbf{e}_1(s) + \kappa_1'(s)\mathbf{e}_2(s) + \kappa_1(s)\kappa_2(s)\mathbf{e}_3(s), \\ \cdots &= \quad \cdots\cdots\cdots \end{aligned} \tag{10}$$

and the expansions (9) inserted in (10) are used to express $\mathbf{f}^{(j)}(0)$ in the required expansion of $\mathbf{f}$,

$$\begin{aligned} \mathbf{f}(s) &= \mathbf{f}(0) + \mathbf{f}'(0)s + \tfrac{1}{2!}\mathbf{f}''(0)s^2 + \tfrac{1}{3!}\mathbf{f}'''(0)s^3 + \cdots \\ &= \mathbf{f}(0) + (s - \tfrac{1}{6}\kappa_{1,0}^2 s^3 + \cdots)\mathbf{e}_1(0) \\ &\quad + (\tfrac{1}{2}\kappa_{1,0}s^2 + \tfrac{1}{6}\kappa_{1,1}s^3 + \cdots)\mathbf{e}_2(0) + (\tfrac{1}{6}\kappa_{1,0}\kappa_{2,0}s^3 + \cdots)\mathbf{e}_3(0) + \cdots \end{aligned} \tag{11}$$

Note that the compound matrix $\overset{2}{\underset{i=1}{\wedge}} A$ is nonsingular iff the matrix A is. So different nonsigular matrices A could be used to simplify the expanded equations (5) and (6) by using the fact

$$\mathbf{x} \wedge \mathbf{y} = 0 \iff \overset{2}{\underset{i=1}{\wedge}} A(\mathbf{x} \wedge \mathbf{y}) = A\mathbf{x} \wedge A\mathbf{y} = 0, \tag{12}$$

and similarly for the curvature continuity equation. In particular, A may be a rotation that brings the Frenet frame at the particular breakpoint ζ_ℓ to $\left((\delta_{ij})_{j=1}^d\right)_{i=1}^d$.

In order to tackle the asymptotic approximation order for a smooth curve $\mathbf{f}$, it is usually enough to consider the single component case, i.e., $N = r + 1$, and follow [7] to estimate the parametric approximation order. This requires a regular reparametrisation of $\mathbf{B}_n$ and $\mathbf{f}$ with the same parameter that satisfies certain additional requirements. The number of the data values, $r + 4 = n+1+(n-1)/(d-1)$ is the expected best approximation order. A particularly simple local reparametrisation can be traced to [2], and works in the more general setup too. Let $\xi_0 = 0$, $\xi_N = h$, and $x_i := \mathbf{f}\cdot\mathbf{e}_i(0)$. The expansion (11) reveals that $\mathbf{f}'(s)\cdot\mathbf{e}_1(0) = 1 + \mathcal{O}(s^2)$. So x_1 is invertible on $[x_1(0), x_1(h)]$ for sufficiently small h. But then the curve $\mathbf{f}$ may be reparametrised by $u = x_1(s)$, with $x_i(s) = x_i(x_1^{-1}(u))$. Note that $\frac{d}{du}x_i(s) = \dfrac{x_i'(s)}{x_1'(s)}$, so the derivative with respect to u is independent of the length of $\mathbf{f}(s)$. But $\mathbf{B}_n$ interpolates $\mathbf{f}$ at boundary points. So $\mathbf{B}_n(0)\cdot\mathbf{e}_1(0) = x_1(0)$, $\mathbf{B}_n(1)\cdot\mathbf{e}_1(0) = x_1(h)$. Thus if $\mathbf{B}_n\cdot\mathbf{e}_1(0)$ is also invertible, $\mathbf{B}_n$ and $\mathbf{f}$ can be reparametrised by the same parameter u. This implies that $\mathbf{B}_n$ and $\mathbf{f}$ agree $(r+4)$-fold. So for a smooth $\mathbf{f}$ the approximation order $r+4$ would be achieved if $\dfrac{d^{r+4}}{du^{r+4}}\mathbf{B}_n$ could be bounded independently of h. The chain rule gives $\dfrac{d^\ell}{du^\ell}x_i(s)$ as the sum of terms

$$x_i^{(k)} \prod_\nu x_1^{(j_\nu)}/(x_1')^{k+\sum j_\nu}, \qquad k + \sum_\nu j_\nu = \ell.$$

Since the derivatives of the reparametrised $\mathbf{B}_n$ are computed in the same manner it is enough to establish

$$\dot{\mathbf{B}}_n \cdot \mathbf{e}_1(0) = \text{const } h + \mathcal{O}(h^2), \text{ const } \neq 0, \quad \mathbf{B}_n^{(\ell)} = \mathcal{O}(h^\ell), \quad \ell = 2, 3, \ldots, n. \tag{13}$$

The general approach outlined here is followed in detail for the case $d = 3, n = 3, r = 1$, i.e., cubic spline interpolation of the space data, and practical examples given.

§2. Cubic G^2 Interpolation in $\mathbb{R}^3$

Let us consider now a particular case, i.e., the cubic G^2 interpolation of the space data. In this case one has $d = 3$, and $r = 1 \leq n = 3$. So one control point can be expressed by the other $2(d-1) = 4$ free parameters. This gives $\mathbf{B}^\ell := \mathbf{B}_3^\ell$ that interpolates $T_{2\ell-1}$ at $t_\ell := t_{\ell,1}$ as

$$\begin{aligned} \mathbf{B}^\ell(t) &= \frac{(1-t)\left((1-t)^2 t_\ell - t(1-t_\ell)^2\right)}{t_\ell} \mathbf{T}_{2(\ell-1)} + \frac{(1-t)\,t}{(1-t_\ell)\,t_\ell} \mathbf{T}_{2\ell-1} \\ &+ \frac{t\left(t^2(1-t_\ell) - (1-t)\,t_\ell^2\right)}{1-t_\ell} \mathbf{T}_{2\ell} + 3(1-t)\,t\,(t-t_\ell)\,\Delta\mathbf{b}_{3\ell-2}. \end{aligned} \tag{14}$$

From here

$$\begin{aligned} \Delta\mathbf{b}_{3\ell-3} &= g_1(t_\ell)\Delta\mathbf{T}_{2\ell-2} - g_2(t_\ell)\Delta\mathbf{T}_{2\ell-1} - t_\ell\Delta\mathbf{b}_{3\ell-2}, \\ \Delta\mathbf{b}_{3\ell-1} &= -g_2(1-t_\ell)\Delta\mathbf{T}_{2\ell-2} + g_1(1-t_\ell)\Delta\mathbf{T}_{2\ell-1} - (1-t_\ell)\Delta\mathbf{b}_{3\ell-2}, \end{aligned} \tag{15}$$

with functions g_1 and g_2 given as

$$g_1(t) := \frac{1}{3}\Big(1 + \frac{1}{t} + t\Big), \quad g_2(t) := \frac{1}{3}\frac{t^2}{1-t}. \tag{16}$$

By inserting (15) in (5) and (6) one obtains a system of $6m$ nonlinear equations $\mathbf{F}(\mathbf{t}, \Delta\mathbf{b}) = 0$ for the unknowns $\mathbf{t} := (t_\ell)_{\ell=1}^m$, $\Delta\mathbf{b} := (\Delta\mathbf{b}_{3\ell-2})_{\ell=1}^m$. Only $4m$ equations are independent.

Let us consider the case $m = 1, \ell = 1$. The elimination of $\Delta\mathbf{b}_1$ from (15) gives

$$\frac{1}{t_1}\Delta\mathbf{b}_0 - \frac{1}{1-t_1}\Delta\mathbf{b}_2 = g(t_1)\Delta\mathbf{T}_0 - g(1-t_1)\Delta\mathbf{T}_1, \tag{17}$$

with

$$g(t) := \frac{1}{3}\Big(\frac{2}{t} + \frac{1}{t^2}\Big).$$

Further, let us multiply (17) by $\mathbf{d}_0\cdot$ and $\wedge\mathbf{d}_2$. By (5), the left hand side vanishes, what reveals the equation

$$g(t_1)\det(\mathbf{d}_0, \Delta\mathbf{T}_0, \mathbf{d}_2) = g(1-t_1)\det(\mathbf{d}_0, \Delta\mathbf{T}_1, \mathbf{d}_2) \tag{18}$$

for the unknown t_1. The function g is strictly decreasing on $(0, \infty)$, and unbounded at 0. Thus if $\det(\mathbf{d}_0, \Delta\mathbf{T}_0, \mathbf{d}_2)\det(\mathbf{d}_0, \Delta\mathbf{T}_1, \mathbf{d}_2) > 0$ the equation (18) has a unique solution $t_1 \in (0,1)$. Further, (17) and (5) then yield

$$\Delta\mathbf{b}_0 = t_1(1,0)(\mathbf{d}_0\mathbf{d}_2)^+(g(t_1)\Delta\mathbf{T}_0 - g(1-t_1)\Delta\mathbf{T}_1)\mathbf{d}_0, \tag{19}$$

and consequently $\Delta\mathbf{b}_1$ if $\mathbf{d}_0 \wedge \mathbf{d}_2 \neq 0$. This proves the following observation.

Theorem 1. *The cubic parametric curve that interpolates the data* $\mathbf{d}_0$, $\mathbf{T}_0$, $\mathbf{T}_1$, $\mathbf{T}_2$, $\mathbf{d}_2$ *is determined uniquely iff* $\det(\mathbf{d}_0, \Delta\mathbf{T}_0, \mathbf{d}_2)\det(\mathbf{d}_0, \Delta\mathbf{T}_1, \mathbf{d}_2) > 0$.

In order to study the asymptotic behaviour of the unknowns let us recall (8), (10), and (11) with the curvature $\kappa := \kappa_1$, and the torsion $\tau := \kappa_2$. By (12) we may assume that $\mathbf{T}_0$ is at origin, and that the Frenet frame at $\mathbf{T}_0$ is the identity matrix. Let $\mathbf{f}$ be parametrised by the arclength, with

$$\mathbf{f}(0) = \mathbf{T}_0,\ \mathbf{f}(\eta_1 h) = \mathbf{T}_1,\ 0 < \eta_1 < 1,\ \mathbf{f}(h) = \mathbf{T}_2.$$

Then

$$\mathbf{f}(s) = \begin{pmatrix} s - \frac{1}{6}\kappa_0^2 s^3 - \frac{1}{8}\kappa_0\kappa_1 s^4 \\ \frac{1}{2}\kappa_0 s^2 + \frac{1}{6}\kappa_1 s^3 - \frac{1}{24}(\kappa_0^3 + \kappa_0\tau_0^2 - \kappa_2)s^4 \\ \frac{1}{6}\kappa_0\tau_0 s^3 + \frac{1}{24}(2\tau_0\kappa_1 + \kappa_0\tau_1)s^4 \end{pmatrix} + \mathcal{O}(s^5). \tag{20}$$

This produces the expansions

$$\begin{aligned}\frac{1}{h^4}\det(\mathbf{d}_0, \Delta\mathbf{T}_0, \mathbf{d}_2) \;=\; & \frac{1}{12}(3 - 2\eta_1)\eta_1^2\kappa_0^2\tau_0 \\ & + \frac{1}{24}\eta_1^2(2 - \eta_1^2)\kappa_0(2\tau_0\kappa_1 + \tau_1\kappa_0)h + \mathcal{O}(h^2),\end{aligned}$$

and

$$\begin{aligned}\frac{1}{h^4}\det(\mathbf{d}_0, \Delta\mathbf{T}_1, \mathbf{d}_2) \;=\; & \frac{1}{12}(1 + 2\eta_1)(1 - \eta_1)^2\kappa_0^2\tau_0 \\ & + \frac{1}{24}(1 - \eta_1^2)^2\kappa_0(2\tau_0\kappa_1 + \tau_1\kappa_0)h + \mathcal{O}(h^2).\end{aligned}$$

From here and (18), the asymptotic expansion of t_1 reads

$$t_1 = \eta_1 - \frac{1}{12}(1 - \eta_1)\eta_1\left(2\frac{\kappa_1}{\kappa_0} + \frac{\tau_1}{\tau_0}\right)h + \mathcal{O}(h^2), \tag{21}$$

and further from (19), and the expansions (20), (21)

$$\Delta\mathbf{b}_1 = \begin{pmatrix} \frac{1}{3}h \\ \frac{1}{6}\kappa_0 h^2 \\ 0 \end{pmatrix} + \mathcal{O}(h^3),\ \mathbf{b}_1 = \begin{pmatrix} \frac{1}{3}h + \frac{1}{36}(2\frac{\kappa_1}{\kappa_0} + \frac{\tau_1}{\tau_0})h^2 \\ 0 \\ 0 \end{pmatrix} + \mathcal{O}(h^3). \tag{22}$$

From here and (20)

$$\Delta\mathbf{b}_i = \begin{pmatrix} \frac{1}{3}h \\ 0 \\ 0 \end{pmatrix} + \mathcal{O}(h^2), \qquad i = 0, 1, 2,$$

what confirms that the second and the third component of the Bézier curve $\mathbf{B}^1$ can be reparametrised by the first one for h small enough, and

$$\Delta^2\mathbf{b}_i = \begin{pmatrix} -\frac{1}{36}(2\frac{\kappa_1}{\kappa_0} + \frac{\tau_1}{\tau_0})h^2 \\ \frac{1}{6}\kappa_0 h^2 \\ 0 \end{pmatrix} + \mathcal{O}(h^3), \quad i = 0, 1, \quad \Delta^3\mathbf{b}_0 = \mathcal{O}(h^3).$$

This verifies (13) in this particular case. In view of more general discussion in the introduction we have thus proved the following.

Theorem 2. *The parametric approximation order of the interpolation sheme considered is optimal, i.e.,* $r + 1 = 5$.

We shall omit the existence and error analysis of the composite case, and illustrate it by numerical examples only.

§3. Numerical Examples

The main part of the numerical procedure is to provide an efficient and reliable way to solve the nonlinear system $\mathbf{F}(\mathbf{t}, \Delta\mathbf{b}) = 0$. As in [4] and [5] the continuation method ([1]) was applied. The solution of the homotopy equation

$$\mathbf{H}(\mathbf{t}, \Delta\mathbf{b}, \lambda) := \mathbf{F}(\mathbf{t}, \Delta\mathbf{b}) - \lambda\mathbf{F}(\mathbf{t}^0, \Delta\mathbf{b}^0) = 0$$

was followed from $\lambda = 1$ to $\lambda = 0$, starting with initial guess obtained by solving the single component problems for data $\mathbf{d}_{2(\ell-1)}, \mathbf{T}_{2(\ell-1)}, \mathbf{T}_{2\ell-1}, \mathbf{T}_{2\ell}, \mathbf{d}_{2\ell}$. Here, $\mathbf{d}_{2\ell}$ is the difference approximation of the direction of the curve $\mathbf{f}$ at $\mathbf{T}_{2\ell}$. In order to assure stability, at each Euler predictor step of the method two equations with the largest residuals were selected (out of three) to represent the wedge product equation in (3) and (4).

As the first example let us consider the interpolation of the spiral

$$\mathbf{f}_1 : [0, 15] \to \mathbb{R}^3 : \xi \mapsto \mathbf{f}_1(\xi) := \begin{pmatrix} \sin\xi \\ \ln(2+\xi)\cos\xi \\ \ln(1+\xi) \end{pmatrix}$$

with positive curvature and negative torsion. Tab. 2 shows that the computed error and approximation order (estimated from two consecutive m). The error column is an estimated parametric distance ([7]), obtained by reparametrisation of $\mathbf{B}_n$ as suggested in [3]: the point $\mathbf{B}_n(t(\xi))$ is the intersection of the curve $\mathbf{B}_n$ and the normal plane for $\mathbf{f}$ at $\mathbf{f}(\xi)$. This implies $(\mathbf{B}_n(t(\xi)) - \mathbf{f}(\xi))\cdot\dot{\mathbf{f}}(\xi) = 0$, so one has to solve a cubic equation to produce $t(\xi)$ for each ξ. The estimated approximation order clearly tends to 5, and the numbers of function and Jacobian matrix evaluations ($\#F$, $\#H'$) are rather small.

m	Error	Rate	$\#F$	$\#H'$
6	$0.72818 * 10^{-1}$	-	14	10
8	$0.17944 * 10^{-1}$	4.87	11	9
10	$0.67640 * 10^{-2}$	4.37	11	10
12	$0.30128 * 10^{-2}$	4.44	10	8
14	$0.14750 * 10^{-2}$	4.63	9	8
16	$0.77685 * 10^{-3}$	4.80	9	8
18	$0.43515 * 10^{-3}$	4.92	9	8
20	$0.25687 * 10^{-3}$	5.00	9	8

Tab. 2. The results for $\mathbf{f}_1$

m	Error	$\#F$	$\#H'$
8	$0.56932*10^{-4}$	15	11
10	$0.46259*10^{-2}$	63	63
12	$0.73456*10^{-5}$	13	8
16	$0.18523*10^{-4}$	24	10
18	$0.14192*10^{-4}$	34	22
22	$0.43003*10^{-5}$	14	9

Tab. 3. The results for $\mathbf{f}_2$

The second example is the interpolation of the part of the closed curve

$$\mathbf{f}_2 : [\pi, 2\pi] \to \mathbb{R}^3 : \xi \mapsto \mathbf{f}_2(\xi) := \begin{pmatrix} (2+\cos\xi)\cos(\frac{2}{3}\xi) \\ (2+\cos\xi)\sin(\frac{2}{3}\xi) \\ \sin\xi \end{pmatrix}.$$

Here, the torsion of $\mathbf{f}_2$ changes sign. Tab. 3 shows that some m produce better results that the others, and the estimated rate would be meaningless. The algorithm even failed to obtain the solution for some m. This is due to the fact that if the tension changes sign, the data may be planar even if they are sampled very closely.

Acknowledgments. Jernej Kozak is supported by the Ministry of Science and Technology of Slovenija.

References

1. Allgower, E. L., and K. Georg, *Numerical Continuation Methods*, Springer-Verlag, 1990.
2. de Boor, C., K. Höllig, and M. Sabin, High accuracy geometric Hermite interpolation, Comput. Aided Geom. Design **4** (1987), 269–278.
3. Degen, W. L. F., Best approximations of parametric curves by splines, in *Mathematical Methods in Computer Aided Geometric Design II*, T. Lyche and L. L. Schumaker (eds.), Academic Press, New York, 1992, 171–184.
4. Feng Y. Y., and J. Kozak, On G^2 continuous interpolatory composite quadratic Bézier curve, J. Comput. Appl. Math. **72** (1996), 141–159.
5. Feng Y. Y., and J. Kozak, On G^2 continuous cubic spline interpolation, BIT **37** (1997), 312–332.
6. Goldman R. N., The method of resolvents: a technique for the implicitization, inversion, and intersection of non-planar parametric, rational cubic curves, Comput. Aided Geom. Design **2** (1985), 237–255.
7. Lyche T., and K. Mørken, A metric for parametric approximation, in *Curves and Surfaces in Geometric Design*, P. J. Laurent, A. Le Méhauté, and L. L. Schumaker (eds.), A. K. Peters, Wellesley MA, 1994, 311–318.
8. Schaback R., Interpolation in $\mathbb{R}^2$ by piecewise quadratic visually C^2 Bézier polynomials, Comput. Aided Geom. Design **6** (1989), 219–233.
9. Schaback R., Planar curve interpolation by piecewise conics of arbitrary type, Constr. Approx. **9** (1993), 373–389.

Yu Yu Feng
Department of Mathematics, University of Science and Technology
230026 Hefei, Anhui, PEOPLE'S REPUBLIC OF CHINA
`FengYY@NSC.USCT.Edu.Cn`

Jernej Kozak
Department of Mathematics, University of Ljubljana
Jadranska 19, 1000 Ljubljana, SLOVENIJA
`Jernej.Kozak@FmF.Uni-Lj.Si`

The Smooth G^2 Filling of a Rectangular Patch Hole Joining Four Heterogeneous Surfaces

J-C. Fiorot and O. Gibaru

Abstract. We propose a method for filling a four-sided hole that interpolates four connected boundary curves in a net of patches. The surfaces defining these given boundary curves can be of different kinds with their suitable representations : Bézier - de Casteljau polynomial patches, (SBR) rational patches, or other parametric surfaces. Pole-functions are introduced as they extend the Gregory square-functions. The pole-functions result from appropriate combinations of order 1 and 2 derivatives at points belonging to the boundary of $[0,1]^2$ of functions that define the surfaces to be joined.

§1. Introduction

We denote $\Delta^{kl} f_{ij}$ the forward difference operator of the function $f_{ij}(u,v)$, for example $\Delta^{10} f_{ij} = f_{i+1,j} - f_{ij}$. A function $\Upsilon : P \subset \mathbb{R}^2 \to \mathbb{R}^3$ is said to be regular if $\forall (u,v) \in P$, $\dfrac{\partial \overrightarrow{\Upsilon}}{\partial u}(u,v) \wedge \dfrac{\partial \overrightarrow{\Upsilon}}{\partial v}(u,v) \neq \vec{0}$.

Let four surfaces S_1, S_2, S_3, S_4 be respectively parametrized by the regular functions Φ_0, Ψ_1, Φ_1, Ψ_0 defined on $[0,1]^2 \to \mathbb{R}^3$ of class C^m $(m \geq 2)$ which satisfy : $\Phi_0(0,1) = \Psi_0(1,0)$, $\Phi_0(1,1) = \Psi_1(1,0)$, $\Psi_1(1,1) = \Phi_1(1,1)$, $\Psi_0(1,1) = \Phi_1(0,1)$. The four corresponding points are denoted A_1, A_2, A_3, A_4 (Fig. 1).

Definition 1. *A filling surface S is a surface parametrized by $f : [0,1]^2 \to \mathbb{R}^3$*

$$f(u,v) = \sum_{i=0}^{n} \sum_{j=0}^{p} B_i^n(u)\, B_j^p(v)\, f_{ij}(u,v), \tag{1}$$

where $f_{ij} : [0,1]^2 \to \mathbb{R}^3$ are functions of class C^m ($m > 1$ except at the corners of the parametric domain where they are at least of class C^0). These functions

Mathematical Methods for Curves and Surfaces II
Morten Dæhlen, Tom Lyche, and Larry L. Schumaker (eds.), pp. 175–182.

ISBN 0-8265-1315-8.

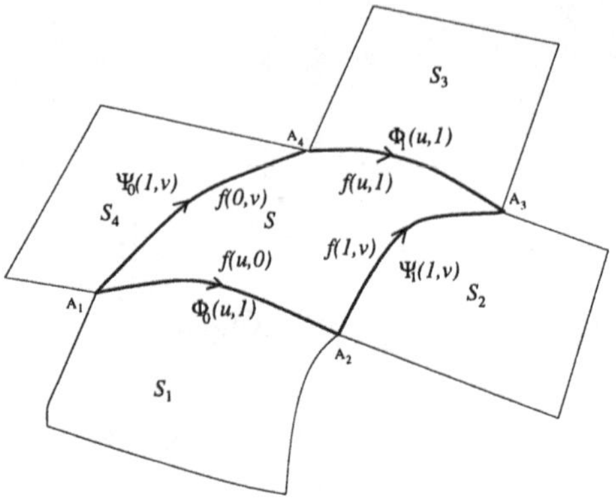

Fig. 1. Filling of a rectangular patch hole joining four heterogenous surfaces.

f_{ij} *are called* pole-functions. *The four curves* $\Phi_0(u,1)$, $\Psi_0(1,v)$, $\Phi_1(u,1)$, $\Psi_1(1,v)$ *define the boundaries of the surface* S *such that*

$$\begin{array}{ll} f(u,0)=\Phi_0(u,1), & f(0,v)=\Psi_0(1,v), \\ f(u,1)=\Phi_1(u,1), & f(1,v)=\Psi_1(1,v). \end{array} \quad \forall (u,v)\in[0,1]^2 .$$

§2. A Class of Pole-functions

We introduce a class of pole-functions where the following functions f_1, f_2, f_3, f_4 extend "Gregory's square-functions" ([4,5]). As we can see from Prop. 1, the two constants of the numerator of "Gregory's square-functions" are replaced by functions of two variables.

Proposition 1. *Let* $\alpha_1, \alpha_2, \alpha_3, \alpha_4, \beta_1, \beta_2, \beta_3, \beta_4 : [0,1]^2 \to \mathbb{R}^3$ *be regular functions of class* C^1 *such that* $\alpha_1(0,0)=\beta_1(0,0)$, $\alpha_2(1,0)=\beta_2(1,0)$, $\alpha_3(0,1)=\beta_3(0,1)$, $\alpha_4(1,1)=\beta_4(1,1)$. *Then the following functions defined on* $[0,1]^2$ *with* $n\in\mathbb{N}$ $(n\geq 2)$ *are of class* C^0 *on* $[0,1]^2$*:*

$$f_1(u,v)=\begin{cases}\alpha_1(0,0), & (u,v)=(0,0)\\ \dfrac{u^n\alpha_1(u,v)+v^n\beta_1(u,v)}{u^n+v^n}, & \text{otherwise,}\end{cases}$$

$$f_2(u,v)=\begin{cases}\alpha_2(1,0), & (u,v)=(1,0)\\ \dfrac{(1-u)^n\alpha_2(u,v)+v^n\beta_2(u,v)}{(1-u)^n+v^n}, & \text{otherwise,}\end{cases}$$

$$f_3(u,v)=\begin{cases}\alpha_3(0,1), & (u,v)=(0,1)\\ \dfrac{u^n\alpha_3(u,v)+(1-v)^n\beta_3(u,v)}{u^n+(1-v)^n}, & \text{otherwise,}\end{cases}$$

$$f_4(u,v)=\begin{cases}\alpha_4(1,1), & (u,v)=(1,1)\\ \dfrac{(1-u)^n\alpha_4(u,v)+(1-v)^n\beta_4(u,v)}{(1-u)^n+(1-v)^n}, & \text{otherwise.}\end{cases}$$

Proposition 2. *With the notations and hypothesis of Prop. 1, f_1 is $\mathcal{C}^1$ on $[0,1]^2$ if*

$$\frac{\partial \alpha_1}{\partial u}(0,0) = \frac{\partial \beta_1}{\partial u}(0,0), \qquad \frac{\partial \alpha_1}{\partial v}(0,0) = \frac{\partial \beta_1}{\partial v}(0,0).$$

Moreover

$$\frac{\partial f_1}{\partial u}(0,0) = \frac{\partial \alpha_1}{\partial u}(0,0), \qquad \frac{\partial f_1}{\partial v}(0,0) = \frac{\partial \alpha_1}{\partial v}(0,0).$$

As we can see in Props. 3 and 4, the originality of these functions is that functions α_1, α_2, α_3, α_4, β_1, β_2, β_3, β_4 are made up of appropriate combinations of order 1 and 2 derivatives at points belonging to the boundary of $[0,1]^2$ of functions that define the surfaces to be joined. They will allow us to create a G^1 or a G^2 filling surface even if the partial derivatives at corners are not compatible.

§3. A G^1 Filling Surface

In the following proposition we give sufficient conditions that pole functions have to satisfy in order to obtain a G^1 filling surface.

Proposition 3. *Let S be a surface parametrized as in (1), and suppose the pole-functions f_{ij} of class $\mathcal{C}^1$ (except at the corners) satisfy the conditions*

(a) for all $(u,v) \in [0,1]^2$,

$$\begin{aligned} f_{i0}(u,0) &= \Phi_0(u,1), \quad f_{ip}(u,1) = \Phi_1(u,1), \quad && i \in [0,n] \\ f_{0j}(0,v) &= \Psi_0(1,v), \quad f_{nj}(1,v) = \Psi_1(1,v), \quad && j \in [0,p], \end{aligned}$$

$$\begin{aligned} f_{i+1,0}(u,0) - f_{i0}(u,0) &= f_{i+1,p}(u,1) - f_{ip}(u,1) = 0, \quad && i \in [0,n-1] \\ f_{0,j+1}(0,v) - f_{0j}(0,v) &= f_{n,j+1}(1,v) - f_{nj}(1,v) = 0, \quad && j \in [0,p-1] \\ f_{1j}(0,v) - f_{0j}(0,v) &= f_{nj}(1,v) - f_{n-1,j}(1,v) = 0, \quad && j \in [0,p] \\ f_{i1}(u,0) - f_{i0}(u,0) &= f_{ip}(u,1) - f_{i,p-1}(u,1) = 0, \quad && i \in [0,n] \end{aligned}$$

(b) there exist a_0, a_1, d_0, $d_1 : [0,1] \to \mathbb{R}$ and b_0, b_1, c_0, $c_1 : [0,1] \to \mathbb{R}^*$ of class $\mathcal{C}^1$ such that for all $i \in [0,n]$ and $u \in]0,1[$,

$$\begin{aligned} \frac{\partial f_{i0}}{\partial u}(u,0) &= \frac{\partial \Phi_0}{\partial u}(u,1), \quad \frac{\partial f_{i0}}{\partial v}(u,0) = a_0(u)\frac{\partial \Phi_0}{\partial u}(u,1) + b_0(u)\frac{\partial \Phi_0}{\partial v}(u,1) \\ \frac{\partial f_{ip}}{\partial u}(u,1) &= \frac{\partial \Phi_1}{\partial u}(u,1), \quad \frac{\partial f_{ip}}{\partial v}(u,1) = a_1(u)\frac{\partial \Phi_1}{\partial u}(u,1) + b_1(u)\frac{\partial \Phi_1}{\partial v}(u,1). \end{aligned}$$

(c) for all $j \in [0,p]$ and $v \in]0,1[$,

$$\begin{aligned} \frac{\partial f_{0j}}{\partial v}(0,v) &= \frac{\partial \Psi_0}{\partial v}(1,v), \quad \frac{\partial f_{0j}}{\partial u}(0,v) = c_0(v)\frac{\partial \Psi_0}{\partial u}(1,v) + d_0(v)\frac{\partial \Psi_0}{\partial v}(1,v) \\ \frac{\partial f_{nj}}{\partial v}(1,v) &= \frac{\partial \Psi_1}{\partial v}(1,v), \quad \frac{\partial f_{nj}}{\partial u}(1,v) = c_1(v)\frac{\partial \Psi_1}{\partial u}(1,v) + d_1(v)\frac{\partial \Psi_1}{\partial v}(1,v). \end{aligned}$$

Then the filling surface S (see Definition 1) joins G^1 with S_1, S_2, S_3, S_4.

Proof: From (a) we obtain the G^0-conditions straighforwardly. For instance on $(u,0)$,

$$\frac{\partial f}{\partial u}(u,0) = n\sum_{i=0}^{n-1} B_i^{n-1}(u)\,\Delta^{10} f_{i0}(u,0) + \sum_{i=0}^{n} B_i^{n}(u)\,\frac{\partial f_{i0}}{\partial u}(u,0)$$

$$\frac{\partial f}{\partial v}(u,0) = p\sum_{i=0}^{n} B_i^{n}(u)\,\Delta^{01} f_{i0}(u,0) + \sum_{i=0}^{n} B_i^{n}(u)\,\frac{\partial f_{i0}}{\partial v}(u,0).$$

By (a), the forward differences are equal to zero. Therefore by (b)-(c), we can conclude that $\frac{\partial f}{\partial u}(u,0) = \frac{\partial \Phi_0}{\partial u}(u,1)$, $\frac{\partial f}{\partial v}(u,0) = a_0(u)\frac{\partial \Phi_0}{\partial u}(u,1) + b_0(u)\frac{\partial \Phi_0}{\partial v}(u,1)$. Thus S_1, S_2, S_3, S_4 are joined G^1-continuously with S. □

In the following proposition (which follows from Prop. 3 by direct calculation), we give a model of pole-functions which allow us to create a G^1 filling surface.

Proposition 4. *Let S be a surface parametrized as in (1) with $n = p = 3$. For any given functions a_0,a_1,d_0, $d_1 : [0,1] \to \mathbb{R}$ and b_0, b_1, c_0, $c_1 : [0,1] \to \mathbb{R}^*$ of class C^1, the following pole-functions satisfy the conditions of Prop. 3:*

$$f_{ij}(u,v) = \begin{cases} \dfrac{u^2\left[\Phi_0(u,1)+v\alpha_0^{[1]}(u)\right]+v^2\left[\Psi_0(1,v)+u\beta_0^{[1]}(v)\right]}{u^2+v^2}, & 0\leq i,j\leq 1 \\ \dfrac{(1-u)^2\left[\Phi_0(u,1)+v\alpha_0^{[1]}(u)\right]+v^2\left[\Psi_1(1,v)+(1-u)\beta_1^{[1]}(v)\right]}{(1-u)^2+v^2}, & \begin{matrix}2\leq i\leq 3\\ 0\leq j\leq 1\end{matrix} \\ \dfrac{u^2\left[\Phi_1(u,1)+(1-v)\alpha_1^{[1]}(u)\right]+(1-v)^2\left[\Psi_0(1,v)+u\beta_0^{[1]}(v)\right]}{u^2+(1-v)^2}, & \begin{matrix}0\leq i\leq 1\\ 2\leq j\leq 3\end{matrix} \\ \dfrac{(1-u)^2\left[\Phi_1(u,1)+(1-v)\alpha_1^{[1]}(u)\right]+(1-v)^2\left[\Psi_1(1,v)+(1-u)\beta_1^{[1]}(v)\right]}{(1-u)^2+(1-v)^2}, & 2\leq i,j\leq 3 \end{cases}$$

where

$$\alpha_0^{[1]}(u) = a_0(u)\frac{\partial \Phi_0}{\partial u}(u,1) + b_0(u)\frac{\partial \Phi_0}{\partial v}(u,1),$$

$$\beta_0^{[1]}(v) = c_0(v)\frac{\partial \Psi_0}{\partial u}(1,v) + d_0(v)\frac{\partial \Psi_0}{\partial v}(1,v),$$

$$\alpha_1^{[1]}(u) = a_1(u)\frac{\partial \Phi_1}{\partial u}(u,1) + b_1(u)\frac{\partial \Phi_1}{\partial v}(u,1),$$

$$\beta_1^{[1]}(v) = c_1(v)\frac{\partial \Psi_1}{\partial u}(1,v) + d_1(v)\frac{\partial \Psi_1}{\partial v}(1,v).$$

Thus, S is joined G^1 with S_1, S_2, S_3, S_4.

In the example below (see Figs. 2 and 3), S_1, S_4 are defined by a polynomial Bézier - de Casteljau surface, and S_2, S_3 are defined by a rational surface (SBR) [3]. We remark that at point A_3 there doesn't exist a common tangent plane, and hence the filling surface cannot join G^1-continuously with S_2 and S_3 at A_3. Nevertheless, we obtain the G^1-conditions along the curves $\Psi_1(1,v)$

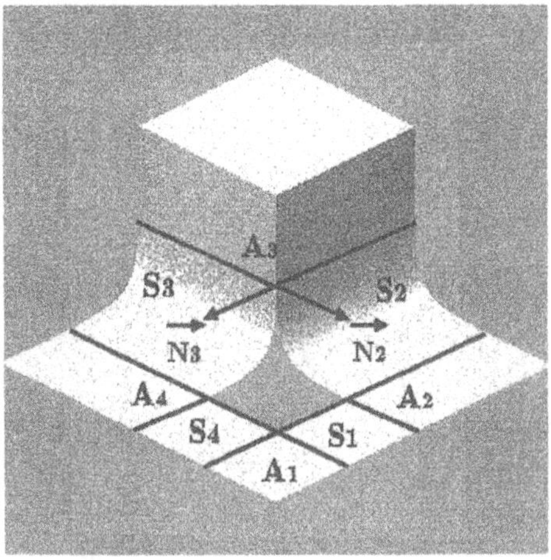

Fig. 2. Four given surfaces.

Fig. 3. A G^1 filling surface.

and $\Phi_1(u,1)$, except at A_3 where the filling surface joins G^0 with S_2 and S_3. Moreover, for instance at point A_1 the strong relations [5]

$$\frac{\partial \Phi_0}{\partial u}(0,1) = \frac{\partial \Psi_0}{\partial u}(1,0), \qquad \frac{\partial \Phi_0}{\partial v}(0,1) = \frac{\partial \Psi_0}{\partial v}(1,0)$$

are not necessarily satisfied. Since at A_1 these derivatives are on the same plane, we can define the free functions a_0, b_0, c_0, d_0 such that the filling surface becomes differentiable at $(u,v) = (0,0)$. This can be proved with Prop. 2. For instance, at A_1 we calculate the free functions a_0, b_0, c_0, d_0 such that the required conditions are satisfied, i.e., $\frac{\partial \alpha_1}{\partial u}(0,0) = \frac{\partial \beta_1}{\partial u}(0,0)$,

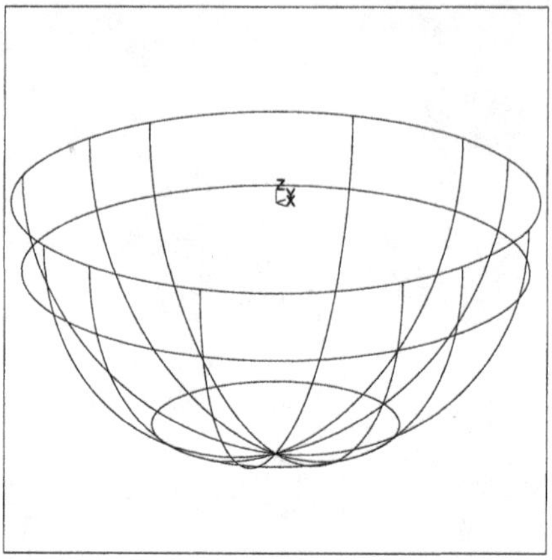

Fig. 4. Four given surfaces.

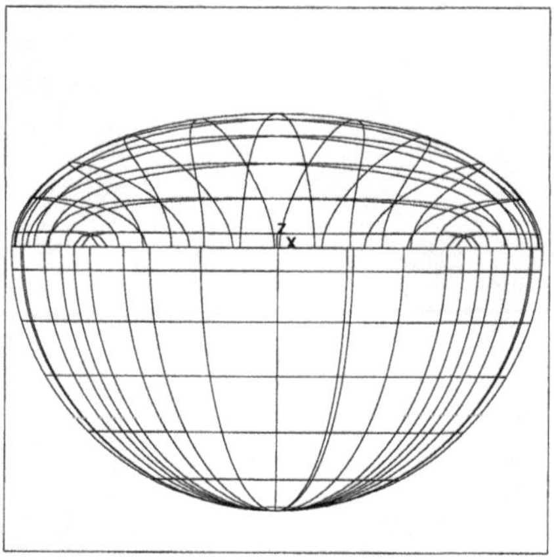

Fig. 5. A G^2 filling surface.

$\dfrac{\partial \alpha_1}{\partial v}(0,0) = \dfrac{\partial \beta_1}{\partial v}(0,0)$. From Prop. 4 we obtain the linear system

$$\frac{\partial \Phi_0}{\partial u}(0,1) = c_0(0)\,\frac{\partial \Psi_0}{\partial u}(1,0) + d_0(0)\,\frac{\partial \Psi_0}{\partial v}(1,0)$$
$$\frac{\partial \Psi_0}{\partial v}(1,0) = a_0(0)\,\frac{\partial \Phi_0}{\partial u}(0,1) + b_0(0)\,\frac{\partial \Phi_0}{\partial v}(0,1)$$

which give one solution for $a_0(0)$, $b_0(0)$, $c_0(0)$, $d_0(0)$ since the given functions are regular.

§4. A G^2 Filling Surface

In the following proposition (which can be verified by direct calculations) we give a model of pole-functions which allow us to create a G^2 filling surface.

Proposition 5. *Let S be a surface parametrized as in (1) with $n = p = 5$. For any given functions $a_0, a_1, d_0, d_1, e_0^*, e_1^*, f_0^*, f_1^*, g_0^*, g_1^*, h_0^*, h_1^* : [0,1] \to \mathbb{R}$, $b_0, b_1, c_0, c_1 : [0,1] \to \mathbb{R}^*$ of class C^2, we define the following pole-functions:*

$$f_{ij} = \begin{cases} \dfrac{u^2\left[\Phi_0(u,1)+v\alpha_0^{[1]}(u)+\frac{v^2}{2}\alpha_0^{[2]}(u)\right]+v^2\left[\Psi_0(1,v)+u\beta_0^{[1]}(v)+\frac{u^2}{2}\beta_0^{[2]}(v)\right]}{u^2+v^2}, & 0\le i,j\le 2, \\[2ex] \dfrac{(1-u)^2\left[\Phi_0(u,1)+v\alpha_0^{[1]}(u)+\frac{v^2}{2}\alpha_0^{[2]}(u)\right]}{(1-u)^2+v^2} \\ \quad + \dfrac{v^2\left[\Psi_1(1,v)+(1-u)\beta_1^{[1]}(v)+\frac{(1-u)^2}{2}\beta_1^{[2]}(v)\right]}{(1-u)^2+v^2}, & 3\le i\le 5,\ 0\le j\le 2, \\[2ex] \dfrac{u^2\left[\Phi_1(u,1)+(1-v)\alpha_1^{[1]}(u)+\frac{(1-v)^2}{2}\alpha_1^{[2]}(u)\right]}{u^2+(1-v)^2} \\ \quad + \dfrac{(1-v)^2\left[\Psi_0(1,v)+u\beta_0^{[1]}(v)+\frac{u^2}{2}\beta_0^{[2]}(v)\right]}{u^2+(1-v)^2}, & 0\le i\le 2,\ 3\le j\le 5 \\[2ex] \dfrac{(1-u)^2\left[\Phi_1(u,1)+(1-v)\alpha_1^{[1]}(u)+\frac{(1-v)^2}{2}\alpha_1^{[2]}(u)\right]}{(1-u)^2+(1-v)^2} \\ \quad + \dfrac{(1-v)^2\left[\Psi_1(1,v)+(1-u)\beta_1^{[1]}(v)+\frac{(1-u)^2}{2}\beta_1^{[2]}(v)\right]}{(1-u)^2+(1-v)^2}, & 3\le i,j\le 5 \end{cases}$$

where for $k = 0, 1$,

$$\begin{aligned} \alpha_k^{[1]}(u) &= a_k(u)\frac{\partial\Phi_k}{\partial u}(u,1) + b_k(u)\frac{\partial\Phi_k}{\partial v}(u,1), \\ \alpha_k^{[2]}(u) &= a_k^2(u)\frac{\partial^2\Phi_k}{\partial u^2}(u,1) + 2a_k(u)b_k(u)\frac{\partial^2\Phi_k}{\partial u\partial v}(u,1) + b_k^2(u)\frac{\partial^2\Phi_k}{\partial v^2}(u,1) \\ &\quad + e_k^*(u)\frac{\partial\Phi_k}{\partial u}(u,1) + f_k^*(u)\frac{\partial\Phi_k}{\partial v}(u,1), \\ \beta_k^{[1]}(v) &= c_k(v)\frac{\partial\Psi_k}{\partial u}(1,v) + d_k(v)\frac{\partial\Psi_k}{\partial v}(1,v), \\ \beta_k^{[2]}(v) &= c_k^2(v)\frac{\partial^2\Psi_k}{\partial u^2}(1,v) + 2c_k(v)d_k(u)\frac{\partial^2\Psi_k}{\partial u\partial v}(1,v) + d_k^2(v)\frac{\partial^2\Psi_k}{\partial v^2}(1,v) \\ &\quad + g_k^*(v)\frac{\partial\Psi_k}{\partial u}(1,v) + h_k^*(v)\frac{\partial\Psi_k}{\partial v}(1,v). \end{aligned}$$

Then S is joined G^2 with S_1, S_2, S_3, S_4.

In the example (see Figs. 4 and 5) the four given surfaces are (SBR) surfaces [3]. Each surface has $\frac{1}{8}$ of a sphere as a support. The four boundary curves taken together make up the horizontal circle.

References

1. Bézier P., Procédé de définition numérique des courbes et des surfaces non mathématiques, Système Unisurf, Automatisme, 13, 1968.
2. de Casteljau, P., Outillage, méthode de calcul, André Citroën Automobiles S.A, Paris, 1959.
3. Fiorot J. C. and P. Jeannin, *Courbes et Surfaces Rationnelles. Applications à la CAO*, RMA12, Masson, Paris, 1989. English version: *Rational Curves and Surfaces. Applications to CAD*, Wiley, Chichester, 1992.
4. Gregory J., Smooth interpolation without twist constraints, in *Computer Aided Geometric Design*, R. Barnhill and R. Riesenfeld (eds.), Academic Press, New York, 1974, 71-87.
5. Gregory J.A., C^1 rectangular and non-rectangular surface patches, in *Surfaces in CAGD*, R. E. Barnhill and W. Boehm (eds.), North-Holland, Amsterdam, 1983.

Jean-Charles Fiorot
ENSIMEV
Université Valenciennes
Le Mont Houy BP311
59304 Valenciennes Cedex - FRANCE
`fiorot@univ-valenciennes.fr`

Olivier Gibaru
C.R.M.A.O.
8 Bld Louis XIV
59046 Lille Cedex - FRANCE
`gibaru@lille.ensam.fr`

Local and Global Convexity Preservation

Michael S. Floater

Abstract. This paper concerns systems of functions which generate convex curves from convex control polygons. We discuss necessary and sufficient conditions for a system to preserve local convexity, global convexity and π-convexity.

§1. Introduction

Convexity is an important property of planar curves in geometric modelling but it has several interpretations. In this paper we emphasize locally convex curves and explore systems of functions on an interval $[a, b]$ which generate locally convex curves from locally convex control polygons. We also treat and compare globally convex and π-convex curves.

By locally convex we mean, roughly speaking, a planar parametric curve which has no inflections and keeps turning anticlockwise (or clockwise) with respect to its parameter. A locally convex curve is free to turn through any angle. In contrast, a globally convex curve is limited to turn through an angle of not more than 2π as it forms part of a closed convex curve. Global convexity preservation has recently been studied in [6] and by Carnicer and García-Esnaola [3]. We will call curves which are globally convex and which turn through an angle of at most π, π-convex. Such curves occur frequently, for example as the graphs of convex functions on an interval, or more generally as such graphs after a suitable rotation of axes. The preservation of convex functions and π-convex curves has been studied by Carnicer, García-Esnaola, and Peña [4,5].

In Section 2 we define what we mean by systems which preserve local convexity of order k, and show that totally positive blending systems with a local support property belong to this category. A B-spline basis of degree $\leq k$ provides an example of such systems. In Section 3 we define globally convex curves and polygons and discuss conditions for global convexity preservation. Finally in Section 4 we propose a simple definition of π-convex curves and we characterize π-convexity preservation.

Mathematical Methods for Curves and Surfaces II
Morten Dæhlen, Tom Lyche, and Larry L. Schumaker (eds.), pp. 183–190.

ISBN 0-8265-1315-8.

§2. Local Convexity

Denoting the determinant of two vectors $p = (p_1, p_2)$ and $q = (q_1, q_2)$ in $\mathbb{R}^2$ by $[p, q] := p_1 q_2 - p_2 q_1$, the usual way to specify local convexity of a twice differentiable planar curve $c : [a, b] \to \mathbb{R}^2$ is by the condition

$$[c'(t), c''(t)] \geq 0, \qquad a \leq t \leq b. \tag{1}$$

However, we prefer to make a definition which does not require differentiability. We will say that an arbitrary curve $c : [a, b] \to \mathbb{R}^2$ is locally convex if there exists $\epsilon > 0$ such that

$$[c(s) - c(r), c(t) - c(s)] \geq 0, \qquad a \leq r < s < t \leq b, \quad t - r \leq \epsilon. \tag{2}$$

A frequently occurring kind of curve $p : [a, b] \to \mathbb{R}^2$ in geometric modelling takes the form

$$p(t) = \sum_{i=0}^{n} P_i u_i(t), \tag{3}$$

where $P_0, \ldots, P_n$ are points in $\mathbb{R}^2$, and $u_0, \ldots, u_n$ are functions defined on the interval $[a, b]$. The polygonal arc $P_0, \ldots, P_n$ is the control polygon of p and the sequence $u_0, \ldots, u_n$ is often referred to as a system, denoted by $(u_0, \ldots, u_n)$. We say that $(u_0, \ldots, u_n)$ is a blending system if for $t \in [a, b]$, $u_i(t) \geq 0$, for $i = 0, \ldots, n$ and $\sum_{i=0}^{n} u_i(t) = 1$. Two common examples of blending systems are the Bernstein and B-spline bases. These two bases are also totally positive [1,7]. The system $(u_0, \ldots, u_n)$ is said to be totally positive if all minors of its collocation matrices

$$M \begin{pmatrix} u_0, \ldots, u_n \\ t_0, \ldots, t_n \end{pmatrix} = \left(u_j(t_i)\right)_{0 \leq i \leq n,\ 0 \leq j \leq n}$$

are nonnegative for $a \leq t_0 < \cdots < t_n \leq b$. As observed by Goodman [7], totally positive blending systems enjoy several shape-preserving properties; the curves they generate (Bézier and B-spline curves in the two examples above) tend to mimic the shapes of their control polygons.

If we identify the control polygon $P_0, \ldots, P_n$ with the piecewise linear curve $P : [0, 1] \to \mathbb{R}^2$ defined as

$$P(t) = (i + 1 - nt)P_i + (nt - i)P_{i+1}, \qquad t \in [\frac{i}{n}, \frac{i+1}{n}], \tag{4}$$

for $i = 0, 1, \ldots, n - 1$, then it is not difficult to show from (2) that P is locally convex if and only if

$$[P_i - P_{i-1}, P_{i+1} - P_i] \geq 0, \qquad 1 \leq i \leq n - 1. \tag{5}$$

This leads us to a natural generalization. Let us say that the control polygon $P_0, \ldots, P_n$ is locally convex of order k, $k \geq 1$, if

$$[P_i - P_{i-1}, P_j - P_{j-1}] \geq 0, \qquad 1 \leq i < j \leq n, \qquad j - i \leq k. \tag{6}$$

When $k = 1$, condition (6) reduces to condition (5). We also note that local convexity of order k implies local convexity of order $k-1$.

Correspondingly, let us say that a system $(u_0, \ldots, u_n)$ preserves local convexity of order k if the curve $p(t)$ in (3) is locally convex when its control polygon is locally convex of order k. We note that local convexity preservation of order k implies local convexity preservation of order $k+1$.

Next we characterize local convexity preservation in terms of the signs of minors of collocation matrices. Let

$$v_i(t) := \sum_{j=i}^{n} u_j(t), \qquad i = 0, 1, \ldots, n. \tag{7}$$

We note that $v_0 \equiv 1$ in the event that $(u_0, \ldots, u_n)$ is a blending system.

Proposition 2.1. *The blending system $(u_0, \ldots, u_n)$ (i) preserves local convexity of order k if and only if (ii) $\exists \epsilon > 0$ such that for $a \le r < s < t \le b$, $t - r \le \epsilon$, and $1 \le i < j \le n$,*

$$\det M \begin{pmatrix} 1, v_i, v_j \\ r, s, t \end{pmatrix} \begin{cases} \ge 0 & j - i \le k, \\ = 0 & j - i > k. \end{cases}$$

Let us remark that from standard properties of determinants,

$$\det M \begin{pmatrix} 1, v_i, v_j \\ r, s, t \end{pmatrix} = \sum_{\alpha=0}^{i-1} \sum_{\beta=i}^{j-1} \sum_{\gamma=j}^{n} \det M \begin{pmatrix} u_\alpha, u_\beta, u_\gamma \\ r, s, t \end{pmatrix}. \tag{8}$$

So the determinant in Proposition 2.1 is nonnegative if $(u_0, \ldots, u_n)$ is totally positive.

Proof: If $p(t)$ is the curve in (3), then using the fact that $\sum_i u_i = 1$,

$$p(t) = P_0 + \sum_{i=1}^{n} v_i(t)(P_i - P_{i-1}) \tag{9}$$

and it follows after a couple of lines of algebra (c.f. Theorem 5.2 of [2]) that

$$[p(s) - p(r), p(t) - p(s)] = \sum_{1 \le i < j \le n} \det M \begin{pmatrix} 1, v_i, v_j \\ r, s, t \end{pmatrix} [P_i - P_{i-1}, P_j - P_{j-1}]. \tag{10}$$

Suppose that $(u_0, \ldots, u_n)$ satisfies property (ii) and choose ϵ as in (ii). Then if the control polygon of p is locally convex of order k, we find from (10) that $[p(s) - p(r), p(t) - p(s)] \ge 0$ for $a \le r < s < t \le b$ and $t - r \le \epsilon$ and so p is locally convex.

Conversely, suppose $(u_0, \ldots, u_n)$ preserves local convexity of order k. For $1 \le i < j \le n$, choose $P_0 = \cdots = P_{i-1} = (0, 0)$, $P_i = \cdots = P_{j-1} = (1, 0)$, and $P_j = \cdots = P_n = (1, 1)$. Since the control polygon of p is locally convex of order

k, p must be locally convex and so $\exists \epsilon_{ij} > 0$ such that for $a \le r < s < t \le b$ and $t - r \le \epsilon_{ij}$,

$$0 \le [p(s) - p(r), p(t) - p(s)] = \det M \begin{pmatrix} 1, v_i, v_j \\ r, s, t \end{pmatrix}.$$

Moreover, if $j - i > k$, the control polygon defined by $P_0 = \cdots = P_{i-1} = (0,0)$, $P_i = \cdots = P_{j-1} = (1,0)$, and $P_j = \cdots = P_n = (1,-1)$ is also locally convex of order k. Thus again p is locally convex and so $\exists \hat{\epsilon}_{ij} > 0$ such that for $a \le r < s < t \le b$ and $t - r \le \hat{\epsilon}_{ij}$,

$$0 \le [p(s) - p(r), p(t) - p(s)] = -\det M \begin{pmatrix} 1, v_i, v_j \\ r, s, t \end{pmatrix}.$$

Hence, letting $\epsilon = \min\{\epsilon_{ij}, \hat{\epsilon}_{ij}\}_{i,j}$ proves (ii). □

Using the characterization of Proposition 2.1, we can show that a totally positive blending system preserves local convexity provided that the supports of the functions do not overlap too much.

Corollary 2.2. *If $(u_0, \ldots, u_n)$ is a totally positive blending system and if $\mathrm{supp}(u_\alpha) \cap \mathrm{supp}(u_\beta) = \emptyset$ for $0 \le \alpha, \beta \le n$ and $|\beta - \alpha| > k+1$, then $(u_0, \ldots, u_n)$ preserves local convexity of order k.*

Proof: Let

$$\epsilon = \min\{d(\mathrm{supp}(u_\alpha), \mathrm{supp}(u_\beta)) : |\beta - \alpha| > k + 1\},$$

where $d(A, B) := \min\{|s - t| : s \in A, t \in B\}$ for compact subsets A and B of $\mathbb{R}$. Then for $a \le r < s < t \le b$ and $t - r \le \epsilon$,

$$\det M \begin{pmatrix} u_\alpha, u_\beta, u_\gamma \\ r, s, t \end{pmatrix} = 0, \qquad 0 \le \alpha < \beta < \gamma \le n, \qquad \gamma - \alpha > k + 1. \quad (11)$$

Indeed, since $t - r \le d(\mathrm{supp}(u_\alpha), \mathrm{supp}(u_\gamma))$, by hypothesis, either $[r, t]$ and $\mathrm{Int}(\mathrm{supp}(u_\alpha))$ are disjoint or $[r, t]$ and $\mathrm{Int}(\mathrm{supp}(u_\gamma))$ are. So either the first or third column of the matrix in (11) is zero.

Due to identity (8) it follows that $(u_0, \ldots, u_n)$ satisfies property (ii) of Proposition 2.1. □

Proposition 2.1 and Corollary 2.2 enable us to determine precisely which B-splines bases preserve local convexity of a given order. We note that Goodman showed in Theorem 1 of [8] that under certain conditions, the number of inflections in a B-spline curve is bounded by the the number of inflections in the control polygon. Local convexity preservation of B-splines is clearly related to that theorem because locally convex curves are curves with no inflections. Letting

$$\tau_0 \le \tau_1 \le \cdots \le \tau_d < \tau_{d+1} < \cdots < \tau_{n+1} \le \cdots \le \tau_{n+d+1}$$

be a sequence of knots (any interior ones being simple), $n \ge d$, the B-splines

$$N_{i,d}(t) = (\tau_{i+d+1} - \tau_i)[\tau_i, \ldots, \tau_{i+d+1}](\cdot - t)_+^d, \qquad i = 0, 1, \ldots, n, \qquad (12)$$

of degree d constitute a totally positive blending system $(N_{0,d}, \ldots, N_{n,d})$ on the interval $[\tau_d, \tau_{n+1}]$. Moreover, recalling the fact that the support of $N_{i,d}$ is the interval $[\tau_i, \tau_{i+d+1}]$, we are able to deduce the following statement.

Corollary 2.3. *The system $(N_{0,d}, \ldots, N_{n,d})$ preserves local convexity of order k if and only if either (i) $k \geq d$ or (ii) $k = d-1$ and $n = d$.*

Proof: Let $u_i = N_{i,d}$, $i = 0, \ldots, n$. We know that $\text{supp}(u_i) \cap \text{supp}(u_j) = \emptyset$ when $|i-j| > d+1$. So by Corollary 2.2, $(u_0, \ldots, u_n)$ preserves local convexity of order d. Furthermore, if $n = d$ then $|i-j| \leq d$ for all $i, j \in \{0, \ldots, n\}$ and so similarly, by Corollary 2.2, $(u_0, \ldots, u_n)$ preserves local convexity of order $d-1$.

Conversely, we first show that $(u_0, \ldots, u_n)$ does not preserve local convexity of order $d-2$. For, if $\tau_d < r < s < t < \tau_{d+1}$, the Schoenberg-Whitney conditions imply that the determinant

$$\det M \begin{pmatrix} u_0, u_\beta, u_d \\ r, s, t \end{pmatrix}$$

is (strictly) positive for all integers β, $0 < \beta < d$ and therefore, from (8),

$$\det M \begin{pmatrix} 1, v_1, v_d \\ r, s, t \end{pmatrix} > 0.$$

Therefore, due to Proposition 2.1, the system $(u_0, \ldots, u_n)$ does not preserve local convexity of order $d-2$ and therefore neither of orders $\leq d-2$.

If $n > d$ and if $\tau_d < r < \tau_{d+1} < t < \tau_{d+2}$ then the Schoenberg-Whitney conditions imply that the determinant

$$\det M \begin{pmatrix} u_0, u_\beta, u_{d+1} \\ r, \tau_{d+1}, t \end{pmatrix}$$

is positive for $0 < \beta < d+1$ and so

$$\det M \begin{pmatrix} 1, v_1, v_{d+1} \\ r, \tau_{d+1}, t \end{pmatrix} > 0.$$

Therefore, we deduce from Proposition 2.1 that $(u_0, \ldots, u_n)$ does not preserve local convexity of order $d-1$. □

Thus a quadratic Bézier curve, which is a special case of a B-spline curve with $n = d = 2$, is locally convex when its control polygon is. However, a quadratic B-spline curve is in general not. This is due to the singular situation at the the internal knot τ_i when $P_{i-2} = P_{i-1}$.

§3. Global Convexity

In this section we discuss global convexity, a stronger property than local convexity. We say that a curve $c : [a, b] \to \mathbb{R}^2$ is globally convex if

$$[c(s) - c(r), c(t) - c(s)] \geq 0, \qquad a \leq r < s < t \leq b.$$

Clearly, from (2), a globally convex curve is also locally convex. It was shown in [6] that a control polygon $P_0, \ldots, P_n$ in $\mathbb{R}^2$, regarded as the piecewise linear curve P in (4), is globally convex if and only if

$$[P_j - P_i, P_k - P_j] \geq 0, \qquad 0 \leq i < j < k \leq n.$$

There are other definitions of global convexity which are equivalent except in pathological cases. For a discussion of several definitions see [3]. In contrast to local convexity preservation, it turns out that *all* totally positive blending systems preserve global convexity. Indeed, by substituting the expression for the determinant in (8) into (10) and interchanging the summations over i, j with those over α, β, γ, we find

$$[p(s) - p(r), p(t) - p(s)] = \sum_{0 \leq \alpha < \beta < \gamma \leq n} \det M \begin{pmatrix} u_\alpha, u_\beta, u_\gamma \\ r, s, t \end{pmatrix} [P_\beta - P_\alpha, P_\gamma - P_\beta]$$

for $r, s, t \in [a, b]$. This identity can alternatively be deduced from a simple application of the Cauchy-Binet theorem as in [6].

A precise characterization for global convexity preservation is as yet unknown when $n \geq 4$. However, it was shown in [6] that a blending system (u_0, u_1, u_2) preserves global convexity if and only if

$$\det M \begin{pmatrix} u_0, u_1, u_2 \\ r, s, t \end{pmatrix} \geq 0, \qquad a \leq r < s < t \leq b,$$

and a blending system (u_0, u_1, u_2, u_3) preserves global convexity if and only if for $a \leq r < s < t \leq b$,

$$\det M \begin{pmatrix} u_0, u_1, u_2 + u_3 \\ r, s, t \end{pmatrix} \geq 0, \quad \det M \begin{pmatrix} u_0, u_1 + u_2, u_3 \\ r, s, t \end{pmatrix} \geq 0,$$
$$\det M \begin{pmatrix} u_0 + u_1, u_2, u_3 \\ r, s, t \end{pmatrix} \geq 0, \quad \det M \begin{pmatrix} u_3 + u_0, u_1, u_2 \\ r, s, t \end{pmatrix} \geq 0.$$

§4. π-convexity

In this last section we analyze an even stronger property of a planar curve. Let us say that a differentiable curve $c : [a, b] \to \mathbb{R}^2$ is π-convex if

$$[c'(s), c'(t)] \geq 0, \qquad a \leq s < t \leq b.$$

Using the fact that

$$[c(s) - c(r), c(t) - c(s)] = \int_r^s \int_s^t [c'(x), c'(y)]\, dx\, dy,$$

we see that a differentiable π-convex curve is also globally convex and indeed

$$\pi\text{-convex} \quad \Rightarrow \quad \text{globally convex} \quad \Rightarrow \quad \text{locally convex}.$$

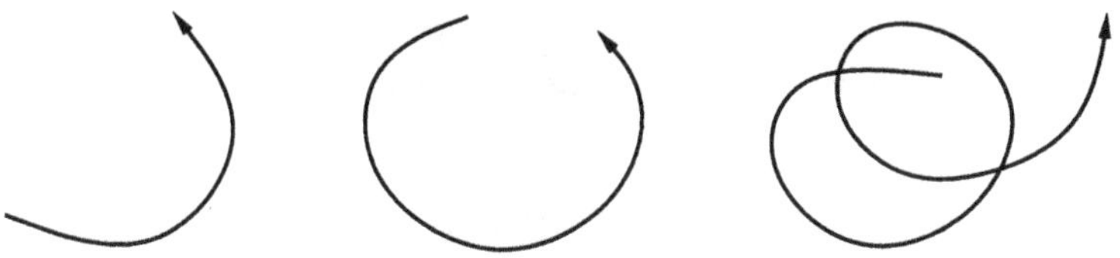

Fig. 1. π-convex, globally convex, and locally convex curves.

Figure 1 shows from left to right, π-convex, globally convex, and locally convex curves. A natural corresponding definition of π-convexity for a control polygon $P_0, \ldots, P_n$ is

$$[P_i - P_{i-1}, P_j - P_{j-1}] \geq 0, \qquad 0 \leq i < j \leq n.$$

We remark that from (6), this condition is equivalent to local convexity of order n. On the other hand, local convexity of order k is equivalent to every subpolygon $P_i, \ldots, P_{i+k}$ being π-convex, for $i = 0, \ldots, n-k$. Making now the obvious definition of π-convexity preservation, we find the following:

Proposition 4.1. *A differentiable blending system $(u_0, \ldots, u_n)$ on $[a, b]$ preserves π-convexity if and only if*

$$\det M \begin{pmatrix} v_i', v_j' \\ s, t \end{pmatrix} \geq 0, \qquad a \leq s < t \leq b, \quad 1 \leq i < j \leq n, \tag{13}$$

where the functions v_i are defined in (7).

Proof: Differentiating (9), we find

$$\begin{aligned} [p'(s), p'(t)] &= \sum_{i,j=1}^{n} v_i'(s) v_j'(t) [P_i - P_{i-1}, P_j - P_{j-1}] \\ &= \sum_{1 \leq i < j \leq n} \det M \begin{pmatrix} v_i', v_j' \\ s, t \end{pmatrix} [P_i - P_{i-1}, P_j - P_{j-1}]. \end{aligned}$$

Thus condition (13) is sufficient for π-convexity preservation. Conversely, if $(u_0, \ldots, u_n)$ preserves π-convexity and $1 \leq i < j \leq n$, let $P_0, \ldots, P_n$ be the control polygon defined by $P_0 = \cdots = P_{i-1} = (0,0)$, $P_i = \cdots = P_{j-1} = (1,0)$, and $P_j = \cdots = P_n = (1,1)$. Since $P_0, \ldots, P_n$ is π-convex we deduce that

$$0 \leq [p'(s), p'(t)] = \det M \begin{pmatrix} v_i', v_j' \\ s, t \end{pmatrix}$$

and so condition (13) is fulfilled. □

Proposition 4.1 is closely related to Corollary 3.4 of [5] since π-convexity is similar to the concept of 'geometric convexity' discussed in [5].

At first, like condition (ii) of Proposition 2.1, condition (13) may seem obscure as it is defined in terms of the auxiliary functions v_i rather than u_i. Nevertheless, in the case of B-splines of degree at least two (and therefore C^1), condition (13) holds. For if $u_i = N_{i,d}$, $i = 0, \ldots, n$, $d \geq 2$, and $(N_{0,d}, \ldots, N_{n,d})$ is the B-spline basis in (12), the well-known identity for derivatives of B-splines (see de Boor [1], page 138) implies

$$v_i'(t) = \sum_{j=i}^{n} N_{i,d}'(t) = \frac{d}{\tau_{i+d} - \tau_i} N_{i,d-1}(t)$$

and so

$$\begin{vmatrix} v_i'(s) & v_j'(s) \\ v_j'(t) & v_j'(t) \end{vmatrix} = \frac{d^2}{(\tau_{i+d} - \tau_i)(\tau_{j+d} - \tau_j)} \begin{vmatrix} N_{i,d-1}(s) & N_{j,d-1}(s) \\ N_{j,d-1}(t) & N_{j,d-1}(t) \end{vmatrix}.$$

The latter determinant is nonnegative when $1 \leq i < j \leq n$ and $a \leq s < t \leq b$ because the B-splines $(N_{1,d-1}, \ldots, N_{n,d-1})$ are totally positive.

References

1. de Boor, C., *A Practical Guide to Splines*, Springer-Verlag, New York, 1978.
2. Carnicer, J. M., M. S. Floater, and J. M. Peña, On systems of functions satisfying hodograph properties, to appear in *Mathematics of Surfaces VII*, T. N. T. Goodman (ed.), Information Geometers, Winchester, 1997.
3. Carnicer, J. M., and M. García-Esnaola, Global convexity of curves and polygons, preprint.
4. Carnicer, J. M., M. García-Esnaola, and J. M. Peña, Convex curves from convex control polygons, in *Mathematical Methods for Curves and Surfaces* M. Dæhlen, T. Lyche & L. L. Schumaker (eds.), Vanderbilt University Press, Nashville, 1995, 63–72.
5. Carnicer, J. M., M. García-Esnaola, and J. M. Peña, Convexity of Rational Curves and Total Positivity, J. Comp. Appl. Math. **71** (1996), 365–382.
6. Floater, M. S., Total positivity and convexity preservation, to appear in J. Approx. Th.
7. Goodman, T. N. T., Shape preserving representations, in *Mathematical methods in Computer Aided Geometric Design*, T. Lyche and L.L. Schumaker (eds.), Academic Press, New York, 1989, 333–357.
8. Goodman, T. N. T., Inflections on curves in two and three dimensions, Comp. Aided Geom. Design **8** (1991), 37–50.

Michael S. Floater
SINTEF Applied Mathematics
Postbox 124, Blindern
0314 Oslo, NORWAY
`Michael.Floater@math.sintef.no`

Convergence of Cascade Algorithms

T. N. T. Goodman and S. L. Lee

Abstract. A nonstationary cascade algorithm for the nonstationary refinement equation $\Phi_k(x) = |M|\sum_{j\in\mathbb{Z}^s} h_{k+1}(j)\Phi_{k+1}(Mx-j)$, $x \in \mathbb{R}^s$, is an iterative process, $\Phi_{k,n}(x) = |M|\sum_{j\in\mathbb{Z}^s} h_{k+1}(j)\Phi_{k+1,n-1}(Mx-j)$, that generates a family of sequences of vector functions $\Phi_{k,n} \in L^2(\mathbb{R}^s)^r$, $k, n = 0, 1, \ldots$, from an initial sequence of vectors $\Phi_{k,0} \in L^2(\mathbb{R}^2)^r$, where each h_k is a finitely supported sequence of $r \times r$ matrices, and M is an $s \times s$ dilation matrix, i.e. an integer matrix with all eigenvalues of modulus > 1 and $\det(M) \geq 2$. We assume that as $k \to \infty$, $h_k \to h$ which defines an ideal or stationary refinement equation $\Phi(x) = |M|\sum_{j\in\mathbb{Z}^s} h(j)\Phi(Mx-j)$. The corresponding cascade algorithm, $\tilde{\Phi}_n(x) = |M|\sum_{j\in\mathbb{Z}^s} h(j)\tilde{\Phi}_{n-1}(Mx-j)$, generates a sequence of vector functions $\tilde{\Phi}_n$ from an initial vector $\tilde{\Phi}_0 \in L^2(\mathbb{R}^s)^r$. The spectrum of the associated transition operator, $Tb(j) := |M|\sum_k\sum_\ell h(k)b(2j-\ell)h(k-\ell)^*$, defined on sequences of $r \times r$ matrices, plays an important role in the convergence of the nonstationary as well as the ideal cascade algorithms. This paper surveys results on various forms of convergence of cascade algorithms, from the scalar case in one dimension, $s = r = 1$ and $M = (2)$, to the latest developments on nonstationary cascade algorithms. Even in the scalar case in one dimension some of the results presented are new.

§1. Introduction

The equation

$$\phi(x) = \sum_{k=0}^{N} 2h(k)\phi(2x-k), \quad x \in \mathbb{R}, \tag{1}$$

is called a refinement equation. It is central to wavelet theory. A solution of (1) is a scaling function or a refinable function. The sequence $h(k)$, $k = 0, 1, \ldots, N$, is the mask for ϕ. We shall restrict our study to real masks. If $\phi \in L^2(\mathbb{R})$ is a compactly supported solution of (1) and is stable, i.e. its integer shifts $\phi(\cdot - j)$, $j \in \mathbb{Z}$, form a Riesz basis of their closed linear span, then it generates a multiresolution $(V_m)_{m\in\mathbb{Z}}$ of $L^2(\mathbb{R})$. Let W_0 be the orthogonal complement

Mathematical Methods for Curves and Surfaces II

Morten Dæhlen, Tom Lyche, and Larry L. Schumaker (eds.), pp. 191–212.

ISBN 0-8265-1315-8.

of V_0 in V_1. If $\psi \in W_0$ is stable, then $\psi(2^m \cdot - j)$, $j, m \in \mathbb{Z}$, form a Riesz basis of $L^2(\mathbb{R})$. The function ψ is called a wavelet, and $\psi(x) = \sum_j 2g(j)\phi(2x - j)$, $x \in \mathbb{R}$, for some finitely supported sequence g, see [3,7,26].

Since a compactly supported stable L^2-solution of a refinement equation generates a multiresolution analysis from which the wavelets are constructed, it is of interest to study the existence of compactly supported L^2-solutions of refinement equations, their stability and how to compute them. To this end we choose a function $\phi_0 \in L^2(\mathbb{R})$ with $\mathrm{supp}(\phi_0) \subset [0, N]$, and generate an L^2-sequence

$$\phi_n(t) = \sum_{k=0}^{N} 2h(k)\phi_{n-1}(2t - k), \quad n = 1, 2, \ldots . \tag{2}$$

Then $\mathrm{supp}(\phi_n) \subset [0, N]$ for all n. Equation (2) is called the cascade algorithm for the refinement equation (1) (or for the mask h). The cascade algorithm is very much like Picard's iteration for the solution of differential equations. In the Fourier domain, (2) becomes

$$\hat{\phi}_n(u) = H(u/2)\hat{\phi}_{n-1}(u/2), \quad u \in \mathbb{R}, \tag{3}$$

where $H(u) := \sum_{k=0}^{N} h(k)e^{-iku}$. Iterating (3) leads to

$$\hat{\phi}_n(u) = \prod_{j=1}^{n} H(u/2^j)\hat{\phi}_0(u/2^n), \quad n = 1, 2, \ldots . \tag{4}$$

If $\sum_{k=0}^{N} h(k) = 1$, the infinite product $\prod_{j=1}^{\infty} H(u/2^j)$ converges locally uniformly in u. Since $\hat{\phi}_0$ is continuous at 0, it follows from (4) that $\hat{\phi}_n(u) \to P(u)\hat{\phi}_0(0)$ locally uniformly in u as $n \to \infty$, where $P(u) := \prod_{j=1}^{\infty} H(u/2^j)$. The function P, being analytic and of polynomial growth as $|u| \to \infty$, defines a compactly support tempered distribution ϕ, and the cascade sequence ϕ_n converges in distribution to ϕ. It also follows that if ϕ_0 is normalized so that $\hat{\phi}_0(0) = 1$, then ϕ is the unique compactly supported solution of the refinement equation (1) and is a tempered distribution whose Fourier transform is $P(u)$. We shall assume throughout this paper that $\sum_k h(k) = 1$ and $\hat{\phi}_0(0) = 1$.

For $n = 0, 1, \ldots$, define

$$b_n(j) := \int_{-\infty}^{\infty} \phi_n(t)\phi_n(t - j)dt, \quad j \in \mathbb{Z}. \tag{5}$$

Then $\|\phi_n\|^2 = b_n(0)$, and $b_n(j) = 0$ for $j \notin [-N + 1, ..., N - 1]$. Further, (2) and (5) lead to

$$b_n(j) = \sum_{k=-N}^{N} 2c(2j - k)b_{n-1}(k), \tag{6}$$

where $c(k) := \sum_\ell h(\ell)h(\ell-k)$ is the autocorrelation of h. Expressed in matrix form, (6) becomes

$$b_n = Tb_{n-1}, \quad n = 1, 2, \ldots, \tag{7}$$

where $b_n = (b_n(-N+1), ..., b_n(N-1))^T$ and $T := (c(2i-j))_{i,j=-N+1}^{N-1}$. The matrix operator T is called the transition matrix for h. In the frequency domain, the transition operator, denoted by $\mathcal{T}$, acts on the space $\mathcal{P}_N := \{\sum_{j=-N+1}^{N-1} b(j)e^{-iju} : b(j) \in \mathbb{R}\}$ of trigonometric polynomials of degree $N-1$.

The matrix operator T is the restriction to $\ell(\{-N+1, ..., N-1\}) \equiv \mathbb{R}^{2N-1}$ of the operator $T : \ell^p(\mathbb{Z}) \to \ell^p(\mathbb{Z})$ defined by

$$(Tb)(j) = \sum_{k \in \mathbb{Z}} c(2j-k)b(k), \; b \in \ell^p(\mathbb{Z}).$$

The operator T is called the transition operator for h. It was also known as the Wavelet-Galerkin operator [23]. Transition operators also arise in the study of uniform subdivision in computer aided geometric design [2,11,14]. Indeed, the associated subdivision process $S_c : \ell^q(\mathbb{Z}) \to \ell^q(\mathbb{Z})$, $\frac{1}{p}+\frac{1}{q} = 1$, is the conjugate of T, i.e. $S_c = T^*$. More generally, for any finitely supported sequence a the corresponding subdivision process $S_a : \ell^q(\mathbb{Z}) \to \ell^q(\mathbb{Z})$ is defined by

$$(S_a b)(j) = \sum_{k \in \mathbb{Z}} c(j-2k)b(k), \; b \in \ell^q(\mathbb{Z}) \; .$$

Lawton [22] characterized the orthonormality of shifts of the solution of a refinement equation in terms of the spectral properties of its transition operator. Indeed, many properties of the solution of a refinement equation, such as the convergence of the cascade algorithm, stabilty and regularity, are embodied in the spectrum of its transition operator. Here, we shall be concerned mainly with the convergence of cascade algorithms. The rest of the paper is divided into two sections. Section 2 covers the main results on convergence of the stationary cascade sequence and its derivative in one dimension. It begins with a friendly exposition where the results and brief proofs are aimed at helping readers understand the subsequent subsections. Section 3 deals with the full generality of nonstationary cascade algorithms.

§2. Convergence of Cascade Algorithms in $L^2(\mathbb{R})$

Iterating (7) leads to

$$b_n = T^n b_0, \quad n = 0, 1, \ldots \; , \tag{8}$$

which is a matrix iteration problem. This formulation is due to Strang [31], where he characterized the strong convergence in $L^2(\mathbb{R})$ of the cascade algorithm in terms of what he called Condition E of the transition matrix. A matrix A is said to satisfy Condition E if 1 is a simple eigenvalue of A and all its other eigenvalues lie inside the unit circle. In order to characterize weak convergence of the cascade algorithm, a more general condition is required. We say that a matrix A satisfies Condition E^{**} if its spectral radius is 1 and all its eigenvalues on the unit circle are nondegenerate. If A satisfies Condition E^{**} and if in addition, 1 is the only eigenvalue on the unit circle, then we say that A satisfies Condition E^*.

Theorem 1. *The following are equivalent:*
(a) *The transition matrix T satisfies Condition E^{**}.*
(b) *The cascade sequence ϕ_n converges weakly in $L^2(\mathbb{R})$ to the compactly supported solution $\phi \in L^2(\mathbb{R})$ of the refinement equation (1), for any starting function ϕ_0.*

Theorem 2. [Strang [31]] *Suppose that h is fundamental, i.e.*

$$\sum_j h(2j) = \sum_j h(2j+1) = \frac{1}{2}.$$

The cascade sequence ϕ_n converges strongly in $L^2(\mathbb{R})$ for any compactly supported initial function ϕ_0 satisfying

$$\sum_{j\in\mathbb{Z}} \phi_0(x-j) = 1, \quad x \in \mathbb{R}, \tag{9}$$

if and only if the transition matrix T satisfies Condition E.

Theorem 1 shows that Condition E^{**} for the transition matrix is sufficient for the solution of a refinement equation to be in $L^2(\mathbb{R})$. The direction (a) implies (b) was established in [25] for the multidimensional scalar case. Theorem 1 was established in its full generality for nonstationary cascade algorithms in higher dimension [15]. Theorem 2, in the above form, was due to Strang [31] (see also [32]). Similar results in a different form, embedded in the theory of L^p-convergence of uniform subdivision processes, were obtained slightly earlier in [14] and [18].

We sketch our proofs of Theorem 1 and the following slightly strongly version of Theorem 2. The proofs below are different from those in [31,14] and [18]. The ideas in the proofs carry over, with some modifications, to higher dimensions for the scalar case [25], the vector case [30] and the nonstationary vector cascade algorithms [15].

Theorem 3. *The following are equivalent:*
(a) *The transition matrix T satisfies Condition E and h is fundamental.*
(b) *The cascade sequence ϕ_n converges strongly in $L^2(\mathbb{R})$, to the compactly solution $\phi \in L^2(\mathbb{R})$ of the refinement equation (1), for any starting function ϕ_0 satisfying (9).*

Proof of Theorem 1: If the transition matrix T satisfies Condition E^{**}, then $\|T^n\|$ is bounded. By (8), $\|b_n\|$ is bounded. In particular, $\|\phi_n\|^2 = b_n(0)$ is bounded. Hence there is a subsequence (ϕ_{n_j}) of (ϕ_n) such that $\phi_{n_j} \to \phi$ weakly in $L^2(\mathbb{R})$. Since ϕ_n converges in distribution to the solution of the refinement equation (1), it follows that ϕ is an L^2-solution of (1) and that the entire sequence ϕ_n converges weakly to ϕ.

Conversely, if (b) is satisfied, let $\phi \in L^2(\mathbb{R})$ be the solution of (1), and define $b(j) := \langle \phi, \phi(\cdot - j)\rangle$, $j = -N+1, \ldots, N-1$. By (1) $Tb = b$. Thus the the spectral radius $\rho(T) \geq 1$. Choose any function ψ in $L^2(\mathbb{R}^s)$ with support in $[0,1]$ and $\hat{\psi}(0) = 1$. Take an arbitrarily fixed $\ell = -N+1,\dots,N+1$, and define $\psi_0 = \psi$, $\phi_0 = \psi(\cdot + \ell)$. For $n = 1,2,\dots$, we define ϕ_n by (2) and similarly we define ψ_n by (2) with ϕ_0 replaced by ψ_0. We then define $c_n(j)$ by $c_n(j) = \langle \psi_n, \phi_n(\cdot - j)\rangle$. Observe that $c_0(j) = \delta(j-\ell)$, $j = -N+1,\dots,N-1$, where δ is the sequence with unit mass concentrated at 0. Thus from (2),

$$c_n = T^n \delta(\cdot - \ell), \quad n = 0,1,2,\dots. \tag{10}$$

By (b), ϕ_n and ψ_n converge to ϕ weakly in $L^2(\mathbb{R})$ as $n \to \infty$. By the Uniform Boundedness Theorem, $\|\phi_n\|_2$ and $\|\psi_n\|_2$ are uniformly bounded in n. Since $|c_n(j)| \le \|\phi_n\|_2\, \|\psi_n\|_2$, for all $j = -N+1,\dots,N-1$, it follows that $c_n(j)$ is uniformly bounded in n. Recalling (10), it follows that $\|T^n\|$ is bounded in n. Since the spectral radius $\rho(T) \ge 1$, T must satisfy Condition E^{**}. □

Proof of Theorem 3: Suppose that T satisfies Condition E. Then the cascade sequence ϕ_n converges weakly in $L^2(\mathbb{R})$ to the solution ϕ of (1). It remains to show that $\|\phi_n\|_2 \to \|\phi\|_2$. Define b_n and b as above. Then $b = Tb$ and $b_n = T^n b_0$, $n = 0,1,\dots$.

We shall show that $b_n \to b$ as $n \to \infty$. Since 1 is a simple eigenvalue of $T : \mathbb{R}^{2N-1} \to \mathbb{R}^{2N-1}$, and all its other eigenvalues are of modulus < 1, we can decompose

$$\mathbb{R}^{2N-1} = M \oplus Q\ ,$$

where $M = \{\alpha b : \alpha \in \mathbb{R}\}$ and $\rho\left(T|_Q\right) < 1$. Let $b_0 = \alpha b + q$, for some $\alpha \in \mathbb{R}$, and $q \in Q$. The crucial step is to show that $\alpha = 1$. To this end, observe that (9) implies $\sum_j b_0(j) = 1$. The assumption that h is fundamental implies that ϕ satisfies the same condition (9) as ϕ_0. Thus $\sum_j b(j) = 1$. Further, for any $v \in \mathbb{R}^{2N-1}$,

$$\sum_j Tv(j) = \sum_j v(j). \tag{11}$$

Therefore, $\sum_j b_n(j) = \sum_j T^n b_0(j) = \sum_j b_0(j) = 1$, $n = 1,2,\dots$. Since $T^n q \to 0$ as $n \to \infty$, (11) again implies that $\sum_j q(j) = 0$. Taken together, we have $\alpha = 1$. Thus $b_0 = b + q$, and $b_n = b + T_h^n q \to b$ as $n \to \infty$. In particular, $\|\phi_n\|_2^2 = b_n(0) \to \|\phi\|_2^2$, as $n \to \infty$.

Conversely, suppose that ϕ_n converges strongly to ϕ in $L^2(\mathbb{R})$ for any compactly supported function $\phi_0 \in L^2(\mathbb{R})$ satisfying (9). Consider the vector b, the initial functions ϕ_0 and ψ_0 as in the proof of Theorem 1, the corresponding cascade sequences ϕ_n and ψ_n and the sequence $c_n(j) = \langle \psi_n, \phi_n(\cdot - j)\rangle$, $j = -N+1,\dots,N-1$. As above, (10) holds. The strong convergence of ϕ_n and ψ_n to ϕ, implies that $c_n \to b$ as $n \to \infty$, where b is an eigenvector of T corresponding to the eigenvalue 1. By (10)

$$T^n \delta(\cdot - \ell) \to b \ \text{ as } n \to \infty,\ \ell = -N+1,\dots,N-1. \tag{12}$$

Thus T^n converges. this means that the spectral radius $\rho(T) \le 1$, 1 is the only eigenvalue on the unit circle and is nondegenerate. Further, it is easy to see from (12) that b is the unique eigenvector, establishing the fact that 1 is a simple eigenvalue.

Finally, to show that h is fundamental, we need only show that c is fundamental. Let $v \in \mathbb{R}^{2N-1}$ be a left eigenvector of T with eigenvalue 1. Then $v^T T^n = v^T$ for all $n = 1, 2, \ldots$. Hence, for each $\ell = -N+1, \ldots, N-1$, $v(j) = v^T T^n \delta(j - \ell) \to v^T b$ for all $j = -N+1, \ldots, N-1$. This means that v is a multiple of $(1, 1, \ldots, 1)^T$. Thus, the relation $v^T T = v^T$ is equivalent to $2\sum_j c(2j - k) = 1$ for all $k = -N+1, \ldots, N-1$. □

2.2. Condition E, Strong Convergence and Stability

Suppose that h is fundamental. Then a necessary and sufficient condition for the strong convergence in $L^2(\mathbb{R})$ of the cascade algorithm for h for any initial compactly supported function $\phi_0 \in L^2(\mathbb{R})$ satisfying (9) is that the transition matrix T satisfies Condition E. Suppose that ϕ is a compactly supported refinable function with mask h. If ϕ is stable, then the cascade algorithm converges strongly in $L^2(\mathbb{R})$ for any compactly supported initial function $\phi_0 \in L^2(\mathbb{R}^d)$ satisfying (9). This result which can be found in [18] follows from the following

Theorem 4. *The scaling function ϕ with mask h is stable if and only if*

(i) T satisfies Condition E, and

(ii) the eigenvector v of T with eigenvalue 1 satisfies $\sum_k v(k)e^{-iku} \neq 0$, $\forall u \in [0, 2\pi)$.

Stability of biorthogonal refinable functions in terms of Condition E of the corresponding transition matrices has been studied in [4] in one dimension and in [27] in higher dimension with dilation matrix $2I$. Theorem 4 was proved in [24] for the multivariate scalar case with a general dilation matrix M. The result was extended to the vector case in [30] with dilation matrix $M = 2I$.

2.3. Convergence of Derivatives of Cascade Sequences

Recall that a mask h whose sum is 1 is fundamental if $\sum_k h(2k) = \sum_k h(2k+1)$. To study the convergence of derivatives of a cascade sequence we need to extend the concept of fundamentality of h to higher order. Since the convergence of derivatives of the cascade sequence for a refinement equation is closely related to the regularity of its solution, the conditions on h for convergence of derivatives of cascade sequences are similar to those for regularity of refinable functions. Results on regularity of refinable functions are given in [9,12] and [10] for the scalar case in one dimension, in [28] and [5] for the vector case in one dimension, in [30,21] and [19] in higher dimensions.

To define the concept of fundamentality of higher order for h let π_m be the space of all polynomials on $\mathbb{R}$ of degree $\leq m$, and define

$$V_{\pi_m} := \{v = (v(j))_{j=-N+1}^{N-1} : \sum_{j\in\mathbb{Z}} p(j)v(j) = 0, \ \forall p \in \pi_m\}.$$

We say that h satisfies the sum rules for π_m if for all $p \in \pi_m$,

$$\sum_{j\in\mathbb{Z}} h(2j)p(2j) = \sum_{j\in\mathbb{Z}} h(2j+1)p(2j+1).$$

If h satisfies the sum rules for π_{m-1} then $W := V_{\pi_{2m-1}}$ is an invariant subspace of T.

Theorem 5. *Suppose that h satisfies the sum rules for π_{m-1}. Then the following are equivalent:*

(a) $2^{2m}T|_W$ *satisfies Condition* E^{**}.

(b) *For any ϕ_0 with support in $[0,N]$, $\phi_0^{(m)} \in L^2(\mathbb{R})$ and satisfying*

$$\sum_{\ell\in\mathbb{Z}} p(\ell)\phi_0(\cdot - \ell) \in \pi_{m-1}, \qquad \text{for all } p \in \pi_{m-1}, \tag{13}$$

the derivatives of the cascade sequence $\phi_n^{(j)}$ converges weakly in $L^2(\mathbb{R})$ to $\phi^{(j)}$ as $n \to \infty$, for $j = 0, 1, \ldots, m$, where ϕ is the solution of (1).

Proof: We first assume (a) and prove (b). Define

$$b_n(\ell) = \langle \phi_n^{(m)}, \phi_n^{(m)}(\cdot - \ell)\rangle, \ \ell \in \mathbb{Z}.$$

Differentiating (2) gives

$$\phi_n^{(m)}(x) = 2^{m+1} \sum_{j=0}^{N} h(j)\phi_{n-1}^{(m)}(2x - j), \ x \in \mathbb{R},$$

and hence $b_n = (2^{2m}T)b_{n-1}$. It follows that $b_n = (2^{2m}T)^n b_0$. Condition (13) shows that $\sum_{\ell\in\mathbb{Z}} p(\ell)\phi_0^{(m)}(\cdot - \ell) = 0$ for all $p \in \pi_{m-1}$. A direct calculation then shows that $b_0 \in W$. Therefore for $n = 0, 1, 2 \ldots$, $b_n \in W$.

By (a), $(2^{2m}T)^n$ restricted to W is bounded in n. Thus b_n is bounded. In particular, $b_n(0) = \|\phi_n^{(m)}\|_2^2$ is bounded. By the same argument as in the proof of Theorem 1, $\phi_n^{(m)}$ converges weakly to $\phi^{(m)}$ as $n \to \infty$. For $0 \le j < m$, $(\phi_n^{(m)})\widehat{\ }(u) = (iu)^{m-j}(\phi_n^{(j)})\widehat{\ }$. Further, $(\phi_n^{(j)})\widehat{\ }$ converges locally uniformly. It follows that $\|(\phi_n^{(j)})\widehat{\ }\|_2$ and hence $\|\phi_n^{(j)}\|_2$ is bounded in n. By the same argument as in the proof of Theorem 1, $\phi_n^{(j)} \to \phi^{(j)}$ weakly as $n \to \infty$, for all $j = 0, 1, \ldots, m$.

Now we assume (b) and deduce (a). Let ϕ_0 be the B-spline of degree m and for some $\alpha \in \{-N+1, \ldots, N-1\}$ define $\psi_0 = \phi_0(\cdot - \alpha)$. Then $\langle \phi_0^{(m)}, \psi_0^{(m)}(\cdot - \ell)\rangle = (-1)^m \nabla^{2m}\delta(\ell - \alpha)$, where $\nabla v(i) := v(i) - v(i-1)$ for any sequence v. Now we define ϕ_n, ψ_n as in (2), and $c_n(\ell) := \langle \phi_n^{(m)}, \psi_n^{(m)}(\cdot - \ell)\rangle$. Thus $c_0 = (-1)^m\nabla^{2m}\delta(\cdot - \alpha) \in W$, for all $\alpha = -N+1, \ldots, N-1$, and $c_n = \left(2^{2m}T\right)^n c_0$, $n = 1, 2, \ldots$. By (b) we know $\phi_n^{(m)}$ and $\psi_n^{(m)}$ converge weakly in $L^2(\mathbb{R})$ to $\phi^{(m)}$ as $n \to \infty$. By the same argument as in the proof of Theorem 1 we show that c_n is bounded. Thus $\left(2^{2m}T\right)^n v$ is bounded for each $v \in W$. Further $b(\ell) := \langle \phi^{(m)}, \phi^{(m)}(\cdot - \ell)\rangle$ is an eigenvector in W of $2^{2m}T$ with eigenvalue 1. Thus $2^{2m}T|_W$ satisfies Condition E^{**}. □

Theorem 6. *Suppose that h satisfies the sum rules for π_m. Then the following are equivalent.*

(a) $(2^{2m}T)|_W$ *satisfies Condition E.*

(b) *For any ϕ_0 with compact support in $[0, N]$, $\phi_0^{(m)} \in L^2(\mathbb{R})$,*

$$\sum_{\ell \in \mathbb{Z}} p(\ell)\phi_0(\cdot - \ell) \in \pi_m, \quad \text{for all } p \in \pi_m, \tag{14}$$

and

$$\sum_{\ell \in \mathbb{Z}} \phi_0(\cdot - \ell) = 1, \tag{15}$$

the derivatives of the cascade sequence, $\phi_n^{(j)}$, converges strongly in $L^2(\mathbb{R})$ to $\phi^{(j)}$ as $n \to \infty$, for $j = 0, 1, \ldots, m$, where ϕ is the solution of (1).

Proof: Define b_n and b as in the proof of Theorem 5. Since $(2^{2m})T|_W$ satisfies Condition E, and $b \in W$ is its eigenvector with eigenvalue 1, we decompose $W = M \bigoplus Q$ as in the proof of Theorem 3, where M is the eigenspace of b and $\rho(2^{2m}T|_Q) < 1$, and write

$$b_0 = \alpha b + q, \quad \text{for some } \alpha \in \mathbb{C},\ q \in Q. \tag{16}$$

As in the proof of Theorem 1, we need only to show that $\alpha = 1$. By (14)

$$\sum_{\ell \in \mathbb{Z}} \ell^m \phi_0(x - \ell) = ax^m + p, \quad \text{for some } a \in \mathbb{C},\ p \in \pi_{m-1}.$$

Defining $\Delta f(x) := f(x) - f(x-1)$, and applying Δ^m to the above equation gives $\sum_{\ell \in \mathbb{Z}} \phi_0(x - \ell) = a$. By (15), $a = 1$. Thus

$$\sum_{\ell \in \mathbb{Z}} \ell^m \phi_0^{(m)}(x - \ell) = m!. \tag{17}$$

Now (14) and (17) imply that

$$\sum_{\ell \in \mathbb{Z}} p(\ell)\phi_0^{(m)}(x - \ell) = 0, \quad \text{for all } p \in \pi_{m-1}\,.$$

By Lemma 3.1 of [15], $\phi_0^{(m)} = \Delta^m \psi$ for some compactly supported $\psi \in L^2(\mathbb{R})$. Therefore,

$$\begin{aligned}
\sum_{\ell \in \mathbb{Z}} \ell^{2m} b_0(\ell) &= \sum_{\ell \in \mathbb{Z}} \ell^{2m} \langle \Delta^m \psi, \Delta^m \psi(\cdot - \ell)\rangle \\
&= (-1)^m \langle \psi(\cdot - m), \sum_{\ell \in \mathbb{Z}} \ell^{2m} \Delta^{2m} \psi(\cdot - \ell)\rangle \\
&= (-1)^m \langle \psi(\cdot - m), \sum_{\ell \in \mathbb{Z}} (2m)!\psi(\cdot - \ell)\rangle.
\end{aligned}$$

From (17) we have

$$m! = \sum_{\ell\in\mathbb{Z}} \ell^m \Delta^m \psi(\cdot - \ell) = \sum_{\ell\in\mathbb{Z}} m!\psi(\cdot - \ell),$$

so that

$$\sum_{\ell\in\mathbb{Z}} \psi(\cdot - \ell) = 1 \text{ and } \int_{\mathbf{R}} \psi(x)dx = 1.$$

Thus

$$\langle \psi(\cdot - m), \sum_{\ell\in\mathbb{Z}} (2m)!\psi(\cdot - \ell)\rangle = (2m)! \int_{\mathbf{R}} \psi(x-m)dx = (2m)!,$$

so that

$$\sum_{\ell\in\mathbb{Z}} \ell^{2m} b_0(\ell) = (-1)^m (2m)!. \tag{18}$$

We now show that if $w = (2^{2m}T)v$, $v \in W$, then

$$\sum_{\ell\in\mathbb{Z}} \ell^{2m} w(\ell) = \sum_{\ell\in\mathbb{Z}} \ell^{2m} v(\ell). \tag{19}$$

The left-hand side of (19) is equal to

$$\begin{aligned}\sum_{\ell\in\mathbb{Z}} \ell^{2m} w(\ell) &= \sum_{k\in\mathbb{Z}} v(k) \sum_{\ell\in\mathbb{Z}} (2\ell)^{2m} 2c(2\ell - k) \\ &= \sum_{k\in\mathbb{Z}} v(2k) \sum_{\ell\in\mathbb{Z}} (2\ell)^{2m} 2c(2\ell - 2k) \\ &+ \sum_{k\in\mathbb{Z}} v(2k+1) \sum_{\ell\in\mathbb{Z}} (2\ell)^{2m} 2c(2\ell - 2k - 1) \\ &= \sum_{k\in\mathbb{Z}} v(2k) \sum_{j\in\mathbb{Z}} (2j+2k)^{2m} 2c(2j) \\ &+ \sum_{k\in\mathbb{Z}} v(2k+1) \sum_{j\in\mathbb{Z}} (2j+2k)^{2m} 2c(2j-1),\end{aligned} \tag{20}$$

where c is the autocorrelation of h. Expanding $(2j+2k)^{2m}$ in the first sum in (20) in powers of $(2j)$ and $(2k)$, and in the second sum in powers of $(2j-1)$ and $(2k+1)$, and noting that c satisfies the sum rules for π_{2m+1} with $\sum_k c(k) = 1$, we have

$$\begin{aligned}\sum_{\ell\in\mathbb{Z}} \ell^{2m} w(\ell) &= \sum_{k\in\mathbb{Z}} k^{2m} v(k) + 2m \sum_{j\in\mathbb{Z}} (2j) 2c(2j) \sum_{k\in\mathbb{Z}} k^{2m-1} v(k) \\ &+ \cdots + \sum_{j\in\mathbb{Z}} (2j)^{2m} 2c(2j) \sum_{k\in\mathbb{Z}} v(k),\end{aligned}$$

which gives (19) since $v \in W$. From (18) and (19),

$$\sum_{\ell \in \mathbb{Z}} \ell^{2m} b_n(\ell) = (-1)^m (2m)!, \quad n = 0, 1, \ldots,$$

and hence $\sum_{\ell \in \mathbb{Z}} \ell^{2m} b(\ell) = (-1)^m (2m)!$. It also follows from (19) that if $v \in W$ is an eigenvector of $(2^{2m}T)$ with eigenvalue < 1, then $\sum_{\ell \in \mathbb{Z}} \ell^{2m} v(\ell) = 0$. Therefore for $q \in Q$, $\sum_{\ell \in \mathbb{Z}} \ell^{2m} q(\ell) = 0$. From (16), $\alpha = 1$. A similar argument as in the proof of Theorem 1 shows that $\|\phi_n^{(m)}\|_2^2 \to \|\phi^{(m)}\|_2^2$ as $n \to \infty$. Thus $\phi_n^{(m)}$ converges strongly to $\phi^{(m)}$ in $L^2(\mathbb{R})$ as $n \to \infty$.

Let $0 \le j < m$. By Theorem 5, $\phi^{(j)} \in L^2(\mathbb{R})$. For $|u| \ge 1$,

$$\begin{aligned} |(\phi_n^{(m)})\widehat{\ }(u) - (\phi^{(m)})\widehat{\ }(u)|^2 &= |u|^{2(m-j)} |(\phi_n^{(j)})\widehat{\ }(u) - (\phi^{(j)})\widehat{\ }(u)|^2 \\ &\ge |(\phi_n^{(j)})\widehat{\ }(u) - (\phi^{(j)})\widehat{\ }(u)|^2. \end{aligned}$$

On the other hand,

$$|(\phi_n^{(j)})\widehat{\ }(u) - (\phi^{(j)} k)\widehat{\ }(u)| \to 0, \quad \text{as } n \to \infty,$$

uniformly for $|u| \le 1$. It follows that $\|(\phi_n^{(j)})\widehat{\ } - (\phi^{(j)})\widehat{\ }\|_2 \to 0$, as $n \to \infty$.

The proof that (b) implies (a) is the same as that in Theorem 5. □

For a nonnegative integer m, define the Sobolev space $L_m^2(\mathbb{R})$ that comprise all real functions f on $\mathbb{R}$ for which $f^{(j)} \in L^2(\mathbb{R})$, for $j = 0, 1, \ldots, m$, with norm

$$\|f\|_{2,m} := \{\sum_{j=0}^{m} \|f^{(j)}\|_2^2\}^{1/2}.$$

Theorem 6 can be recast in terms of Sobolev convergence.

Corollary 7. *Suppose that h satisfies the sum rules for π_m. Then the following are equivalent.*

(a) $(2^{2m}T)|_W$ *satisfies Condition E.*

(b) *For any $\phi_0 \in L_m^2(\mathbb{R})$ with compact support in $[0, N]$, satisfying*

$$\sum_{\ell \in \mathbb{Z}} p(\ell) \phi_0(\cdot - \ell) \in \pi_m, \qquad \text{for all } p \in \pi_m, \tag{21}$$

and

$$\sum_{\ell \in \mathbb{Z}} \phi_0(\cdot - \ell) = 1, \tag{22}$$

the cascade sequence ϕ_n converges strongly in $L_m^2(\mathbb{R})$, as $n \to \infty$, to ϕ which is the solution of (1).

2.4. Matrix Cascade Algorithms and Transition Operators

We now consider the matrix refinement equation (MRE) of the form

$$\Phi(x) = \sum_{k=0}^{N} 2h(j)\Phi(2x - j), \quad x \in \mathbb{R}, \tag{23}$$

where $\Phi = (\phi_1, \ldots, \phi_r)^T$ and h is a finitely supported sequence of $r \times r$ matrices. For a starting vector function $\Phi_0(t) = (\phi_{0,1}, \ldots, \phi_{0,r})^T$ in $L^2(\mathbb{R})^r$ with support in $[0, N]$, i.e. $\text{supp}(\phi_{0,j}) \subset [0, N]$ for $j = 1, \ldots, r$, the cascade algorithm defines a sequence

$$\Phi_n(x) = \sum_{k=0}^{N} 2h(j)\Phi_{n-1}(2x - j), \quad x \in \mathbb{R},\ n = 1, 2, \ldots . \tag{24}$$

Each $\Phi_n \in L^2(\mathbb{R})^r$ and has support in $[0, N]$. The $r \times r$ matrix

$$b_n(x) = \int_{\mathbb{R}} \Phi_n(t)\Phi_n(t - x)^* dt \tag{25}$$

defines the Gramian of Φ_n. Equations (24) and (25) lead to

$$b_n(x) = \sum_{k=0}^{N} \sum_{\ell=k-N}^{k} 2h(k)b_{n-1}(2x - \ell)h(k - \ell)^* . \tag{26}$$

Setting $t = \nu$ in (26) gives

$$b_n(\nu) = \sum_{k=0}^{N} \sum_{\ell=k-N}^{k} 2h(k)b_{n-1}(2\nu - \ell)h(k - \ell)^* . \tag{27}$$

Defining the *transition operator* T on all sequences $(b(n))_{n \in [-N+1, N-1]}$ of $r \times r$ matrices by

$$(Tb)(\nu) = \sum_{k=0}^{N} \sum_{\ell=k-N}^{k} 2h(k)b(2\nu - \ell)h(k - \ell)^T, \tag{28}$$

the relation (27) becomes

$$b_n = Tb_{n-1}, \quad n = 1, 2 \ldots . \tag{29}$$

To describe the transition operator T in the Fourier domain which is easier to deal with at times, we introduce the space $\mathcal{V}_N$ of all $r \times r$ matrices with trigonometric polynomial entries whose Fourier coefficients are real and supported in $[-N + 1, N - 1]$. Let

$$\mathcal{T}B(u) := \sum_{k=-N+1}^{N-1} Tb(k)e^{-iku},$$

where $B(u) := \sum_{k=-N+1}^{N-1} b(k)e^{-iku}$. Then $\mathcal{T}$ is a linear operator on $\mathcal{V}_n$ given by

$$\mathcal{T}B(u) = H(\frac{u}{2})B(\frac{u}{2})H^*(\frac{u}{2}) + H(\frac{u}{2}+\pi)B(\frac{u}{2}+\pi)H^*(\frac{u}{2}+\pi), \quad B \in \mathcal{V}_N.$$

Equation (29) becomes $B_n = T_H B_{n-1}$, $n = 1, 2\ldots$, where B_n is the Fourier series with vector Fourier coefficients $(b_n(\nu))_{\nu=-N+1}^{N-1}$.

Also, in the Fourier domain the cascade algorithm (24) becomes

$$\hat{\Phi}_n(u) = H(u/2)\hat{\Phi}_{n-1}(u/2), \quad u \in \mathbb{R}, \tag{30}$$

where $H(u) := \sum_{k=0}^{N} h(k)e^{-iku}$. Iterating (30) gives

$$\hat{\Phi}_n(u) = \prod_{j=1}^{n} H(u/2^j)\hat{\Phi}_0(u/2^n), \quad n = 1, 2, \ldots . \tag{31}$$

Since Φ_0 is compactly supported, its Fourier transform $\hat{\Phi}_0$ is continuous at 0. Thus $\lim_{n\to\infty} \hat{\Phi}_0(u/2^n) = \Phi_0(0) =: v$ for some $v \in \mathbb{R}^r$. From (30), v is an eigenvector of $H(0)$ with eigenvalue 1. For a finitely supported mask, the infinite product $P(u) = \prod_{j=1}^{\infty} H(u/2^j)$ converges locally uniformly if and only if $H(0)$ satisfies Condition E^*.

Example: Consider the sequence of 2×2 matrices $h(k)$ with 3 nonzero terms

$$h(0) = \frac{1}{16}\begin{pmatrix} 4 & 6 \\ -1 & -1 \end{pmatrix}, \quad h(1) = \frac{1}{4}\begin{pmatrix} 2 & 0 \\ 0 & 1 \end{pmatrix}, \quad h(2) = \frac{1}{16}\begin{pmatrix} 4 & -6 \\ 1 & -1 \end{pmatrix}.$$

Then $H(u) = h(0) + h(1)^{-iu} + h(2)e^{-i2u}$, and

$$H(0) = h(0) + h(1) + h(2) = \begin{pmatrix} 1 & 0 \\ 0 & 1/8 \end{pmatrix},$$

which satisfies Condition E^*. In fact $H(0)$ satisfies Condition E.

Suppose $H(0)$ satisfy Condition E^*. Then $P(u) := \prod_{j=1}^{\infty} H(u/2^j)$, $u \in \mathbb{R}$ defines an $r \times r$ matrix of continuous functions which are of polynomial growth as $|u| \to \infty$. The proof is standard and can be found in [7] for the scalar case and [16] for the vector case. In this case, (31) shows that

$$\lim_{n\to\infty} \hat{\Phi}_n(u) = P(u)v, \quad u \in \mathbb{R}^s,$$

where $P(u)v$ is a distributional solution of the MRE (23). Suppose further that the eigenvalue 1 of $H(0)$ has multiplicity d with d linearly independent eigenvectors $v_1, \ldots, v_d$. Then Φ_j defined by $\hat{\Phi}_j := P(u)v_j$ $j = 1, 2, \cdots, d$, are distributional solutions of the MRE (23) since they satisfy $\widehat{\Phi}_j(u) = H(u/2)\widehat{\Phi}_j(u/2)$, and any solution of (23) lies in the linear span of $\{\hat{\Phi}_1, \hat{\Phi}_2, \ldots, \hat{\Phi}_d\}$.

We shall say T satisfies Condition E^{**} (Condition E^* or Condition E) if its representation matrix satisfies Condition E^{**} (Condition E^* or Condition E, respectively).

The following theorems extend Theorem 1 to the vector case.

Theorem 8. *The cascade sequence Φ_n converges weakly in $L^2(\mathbb{R})^r$ for any starting vector Φ_0 if and only if T satisfies Condition E^{**}.*

Another extension of the concept of fundamentality of a mask is required for the strong convergence of vector cascade algorithms in $L^2(\mathbb{R})^r$. We say that a matrix mask h is fundamental with respect to a nonzero vector $v \in \mathbb{R}^r$ if

$$v^T \sum_{j\in\mathbb{Z}} h(2j) = v^T \sum_{j\in\mathbb{Z}} h(2j+1) = \frac{v^T}{2}.$$

Theorem 9. *Suppose that*

(a) *T satisfies Condition E and h is fundamental with respect to v.*

Then

(b) *For any initial vector Φ_0 satisfying*

$$v^T \sum_{j\in\mathbb{Z}} \Phi_0(\cdot - j) = 1, \quad \textit{almost everywhere}, \tag{32}$$

the cascade sequence Φ_n converges strongly in $L^2(\mathbb{R})^r$ to a solution of the MRE (23).

Conversely, if $H(0)$ satisfies Condition E, then (b) implies (a).

We note that if $H(0)$ satisfies Condition E, the MRE (23) has a unique solution. Under this assumption the statements (a) and (b) in Theorem 9 are equivalent, a result obtained earlier in [30].

In order to check for Conditions E, E^* and E^{**} for the transition operator, it is necesssary to compute the matrix representation of the operator. This has been done using Kronecker products, see [29,20]. The Kronecker product $B \otimes C$ of two $r \times r$ matrices $B = (b_{ij})_{i,j=1}^r$ and C is the $r^2 \times r^2$ matrix $(b_{i,j}C)_{i,j=1}^r$. Indeed the matrix representation for the transition operator for the matrix mask h is $(2a(2i-j))_{1-N\le i,j\le N-1}$, where $a(j)$ is the $r^2 \times r^2$ matrix defined by $a(j) := \sum_{\ell=0}^N h(\ell - j) \otimes h(\ell)$.

§3. Nonstationary Cascade Algorithms in $L^2(\mathbb{R}^s)$

In a stationary multiresolution $(V_k)_{k\in\mathbb{Z}}$ of $L^2(\mathbb{R}^s)$ the spaces V_k are generated by the shifts and dilations of one scaling function ϕ that is a solution of a stationary refinement equation of the form

$$\phi(x) = |M| \sum_{j\in\Omega} 2h(j)\phi(Mx - j), \quad x \in \mathbb{R}^s, \tag{33}$$

where h is a finitely supported sequence with support in a finite set $\Omega \subset \mathbb{Z}^s$ and M is an $s \times s$ dilation matrix. Stationary multiresolutions do not exist in Hilbert spaces, such as the Sobolev spaces and periodic L^2-spaces, which do not have a unitary dilation operator. Therefore, in the multiresolution decomposition of such a space, it is natural to consider nonstationary multiresolution. In a nonstationary multiresolution $(V_k)_{k\in\mathbb{Z}}$, each V_k is generated by the 2^{-k}-shifts of a function ϕ_k which depends on the multiresolution level k. These functions constitute a sequence (ϕ_k) that is a solution of a nonstationary refinement equation

$$\phi_k(x) = |M| \sum_{j\in\Omega} h_{k+1}(j)\phi_{k+1}(Mx-j), \quad x \in \mathbb{R}^s,\ k = 0,1,2,\ldots, \tag{34}$$

where h_k is a finitely supported sequence with support in Ω for each k. Here, we have assumed that $\phi_k = \phi_0$ for all $k < 0$.

The equation (34) is the scalar case of the nonstationary matrix refinement equations of the form

$$\Phi_k(x) = |M| \sum_{j\in\Omega} h_{k+1}(j)\Phi_{k+1}(Mx-j), \quad x \in \mathbb{R}^s,\ k = 0,1,2,\ldots, \tag{35}$$

where $\Phi = (\phi_1, \ldots, \phi_r)^T$ and h_k is a finitely supported sequence of $r \times r$ matrices with support in Ω for each k. We shall also assume throughout that there is a sequence of $r \times r$ matrices h supported on Ω such that for each $j \in \Omega$

$$\sum_{k=0}^{\infty} \|h_k(j) - h(j)\| < \infty. \tag{36}$$

In the Fourier domain, (35) is equivalent to

$$\hat{\Phi}_k(u) = H_{k+1}(\tilde{M}u)\hat{\Phi}_{k+1}(\tilde{M}u), \quad u \in \mathbb{R}^s, \tag{37}$$

where $H_k(u) = \sum_{j\in\mathbb{Z}^s} h_k(j)e^{-iju}$, $u \in \mathbb{R}^s$ and $\tilde{M} = (M^{-1})^T$. A sequence (Φ_k) is a *solution* of (35) if it satisfies (35) or (37) and

$$\lim_{n\to\infty} \hat{\Phi}_{k+n}(\tilde{M}^n u) = v, \quad \text{for some } v \in \mathbb{R}^r, \tag{38}$$

uniformly in k and locally uniformly in $u \in \mathbb{R}^s$.

If $h_k = h$ for all k, then $\Phi_k = \Phi$ for all k, and equation (35) reduces to the stationary refinement equation

$$\Phi(x) = |M| \sum_{j\in\Omega} 2h(j)\Phi(Mx-j), \quad x \in \mathbb{R}^s. \tag{39}$$

Equation (39) can also be viewed as the limiting case of the nonstationary refinement equation (35), and we shall also refer to it as the ideal refinement equation associated with (35). Equation (39) is equivalent to

$$\widehat{\Phi}(u) = H(\tilde{M}u)\widehat{\Phi}(\tilde{M}u),$$

where $H(u) := \sum_j h(j)e^{-iju}$. This leads to

$$\widehat{\Phi}(u) = \prod_{\ell=1}^{n} H(\tilde{M}^{\ell}u)\widehat{\Phi}(\tilde{M}^{n}u).$$

As in the univariate case, the infinite product $\prod_{\ell=1}^{n} H(\widetilde{M}^{\ell}u)$ converges locally uniformly if and only if $H(0)$ satisfies Condition E^*. We shall assume throughout this section that $H(0) = \sum_j h(j)$ satisfies Condition E^*. By (36) the sequence (h_k) satisfies

$$\sum_{k=1}^{\infty} \| \sum_{j} h_k(j) - H(0) \| < \infty. \tag{40}$$

Example: Take $r = s = 1$. Let $\lambda_0, \ldots, \lambda_N \in \mathbb{R}$, and for $k = 0, 1, \ldots,$ define a function ϕ_k by

$$\hat{\phi}_k(u) := \frac{\prod_{j=0}^{N}(e^{-2^{-k}\lambda_j} - e^{-iu})}{\prod_{j=0}^{N}(iu - 2^{-k}\lambda_j)}.$$

Then

$$\hat{\phi}_k(u) = H_{k+1}(u/2)\hat{\phi}_{k+1}(u/2),$$

where

$$H_{k+1}(u) := \prod_{j=0}^{N}\left(\frac{e^{-2^{-k-1}\lambda_j} + e^{-iu}}{2}\right).$$

Clearly,

$$\lim_{k\to\infty} H_k(u) = \left(\frac{1+e^{iu}}{2}\right)^{N+1},$$

and it is easy to see that h_k and h satisfy (40).

We choose a bounded set $S \subset \mathbb{R}^s$ satisfying $\cup_{j\in\Omega} M^{-1}(S+j) \subset S$, and choose a sequence $\Phi_{k,0}$, $k = 0, 1, 2, \ldots,$ in $L^2(\mathbb{R}^s)^r$ with support in S satisfying

$$\lim_{k\to\infty} \Phi_{k,0} = \tilde{\Phi}_0 \ \text{ strongly in } L^2(\mathbb{R}^s), \tag{41}$$

and

$$\lim_{n\to\infty} \hat{\Phi}_{k+n,0}(\tilde{M}^n u) = v, \tag{42}$$

uniformly in k and locally uniformly in u, where v is the same as in (38).

Now for $n = 1, 2, \ldots,$ $k = 0, 1, 2, \ldots,$ we define $\Phi_{k,n}$ in $L^2(\mathbb{R}^s)^r$ by

$$\Phi_{k,n}(x) = |M| \sum_{j\in\Omega} h_{k+1}(j)\Phi_{k+1,n-1}(Mx - j), \quad x \in \mathbb{R}^s. \tag{43}$$

We call (43) a nonstationary cascade algorithm for the refinement equation (35) (or for the mask (h_k)). The nonstationary cascade sequence $\Phi_{k,n}$ is a double sequence. The supports of $\Phi_{k,n}$ are in S for all k and n. We are interested in its limits as $n \to \infty$ and $k \to \infty$.

As in the stationary case, for $k, n = 0, 1, \ldots$, $j \in \mathbb{Z}^s$, we define a $r \times r$ matrix

$$b_{k,n}(j) := \int_{\mathbb{R}^s} \Phi_{k,n}(t)\Phi_{k,n}(t-j)^*.$$

Then for any k and n, $b_{k,n}$ is a finitely supported sequence of $r \times r$ matrices with support in some finite set $I \subset \mathbb{Z}^s$. Indeed, I can be chosen such that $S \cap (S + j) = \emptyset$ for $j \in I$. Further (43) gives

$$b_{k,n}(j) = |M| \sum_{\mu,\nu \in \mathbb{Z}^s} h_{k+1}(\mu) b_{k+1,n-1}(Mj - \nu) h_{k+1}(\mu - \nu)^T, \quad j \in I. \tag{44}$$

We shall denote by $\ell(I)$ the set of all sequences supported on I. Note that in the scalar case ($r = 1$) (44) simplifies to

$$b_{k,n}(j) = |M| \sum_{\nu \in \mathbb{Z}^s} c_{k+1}(Mj - \nu) b_{k+1,n-1}(\nu), \quad j \in I, \tag{45}$$

where c_{k+1} is the autocorrelation of h_{k+1}. Expressed in matrix form (44) becomes

$$b_{k,n} = T_{k+1} b_{k+1,n-1}, \quad k = 0, 1, \ldots, \ n = 1, 2, \ldots, \tag{46}$$

where T_k is the transition operator for the mask (h_k), i.e.

$$T_k b(j) = |M| \sum_{\mu,\nu \in \mathbb{Z}^s} h_k(\mu) b(Mj - \nu) h_k(\mu - \nu)^T, \quad j \in I. \tag{47}$$

In the scalar case (r=1) the corresponding transition operator is defined according to the relation (45). Iterating (46) leads to

$$b_{k,n} = \left(\prod_{\ell=k+1}^{k+n} T_\ell \right) b_{k+n,0}, \quad k, n = 0, 1, 2, \ldots. \tag{48}$$

Let T be the transition operator for the matrix mask h. The assumption (36) on h_k and h carries over to T_k and T, i.e.

$$\sum_{k=0}^{\infty} \|T_k - T\| < \infty, \tag{49}$$

a condition that is crucial for the boundedness and convergence of the product on the right of (48). The analysis on the convergence of nonstationary cascade algorithms depends on the following results on infinite products of matrices.

Theorem 10. *Suppose that (49) holds and the spectral radius $\rho(T) \geq 1$. Then $\left\|\prod_{\ell=k+1}^{k+n} T_\ell\right\|$ is bounded for all n and k if and only if T satisfies Condition E^{**}.*

Theorem 11. *Suppose that (49) holds. Then the following are equivalent.*

(a) *T satisfies Condition E^*.*

(b) *The matrix product $\prod_{\ell=k+1}^{k+n} T_\ell$ converges as $n \to \infty$, uniformly in k.*

Moreover, if (a) or (b) holds then

$$\lim_{k\to\infty}\lim_{n\to\infty}\prod_{\ell=k+1}^{k+n} T_\ell = \lim_{n\to\infty}\lim_{k\to\infty}\prod_{\ell=k+1}^{k+n} T_\ell = \lim_{n\to\infty} T^n.$$

Theorem 11 was proved in [13]. The proof in [13] can be adapted for Theorem 10 (see also [15]).

Let us return to the nonstationary refinement equation and the cascade algorithms. Iterating (37) gives

$$\hat{\Phi}_k(u) = P_{k,n}(u)\hat{\Phi}_{k+n}(\tilde{M}^n u), \quad u \in \mathbb{R}^s, \tag{50}$$

where

$$P_{k,n}(u) := \prod_{\ell=1}^{n} H_{k+\ell}(\tilde{M}^\ell u), \quad u \in \mathbb{R}^s. \tag{51}$$

The assumption (40) leads to

$$\sum_{\ell=1}^{\infty} \|H_{k+\ell}(\tilde{M}^\ell u) - H(0)\| < \infty,$$

where the convergence is locally uniform in u and uniform in k. Theorem 11 implies (not directly) that the infinite product $P_k(u) := \prod_{\ell=1}^{\infty} H_{k+\ell}(\tilde{M}^\ell u)$, $u \in \mathbb{R}^s$, converges uniformly in k and locally uniformly in u. A similar argument as before shows that $P_k(u)v$ is the Fourier transform of a distributional solution Φ_k of (34). As in the scalar case, since $H(0)$ satisfies Condition E^* all the solutions of (35) are in the linear span of $\{\Phi_{k,1}, \dots, \Phi_{k,d}\}$ where $\hat{\Phi}_{k,j} = P_k v_j$, and v_j, $j = 1, \dots, d$, are the linearly independent eigenvectors of $H(0)$ with eigenvalue 1.

Similarly, (43) is equivalent to

$$\hat{\Phi}_{k,n}(u) = H_{k+1}(\widetilde{M}u)\hat{\Phi}_{k+1,n-1}(\widetilde{M}u),$$

which leads to

$$\hat{\Phi}_{k,n}(u) = \prod_{\ell=1}^{n} H_{k+\ell}(\tilde{M}^\ell u)\hat{\Phi}_{k+n,0}(\tilde{M}^n u), \quad u \in \mathbb{R}^s,$$

for $k, n = 0, 1, 2, \dots$. From the convergence of the product on the right and and from (42), we have $\lim_{n\to\infty} \hat{\Phi}_{k,n}(u) = P_k(u)v$, where the convergence is uniform over k and locally uniform in u.

3.1. The Scalar Case in $\mathbb{R}^s$

For $r = 1$, we write $\Phi = \phi$, and $\Phi_{k,n} = \phi_{k,n}$. In this case, h_k and h are scalar sequences. The vector v in (42) is now a scalar and is taken to be 1. Extending Theorems 1 and 3 we have the following results on the convergence of nonstationary cascade algorithms in $L^2(\mathbb{R}^s)$.

Theorem 12. *The following are equivalent.*

(a) *T satisfies Condition E^{**}.*

(b) *For any compactly support $(\phi_{k,0}) \in L^2(\mathbb{R}^s)$ satisfying (41) and (42), $\phi_{k,n}$ converges weakly to ϕ_k in $L^2(\mathbb{R})$ as $n \to \infty$, uniformly in k, where (ϕ_k) is the L^2-solution of (34), and ϕ_k converges weakly in $L^2(\mathbb{R})$ to ϕ which is the solution of the ideal refinement equation (33).*

(c) *For any compactly supported $\phi_{k,0} \in L^2(\mathbb{R}^s)$ satisfying (41) and (42), $\phi_{k,n}$ converges weakly to $\tilde{\phi}_n$ in $L^2(\mathbb{R})$ as $k \to \infty$, uniformly in n, where $(\tilde{\phi}_n)$ is the cascade sequence for h with starting function $\tilde{\phi}_0$, and $\tilde{\phi}_n$ converges weakly in $L^2(\mathbb{R})$ to ϕ which is the solution of (33).*

Theorem 13. *The following are equivalent.*

(a) *T satisfies Condition E and h is fundamental.*

(b) *For any compactly supported $\phi_{k,0} \in L^2(\mathbb{R}^s)$ satisfying (41), (42) and*

$$\sum_{j\in\mathbb{Z}^s} \tilde{\phi}_0(\cdot - j) = 1, \qquad \textit{almost everywhere,}$$

the cascade sequence $\phi_{k,n}$ converges strongly to ϕ_k in $L^2(\mathbb{R})$ as $n \to \infty$, uniformly in k, where (ϕ_k) is the solution of (34), and ϕ_k converges strongly in $L^2(\mathbb{R})$ to ϕ which is the solution of (33).

(c) *For any $(\phi_{k,0})$ as in (b), the cascade sequence $\phi_{k,n}$ converges strongly to $\tilde{\phi}_n$ in $L^2(\mathbb{R})$ as $k \to \infty$, uniformly in n, where $(\tilde{\phi}_n)$ is the cascade sequence for h with starting function $\tilde{\phi}_0$, and $\tilde{\phi}_n$ converges strongly in $L^2(\mathbb{R})$ to ϕ which is the solution of (33).*

Complete proofs of Theorems 12 and 13 are given in [15]. The proofs combine results on the boundedness and convergence of infinite products of matrices in Theorems 10 and 11 and the ideas in the proof of Theorems 1 and 3. These results can be partially extended to the vector case of nonstationary cascade algorithms in the same way as Theorem 1 and Theorem 3 are extended to Theorems 8 and 9 respectively. They will be considered in Subsection 3.2. We remark that the conditions on $(\phi_{k,0})$ in Theorems 12 and 13 are easily satisfied if $\phi_{k,0}$ is independent of k.

Theorems 5 and 6 characterize weak and strong convergence respectively, of derivatives of cascade sequences in one dimension. The corresponding results in higher dimensions with a general dilation matrix are more involved and take many forms. In [15] only weak convergence of derivatives is considered, and the results are not as complete as in the scalar case in one dimension. More has to be done. We shall state here only one of the results in [15].

First we extend further the concept of fundamentality of a scalar mask h. Recall that M is an integer matrix with all eigenvalues of modulus > 1 and $\det(M) \geq 2$. Let Γ comprise coset representatives of $\mathbb{Z}^s/M\mathbb{Z}^s$. For any finite dimensional space π of polynomials on $\mathbb{R}^s$, we say that a mask h satisfies the *sum rules for* π if for any $\gamma \in \Gamma$ and $p \in \pi$,

$$\sum_j h(Mj)p(Mj) = \sum_j h(Mj+\gamma)p(Mj+\gamma).$$

When $\pi = \pi_{m-1}$, the space of all polynomials in $\mathbb{R}^s$ of total degree $\leq m-1$, the above definition coincides with the sum rules of order m as defined in [19]. We also write $V_\pi = \{v \in \ell(I) : \sum_{j\in I} p(j)v(j) = 0 \text{ for all } p \in \pi\}$.

Theorem 14. *Suppose that M has s linearly independent eigenvectors in $\mathbb{Z}^s$ with eigenvalues $\lambda_1, \ldots, \lambda_s$, h_k satisfies the sum rules for π_{m-1} for some $m \geq 1$, for all $k \geq 0$, and T satisfies Condition E^{**}. Then $W = V_{\pi_{2m-1}}$ is an invariant subspace of T_k, and (a) implies (b), where (a) and (b) are as follows:*

(a) *$\rho(M)^{2m}T$ restricted to W satisfies Condition E^{**}.*

(b) *For any eigenvectors $u_1, \ldots, u_m$ of M and any sequence of functions $\phi_{k,0}$ with support in S satisfying (42) with $v = 1$, $D_{u_1} \cdots D_{u_m}\phi_{k,0} \in L^2(\mathbb{R}^s)$, and*

$$\sum_{\ell\in\mathbb{Z}^s} p(\ell)D_{u_1}\cdots D_{u_m}\phi_{k,0}(\cdot - \ell) = 0, \quad \text{for all } p \in \pi_{m-1},$$

we have $D_{u_1}\cdots D_{u_m}\phi_{k,n} \to D_{u_1}\cdots D_{u_m}\phi_k$ weakly in $L^2(\mathbb{R}^s)$ as $n \to \infty$, uniformly in k, where (ϕ_k) is the solution of (34), and $D_{u_1}\cdots D_{u_m}\phi_k \to D_{u_1}\cdots D_{u_m}\phi$ weakly in $L^2(\mathbb{R}^s)$ as $k \to \infty$, where ϕ is the solution of (33).

3.2. Nonstationary Matrix Cascade Algorithms in $\mathbb{R}^s$

We end this paper with some results on the convergence of nonstationary matrix cascade algorithms extending Theorems 8 and 9 in Subsection 2.4. The cascade algorithm for (h_k) is defined in (43), where h and h_k are finitely supported sequences of $r \times r$ matrices. We say that h is fundamental with respect to a non-zero vector v in $\mathbb{R}^r$ if for any $\gamma \in \Gamma$,

$$v^T \sum_{j\in\mathbb{Z}^s} h(Mj+\gamma) = \frac{v^T}{|M|}.$$

Theorem 15. *The following are equivalent:*

(a) T satisfies condition E^{**}.

(b) For any initial sequence $(\Phi_{k,0})$ satisfying (41) and (42), the cascade algorithm $\Phi_{k,n}$ converges weakly to Φ_k in $L^2(\mathbb{R}^s)^r$ as $n \to \infty$, uniformly in k, where (Φ_k) is a solution of (35), and Φ_k converges weakly in $L^2(\mathbb{R}^s)^r$ to a solution Φ of (33).

Theorem 16. *Suppose that*
(a) *T satisfies Condition E and h is fundamental with respect to v. Then*
(b) *for any initial sequence $(\Phi_{k,0})$ satisfying (41), (42) and*

$$v^T \sum_{j\in\mathbb{Z}^s} \tilde{\Phi}_0(\cdot - j) = 1 \text{ almost everywhere },$$

the cascade algorithm $\Phi_{k,n}$ converges strongly in $L^2(\mathbb{R}^s)^r$ to a solution Φ_k of (35) as $n \to \infty$, uniformly in k. Moreover as $k \to \infty$, ϕ_k converges strongly in $L^2(\mathbb{R}^s)^r$ to a solution Φ of (33) which satisfies

$$v \sum_{j\in\mathbb{Z}^s} \Phi(\cdot - j) = 1 \text{ almost everywhere }.$$

Further if $H(0)$ satisfies Condition E, then (b) implies (a).

Acknowledgments. Supported by the Wavelets Strategic Research Programme, National University of Singapore, under a grant from the National Science and Technology Board and the Ministry of Education, Singapore.

References

1. deBoor, C., R. DeVore, and A. Ron On the construction of multivariate (pre) wavelets, Constr. Approx. **9** (1993), 123–166.
2. Cavaretta, A. S., W. Dahmen, and C. A. Micchelli, *Stationary Subdivision*, Memoir Amer. Math. Soc. Vol. 93 , 1991, 1–186.
3. Chui, Charles, K., *An Introduction to Wavelets,* Academic Press, Boston, 1992.
4. Cohen, A., I. Daubechies, and J. C. Feauveau, Biorthogonal basis of compactly supported wavelets, Comm. Pure and Appl. Math. **45** (1992) 485–560.
5. Cohen, A., I. Daubechies, and G. Plonka, Regularity of refinable functions, J. Fourier Anal. and Appl. (1996).
6. Cohen, A., K. Gröchenig, and L. Villemoes Regularity of multivariate refinable functions, preprint, 1996.
7. Daubechies, I., *Ten Lectures on Wavelets,* CBMS-NSF Series in Applied Mathematics #61, SIAM Publ., Philadelphia, 1992.
8. Daubechies, I., Orthonormal bases of compactly supported wavelets, Com. Pure and Appl. Math. **41** (1988), 909–996.
9. Daubechies, I., and J. C. Lagaria, Two-scale difference equations I. Existence and global regularity of solutions, SIAM J. Math. Anal. **22** (1991), 1388–1410.

10. Daubechies, I., and J. C. Lagaria, Two-scale difference equations II. Local regularity, infinite products of matrices and fractals, SIAM J. Math. Anal. **23** (1992), 1031–1079.

11. Dyn, N., D. Levin, and C. A. Micchelli, Using parameters to increase smoothness of curves and surfaces generated by subdivision, Computer Aided Geom. Design **7** (1990), 91–106.

12. Eirola, T., Sobolev characterization of solutions of dilation equations, SIAM J. Math. Anal. **23** (1992), 1015–1030.

13. Goh Say Song, S. L. Lee, and K. M. Teo, Multidimensional periodic multiwavelets, preprint, National University of Singapore, 1997.

14. Goodman, T. N. T., C. A. Micchelli, and J. Ward, Spectral radius formulas for subdivision operators , in *Recent Advances in Wavelet Analysis*, L. L. Schumaker and G. Webb (eds), Academic Press, 1995, 335–360.

15. Goodman, T. N. T., and S. L. Lee, Convergence of nonstationary cascade algorithms, preprint, National University of Singapore, 1997.

16. Heil, C., and D. Colella, Matrix refinement equations: Existence and Uniqueness, J. Fourier Anal. and Appl. (1996).

17. Hervé, L., Construction et regularite des fonctions d' echelle, SIAM J. Math. Anal. **26** (1995), 1361–1385.

18. Jia, R. Q., Subdivision schemes in L^p spaces, Advances in Computational Mathematics **3** (1995), 309–341.

19. Jia, R. Q., Characterization of smoothness of multivariate refinable functions in Sobolev spaces, preprint, University of Alberta, 1996.

20. Jiang, Q., Orthogonal multiwavelets with optimum time-frequency resolution, IEEE Trans. Signal Processing, to appear.

21. Jiang Q., On the regularity of matrix refinable functions, SIAM J. Math. Anal., to appear.

22. Lawton, W., Necessary and sufficient conditions for constructing orthonormal wavelets, J. Math. Phys. **32** (1991), 52–61.

23. Lawton, W., Multilevel properties of the wavelet-Galerkin operator, J. Math. Phys. **32** (1991), 1440–1443.

24. Lawton, W., S. L. Lee and Zuowei Shen, Stability and orthonormality of multivariate refinable functions, SIAM J. Math. Anal. **28** (1997), 999–1014.

25. Lawton, W., S. L. Lee, and Zuowei Shen, Convergence of multidimensional cascade algorithm, Numer. Math., to appear.

26. Lee, S. L., H. H. Tan, and W. S. Tang, Wavelet bases for a unitary operator, Proc. Edinburgh Math. Soc. **38** (1995), 233–260.

27. Long, R., and D. R. Chen, Biorthogonal wavelets bases on $\mathbb{R}^d$, Appl. Comp. Harmonic Anal. **2** (1995), 230–242.

28. Micchelli, C. A., and Thomas Sauer, Regularity of multiwavelets, Advances in Comput. Math. **7** (1997), 455–545.

29. Plonka, G., On stability of scaling vectors, preprint.

30. Shen, Z. Refinable function vectors, SIAM J. Math. Anal. (1997).

31. Strang, G., Eigenvalues of $(\downarrow 2)H$ and convergence of cascade algorithm, IEEE Trans. Signal Proc. **44** (1996).

32. Strang, G., and T. Nguyen, *Wavelets and Filter Banks*, Wellesley- Cambridge Press, 1996.

33. Villemoes, L., Wavelet analysis of refinable equations, SIAM J. Math. Anal. **25** (1994), 1433–1460.

T. N. T. Goodman
Department of Mathematical and Computer Sciences
University of Dundee
Dundee DD1 4HN, SCOTLAND
`tgoodman@mcs.dundee.ac.uk`

S. L. Lee
Department of Mathematics
National University of Singapore
SINGAPORE 119260
`matleesl@math.nus.sg`

Refinable Vectors of Spline Functions

T. N. T. Goodman and S. L. Lee

Abstract. In this paper we study refinable vectors of spline functions for general knots. A characterisation is given of refinable spline spaces, i.e. spaces $\mathcal{S}$ of spline functions such that $f \in \mathcal{S}$ implies $f(\frac{\cdot}{2}) \in \mathcal{S}$. By considering bases for such spaces we characterise refinable splines whose integer translates are linearly independent and for which the number of spline functions is 'maximal'. Further results are also mentioned.

§1. Introduction

For functions $\phi_i : \mathbb{R} \to \mathbb{R}$, $i = 1, \ldots, r$, with compact support, we say $\phi = (\phi_1, \ldots, \phi_r)^T$ is refinable with dilation 2 if

$$\phi(x) = \sum_{j=-N}^{N} A(j)\phi(2x - j), \quad x \in \mathbb{R}, \tag{1}$$

for $N > 0$ and $r \times r$ matrices $A(j)$. Refinable functions play a central role in multiresolution methods, in particular in the construction of scaling functions and wavelets [1,3,4,5] and in the construction of scaling vectors and multiwavelets [2,7,9,12,14].

We are concerned with the case when ϕ comprises spline functions, i.e. piecewise polynomials. The work is largely motivated by the results in [15] where it was shown that for $r = 1$, the only refinable splines whose integer translates form Riesz bases are B-splines with simple knots at the integers $\mathbb{Z}$. For $r \geq 2$ the situation is more complex. In [8], Donovan, Geronimo and Hardin have constructed refinable splines for $r = 3$ whose integer translates are orthogonal. In [10], refinable splines with knots at $\mathbb{Z}$ and which form Riesz bases are characterised for $r = 2$. In this paper we continue this study for general knots and any r. In Section 2 we characterise refinable spline spaces, i.e., spaces $\mathcal{S}$ of spline functions such that $f \in \mathcal{S}$ implies $f(\frac{\cdot}{2}) \in \mathcal{S}$. By considering bases for such spaces, we characterise in Section 3 refinable splines whose integer translates are linearly independent and for which r is 'maximal'. Further results are also mentioned.

Mathematical Methods for Curves and Surfaces II
Morten Dæhlen, Tom Lyche, and Larry L. Schumaker (eds.), pp. 213–220.

ISBN 0-8265-1315-8.

§2. Refinable Spline Spaces

In order to study our spaces of spline functions, we shall introduce some spline functions of compact support which will act as basis functions. These were introduced in [10], using the Fourier transform

$$\hat{f}(u) := \int_{-\infty}^{\infty} f(x)e^{-ixu}dx.$$

For $n \geq 1$ and $0 \leq k < n$, define the function $B_{n,k}$ by

$$\widehat{B}_{n,k}(u) = \frac{(iu)^{n-k} - (-1)^{n-k} P_{n,k}(e^{-iu})}{(iu)^{n+1}}, \quad u \in \mathbb{R}, \tag{2}$$

where $P_{n,k}(z)$ is the Taylor polynomial of degree n at $z = 1$ for $(\log z)^{n-k}$. For $z = e^{-iu}$, the numerator in (2) is $(-1)^{n-k}((\log z)^{n-k} - P_{n,k}(z))$ near $z = 1$ and so has a zero of order $n+1$ at $z = 1$. Thus (2) is well-defined at $u = 0$. By integrating by parts in the inverse Fourier transform for $B_{n,k}$ we can derive the following.

(a) The function $B_{n,k}$ is a piecewise polynomial of degree n with support on $[0, n]$.

(b) The only discontinuities of derivatives of $B_{n,k}$ are in $B_{n,k}^{(n)}$ at $0, \ldots, n$ and in $B_{n,k}^{(k)}$ at 0, where $B_{n,k}^{(k)}(0^+) = 1$.

(c) Moreover, these properties determine $B_{n,k}$ uniquely.

We now define our spline spaces. Take $n \geq 1$ and let S_n be a finite subset of $[0, 1)$. For $k = 0, 1, \ldots, n-1$, we let S_k be a subset of S_n. We then denote by $\mathcal{S}$ the space of all piecewise polynomials f of degree n on $\mathbb{R}$ whose only discontinuities of derivatives are in $f^{(k)}$ at $S_k + \mathbb{Z}$, $k = 0, 1 \ldots, n$.

Example 1. If for some r, $1 \leq r \leq n+1$, for $S_k = S_n$, $n - r + 1 \leq k \leq n$, and $S_k = \emptyset$, for $0 \leq k \leq n - r$, then $\mathcal{S}$ comprises all spline functions of degree n with knots of multiplicity r at $S_n + \mathbb{Z}$.

For α in S_n we denote by B_α the usual B-spline of degree n with consecutive simple knots in $S_n + \mathbb{Z}$, starting at α. To be precise, let $S_n = \{\alpha_0, \ldots, \alpha_{r-1}\}$ with $0 \leq \alpha_0 < \cdots < \alpha_{r-1} < 1$, and define α_j, $j \in \mathbb{Z}$, by $\alpha_{\ell r+m} = \ell + \alpha_m$, $\ell \in \mathbb{Z}$, $m = 0, \ldots, r-1$. Then for $\alpha = \alpha_\ell$, $0 \leq \ell \leq r-1$, B_α is the B-spline with simple knots at $\alpha_\ell, \ldots, \alpha_{\ell+n+1}$. The normalisation of the B-spline is not important.

Theorem 1. *Any function f in $\mathcal{S}$ can be written uniquely as*

$$f = \sum_{\alpha \in S_n} \sum_{j \in \mathbb{Z}} a_{j,\alpha} B_\alpha(\cdot - j) + \sum_{k=0}^{n-1} \sum_{\alpha \in S_k} \sum_{j \in \mathbb{Z}} b_{j,k,\alpha} B_{n,k}(\cdot - \alpha - j), \tag{3}$$

for constants $a_{j,\alpha}$, $b_{j,k,\alpha}$ in $\mathbb{R}$. Moreover, f has compact support if and only if only a finite number of these constants are non-zero.

Proof: Take $f \in \mathcal{S}$. For $0 \le k \le n-1$, $\alpha \in S_k$, $j \in \mathbb{Z}$, let

$$b_{j,k,\alpha} = f^{(k)}((\alpha + j)^+) - f^{(k)}((\alpha + j)^-), \tag{4}$$

and define

$$g = \sum_{k=0}^{n-1} \sum_{\alpha \in S_k} \sum_{j \in \mathbb{Z}} b_{j,k,\alpha} B_{n,k}(\cdot - \alpha - j). \tag{5}$$

Then for $\alpha \in S_k$, and $j \in \mathbb{Z}$,

$$\begin{aligned} &g^{(k)}((\alpha + j)^+) - g^{(k)}((\alpha + j)^-) \\ &= \sum_{k=0}^{n-1} \sum_{\beta \in S_k} \sum_{\ell \in \mathbb{Z}} b_{j,k,\beta}(B_{n,k}^{(k)}((\alpha - \beta + j - \ell)^+) - B_{n,k}^{(k)}((\alpha - \beta + j - \ell)^-)) \\ &= b_{j,k,\alpha} = f^{(k)}((\alpha + j)^+) - f^{(k)}((\alpha + j)^-). \end{aligned}$$

Thus $f - g$ has discontinuities only in the nth order derivatives at $S_n + \mathbb{Z}$ and so we can write

$$f - g = \sum_{\alpha \in S_n} \sum_{j \in \mathbb{Z}} a_{j,\alpha} B_\alpha(\cdot - j) \tag{6}$$

for constants $a_{j,\alpha}$, which gives (3).

If f has compact support, then by (4) only a finite number of the constants $b_{j,k,\alpha}$ are non-zero. So by (5), g has compact support. Since $f - g$ has compact support, (6) shows that only a finite number of the constants $a_{j,\alpha}$ are non-zero. □

We next consider a spline space $\mathcal{S}$ that is refinable with dilation by 2, i.e. $f \in \mathcal{S} \Rightarrow f(\frac{\cdot}{2}) \in \mathcal{S}$. Since f has a discontinuity in a derivative at α if and only if $f(\frac{\cdot}{2})$ has such a discontinuity at 2α, we see that $\mathcal{S}$ is refinable if and only if for $k = 0, \ldots, n$, $\alpha \in S_k \Rightarrow 2\alpha (\mathrm{mod}\, \mathbb{Z}) \in S_k$.

Lemma 2. *If $\mathcal{S}$ is refinable, then every element of S_n is rational and can therefore be written in the form*

$$\frac{\beta}{2^\gamma(2^\delta - 1)} \quad \text{for } \alpha, \beta, \gamma \in \mathbb{Z}, \quad \gamma \ge 0,\ \delta \ge 1. \tag{7}$$

Proof: Take $\alpha \in S_n$. Then $\{2^j \alpha (\mathrm{mod}\, \mathbb{Z}) : j = 0, 1, 2, \ldots\} \subset S_n$ and so is finite. So for some $j \ne k$, $2^j \alpha = 2^k \alpha + \ell$, for some $\ell \in \mathbb{Z}$. Therefore $\alpha = \frac{\ell}{2^j - 2^k}$. We now show that any rational can be written in the form (7). Take any integer $N \ge 1$. Then $\{2^j (\mathrm{mod}\, N) : j = 0, \ldots, N\}$ has at most N distinct elements and so there exist $i, j \in \{0, \ldots, N\}$, $i < j$, with $2^j = 2^i + kN$, for some integer k. So

$$\frac{1}{N} = \frac{k}{2^i(2^{j-i} - 1)}.$$

Hence any rational can be written in the form (7). □

Clearly in the representation (7) we can choose β to be odd, if $\gamma > 0$. Even so this representation is not unique, e.g. $\frac{1}{2^2-1} = \frac{5}{2^4-1}$.

Lemma 3. *Suppose that $\alpha = \frac{\beta}{2^\delta - 1}$, where δ is the smallest positive integer for such a representation. Then $\{2^j\alpha \,(\mathrm{mod}\, \mathbb{Z}) : j = 0, 1, 2, \ldots\}$ comprises the distinct elements $2^j\alpha(\mathrm{mod}\, \mathbb{Z})$, $j = 0, \ldots, \delta - 1$.*

Proof: For α as above, $2^\delta \alpha = \beta + \frac{\beta}{2^\delta - 1} = \alpha (\mathrm{mod}\, \mathbb{Z})$. So

$$\{2^j\alpha \,(\mathrm{mod}\, \mathbb{Z}) : j = 0, 1, \ldots\} = \{2^j\alpha \,(\mathrm{mod}\, \mathbb{Z}) : j = 0, 1, 2, \ldots, \delta - 1\}.$$

Now suppose $2^\ell \alpha = 2^j \alpha + m$, $0 \le j < \ell \le \delta - 1$, $m \in \mathbb{Z}$. Then $\alpha = \frac{m}{2^j(2^{\ell - j} - 1)}$ and since α can be written with odd denominator, $\alpha = \frac{p}{2^{\ell-j}-1}$, for some $p \in \mathbb{Z}$. As $1 \le \ell - j < \delta$, this contradicts our assumptions on α. So $\{2^j\alpha \,(\mathrm{mod}\, \mathbb{Z}) : j = 0, \ldots, \delta - 1\}$ comprises δ distinct elements. □

Note that for any α with odd denominator (possibly 1), Lemma 2 shows that it can be written in the form in Lemma 3. With such an α we shall call the set $\{2^j\alpha(\mathrm{mod}\, \mathbb{Z}) : j = 0, \ldots, \delta - 1\}$ a cycle. Note that if a cycle has δ elements, then its elements are in $\mathbb{Z}/(2^\delta - 1)$. By a tail of a cycle C we mean any finite set $T \subset [0, 1)$ disjoint from C which satisfies the following. For any $t \in T$, there is some $\gamma \ge 1$ such that $2^j t$ (mod $\mathbb{Z}$), $j = 0, \ldots, \gamma - 1$, are distinct elements of T and $2^\gamma t$ (mod $\mathbb{Z}$) is in C.

Example 2. A cycle C is given by $\{\frac{1}{3}, \frac{2}{3}\}$ and a tail for C is $\{\frac{1}{24}, \frac{1}{12}, \frac{1}{6}, \frac{7}{12}\}$.

Theorem 4. *A space S is refinable if and only if for $k = 0, 1, \ldots, n$, S_k comprises a disjoint union of cycles with tails.*

Proof: If S_k is a union of cycles with tails, then clearly $\alpha \in S_k \Rightarrow 2\alpha$ (mod $\mathbb{Z}$) $\in S_k$ and so S is refinable.

Next suppose that S is refinable. So for $k = 0, 1, \ldots, n$, $\alpha \in S_k$ implies 2α (mod $\mathbb{Z}$) $\in S_k$. Take $\alpha \in S_k$. If α has odd denominator (possibly 1), then it lies in a cycle. Otherwise we can write it as in (7) with β odd, $\gamma \ge 1$. Then $2^\gamma \alpha = \frac{\beta}{2^\delta - 1}$, which has odd denominator and so lies in a cycle C. For $j = 0, \ldots, \gamma - 1$, $2^j\alpha$ cannot be written with an odd denominater and so does not lie in any cycle. Also $2^j\alpha$ (mod $\mathbb{Z}$), $j = 0, \ldots, \gamma - 1$, are distinct, since otherwise $2^j\alpha = 2^i\alpha + m$ for some $m \in \mathbb{Z}$, $0 \le i < j \le \gamma - 1$, and so $2^i\alpha = \frac{m}{2^{j-i}-1}$, which has odd denominator. Thus α is in a tail of C. So S_k is a union of cycles with tails.

Further, if α is in a cycle C, then $C = \{2^j\alpha(\mathrm{mod}\, \mathbb{Z}) : j = 0, 1, \ldots\}$. Thus distinct cycles are disjoint. Next suppose that α is in a tail for C and express α in the form (7) above. Then we have seen that for $j = 0, \ldots, \gamma - 1$, $2^j\alpha$ does not lie in any cycle, while for $j \ge \gamma$, $2^j\alpha \in \mathbf{C}$. Thus α cannot lie in the tail of any other cycle. So S_k is a disjoint union of cycles with tails. □

We now list all refinable spaces with two knots (mod $\mathbb{Z}$), i.e. $\sum_{k=0}^{n} |S_k| = 2$. Note that such a refinable spline space is completely determined by the knots.

1) $S_n = S_j = \{0\}$, for some j, $0 \le j \le n-1$.
2) $S_n = \{\frac{1}{2}, 0\}$. 3) $S_n = \{\frac{1}{3}, \frac{2}{3}\}$.

To finish the section we list all refinable spaces $\mathcal{S}$ with 3 knots (mod $\mathbb{Z}$).

1) $S_n = S_j = S_\ell = \{0\}$, for some j, ℓ, $0 \le j < \ell \le n-1$.
2) $S_n = \{\frac{1}{2}, 0\}$, $S_j = \{0\}$, for some j, $0 \le j \le n-1$.
3) $S_n = \{0, \frac{1}{3}, \frac{2}{3}\}$. 4) $S_n = \{\frac{1}{4}, \frac{1}{2}, 0\}$.
5) $S_n = \{\frac{3}{4}, \frac{1}{2}, 0\}$. 6) $S_n = \{\frac{1}{6}, \frac{1}{3}, \frac{2}{3}\}$.
7) $S_n = \{\frac{5}{6}, \frac{1}{3}, \frac{2}{3}\}$. 8) $S_n = \{\frac{1}{7}, \frac{2}{7}, \frac{4}{7}\}$.
9) $S_n = \{\frac{3}{7}, \frac{6}{7}, \frac{5}{7}\}$.

§3. Refinable Splines

We now return to our original problem of finding refinable vectors of spline functions $\phi = (\phi_1, \ldots, \phi_n)^T$ satisfying (1). We first recall some results of Jia and Micchelli [13] on the linear independence of integer translates of integrable functions ϕ with compact support. The integer translates of ϕ form a Riesz basis in $L^2(\mathbb{R})$ if and only if for each u in $\mathbb{R}$ there are integers $j_1, \ldots, j_r$, for which the matrix $[\hat{\phi}(u + 2\pi j_1) \ldots \hat{\phi}(u + 2\pi j_r)]$ is non-singular. Since $\hat{\phi}$ is analytic we can extend it to the complex plane. Then the integer translates of ϕ are linearly independent if and only if for each z in $\mathbb{C} \setminus \{0\}$, there are integers $j_1, \ldots, j_r$, for which the matrix $[\hat{\phi}(z + 2\pi j_1) \ldots \hat{\phi}(z + 2\pi j_r)]$ is non-singular.

Theorem 5. *Let $\mathcal{S}$ be a refinable spline space with r knots (mod $\mathbb{Z}$), i.e. $\sum_{k=0}^{n} |S_k| = r$. Let $\psi_1, \ldots, \psi_r$ denote the functions $B_\alpha, B_{n,k}$ as in Theorem 1, and write $\psi = (\psi_1, \ldots, \psi_r)^T$. Then $\phi = (\phi_1, \ldots, \phi_r)^T$ is a refinable vector of functions in $\mathcal{S}$ whose integer translates are linearly independent if and only if*

$$\hat{\phi}(u) = M(e^{-iu})\hat{\psi}(u), \quad u \in \mathbb{R}, \tag{8}$$

where $M(z)$ is a $r \times r$ matrix of Laurent polynomials with $\det M(z) = cz^\ell$ for some $\ell \in \mathbb{Z}$ and non-zero constant c.

Proof: Suppose that (8) is satisfied. Then we can extend (8) to z in $\mathbb{C}$, and for integers $j_1, \ldots, j_r$,

$$[\hat{\phi}(z + 2\pi j_1) \ldots \hat{\phi}(z + 2\pi j_r)] = M(e^{-iz})[\hat{\psi}(z + 2\pi j_1,) \ldots \hat{\psi}(z + 2\pi j_r)]. \tag{9}$$

Note that $M(e^{-iz})$ is non-singular for all z in $\mathbb{C} \setminus \{0\}$. By Theorem 1, the integer translates of ψ are linearly independent. So for each $z \in \mathbb{C} \setminus \{0\}$ we can choose $j_1, \ldots, j_r$ so that the right-hand side of (9) is non-singular, and from the left-hand side we see that the integer translates of ϕ are linearly independent.

Taking Fourier transforms of (3) shows that for any f in $\mathcal{S}$, there is a $1 \times r$ matrix $P(z)$ of Laurent polynomials such that for u in $\mathbb{R}$,

$$\hat{f}(u) = P(e^{-iu})\hat{\psi}(u) = P(e^{-iu})M^{-1}(e^{-iu})\hat{\phi}(u),$$

by (8). Since $\det M(z) = cz^{\ell}$, $M^{-1}(z)$ is a matrix of Laurent polynomials. Taking inverse Fourier transforms shows that for some $N \geq 0$,

$$f(x) = \sum_{j=-N}^{N} A(j)\phi(x-j), \quad x \in \mathbb{R},$$

where $A(j)$ are $1 \times r$ matrices. Since $\mathcal{S}$ is refinable, $\phi(\frac{\cdot}{2}) \in \mathcal{S}$ and so for some $N \geq 0$,

$$\phi(\frac{x}{2}) = \sum_{j=-N}^{N} B(j)\phi(x-j), \quad x \in \mathbb{R},$$

where $B(j)$ are $r \times r$ matrices. Hence ϕ is refinable.

Now suppose that $\phi = (\phi_1, \ldots, \phi_r)^T$ are functions in $\mathcal{S}$ with linearly independent integer translates. Then by Theorem 1 we can express ϕ as a finite linear combination of integer translates of ψ. Taking Fourier transforms gives (8), where $M(z)$ is a $r \times r$ matrix of Laurent polynomials. Extending to z in $\mathbb{C} \setminus \{0\}$ gives (9) for any integers $j_1, \ldots, j_r$. Since the integer translates of ϕ are linearly independent, the left-hand side of (9) must be non-singular for each z in $\mathbb{C} \setminus \{0\}$ and some choice of $j_1, \ldots, j_r$. Thus $M(z)$ must be non-singular for all $z \neq 0$, i.e. $\det M(z) = cz^{\ell}$ for some $\ell \in \mathbb{Z}$ and a non-zero constant c. □

Results related to Theorem 5 are given in [18]. Examples of such refinable splines ϕ are given in [10] for $r = 2$ and $S_n = S_k = \{0\}$. For $\mathcal{S}$ with r knots we can certainly have more than r refinable functions, e.g. by simply repeating functions in ϕ. However more than r functions cannot generate a Riesz basis, regardless of refinability properties, as we see next.

Theorem 6. *Let $\mathcal{S}$ be a spline space with s knots. If $\phi_1, \ldots, \phi_r$ in $\mathcal{S}$ have integer translates which form a Riesz basis, then $r \leq s$.*

Proof: Take $\phi_1, \ldots, \phi_r$ in $\mathcal{S}$ and let $\psi_1, \ldots, \psi_s$ denote the functions B_α, $B_{n,k}$ as in Theorem 1. As in (3) we can express ϕ as a finite linear combination of integer translates of ψ, and taking Fourier transforms gives (8), where $M(z)$ is an $r \times s$ matrix of Laurent polynomials. So for any $u \in \mathbb{R}$ and integers $j_1, \ldots, j_r$,

$$[\hat{\phi}(u + 2\pi j_1) \ldots \hat{\phi}(u + 2\pi j_r)] = M(e^{-iu})[\hat{\psi}(u + 2\pi j_1) \ldots \hat{\psi}(u + 2\pi j_r)].$$

This $r \times r$ matrix has rank at most s. So for $r > s$ it must be singular and so the integer translates of $\phi_1, \ldots, \phi_r$ cannot form a Riesz basis. □

In a refinable space $\mathcal{S}$ with s knots, we may have less than s linearly independent refinable functions as the following example shows.

Example 3. Take $n = 0$, $S_0 = \{0, \frac{1}{4}, \frac{1}{2}, \frac{3}{4}\}$, and

$$\phi_1 = \chi_{[0,\frac{1}{2}]}, \quad \phi_2 = \chi_{[\frac{1}{4},\frac{3}{4}]}, \quad \phi_3 = \chi_{[\frac{1}{2},1]}.$$

Here $s = 4$ but the three functions ϕ_1, ϕ_2, ϕ_3 are easily seen to be refinable and linearly independent.

Our next example shows that in Theorem 5 it is not possible to relax the condition of being linearly independent to being a Riesz basis.

Example 4. Take $n = 0$, $S_0 = \{0, \frac{1}{2}\}$,

$$\phi_1 = \chi_{[0,1]} \text{ and } \phi_2(x) = \begin{cases} c, & 0 \le x \le \frac{1}{2}, \\ 1, & 1 \le x \le \frac{3}{2}, \end{cases}$$

where $c < 0$, $c \neq -1$. Here $s = 2$ and it is easily seen that ϕ_1, ϕ_2 are refinable. It can be checked from the Jia and Micchelli conditions [13] that the integer translates of ϕ_1, ϕ_2 form a Riesz basis but are not linearly independent.

In another paper we shall investigate these possibilities and, in particular, show the following.

1) Suppose that $\mathcal{S}$ has s knots (mod $\mathbb{Z}$) comprising only cycles and a single tail T of the form $\{2^j\alpha : j = 0, 1, \ldots, \gamma - 1\}$. Then there cannot be less than s refinable functions.
2) Suppose that $\mathcal{S}$ has r knots (mod $\mathbb{Z}$) comprising only cycles, i.e. all knots have odd denominators. If the integer translates of the refinable functions $\phi_1, \ldots, \phi_r$ form a Riesz basis, then they are linearly independent.

Acknowledgments. Supported by the Wavelets Strategic Research Programme, National University of Singapore, under a grant from the National Science and Technology Board and the Ministry of Education, Singapore.

References

1. Chui, C. K., *An Introduction to Wavelets,* Academic Press, Boston, 1992.
2. Chui, C. K., and J. Lian, A study on orthonormal multi-wavelets, J. Appl. Numer. Math. **20** (1996), 272–298.
3. Chui, C. K., and J. Z. Wang, On compactly supported spline wavelets and a duality principle, Trans. Amer. Math. Soc. **330** (1992), 903–915.
4. Cohen, A., I. Daubechies, and J. C. Feauveau, Biorthogonal basis of compactly supported wavelets, Comm. Pure and Appl. Math. **45** (1992), 485–560.
5. Daubechies, I., Orthonormal bases of compactly supported wavelets, Comm. Pure and Appl. Math. **41** (1988), 909–996.
6. Daubechies, I., *Ten Lectures on Wavelets,* CBMS-NSF Series in Applied Mathematics #61, SIAM Publ., Philadelphia, 1992.

7. Donovan, G., J. S. Geronimo, D. P. Hardin, and P. R. Massopust, Construction of orthogonal wavelets using fractal interpolation functions, SIAM J. Math. Anal. **27** (1996), 1150–1192.
8. G. Donovan, J. S. Geronimo, and D. P. Hardin, Intertwining multiresolution analysis and the construction of piecewise polynomial wavelets, SIAM J. Math. Anal. **27** (1996), 1791–1815.
9. Geronimo, J. S., D. H. Hardin, and P. R. Massopust, Fractal functions and wavelet expansions based on several scaling functions, J. Approx. Theory **78** (1994), 373–401.
10. Goodman, T. N. T., Characterising pairs of refinable splines, J. Approx. Theory, to appear.
11. Goodman, T. N. T., Properties of bivariate refinable spline pairs, in *Multivariate Approximation, Recent Trends and Results*, W. Haußmann, K. Jetter and M. Reimer (eds.), Akademie Verlag, Berlin, 1997, 63–82.
12. Goodman, T. N. T., and S. L. Lee, Wavelets of multiplicity r, Trans. Amer. Math. Soc. **342** (1994), 307–324.
13. Jia, R. Q., and C. A. Micchelli, On linear independence of integer translates of a finite number of functions, Proc. Edinburgh Math. Soc. **36** (1992), 69–85.
14. Jiang, Q., Orthogonal multiwavelets with optimum time-frequency resolution, IEEE Trans. Signal Processing (Special Issue), to appear.
15. Lawton, W., S. L. Lee, and Z. Shen, Complete characterisation of refinable splines, Advances in Comp. Math. **3** (1995), 137–145.
16. Micchelli, C. A., and T. Sauer, Regularity of multiwavelets, Advances in Comp. Math. **7** (1997), 455–545.
17. Shen, Z. Refinable function vectors, SIAM J. Math. Anal., to appear.
18. Wang, J., Stability and linear independence associated with scaling vectors, SIAM J. Math. Anal., to appear.

T. N. T. Goodman
Department of Mathematics
University of Dundee
Dundee DD1 4HN, SCOTLAND
`tgoodman@mcs.dundee.ac.uk`

S. L. Lee
Department of Mathematics
National University of Singapore
SINGAPORE 119260
`matleesl@math.nus.sg`

De Casteljau's Algorithm Revisited

Jens Gravesen

Abstract. It is demonstrated how all the basic properties of Bézier curves can be derived swiftly and efficiently without any reference to the Bernstein polynomials and essentially with only geometric arguments. This is achieved by viewing one step in de Casteljau's algorithm as an operator (the de Casteljau operator) acting on a sequence of points, producing a sequence with one point less. The properties of Bézier curves are then derived by analysing de Casteljau's operator.

§1. Introduction

We consider one step in de Casteljau's algorithm as an *operator* (the de Casteljau operator), acting on a sequence of points, producing a new sequence with one point less. We produce a Bézier curve by applying de Casteljau's operator a number of times, so it is clear that the properties of a Bézier curve can be found by studying products of de Casteljau's operator with itself. The final ingredient is the observation that de Casteljau's operator is built from the two extremely simple operators, which remove the first and the last points from a sequence, respectively.

A slight modification of the idea has been used by M. Hosaka and M. Kuroda, cf. [3,4], and by J. Hoschek and D. Lasser cf. [5]. Instead of finite sequences they use infinite sequences.

§2. The de Casteljau Algorithm

The de Casteljau algorithm can be described by the scheme

$$\begin{array}{ccccccccc} P_0 & \to & P_0^1 & \to & \cdots & \to & P_0^{n-1} & \to & P_0^n \\ & \nearrow & & \nearrow & & \nearrow & & \nearrow & \\ P_1 & \to & P_1^1 & \to & \cdots & \to & P_1^{n-1} & & \\ & \nearrow & & \nearrow & & \nearrow & & & \\ \vdots & & \vdots & & & & & & \\ & \nearrow & & \nearrow & & & & & \\ P_{n-1} & \to & P_{n-1}^1 & & & & & & \\ & \nearrow & & & & & & & \\ P_n & & & & & & & & \end{array} \qquad (1)$$

Mathematical Methods for Curves and Surfaces II
Morten Dæhlen, Tom Lyche, and Larry L. Schumaker (eds.), pp. 221–228.

ISBN 0-8265-1315-8.

Using matrix notation one step in the algorithm (going from one column to the next) can be written as

$$\begin{bmatrix} P_0^{k+1} \\ P_1^{k+1} \\ \vdots \\ P_{n-k-1}^{k+1} \end{bmatrix} = \begin{bmatrix} 1-t & t & 0 & \dots & 0 \\ 0 & 1-t & t & \ddots & \vdots \\ \vdots & \ddots & \ddots & \ddots & 0 \\ 0 & \dots & 0 & 1-t & t \end{bmatrix} \begin{bmatrix} P_0^k \\ P_1^k \\ \vdots \\ P_{n-k}^k \end{bmatrix}$$

$$= \left((1-t) \begin{bmatrix} 1 & 0 & 0 & \dots & 0 \\ 0 & 1 & 0 & \ddots & \vdots \\ \vdots & \ddots & \ddots & \ddots & 0 \\ 0 & \dots & 0 & 1 & 0 \end{bmatrix} + t \begin{bmatrix} 0 & 1 & 0 & \dots & 0 \\ 0 & 0 & 1 & \ddots & \vdots \\ \vdots & \ddots & \ddots & \ddots & 0 \\ 0 & \dots & 0 & 0 & 1 \end{bmatrix} \right) \begin{bmatrix} P_0^k \\ P_1^k \\ \vdots \\ P_{n-k}^k \end{bmatrix}.$$

Inspired by this, we introduce two basic operators which act on a sequence of points.

Definition 1. *The* right truncation operator R *and the* left truncation operator L *is defined by*

$$R : (P_0, \dots, P_m) \mapsto (P_0, \dots, P_{m-1}),$$
$$L : (P_0, \dots, P_m) \mapsto (P_1, \dots, P_m).$$

Remark 1. *To be precise we should indicate the length of the sequence on which the operators acts, so the definition reads*

$$R_m : (P_0, \dots, P_m) \mapsto (P_0, \dots, P_{m-1}),$$
$$L_m : (P_0, \dots, P_m) \mapsto (P_1, \dots, P_m).$$

Remark 2. *In this paper all sequences are finite. It is possible to work with infinite sequences, $(P_0^i, P_1^i, \dots)$, then the scheme (1) becomes infinite downward and to the right, and our scheme is just the top left triangle of this infinite scheme. The operator R is then the identity and the operator L is left-shift. We may even use bi-infinite sequences, $(\dots, P_{-1n}, P_0^i, P_1^i, \dots)$ in which case the left-shift operator L is invertible.*

Even though the definition of R and L is as simple as you could possibly want, they are all we need for the development of the Bézier curve theory. We start by stating some basic properties of R and L.

Lemma 2. *The operators R and L "commute", i.e.*

$$LR = RL.$$

Proof: This is clear, both LR and RL removes the first and the last point of a sequence. □

Remark. *Strictly speaking the word "commute" is a misnomer, because the L's and the R's on the two sides of the equation is different. The equation should read $L_{m-1}R_m = R_{m-1}L_m$.*

Lemma 3. *If we identify a point P and the sequence (P) consisting of just that point, then,*

$$P_k = L^k R^{m-k}(P_0, \dots, P_m), \quad \text{for } 0 \le k \le m.$$

Proof: The operator $L^k R^{m-k}$ removes the first k points $P_0, \dots, P_{k-1}$ and the last $m-k$ points $P_{k+1}, \dots, P_m$, so we are left with the sequence consisting of just one point P_k. □

From the basic operators R and L we can define two new operators.

Definition 4. *The (well known)* forward difference operator Δ *is*

$$\Delta = L - R.$$

Definition 5. *If $t \in \mathbb{R}$, then the* de Casteljau operator $C(t)$ *is*

$$C(t) = (1-t)R + tL = R + t\Delta.$$

By definition

$$C(t) : (P_0, \dots, P_m) \mapsto ((1-t)P_0 + tP_1, \dots, (1-t)P_{m-1} + tP_m),$$

and we see that $C(t)$ describes one step in de Casteljau's algorithm. By Lemma 2 we have the following corollary.

Corollary 6. *Let $s, t \in \mathbb{R}$, then the operators R, L, Δ, $C(s)$, $C(t)$ "commute".*

Again the word "commute" should not be taking literally. The operators we have introduced are affine operators:

Lemma 7. *Let $t \in \mathbb{R}$ and let M be any of the operators R, L, Δ, $C(t)$. If $\alpha + \beta = 1$, then*

$$M\big(\alpha(P_0, \dots, P_n)) + \beta(Q_0, \dots, Q_n)\big) = \alpha M(P_0, \dots, P_n)) + \beta M(Q_0, \dots, Q_n).$$

Proof: As linear combinations of affine operators are affine, we only have to show that the lemma holds for the operators R and L, but this is obvious. □

Finally we have that de Casteljau's operator is affine in it's argument:

Lemma 8. *Let $a, b, t \in \mathbb{R}$ then*

$$C((1-t)a + tb) = (1-t)C(a) + tC(b).$$

Proof: The proof is a straightforward calculation:

$$\begin{aligned} C((1-t)a + tb) &= R + ((1-t)a + tb)(L - R) \\ &= (1-t)R + (1-t)a(L-R) + tR + tb(L-R) \\ &= (1-t)(R + a(L-R)) + t(R + tb(L-R)) \\ &= (1-t)C(a) + tC(b). \quad \square \end{aligned}$$

The de Casteljau's algorithm is encoded in the operator $C(t)$, so we can use it to define a Bézier curve.

Definition 9. *A Bézier curve of degree n with control points $P_0, \ldots, P_n$ is defined by*

$$\mathbf{b}(t) = C(t)^n (P_0, \ldots, P_n).$$

Remark. *The symbol $C(t)^n$ is a misuse of notation. The definition should read $\mathbf{b}(t) = C_2(t)C_3(t)\ldots C_{n+1}(t)(P_0, \ldots, P_n)$, but as it is clear from the context what the interpretation of $C(t)^n$ is, we will stick to the simple notation. The same remark holds each time we compose any of our operators and in the following we will use the abusive but simple notation without any further comments.*

§3. Properties of Bézier Curves

As a composition of affine operators gives an affine operator, Lemma 7 immediately give us that the operator $C(t)^n$ is affine. This can be formulated as:

Corollary 10. *Let $(P_0, \ldots, P_n)$ be control points for the Bézier curve $\mathbf{b}(t)$ and let $(Q_0, \ldots, Q_n)$ be control points for the Bézier curve $\mathbf{c}(t)$. If $\alpha + \beta = 1$ then the Bézier curve $\alpha\mathbf{b}(t) + \beta\mathbf{c}(t)$ has control points $(\alpha P_0 + \beta Q_0, \ldots, \alpha P_n + \beta Q_n)$.*

Next we have the *symmetry* property of Bézier curves:

Theorem 11. *If $\mathbf{b}(t)$ is a Bézier curve with control points $P_0, \ldots, P_n$, and $\widetilde{\mathbf{b}}(t)$ is the Bézier curve with control points $P_n, \ldots, P_0$, then*

$$\widetilde{\mathbf{b}}(t) = \mathbf{b}(1 - t).$$

Proof: Let J denote the operator which reverse the order of a sequence, i.e.

$$J : (P_0, \ldots, P_n) \mapsto (P_n, \ldots, P_0).$$

We clearly have $RJ = JL$ and $JR = LJ$, and hence

$$\begin{aligned} C(t)J &= ((1-t)R + tL)J = (1-t)RJ + tLJ \\ &= (1-t)JL + tJR = J(tR + (1-t)L) = JC(1-t). \end{aligned}$$

By induction we have

$$C(t)^n J = JC(1-t)^n,$$

and as J is the identity on a one point sequence, the theorem follows. □

The *derivative* of a Bézier curve is easily found.

Theorem 12. *If $\mathbf{b}(t)$ is a Bézier curve with control points $P_0, \ldots, P_n$, then the derivative is given by*

$$\frac{d}{dt}\mathbf{b}(t) = nC(t)^{n-1}\Delta(P_0, \ldots, P_n) = n\Delta C(t)^{n-1}(P_0, \ldots, P_n).$$

Proof: As $C(t) = R + t\Delta$ we have $\frac{d}{dt}C(t) = \Delta$ and as Δ and $C(t)$ commute,

$$\frac{d}{dt}\left(C(t)^n\right) = \sum_{k=0}^{n-1} C(t)^k \left(\frac{d}{dt}C(t)\right) C(t)^{n-1-k} = \sum_{k=0}^{n-1} C(t)^k \Delta C(t)^{n-1-k}$$
$$= nC(t)^{n-1}\Delta = n\Delta C(t)^{n-1}. \quad \square$$

Remark. *The expression*

$$\frac{d}{dt}\mathbf{b}(t) = nC(t)^{n-1}\Delta(P_0, \ldots, P_n)$$

tells us that the derivative of a Bézier curve is a new Bézier curve where the control points are the forward differences of the original control points multiplied by the degree. The expression

$$\frac{d}{dt}\mathbf{b}(t) = n\Delta C(t)^{n-1}(P_0, \ldots, P_n)$$

shows us that the derivative of a Bézier curve is the forward difference of the two points in the second last step in de Casteljau's algorithm multiplied by the degree.

Corollary 13. *If* $\mathbf{b}(t)$ *is a Bézier curve with control points* $P_0, \ldots, P_n$, *then*

$$\frac{d^k}{dt^k}\mathbf{b}(t) = \frac{n!}{(n-k)!}C(t)^{n-k}\Delta^k(P_0, \ldots, P_n)$$
$$= \frac{n!}{(n-k)!}\Delta^k C(t)^{n-k}(P_0, \ldots, P_n).$$

Proof: The corollary follows by induction, using theorem 12. $\square$
As another corollary we get linear precision:

Corollary 14. *If* P *and* Q *are two points and the Bézier curve* $\mathbf{b}(t)$ *has the control points*

$$P_i = \frac{(n-i)}{n}P + \frac{i}{n}Q, \qquad i = 0, \ldots, n,$$

then

$$\mathbf{b}(t) = (1-t)P + tQ.$$

Proof: We only need to show that the derivative of $\mathbf{b}(t)$ is constant $P - Q$, but this is easy. The derivative $\mathbf{b}'(t)$ has control points $n(P_{i+1} - P_i) = P - Q$, and if all the control points of a Bézier curve are one and the same point, then the curve is constant equal to that point. $\square$

The degree elevation formula is the most complicated in the this context, and it is the one instance where it is better to use the Bernstein representation.

Definition 15. *The* degree elevation operator *is defined by*

$$U : (P_0, \ldots, P_n) \mapsto \left(P_0, \frac{nP_1 + P_0}{n+1}, \ldots, \frac{P_n + nP_{n-1}}{n+1}, P_n\right),$$

i.e., the k'th element of $U(P_0, \ldots, P_n)$ *is*

$$\frac{(n+1-k)P_k + kP_{k-1}}{n+1}.$$

Before we prove the degree elevation formula we need a lemma.

Lemma 16. *Acting on sequences of length* $n+1$ *we have,*

$$\Delta U = \frac{n}{n+1} U \Delta.$$

Proof: This is just a calculation. Starting with the k-th element of the sequence $\Delta U(P_0, \ldots, P_n)$, we get

$$\begin{aligned}
&\frac{(n+1-(k+1))P_{k+1} + (k+1)P_k}{n+1} - \frac{(n+1-k)P_k + kP_{k-1}}{n+1} \\
&= \frac{n+1-k-1}{n+1} P_{k+1} - \frac{n+1-k}{n+1} P_k + \frac{k+1}{n+1} P_k - \frac{k}{n+1} P_{k-1} \\
&= \frac{n-k}{n+1} P_{k+1} - \frac{n-k}{n+1} P_k + \frac{k}{n+1} P_k - \frac{k}{n+1} P_{k-1} \\
&= \frac{n}{n+1} \left(\frac{n-k}{n} (P_{k+1} - P_k) + \frac{k}{n} (P_k - P_{k-1}) \right).
\end{aligned}$$

and this is exactly the k-th element of the sequence $\frac{n}{n+1} U \Delta(P_0, \ldots, P_n)$. □
We can now formulate the degree elevation formula as

Theorem 17. *Acting on sequences of length* n, *we have*

$$C(t)^n U = C(t)^{n-1}. \tag{2}$$

Proof: This is done by induction on n. If $n = 1$, (2) is trivial. So assume $C(t)^n U = C(t)^{n-1}$. We clearly have

$$C(0)^{n+1} U = R^{n+1} U = R^n = C(0)^n,$$

and hence it is sufficient to show that (2) holds after differentiation, i.e. we need to show that

$$(n+1) C(t)^n \Delta U = n C(t)^{n-1} \Delta.$$

By Lemma 16 the left-hand side is $nC(t)^n U \Delta$, but this is the same as the right-hand side by the induction assumption. □

Remark. *Even though both sides of equation (2) can act on sequences of length greater than* n, *it is only valid when acting on sequences of length* n.

An important property of a Bézier curve is given in the *subdivision theorem*, which states that the part of a Bézier curve, which is defined for $t \in [0,c]$, is a Bézier curve with control points $P_0, P_0^1, \dots, P_0^n$, and the part, which is defined for $t \in [c,1]$, is a Bézier curve with control points $P_0^n, P_1^{n-1}, \dots, P_n$, where P_i^k are the intermediate points in de Casteljau's algorithm, evaluated for the parameter value c. We obtain the subdivision theorem by letting $(a,b) = (0,c)$ and $(a,b) = (c,1)$ in the following theorem.

Theorem 18. *Let* $\mathbf{b}(t)$ *be a Bézier curve with control points* $P_0, \dots, P_n$, *and let* a, b *be two real numbers, then the restriction of* $\mathbf{b}$ *to the interval* $[a,b]$ *is a Bézier curve with control points* $\widetilde{P}_0, \dots, \widetilde{P}_n$, *where*

$$\widetilde{P}_k = C(a)^{n-k} C(b)^k (P_0, \dots, P_n).$$

Proof: If $0 \le k \le n-1$, then the k'th element of the sequence

$$C(t)\big(C(a)^n, C(a)^{n-1}C(b), \dots, C(b)^n\big) = \\ \big((1-t)R + tL\big)\big(C(a)^n, C(a)^{n-1}C(b), \dots, C(b)^n\big),$$

is given by

$$\begin{aligned}(1-t)C(a)^{n-k}C(b)^k &+ tC(a)^{n-k-1}C(b)^{k+1} \\ &= C(a)^{n-k-1}\big((1-t)C(a) + tC(b)\big)C(b)^k \\ &= C(a)^{n-k-1}C\big((1-t)a + tb\big)C(b)^k \\ &= C(a)^{n-k-1}C(b)^k C\big((1-t)a + tb\big).\end{aligned}$$

Thus, we have

$$\begin{aligned}C(t)\big(C(a)^n, &C(a)^{n-1}C(b), \dots, C(b)^n\big) \\ &= \big(C(a)^{n-1},\ C(a)^{n-2}C(b), \dots, C(b)^{n-1}\big)C\big((1-t)a + tb\big),\end{aligned}$$

and by induction we get

$$C(t)^n\big(C(a)^n, C(a)^{n-1}C(b), \dots, C(b)^n\big) = C\big((1-t)a + tb\big)^n.$$

Hence $C(t)^n\big(\widetilde{P}_0, \dots, \widetilde{P}_n\big) = C\big((1-t)a + tb\big)^n (P_0, \dots, P_n)$ and the proof is complete. □

Finally, we demonstrate how the Bernstein form is a consequence of the binomial formula.

Theorem 19. *Let* $\mathbf{b}(t)$ *be a Bézier curve with control points* $P_0, \dots, P_n$. *Then*

$$\mathbf{b}(t) = \sum_{k=0}^{n} B_k^n(t) P_k,$$

where

$$B_k^n(t) = \binom{n}{k}(1-t)^{n-k}t^k, \quad k = 0, \dots, n,$$

are the Bernstein polynomials of degree n.

Proof: We have

$$C(t)^n = ((1-t)R + tL)^n = \sum_{k=0}^{n} \binom{n}{k} ((1-t)R)^{n-k} (tL)^k$$
$$= \sum_{k=0}^{n} \binom{n}{k} (1-t)^{n-k} t^k R^{n-k} L^k = \sum_{k=0}^{n} B_k^n(t) R^{n-k} L^k,$$

so

$$\mathbf{b}(t) = C(t)^n (P_0, \dots, P_n)$$
$$= \sum_{k=0}^{n} B_k^n(t) R^{n-k} L^k (P_0, \dots, P_n) = \sum_{k=0}^{n} B_k^n(t) P_k. \quad \square$$

It has recently been popular to view a Bézier curve as a symmetric multiaffine map (the so called *blossom*) evaluated at the diagonal. In our notation this amounts to the following:

Definition 20. *The blossom of a Bézier curve of degree n with control points $P_0, \dots, P_n$ is defined by*

$$\mathbf{b}(t_1, \dots, t_n) = C(t_1) \circ \dots \circ C(t_n)(P_0, \dots, P_n).$$

Once again we can derive the basic properties of the blossom by utilizing the commutativity of R and L, but we leave this as an exercise for the reader.

Acknowledgments. I am indebted to Lars-Erik Andersson who presented the proof of Theorem 18 (using infinite sequences) at a summer school in computational geometry, held at Linköping University, May 7–11, 1990.

References

1. Andersson, L-E., Private Notes, from a summer school in Computational Geometry, Linköping, 1990.
2. Farin, G., *Curves and Surfaces for Computer Aided Geometric Design*, Academic Press, NY, 1988.
3. Hosaka, M. and M. Kuroda, (in Japanese), J. of Information Processing Society of Japan **17** (1976), 1120–1127.
4. Hosaka, M. and M. Kuroda, A Synthesis Theory of Curves and Surfaces for CAD, Information Processing in Japan **17** (1977), 75–79.
5. Hoschek, J. and D. Lasser, *Fundamentals of Computer Aided Geometric Design*, A. K. Peters, Boston MA, 1993.

Jens Gravesen
Department of Mathematics
Building 303
Technical University of Denmark
DK-2800 Lyngby, DENMARK
`J.Gravesen@mat.dtu.dk`

Triangulations from Repeated Bisection

Bernd Hamann and Benjamin W. Jordan

Abstract. We present a method for the iterative refinement of triangulations. Given a coarse triangulation of the compact domain of a bivariate function, we present a refinement strategy based on approximation error. The triangulation is used to compute a best linear spline approximation, using the term best approximation in an integral least squares sense. We improve an approximation by identifying the triangle with largest error and refine the triangulation by bisecting this triangle.

§1. Introduction

In the context of visualizing very large data sets, it is imperative to use hierarchical data representations that allow us to study physical phenomena at multiple levels of detail. General, robust, and efficient methodologies are needed to support hierarchical data representation and visualization.

Triangulations are a natural choice when complicated regions in two – or three – dimensions must be represented. We present a construction of hierarchies of triangulations, constructed as best linear spline approximations. The basic idea is repreated bisection. The coefficients associated with each vertex in a triangulation are computed in a best approximation sense. Whenever one bisects a triangle one needs to compute new spline coefficients for all vertices. One can perform the matrix inversions efficiently due to the fact that the matrices are very sparse. This sparseness allows us to reduce matrix bandwidth significantly. The main principles of our approach become evident from the discussion of the univariate case.

Three main objectives have influenced the design of our construction. *The construction should be simple.* The number of special cases to be considered should be small; a refinement step should cause minimal topological change; and the computation of a best linear spline approximation should be robust. *The construction should be general.* The approach should be applicable to multivariate and multi-valued functions as well; and the approach should handle arbitrary, complex-shaped domains. *The construction should*

Mathematical Methods for Curves and Surfaces II
Morten Dæhlen, Tom Lyche, and Larry L. Schumaker (eds.), pp. 229–236.

ISBN 0-8265-1315-8.

be efficient. The computation of the triangulation hierarchy should be reasonably efficient. The single steps of our construction are:

1) **Initial approximation.** Define an initial, coarse triangulation of the function's domain and, for all vertices, compute the spline coefficients defining the initial best linear spline approximation.

2) **Error estimation.** Compute appropriate global error and local error (=triangle-specific error) estimates.

3) **Bisection.** Identify the triangle with largest local error and bisect it by splitting its longest edge into two segments – thereby also bisecting all triangles sharing this edge.

4) **Computation of new approximation.** Based on the new vertex set and new triangulation, compute a new best linear spline approximation. Iterate steps 2), 3) and 4) until a termination condition is met.

Remark 1.1. We assume that the function to be approximated is known analytically. Should this not be the case, one needs to first determine such a function, possibly from a finite set of scattered data.

From the set of all best linear spline approximations of F we select a subset consisting of those that we can associate with a particular level in an approximation hierarchy. What does the notion of "level" mean? Given an error tolerance tuple $\mathbf{E} = (\epsilon_0, ..., \epsilon_{m-1})$, $\epsilon_j > \epsilon_{j+1}$, we call an approximation a level-j approximation when its global error (see below) lies in the interval $(\epsilon_{j+1}, \epsilon_j]$. A spline approximation associated with level j is called the representative of level j if its number of knots is minimal among all level-j approximations.

§2. The Univariate Case

Our construction of best approximations requires a few, simple notions from linear algebra and approximation theory. We use an interval-weighted scalar product $\langle f, g \rangle$ for two functions $f(x)$ and $g(x)$, defined over $[a, b]$,

$$\langle f, g \rangle = \frac{1}{b-a} \int_a^b f(x)\, g(x)\, dx, \tag{1}$$

and an interval-weighted L^2 norm to measure a function $f(x)$,

$$\|f\| = \langle f, f \rangle^{1/2} = \left(\frac{1}{b-a} \int_a^b (f(x))^2\, dx \right)^{1/2}. \tag{2}$$

We include the factor $\frac{1}{b-a}$ to eliminate the influence of interval length when computing interval-specific error estimates for the approximation process.

This ensures "normalization" – we are using *error per unit* to determine which segment to bisect next.

The best approximation of a function F, when approximating it by a linear combination $\sum_{i=0}^{n-1} c_i f_i$ of independent functions $f_0, ..., f_{n-1}$, is defined by the solution of the normal equations

$$\begin{pmatrix} \langle f_0, f_0 \rangle & \cdots & \langle f_{n-1}, f_0 \rangle \\ \vdots & & \vdots \\ \langle f_0, f_{n-1} \rangle & \cdots & \langle f_{n-1}, f_{n-1} \rangle \end{pmatrix} \begin{pmatrix} c_0 \\ \vdots \\ c_{n-1} \end{pmatrix} = \begin{pmatrix} \langle F, f_0 \rangle \\ \vdots \\ \langle F, f_{n-1} \rangle \end{pmatrix}, \quad (3)$$

see, [4]. We will also write this linear system as $M^{[n-1]}\mathbf{c}^{[n-1]} = \mathbf{F}^{[n-1]}$. For an arbitrary set of basis functions f_i one has to investigate means for the efficient solution of this system.

We are concerned with the construction of a hierarchy of best linear spline approximations to a function F by increasing the number of basis functions, or in other words, the number of knots. We increase the number of knots, generally one-by-one, until a best approximation is obtained whose associated error is smaller than some threshold.

In general, the computation of a best linear spline approximation to a given function F is an optimization problem, which, when allowing variable knots, is quite involved, see [2]. Our current approach does not permit variable knots, it is based on repeated bisection using the midpoints of intervals. This approach is computationally less expensive than the general optimization problem. In each step, we simply determine the interval with largest error and bisect it. Bisection implies the insertion of the midpoint as a new new knot.

We start by approximating a function F using a single linear segment by computing $M^{[1]}\mathbf{c}^{[1]} = \mathbf{F}^{[1]}$. We assume that F is defined over $[0, 1]$ and that the basis functions f_i are hat functions, *i.e.*, $f_i(x_j) = \delta_{i,j}$ (see Fig. 1). The initial error is $E^{[1]} = ||F - (c_0 f_0 + c_1 f_1)||$, and if this value is larger than the given threshold we insert an additional knot at $x = \frac{1}{2}$. (Inserting this additional knot changes the knot sequence, and, consequently, one obtains three new hat basis functions.) Thus, the next best approximation, using the updated knot sequence, is obtained by solving $M^{[2]}\mathbf{c}^{[2]} = \mathbf{F}^{[2]}$, and the new error value is given by $E^{[2]} = \left|\left|F - \sum_{i=0}^{2} c_i f_i\right|\right|$. Since our method is based on repeated bisection, we compute segment-specific, local errors for the intervals $\left[0, \frac{1}{2}\right]$ and $\left[\frac{1}{2}, 1\right]$ which determine the segment to be bisected next.

Assuming that an intermediate approximation is based on the knot sequence $0 = x_0 < x_1 < x_2 < \cdots < x_{k-2} < x_{k-1} = 1$ and that its coefficient vector is $(c_0, ..., c_{k-1})^T$, we define the global error as

$$E^{[k-1]} = \| F - \sum_{i=0}^{k-1} c_i f_i \| \quad (4)$$

and local (segment-specific) errors as

$$e_i^{[k-1]} = \left(\frac{1}{x_{i+1} - x_i} \int_{x_i}^{x_{i+1}} \left(F - (c_i f_i + c_{i+1} f_{i+1})\right)^2 dx \right)^{1/2}, \quad i = 0, ..., k-2. \quad (5)$$

We compute the local error values for each segment and bisect the segment with largest local error value. If there are multiple segments with the same maximal error value, we randomly pick one of them to be bisected. (It is of course possible to bisect all segments with the same maximal error at the same time, thus leading to a unique answer.)

When using hat functions as basis functions, the only non-zero elements of $M^{[k-1]}$, for a particular row i, are the elements $\langle f_{i-1}, f_i \rangle = \Delta_{i-1}/6$, $\langle f_i, f_i \rangle = (\Delta_{i-1} + \Delta_i)/3$, and $\langle f_{i+1}, f_i \rangle = \Delta_i/6$, where $\Delta_j = x_{j+1} - x_j$. Thus, $M^{[k-1]}$ is the tridiagonal matrix

$$M^{[k-1]} = \frac{1}{6} \begin{pmatrix} 2\Delta_0 & \Delta_0 & & & \\ \Delta_0 & 2(\Delta_0 + \Delta_1) & \Delta_1 & & \\ & \Delta_1 & 2(\Delta_1 + \Delta_2) & \Delta_2 & \\ & & \ddots & \ddots & \ddots \\ & & & \Delta_{k-2} & 2\Delta_{k-2} \end{pmatrix}. \tag{6}$$

It is necessary, due to the global nature of the problem, to re-compute all components of the coefficient vectors $(c_0, c_1, c_2, \ldots)^T$ whenever one inserts a knot.

Remark 2.1. A drawback when inserting knots one-by-one is the fact that the method is not local, *i.e.*, inserting a single knot leads to a new system of normal equations. One can significantly enhance the efficiency of the method by bisecting multiple intervals in one step, thereby refining an approximation in different regions in one step.

§3. The Bivariate Case

We proceed similarly in the bivariate case. Given an initial best linear spline approximation based on a small number of triangles, we compute the global approximation error of the associated best linear spline approximation and, should this error be too large, insert the midpoint of the longest edge of the triangle with largest local error as a new knot. Bisection leads to the split of one or two triangles – depending on whether the bisected edge is shared by a second triangle or not.

Bisection terminates when a certain global error condition is satisfied. If the triangle with largest local error has more than one edge with maximal length, we bisect one of the edges, chosen randomly, with maximal length. (Alternatively, one could split multiple edges simultaneously to guarantee uniqueness.) If multiple triangles share the same maximal local error value, we choose, randomly, one of these triangles to be bisected. (One could choose to bisect larger triangles first – or vice versa – or choose to simultaneously split all triangles sharing the same maximal local error value.)

The union of the triangles in the initial triangulation defines the region to which we apply our refinement scheme to obtain improved approximations of a bivariate function $F(x, y)$. We assume that the boundary of F's domain is approximated well enough by the coarse, initial triangulation.

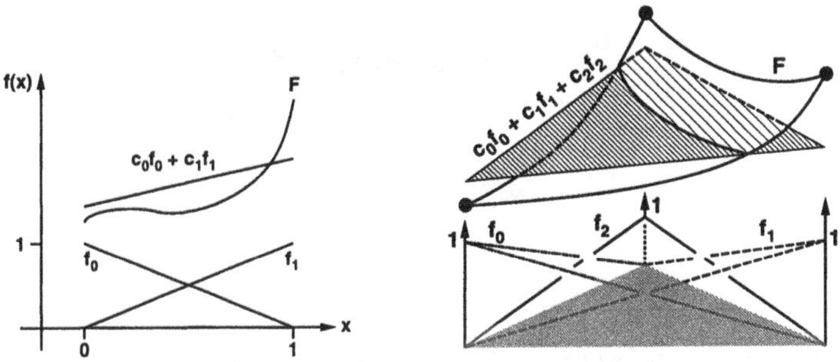

Fig. 1. Basis functions f_i, function F, and F's best approximation.

Some definitions are required for the bivariate setting. Denoting the vertex (knot) set of an intermediate approximation by $\{\mathbf{v}_i = (x_i, y_i)^T | i = 0, ..., k-1\}$ and the coefficient vector associated with the hat basis functions $f_i(x, y)$ by $(c_0, ..., c_{k-1})^T$, the global error is

$$E^{[k-1]} = \| F - \sum_{i=0}^{k-1} c_i f_i \|. \tag{7}$$

and the triangle-specific, local error for triangle T_j is

$$e_j^{[k-1]} = \left(\frac{1}{\text{area\_of\_}T_j} \int_{T_j} \left(F - \text{spline\_over\_}T_j)\right)^2 dxdy \right)^{1/2}, \quad j = 0, ..., n_T - 1, \tag{8}$$

where n_T is the number of triangles.

Considering the m vertices in the initial triangulation, the initial best linear spline approximation is defined by $M^{[m-1]}\mathbf{c}^{[m-1]} = \mathbf{F}^{[m-1]}$. The initial error is $E^{[m-1]} = \left|\left|F - \sum_{i=0}^{m-1} c_i f_i\right|\right|$. If $E^{[m-1]}$ is larger than a prescribed tolerance, we insert the midpoint of the longest edge of the triangle with largest local error as a new knot. We continue to insert knots until an approximation is good enough. In the bivariate setting the local errors are determined by the difference between $F(x, y)$ and an intermediate best linear approximation over each triangle. Fig. 1 illustrates F, hat basis functions, and a best approximation for the univariate and the bivariate cases.

Considering k knots, the solution of the normal equations requires the inversion of a matrix $M^{[k-1]}$, where the number of non-zero entries per column in this matrix is defined by the valences of the vertices in the triangulation. A vertex $\mathbf{v}_i$ with valence v_i causes $v_i + 1$ non-zero entries in column i. These valence values are not limited, and it is possible to obtain large numbers of non-zero elements in certain columns of $M^{[k-1]}$. Thus, the computation of the coefficients in the bivariate case is much more expensive than in the univariate case.

Remark 3.1. Arbitrarily skinny triangles can result from repeated bisection. Therefore, one should incorporate a constraint into the algorithm that prohibits the generation of triangles whose shapes are not acceptable.

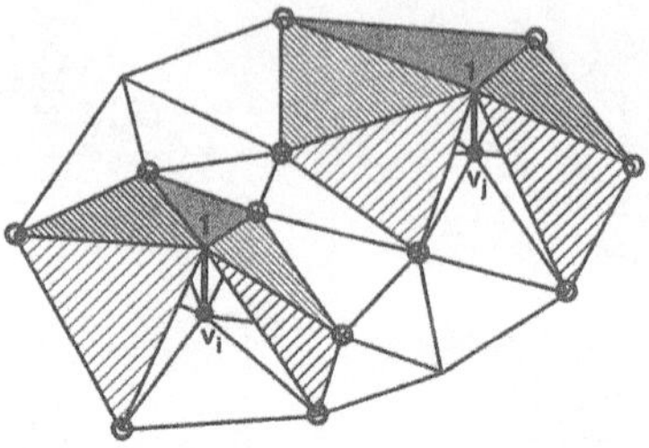

Fig. 2. Platelets of v_i and v_j and associated hat basis functions.

The change-of-variables theorem allows us to effectively compute scalar products of hat functions in the bivariate case. A hat function $f_i(x, y)$ associated with vertex $\mathbf{v}_i$ is the spline basis function whose function value is one at $\mathbf{v}_i$ and zero at all other vertices. The function f_i varies linearly between zero and one over the triangles defining $\mathbf{v}_i$'s platelet, see Fig. 2. (The platelet of $\mathbf{v}_i$ is the set of triangles sharing $\mathbf{v}_i$ as a common vertex.)

Applying the change-of-variables theorem, the scalar product $\langle f_i, f_i \rangle$ is given by

$$\langle f_i, f_i \rangle = \sum_{j=0}^{n_i-1} \int_{T_j} f_i \, f_i \, dxdy = \frac{1}{12} \sum_{j=0}^{n_i-1} J_j, \tag{9}$$

where n_i is the number of platelet triangles associated with vertex $\mathbf{v}_i$, T_j is the j-th platelet triangle, and J_j is the Jacobian of the j-th platelet triangle,

$$J_j = \det \begin{pmatrix} x_1^j - x_0^j & x_2^j - x_0^j \\ y_1^j - y_0^j & y_2^j - y_0^j \end{pmatrix}. \tag{10}$$

Here, $(x_0^j, y_0^j)^T$, $(x_1^j, y_1^j)^T$, and $(x_2^j, y_2^j)^T$ are the coordinate vectors of T_j's counterclockwise-oriented vertices. (The platelet of $\mathbf{v}_i$ is the set of triangles $\mathcal{T}_i = \{T_j\}_{j=0}^{n_i-1}$.) The scalar product $\langle f_j, f_k \rangle$ of two basis functions whose associated vertices $\mathbf{v}_j$ and $\mathbf{v}_k$ are connected by an edge is

$$\langle f_j, f_k \rangle = \sum_{i=0}^{n_{j,k}-1} \int_{T_i} f_j \, f_k \, dxdy = \frac{1}{24} \sum_{i=0}^{n_{j,k}-1} J_i. \tag{11}$$

(The set $\mathcal{T}_{j,k} = \{T_i\}_{i=0}^{n_{j,k}-1}$ is the union of the triangles in $\mathbf{v}_j$'s and $\mathbf{v}_k$'s platelets.)

Remark 3.2. The matrices $M^{[k-1]}$ are sparse, *i.e.*, relatively few matrix entries are different from zero, and we utilize the Cuthill-McKee algorithm to reduce the bandwidths of the matrices before inverting them, see [3]. A lower bound for the bandwidth that one can achieve when applying a bandwidth-reducing strategy is given by the vertex with maximal valence. One might therefore consider imposing a threshold on maximal vertex valence. Efficiency

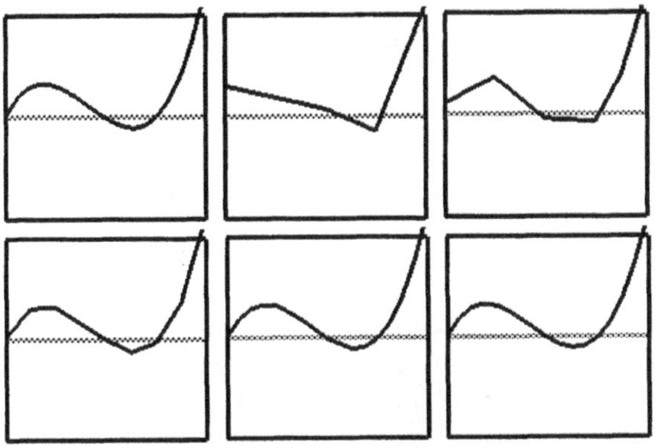

Fig. 3. $F = 10x\left(x - \frac{1}{2}\right)\left(x - \frac{3}{4}\right)$; 4-, 6-, 8-, 14-, and 19-knot approximations.

is a concern when performing bisection for a triangulation consisting of a large number of vertices, since each bisection step requires the re-computation of all coefficients. The Cuthill-McKee algorithm, a graph-based vertex indexing scheme, allows one to efficiently insert a new vertex; since the insertion of a vertex is a "local" graph operation one can also efficiently determine the new vertex indices leading to a new small-bandwidth matrix.

§4. Examples

We have tested our approach for univariate and bivariate test functions, using Romberg integration for the computation of the scalar products $\langle F, f_i \rangle$ and the errors, see [1]. We have inserted knots one-by-one.

Figs. 3 and 4 show a univariate and a bivariate example. The upper-left corner image in both figures is an "exact" rendering of the original function. Univariate functions are rendered as graphs $\left(x, f(x)\right)^T$, where the squares in Fig. 3 indicate the region $[0,1] \times [-1,1]$. Bivariate functions are rendered as graphs $\left(x, y, f(x,y)\right)^T$ using flat-shaded surface triangulations, where the cubes in Fig. 4 indicate the region $[0,1] \times [0,1] \times [0,1]$. (The initial triangulation is defined by splitting the unit square into the two triangles obtained by connecting $(0,0)^T$ and $(1,1)^T$. Global approximation errors are based on equations (4) and (7). Using a vector notation, the global errors of the five approximations shown in Fig. 3 are given by $(0.816, 0.241, 0.073, 0.022, 0.010)$, and the errors for the three approximations shown in Fig. 4 are $(0.062, 0.009, 0.001)$.

§5. Conclusions

We have presented a method for the refinement of best linear spline approximations for univariate and bivariate functions. The spline basis functions are defined over collections of triangles. The approach is sound, robust, general, and fairly efficient. We are extending the scheme to trivariate functions in the context of hierarchical volume visualization. We plan to improve the efficiency

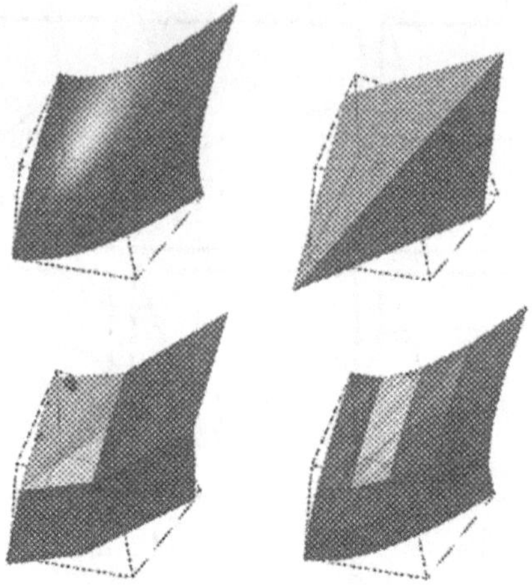

Fig. 4. $F = x^2 + y^2$; 4-, 11-, and 41-knot approximations.

of our approach and generalize it to allow more general knot placement. We will extend the approach to multi-valued functions, in particular to vector fields.

Acknowledgments. This work was supported by NSF grants ASC-9210439 (Research Initiation Award) and ASC-9624034 (CAREER Award), by ONR grant N00014-97-1-0222, and by ARO grant ARO-36598-MA-RIP.

References

1. Boehm, W. and H. Prautzsch, *Numerical Methods*, A K Peters, Wellesley, 1993.
2. Cantoni, A., Optimal curve fitting with piecewise linear functions, IEEE Transactions on Computers **20** (1971), 59–67.
3. Cuthill, E. and J. McKee, Reducing the bandwidth of sparse symmetric matrices, in *Proc. ACM National Conference*, Association for Computing Machinery, New York, 1969, 157–172.
4. Davis, P. J., *Interpolation and Approximation*, Dover Publications, New York, 1975.

Bernd Hamann, Co-Director
Center for Image Processing
and Integrated Computing (CIPIC)
Department of Computer Science
University of California, Davis
Davis, CA 95616-8562 USA
`hamann@cs.ucdavis.edu`

Benjamin W. Jordan
Pixar Animation Studios
1001 West Cutting Boulevard
Richmond, CA 94804 USA
`bjordan@pixar.com`

The Normal Form of a Planar Curve and Its Application to Curve Design

Erich Hartmann

Abstract. Analogously to the Hessian normal form of a line, the normal form $h(\boldsymbol{x}) = 0$ of a planar curve is defined. Thus any planar curve for which the normal form exists can be considered as an implicit curve and all such curves can be treated in a uniform way. Hence the simple implicit blending methods are applicable. Offsets are just level curves of the oriented distance function h. Bisectors of two curves have simple implicit representations.

§1. Introduction

In CAGD a planar curve is usually defined as a parametric curve $\boldsymbol{x} = \gamma(t), t \in [a, b]$ or as an implicit curve $f(\boldsymbol{x}) = 0$. But there are important curves which have no such standard representations, for example (except for special cases) the offset curves of an implicit curve, the bisectors of two curves, and blend curves of two parametric curves. The idea of the normal form (introduced below) allows us to consider any curve as an implicit one, and all the powerful methods defined for implicit curves are applicable. Furthermore, the simple geometric meaning of the oriented distance function h yields a *uniform treatment of all curves independent of their definitions.*

In Section 2 we define the normal form for a planar parametric curve and generalize the idea to arbitrary (planar) curves. Special attention is paid to implicit curves. Section 3 shows that the normal form is a powerful tool for solving basic problems in CAGD. In [3] the idea of the normal form is extended to surfaces.

§2. The Normal Form of a Planar Curve

We introduce first the normal form for a parametric curve, generalize it to arbitrary curves, and show how to handle implicit curves.

Mathematical Methods for Curves and Surfaces II
Morten Dæhlen, Tom Lyche, and Larry L. Schumaker (eds.), pp. 237–244.

ISBN 0-8265-1315-8.

2.1 The Normal Form of a Parametric Curve

Let

- $\Gamma : \boldsymbol{x} = \boldsymbol{\gamma}(t) = (\gamma_1(t), \gamma_2(t)),\ t \in [a, b]$ be a smooth plane curve,
- $\boldsymbol{n}(t) := (\gamma_2(t), -\gamma_1(t))/\|...\|$ the unit normal,
- $D \subset \mathbb{R}^2$ such that for any $\boldsymbol{x} \in D$ the function $d_{\boldsymbol{x}}(t) := \|\boldsymbol{x} - \boldsymbol{\gamma}(t)\|$ has exactly one minimum,
- $t_m(\boldsymbol{x})$ the parameter of the minimum, $\boldsymbol{\gamma}(t_m(\boldsymbol{x}))$ the footpoint (of point $\boldsymbol{x}$).
- $h(\boldsymbol{x}) := sign((\boldsymbol{x} - \boldsymbol{\gamma}(t_m(\boldsymbol{x}))) \cdot \boldsymbol{n}(t_m(\boldsymbol{x})))\ d_{\boldsymbol{x}}(t_m(\boldsymbol{x}))$ the oriented distance of $\boldsymbol{x}$ to curve Γ.

The equation $h(\boldsymbol{x}) = 0$ is an implicit representation of curve Γ and is called the normal form of Γ.

Remark: The normal form of a line is the well known Hessian normal form: $h(\boldsymbol{x}) := \boldsymbol{n} \cdot (\boldsymbol{x} - \boldsymbol{x}_1) = 0$ where $\boldsymbol{n}$ is a unit normal and $\boldsymbol{x}_1$ a point of the line.

Essential Properties of the Oriented Distance Function:

(O) Level curves of the oriented distance function h are offset curves of curve Γ.

(D) h is differentiable: $\nabla h(\boldsymbol{x}) = \boldsymbol{n}(t_m(\boldsymbol{x}))$ (unit normal at the footpoint belonging to $\boldsymbol{x}$).

The second property can be derived by considering the offset curves Γ_h : $\boldsymbol{x} = \boldsymbol{\gamma}(t) + h\boldsymbol{n}(t)$ of Γ. Applying the implicit function theorem to the equation $\boldsymbol{x} - \boldsymbol{\gamma}(t) + h\boldsymbol{n}(t) = 0$ for given $\boldsymbol{x}$ yields the existence of the real function $h(\boldsymbol{x})$ and statement (D).

Evaluation of the Distance Function:

The usage of the normal form depends on the possibility of evaluating the oriented distance function h. This means solving for a given $\boldsymbol{x}$ the equation

$$(\boldsymbol{x} - \boldsymbol{\gamma}(t)) \cdot \dot{\boldsymbol{\gamma}}(t) = 0.$$

This can be done either by

- Newton iteration (the second derivative of $\boldsymbol{\gamma}$ is necessary) or
- repeatedly determining footpoints on tangents (only the first derivative of $\boldsymbol{\gamma}$ is used): $t_{i+1} = t_i + \frac{(\boldsymbol{x} - \boldsymbol{\gamma}(t_i)) \cdot \dot{\boldsymbol{\gamma}}(t_i)}{\dot{\boldsymbol{\gamma}}(t_i)^2}$.

The necessary starting parameter can be determined from the minimum of $\{(\boldsymbol{x} - \boldsymbol{\gamma}(t_1))^2, ..., (\boldsymbol{x} - \boldsymbol{\gamma}(t_n))^2\}$ for an equally spaced polygon $\{\boldsymbol{\gamma}(t_1), ..., \boldsymbol{\gamma}(t_n)\}$ on Γ or from the footpoint of a point $\boldsymbol{x}'$ in the vicinity of $\boldsymbol{x}$.

2.2 Generalization

It is not necessary that the curve Γ from the previous subsection be given by a parametric representation. The only necessary properties are

(NF1) Γ has to be *smooth* and

(NF2) there is a region $D \subset \mathbb{R}^2$ in the vicinity of Γ such that for any point $\boldsymbol{x} \in D$ it is possible to determine a (unique) *footpoint* (point on Γ with minimal distance to $\boldsymbol{x}$).

Conclusion: Any planar curve Γ with properties (NF1) and (NF2) can be considered as an implicit curve $h(\boldsymbol{x}) = 0$, where $h(\boldsymbol{x})$ is the oriented distance of point $\boldsymbol{x} \in D$ to the curve Γ and $\nabla h(\boldsymbol{x})$ is the unit normal at the footpoint belonging to $\boldsymbol{x}$. The equation $h(\boldsymbol{x}) = 0$ is called the normal form of Γ.

So any algorithm for implicit curves which only needs $h(\boldsymbol{x})$ and $\nabla h(\boldsymbol{x})$ for a point $\boldsymbol{x}$ of consideration are applicable to Γ. For example, the tracing algorithm or Newton iteration for determining intersection points (see Section 3).

Remarks: a) The determination of the region D, for which a unique footpoint exists, is usually a difficult task and subject to further investigation.

b) The convergence of the footpoint algorithm becomes critical for points near the boundary of D. Statements on the quality of convergence are difficult.

2.3 The Normal Form of an Implicit Curve

Let $\Gamma_0 : \ f = 0$ be a smooth implicit curve. Usually the function value $f(\boldsymbol{x})$ has no geometric meaning. The advantage of the normal form $h = 0$ of a curve is the simple geometric meaning of the function value $h(\boldsymbol{x})$ (oriented distance) and the level curves $h(\boldsymbol{x}) = c$ (offset curves). So, whenever the distance of a point or an offset curve of an implicit curve is needed, one should apply the normal form of the curve.

Evaluation of the Distance Function:

Let $\Gamma_0 : \ f = 0$ be an implicit curve and $\boldsymbol{x}$ a point within the vicinity of Γ_0. A footpoint $\boldsymbol{x}_0$ of $\boldsymbol{x}$ fulfills the following two conditions:

1) $f(\boldsymbol{x}_0) = 0$ and 2) $\nabla f(\boldsymbol{x}_0)$ is parallel to $\boldsymbol{x} - \boldsymbol{x}_0$.

For the determination of $\boldsymbol{x}_0$ we use the following procedure curvepoint which calculates a *curve point* $\mathbf{c}$ along the steepest way for a given point $\boldsymbol{p}$:

(CP0) $\boldsymbol{q}_0 = \boldsymbol{p}$

(CP1) repeat $\boldsymbol{q}_{k+1} := \boldsymbol{q}_k - \frac{f(\boldsymbol{q}_k)}{\nabla f(\boldsymbol{q}_k)^2} \nabla f(\boldsymbol{q}_k)$ (Newton step)

until $\|\boldsymbol{q}_{k+1} - \boldsymbol{q}_k\|$ is "sufficiently" small.

$\mathbf{c} = \boldsymbol{q}_{k+1}$.

Determination of *footpoint* $\boldsymbol{x}_0$ (belonging to $\boldsymbol{x}$):

(FP0) $\mathbf{c}_0 = \mathtt{curvepoint}(\boldsymbol{x})$

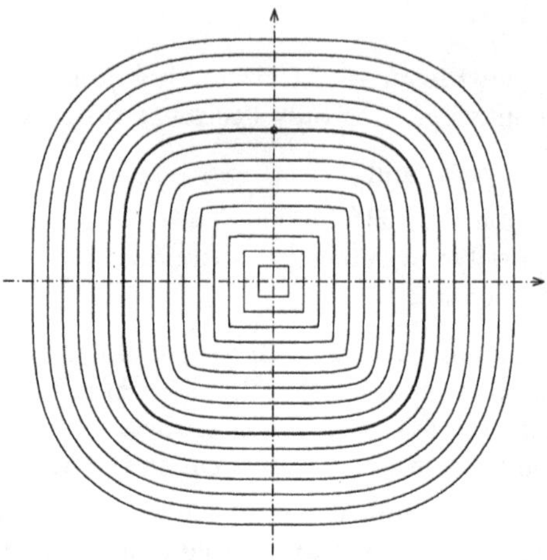

Fig. 1. Offset curves of the implicit curve $f(\boldsymbol{x}) = x^4 + y^4 - 1 = 0$ (bold curve).

(FP1) repeat $\boldsymbol{t}_{i+1} = \boldsymbol{x} - \frac{(\boldsymbol{x}-\boldsymbol{c}_i)\cdot\nabla f(\boldsymbol{c}_i)}{\nabla f(\boldsymbol{c}_i)^2}\nabla f(\boldsymbol{c}_i)$ (near point on tangent line),

$\boldsymbol{c}_{i+1} = \texttt{curvepoint}(\boldsymbol{t}_{i+1})$.

until $\|\boldsymbol{c}_{i+1} - \boldsymbol{c}_i\|$ is "sufficiently" small.

$\boldsymbol{x}_0 = \boldsymbol{c}_{i+1}$.

Displaying an Implicit Curve:

All curves in Section 3 are either implicit curves or represented implicitly by their normal forms and *traced* by the following algorithm for regular implicit curves:

Let $\Gamma_0 :\ f = 0$ be an implicit curve.

(IC0) Choose a starting point $\boldsymbol{q}_1$ in the vicinity of Γ

(IC1) $\boldsymbol{p}_i := \texttt{curvepoint}(\boldsymbol{q}_i)$ (see algorithm **curvepoint** above)

$\boldsymbol{t}_i := (-f_y(\boldsymbol{p}_i), f_x(\boldsymbol{p}_i))/\|...\|$ (unit tangent)

$\boldsymbol{q}_{i+1} := \boldsymbol{p}_i + \delta \boldsymbol{t}_i$ (δ: steplength)

The tracing algorithm stops if $\boldsymbol{p}_i$ is "near" a prescribed endpoint (or another termination).

§3. Applications of the Normal Form

3.1 Offset Curves of Implicit Curves

Offset curves of an implicit curve $f = 0$ are level curves of the oriented distance function: $h = c$.

Example: Figure 1 shows the curve $f(\boldsymbol{x}) = x^4 + y^4 - 1 = 0$ (bold curve) and offset curves. One recognizes that some of the inner level curves possess singularities at lines of symmetry.

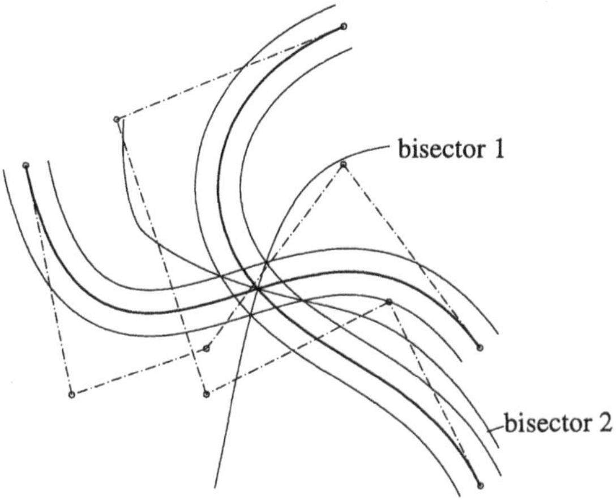

Fig. 2. Offset curves and bisectors of two Bezier curves (bold curves).

3.2 Bisectors of Two Curves

Let Γ_1, Γ_2 be two smooth parametric or implicit curves, and $h_1 = 0, h_2 = 0$ their normal forms. Hence the equations

$$h_1(\boldsymbol{x}) - h_2(\boldsymbol{x}) = 0, \quad h_1(\boldsymbol{x}) + h_2(\boldsymbol{x}) = 0$$

are implicit representations of the bisector curves (points, which have equal distance to Γ_1 and Γ_2).

The bisectors can be traced by any marching algorithm that uses only the first derivative (see Section 2.3).

Example 1: Figure 2 shows two Bezier curves (bold curves) together with two offset curves and their bisectors.

Example 2: Figure 3 shows the curves

$x^2 + y^2 - r = 0$ (circle), $\quad (x - 1.5)^4 + (y - 0.5)^4 - 1 = 0$ (hypercircle)

and one bisector for the cases: $r = 1$, $r = 0.5$, $r = 0.2$, and $r = 0.002$.

Example 3: Figure 4 shows the bisector curve of

a) the ellipse $x^2 + \frac{y^2}{4} = 1$ and the point $(-0.5, 1)$

b) the hypercircle $x^4 + y^4 = 1$ and the point $(-0.4, 0.6)$.

While tracing the bisectors one should pay attention to singularities on the lines of symmetry of the ellipse and the hypercircle respectively. Further information on bisectors is contained in [1,2].

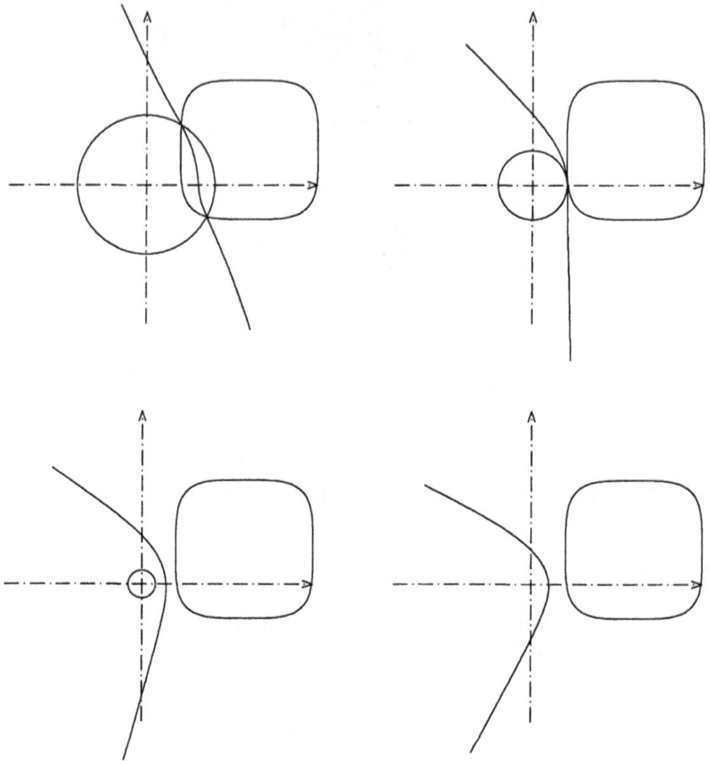

Fig. 3. One bisector of a circle with various radii and a hypercircle .

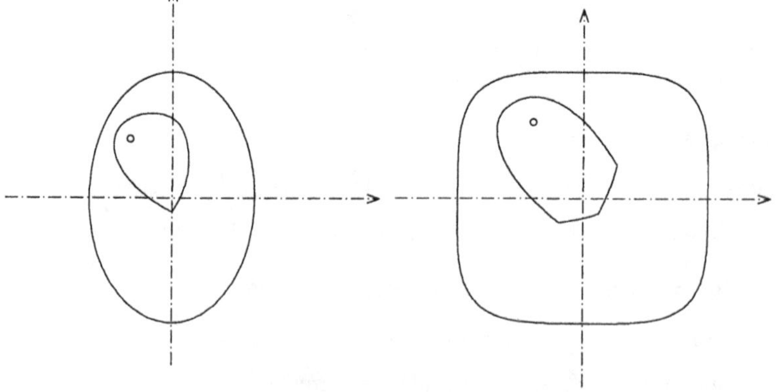

Fig. 4. The bisector of an ellipse/hypercircle and a point.

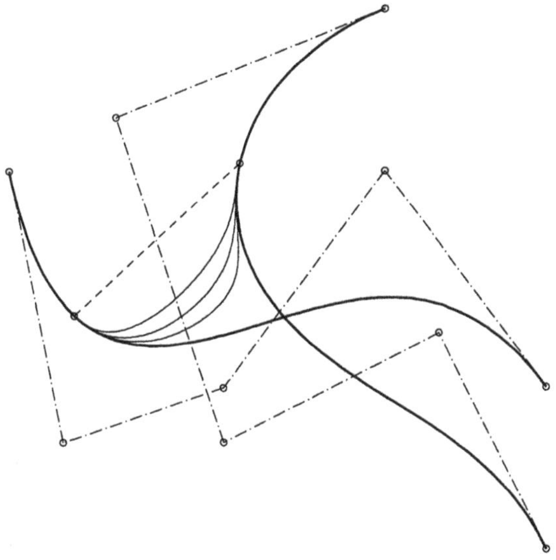

Fig. 5. G^2–Blending curves of two Bezier curves.

3.3 G^n–Blending of Curves

Let $\Gamma_1 : f_1(\boldsymbol{x}) = 0$, $\Gamma_2 : f_2(\boldsymbol{x}) = 0$ be two implicit curves and $\Gamma_0 : f_0(\boldsymbol{x}) = 0$ a line which intersects Γ_1 and Γ_2 (f_i differentiable enough). Then

$$\Phi_\mu : (1-\mu) f_1 f_2 - \mu f_0^{n+1} = 0, \;\; 0 < \mu < 1,$$

is for any μ a curve with G^n-contact to the curves Γ_1 and Γ_2 at the intersection points $\Gamma_1 \cap \Gamma_0$, $\Gamma_2 \cap \Gamma_0$. The blending curves Φ_μ are called parabolic functional splines (see [4]).

Example: Using the normal form this (and any other) implicit blending method is applicable to almost arbitrary curves, especially to parametric curves. Figure 5 shows G^2–blending curves of two Bezier curves for $\mu = 0.01, 0.03$ and 0.1, respectively.

3.4 Intersection Points of Curves

Using normal forms of curves we are able to determine intersection points by an algorithm for implicit curves (Newton iteration).

Example: The intersection points of a blending curve (normal form: $h(\boldsymbol{x}) = 0$) of the previous subsection with a bisector curve (normal form: $b(\boldsymbol{x}) = 0$) of the given Bezier curves can be determined by solving the nonlinear system

$$h(\boldsymbol{x}) = 0, \quad b(\boldsymbol{x}) = 0$$

with a Newton iteration.

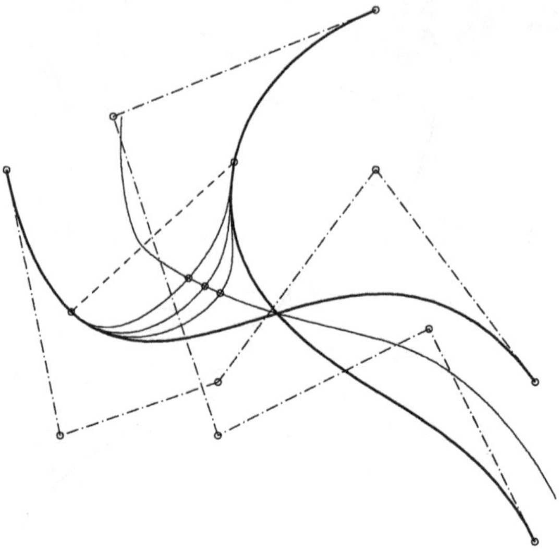

Fig. 6. Intersection points of blending curves and a bisector.

References

1. Farouki, R. T. and J. K. Johnstone, The bisector of a point and a plane parametric curve, Comput. Aided Geom. Design **11** (1994), 117–151.
2. Farouki, R. T. and J. K. Johnstone, Computing point/curve and curve/curve bisectors, in *Design and Application of Curves and Surfaces: Mathematics of Surfaces V*, R. B. Fisher (ed.), Oxford University Press, 1994, 327–354.
3. Hartmann, E., Numerical Implicitization for Intersection and G^n-continuous Blending of Surfaces, Comput. Aided Geom. Design, to appear.
4. Li, J., J. Hoschek, and E. Hartmann, G^{n-1}-functional splines for interpolation and approximation of curves, surfaces and solids, Comput. Aided Geom. Des. **7** (1990), 209–220.

Erich Hartmann
Fachbereich Mathematik
Technische Universität Darmstadt
Schlossgartenstr. 7
64289 Darmstadt, GERMANY
`ehartmann@mathematik.tu-darmstadt.de`

Finite Element Thin Plate Splines for Data Mining Applications

Markus Hegland, Stephen Roberts, and Irfan Altas

Abstract. Thin plate splines have been used successfully to model curves and surfaces. A new application is in data mining where they are used to model interaction terms. These interaction splines break the "curse of dimensionality" by reducing the high-dimensional nonparametric regression problem to the determination of a set of interdependent surfaces. However, the determination of the corresponding thin plate splines requires the solution of a dense linear system of equations of order n where n is the number of observations. For data mining applications n can be in the millions, and so standard thin plate splines, even using fast algorithms may not be practical. A finite element approximation of the thin plate splines will be described. The method uses H^1 elements in a formulation which only needs first order derivatives. The resolution of the method is chosen independently of the number of observations which only need to be read from secondary storage once and do not need to be stored in memory. The formulation leads to a saddle point problem. Convergence and solution of the method and its relationship to the standard thin plate splines will be discussed.

§1. Introduction

Recent years have seen a considerable effort to develop reliable and efficient data mining tools to discover hidden knowledge in very large data bases. Data mining methods have been applied to a number of application areas, for instance the calculation of profit according to customer, product and acquisition channel; trend and response prediction and identification; the development of tools to examine corporate intelligence reports, tax/insurance frauds and patterns in airline crashes. The interested reader can refer to [5] for detailed information.

The fundamental issue is the development and adaptation of algorithms to extract some useful information from very large databases. Often the number of observations can be of the order of millions, where there may be thousands of variables recorded.

Mathematical Methods for Curves and Surfaces II

Morten Dæhlen, Tom Lyche, and Larry L. Schumaker (eds.), pp. 245–252.

ISBN 0-8265-1315-8.

In this work we study the development of nonparametric models $y = f(\xi_1, \dots, \xi_m)$ to fit very large data sets. Here y is the expectation of an observation and $\xi_1, \dots, \xi_m$ represent the dimensionality of the data. If the data set is complex (i.e., m is large) one will have to deal with the *curse of dimensionality* when one tries to estimate the model f. To overcome this curse, additive models or interaction splines have been used [8]. In generalised additive models, if interaction terms are limited to order two interactions, this leads to a problem of determination of coupled surfaces. Thus, an important part of any data analysis algorithm for this problem is the determination of an approximating surface for extremely large data sets.

Thin plate splines are very popular for the estimation of surfaces $y = f(\xi_1, \xi_2)$ from observed data. A variational characterisation of the thin plate splines was proposed by Duchon [7]. He showed that an explicit formulation of the solution can be obtained in terms of radial basis functions which leads to a symmetric indefinite dense linear system of equations. It was seen that this system can be reduced to a positive definite system of equations which can be solved by a conjugate gradient method [11]. Further improvements using ideas from multipole expansions and Lagrange functions lead to methods which are of $O(n)$ or $O(n\log(n))$ in complexity [2,3,4], where n is the number of observations. Though these methods provide fast algorithms, the amount of storage still depends on the number of observations.

In this work, we introduce a new smoothing method which can be viewed as a discrete thin plate spline like those proposed in [1,12]. This new spline has similar properties to ordinary thin plate splines, but combines the favourable properties of finite element smoothing with those of thin plate splines. In Section 2, we introduce the method which is based on formulating the thin plate spline functional in terms of first order derivatives. This is similar to mixed finite element techniques for the biharmonic equation. In Section 3 we describe the discrete problem and explain the assembly and solution of equations. In Section 4, we provide some computational examples comparing our method with Duchon's thin plate splines and the method of finite element least squares fitting. Finally we end with a short conclusion.

§2. First Order Thin Plate Splines

The thin plate spline for a general domain Ω is the function f_α for which all the partial derivatives of total order 2 are in $L^2(\Omega)$ and which minimises the functional

$$\frac{1}{n}\sum_{i=1}^{n}(f(\boldsymbol{x}_i) - y_i)^2 + \alpha \int_\Omega \left(f_{\xi_1\xi_1}(\boldsymbol{x})^2 + 2f_{\xi_1\xi_2}(\boldsymbol{x})^2 + f_{\xi_2\xi_2}(\boldsymbol{x})^2\right) d\boldsymbol{x},$$

where $\boldsymbol{x} = (\xi_1, \xi_2)^T$ and $\boldsymbol{x}_i = (\xi_{1i}, \xi_{2i})^T \in \Omega$, $y_i \in \mathbb{R}$, $i = 1, \dots, n$ denote the measurements. The case $\Omega = \mathbb{R}^2$ was extensively investigated by Duchon [7]. For compact Ω, finite element approximations, called D^m splines, have been investigated by Arcangeli [1], where the function f_α is approximated by functions in a discrete subspace of $H^2(\Omega)$.

In this work we investigate a mixed method in which the need to work with $H^2(\Omega)$ is removed. We replace the second derivatives in the above functional with the first derivatives of the gradient $\boldsymbol{u}$ of the function f. Now the condition $\nabla \boldsymbol{u} = f$ is replaced with the condition $\Delta f = \mathrm{div}(u)$ with appropriate boundary conditions, together with the condition $\mathrm{curl}(u) = 0$.

This first order derivative formulation allows us to work with simple continuous piecewise polynomial finite element spaces. In addition the underlying equations that need to be solved are Poisson equations, for which robust fast solvers exist. In addition we have found that the condition $\mathrm{curl}(u) = 0$ does not need to be enforced to lead to good data fitting.

To describe our reformulation we introduce the space $V = \bar{H}^1(\Omega)^2 \times \mathbb{R}^3$, where $\bar{H}^1(\Omega)$ is the closed subspace of the Sobolev space $H^1(\Omega)$ consisting of functions with zero mean. It is seen that $|v|_1^2 := \int_\Omega |\nabla v|^2 dx$ defines a norm on $\bar{H}^1(\Omega)$ and so $\|(\boldsymbol{u}, \boldsymbol{c})\|_V^2 := |u_1|_1^2 + |u_2|_1^2 + \boldsymbol{c}^T \boldsymbol{c}$ defines a norm on V.

For $\boldsymbol{u} \in \bar{H}^1(\Omega)^2$ let $u_0 \in \bar{H}^1(\Omega)$ be the solution of $(\nabla v, \nabla u_0) = (\nabla v, \boldsymbol{u})$ for all $v \in \bar{H}^1(\Omega)$, where $(\cdot, \cdot)$ denotes the $L^2(\Omega)$ inner product. Let the function values of u_0 on the measurement points be denoted by

$$P\boldsymbol{u} := (u_0(\boldsymbol{x}_1), \ldots, u_0(\boldsymbol{x}_n))^T.$$

Furthermore, let

$$X := \begin{bmatrix} 1 & \cdots & 1 \\ \boldsymbol{x}_1 & \cdots & \boldsymbol{x}_n \end{bmatrix}^T \in \mathbb{R}^{n,3}$$

and $\boldsymbol{y} := (y_1, \ldots, y_n)^T$. Using all of this one can introduce the functional

$$J_\alpha(\boldsymbol{u}, \boldsymbol{c}) = n^{-1}(P\boldsymbol{u} + X\boldsymbol{c} - \boldsymbol{y})^T(P\boldsymbol{u} + X\boldsymbol{c} - \boldsymbol{y}) + \alpha(|u_1|_1^2 + |u_2|_1^2)$$

and the associated bilinear form a and linear form F given by

$$\begin{aligned} a(\boldsymbol{u}, \boldsymbol{c}; \boldsymbol{v}, \boldsymbol{d}) &= n^{-1}(P\boldsymbol{u} + X\boldsymbol{c})^T(P\boldsymbol{v} + X\boldsymbol{d}) + \alpha((u_1, v_1)_1 + (u_2, v_2)_1) \\ F(\boldsymbol{v}, \boldsymbol{d}) &= n^{-1}(P\boldsymbol{v} + X\boldsymbol{d})^T \boldsymbol{y}. \end{aligned}$$

Now we can define our new spline in terms of the minimiser of J_α over the space V. The following theorem shows that such a minimiser exists and is unique.

Theorem 1. *If the data points $\{\boldsymbol{x}_i\}$ are not collinear then there is exactly one $(\boldsymbol{u}, \boldsymbol{c}) \in V$ such that*

$$J_\alpha(\boldsymbol{u}, \boldsymbol{c}) \leq J_\alpha(\boldsymbol{v}, \boldsymbol{d}), \text{ for all } (\boldsymbol{v}, \boldsymbol{d}) \in V,$$

and this $(\boldsymbol{u}, \boldsymbol{c}) \in V$ is also determined by the variational equations

$$a(\boldsymbol{u}, \boldsymbol{c}; \boldsymbol{v}, \boldsymbol{d}) = F(\boldsymbol{v}, \boldsymbol{d}), \text{ for all } (\boldsymbol{v}, \boldsymbol{d}) \in V.$$

Proof: The proof of this theorem is based on the establishment of the continuity of a and F and the V-ellipticity of a.

The continuity of a and F follows immediately from the fact that the mapping $\boldsymbol{u} \mapsto P\boldsymbol{u}$ is continuous from $\bar{H}^1(\Omega)^2$ to $\mathbb{R}^n$. This follows from standard regularity theorems for elliptic problems which imply that if $\boldsymbol{u} \in H^1(\Omega)^2$ then $u_0 \in H^2(\Omega)$. The pointwise control is provided by the Sobolev imbedding theorem which states that $H^2(\Omega) \subset C^0(\bar{\Omega})$. Indeed, there exists a constant C_Ω such that $\|u_0\|_{L^\infty} \le C_\Omega \|\boldsymbol{u}\|_1$.

To prove that a is V-elliptic, we note that

$$(P\boldsymbol{u} + X\boldsymbol{c})^T(P\boldsymbol{u} + X\boldsymbol{c}) \ge (1-\gamma)\boldsymbol{c}^T X^T X\boldsymbol{c} + (1-\gamma^{-1})P\boldsymbol{u}^T P\boldsymbol{u}$$

for any $\gamma < 1$, and so

$$a\,(\boldsymbol{u},\boldsymbol{c};\boldsymbol{u},\boldsymbol{c}) \ge n^{-1}(1-\gamma)\boldsymbol{c}^T X^T X\boldsymbol{c} + n^{-1}(1-\gamma^{-1})P\boldsymbol{u}^T P\boldsymbol{u} + \alpha(|u_1|_1^2 + |u_2|_1^2).$$

Provided $\alpha > 0$ and $X^T X$ is positive definite, that is the data points are not collinear, we can make a choice of γ close to 1 which ensures that $n^{-1}(1-\gamma^{-1})P\boldsymbol{u}^T P\boldsymbol{u}$ is smaller than $\alpha(|u_1|_1^2 + |u_2|_1^2)$. Consequently there exists a positive constant C_E so that $a\,(\boldsymbol{u},\boldsymbol{c};\boldsymbol{u},\boldsymbol{c}) \ge C_E\,\|(\boldsymbol{u},\boldsymbol{c})\|_V^2$. In particular the bilinear form a is V-elliptic. □

Zero Curl Condition

With this new method we have not guaranteed that $\boldsymbol{u}$ is the gradient of a function. To ensure this we need to apply the condition $\mathrm{curl}(\boldsymbol{u}) = 0$, where $\mathrm{curl}(\boldsymbol{u}) = u_{2,\xi_1} - u_{1,\xi_2}$. The space $W = \{(\boldsymbol{u},\boldsymbol{c}) \in V \mid \mathrm{curl}(\boldsymbol{u}) = 0\}$ is a closed subspace of V, and so one has

Theorem 2. *If the data points $\{\boldsymbol{x}_i\}$ are not collinear, then there is exactly one $(\boldsymbol{u},\boldsymbol{c}) \in W$ such that*

$$J_\alpha(\boldsymbol{u},\boldsymbol{c}) \le J_\alpha(\boldsymbol{v},\boldsymbol{d}), \quad \textit{for all } (\boldsymbol{v},\boldsymbol{d}) \in W.$$

If $u_0 \in \bar{H}^1(\Omega)$ such that

$$(\nabla v, \nabla u_0) = (\nabla v, \boldsymbol{u}), \quad \textit{for all } v \in \bar{H}^1(\Omega),$$

then

$$f_{\alpha,\Omega}(\boldsymbol{x}) = u_0(\boldsymbol{x}) + \boldsymbol{c}^T\boldsymbol{x}$$

is the thin-plate spline over Ω, i.e., the minimiser of the original thin plate spline functional defined over Ω.

Proof: The existence and uniqueness of $(\boldsymbol{u},\boldsymbol{c})$ follows from the closedness of W in V [6]. For any $(\boldsymbol{u},\boldsymbol{c}) \in W$ with u_0 as defined above one gets $\boldsymbol{u} = \nabla u_0$. With $f(\boldsymbol{x}) := u_0(\boldsymbol{x}) + \boldsymbol{c}^T\boldsymbol{x}$ one observes that the value of $J_\alpha(\nabla u_0, \boldsymbol{c})$ is equal to the value of the original thin plate spline functional in terms of f. □

Thus one gets a first order characterisation of the thin-plate smoothing spline for a compact domain Ω. In practical tests it was seen that the difference between the minima over V and W is small. Thus in the following we will discuss the minimisation over V which provides an alternative to Duchon's thin plate spline $f_{\alpha,\Omega}$.

§3. Discretisation of the First Order Thin Plate Spline

The discretisation of the problem follows the standard finite element framework. We consider a finite dimensional subspace $X_h \subset H^1(\Omega)$, e.g., continuous piecewise bilinear functions on quadrilaterals or piecewise linear function on triangles. The space V_h approximating V is simply $\bar{X}_h^2 \times \mathbb{R}^3$, where $\bar{X}_h = X_h \cap \bar{H}^1(\Omega)$. In these numerical experiments we are not enforcing the zero curl condition, and so can use simple H^1 elements, in particular continuous piecewise bilinear functions on rectangles.

Of course to actually implement this method we need to be able to solve the Poisson equation for u_0 which we can only do approximately. For this reason we introduce the following discrete functional

$$J_{\alpha,h}(\boldsymbol{u},\boldsymbol{c}) = n^{-1}(P_h\boldsymbol{u} + X\boldsymbol{c} - \boldsymbol{y})^T(P_h\boldsymbol{u} + X\boldsymbol{c} - \boldsymbol{y}) + \alpha(|u_1|_1^2 + |u_2|_1^2)$$

and the associated bilinear form a_h and the linear form F_h

$$\begin{aligned} a_h(\boldsymbol{u},\boldsymbol{c};\boldsymbol{v},\boldsymbol{d}) &= n^{-1}(P_h\boldsymbol{u} + X\boldsymbol{c})^T(P_h\boldsymbol{v} + X\boldsymbol{d}) + \alpha((u_1,v_1)_1 + (u_2,v_2)_1) \\ F_h(\boldsymbol{v},\boldsymbol{d}) &= n^{-1}(P_h\boldsymbol{v} + X\boldsymbol{d})^T\boldsymbol{y}. \end{aligned}$$

Here $P_h\boldsymbol{u}$ is the vector of pointwise values of the function $u_{0,h}$ which solves the discrete Poisson problem $(\nabla v_h, \nabla u_{0,h}) = (\nabla v_h, \boldsymbol{u})$, for all $v_h \in \bar{X}_h$. Using the methods of the last section, together with a standard L^∞ convergence bound for finite element approximations of the Poisson equation, it is possible to show that a_h and F_h are continuous and that a_h is V-elliptic.

The discrete spline is defined as the minimiser of $J_{\alpha,h}$ over V_h. This is a non-conforming method since $a_h \neq a$ and $F_h \neq F$, but we can show that

Theorem 3. *If the data points $\{\boldsymbol{x}_i\}$ are not collinear, then there is exactly one $(\boldsymbol{u}_h,\boldsymbol{c}_h) \in V_h$ such that*

$$J_{\alpha,h}(\boldsymbol{u}_h,\boldsymbol{c}_h) \leq J_{\alpha,h}(\boldsymbol{v}_h,\boldsymbol{d}_h), \text{ for all } (\boldsymbol{v}_h,\boldsymbol{d}_h) \in V_h$$

and this $(\boldsymbol{u}_h,\boldsymbol{c}_h) \in V_h$ is also determined by the variational equations

$$a_h(\boldsymbol{u}_h,\boldsymbol{c}_h;\boldsymbol{v}_h,\boldsymbol{d}_h) = F_h(\boldsymbol{v}_h,\boldsymbol{d}_h), \text{ for all } (\boldsymbol{v}_h,\boldsymbol{d}_h) \in V_h.$$

In addition, for any $h_0 > 0$ and $\epsilon > 0$ there exists a positive constant C such that

$$\|(\boldsymbol{u},\boldsymbol{c}) - (\boldsymbol{u}_h,\boldsymbol{c}_h)\|_V \leq Ch^{1-\epsilon}n^{-1/2}\|\boldsymbol{y}\|_{\mathbb{R}^n}$$

for all $h < h_0$. Here $(\boldsymbol{u},\boldsymbol{c})$ is the solution of the continuous problem described in Theorem 1.

Hence our method is essentially first order accurate in the V norm. The ϵ appearing in this theorem is due to the fact that $H^{1+\epsilon}(\Omega) \subset C^0(\bar{\Omega})$ for any $\epsilon > 0$. The bound on $\|(\boldsymbol{u},\boldsymbol{c}) - (\boldsymbol{u}_h,\boldsymbol{c}_h)\|_V$ follows from the fact that

$$\begin{aligned} |a(\boldsymbol{u},\boldsymbol{c},\boldsymbol{v},\boldsymbol{d}) - a_h(\boldsymbol{u},\boldsymbol{c},\boldsymbol{v},\boldsymbol{d})| &\leq C_\epsilon h^{1-\epsilon}\|(\boldsymbol{u},\boldsymbol{c})\|_V\|(\boldsymbol{v},\boldsymbol{d})\|_V \\ \|F(\boldsymbol{v},\boldsymbol{d}) - F_h(\boldsymbol{v},\boldsymbol{d})\| &\leq C_\epsilon h^{1-\epsilon}n^{-1/2}\|\boldsymbol{y}\|_{\mathbb{R}^n}\|(\boldsymbol{v},\boldsymbol{d})\|_V \end{aligned}$$

for $(\boldsymbol{u},\boldsymbol{c}),(\boldsymbol{v},\boldsymbol{d}) \in V$.

The Equations

Now we can describe the equations we must solve. We need to find $(\boldsymbol{u}_h, \boldsymbol{c}_h) \in V_h$ and $u_{0,h} \in \bar{X}_h$. Let $b_0, b_1, \ldots, b_m$ be a basis of X_h such that $b_0(\boldsymbol{x}) = 1$ for all $\boldsymbol{x} \in \Omega$. Then one can uniquely represent $v \in X_h$ by $v = c + \sum_{i=1}^{m} v_i b_i$.

Now set $\boldsymbol{v}_j = (v_{j1}, \ldots, v_{jm})^T$, $j = 0, 1, 2$. Then one gets the matrix representation of J as

$$J(\boldsymbol{v}_0, \boldsymbol{v}_1, \boldsymbol{v}_2, \boldsymbol{c}) = n^{-1}\|N\boldsymbol{v}_0 + X\boldsymbol{c}\|^2 + \alpha(\boldsymbol{v}_1^T A \boldsymbol{v}_1 + \boldsymbol{v}_2^T A \boldsymbol{v}_2),$$

where $\|\cdot\|$ denotes the Euclidian norm in $\mathbb{R}^n$, $[N]_{ij} = b_j(\boldsymbol{x}_i)$ and $[A]_{ij} = (\nabla b_i, \nabla b_j)$, $i, j = 1, \ldots, m$. Note that A is positive definite as b_0 (which corresponds to the nullspace of the Laplacian) has been treated separately. This representation has also been chosen so that the terms with c_i drop out and one gets the constraint $A\boldsymbol{v}_0 - B_1\boldsymbol{v}_1 - B_2\boldsymbol{v}_1 = 0$, where $[B_k]_{ij} = (\partial b_i / \partial \xi_k,\, b_j)$.

This is a quadratic optimisation problem, and one gets the equivalent linear system of equations [10]

$$\begin{bmatrix} \alpha A & 0 & 0 & 0 & -B_1^T \\ 0 & \alpha A & 0 & 0 & -B_2^T \\ 0 & 0 & n^{-1}N^T N & n^{-1}N^T X & A \\ 0 & 0 & n^{-1}X^T N & n^{-1}X^T X & 0 \\ -B_1 & -B_2 & A & 0 & 0 \end{bmatrix} \begin{bmatrix} \boldsymbol{u}_1 \\ \boldsymbol{u}_2 \\ \boldsymbol{u}_0 \\ \boldsymbol{c} \\ \boldsymbol{w} \end{bmatrix} = \begin{bmatrix} 0 \\ 0 \\ n^{-1}N^T \boldsymbol{y} \\ n^{-1}X^T \boldsymbol{y} \\ 0 \end{bmatrix},$$

where $\boldsymbol{w}$ is the Lagrange multiplier vector. It is seen that the system matrix is symmetric indefinite and sparse as the blocks A, $N^T N$ and B_k are sparse. The equation is solved using a variant of Uzawa's method. The details of the solution method are described in [9].

All matrices not involving X or N can be treated as standard finite element matrices and are independent of the data size, n. The building up of the matrices $N^T N$, $N^T X$ and $X^T X$ requires that each measurement is visited. However, the storage of the information and the further processing only depends on the resolution of the elements determined by m.

§4. Examples

Finally we would like to demonstrate our method on an example problem. For the following example we use the "peaks" function from `Matlab`. The peaks function consists of a linear combination of several scaled and translated Gaussian distributions. All the computations were done with `Matlab` on a Sun Sparc workstation at ANU.

We compare our method with Duchon's standard thin plate splines and with finite element least square fitting. We use 500 random equally distributed data points on $[0, 1]^2$. To the data which is in the range of $[-10, 10]$ a data error normally distributed with expectation 0 and standard deviation 0.1 was added. As the choice of the smoothing parameter is not discussed here we choose $\alpha = 10^{-5}$ fixed.

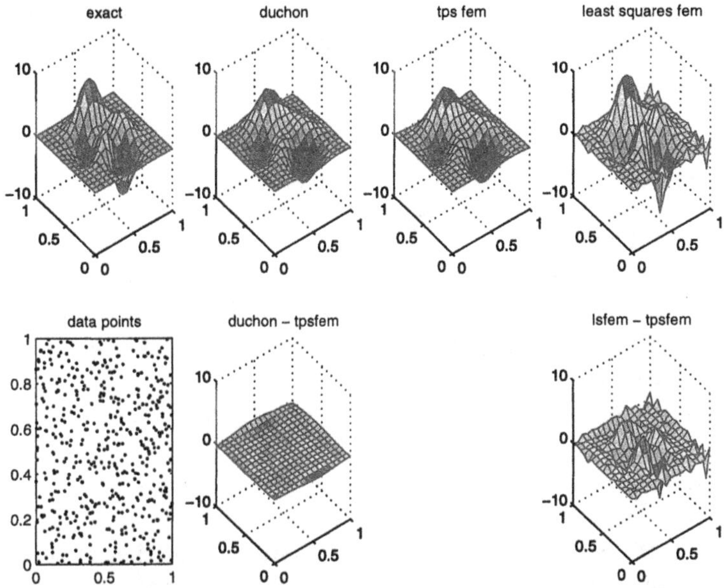

Fig. 1. Comparison of thin plate splines (duchon), First order Thin plate splines (tps fem) and Finite element least square fit (lsfem).

As can be seen in Figure 1, the thin plate spline (duchon) and the first order thin plate spline (tpsfem) produce very similar results except in the region close to the boundary. Remember that the thin plate splines are computed for the domain $\mathbb{R}^2$ where the first order thin plate spline is computed for the square. Furthermore, the curl condition is not enforced.

The finite element least squares solution has too few points and produces an ill-conditioned linear system and poor results.

§5. Conclusions

Let us summarise the main points of our paper. We have described a new discrete thin plate spline based on a first order formulation of the variational problem. This provides a method which can deal with large data sets, and produces results similar to Duchon's standard thin plate splines.

Acknowledgments. This work was supported by the Cooperative Research Centre for Advanced Computational Systems at the Australian National University.

References

1. Arcangéli, R., Some applications of discrete D^m splines, in *Mathematical Methods in Computer Aided Geometric Design*, T. Lyche and L. L. Schumaker (eds.), Academic Press, New York, 1989, 35–44.

2. Beatson, R. K. and G. N. Newsam, Fast evaluation of radial basis functions. I, Comput. Math. Appl. **24** (1992), 7–19.
3. Beatson, R. K. and M. J. D. Powell, An iterative method for thin plate spline interpolation that employs approximations to Lagrange functions, Pitman Res. Notes Math. Ser. **303** (1994), 17–39.
4. Beatson, R. K., G. Goodsell, and M. J. D. Powell, On multigrid techniques for thin plate spline interpolation in two dimensions, Lectures in Appl. Math. **32** (1996), 77–97.
5. Berry, M. J. A. and G. S. Linoff, *Data Mining Techniques for Marketing, Sales and Customer Support,* John Wiley and Sons, New York, 1997.
6. Ciarlet, P. G., *The Finite Element Method for Elliptic Problems,* North-Holland, Netherlands, 1978.
7. Duchon, J., Splines minimizing rotation-invariant semi-norms in Sobolev spaces, Lecture Notes in Math. **571** (1977), 85–100.
8. Hastie, T. J. and R. J. Tibshirani, *Generalized Additive Models,* Chapman and Hall, London, 1990.
9. Hegland, M., S. Roberts, and I. Altas, Finite element thin plate splines for smoothing applications, submitted to *Computational Techniques and Applications: CTAC97.*
10. Golub, G. H. and C. F. Van Loan, *Matrix Computations,* Johns Hopkins University Press, Baltimore, 1989.
11. Sibson, R. and G. Stone, Computation of thin-plate splines, SIAM J. Sci. Statist. Comput. **12** (1991), 1304–1313.
12. Torrens, J. J., F. J. Serón, M. C. López de Silanes, R. Arcangéli, and D. Apprato, Finite-element interpolation of nonregular parametric surfaces, (Contributions to the computation of curves and surfaces), (Puerto de la Cruz), 1989, 101–108.

Markus Hegland
Computer Sciences Laboratory
The Australian National University
Canberra ACT 0200, AUSTRALIA
markus.hegland@anu.edu.au

Stephen Roberts
School of Mathematical Sciences
The Australian National University
Canberra ACT 0200, AUSTRALIA
stephen.roberts@anu.edu.au

Irfan Altas, School of Information Studies
Charles Sturt University
Wagga Wagga, NSW 2678, AUSTRALIA
ialtas@csu.edu.au

A Geometric Concept of Reverse Engineering of Shape: Approximation and Feature Lines

Josef Hoschek, Ulrich Dietz and Wilhelm Wilke

Abstract. Reverse Engineering is the reconstruction process of surfaces of physical objects into geometric surface models. The objects have to be digitized and these discrete data have to be converted into smooth surface models. In the presented paper we will improve overall approximations of digitized set of points with help of feature lines. These feature lines are determined by the angular variation of estimated normal vectors.

§1. A Geometric Concept

In the computer assisted manufacturing process there often remain objects which are not originally described in a CAD–System: reasons are modifications (grinding down or putting on material) of existing parts, which are required to improve the quality of the product, real-scale clay or wood models are needed as stylists, and management often rely more on evaluating real 3D objects than on viewing projection of objects on 2D screens at reduced scale. These physical objects are measured by mechanical digitizing machines or laser scanners or Moiré based optical sensors. The resulting data set may be partially ordered (digitizing machines) or unorganized (laser scanners), and have to be transformed into smooth surfaces. In [16] it is mentioned, that the main purpose of reverse engineering is to convert discrete data sets into a piecewise smooth, continuous model.

In this paper we will focus on the geometric part of reverse engineering, and will assume that the data acquiring process is finished and a (unorganized) cloud of points is given, describing the shape of the considered objects. The crucial points of geometric reverse engineering are

- segmentation (determing of feature lines or key curves of an object)
- surface fitting (approximation of the cloud of points by a suitable surface construction and smoothing the geometric shape).

Mathematical Methods for Curves and Surfaces II
Morten Dæhlen, Tom Lyche, and Larry L. Schumaker (eds.), pp. 253–262.

ISBN 0-8265-1315-8.

Up to now, these steps are carried out manually in CAD–Systems: segmentation is done interactively by experience of the user; afterwards, these manually chosen patches (parts of the given object) are approximated by the CAD–system–software more or less effectively [6,9].
Therefore, the goal of our research is

- to avoid manual interaction of the user as far as possible
- to approximate the whole given cloud of points with an overall surface and improve parts of this overall approximation, in case those parts have less shape quality. For quality improvment we will automatically determine segmentation or feature lines as far as possible.

Our concept on geometric Reverse Engineering is visualized in Fig. 1: One starts with an overall approximation surface and gets the feature lines with help of triangulations. Surface parts with less quality can be improved with the help of these feature lines. For surface improvement we need closed loops — if the feature lines determined open regions, we close the region with suitable choosen curves.

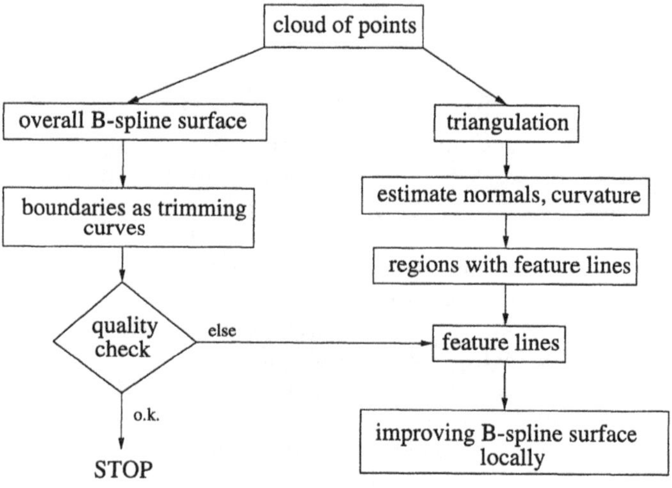

Fig. 1. Geometric Concept of Reverse Engineering.

§2. B–Spline Surface Approximation

Given is the cloud of points $\mathbf{P}_i$ $(i = 1...N)$, without any structure, with arbitrary boundaries and holes in the interior. As tensor–product B–spline (or Bézier) surface representations are used in CAD–systems, we will approximate the cloud of points by a B–spline surface (for other strategies like α–shapes see [1])

$$\mathbf{X}(u,v) = \sum_{i=0}^{n}\sum_{j=0}^{m} \mathbf{d}_{ij} N_{ik}(u) N_{jk}(v) \tag{1}$$

with suitable chosen n, m, the order k of the B–spline basis functions, and the unknown control points $\mathbf{d}_{ij}$. The parameters (u, v) are defined over suitable knot sequences [8].

To determine an initial parametrization of the given points $\mathbf{P}_i$, we project orthogonally the points on a least squares plane ε of all $\mathbf{P}_i$, and choose a suitable rectangle in ε as a parametric domain for the parameter values of the points $\mathbf{P}_i$. The only requirement for the rectangle is that: it should contain all the pre-images of points $\mathbf{P}_i$. If projection in ε is not unique, the parameter domain must be subdivided into parts.

We can interpret the plane ε as a metal sheet and try to deform it to an approximation surface of the cloud of points [4]. This deformation can be described by the minimization problem

$$\tilde{E} = \frac{1}{N}\sum_i (\mathbf{X}(u_i, v_i) - \mathbf{P}_i)^2 + \lambda \sum_p \alpha_p E_p \longrightarrow \min \tag{2}$$

with the design parameters $\alpha_p (\sum_p \alpha_p = 1)$, some energy constraints E_p, and a penalty factor $\lambda \in \mathbb{R}$. To get linear and fast schemes, we assume that the energy terms E_p are quadratic in the unknown control points. A very effective energy term is an approximation of the thin plate energy [10]

$$E_1 = \int\int (\mathbf{X}_{uu}^2 + 2\mathbf{X}_{uv}^2 + \mathbf{X}_{vv}^2) dudv. \tag{3}$$

The use of energy constraints closes gaps in the coefficient matrix for solving the minimization problem (2). Such gaps appear in regions of the given data set with few or no points [3,10]. Problem (2) leads to a coefficient matrix with band structure which can be solved very fast.

The most important thing for the whole approximation process is the parametrization of the given points. Therefore, we determine a first solution of (2) on the basis of the initial parametrization and use the first approximation surface to reparametrize the given points with methods proposed in [8]. Then, we solve (2) and repeat the reparametrization. This iterative sequence is repeated. Additionally, the penalty factor is reduced by $\lambda_{i+1} = \frac{\lambda_i}{2}$ in order to reduce the influence of total energy, thus the surface can move closer to the points and the distance tolerance can be fulfilled.

Now, we change the approximation strategy (2) and introduce local working energy constraints to improve the approximation surface in critical parts. The new objective function may be [4]

$$\tilde{E} = \sum_i (\mathbf{P}_i - \mathbf{X}(u_i, v_i))^2 + \int\int \lambda(u, v) E_p dudv +$$

$$\gamma \int\int (\lambda(u, v) - \lambda_0)^2 dudv \longrightarrow \min. \tag{4}$$

with $\lambda(u, v) = \sum_{i,j} a_{ij} N_{ik}(u) N_{jk}(v)$ and unknown scalar coefficients a_{ij}. λ_0 is a constant to control $\lambda(u, v)$, γ is a penalty factor. To get a linear scheme, we have minimized at first (4) with respect to the control points $\mathbf{d}_{ik}$ as unknowns, then the points $\mathbf{P}_i$ are reparametrized and (4) is minimized with respect to a_{ij}. This procedure is repeated iteratively whilst $\gamma_{j+1} = \frac{\gamma_j}{2}$ until the required tolerance is obtained.

Fig. 2 shows a flat object with an extreme varying point density. The first figure gives points and the overall approximation surface, the second one gives the shape of the local energy function. One can see that this energy constraint is non–active in areas with larger curvatures.

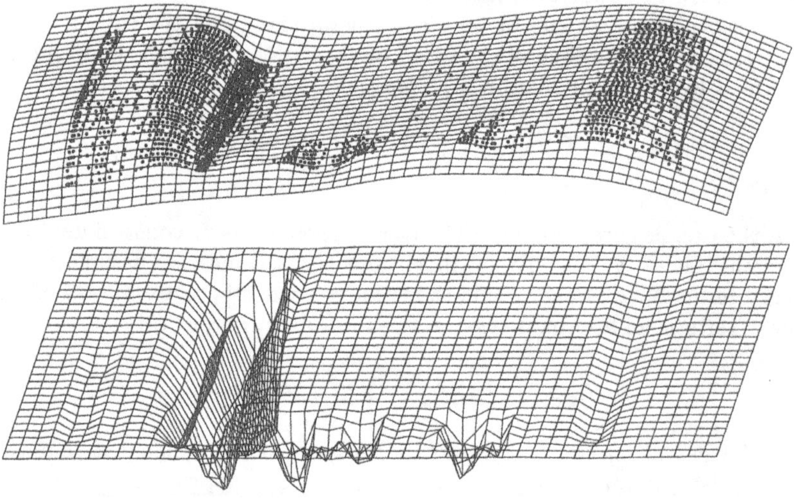

Fig. 2. Overall bicubic B–spline approximation (20 x 40 segments) of a cloud of points: extension 380 mm, max. deviation 0.18 mm and the plot of the local energy $\lambda(u, v)$ (scaled in z–direction).

If only the part of this surface covered by the cloud of points is required (as in Fig. 2), we can thus determine the boundary points with the help of those in a triangulation of the data set: we project these points perpendicular onto the approximation surface and calculate for each projected point the pre–image (u_i, v_i) in the parameter domain. These points (u_i, v_i) are approximated by a B–spline curve

$$\mathbf{c}(t) = \begin{pmatrix} u(t) \\ v(t) \end{pmatrix} = \sum_{\mathbf{i}} \mathbf{d}_{\mathbf{i}} N_{ik}(t)$$

with least squares techniques and parameter correction [11]. The final trimming curves are obtained by inserting $\mathbf{c}(t)$ into the surface representation.

In spite of these two approximation stages, it could happen that some parts of the approximation surface do not satisfy the required quality criteria (small error tolerance, good shape of reflection lines). To improve the shape of these critical parts we will determine feature lines, cutting off the approximation surface at these lines and approximate those critical parts with the help of now given feature lines as boundaries of the approximation problem.

§3. Feature Lines of a Cloud of Points

To determine feature lines, a lot of methods have been developed. Unfortunately, all of these strategies give complete solutions only for very special cases [2,5,13]. In general, two approaches to the segmentation are used: edge–based or face–based methods. The first tries to find boundaries in the data set by curvature estimation and rapid changes of those, the second starts from seed points by growing regions with the same properties until these are changing [16]. Recently in [12], a differential geometric approach for C^3 surfaces was proposed: the feature lines are determined by the vertices of the lines of curvature.

Summarizing, we will use an edge–based technique combined with face–based properties: at first, we will use the given cloud of points to get feature lines independently from the approximation surface, then the feature lines are projected on the approximation surface to improve its critical parts.

To determine the feature lines, we will triangulate the given cloud of points: in the past a lot of algorithms have been developed for 3D–triangulations of unorganized data. Well-known examples are: 3D–Delaunay triangulations or marching–cube based algorithms.

For surface reconstruction with a large data set (more than 100,000 points), these algorithms are mostly too slow and often lead only to the convex hull of the given data set. To get a fast algorithm in [17] a multistep algorithm is developed

- At first, the data set is subdivided in regular solids (voxels) of a basis extention chosen by the user, then the points in a voxel are projected on its least squares plane. If this projection is not unique, the voxels are refined. With the help of an octree data structure the topology of the object can be reconstructed. The given data cloud can be an open set or a closed set.
- In each least squares plane the points are triangulated by a planar Delaunay triangulation and then lifted to a 3D–triangulation of the points in a voxel.
- Now, the edges of each voxel triangulation are connected with neighbouring voxels using algorithms introduced in [15]. Due to measurement errors some gaps may remain; these are closed interactively.

On a Pentium Pro 200 MHz this multistep algorithm for planar Delaunay triangulation needs five seconds for 100,000 points, and one minute for 100,000 triangles (see Fig. 3).

We assume that feature lines appear where the oriented normal vector field of the required surface changes rapidly. To get an estimate of the normal vector in a point $\mathbf{P}_k$, we consider the normal vectors $\mathbf{D}_i$ (with same orientation) of the neighbouring triangles $\{T_i\}$, and introduce as normal vector $\mathbf{N}_k$ in $\mathbf{P}_k$ the mean value

$$\mathbf{N}_k = \frac{\sum w_i \mathbf{D}_i}{|\sum w_i \mathbf{D}_i|}, \qquad i \in \{T_i\}$$

in which w_i are suitable weights, like the inverse area of each triangle T_i

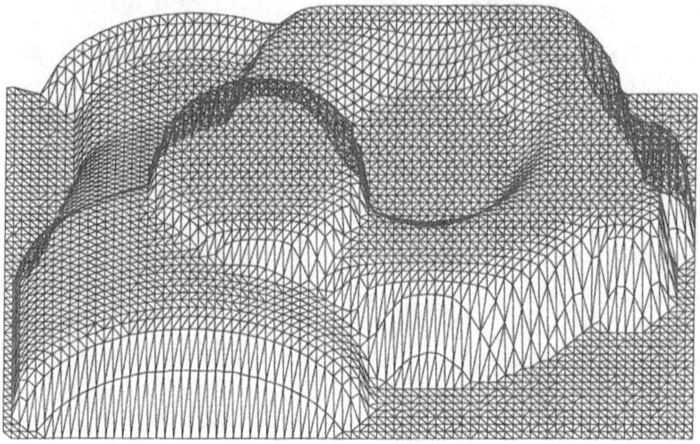

Fig. 3. Triangulation of a set of points digitized from a mechanical model.

$$w_i = \frac{2}{|(\mathbf{P}_i - \mathbf{P}_k) \times (\mathbf{P}_{i+1} - \mathbf{P}_k)|}$$

To get an estimate of the variation of the normal vector in $\mathbf{P}_k$, we use the angles

$$\alpha_{j,k} := \arccos(\mathbf{N}_j \cdot \mathbf{N}_k)$$

between $\mathbf{N}_k$ and a neighbouring normal vector $\mathbf{N}_j$ in $\mathbf{P}_j$, and the euclidean distance

$$l_{j,k} := |\mathbf{P}_j - \mathbf{P}_k|,$$

and introduce as an estimate of the *angular variation* of the normal vector $\mathbf{N}_k$ in point $\mathbf{P}_k$

$$\beta_k = \frac{\sum \frac{\alpha_{j,k}}{l_{j,k}}}{\sum \frac{1}{l_{j,k}}}.$$

If all $l_{j,k}$ are equal, β_k is the arithmetic mean of $\alpha_{j,k}$.

Firstly, the angular variations can be used to recognize flat regions in the sense of face–based technique: regions with very small or no changes of β could be expected to be flat. The corresponding triangles are cancelled and the arising gaps are closed suitably. Additionally, this step reduces the data volume.

Feature lines can be expected in regions in which the angular variation changes rapidly. We can deduct these variations with an erosion operator well-known in image analysis [7]. In [14] a Laplace operator was used but in our experience the erosion operator leads to much better results.

To indicate the changes of the angular variation, we assign a grey level to each point $\mathbf{P}_i$ (light colors indicate high variation and dark colors low

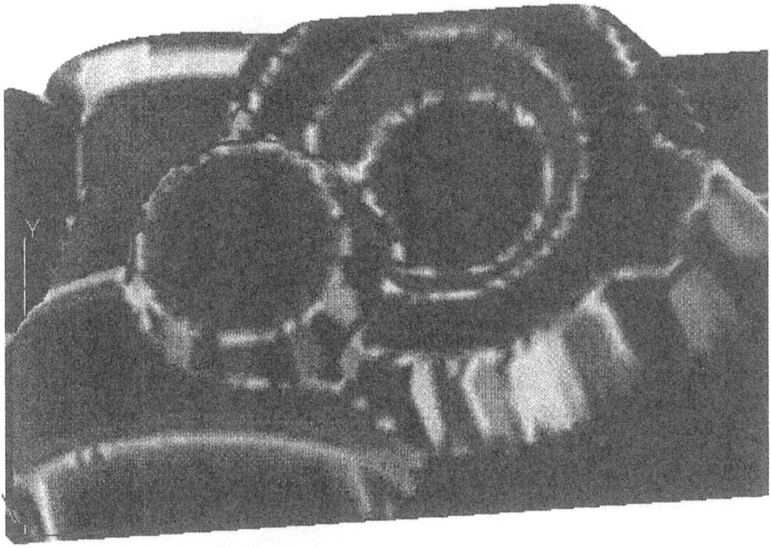

Fig. 4. Changes of angular variation are the loci of the expected feature lines, they seperate planar and non planar parts of the object.

variation), and then interpolate linearly in each triangle. Feature lines are visible as jumps in the grey scale.

To obtain points on the feature lines automatically, we use an algorithm based on the principle of region growing: In order to start an algorithm we choose an initial triangle in the neighbourhood of a potentional feature line and prescribe a threshhold value σ. An estimate of σ can be extracted from the plot of angular variation in Fig. 4. We choose the value in which the colour changes from dark grey to medium grey (in Fig. 5 the value of $\sigma = 4$ degree). The algorithm connects neighbouring triangles at the boundaries of a feature region with the following criteria:

- each triangle is only inspected once.
- the minimal angular variations in the vertices of the neighbouring triangle are less than σ. The triangles which fulfill this criteria are also the candidates for the initial triangles of the first step in the region growing process.
- the changes between minimal angular variation in the vertices of the triangle and the maximal angular variation in the vertices of the neighbouring triangle are less σ.

As a result we get a polygonial line as the boundary of the feature region. This polygonal line (see Fig. 5) is interpreted as the pre-image of a feature line (see Fig. 6). Now we select points from these boundaries and project them on the approximation surface obtained in Chapter 2. These points determine the required feature lines on the approximation surface.

Of interest are only points of potentional feature lines in parts of the

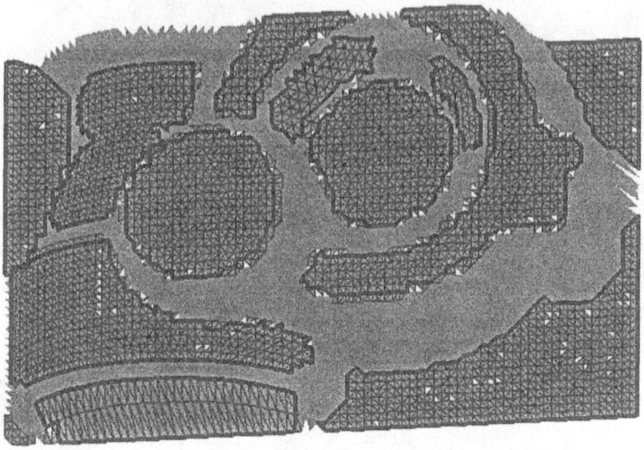

Fig. 5. Polygonal feature lines for the model in Fig. 3, the graet areas are not used for feature line detection.

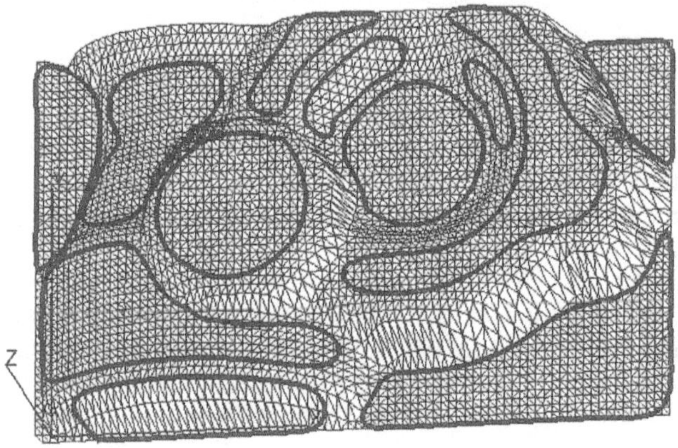

Fig. 6. B–Spline representation of the feature lines.

approximation surface with less surface quality (for instance planar parts in the model surface of Figs. 3, 4). We approximate the points of the feature lines by B–Spline curves in the parametric domain of the approximation surface. For smooth approximation of the polygonal feature lines energy constraints are used (see Fig. 6). Thus, we get a structure of curves in the critical parts of the approximation surface. Suitable feature lines are combined with rectangular parts to n–sided regions or closed loops. These closed figures are interpreted as trimming curves of the approximation surface cutting off its undesired parts. To fill these holes we can use an approximation process preserving the boundary areas of trimming curves: in [4] a method is developed to use points and normal vector fields out of such boundary areas in order to fill the holes.

Additionally, in our case, we can use the points of the given data set which lie inside the trimming holes.

§4. Conclusion

The algorithms for determing feature lines are intependent on the used approximation strategies for the given set of points. Thus in CAD–systems the usual interactive segmentation of the data set may be replaced by our approach for automatic segmentation.

References

1. Bajaj, Ch. L., F. Bernardini, and G. Xu, Automatic reconstruction of surfaces and scalar fields from 3D scans, Computer Graphics (SIGGRAPH'95 Proceedings) (1995), 109–118.
2. Besl, P. J., and R. C. Jain, Segmentation through variable–order surface fitting, IEEE PAMI **10** (1988), 167–192.
3. Dietz, U., B–Spline approximation with energy constraints, in *Advanced Course on Fairshape*, J. Hoschek and P. Kaklis (eds.), Teubner, Stuttgart 1996, 229–240.
4. Dietz, U., Fair surface reconstruction from point clouds, in *Mathematical Methods for Curves and Surfaces II*, M. Dæhlen, T. Lyche, and L. L. Schumaker (eds.), Vanderbilt Univ. Press, Nashville, 1998, 79–86.
5. Dong, J., and S. Vijayan, Manufacturing feature determination and extraction — Part I / Part II: Computer–Aided Design **29** (1997), 427–440, 475–484.
6. Eck, M., P. Bonitz, R. Luehrs, and R. Schmidt, High quality surface reconstruction, preprint.
7. Haberäcker, P.: *Digitale Bildverarbeitung*, Hansa Verlag, 1991.
8. Hoschek, J., and D. Lasser, *Fundamentals of Computer Aided Geometric Design*, AK Peters, Wellesley, 1993.
9. Hoschek, J., and W. Dankwort (eds.), *Reverse Engineering*, Teubner, Stuttgart, 1996.
10. Hoschek, J., and U. Dietz, Smooth B–spline surface approximation to scattered data, in *Reverse Engineerung*, J. Hoschek and W. Dankwort (eds.), Teubner Stuttgart, 1996, 143–152.
11. Hoschek, J., and F.-J. Schneider, Approximate spline conversion for integral and rational Bezier and B-spline surfaces, in *Geometry Processing for Design and Manufacturing*, R.E. Barnhill (ed.), SIAM, Philadelphia, 1992, 45–86.
12. Lukacs, G., and L. Andor, Computing natural division lines on free–form surfaces based on measured data, in *Mathematical Methods for Curves and Surfaces II*, M. Dæhlen, T. Lyche, and L. L. Schumaker (eds.), Vanderbilt Univ. Press, Nashville, 1998, 319–326.

13. Milroy, M. J., C. Bradley, and G. W. Vickers, Segmentation of a wrap-around model using an active contour, *Computer–Aided Design* **29** (1997), 299–320.

14. Sarkar, B., and C. H. Menq, Smooth–surface approximation and reverse engineering, Computer–Aided Design **23** (1991), 623–628.

15. Shewchuck, J. R., Triangle –A two dimensional quality mesh generator and Delaunay Triangulator. Tech. Report Carnegie Mellon Univ. Pittsburgh, 1995.

16. Varady, T., R. R. Martin, and J. Cox, Reverse engineering of Geometric models — an introduction, Computer–Aided Design **29** (1997), 255–268.

17. Wilke, W.: Private communication, Daimler Research Center Ulm, 1997.

Josef Hoschek and Ulrich Dietz
Dept. of Mathematics
Darmstadt University of Technology
Schlossgartenstr. 7
6428 Darmstadt, GERMANY
`hoschek@mathematik.tu-darmstadt.de`

Wilhelm Wilke
Daimler Research Center Ulm
Ulm, GERMANY
`wilke@dbag.ulm.daimlerbenz.com`

Convex Surface Fitting with Parametric Bézier Surfaces

Bert Jüttler

Abstract. We present a method for approximating scattered data with a convex parametric Bézier surface patch. In the first step we construct a reference surface which roughly specifies the expected shape of the approximating surface. Based on this reference surface we generate linearized convexity conditions. The approximating surface is then found by solving a quadratic programming problem.

§1. Introduction

Convexity conditions for bivariate piecewise polynomial functions have been studied in a number of publications, see the survey articles by Goodman [7] and Dahmen [2]. In the case of bivariate polynomials in Bernstein–Bézier representation with respect to a basis triangle, Chang and Davis [1] observed that convexity of the control net implies convexity of the polynomial. This result was later generalized to the multivariate case by Dahmen and Micchelli [3]. For tensor–product spline functions, weak convexity conditions were developed by Floater [5]. These conditions lead to a system of quadratic inequalities for the spline coefficients, where each inequality involves only relatively few coefficients.

Willemans and Dierckx [13] developed a method for convex surface fitting with Powell–Sabin spline functions. Based on the quadratic convexity conditions by Chang and Feng (see [2]) they were able to formulate this task as a quadratic optimization problem with quadratic inequality constraints.

In the case of *parametric* surface patches, only a few related results seem to exist. A very strong sufficient convexity condition for parametric tensor–product Bézier surfaces has been formulated by Schelske in 1984, see [8]. This condition is fulfilled only by convex translational surfaces. Zhou [14] derived convexity conditions for parametric triangular Bézier surfaces. These conditions lead to a system of inequalities whose left–hand sides are polynomials

Mathematical Methods for Curves and Surfaces II
Morten Dæhlen, Tom Lyche, and Larry L. Schumaker (eds.), pp. 263–270.

ISBN 0-8265-1315-8.

of degree 6 in the components of the control points. Recently, similar conditions for parametric tensor–product surfaces were developed by Koras and Kaklis [12]. An approximate method for removing shape flaws from tensor–product B–spline surfaces is described in [11]. This method is based on local modifications of the control net.

We present a new method for convex surface fitting with parametric tensor–product Bézier surfaces. Based on linearized convexity conditions we formulate this task as a quadratic optimization problem with linear inequality constraints. This problem can be solved with the help of standard algorithms from optimization theory. The method is illustrated by an example.

§2. Outline of the Method

We present a method for solving the following approximation problem. A cloud of data $\boldsymbol{p}_i \in \mathbb{R}^3$ $(i = 0, \ldots, P)$ is given. In addition to the data, we assume that associated parameter values $(u_i, v_i) \in [0,1]^2$ are given. If these parameters are unknown, then they can be estimated from the data, e.g. by projecting them into a suitably chosen plane. A more sophisticated scheme for assigning the parameter values has been developed by Floater [6].

The given data are to be approximated by the parametric tensor–product Bézier surface patch (see [8])

$$\boldsymbol{x}(u,v) = \sum_{r=0}^{m} \sum_{s=0}^{n} B_r^m(u)\, B_s^n(v)\, \boldsymbol{b}_{r,s}, \quad (u,v) \in [0,1]^2, \tag{1}$$

with the unknown control points $\boldsymbol{b}_{r,s} = (\, b_{r,s,1}\ b_{r,s,2}\ b_{r,s,3}\,)^\top \in \mathbb{R}^3$ and with the well–known Bernstein polynomials $B_q^p(t) = \binom{p}{q} t^q (1-t)^{p-q}$. The control points $\boldsymbol{b}_{r,s}$ are found by the following procedure.

1) Find a *reference surface.* This surface is used in order to specify the expected shape of the approximating surface (1). Its construction is described in Section 3.
2) Generate *linearized convexity conditions.* Based on the reference surface we generate a system of linear inequalities for the components of the control points $\boldsymbol{b}_{r,s}$ which guarantee the convexity of the surface patch (1), see Section 4.
3) Compute the control points $\boldsymbol{b}_{r,s} \in \mathbb{R}^3$. The control points are found by solving a quadratic programming problem as outlined in Section 5.

This procedure can be iterated several times; one may use the first result as a new reference surface. The new reference surface leads to linear convexity constraints which are better suited for approximating the given data in the third step.

In addition, one may use the idea of *parameter correction* in each cycle: the parameter values (u_i, v_i) are replaced by new values $(\bar{u}_i, \bar{v}_i) \in [0,1]^2$ such that the new error vectors $\boldsymbol{x}(\bar{u}_i, \bar{v}_i) - \boldsymbol{p}_i$ are perpendicular to the surface, see [8] for details. By using the new parameters we can improve the result of the approximation procedure.

§3. The Reference Surface

An approximating parametric Bézier surface for scattered data can be found by minimizing the least–squares sum

$$L = \sum_{i=0}^{P} \|\boldsymbol{x}(u_i, v_i) - \boldsymbol{p}_i\|^2. \tag{2}$$

The minimum can easily be computed by solving the system of normal equations, see [8]. In general, however, minimizing the sum (2) leads to a non–convex surface.

In order to find a suitable convex reference surface, we modify the least squares sum by adding a *tension term*, $F = L + w\,T$, with

$$T = \int_0^1 \int_0^1 \|\boldsymbol{x}_{uuu}\|^2 + \|\boldsymbol{x}_{vvv}\|^2 + \|\boldsymbol{x}_{uv}\|^2 \,\mathrm{d}u\,\mathrm{d}v. \tag{3}$$

The subscripts denote the partial derivatives of the surface $\boldsymbol{x}(u, v)$. The tension term is introduced in order to increase the "stiffness" of the reference surface. Its influence is controlled by the weight w. The value of the tension term is zero if and only if the surface is a biquadratic translational surface, i.e. the parameter lines of both systems u = constant and v = constant are translated copies of the curves $\boldsymbol{x}(u, 0)$ and $\boldsymbol{x}(0, v)$.

Figure 1 shows a biquadratic translational surface and its Bézier control net (dashed lines). Three samples of the congruent parameter lines u = constant have been drawn as solid black curves. All faces of the control net are parallelograms; this property characterizes translational Bézier surfaces.

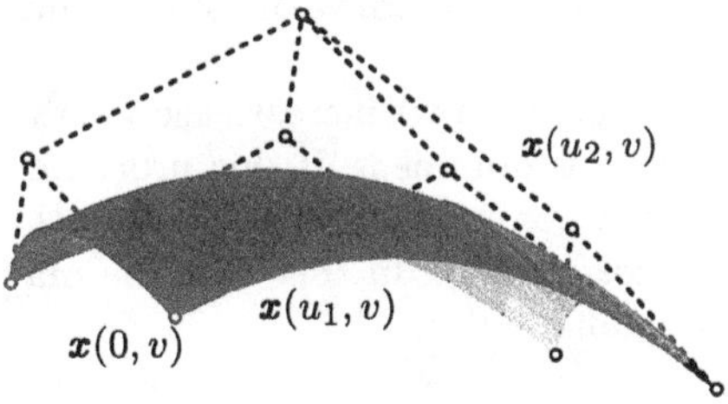

Fig. 1. A translational surface.

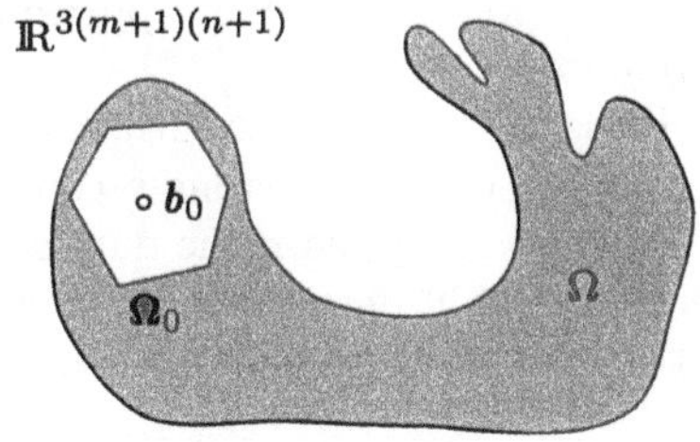

Fig. 2. Linearized convexity conditions.

If the weight w is increased, then the surface which minimizes the modified functional $F + w\,T$ converges to a biquadratic translational surface. Note that this limit translational surface is not guaranteed to be convex. In all practical examples, however, one can expect that this surface is convex, provided that the data stem from a roughly convex surface. Otherwise the given scattered data is unsuitable for a convex approximation, see also the comments at the end of this section.

The above tension term (3) is just one possible choice. One may choose any other functional which increases the stiffness of the resulting surface. The kernel of the tension term, however, should not only contain the linear or bilinear Bézier surfaces (this would happen if we choose the tension term as an integral of squared second derivative vectors), as these surfaces can never be strongly convex. (A surface is said to be strongly convex if it is convex and the Gaussian curvature is positive everywhere.) The tension term (3) seems to be a reasonable choice as its kernel (biquadratic translational surfaces) offers enough degrees of freedom for providing a reasonable approximation.

The control points of the reference surface are found by solving the system of normal equations

$$\frac{\partial F}{\partial b_{r,s,k}} = 0 \qquad (r = 0, \dots, m;\ s = 0, \dots, n;\ k = 1, 2, 3). \tag{4}$$

They form a system of linear equations for the components of the control points.

Still, the weight w has to be chosen. On the one hand, the influence of the tension term should so big enough so that solving (4) leads to a convex reference surface. On the other hand, the weight w should be as small as possible in order to get linearized convexity conditions which are well adapted to the specific data.

We compute an appropriate value for the weight w with the help of simple binary search. Let $w = 1.0$ be the initial weight and compute the resulting reference surface. If this surface is strongly convex but it does not fit very well to the data, then we decrease the weight, $w_{\text{new}} = \frac{1}{2} w_{\text{old}}$, until the approximation gets good enough. If the reference surface becomes non–convex, however, than we go back to the previously used value of w. It may happen that the reference surface never gets non–convex. In this case the reference surface is already the final approximating surface; no convexity constraints are required.

If the initial reference surface is non–convex, then we increase the weight, $w_{\text{new}} = 2w_{\text{old}}$, until a strongly convex surface is obtained. If this procedure fails (this may happen if the data stem from a non–convex surface), then the data is probably unsuitable for convex approximation. In this case one may use the best fitting plane as a convex approximation.

§4. Linearized Convexity Conditions

With the help of the reference surface it is now possible to find linear constraints which guarantee the convexity of the surface (1). The situation is illustrated by the schematic Figure 2. Each surface patch (1) is associated with the point $\boldsymbol{b} \in \mathbb{R}^{3(m+1)(n+1)}$ with the coordinates

$$\boldsymbol{b} = (\, \boldsymbol{b}_{0,0}^{\top}\ \boldsymbol{b}_{0,1}^{\top}\ \boldsymbol{b}_{0,n}^{\top}\ \boldsymbol{b}_{1,0}^{\top}\ \cdots\ \boldsymbol{b}_{m,n}^{\top}\,)^{\top}. \tag{5}$$

The convex surfaces correspond to a certain subset Ω of this space. The reference surface is associated with an interior point $\boldsymbol{b}_0$ of this subset. By generating linear sufficient convexity conditions we construct a circumscribed polyhedron $\Omega_0 \subset \Omega$ for the point $\boldsymbol{b}_0$.

The linearized convexity conditions are found with the help of the following procedure.

1) Use the reference surface for specifying bounding polyhedral cones for the first derivative vectors of the approximating surface. Let $(\vec{\boldsymbol{r}}_i)_{i=1,\ldots,R}$ and $(\vec{\boldsymbol{s}}_j)_{j=1,\ldots,S}$ be the spanning vectors of these cones. We generate linear inequalities for the unknown control points $\boldsymbol{b}_{r,s}$ which guarantee that the first derivative vectors are contained within these cones.
2) Find a bounding polyhedral cone for the cross product $\boldsymbol{x}_u \times \boldsymbol{x}_v$. Let $(\vec{\boldsymbol{t}}_k)_{k=1,\ldots,T}$ be the spanning vectors of this cone. They are a certain subset of $\{\vec{\boldsymbol{r}}_i \times \vec{\boldsymbol{s}}_j \mid i = 1,\ldots,R;\ j = 1,\ldots,S\}$. Then,

$$\boldsymbol{x}_u(u,v) \times \boldsymbol{x}_v(u,v) = \sum_{k=1}^{T} \tau_k(u,v)\, \vec{\boldsymbol{t}}_k, \quad (u,v) \in [0,1]^2, \tag{6}$$

holds with some non–negative functions $\tau_k(u,v)$.
3) Generate linear constraints which guarantee that the second fundamental form of the surface is either non–negative or non–positive definite for all $(u,v) \in [0,1]^2$. As a sufficient condition, we guarantee that the T matrices

$$H_k = \begin{pmatrix} \boldsymbol{x}_{uu} \cdot \vec{\boldsymbol{t}}_k & \boldsymbol{x}_{uv} \cdot \vec{\boldsymbol{t}}_k \\ \boldsymbol{x}_{uv} \cdot \vec{\boldsymbol{t}}_k & \boldsymbol{x}_{vv} \cdot \vec{\boldsymbol{t}}_k \end{pmatrix} \qquad (k = 1,\ldots,T) \tag{7}$$

are either all non–negative or non–positive definite. This is sufficient for (local) convexity as the second fundamental form

$$\frac{1}{\|\boldsymbol{x}_u \times \boldsymbol{x}_v\|} \begin{pmatrix} \boldsymbol{x}_{uu} \cdot (\boldsymbol{x}_u \times \boldsymbol{x}_v) & \boldsymbol{x}_{uv} \cdot (\boldsymbol{x}_u \times \boldsymbol{x}_v) \\ \boldsymbol{x}_{uv} \cdot (\boldsymbol{x}_u \times \boldsymbol{x}_v) & \boldsymbol{x}_{vv} \cdot (\boldsymbol{x}_u \times \boldsymbol{x}_v) \end{pmatrix} \tag{8}$$

is a non–negative linear combination of these matrices, cf. (6). The matrices (7) are non–negative (non–positive) definite if and only if the $2T$ quadratic polynomials

$$(\ \xi \quad \pm(1-\xi)\ )\ H_k \begin{pmatrix} \xi \\ \pm(1-\xi) \end{pmatrix} \qquad (k = 1,\ldots,T) \tag{9}$$

are non–negative (non–positive) for $\xi \in [0,1]$. These polynomials are obtained if we restrict the quadratic form $\boldsymbol{z}^\top H_k \boldsymbol{z}$ ($\boldsymbol{z} \in \mathbb{R}^2$) to the linearly parameterized edges of the square with the vertices $(\pm 1, 0)$ and $(0, \pm 1)$. Their coefficients depend linearly on the control points $\boldsymbol{b}_{r,s}$. Based on this fact we are able to generate linear inequalities for the control points which imply convexity.

This procedure leads to a system $\mathcal{I}$ of linear inequalities for the components of the control points $\boldsymbol{b}_{r,s}$. It may be necessary to subdivide the reference surface in order to find a suitable bounding cone for the cross product $\boldsymbol{x}_u \times \boldsymbol{x}_v$

(for instance, if the reference surface cannot be considered as the graph of a function). Due to space limitations we cannot present any details of the procedure. See [9] for more information. As observed in [9], the linearized convexity conditions can be adapted to any strongly convex reference surface. This is also obvious from Figure 2; each inner point of Ω possesses a circumscribed polyhedron.

§5. Quadratic Programming

The control points $\boldsymbol{b}_{r,s}$ of the approximating surface patch (1) are found by minimizing the least–squares sum (2) subject to the linearized convexity conditions $\mathcal{I}$. This is a quadratic programming (qp) problem; a quadratic objective function is to be minimized subject to linear equality and inequality constraints.

A number of fast and efficient solvers for solving problems of this type have been developed in optimization theory. For instance, an approximate solution can be found with the help of the LOQO package by Vanderbei (available from `http://www.princeton.edu/~rvdb/`). Alternatively one may use an active set strategy (which works similarly to the simplex algorithm) as described in the textbook by Fletcher [4]. The example in the next section has been computed by using LOQO.

§6. An Example

We sampled 51 points from an ellipsoid and perturbed them by using random numbers. These data are to be approximated by a bicubic Bézier surface patch (1). Figure 3 compares the unconstrained approximating surface (a) and the convex approximating surface (b). Whereas the unconstrained surface possesses a huge number of oscillations, the constrained approximation possesses a convex shape. The least squares sums of both surfaces are equal to 0.153 and 0.244, respectively (reference surface: 0.422). The convex approximation has been obtained after 2 iterations (with the use of parameter correction) of the procedure from Section 2. In the first (second) iteration we had to solve a qp problem with 1761 (9674) linear inequalities for 48 (48) unknowns. In addition to the surfaces, both Figures 3 a,b show the Bézier control nets and some level curves x_3 = constant. The ellipses in Figure 3b visualize the curvature distribution. Their principal axes are the principal curvature directions, the diameters are proportional to the principal curvatures.

The two plots in Figure 4 visualize the distribution of the Gaussian curvature $K(u,v)$ for both approximating surfaces. The Gaussian curvature of the unconstrained approximation indicates huge hyperbolic surface regions, $-63.33 \leq K_{\text{unc}}(u,v) \leq 12.44$. In contrast to this, the convex surface has non–negative Gaussian curvature values, $0.01 \leq K_{\text{conv}}(u,v) \leq 2.32$.

§7. Final Remark

In this article we described a method for convexity–preserving surface fitting with parametric tensor–product Bézier surface patches. An analogous method

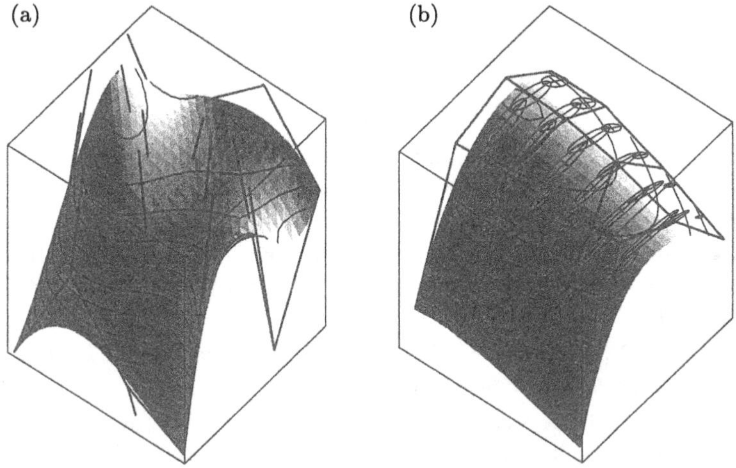

Fig. 3. The unconstrained approximation (a) and the convex approximation (b).

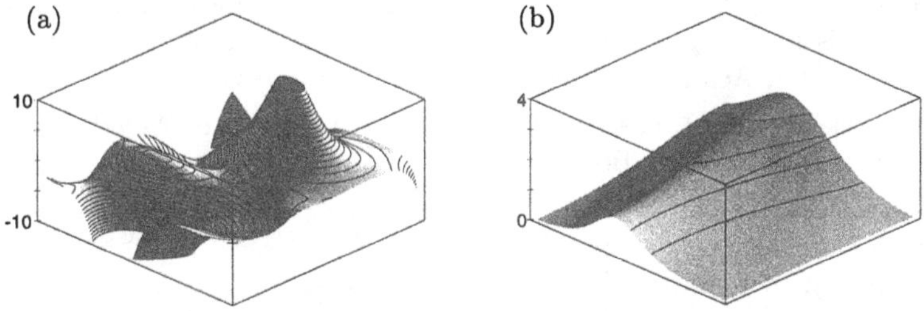

Fig. 4. The Gaussian curvature plots of the surfaces shown in Fig. 3.

can be formulated for approximating scattered data with a tensor–product spline *function* subject to piecewise convexity/concavity constraints, see [10]. Similar to the parametric case, the convexity of the approximating surface can be guaranteed by linear inequalities for the spline coefficients. In the functional case, however, the approximation scheme is much simpler as no reference surface is required. The third step of the algorithm from Section 4 can be applied directly to the Hessian matrix of the spline surface. Moreover, the linear inequalities can be shown to be *asymptotically necessary:* if the number of inequalities is increased in a suitable manner, the feasible set of spline functions approximates the set of all convex spline functions as accurately as desired. This property can be achieved as the convex spline functions form a convex set.

The set Ω of all convex *parametric* surface patches, by contrast, is non–convex, see Section 4. Hence, no asymptotically necessary linearized convexity conditions for parametric Bézier surfaces can be found. As outlined in Section 4, is possible to find linear sufficient convexity conditions for each strongly convex surface, i.e. for each point from the interior of Ω.

References

1. Chang, G.-Z. and P. J. Davis, The convexity of Bernstein polynomials over triangles, J. Approx. Theory **40** (1984), 11–28.
2. Dahmen, W., Convexity and Bernstein–Bézier polynomials, in *Curves and Surfaces,* P.-J. Laurent, A. Le Méhauté, and L. L. Schumaker (eds), Academic Press, New York, 1991, 107–134.
3. Dahmen, W. and C. A. Micchelli, Convexity of multivariate Bernstein polynomials and box spline surfaces, Stud. Sci. Math. Hung. **23** (1988), 265–287.
4. Fletcher, R., *Practical Methods of Optimization,* Second Edition, John Wiley & Sons, Chichester, 1987.
5. Floater, M. S., A weak condition for the convexity of tensor–product Bézier and B–spline surfaces, Advances in Comp. Math. **2** (1994), 67–80.
6. Floater, M. S., Parametrization and smooth approximation of surface triangulations, Comput. Aided Geom. Design **14** (1997), 231–250.
7. Goodman, T. N. T., Shape preserving representations, in *Mathematical Methods in Computer Aided Geometric Design,* T. Lyche and L. Schumaker (eds), Academic Press, New York, 1989, 333–351.
8. Hoschek, J. and D. Lasser, *Fundamentals of Computer Aided Geometric Design,* A. K. Peters, Boston MA, 1993.
9. Jüttler, B., Linear convexity conditions for parametric tensor–product Bézier surface patches, in *Mathematics of Surfaces VII,* T. N. T. Goodman (ed), Information Geometers, Winchester, 1997, in press.
10. Jüttler, B., Surface fitting using convex tensor–product splines, J. Comput. Appl. Math. **84** (1997), 23–44.
11. Kaklis, P. D. and G. D. Koras, A quadratic–programming method for removing shape-failures from tensor-product B-spline surfaces, NTU Athens, Dept. of Naval Architecture, Report TR–SDL–CAGD–1/97.
12. Koras, G. D. and P. D. Kaklis, Convexity conditions for parametric tensor–product B-spline surfaces, NTU Athens (Greece), Dept. of Naval Architecture, manuscript, 1997.
13. Willemans, K. and P. Dierckx, Surface fitting using convex Powell–Sabin Splines, J. Comput. Appl. Math. **56** (1994), 263–282.
14. Zhou, C.-Z., On the convexity of parametric Bézier triangular surfaces, Comput. Aided Geom. Design **7** (1990), 459–463.

Bert Jüttler
Technische Universität Darmstadt
Fachbereich Mathematik
Schloßgartenstraße 7
64289 Darmstadt, GERMANY
`juettler@mathematik.tu-darmstadt.de`

A Natural Geometric Norm for Multi-Channel Image Processing

Ron Kimmel

Abstract. The geometrical framework in which images are considered as surfaces is shown to be a natural selection for color image processing. The steepest descent flow associated with the first variation of the area functional is a significant selective smoothing procedure. Generally, the steepest descent flow for multi-channel variational methods smoothes the different channels of the image. The functional should capture the way the smoothing process acts on the different color channels while exploring the coupling between them. Here we justify the usage of the area norm obtained by the geometric framework, and the Beltrami steepest descent flow as its natural scale-space. We list the requirements, relate to other recent norms, and present simulation results.

§1. Introduction

Recently, [17,10,9], a geometrical framework for image diffusion was introduced. It was claimed that a natural norm for image processing is given by minimizing the area of the 'image surface' in a special way. In this note we justify the usage of the area norm obtained by the geometric framework and the Beltrami flow as its natural scale-space. We limit our discussion to variational methods in nonlinear scale space image processing, and to Euclidean spaces: We consider the given color space (multi-channel space, or feature space) to be Euclidean; the flow is invariant to any Euclidean change of the color coordinates, and is obviously invariant to Euclidean transformations in the spatial domain (translations and rotations of the xy coordinates). Note that given any significant group of transformations in color space, one could design the invariant flow with respect to that group based on the philosophy of images as surfaces: The question then is the meaningful definition of an invariant arclength in the (x, y, R, G, B) hybrid space.

The structure of this note is as follows: Sect. 2 is a brief overview of the geometric framework for image processing and the Beltrami flow. Sect. 3 lists the coupling requirements for the multi-channel case. We show that for a simple 'color image formation' model, the natural order of events is captured by the area norm. We conclude with experimental results of the Beltrami flow in color.

Mathematical Methods for Curves and Surfaces II
Morten Dæhlen, Tom Lyche, and Larry L. Schumaker (eds.), pp. 271–278.

ISBN 0-8265-1315-8.

§2. The Geometric Framework: Brief Overview

A new geometrical framework for image processing was introduced in [17,10,9]. This framework finds a seamless link between the TV-L_1 ($\int |\nabla I|$), [13], and the L_2 ($\int |\nabla I|^2$) norms that are often used in image processing, based on the geometry of the image and its interpretation as a surface. It unifies most of the current 'scale space' models for images by a simple selection of one parameter, yet more importantly, it enables the introduction of new methods to deal with images in a simple and natural way.

A functional called the Polyakov action borrowed from high energy physics was shown to be useful for image enhancement in color, texture, volumetric medical data, movies, and more. The idea is to consider images as surfaces rather than functions. Then, minimize the area of the surface in a special way; e.g. a gray level image is considered to be a 2D surface given by the graph $I(x,y)$ in the 3D space (x,y,I). Similarly, a color image is a 2D surface that is given by the three graphs: $R(x,y), G(x,y)$, and $B(x,y)$, in the 5D space (x,y,R,G,B).

Consider a gray level image as a map from a two dimensional surface to a three dimensional space $\mathbb{R}^3$. We have at each point of the xy coordinate plane an intensity $I(x,y)$. Our space has Cartesian coordinates (x,y,I) where x and y are the *spatial* coordinates and I is the *feature* coordinate. Now, assume the image is corrupted by an unknown noise and should be 'denoised', or a 'clean' image should be produced for further processing. The idea is to invent a geometric flow that minimizes the area of the image as a surface in a way that preserves the edges.

An important question is how to treat multi-channel images. A color image is a good example since one actually considers 3 images Red, Green, and Blue, that are composed into one. To answer this question, we view images as imbedding maps, that flow towards minimal surfaces.

Let us draw a rough sketch of the method: As a first step define an arc-length in the relevant space. For example, an arclength in the (x,y,I) Euclidean space is given by $ds^2 = dx^2 + dy^2 + dI^2$.

Next, the induced metric of the image surface given by the graph surface $(x,y,I(x,y))$ is 'pulled back' from the arclength equation. By applying the chain rule $dI = I_x dx + I_y dy$, the metric is obtained by rearranging the terms into a bilinear structure that measures distance on the surface via the arclength

$$ds^2 = g_{11}dx^2 + 2g_{12}dxdy + g_{22}dy^2,$$

where $g_{11} = 1 + I_x^2$, $g_{22} = 1 + I_y^2$, and $g_{12} = I_x I_y$ are the induced metric coefficients. In a similar way, the distance measure and the induced metric are pulled back from the arclength definition for the 2D surface described by a color image in the 5D (x,y,R,G,B) space, where now the arclength is given by $ds^2 = dx^2 + dy^2 + dR^2 + dG^2 + dB^2$. See [18] for a related effort.

The induced metric $g_{\mu\nu}$ is plugged into an action which is the most general form for measuring area. For two dimensional surfaces, this functional, was first proposed by Polyakov in the context of high energy physics. In the next section we will further elaborate on the selection of area as a proper measure for color images.

Denote by (Σ, g) the image manifold and its metric and by (M, h) the space-feature manifold and its metric. For 2D surfaces $\Sigma = (\sigma_1, \sigma_2)$ is the parametrization that we later identify with the image plane, i.e., $\sigma_1 = x$, $\sigma_2 = y$. $(g_{\mu\nu})$ is the surface metric, and can be written as a 2×2 matrix:

$$(g_{\mu\nu}) = \begin{pmatrix} g_{11} & g_{12} \\ g_{12} & g_{22} \end{pmatrix}.$$

M for our color case stands for the (x, y, R, G, B) space, and its metric $(h_{\mu\nu})$ is a 5×5 matrix that describes the way we measure distances in this space. We can consider the simple Euclidean color space in which $h_{\mu\nu} = \delta_{\mu\nu}$, i.e. the identity matrix. However, other selections that describe different measures in the color space are possible. The map $\mathbf{X} : \Sigma \to M$ has the weight

$$S[X^i, g_{\mu\nu}, h_{ij}] = \int d^m\sigma \sqrt{g} g^{\mu\nu} \partial_\mu X^i \partial_\nu X^j h_{ij}(\mathbf{X}), \tag{1}$$

where m is the dimension of Σ, g is the determinant of the image metric, $g^{\mu\nu}$ is the inverse of the image metric, the range of indices is $\mu, \nu = 1, \ldots, \dim \Sigma$, and $i, j = 1, \ldots, \dim M$, and h_{ij} is the metric of the embedding space. For our color case $\mathbf{X} = (x(\sigma_1, \sigma_2), y(\sigma_1, \sigma_2), R(\sigma_1, \sigma_2), G(\sigma_1, \sigma_2), B(\sigma_1, \sigma_2))$. We used the Einstein summation convention: The summation is applied on each index that appears twice, once as a subscript and once as a superscript. Let us consider the simple example of a gray level image $\mathbf{X} = (x(\sigma_1, \sigma_2), y(\sigma_1, \sigma_2), I(\sigma_1, \sigma_2))$. If we identify the xy plane with the parametrization manifold Σ, and consider a Euclidean space $h_{\mu\nu} = \delta_{\mu\nu}$, we get the area element $\sqrt{g} = \sqrt{1 + I_x^2 + I_y^2}$, and the area measure is then given by $S = \int dx dy \sqrt{g}$. Given the above functional, we have to choose the minimization. In [17] it was shown how different choices yield different flows. Some flows are recognized as existing methods like the heat flow, with passive coordinate transformation [6], the Perona-Malik flow [12], the minimal surface segmentation [2], the color flow [15,3], the mean-curvature flow [11] and its variants [4]. The new result in [17,9] is the steepest descent flow that results by minimizing with respect to the feature coordinates.

The minimization of Polyakov action yields the steepest descent direction for area minimization. If we vary with respect to the feature coordinate (fixing the x and y coordinates for the gray level and color images), we obtain the area minimization direction given by Beltrami operator operating on the feature coordinate(s). Evolving the image using this result, yields an efficient geometric flow for smoothing the image while preserving the edges. It is written as

$$\mathbf{I}_t = \Delta_g \mathbf{I}, \tag{2}$$

where for the color case $\mathbf{I} = (R, G, B)$. The operator that is acting on $\mathbf{I}$ is the natural generalization of the Laplacian from flat spaces to manifolds and is called *the second order differential operator of Beltrami*, or for short *Beltrami operator*, and is denoted by Δ_g. It is defined by

$$\Delta_g \mathbf{I} \equiv \frac{1}{\sqrt{g}} \partial_\mu (\sqrt{g} g^{\mu\nu} \partial_\nu \mathbf{I}). \tag{3}$$

Explicitly, for multi-channel $2D$ surfaces, the flow is given by

$$I_t^i = \frac{1}{g}(p_x^i + q_y^i) - \frac{1}{2g^2}(g_x p^i + g_y q^i), \tag{4}$$

where $g_x = \partial_x g$ ($g_y = \partial_y g$), $g_{\mu\nu} = \delta_{\mu\nu} + \sum_i I_\mu^i I_\nu^i$, $g = g_{11}g_{22} - g_{12}^2$, $p^i = g_{22}I_x^i - g_{12}I_y^i$, and $q^i = -g_{12}I_x^i + g_{11}I_y^i$.

Geometrically, for the gray level case, the above evolution equation is the mean curvature flow of the image surface divided by the induced metric $g = \det(g_{\mu\nu})$. Equivalently, it is the evolution via the $\mathbf{I}$ components of the mean curvature vector $\mathbf{H}$. I.e. for the surface $(\mathbf{x}(\sigma_1, \sigma_2), \mathbf{I}(\sigma_1, \sigma_2))$ in the Euclidean space $(\mathbf{x}, \mathbf{I})$, the curvature vector is given by $\mathbf{H} = \Delta_g(\mathbf{x}(\sigma_1, \sigma_2), \mathbf{I}(\sigma_1, \sigma_2))$. If we identify $\mathbf{x}$ with σ then $\Delta_g I^i(\mathbf{x}) = \mathbf{H} \cdot \hat{I}^i$, i.e., the $\hat{I}^i$ component of the mean curvature vector. Observe that this direct computation applies for co-dimensions > 1. The determinant of the induced metric matrix $g = \det(g_{ij})(= 1 + I_x^2 + I_y^2$ for the gray level case) may be considered as a generalized form of an edge indicator. Therefore, (3) is a selective smoothing flow that preserves edges and can be generalized to any dimension. For gray level images $\mathbf{I} = I$, the Beltrami flow is given explicitly by

$$I_t = \Delta_g I = \frac{1}{\sqrt{g}} div \left(\frac{\nabla I}{\sqrt{g}} \right) = \frac{(1 + I_x^2)I_{yy} - 2I_x I_y I_{xy} + (1 + I_y^2)I_{xx}}{(1 + I_x^2 + I_y^2)^2}. \tag{5}$$

For the Euclidean multi-channel case, the norm we consider is $\int \sqrt{g}$. Here g is the determinant of the metric matrix $g = \det(g_{ij}) = g_{11}g_{22} - g_{12}^2$ given by its components $g_{\mu\nu} = \delta_{\mu\nu} + \sum_i I_\mu^i I_\nu^i$. This action functional is given explicitly by

$$S = \int \sqrt{1 + \sum_i |\nabla I^i|^2 + \frac{1}{2} \sum_{ij} (\nabla I^i, \nabla I^j)^2} dx dy, \tag{6}$$

where $(\nabla R, \nabla G) \equiv R_x G_y - R_y G_x$ stand for the magnitude of the vector product (cross) of the vectors ∇R and ∇G. The action in (6) is simply the area of the image as a surface.

Let us explore the effects of scaling the intensity axis. If we multiply the intensities by a constant β, the above norm may be read as

$$S = \int \sqrt{1 + \beta^2 \sum_i |\nabla I^i|^2 + \beta^4 \frac{1}{2} \sum_{ij} (\nabla I^i, \nabla I^j)^2} dx dy. \tag{7}$$

The steepest descent flow for this functional depends on the value of β. For $\beta \gg sup_{i,\mathbf{x}} |\nabla I^i|$, the order of events along the scale of the flow is as follows: First the channels are aligned together, and only then starts the selective smoothing geometric flow (similar to the single channel TV-L_1). On the other limit, as $\beta \ll \sup_{i,\mathbf{x}} |\nabla I^i|$, the smoothing will tend to occur uniformly in all directions as a multi-channel heat equation (L_2).

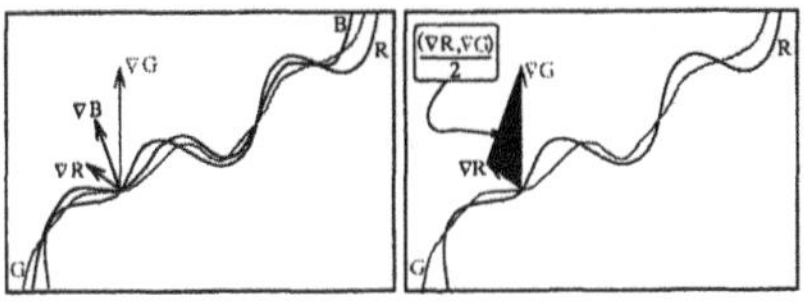

Figure 1: Left: One level set of each channel and their corresponding gradient vectors at one point. Right: $\frac{(\nabla G, \nabla R)}{2}$, the area of the gray triangle, measures the alignment between the two channels,

§3. Coupling Requirements

When considering multi-channel images we need to define the way the channels are to be coupled. Assume that each channel is 'equally important' and thus the measure that links between the different channels should be symmetric in this aspect. Within the scale space philosophy, we want the different channels to get smoother in scale. This requirement leads to the minimization of the different channels gradient magnitudes $|\nabla I^i|$ combined in one way or another that yields coupling.

A different, yet very important demand for multi-channel image processing is the alignment requirement of the different channels in scale. That is, we want the different channels to align together as they become smoother in scale. Fig. 1 shows one level set of each of the three color channels and the corresponding gradient ∇I^i at one point along the level set. The requirement that the different channels align together as they evolve amounts to minimizing the cross products between their gradient vectors $(\nabla I^i, \nabla I^j)^2$.

On inspection of (7), the minimization includes the gradient magnitudes of the channels and the cross products between the different channels, which is a desired norm. Next, we consider an axiomatic approach for the above claims that will set a proper order of events and lead us to the area minimization via the Beltrami flow.

A simplified model for color images is a result of viewing Lambertian surface patches that are not necessarily flat. Such a scene is a generalization of what is known as a 'Mondriaan world'. In this case, each channel may be considered as the projection of the real 3D world surface normal $\hat{\mathbf{N}}(\mathbf{x})$ onto the light source direction $\vec{l}$, multiplied by the albedo $\rho(x,y)$. The albedo captures the characteristics of the 3D object's material, and is different for each spectral channel. That is, the 3 color channels may be written as

$$I^R(\mathbf{x}) = \rho_R(\mathbf{x})\hat{\mathbf{N}}(\mathbf{x}) \cdot \vec{l}, \quad I^G(\mathbf{x}) = \rho_G(\mathbf{x})\hat{\mathbf{N}}(\mathbf{x}) \cdot \vec{l}, \quad I^B(\mathbf{x}) = \rho_B(\mathbf{x})\hat{\mathbf{N}}(\mathbf{x}) \cdot \vec{l}.$$

This means that the different colors capture the change in material via ρ_i (where i stands for R, G, B) that multiplies the normalized shading image $\tilde{I}(\mathbf{x}) = \hat{\mathbf{N}}(\mathbf{x}) \cdot \vec{l}$. The above image formation model [5] was used for color based segmentation [8] and shading extraction from color images [7]. Let us follow the above generalization of this model and assume that the material, and therefore the albedo, are the same within a given object in the image, e.g. $\rho_i(\mathbf{x}) = c_i$, where c_i is a given constant. Thus, $\nabla \rho_i(\mathbf{x}) = 0$ within the interior of a given object. The intensity gradient for each channel within a given object is then given by

$$\nabla I^i(\mathbf{x}) = \tilde{I}(\mathbf{x})\nabla\rho_i(\mathbf{x}) + \rho_i(\mathbf{x})\nabla\tilde{I}(\mathbf{x}) = \tilde{I}(\mathbf{x})\nabla c_i + c_i\nabla\tilde{I}(\mathbf{x}) = c_i\nabla\tilde{I}(\mathbf{x}).$$

Observe that, under the above assumptions, all color channels should have the same gradient direction within a given object.

Now we are ready to deal with the boundaries between objects. Since along the boundaries, both the normalized shading image $\tilde{I}$ and the albedo ρ_i go through a sudden change. The gradient direction should be orthogonal to the boundary for each of the channels.

Following the above claims, the first step should be the alignment of the channels so that their gradient directions agree. Next, comes the diffusion of all the channels simultaneously, while keeping the alignment property. For a large enough β, (7) follows exactly these requirements. For a large β, the area norm (7) is a regularization form of

$$\int\sqrt{\sum_i |\nabla I^i|^2 + \beta^2 \sum_{ij} (\nabla I^i, \nabla I^j)^2 dxdy}, \tag{8}$$

that captures the right order of events as described above. Actually, for a 'larger' β, (8) can be considered as a regularization of the *affine invariant* norm $\int\sqrt{\sum_{ij}(\nabla I^i, \nabla I^j)^2 dxdy}$. If we also add the demand that edges should be preserved and search for the simplest geometric parametrization for the steepest descent flow, we end up with the Beltrami flow as a natural selection.

Experimental Results: The Beltrami flow $\mathbf{I} = \Delta_g \mathbf{I}$ is used to selectively smooth the JPEG compression distortions of an image that was extracted from the net. Fig. 2 shows results for color image denoising via the Beltrami flow. Observe how the color perturbation are smoothed: The cross correlation between the channels holds the edge while selectively smoothing the noncorrelated data. Next, Fig. 3 shows three snapshots of the Beltrami scale space in color.

It is important to note that the existing methods for multi-channel image processing (to name a few [1,3,14,15,16]) do not consider the cross-alignment terms as a correct measure.

§4. Conclusion

The geometric framework of images as surfaces lead us to the norm that resolves the twist (torsion) between the channels via the cross-alignment term. It is important for color image reconstruction after distortion of the different channels. This was demonstrated by our example in which color fluctuations occur along the edges as a result of JPEG lossy compression. In order to preserve the edge and resolve these fluctuations one needs to use the cross alignment within the definition of the norm.

The geometric framework with the area ($\int\sqrt{g}$) norm, yields a natural coupling between the channels via the Beltrami flow that preserves edges in

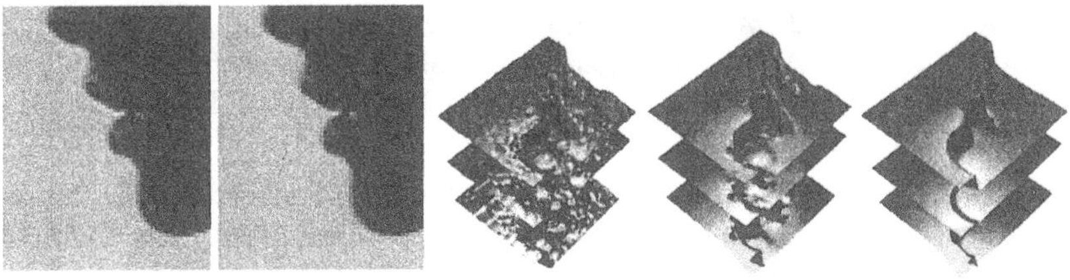

Figure 2: Denoising JEPG lossy compression perturbations. Left to right: The original image, the result of the flow (70 numerical iterations, $\Delta t = 0.21, \Delta x = 1$), the channels presented as surface (original, middle step, and final result). [This is a color figure]

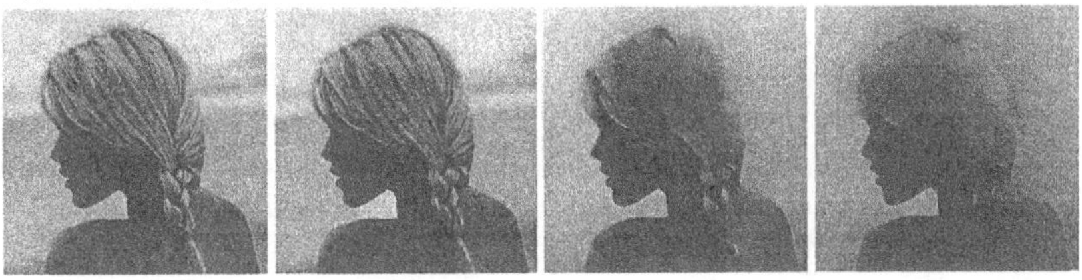

Figure 3: Three snapshots along the scale space of two color images (left most is the original image). [This is a color figure]

a geometrical way. The cross alignment and the gradient magnitude terms appear as proper measures in the definition of the area norm. We have shown that the geometric framework yields the most natural norm with respect to previous norms, and with respect to a list of objective requirements and considerations of the color image formation, and color perception process.

Acknowledgments. I would like to thank Drs. Nir Sochen, Ravi Malladi, and Sherif Makram-Ebeid for interesting discussions and Dr. David Adalsteinsson for his help with Fig. 2. This work is supported in part by the Applied Mathematics Subprogram of the Office of Energy Research under DE-AC03-76SFOOO98, and ONR grant under NOOO14-96-1-0381.

References

1. Blomgren, P., and T. F. Chan, Color TV: Total variation methods for restoration of vector valued images, CAM TR, UCLA, 1996.
2. Caselles, V., R. Kimmel, G. Sapiro, and C. Sbert, Minimal surfaces: A geometric three dimensional segmentation approach, Numerische Mathematik **77** (1997), 423-425.
3. Chambolle A., Partial differential equations and image processing, Proc. IEEE ICIP **1** (1994), 16–20.
4. El-Fallah, A. I., G. E. Ford, V. R. Algazi, and R. R. Estes, The invariance of edges and corners under mean curvature diffusions of images, in Processing III SPIE, Vol. 2421, 1994, 2–14.
5. Eliason, P. T., L. A. Soderblom, and P. S. Chavez, Extraction of topographic and spectral albedo information from multi spectral images, Photogrommetric Engineering and Remote Sensing **48** (1981), 1571–1579.

6. Florack, L. M. J., A. H. Salden, B. M. ter Haar Romeny, J. J. Koendrink, and M. A. Viergever, Nonlinear scale-space, in *Geometric–Driven Diffusion in Computer Vision*, B. M. ter Haar Romeny (ed.), Kluwer Academic Publishers, The Netherlands, 1994.
7. Funt, B. V., M. S. Drew, and M. Brockington, Recovering shading from color images, in Lecture Notes in Comp. Science Vol. 588: Computer Vision: ECCV'92, Springer-Verlag, 1992, 124–132.
8. Healey, G., Using color for geometry-insensitive segmentation, J. Opt. Soc. Am. A **6** (1989), 920–937.
9. Kimmel, R., N. Sochen, and R. Malladi, From high energy physics to low level vision, in Lecture Notes In Comp. Science Vol. 1252, 1st Int. Conf. on Scale-Space Theory in Comp. Vision, Springer-Verlag, 1997, 236–247.
10. Kimmel, R., N. Sochen, and R. Malladi, Images as embedding maps and minimal surfaces: Movies, color, and volumetric medical images, in Proc. of IEEE CVPR'97, 1997, 350–355.
11. Malladi, R., and J. A Sethian, Image processing: Flows under min/max curvature and mean curvature, Graphical Models and Image Processing **58** (1996), 127–141.
12. Perona P., and J. Malik, Scale-space and edge detection using anisotropic diffusion, IEEE-PAMI **12** (1990), 629–639.
13. Rudin, L., S. Osher, and E. Fatemi, Nonlinear total variation based noise removal algorithms, Physica D **60** (1992), 259–268.
14. Sapiro, G., Vector-valued active contours, in Proc. IEEE CVPR'96, 1996, 680–685.
15. Sapiro, G., and D. L. Ringach, Anisotropic diffusion of multivalued images with applications to color filtering, IEEE Trans. Image Proc. **5** (1996), 1582–1586.
16. Shah, J., Curve evolution and segmentation functionals: Application to color images, in Proc. IEEE ICIP'96, 1996, 461–464.
17. Sochen, N., R. Kimmel, and R. Malladi, A general framework for low level vision, IEEE Trans. Image Proc., to appear.
18. Yezzi, A., Modified curvature motion for image smoothing and enhancement, IEEE Trans. Image Proc., to appear.

Ron Kimmel
Lawrence Berkeley National Laboratory
and Dept. of Mathematics
University of California, Berkeley, CA 94720 USA
`ron@math.lbl.gov`
`http://www.lbl.gov/~ron/`

Fairing by Finite Difference Methods

Leif Kobbelt

Abstract. We propose an efficient and flexible scheme to fairly interpolate or approximate the vertices of a given triangular mesh. Instead of generating a piecewise polynomial representation, our output will be a refined mesh with vertices lying densely on a surface with minimum bending energy. To obtain them, we generalize the finite difference technique to parametric meshes. The use of local parameterizations (charts) makes it possible to recast the minimization of non-linear geometric functionals as sparse linear systems. Efficient multi-grid solvers can then be applied, leading to fast algorithms which generate surfaces of high quality.

§1. Introduction

Fairing schemes which construct a surface by solving a constrained optimization problem are traditionally based on piecewise polynomial representations [2,9,16,18]. The major difficulty in this approach is that on one hand efficient (linear) schemes are in general dependent on the specific parameterization and hence fail to be proper models of the physical or geometric intent [7]. On the other hand, more sophisticated non-linear optimization is computationally involved and often unstable [15].

The basic idea of the variational design approach is to measure the quality of a surface in terms of its bending energy. The most common functional is the total curvature

$$\boldsymbol{E}(\mathcal{S}) \;:=\; \int_{\mathcal{S}} \left(\kappa_1^2 + \kappa_2^2\right) d\mathcal{S} \tag{1}$$

which approximates the bending energy of a thin plate. However, since the principal curvatures and the area element depend non-linearly on the surface $\mathcal{S}$, this functional is difficult to minimize. For practical fairing schemes, the total curvature is therefore replaced by the so-called thin-plate functional

$$\boldsymbol{E}(F) \;:=\; \int_{\Omega} \left(F_{uu}^2 + 2\,F_{uv}^2 + F_{vv}^2\right) du\, dv \tag{2}$$

Mathematical Methods for Curves and Surfaces II
Morten Dæhlen, Tom Lyche, and Larry L. Schumaker (eds.), pp. 279–286.

ISBN 0-8265-1315-8.

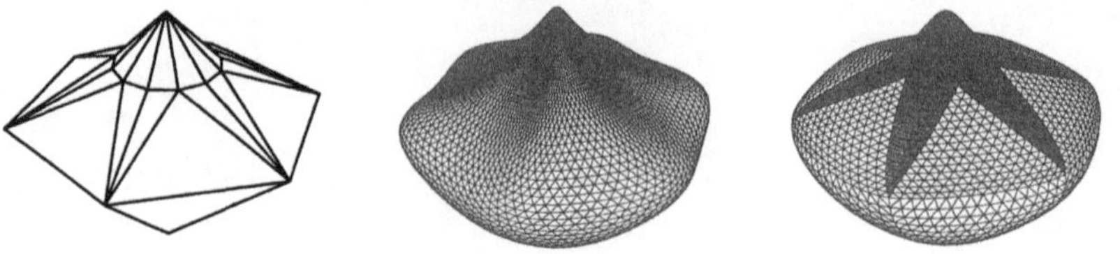

Fig. 1. The triangular mesh on the left is interpolated by surfaces minimizing the thin-plate energy. In the center all patches are parameterized over equilateral triangles; on the right the parameterization is approximately isometric. The shape of the right surface looks 'better' since the true total curvature functional is approximated closer in this case.

which turns out to be identical to the total curvature if the parameterization $F : \Omega \in \mathbb{R}^2 \to \mathcal{S}$ is isometric. Alas, this assumption is far from being satisfied in general. Minimizing (2) instead of (1) hence changes the mathematical model for fairness significantly, and the geometric justification for the approach is no longer valid (cf. Fig 1).

Opposed to the *exact* minimization of the *approximate* energy functional is another class of fairing schemes which *approximate* the minimum of the *true* functional by non-linear optimization techniques. These schemes are mainly based on point evaluation of the integrand function $\kappa_1^2 + \kappa_2^2$, and apply a quadrature formula with respect to an estimated area element [15].

A third way of dealing with the intrinsic non-linearity in the problem of generating fair surfaces is to hide it in the parameterization. In [7] a promising approach to achieve this goal is proposed where the parameterization of the surface $\mathcal{S}$ is defined over a non-planar domain. However, the concept of *data-dependent functionals* still lacks the necessary flexibility to construct surfaces of arbitrary topology since G^k boundary conditions between individual polynomial patches have to be observed.

All the above-mentioned approaches are focused on the generation of *spline surfaces*. In this paper I propose a practical (i.e. simple, fast and robust) scheme to compute refined *triangular meshes* with vertices lying densely on a fair surface. There are no topological restrictions as long as the mesh is locally homeomorphic to a disc. The scheme has linear complexity in the number of generated triangles and works completely automatically.

§2. Classical Fairing

In the classical fairing setting an optimal surface is sought in a space spanned by a finite element basis $\{\Phi_i\}$ while maintaining smoothness conditions across the boundaries between elements (e.g., a spline space). There are two major difficulties in this approach:

• A consistent C^k-parameterization of closed surfaces is not possible in general, and hence *geometric* continuity conditions are introduced. By this, however, we lose the linear structure of the search space which makes the optimization much more difficult.

• Approximating geometric curvatures by a combination of second order partial derivatives is a rather bad mathematical model (cf. Fig 1). A parameter correcting optimization algorithm could reduce the fairness energy without actually modifying the geometric shape of the surface.

The conditions for interpolation or approximation and for the smooth connection between adjacent patches can be imposed either by directly eliminating degrees of freedom, or by Lagrangian multipliers, or by some penalty method that introduces additional energy terms measuring the approximation error and the non-smoothness between adjacent patches.

Once a proper basis $\{\Phi_i\}$ for the search space is chosen, the minimization of (2) is a rather simple task: We formally compute the partial derivatives of the objective functional with respect to the coefficients $\boldsymbol{c}_i$ in the expansion

$$\boldsymbol{E}(F) \;:=\; \boldsymbol{E}\Big(\sum_i c_i\,\Phi_i\Big)$$

and set them to zero. The solvability of the optimization problem is guaranteed if the kernel of the energy functional (2) and the kernel of the approximation constraint (i.e., the space of functions having the value zero at all approximation sites) have a trivial intersection.

§3. Non-classical Fairing

When generating globally fair surfaces, the goal of achieving well-distributed curvature with as few oscillations as possible is much more important than strict differentiability. For instance the penalty methods [15] exploit the fact that in practical applications, tangent plane continuity is established if the jumps of the normal vectors fall below some ε-threshold.

Non-classical fairing schemes use this observation as a justification for no longer requiring C^k smoothness. Surfaces are approximated by piecewise linear C^0 polyhedra, and the shape of these polygonal meshes is optimized. The major advantage provided by those *mesh-smoothing* algorithms is their flexibility with respect to the topology of the surface to be modeled: the C^0 conditions in a triangular mesh are trivially guaranteed.

There are several approaches that apply smoothing operators to meshes in order to improve their fairness. Linear operators have the advantage of being fast and easy to implement [17], but non-linear operators are able to additionally preserve important geometric properties [6,14]. Some schemes increase the fairness of a mesh by changing the positions of the vertices only, others also allow topological changes of the mesh [3,10].

Most of the non-classic schemes are designed to operate on fine meshes where, e.g., noise has to be removed. The task of interpolating a given coarse mesh by a finer one can obviously be reduced to the first problem by taking the original mesh, subdividing all the triangles uniformly to introduce degrees of freedom for the optimization, and then applying a smoothing operator. Most of the proposed schemes, however, will perform very poorly on this type

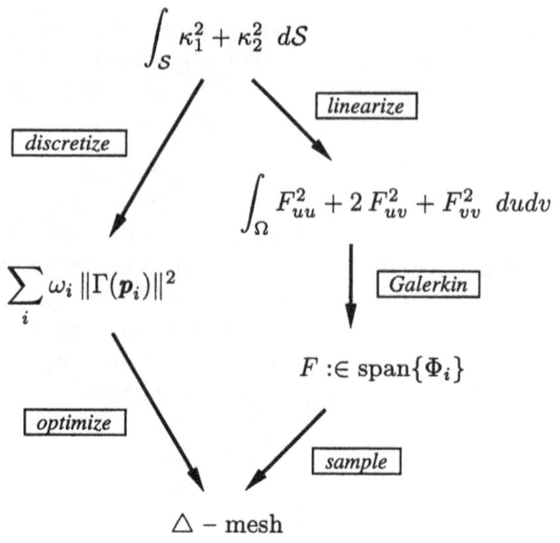
$\int_S \kappa_1^2 + \kappa_2^2 \; dS$
linearize
discretize
$\int_\Omega F_{uu}^2 + 2\,F_{uv}^2 + F_{vv}^2 \; dudv$
$\sum_i \omega_i \, \|\Gamma(\boldsymbol{p}_i)\|^2$
Galerkin
$F :\in \mathrm{span}\{\Phi_i\}$
optimize
sample
$\triangle$ - mesh

Fig. 2. Different paths lead from the original mathematical model of *fairness* (top) to an approximative solution represented by a triangular mesh (bottom).

of data. The reason for this is the typical low-pass characteristics of local smoothing operators. Local high frequency noise is filtered out very quickly, but changes of the global shape are propagated very slowly.

§4. The Discrete Fairing Approach

An algorithmic approach to freeform surface design are *subdivision schemes* [5] where surfaces are no longer defined by an explicit map from a parameter domain into $\mathbb{R}^3$. Generalizing the concept of knot insertion for B-splines, a subdivision surface is defined by a coarse mesh roughly describing its shape and a rule for refining the mesh. By applying this rule recursively, we obtain a sequence of finer and finer meshes which converge to a smooth limit surface [1,4,12]. In practice the refinement is only repeated until the resulting mesh approximates the final surface up to a prescribed tolerance.

In the discrete fairing algorithm [13] we use the topological aspect of this iterative surface generation paradigm to define a nested sequence of meshes. The actual position of the new vertices when refining a given mesh is determined by the minimization of a bending energy functional.

Algorithmically, we are exactly in the situation described at the end of the last section: We refine a given mesh topologically and then apply a smoothing scheme to improve the fairness. However, we can exploit the fact that iterative refinement generates a sequence of meshes which match the requirements for multi-grid solvers [8]. This accelerates our scheme considerably.

Fig. 2 compares the different approaches to fairing. The crucial step in the discrete fairing approach is to discretize the original continuous functional.

The flexibility of this approach stems form the fact that the requirements for the discretization are much less demanding than the set-up for the Galerkin projection based approaches.

For the discretization step we only need two ingredients. First, we have to replace the surface integral by a quadrature formula, i.e., by a weighted sum of samples of the integrand function

$$\sum_i \omega_i \left(\|\Gamma_{i,uu}(\boldsymbol{p}_i)\|^2 + 2\,\|\Gamma_{i,uv}(\boldsymbol{p}_i)\|^2 + \|\Gamma_{i,vv}(\boldsymbol{p}_i)\|^2 \right) \approx \int_S \kappa_1^2 + \kappa_2^2 \, dS.$$

A natural choice for the sampling sites are the vertices $\boldsymbol{p}_i$ of the mesh (it is here where the discrete curvature is actually located). The weight coefficients ω_i have to reflect the local area element, i.e., the triangles' relative size.

To sample the integrand function at each vertex, we have to compute second order partial derivatives. We do this by applying divided difference operators $\Gamma_{i,*}$ which are constructed by finding a quadratic least squares approximant to the direct neighbors of each vertex with respect to a local parameterization

$$\mu_i : \boldsymbol{p}_j - \boldsymbol{p}_i \mapsto (u_j, v_j) \in \mathbb{R}^2.$$

Since we evaluate the derivatives at isolated points only, we can find locally isometric parameterizations and exploit the fact that geometric curvatures coincide with a combination of second order derivatives in this case. Constructing a new set of divided difference operators $\Gamma_{i,*}$ for every vertex $\boldsymbol{p}_i$ means to choose a different parameterization μ_i for each. Hence, we compute derivatives (divided differences) always with respect to *local* parameterizations (*charts*). To define those we estimate the tangent plane at each vertex and get the parameter values for the neighbors by orthogonal projection. Another possibility is to approximate the exponential map by assigning parameter values according to the length of the edges and the angles between them [13,19].

It is tempting to re-estimate the local charts in every step of the iterative optimization. This makes sense given that the current mesh is always the best available approximation to the optimum. However, it turns out that this additional freedom makes the problem severely unstable. In [19] such a smoothing scheme is proposed, but the authors have to introduce additional topological operations in order to balance the instability.

We want to cast the fairing problem into a simple quadratic optimization problem and we want to find the "best" discrete approximation to the original objective functional. The way to satisfy both goals is to estimate the local charts, i.e., the local metric of the final surface by looking at the given data only. In [13], a weighted average of the discrete exponential map and the uniform parameterization is proposed. This initial estimate is kept fixed during the whole optimization process making the fairing energy functional quadratic with respect to the variables (= vertices). It can be shown that the solution of the optimization problem is uniquely defined, i.e., the fairing functional is strictly positive definite if the local charts are adapted to the subdivision connectivity of iteratively refined meshes [13].

We can understand this approach as being *dual piecewise polynomial.* Instead of assigning polynomial patches to the *faces* of a given mesh, we assign them to the *vertices.* This is implicitly done when solving the Vandermonde-system to construct the divided difference operators. For the evaluation of the discrete functional we only have to know these polynomials (and their derivatives) at the vertices but we can assume the virtual existence of a continuous surface consisting of patches around each vertex.

§5. Results

The discrete fairing scheme has been implemented based on a modification of a V-cycle multi-grid scheme [8]. Since appropriate stationary subdivision schemes provide smooth (but not necessarily fair) meshes, we don't have to pre-smooth the mesh before going to a coarser level. In fact, is it enough to compute the back-leg of the V-cycle, i.e., we alternate topological subdivision and iterative smoothing.

The special eigenstructure of the Gauß-Seidel iteration matrix in our case implies fast convergence in high frequency sub-spaces but very slow convergence at low frequencies. As a consequence, the mesh is flattened very quickly but converges slowly to the true solution. This effect can sometimes be observed at the original vertices where interpolation constraints are imposed: Occasionally visible cusps remain and are smoothed out rather slowly.

Since exact interpolation in the *interior* of the surface is an artificial constraint anyway, we avoided this problem by letting the original vertices move by some ε in a post-processing step. This removed the "pimples" but did not have significant effect on the global shape.

An important issue for practical applications is the control of the boundaries. As suggested in [12], we use a univariate scheme [11] to generate boundary curves that are independent of the interior of the mesh. This allows not only to join two surfaces along a common C^0 feature line, but it also gives the possibility to model sharp feature lines within an otherwise smooth surface. Fig. 3 shows an example of a surface generated by the discrete fairing scheme.

§6. Conclusions and Future Work

I have presented an efficient and robust scheme to generate triangular meshes that smoothly interpolate given vertices. There are no restrictions on the topology since no global parameterization has to be constructed. All we need are estimates of the local metric at each vertex. Upon this we can build divided difference operators allowing the approximation of partial derivatives with respect to an isometric parameterization. Together with an estimate of the relative area element this defines a discrete version of the continuous objective functional.

Due to the multi-resolution structure of meshes with subdivision connectivity, we can apply fast multi-grid solvers to compute the positions of the vertices in the optimal mesh. Right now, we generate moderately complex

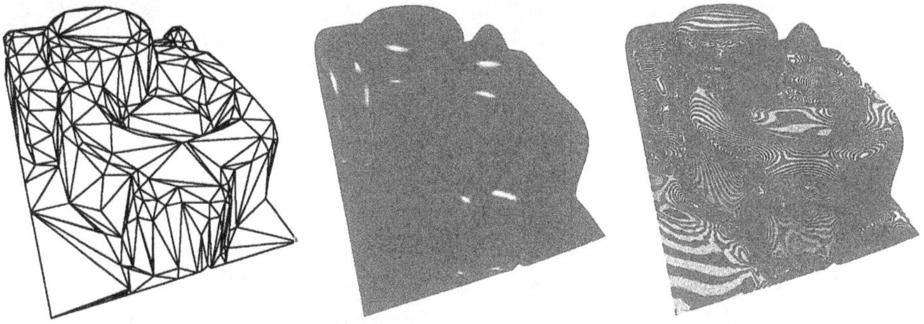

Fig. 3. A fair surface generated by the discrete fairing scheme. The flexibility of the algorithm allows interpolating rather complex data by high quality surfaces. The process is completely automatic and took about 10 sec to compute the refined mesh with 50K triangles.

models within a few seconds — in the future interactive modeling should become possible.

Future research in this field should address the investigation of higher order fairing functionals since those lead to even better results in some cases. However, higher order partial derivatives require local parameterizations that cover a larger neighborhood around each vertex. This increases the number of topological special cases that have to be considered.

Unlike classical fairing schemes, the discrete scheme does not generate spline surfaces. However, if compatibility to some standard CAD data exchange format matters, the obtained meshes can be converted into a spline representation. This can be achieved by any simple least squares fitting scheme with parameter correction since the duty to establish fairness has already been taken care of by the discrete fairing scheme. The philosophy behind this procedure is to do the fairing by some flexible discrete (non-classic) method, and once a sufficient number of points on the optimal surface is computed and the fair shape is recovered, we can use standard spline surfaces for the concluding fitting step.

If the original approximation constraints are considered as local forces that pull or push the surface into the wanted position then the discrete fairing allows to generate many, more densely distributed mini-forces that pull the spline surface more accurately into the optimal position.

References

1. Catmull, E., and J. Clark, Recursively generated B-spline surfaces on arbitrary topological meshes, CAD **10** (1978), 350–355

2. Celniker, G., and D. Gossard, Deformable curve and surface finite elements for free-form shape design, ACM Computer Graphics **25** (1991), 257–265.

3. van Damme, R., and L. Alboul, Tight triangulations, in *Mathematical Methods for Curves and Surfaces*, M. Dæhlen, T. Lyche, and L. L. Schumaker (eds.), Vanderbilt University Press, Nashville, 1995, 517–526.
4. Doo, D., and M. Sabin, Behaviour of recursive division surfaces near extraordinary points, CAD **10** (1978), 356–360
5. Dyn, N., Subdivision schemes in CAGD, in *Advances in Numerical Analysis II*, W.A. Light (ed.), Oxford University Press 1991, 36–104.
6. Dyn, N., D. Levin, and D. Liu, Interpolatory convexity-preserving subdivision schemes for curves and surfaces, CAD **24** (1992), 211–216
7. Greiner, G., Variational design and fairing of spline surfaces, Computer Graphics Forum **13** (1994), 143–154.
8. Hackbusch, W., *Multi-Grid Methods and Applications*, Springer Verlag 1985, Berlin.
9. Hagen, H., and G. Schulze, Automatic smoothing with geometric surface patches, CAGD **4** (1987), 231–235.
10. Hoppe, H., T. de Rose, and T. Duchamp et. al., Mesh optimization, ACM Computer Graphics **27** (1993), 19–26.
11. Kobbelt, L., A variational approach to subdivision, CAGD **13** (1996), 743–761.
12. Kobbelt, L., Interpolatory subdivision on open quadrilateral nets with arbitrary topology, Comp. Graph. Forum **15** (1996), 409–420.
13. Kobbelt, L., Discrete fairing, Proceedings of the Seventh IMA Conference on the Mathematics of Surfaces, 1996.
14. Le Méhauté, A., and F. Utreras, Convexity-preserving interpolatory subdivision, CAGD **11** (1994), 17–37.
15. Moreton, H., and C. Séquin, Functional optimization for fair surface design, ACM Computer Graphics **26** (1992), 167–176.
16. Sapidis, N., (ed.), *Designing Fair Curves and Surfaces*, SIAM, 1994.
17. Taubin, G., A signal processing approach to fair surface design, ACM Computer Graphics **29** (1995), 351–358.
18. Welch, W., and A. Witkin, Variational surface modeling, ACM Computer Graphics **26** (1992), 157–166.
19. Welch, W., and A. Witkin, Free-form shape design using triangulated surfaces, ACM Computer Graphics **28** (1994), 247–256.

Leif Kobbelt
Computer Graphics Group
University of Erlangen–Nürnberg
Am Weichselgarten 9
91058 Erlangen, GERMANY
`kobbelt@informatik.uni-erlangen.de`

Towards Free-Form Fractal Modelling

Ljubiša M. Kocić and Alba Chiara Simoncelli

Abstract. An alternative form of coding Iterated Function Systems, based on a control polygon and a subdivision technique, is offered. Corresponding attractors are more flexible and controllable. Some geometric properties necessary in free-form curve/surface modeling, such as affine invariance, convex hull property, continuity, interpolation, symmetry, can be preserved. The new model can generate smooth shapes, like polynomial or spline curves, as well as classical fractal sets, so it offers an adequate tool to model both in the same context.

§1. Introduction

Features like affine invariance, convex hull property or interpolation capability are basic for free-form modelling systems. Indeed, most classical curve and surface models meant to describe smooth shapes of man-made objects possess them. But those models cannot efficiently describe highly irregular and complex shapes, widely present in nature, such as the shape of clouds or of live tissues. These are suitably modeled by fractal sets, and can be reconstructed by means of different specific algorithms [1,4,12] based upon, or connected with, the theory of Iterated Function Systems (IFS). This theory was developed by M. Barnsley [1] in an arbitrary complete metric space. Since our natural setting is the m-dimensional real space, we deal here with IFS in the metric space $(\mathbb{R}^m, d)$, $m \geq 1$.

Let $\{w_i\}_{i=1}^n$ be a set of contractive Lipschitz mappings with factors $s_i < 1$. Let $\mathcal{H}(\mathbb{R}^m)$ be the set of all nonempty compact subsets of $\mathbb{R}^m$, and let h denote the Hausdorff metric induced by d, so that $\big(\mathcal{H}(\mathbb{R}^m), h\big)$ is a complete metric space. The system

$$\{\mathbb{R}^m; w_1, \ldots, w_n\} \tag{1}$$

is called an (*hyperbolic*) Iterated Function System, or IFS. The associated *Hutchinson operator* $W : \mathcal{H}(\mathbb{R}^m) \to \mathcal{H}(\mathbb{R}^m)$, defined by

$$W(B) = \cup_{i=1}^n w_i(B), \quad B \subset \mathbb{R}^m , \tag{2}$$

Mathematical Methods for Curves and Surfaces II
Morten Dæhlen, Tom Lyche, and Larry L. Schumaker (eds.), pp. 287–294.

ISBN 0-8265-1315-8.

is a contraction of $(\mathcal{H}(\mathbb{R}^m), h)$ with factor $s = \max_i\{s_i\} < 1$, see [1,12]. By the Banach fixed-point theorem, there is a unique set $A \in \mathcal{H}(\mathbb{R}^m)$ being a fixed point of W and, in the Hausdorff metric, for *any* $B \in \mathcal{H}(\mathbb{R}^m)$, see [6,13]

$$A = W(A) = \lim_{k\to\infty} W^{\circ k}(B). \tag{3}$$

The set A is called an attractor of (1), and sometimes a *deterministic fractal.*

The IFS uniquely determines its attractor, namely the fractal set to be reproduced, and (3) provides a tool to generate it in a natural way. But this IFS does not fulfil the basic needs of free-form modelling that we mentioned.

A rendering algorithm can be seen as an operator mapping input data into output data. It has the *free-form property* provided that any change in the input data causes predictable changes of the output. In particular, it is *invariant* under a transformation ω if transforming the input data by ω causes the output data to also be transformed by ω. In this sense, algorithms based on IFS (1) fail to have free-form property. In fact, if ω is an invertible affine map of $\mathbb{R}^m$, and A is the attractor of the IFS given by (1), in order to generate the set $\omega(A)$ the IFS must be changed into $\{\mathbb{R}^m; \hat{w}_1, \dots, \hat{w}_n\}$, where $\hat{w}_i = \omega \circ w_i \circ \omega^{-1}$ and $(f \circ g)(x) = f(g(x))$. So, no algorithm based on the IFS (1) can be affinely invariant. Actually the Initiator-Generator algorithm [12] is invariant under plane similarities, and IFS methods based on fractal interpolation [8,13] offer controllability, but none of them offers affine invariance.

The need to introduce a free-form property into an IFS, so as to be able to deform its attractor, is not completely new: Zair and Tosan [15] used binary subdivision and the deterministic algorithm, and Berger [2] constructed Bézier and spline curves using an IFS and the random iteration approach. In [10] Kocić and Simoncelli investigated affine invariant IFS based on a three point control polygon. In the present paper the authors wish to reestablish these approaches in a more clear, concise, yet comprehensive form. A controllable fractal model, the AIFS or *affine IFS*, is introduced, and its geometrical properties are examined. The AIFS is based on a control polygon and on its refinements expressed by means of subdivision matrices. The control polygon plays a triple role, as it defines the space in which contractions act, takes part in the definition of contractions, and also is a first approximation of the attractor itself.

The concept of AIFS is established in Section 2, where also some geometric properties of the model are proved. In Section 3 it is shown how it can be used for modelling both smooth and fractal shapes, and some problems connected with construction of the AIFS are briefly discussed.

§2. Random Affine Algorithm

Let $m \geq 3$ and $2 \leq d < m$. We call the *control polygon* of length m in $\mathbb{R}^d$ the ordered set of points $\boldsymbol{P} = (\boldsymbol{P}_1, \dots, \boldsymbol{P}_m)$, $\boldsymbol{P}_i \in \mathbb{R}^d$. It can be written in vector form as $\boldsymbol{P} = [\boldsymbol{P}_1 \dots \boldsymbol{P}_m]^T$, and also can be considered as a simplex in

$\mathbb{R}^{m-1}$. Therefore, given a control polygon $\boldsymbol{P} \subset \mathbb{R}^d$, for each $\boldsymbol{x} \in \mathbb{R}^{m-1}$ there exists a vector $\boldsymbol{\rho} = [\rho_1 \dots \rho_m]^T$, such that $\sum_i \rho_i = 1$ and $\boldsymbol{x} = \boldsymbol{\rho}^T \boldsymbol{P}$, whose components are the *barycentric coordinates* of the point $\boldsymbol{x}$ with respect to $\boldsymbol{P}$.

Let S_i denote a square row stochastic matrix of order m, namely $S_i = (s^i_{kj})^m_{k,j=1}$ with $\sum_j s^i_{kj} = 1, \forall k$. Denote by the same name the associated linear transformation $S_i : \mathbb{R}^m \to \mathbb{R}^m$, and denote by $\mathcal{A}_i$ the affine mapping of $\mathbb{R}^{m-1}$ naturally associated with S_i by means of coordinate extension [3]. If $\mathcal{A}_i$ is a contraction, with contractivity factor $s_i < 1$, we call S_i a *refinement matrix*. With this notation, we define an AIFS as follows:

Definition 1. *Let $\boldsymbol{P}$ be a control polygon. Let $S_1, \dots, S_n$ be $n \geq 2$ refinement matrices such that $S_i \neq S_j$, if $i \neq j$. Let $\sum_i p_i = 1$ and $p_i \geq 0$ $(i = 1, \dots, n)$. The system $\sigma = \{\boldsymbol{P}\,; S_1, \dots, S_n; p_1, \dots, p_n\}$ is called a (hyperbolic)* Affine Iterated Function System *(AIFS) with probabilities $p_1, \dots, p_n$.*

The image of $\boldsymbol{P}$ under the map S_i, namely $S_i\boldsymbol{P} = [\boldsymbol{P}^i_1 \dots \boldsymbol{P}^i_m]^T$, is called the i-th polygonlet of σ. The Hutchinson operator associated with σ satisfies

$$W(\boldsymbol{P}) = \cup^n_{i=1} S_i \boldsymbol{P}, \quad n \geq 2\,, \tag{4}$$

so the union of all polygonlets gives the image of the control polygon by the Hutchinson operator, or preattractor of σ. The attractor of σ will be denoted by $\sigma(\boldsymbol{P})$. By (3), $\sigma(\boldsymbol{P}) = W(\sigma(\boldsymbol{P}))$. By virtue of Theorem 1, below, the following algorithm constructs the projection of $\sigma(\boldsymbol{P})$ on the d-dimensional real space (the *modelling space* in [15]), we will denote it by $\hat{\sigma}(\boldsymbol{P})$.

Algorithm (Random AIFS). *Let $\sigma = \{\boldsymbol{P}\,; S_1, \dots, S_n; p_1, \dots, p_n\}$ be an hyperbolic AIFS, and let $\mathbf{e} = [1\,1 \dots 1]^T \in \mathbb{R}^m$.*

(i) *Take $\boldsymbol{\rho}^0 \in \mathbb{R}^m$ such that $S_j^T \boldsymbol{\rho}^0 = \boldsymbol{\rho}^0$ (for some $1 \leq j \leq n$) and $\mathbf{e}^T \boldsymbol{\rho}^0 = 1$;*
(ii) *For $k = 1, 2, \dots,$ with probability p_i $(1 \leq i \leq n)$, calculate*

$$\boldsymbol{\rho}^k = S_i^T \boldsymbol{\rho}^{k-1}, \ \boldsymbol{\rho}^k = [\rho^k_1 \dots \rho^k_m]^T \in \mathbb{R}^m\,, \tag{5}$$

$$\boldsymbol{x}_k = \sum_{\nu=1}^{m} \boldsymbol{P}_\nu \rho^k_\nu \ . \tag{6}$$

Theorem 1. *The point $\hat{\boldsymbol{x}}_k = [x^1_k \dots x^d_k]^T$, a projection on $\mathbb{R}^d$ of the point $\boldsymbol{x}_k$ defined by (6), lies on $\hat{\sigma}(\boldsymbol{P})$.*

Proof: Let $\boldsymbol{x}_0$ be the fixed point of the contraction $\mathcal{A}_j$ (associated with S_j) and let $\boldsymbol{\rho}^0$ be the vector of its barycentric coordinates w.r.t. $\boldsymbol{P}$. Then $\boldsymbol{\rho}^0$ is the right eigenvector of S_j^T corresponding to the unit eigenvalue and it satisfies the condition in step (i) of the Algorithm. Since S_i^T is column stochastic, the iteration (5) produces, starting by $\boldsymbol{\rho}^0$, a sequence of vectors $\{\boldsymbol{\rho}^k\}$ such that $\mathbf{e}^T \boldsymbol{\rho}^k = 1$, i.e. $\boldsymbol{\rho}^k$ contains barycentric coordinates of a point $\boldsymbol{x}_k \in \mathbb{R}^{m-1}$.

Moreover, being a fixed point of $\mathcal{A}_j$, the point $\boldsymbol{x}_0$ lies on the attractor $\sigma(\boldsymbol{P})$ which implies that $\boldsymbol{x}_1, \boldsymbol{x}_2, \ldots$ also are on $\sigma(\boldsymbol{P})$ and thereby, $\hat{\boldsymbol{x}}_k \in \hat{\sigma}(\boldsymbol{P})$. □

It should be noted that the starting point $\boldsymbol{\rho}^0 \in \mathbb{R}^m$ can in fact be an arbitrary point, and in this case the sequence of points $\{\hat{\boldsymbol{x}_k}\}$ will only approximate $\hat{\sigma}(\boldsymbol{P})$, which can also be acceptable for modelling purposes. □

The Random AIFS Algorithm is invariant under affine transformations of the control polygon $\boldsymbol{P}$, which can be used to control the attractor's shape. In fact, the following theorem holds.

Theorem 2. *The Random AIFS Algorithm is affinely invariant, that is: for any affine transformation* $\mathcal{A} : \mathbb{R}^{m-1} \to \mathbb{R}^{m-1}$, $\mathcal{A}(\sigma(\boldsymbol{P})) = \sigma(\mathcal{A}(\boldsymbol{P}))$.

Proof: In step (ii) of the algorithm, (5) produces the same sequence of vectors of barycentric coordinates, regardless of the control polygon $\boldsymbol{P}$. Then (6) reconstructs the affine coordinates of the point being generated as an affine combination of the points in $\boldsymbol{P}$. By definition, affine transformations preserve affine combinations [3], thus $\boldsymbol{x}'_k = \mathcal{A}(\boldsymbol{x}_k) \in \mathcal{A}(\sigma(\boldsymbol{P}))$ will yield $\boldsymbol{x}'_k = \mathcal{A}(\sum_{\nu=1}^{m} \boldsymbol{P}_\nu \rho_\nu^k) = \sum_{\nu=1}^{m} \mathcal{A}(\boldsymbol{P}_\nu)\rho_\nu^k \in \sigma(\mathcal{A}(\boldsymbol{P}))$ and vice versa. □

Fig. 1. Action on a Lévy curve in $\mathbb{R}^2$.

An example of a fractal set, known as a Lévy curve [4,6], is shown in Figure 1 (left). It was constructed using the Random AIFS Algorithm with $\sigma = \{\boldsymbol{P};\ S_1, S_2;\ 0.5, 0.5\}$, where $\boldsymbol{P}$ is the three-points control polygon also shown in the figure, $S_1 = [1\,0\,0;\ 0.5\,1\,-0.5;\ 0\,1\,0]^T$ and $S_2 = [0\,1\,0;\ -0.5\,1\,0.5;\ 0\,0\,1]^T$. Figure 1 (right) illustrates how, acting only on the control polygon, the shape of the attractor can be modified in a predictable way.

Theorem 3. *If all refinement matrices in the hyperbolic AIFS* σ *have non-negative entries, then* $\sigma(\boldsymbol{P})$ *is included in the convex hull of* $\boldsymbol{P}$.

Proof: Let $conv(\boldsymbol{B})$ denote the convex hull of set $\boldsymbol{B}$. The hypothesis that $s_{kj}^i \geq 0\ \forall i, k, j$, along with $\sum_j s_{kj}^i = 1\ \forall i, k$, yields $\boldsymbol{P}_k^i \subset conv(\boldsymbol{P})$, $\forall i, k$. So every polygonlet $S_i\boldsymbol{P} \subset conv(\boldsymbol{P})$ and, by (4), $W(\boldsymbol{P}) \subset conv(\boldsymbol{P})$ also. Iterating will leave $W^{\circ 2}(\boldsymbol{P})$ in the $conv(W(\boldsymbol{P}))$, and subsequent iterations lead to the chain relation $conv(\boldsymbol{P}) \supset conv(W(\boldsymbol{P})) \supset \cdots \supset \sigma(\boldsymbol{P})$. □

Theorem 4. *Let* $\boldsymbol{e}_k = [0 \ldots 0\,1\,0 \ldots 0]^T$ *be the k-th unit vector in* $\mathbb{R}^m$. *If*

$$\boldsymbol{e}_1^T S_1 = \boldsymbol{e}_1^T, \qquad \boldsymbol{e}_m^T S_n = \boldsymbol{e}_m^T, \tag{7}$$

$$\boldsymbol{e}_m^T S_i = \boldsymbol{e}_1^T S_{i+1} \qquad 1 \leq i \leq n-1, \tag{8}$$

$\sigma(\boldsymbol{P})$ is a continuous curve in $\mathbb{R}^d$. If, in addition, for some i, $\boldsymbol{e}_m^T S_i = \boldsymbol{e}_1^T S_{i+1} = \boldsymbol{e}_j^T$, this curve interpolates the j-th vertex of $\boldsymbol{P}$.

Proof: In fact, (7) means that $\boldsymbol{P}_1^1 = \boldsymbol{P}_1$ and $\boldsymbol{P}_m^n = \boldsymbol{P}_m$ while (8) implies $\boldsymbol{P}_m^i = \boldsymbol{P}_1^{i+1}$, $1 \le i \le n-1$. Which proves that the set $W(\boldsymbol{P})$ is a polygon itself, namely a continuous curve in $\mathbb{R}^d$. Further refinement does not spoil this continuity, so $W^{\circ 2}(\boldsymbol{P})$ is also a polygon, and so on. The continuity being preserved by the limiting process, the attractor is also a continuous curve.

The additional conditions ensure that $\boldsymbol{P}_m^i = \boldsymbol{P}_1^{i+1} = \boldsymbol{P}_j$. By (7) and (8) a new iteration will produce refined polygonlets that interpolate both $S_i\boldsymbol{P}$ and $S_{i+1}\boldsymbol{P}$, and therefore interpolate the vertex $\boldsymbol{P}_j$. The interpolation property being preserved by the operator W, the limit curve will interpolate it as well. □

Theorem 5. *Let $\boldsymbol{P}$ be a symmetric polygon. Let π denote the $m \times m$ permutation matrix having unit secondary diagonal and the other items equal to zero. If $\sigma = \{\boldsymbol{P}; S_1, \ldots, S_n; p_1, \ldots, p_n\}$ with $S_i = \pi S_{n-i+1}\pi$ $(i = 1, \ldots, n)$, then $\sigma(\boldsymbol{P})$ possesses the same symmetry property as $\boldsymbol{P}$.*

Proof: By hypothesis there exists a symmetry transformation $\lambda : \mathbb{R}^d \to \mathbb{R}^d$ such that $\lambda(\boldsymbol{P}) = \pi\boldsymbol{P} = [\boldsymbol{P}_m \ldots \boldsymbol{P}_1]^T$. Then, being λ an affine map and S_i a row stochastic matrix, $\forall i$ we have that $\lambda(S_i\boldsymbol{P}) = S_i\lambda(\boldsymbol{P}) = S_i\pi\boldsymbol{P}$ which, being, by supposition, $S_i = \pi S_{n-i+1}\pi$, and being $\pi^2 = I$, leads to $\lambda(S_i\boldsymbol{P}) = \pi S_{n-i+1}\pi^2\boldsymbol{P} = \pi S_{n-i+1}\boldsymbol{P}$. This means that, regardless of the order of points, every polygonlet is the symmetric image of another one, and therefore the preattractor $W(\boldsymbol{P})$ is a symmetric set, namely $\lambda(W(\boldsymbol{P})) = W(\boldsymbol{P})$, which implies symmetry of $W^{\circ 2}(\boldsymbol{P})$, $W^{\circ 3}(\boldsymbol{P})$,..., $\sigma(\boldsymbol{P})$. □

Usual nondegeneracy property is not valid for this model, in fact some examples exist that an AIFS with nondegenerate control polygon and nonsingular refinement matrices can indeed produce a one-point attractor [11].

§3. Smooth Curves and Other Questions

Figure 2 shows how gradual continuous bending of a known attractor, the very famous Barnsley fern, is obtained by means of the Random AIFS Algorithm, simply acting on the control polygon.

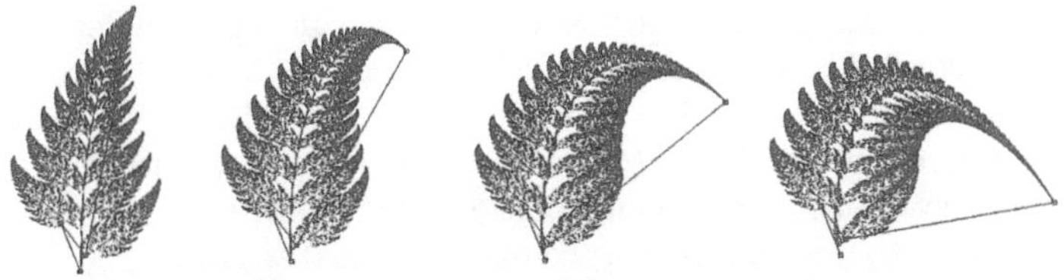

Fig. 2. Gradual changing of an attractor.

The technique by which this well-known attractor was endowed with a control polygon is explained later in this section, and illustrated in Figure 4.

First we want to point out that the Random AIFS Algorithm can produce smooth curves as well: for example Bézier and spline curves, that are a classical tool in free form modelling. Given an m-points control polygon $\boldsymbol{P}$, and the matrices $S_1 = [\binom{i-1}{j-1}2^{-i+1}]_{i,j=1}^{m}$ and $S_2 = \pi S_1 = [\binom{m-i}{m-j}2^{-m+i}]_{i,j=1}^{m}$, the AIFS $\{\boldsymbol{P};\, S_1, S_2;\, 0.5, 0.5\}$ will produce the Bézier curve of degree $m-1$ (Fig. 3, leftmost). These refinement matrices correspond to the so called mid-point subdivision [14]: this is not the only possible choice. In fact an IFS for Bézier curves of any degree can also be obtained by a different but related approach, as used in [7,9]. The AIFS with the same control polygon and refinement matrices given by $S_1 = [\binom{m}{2j-i-1}2^{-m+1}]_{i,j=1}^{m}$ and $S_2 = \pi S_1 = [\binom{m}{m-2j+i}2^{-m+1}]_{i,j=1}^{m}$ will generate the cardinal spline curve of degree $m-1$. Figure 3, middle, represents the cubic case. The cubic spline interpolating the end-points in the rightmost figure was obtained by the same refinement matrices, simply tripling the end-points of $\boldsymbol{P}$.

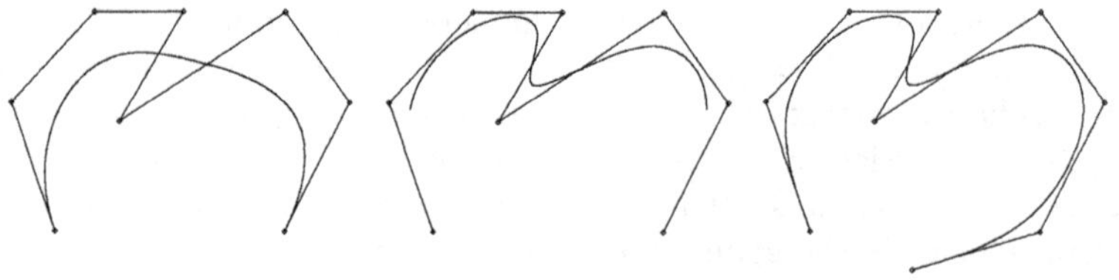

Fig. 3. Bézier curve and spline curves by AIFS.

We will now briefly examine some questions connected with construction of an AIFS. This requires specifying a control polygon $\boldsymbol{P}$, n refinement matrices $S_1, \ldots, S_n$, and the set of probabilities $p_1, \ldots, p_n$.

The most natural choice for probabilities is that the value of p_i be proportional to the ratio of the "size" of the i-th polygonlet to the "size" of $W(\boldsymbol{P})$. Thus, $p_i = L_i / \sum L_j$. As is known from the IFS theory, though, this formula is approximative and can be modified in order to produce different effects.

In the case of classical smooth curves (illustrated above), the refinement matrices are generally known from the literature. For example, the refinement matrices for Lagrange interpolating curve, for Beta-2 spline, Catmull-Rom spline, midpoint cubic and Chaikin curve are given in [5]. Rendering such curves by means of the Random AIFS Algorithm is straightforward. Otherwise, in cases when the matrices are not known, but the polygonlets are, refinement matrices can be derived from these, by solving a linear system. If $m = d + 1$ the system has a unique solution. In the case $m > d + 1$ it is advisable to resort to generalized barycentric coordinates [3,11].

As for determining the control polygon, when this is not known in advance, the points $\boldsymbol{P}_1, \ldots, \boldsymbol{P}_m$, should be chosen on the attractor. One way of doing this is to use the fixed points of contractions in the IFS, i.e. to take

$\boldsymbol{P}_i = \lim_k w_i^{\circ k}(\boldsymbol{x})$, where $\boldsymbol{x}$ is any point from $\mathbb{R}^2$. In fact, this is what was done to create the *flexible Barnsley fern* example in Figures 2 and 4. The control points $\boldsymbol{P}_1$, $\boldsymbol{P}_2$, $\boldsymbol{P}_3$ and $\boldsymbol{P}_4$ are the fixed points of the four affine contractions in the IFS that produces the original image (Figure 4, leftmost). The four polygonlets are labeled as w_1, w_2, w_3 and w_4 in Figure 4.

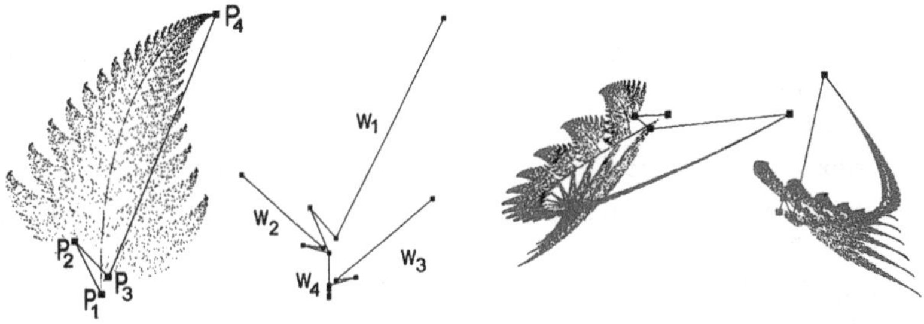

Fig. 4. Modelling of Barnsley fern.

The control polygon selected by this method has the nice property of being uniquely determined by the affine IFS, which makes transformation of a given IFS into an AIFS an automatic process. But the question wether this choice is the best is open. Our experience shows that sometimes this kind of polygon allows less flexibility [11], and this question deserves deeper investigation.

As a last remark we notice that, since convex linear combinations of two matrices preserve the property of being row stochastic, it is possible to blend two AIFSs. As shown, for example, in [11], this gives the possibility of creating sweeping surfaces. Also, other smooth surface patches like triangular patches or tensor product rectangular patches can easily be evaluated by AIFS.

§4. Conclusion

This paper is an attempt to introduce a useful variant of Iterated Function System that is affine invariant, and thus called Affine Iterated Function System (AIFS). The idea is not completely new - it has roots in the theory of stationary subdivision [14,5], and the problem was addressed also in [2,10,15]. But it seems that so far fractal modelling is not incorporated in the practice of CAGD. The AIFS offers the possibility to construct and manipulate fractal sets in a similar way as it is done in classical free form curve and surface modelling. Like the usual schemes, it provides a control polygon as a modelling tool, but it also takes advantage of some special features of known fractal algorithms, like the use of probabilities to implement shading and/or colouring. The ability to produce smooth curves and surfaces as well as very irregular fractal sets, preserving some desirable geometric properties from the classical arsenal like convex hull property, symmetry and interpolating capability, makes the AIFS a suitable tool for versatile modeling tasks.

References

1. Barnsley, M. F., *Fractals Everywhere*, Academic Press, 1988.
2. Berger, P. A., Random affine iterated function systems: curve generation and wavelets, SIAM Review **34** (1992), 361–385.
3. Boehm, W., and H. Prautzsch, *Geometric Concepts for Geometric Design*, A. K. Peters, Wellesley MA, 1993.
4. Dubuc, S., and A. Elqortobi, Approximation of fractal sets, J. Comput. Appl. Math. **29** (1990), 79–89.
5. Goldman, R. N., and T. D. DeRose, Recursive subdivision without the convex hull property, Comput. Aided Geom. Design **3** (1986), 247–265.
6. Hutchinson, J. E., Fractals and self similarity, Indiana Univ. Math. J. **30** (1981), 713–747.
7. Kocić, Lj. M., Fractals and Bernstein polynomials, Periodica Mathematica Hungarica **33** (1996), 185–195.
8. Kocić, Lj. M., Monotone interpolation by fractal functions, in *Approximation and Optimization, Proc. of ICAOR (Cluj-Napoca, Ro. 1996)*, Vol.I, Transilvania Press, Cluj-Napoca RO, 1997, 291–298.
9. Kocić, Lj. M., and A. C. Simoncelli, Bézier curves via fractal algorithm, Proc. I. C. Func. Anal. Approx. Th. (Maratea, It. 1996) in Rend. Circ. Mat. Palermo **47** (1998), (to appear).
10. Kocić, Lj. M., and A. C. Simoncelli, Fractals generated by a triangle, Fractalia **21** (1997), 13–18.
11. Kocić, Lj. M., and A. C. Simoncelli, AIFS: a tool for modelling fractal sets, Pubbl. Dip. Mat. Appl. n.17(1997), Universitá Federico II, Napoli.
12. Mandelbrot, B., *The Fractal Geometry of Nature*, W.H. Freeman and Co., New York, 1983.
13. Massopust, P. R., *Fractal Functions, Fractal Surfaces, and Wavelets*, Academic Press, New York, 1994.
14. Micchelli, C. A., *Mathematical Aspects of Geometric Modeling*, VBMS-NSF series in Applied Math., SIAM, Philadelphia PE, 1995.
15. Zair, C. E., and E. Tosan, Fractal modeling using free form techniques, Computer Graphics Forum **15** (1996), 269–278.

Ljubiša M. Kocić
Dept. of Mathematics, Faculty of Electronic Engineering
P. O. Box 73, 18000 Niš, YUGOSLAVIA
`kole@laplace.elfak.ni.ac.yu`

Alba Chiara Simoncelli
Dipartimento di Matematica ed Applicazioni, Università degli Studi
di Napoli Federico II, Via Cinthia, Monte S. Angelo, 80126 Napoli, ITALY
`simoncel@matna2.dma.unina.it`

Fairness Approximation by Modified Discrete Smoothing D^m-splines

A. Kouibia, M. Pasadas and J. J. Torrens

Abstract. This paper addresses the problem of constructing curves and surfaces from Lagrange data. We modify the classical definition of discrete smoothing D^m-splines in order to handle parametric curves and surfaces and additional fairness criteria. A convergence result is established and some numerical and graphical examples are given.

§1. Introduction

In past few years, variational methods in CAGD have received considerable attention, due to their efficiency and usefulness in the fitting and design of curves and surfaces. The basic idea of these methods is to minimize a functional which typically contains two terms: the first indicates how well the curve or surfaces approximates a given set of data, while the second controls the degree of smoothness or fairness of the curve or surface.

A wide range of fairness measures have been proposed, derived from physical considerations (e.g. stretch energy, bending energy) or geometric entities (e.g. curve length, surface area, curvature). See, for example, [7,8,10,12] and references therein.

Discrete smoothing D^m-splines provide specific examples of variational curves and surfaces. These splines minimize, in a finite element space, a quadratic functional in which the fairness term is a Sobolev semi–norm of order m (see [2,3] or, in this volume, [11] for definitions in the case of real–valued functions). In this paper, we modify the classical definition of this class of splines in order to handle additional fairness criteria.

This paper is organized as follows. In Section 2 we briefly recall some preliminary notations and results. In Section 3 we present a modified class of discrete smoothing D^m-splines, more adapted to fairing processes. In Section 4 we establish a convergence result for this type of splines. Finally, Section 5 contains some numerical and graphical examples.

Mathematical Methods for Curves and Surfaces II
Morten Dæhlen, Tom Lyche, and Larry L. Schumaker (eds.), pp. 295–302.

ISBN 0-8265-1315-8.

§2. Notations and Preliminaries

Let $p, n \in \mathbb{N}^*$. We denote by $\langle \cdot \rangle_{\mathbf{R}^n}$ and $\langle \cdot , \cdot \rangle_{\mathbf{R}^n}$, respectively, the Euclidean norm and inner product in $\mathbb{R}^n$. Likewise, for any nonempty open set Ω in $\mathbb{R}^p$ and any $m \in \mathbb{N}$, we denote by $H^m(\Omega; \mathbb{R}^n)$ the usual Sobolev space of (classes of) functions u which belong to $L^2(\Omega; \mathbb{R}^n)$, together with all their partial derivatives $\partial^\alpha u$, in the distribution sense, of order $|\alpha| \leq m$, where $\alpha = (\alpha_1, \ldots, \alpha_p) \in \mathbb{N}^p$ and $|\alpha| = \alpha_1 + \cdots + \alpha_p$. This space is equipped with the norm

$$\|u\|_{m,\Omega,\mathbf{R}^n} = \left(\sum_{|\alpha| \leq m} \int_\Omega \langle \partial^\alpha u(x) \rangle^2_{\mathbf{R}^n} \, dx \right)^{\frac{1}{2}},$$

the semi–norms

$$|u|_{j,\Omega,\mathbf{R}^n} = \left(\sum_{|\alpha| = j} \int_\Omega \langle \partial^\alpha u(x) \rangle^2_{\mathbf{R}^n} \, dx \right)^{\frac{1}{2}}, \quad 0 \leq j \leq m,$$

and the corresponding inner semi–products

$$(u, v)_{j,\Omega,\mathbf{R}^n} = \sum_{|\alpha| = j} \int_\Omega \langle \partial^\alpha u(x), \partial^\alpha v(x) \rangle_{\mathbf{R}^n} \, dx.$$

When Ω is a bounded, open subset of $\mathbb{R}^p$ with Lipschitz–continuous boundary (in the J. Nečas [9] sense), it follows from Sobolev's imbedding theorem that, for any $k \in \mathbb{N}$ such that $m > k + p/2$, $H^m(\Omega; \mathbb{R}^n)$ is a subset of $C^k(\overline{\Omega}; \mathbb{R}^n)$ with continuous injection, where $C^k(\overline{\Omega}; \mathbb{R}^n)$ stands for the space of functions with values in $\mathbb{R}^n$ which are bounded and uniformly continuous on Ω, together with all their partial derivatives of order $\leq k$.

Finally, for any $m, k \in \mathbb{N}$, we denote by $\mathbb{R}^{m,k}$ and P_k, respectively, the space of real matrices with m rows and k columns and the space of all polynomials of degree $\leq k$.

§3. Modified Discrete Smoothing D^m-splines

3.1. Motivation

Let m, n and p be three positive integers such that $1 \leq p < n \leq 3$ and $m > \frac{p}{2}$. Let $\Upsilon \subset \mathbb{R}^n$ be a curve (if $p = 1$) or a surface (if $p = 2$) parameterized by a function f that belongs to the Sobolev space $H^m(\Omega; \mathbb{R}^n)$, where Ω is a bounded, open subset of $\mathbb{R}^p$. For the sake of simplicity, assume that Ω is polygonal if $p = 2$. We consider the following problem:

$$\begin{cases} \textit{Given the sets } A = \{a_1, \ldots, a_N\} \subset \overline{\Omega} \textit{ and } \{f(a_1), \ldots, f(a_N)\} \subset \Upsilon, \\ \textit{find an "optimal" approximant of } \Upsilon. \end{cases} \tag{1}$$

Obviously, many variational methods can be devised to cope with this problem, depending on the meaning conferred to the word "optimal". A possible choice is to consider that "optimal" approximants are those parameterized by discrete smoothing D^m-splines, whose definition we briefly recall.

Let H be a set of real, positive numbers of which 0 is an accumulation point. Suppose we are given, for any $h \in H$, a tesselation $\mathcal{T}_h$ of $\overline{\Omega}$, made with elements of diameter $\leq h$ (intervals, if $p = 1$, or triangles or rectangles, if $p = 2$), and a finite element space X_h, constructed on $\mathcal{T}_h$ from a generic finite element (K, P_K, Σ_K) of class C^k, with $m \leq k+1$, and such that $P_K \subset H^m(K)$ (see [5] for notations). It is easily shown that, for any $h \in H$, the parametric finite element space $V_h = (X_h)^n$ satisfies

$$V_h \subset H^m(\Omega; \mathbb{R}^n) \cap C^k(\overline{\Omega}; \mathbb{R}^n).$$

Assume that A contains a P_{m-1}-unisolvent subset. Then, for any $h \in H$ and $\varepsilon > 0$, the *V_h-discrete smoothing D^m-spline relative to A, $(f(a_i))_{1 \leq i \leq N}$ and ε* is the unique solution of the problem: find $\sigma_{\varepsilon h}$ such that

$$\sigma_{\varepsilon h} \in V_h \text{ and } J_\varepsilon(\sigma_{\varepsilon h}) = \inf_{v \in V_h} J_\varepsilon(v),$$

where

$$J_\varepsilon(v) = \sum_{i=1}^{N} \langle v(a_i) - f(a_i) \rangle^2_{\mathbf{R}^n} + \varepsilon |v|^2_{m,\Omega,\mathbf{R}^n}. \tag{2}$$

Observe that the first term in $J_\varepsilon(v)$ indicates how well v approaches f in a discrete least squares sense, while the second one is a smoothness measure. The parameter ε weights the importance given to every term. Hence, its value has a strong influence on the global shape of the corresponding curve or surface.

In some cases, it would be desirable to dispose of additional fairness criteria and shape parameters in order to allow a better shape control. Thus, for fairing purposes, it is advantageous to include in the functional J_ε the Sobolev semi–norms of orders 1 to $m-1$, as we shall do in the next subsection. We remark that $| \cdot |_{j,\Omega,\mathbf{R}^n}$ can be seen as a simplified measure of curve length ($p = 1$, $j = 1$), surface area ($p = 2$, $j = 3$), bending energy ($j = 2$), or variation of curvature ($j = 3$).

3.2. Definition

For any m-tuple $\tau = (\tau_1, \ldots, \tau_m)$ of nonnegative real numbers with $\tau_m > 0$, let J_τ be the functional defined in $H^m(\Omega; \mathbb{R}^n)$ by

$$J_\tau(v) = \sum_{i=1}^{N} \langle v(a_i) - f(a_i) \rangle^2_{\mathbf{R}^n} + \sum_{j=1}^{m} \tau_j |v|^2_{j,\Omega,\mathbf{R}^n}.$$

The parameters $\tau_1, \ldots, \tau_m$ act as shape parameters, controlling the relative weight given in J_τ to the least squares term and the different smoothness terms. Of course, for $\tau = (0, \ldots, 0, \varepsilon)$, J_τ is simply the functional J_ε defined in (2).

For any $\tau = (\tau_1, \dots, \tau_m) \in \mathbb{R}^m$, with $\tau_1, \dots, \tau_{m-1} \geq 0$ and $\tau_m > 0$, and for any $h \in H$, it is clear that an "optimal" approximant of Υ is now the curve or surface parameterized by any solution of the problem: find $\sigma_{\tau h}$ such that

$$\sigma_{\tau h} \in V_h \text{ and } J_\tau(\sigma_{\tau h}) = \inf_{v \in V_h} J_\tau(v).$$

From the Lax–Milgram's Lemma, it follows that this problem has a unique solution $\sigma_{\tau h}$, also characterized by the relations

$$\begin{cases} \sigma_{\tau h} \in V_h, \\ \forall v \in V_h, \ \sum_{i=1}^{N} \langle \sigma_{\tau h}(a_i), v(a_i) \rangle_{\mathbf{R}^n} + \sum_{j=1}^{m} \tau_j (\sigma_{\tau h}, v)_{j,\Omega,\mathbf{R}^n} \\ \qquad = \sum_{i=1}^{N} \langle f(a_i), v(a_i) \rangle_{\mathbf{R}^n}. \end{cases} \tag{3}$$

We say that $\sigma_{\tau h}$ is the *modified* V_h*-discrete smoothing* D^m*-spline relative to* A, $(f(a_i))_{1 \leq i \leq N}$ *and* τ.

3.3. Computation

Let $\{w_1, \dots, w_M\}$ be a basis of the finite element space X_h. Thus $\sigma_{\tau h} = \sum_{j=1}^{M} \alpha_j w_j$ for some $\alpha_1, \dots, \alpha_M \in \mathbb{R}^n$. Consider the matrices

$$\begin{aligned} \mathcal{A} &= \big(w_j(a_i)\big)_{\substack{1 \leq i \leq N \\ 1 \leq j \leq M}} \in \mathbb{R}^{N,M}, \\ \mathcal{R}_\ell &= \big((w_j, w_i)_{\ell,\Omega}\big)_{1 \leq i,j \leq M} \in \mathbb{R}^{M,M}, \quad 1 \leq \ell \leq m, \\ \beta &= \big(f(a_i)\big)_{1 \leq i \leq N} \in \mathbb{R}^{N,n}, \end{aligned}$$

where $(\,\cdot\,,\,\cdot\,)_{\ell,\Omega}$ is the inner semi–product of order ℓ in the Sobolev space $H^m(\Omega)$. From (3), we deduce that the matrix $\alpha = (\alpha_j)_{1 \leq j \leq M} \in \mathbb{R}^{M,n}$ satisfies

$$\left(\mathcal{A}^T \mathcal{A} + \sum_{\ell=1}^{m} \tau_\ell \mathcal{R}_\ell \right) \alpha = \mathcal{A}^T \beta.$$

Hence, $\sigma_{\tau h}$ is computed by solving n linear systems, each one of which has the same symmetric, positive definite matrix $\mathcal{A}^T \mathcal{A} + \sum_{\ell=1}^{m} \tau_\ell \mathcal{R}_\ell$. We remark that the order of every linear system is $M = \dim X_h$, independent of the number N of data points.

§4. Convergence

In this section we establish a convergence result for modified discrete smoothing D^m-splines analogous to Theorem 5 in [11] (see also [4]). Due to its similarity, we refer the reader to that paper for remarks on the hypotheses and additional comments.

Suppose we are given a set D of real positive numbers for which 0 is an accumulation point, and, for every $d \in D$, an ordered set A^d of $N = N(d)$ points in $\overline{\Omega}$ such that

$$\sup_{x\in\Omega} \delta(x, A^d) = d,$$

δ being the Euclidean distance in $\mathbb{R}^n$.

Assume that the family of finite element spaces $(X_h)_{h\in H}$ is constructed in such a way that the following result, due to P. Clément (cf. [6]), holds:

> There exist an integer $m' \geq m$ and, for all $h \in H$, a linear operator $\Pi_h : L^2(\Omega) \to X_h$ such that:
> (i) $\exists C > 0$, $\forall h \in H$, $\forall l = 0, \ldots, m' - 1$, $\forall v \in H^{m'}(\Omega)$,
> $$\left(\sum_{K\in\mathcal{T}_h} |v - \Pi_h v|_{l,K}^2\right)^{1/2} \leq C\, h^{m'-l} |v|_{m',\Omega}; \tag{4}$$
> (ii) $\forall v \in H^{m'}(\Omega)$, $\lim_{h\to 0} \left(\sum_{K\in\mathcal{T}_h} |v - \Pi_h v|_{m',K}^2\right)^{1/2} = 0.$

Assume, in addition, that the families $(A^d)_{d\in D}$ and $(\mathcal{T}_h)_{h\in H}$ satisfy the relation

$$\exists C > 0,\ \forall (d, h) \in D \times H,\ \forall K \in \mathcal{T}_h : \frac{\operatorname{card} A^d \cap K}{\operatorname{meas} K} \leq C\, d^{-n},$$

which bounds above the density of data points in every element of every tesselation $\mathcal{T}_h$. Finally, considering h and τ as functions of d, suppose that

$$\tau_m = o(d^{-p}),\ d \to 0,$$

$$\frac{h^{2m'}}{d^p \tau_m} = o(1),\ d \to 0,$$

and that

$$\frac{\tau_1 + \cdots + \tau_{m-1}}{\tau_m} = o(1),\ d \to 0.$$

Under these conditions we have the following result.

Theorem 1. *Let f be a function in $H^{m'}(\Omega; \mathbb{R}^n)$, with m' given in (4). For all $d \in D$, let $\sigma_{\tau h}^d$ be the modified V_h-discrete smoothing D^m-spline relative to A^d, $(f(a))_{a\in A^d}$ and τ. Then*

$$\lim_{d\to 0} \|\sigma_{\tau h}^d - f\|_{m,\Omega} = 0.$$

§5. Numerical and Graphical Examples

Consider the surface (a piece of a cylinder) defined by the parameterization $f : \overline{\Omega} \to \mathbb{R}^3$, with $\Omega = (0,1) \times (0,1)$ and $f(u,v) = (3u, \cos \pi v, \sin \pi v)$. Its area is 3π and, at every point, the total curvature (e.g. the sum of the squares of the principal curvatures) is constant and equal to 1.

We have computed several modified discrete smoothing D^3-splines σ_τ from a set of 400 scattered points. The values of τ are indicated in Table 1. The parametric finite element space has been constructed from the Bogner–Fox–Schmit's rectangle of class C^2 (cf. [1]) on a tesselation of $\overline{\Omega}$ having 4 equal squares.

By evaluation at 10000 random points in $\overline{\Omega}$, we have obtained, for every τ, estimates for the relative error $\|f - \sigma_\tau\|_{0,\Omega,\mathbf{R}^3} / \|f\|_{0,\Omega,\mathbf{R}^3}$, the area of the approximating surface parameterized by σ_τ, and the mean value of its total curvature. They are denoted by $\tilde{E}$, $\tilde{A}$ and $\tilde{T}$, respectively, and showed in Table 1.

Graphically, the three approximating surfaces are very similar. They present almost the same degree of approximation, measured by $\tilde{E}$. Differences are more apparent if $\tilde{A}$, $\tilde{T}$, and the plots of the total curvature shown in Fig. 1 are examined. For $\tau = (0, 10^{-2}, 0.3 \times 10^{-4})$, since $\tau_2 \neq 0$, the semi-norm $|\cdot|_{2,\Omega,\mathbf{R}^3}$, related to the total curvature, is minimized (together with the least squares term and the semi-norm of order 3). Compared with the first case (e.g. $\tau = (0, 0, 0.8 \times 10^{-4})$), this results in a diminution of the total curvature (see the values of $\tilde{T}$ and Fig. 1). Analogously, the surface area, related to the semi-norm $|\cdot|_{1,\Omega,\mathbf{R}^3}$, strongly decreases for $\tau = (0.3, 0, 10^{-4})$, at the price, however, of a considerable increase in the total curvature. Observe that, for such value of τ, one also has the smallest range of variation in the total curvature (see Fig. 1), because, in this case, τ_3 reaches the biggest value. The user has to set the vector parameter τ according to the fairing effect (if any) which he or she is looking for.

τ	$\tilde{E}$	$\tilde{A}$	$\tilde{T}$
$(0, 0, 0.8 \times 10^{-4})$	6.2847×10^{-3}	9.5500	0.89251
$(0, 10^{-2}, 0.3 \times 10^{-4})$	6.2439×10^{-3}	9.3551	0.79840
$(0.3, 0, 10^{-4})$	6.4238×10^{-3}	8.9996	0.99420

Table 1. Values of estimates $\tilde{E}$, $\tilde{A}$, and $\tilde{T}$.

Acknowledgments. This work has been supported in part by the Gobierno de Navarra (Research Project BON nº 33–18/3/94), the European Union (Research Project INTAS 94–4070) and the Diputación General de Aragón (Research Project PCB0994).

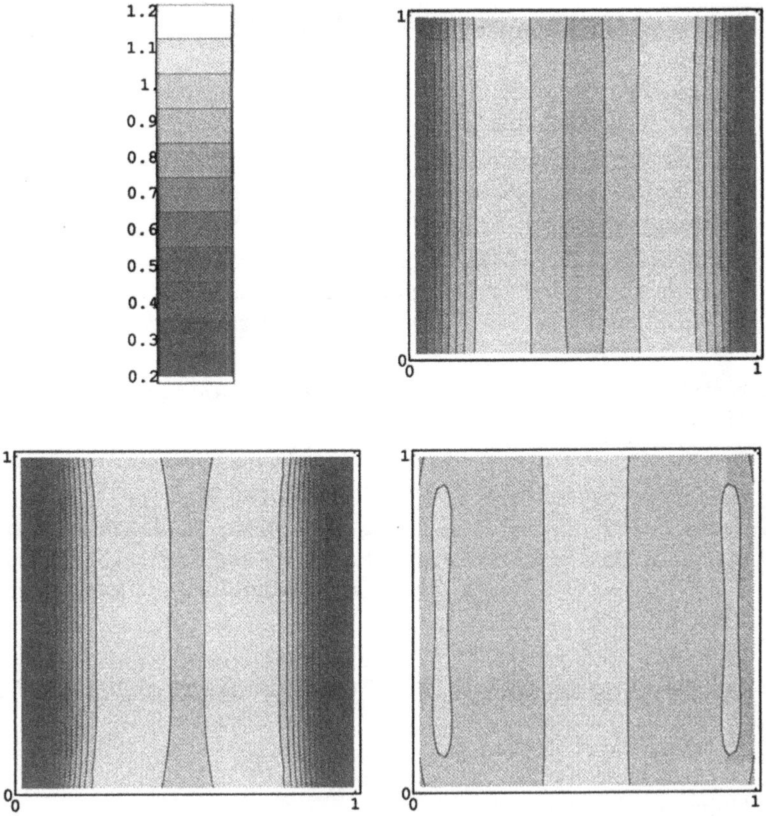

Fig. 1. Total curvature for $\tau = (0, 0, 0.8 \times 10^{-4})$ (top right), $(0, 10^{-2}, 0.3 \times 10^{-4})$ (bottom left), and $(0.3, 0, 10^{-4})$ (bottom right).

References

1. Apprato, D., R. Arcangéli, and R. Manzanilla, Sur la construction de surfaces de classe C^k à partir d'un grand nombre de données de Lagrange, Math. Model. Numer. Anal. **21** (1987), 529–555.
2. Arcangéli, R., D^m-splines sur un domaine borné de $\mathbb{R}^n$, Publication URA CNRS 1204 n° 86/2, Pau, 1986.
3. Arcangéli, R., Some applications of discrete D^m-splines, in *Mathematical Methods in Computer Aided Geometric Design*, T. Lyche and L. L. Schumaker (eds.), Academic Press, New York, 1989, 35–44.
4. Arcangéli, R., R. Manzanilla, and J. J. Torrens, Approximation spline de surfaces de type explicite comportant des failles, Math. Model. Numer.

Anal. **31**(5) (1997), 643–676.

5. Ciarlet, P. G., *The Finite Element Method for Elliptic Problems*, North–Holland, 1978.
6. Clément, P., Approximation by finite element functions using local regularization, RAIRO R–2 (1975), 77–84.
7. Greiner, G., Surface construction based on variational principles, in *Wavelets, Images, and Surface Fitting*, P.-J. Laurent, A. Le Méhauté, and L. L. Schumaker (eds.), A. K. Peters, Wellesley MA, 1994, 277–286.
8. Kaasa, J., and G. Westgaard, Analysis of curvature–related surface properties, in *Curves and Surfaces with Applications in CAGD*, A. Le Méhauté, C. Rabut, and L. L. Schumaker (eds.), Vanderbilt Univ. Press, Nashville, 1997, 211–215.
9. Nečas, J., *Les Méthodes Directes en Théorie des Équations Elliptiques*, Masson, 1967.
10. Nowaki, H., and X. Lü, Fairing composite polynomial curves with constraints, Comput. Aided Geom. Design **11** (1994), 1–5.
11. Torrens, J. J., Discrete smoothing D^m-splines: Applications to surface fitting, in *Mathematical Methods for Curves and Surfaces II*, M. Dæhlen, T. Lyche, and L. L. Schumaker (eds.), Vanderbilt Univ. Press, Nashville, 1998, 477–484.
12. Wesselink, W., and R. C. Veltkamp, Interactive design of constrained variational curves, Comput. Aided Geom. Design **12** (1995), 533–546.

A. Kouibia and M. Pasadas
Departamento de Matemática Aplicada
Universidad de Granada
Severo Ochoa s/n
18071 Granada, SPAIN
`kouibia@goliat.ugr.es`
`mpasadas@goliat.ugr.es`

J. J. Torrens
Departamento de Matemática e Informática
Universidad Pública de Navarra
Campus de Arrosadía s/n
31006 Pamplona, SPAIN
`jjtorrens@upna.es`

Interpolating Curve with B-spline Curvature Function

Mitsuru Kuroda, Masatake Higashi, Tsuyoshi Saitoh, Yumiko Watanabe, and Tetsuzo Kuragano

Abstract. Describing curvature with a B-spline function whose parameter is arc length, we derive a method for obtaining an interpolanting spline curve. By B-spline techniques in curvature space, designers are able to introduce straight lines and circular arcs into an extended clothoidal spline curve with piecewise linear/quadratic curvature, and specify tangents and curvatures at junction points. We solve the relevant nonlinear equations easily by Newton-Raphson method, using a system which manipulates symbolic expressions as well as numerical data.

§1. Introduction

Nutbourne et al. tried a method to specify, blend and integrate a piecewise linear curvature and developed a local scheme to generate an interpolating clothoidal spline curve span-by-span [8,9,11,10]. Meek et al. proposed an interactive method for a clothoidal spline interpolation [4,5]. These two methods have drawbacks because they heavily depend on conditions at the starting point and require interactive assistance or "the sensible way to use" [5].

Mehlum studied a nonlinear spline interpolant and developed a method to approximate the clothoidal splines as its solution with a sequence of circular arcs within a prespecified tolerance [7]. Stoer proposed a method for the interpolating clothoidal spline curve that minimizes the integral of squared curvature [12]. Brunnett et al. developed a method to approximate interpolating minimal-energy splines with curves of a piecewise polynomial curvature function [1]. These variational methods lead to relatively large nonlinear systems. Moreover, Stoer's curve does not always exist depending on the configuration of data points and Brunnett's one is insignificantly G^2 discontinuous.

In this paper, we present a straightforward design method of an interpolating curve whose curvature is a B-spline function of arc length. Using the preferable B-spline techniques such as multiple knots including pseudo knots

Mathematical Methods for Curves and Surfaces II
Morten Dæhlen, Tom Lyche, and Larry L. Schumaker (eds.), pp. 303–310.

ISBN 0-8265-1315-8.

(of multiplicity 0), knot insertion and degree reduction, designers are able to control the curvature profile directly to a certain extent and specify tangents and curvatures at junction points. They are also able to make the curve more flexible under the condition of a limited number of segments by using pseudo knots at data points (see [13,6] in the case of arc splines). The interpolation conditions and these other requirements lead to a system of nonlinear equations. This system is solved easily in the Newton-Raphson method, with a system manipulating both symbolic expressions and numerical data.

We discuss a planar curve with arc length parameter and introduce notation in Section 2. We describe a system of equations in Section 3 and discuss how to solve it in Section 4. Then we obtain some examples of curves satisfying various additional requirements and illustrate the properties of the newly developed curves in Section 5.

§2. Tangent Angle Profile and Unknowns

A planar curve $\boldsymbol{r}(s)$ is expressed as

$$\boldsymbol{r}(s) = \boldsymbol{r}_0 + \int_0^s \begin{pmatrix} \cos\psi(s) \\ \sin\psi(s) \end{pmatrix} ds, \tag{1}$$

where s is the arc length from the starting point $\boldsymbol{r}_0$ and $\psi(s)$ is tangent angle (the angle between the tangent vector and the direction of the x axis). The following relations among $\boldsymbol{r}$, unit tangent vector $\boldsymbol{t}$, unit normal vector $\boldsymbol{n}$ and curvature κ hold:

$$\frac{d^2\boldsymbol{r}}{ds^2} = \frac{d\boldsymbol{t}}{ds} = \frac{d\psi}{ds}\boldsymbol{n} = \kappa\boldsymbol{n}. \tag{2}$$

We express the tangent angle $\psi(s)$ as a cubic B-spline function of parameter s, and so the curve includes straight lines, circular arcs, clothoids or second clothoids with quadratic curvature. Since $\psi(s)$ determines a unique curve shape, we are able to specify and control an interpolating curve by a control polygon (de Boor ordinates and Greville abscissae) of $\psi(s)$ under the conditions

$$\boldsymbol{r}(s_i) = \boldsymbol{p}_i \equiv \begin{pmatrix} x_i \\ y_i \end{pmatrix}, \qquad i = 0, 1, \ldots, n. \tag{3}$$

We are going to derive a set of equations with unknown parameters of $\psi(s)$ based on the conditions (3) and solve it. In this section, we arrange the parameters and choose a set of unknowns. We use the following notation throughout the paper, where Δ is the forward difference operator defined by $\Delta z_i = z_{i+1} - z_i$.

1) Knots: $v_{-2}, v_{-1}, \ldots, v_q$. End knots are of multiplicity 3. Each knot corresponds to a junction point. See Fig. 1.
   - $\Delta w_0, \Delta w_1, \ldots, \Delta w_{k-1}$ are a sequence of all the non-zeros of $\{\Delta v_{-2}, \Delta v_{-1}, \ldots, \Delta v_{q-1}\}$.
   - $S_0, S_1, \ldots, S_{l-1}$ are all the independent ones among $\{\Delta w_0, \Delta w_1, \ldots, \Delta w_{k-1}\}$.

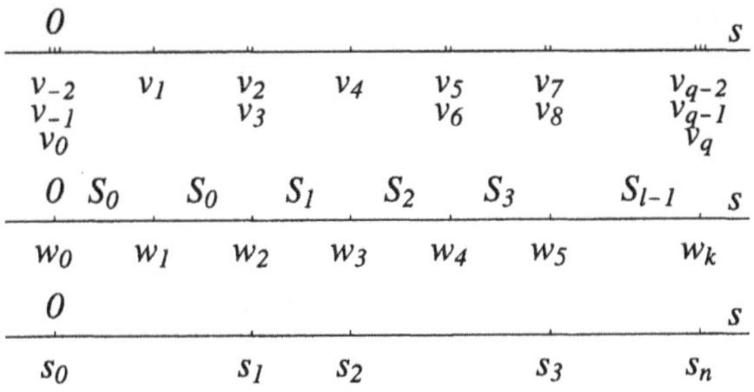

Fig. 1. Relation among parameters $\{v_i\}$, $\{w_i\}$, $\{S_i\}$ and $\{s_i\}$.

- List of differences of knots: which consists of zeros and S_i's. We use this list to specify the structure of the curve compactly, denoting a zero segment due to multiple knots and a multiple segment span by a zero and a sublist, respectively. This list in the case of Fig. 1 is

$$\{0,0,\{S_0,S_0\},0,S_1,\{S_2,0,S_3\},0,\ldots,0,0\}.$$

2) de Boor ordinates [2]: $B_0, B_1, \ldots, B_q$, where B_i corresponds to the Greville abscissa $(v_{i-2}+v_{i-1}+v_i)/3$.
3) Bézier ordinates [2]: $b_0, b_1, \ldots, b_{3k}$. For easy manipulation, we transform $\psi(s)$ into $\{\psi_0(t), \psi_1(t), \ldots, \psi_{k-1}(t)\}$ with local parameter t. We get the Bézier ordinates that correspond to k segments without zero segments due to multiple knots. See Section 4.

$$\begin{aligned}\psi_i(t) \equiv& (b_{3i}, b_{3i+1}, b_{3i+2}, b_{3i+3}) \cdot ((1-t)^3, 3(1-t)^2t, 3(1-t)t^2, t^3)^t,\\ 0 \le t \equiv& \frac{s-w_i}{w_{i+1}-w_i} \le 1, \quad i=0,1,\ldots,k-1.\end{aligned} \tag{4}$$

Since the tangent angle $\psi(s)$ is expressed in terms only of knot sequence and de Boor ordinates, we put together the independent lengths of segments S_i's and the de Boor ordinates B_i's into the following unknown vector $\boldsymbol{u}$ whose elements have aliases for ease of notation.

$$\boldsymbol{u} \equiv (u_0, u_1, \ldots, u_{l+q})^t \equiv (S_0, S_1, \ldots, S_{l-1}, B_0, B_1, \ldots, B_q)^t. \tag{5}$$

§3. System of Equations

The number of unknowns is $(l+q+1)$ in (5). On the other hand, we have $2n$ interpolation conditions that each component x and y of the curve interpolates to each pair of data points:

$$\begin{aligned}\begin{pmatrix} f_{2i} \\ f_{2i+1}\end{pmatrix} \equiv& \int_{s_i}^{s_{i+1}} \begin{pmatrix}\cos\psi(s)\\ \sin\psi(s)\end{pmatrix} ds - \begin{pmatrix}\Delta x_i\\ \Delta y_i\end{pmatrix}\\ =& \sum_{i_{th}\,span} \Delta w_j \int_0^1 \begin{pmatrix}\cos\psi_j(t)\\ \sin\psi_j(t)\end{pmatrix} dt - \begin{pmatrix}\Delta x_i\\ \Delta y_i\end{pmatrix} = 0,\\ & i=0,\ 1, \ldots,\ n-1,\end{aligned} \tag{6}$$

where the summation $\sum_{i_{th} span}$ is over segment(s) of the i-th span. Therefore we are able to add the other $(l+q-2n+1)$ requirements to these conditions. For example, we may specify tangent directions, curvatures or radii of curvature at junction points, using the corresponding mathematical expressions $\{\psi(w_i),\ \kappa(w_i),\ 1/\kappa(w_i)\}$ obtained by symbolic manipulation. In the case of Fig. 4,

$$\psi(w_i) = \frac{B_{2i+1}S_{i-1} + B_{2i}S_i}{S_{i-1} + S_i}, \quad \kappa(w_i) = -3\frac{B_{2i} - B_{2i+1}}{S_{i-1} + S_i}, \quad i = 0,\ 1, \ldots, 16,$$

where $S_{-1} = S_{17} = 0$. We may also require $\kappa_i(t) = 0$ or $\kappa_i(t) = constant$ for degree reduction so that the segment results in straight line or circular arc, respectively. We summarize these independent requirements as

$$f_i(\boldsymbol{u}) = 0, \quad i = 2n,\ 2n+1, \ldots, m, \quad m \le l + q. \tag{7}$$

§4. Curve Derivation

Our problem leads to the following system of equations with the unknowns $\boldsymbol{u}$:

$$\boldsymbol{f}(\boldsymbol{u}) \equiv (f_0,\ f_1, \ldots, f_m)^t = 0, \qquad m \le l + q. \tag{8}$$

We solve this system of equations iteratively by Newton-Raphson method:

$$\boldsymbol{u}^{(i+1)} = \boldsymbol{u}^{(i)} - J^{-1}\boldsymbol{f}(\boldsymbol{u}^{(i)}), \tag{9}$$

where J is the following Jacobian matrix.

$$J \equiv \frac{\partial(f_0, f_1, \ldots, f_m)}{\partial(u_0, u_1, \ldots, u_{l+q})}|_{\boldsymbol{u}=\boldsymbol{u}^{(i)}}. \tag{10}$$

A good initial value $\boldsymbol{u}^{(0)}$ for this iteration is available from the parametric cubic C^2 interpolant.

When $m < l + q$ in (8), we may introduce default constraints from the conventional curves or replace the inverse matrix of J with the Moore-Penrose pseudo inverse [14,15] for non-square matrix. This inverse matrix has the favorable property that the sum of the squares of all the entries in $(JJ^{-1}-I)$, where I is an identity matrix, is minimized.

We have programmed this algorithm with *Mathematica* (Wolfram Research Inc.), which results in compact coding. For example, we are able to get Bézier ordinates $\{b_0,\ b_1, \ldots, b_k\}$ expressed by the unknowns $\boldsymbol{u}$, using the following function *Bezbi* with a list of differences of knots as the argument *S0S0*. Our program is available on `http://www.toyota-ti.ac.jp/people/kuroda`.

```
Bezbi[S0S0_List]:=Module[{n,i,a,b,c,b3,b0,b1,d0,d1,d2,
  S03=Partition[Flatten[S0S0],3,1]},
n=-1;({a[n],b[++n]}={#[[1]],#[[3]]}/Plus@@#;
      c[n]=#[[2]]/Plus@@Rest[#]/.Indeterminate->1/2)&/@S03;
b3=Flatten[Table[{b0=b[i]B[i]+(1-b[i])B[i+1],
                 b1=(1-a[i])B[i+1]+a[i]B[i+2],
                 (1-c[i])b0+c[i]b1}[[{1,3,2}]],{i,0,--n}]];
Simplify[b3]//.{d0___,d1_,d1_,d2___}->{d0,d1,d2}];
```

§5. Some Examples of Newly Developed Curves

We show some examples of the new curves, using two sets of data points in the references [10,3] and the other two sets of ours. Each figure includes curve(s) and the corresponding curvature plot(s). In the figures:

o denotes data point;
• denotes a junction point between a pair of consecutive data points;
ticks of the horizontal (arc length) axis correspond to junction points;
the total length of the curve is at the end of this axis.

Figure 2 is a closed G^2 clothoidal curve that fits data points given in [10].

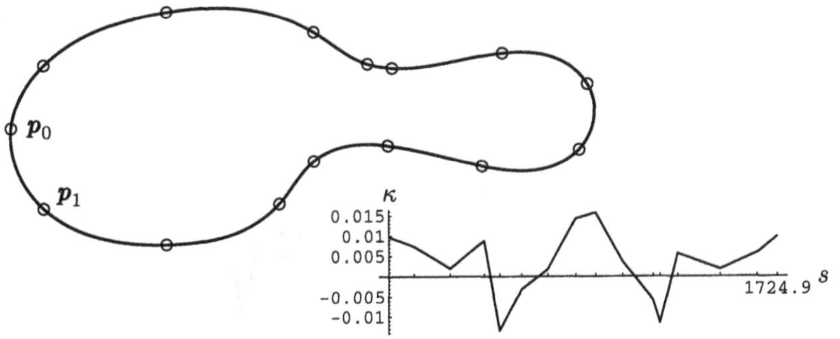

Fig. 2. Closed G^2 clothoidal curve fitting data points given in [10].

Figure 3 is a G^1 interpolating curve that includes straight lines (1st, 3rd, 5th, 8th and 12th spans) and circular arcs (u, v and w spans). Data points are {(9,1), (4,1), (1,3), (1,6), (2,7), (7,7), (10,3.5), (14,3.5), (14,6), (15,7), (17,4.2), (20,3), (22,3)}.

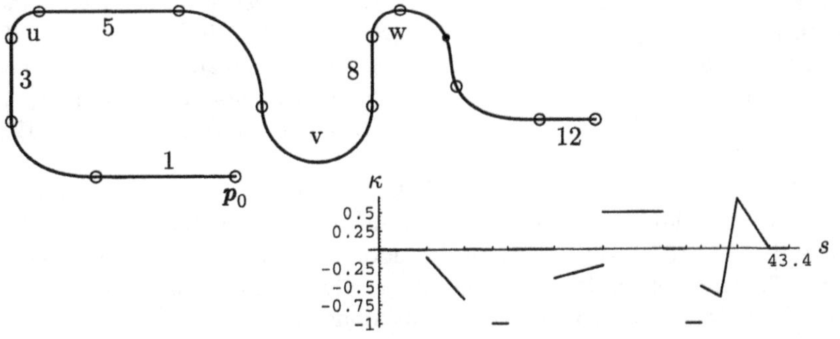

Fig. 3. G^1 clothoidal interpolant including straight lines (1, 3, 5, 8, 12) and circular arcs (u, v, w).

Figure 4 is a G^2 interpolating clothoidal curve that includes circular arcs (u, v, ..., z). The curve is relatively smooth with a simple profile of curvature. The data points of Fig. 4 are from [3]. The list of differences of knots is {0, 0,

S_0, 0, S_1, 0, {S_2, 0, S_3}, 0, S_4, 0, S_5, 0, {S_6, 0, S_7}, 0, {S_8, 0, S_9}, 0, {S_{10}, 0, S_{11}}, 0, S_{12}, 0, S_{13}, 0, {S_{14}, 0, S_{15}}, 0, S_{16}, 0, 0}. The data points $\boldsymbol{p}_5, \boldsymbol{p}_6, \boldsymbol{p}_8$ are not junction points since the corresponding knots have been changed to pseudo knots by full continuity conditions.

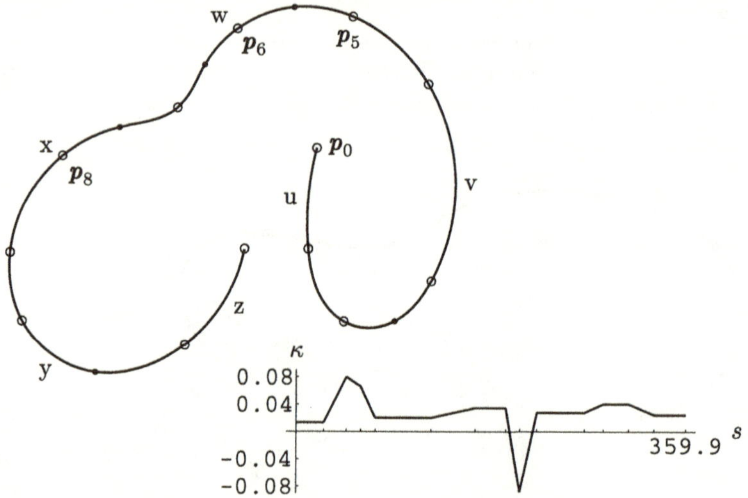

Fig. 4. G^2 clothoidal curve fitting data points given in [3].

In Figure 5, (a) and (b) are a G^3 second clothoidal interpolant and a G^2 clothoidal one, respectively. (c) is a G^2 bi-clothoidal interpolant whose tangents are the mean tangents of the curves (a) and (b) at data points. (d) is a G^3 second clothoidal interpolant that has a junction point between each pair of consecutive data points. (e) is a G^3 hybrid interpolant of the curves (a) and (d) with the same end tangents as (d), whose list of differences of knots is {0, 0, S_0, S_1, S_2, S_3, S_4, {S_5, S_6}, {S_7, S_8}, {S_9, S_{10}}, S_{11}, 0, 0}. The data points $\boldsymbol{p}_6, \boldsymbol{p}_7$ are not junction points. (e) has the shortest curve length among the G^3 curves (a), (d) and (e). The data points of Fig. 5 are {(11.6, 7), (10, 3.33), (8.5, 7.65), (12.5, 10.8), (8, 13.1), (7, 8.05), (2, 8.45), (5, 12.2), (5, 7.57), (5, 3.5)}.

All the curves were obtained with 4 digit precision after 3 to 5 iterations of Newton-Raphson method. Five iterations in Fig. 5(a) took 18.79 sec (Sun Ultra SPARC 167 MHz), using the pseudo inverse matrix with *Mathematica* Ver. 3.0. These examples show that the newly developed method works in a reliable and accurate manner for various kinds of curves.

§6. Concluding Remarks

We have proposed the simple yet flexible method to control a piecewise polynomial curvature of a composite interpolant directly to a certain extent for shape design, using a system which manipulates symbolic expressions as well as numerical data. Now, the commonly used techniques of the B-splines are

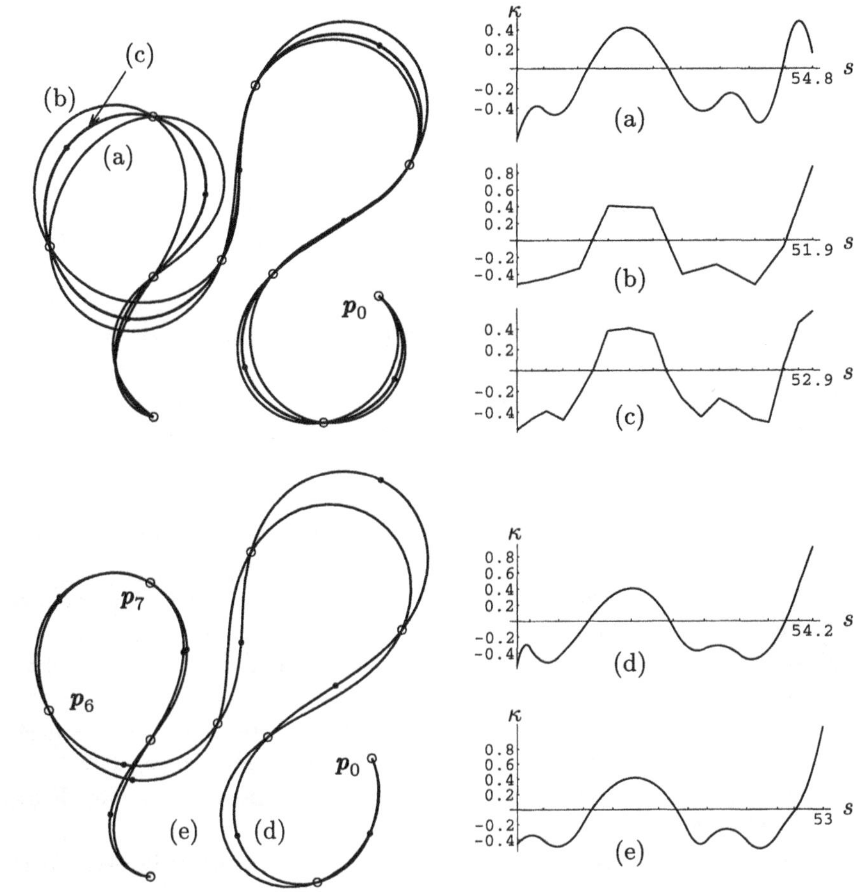

Fig. 5. G^2 clothoidal interpolants and G^3 second clothoidal ones.

not used in curve space, but in curvature space. We are on extending this method to three dimensional curves and surfaces.

Acknowledgments. This research was partly supported by the Ministry of Education, Science, Sports and Culture under Grant-in-Aid for Scientific Research (C) 08650182.

References

1. Brunnett, G. and J. Kiefer, Interpolation with minimal-energy splines, Computer-Aided Design **26** (1994), 137–144.
2. Farin, G., *Curves and Surfaces for Computer Aided Geometric Design: A Practical Guide*, Academic Press, London, 1988.

3. Makino, H., N. Furuya and M. Iwatsuki, Performance tests for CP motion with the SCARA robot, in 15th ISIR, Tokyo, 1985, 1011–1020.
4. Meek, D. S. and D. J. Walton, The use of Cornu spirals in drawing planar curves of controlled curvature, Journal of Computational and Applied Mathematics **25** (1989), 69–78.
5. Meek, D. S. and R. S. D. Thomas, A guided clothoid spline, Comput. Aided Geom. Design **8** (1991), 163–174.
6. Meek, D. S. and D. J. Walton, Planar osculating arc splines, Comput. Aided Geom. Design **13** (1996), 653–671.
7. Mehlum, E., Nonlinear splines, in *Computer Aided Geometric Design*, R. E. Barnhill and R. F. Riesenfeld (eds.), Academic Press, New York, 1974, 173–207.
8. Nutbourne, A. W., P. M. McLellan and R. M. L. Kensit, Curvature profiles for plane curves, Computer-Aided Design **4** (1972), 176–184.
9. Pal, T. K. and A. W. Nutbourne, Two-dimensional curve synthesis using linear curvature elements, Computer-Aided Design **9** (1977), 121–134.
10. Pal, T. K., Intrinsic spline curve with local control, Computer-Aided Design **10** (1978), 19–29.
11. Schechter, A., Synthesis of 2D curves by blending piecewise linear curvature profiles, Computer-Aided Design **10** (1978), 8–18.
12. Stoer, J., Curve fitting with clothoidal splines, Journal of Research of the National Bureau of Standards **87** (1982), 317–346.
13. Su, B.-Q. and D.-Y. Liu, *Computational Geometry: Curve and Surface Modeling*, G.-Z. Chang, Trans., Academic Press, San Diego, 1989.
14. Wolfram, S., *Mathematica: Third Edition*, Cambridge University Press, Cambridge, 1996.
15. Yanai, H. and K. Takeuchi, *Projection Matrix, Generalized Inverse Matrix and Singular Value Decomposition*, University of Tokyo, Tokyo, 1983.

Mitsuru Kuroda and Masatake Higashi
Toyota Technological Institute
2-12 Hisakata, Tempaku, Nagoya 468-8511, JAPAN
`kuroda@toyota-ti.ac.jp`
`higashi@toyota-ti.ac.jp`

Tsuyoshi Saitoh and Yumiko Watanabe
Tokyo Denki University
2-2 Kanda-Nishiki, Chiyoda, Tokyo 101-8457, JAPAN
`saitoh@saitohlab2.c.dendai.ac.jp`
`watanabe@saitohlab2.c.dendai.ac.jp`

Tetsuzo Kuragano
Sony Corp.
4-10-18 Takanawa, Minato, Tokyo 108-8621, JAPAN
`kuragano@sees.sony.co.jp`

Some Families of Triangular Finite Elements Which Can Provide Sequences of Nested Spaces

A. Le Méhauté

Abstract. The present paper is a continuation of the discussion of hierarchical bases of triangular finite elements. It is an attempt to determine in what sense it is possible to define nested sequences of finite elements spaces and, more precisely, what are the triangular finite elements which are relevant to use in such nested sequences.

§1. Introduction

It is the aim of this note to complement and continue the discussion in [19] and try to determine how it is possible to define nested sequences of finite element spaces and more precisely what are the triangular finite elements which are relevant to use in such nested sequences.

Let us first make explicit what we mean by this idea of nested sequences of finite element spaces. Suppose $\Omega \subset \mathbb{R}^2$ is a bounded polygonal domain, *i.e.*, the boundary $\partial\Omega$ of Ω consists of straight line segments. Let Δ be a triangulation of Ω built upon a set $\mathcal{A}$ of given (initial) points, $\mathcal{A} = \{a_1, \dots, a_N\}$. Let $\mathcal{S}_d^k(\Delta)$ be the set of piecewise polynomial spline functions with continuity k and degree d built upon Δ, *i.e.*,

$$\mathcal{S}_d^k(\Delta) = \{f : \; f \in \mathcal{C}^k(\bar{\Omega}), f|_T \in \mathbb{P}_d[\mathbb{R}^2], \forall T \in \Delta\},$$

where $\mathbb{P}_d[\mathbb{R}^2]$ is the set of polynomials with two real variables and degree d.

The initial set of approximating functions will be $V_1 = \mathcal{S}_d^k(\Delta)$ for some k and some d and we want to describe a sequence of spaces $\{V_i\}_{i \geq 1}$ such that

$$V_1 \subset V_2 \subset \cdots \subset V_i \subset \cdots.$$

As we will show there exist different possibilities for the construction of such nested families and at least one could think of the two different approaches that are related to each of the two aspects of finite element: the usual h-version and the p-version (see [4]), without mentioning the use of macro-elements generalizing Hsieh - Clough - Tocher or Fraeijs de Veubeke - Sanders elements (see [8,9,19,22] and references therein). Another possibility is to use rational functions instead of polynomials, see Section 4 below.

Mathematical Methods for Curves and Surfaces II
Morten Dæhlen, Tom Lyche, and Larry L. Schumaker (eds.), pp. 311–318.

ISBN 0-8265-1315-8.

§2. Basic Aspects of Finite Element Approximation

Let us recall here some (well known) basic aspects of the finite element method in $\mathbb{R}^2$, or equivalently of the construction of piecewise polynomial spline approximation of functions of two real variables.

Let Δ be a (proper) triangulation of a polygonal domain Ω, with no splitting of the triangles (which means that we are dealing with a unique polynomial piece defined on each triangle).

The approximation by simplicial polynomial finite elements is based on the following hypothesis:

(h1) Interpolation data are linearly independent;

(h2) Interpolation is local (*i.e.*, any alteration of a single datum should affect only the triangles containing the point supporting this datum);

(h3) Those degrees of freedom that are necessary to ensure k-continuity between adjacent triangles (called *first type data*) are given on the boundary of each triangle. Other data are needed to ensure unisolvency on each triangle T , and are given either on the boundary or in the interior of T (they are called *second type data*);

(h4) The first type data are "symmetric" relatively to the structure of each triangle (*i.e.*, the 3 vertices play the same rôle, just like the three edges);

(h5) First type data are linear functionals of the form $\delta^{\alpha}_{a_i} : f \mapsto \partial^{\alpha} f(a_i)$, where a_i are points on the boundary of T , $\alpha \in \mathbb{N}^n$, or belong to $Span\{\delta^{\alpha}_{a_i}\}$(in order to use derivatives normal to the edges, for instance).

2.1. Morgan-Scott Revisited

As was already noticed in [10] for the C^1 case and more generally in [19], it is not possible to use the Argyris elements and its generalizations which lead to the spaces $\mathcal{Z}^k_d(\Delta)$, [25, 13], because of the *super - spline effect* [7], *i.e.*, the fact that the order of the derivatives involved at the vertices (and for some cases at points corresponding to the second type data) is higher than the order of continuity of the space. Using this space for refining the triangulation will produce a dense subset of points in Ω with a higher continuity than could be achieved by the space itself. Instead, we have to use the full space $\mathcal{S}^k_d(\Delta)$, which is after all possible because the usual Morgan-Scott-Ženíšekresult [13] can be restated as follows [19]:

Proposition 2.1 *Given any integer $k \geq 0$ and in order to obtain C^k continuity between simplices, it is sufficient that for any of the vertices a in the triangulation, the linear forms $v \mapsto \partial^{\alpha} v(a)$ are available from the data set, for all α such that $|\alpha| \leq 2k$.*

Of course, this does not mean that they are necessarily given in the data set, and in the latter case, we will have to pay a price, not only in the evaluation of those degrees of freedom, but also because of the fact that the hypotheses ($h2, h3$) have to be enlarged: the evaluation of the interpolant on one triangle will involve data from the adjacent ones.

2.2. A basis for $\mathcal{S}_d^k(\Delta)$

In order to get a basis of $\mathcal{S}_d^k(\Delta)$, we use the following data ([19]):

(2.1) On each edge $e = \sigma_s$ of Δ, the normal derivatives $\frac{\partial^i}{\partial \nu_s^i} f(Q_{r,s}^i)$, for $i = 1, \ldots, k$ and $r = 1, \ldots, i$, where Q_{rs}^i are points on e such that Q_{rs}^i is not a vertex and $Q_{rs}^i \neq Q_{ts}^i$ whenever $t \neq r$.

(2.2) Inside each triangle T, the second type degrees of freedom needed to ensure the $(d = 4k+1)$–unisolvency, which could be $\partial^\alpha f(a_0), |\alpha| \leq k-2$, where a_0 is a strictly interior point, or any equivalent set.

(2.3) Around each vertex a, two possibilities according to the fact that a is an interior or a boundary vertex.

(2.3a) For an interior vertex, let $e_1, e_2, \ldots, e_n$ be the different edges emanating from a, ordered clockwise for instance. If a is not a defective vertex [2], all the slopes are different and we take the data set

$$\partial^\alpha f(a), |\alpha| \leq k, \qquad D^{i+j} f(a) \cdot (e_s^i, e_{s+1}^j),\ i, j = 1, \ldots, k; s = 1, \ldots, n.$$

When a is a defective vertex , we have to add another set of data, corresponding to the directional derivatives $D^{i+j} f(a) \cdot (e_s^i, e_{s+1}^j),\ \ i, j = 1, \ldots, 2k$ which cannot be obtained from the previous set . (This corresponds to the terms in the sum σ in [2]).

(2.3b) For a boundary vertex a, let $e_1, e_2, \ldots, e_n$ be the different edges emanating from a, ordered clockwise for instance in such a way that the boundary edges are e_1 and e_n. If a is not a defective vertex, all the slopes are different and we take the data set

$$\begin{aligned} &\partial^\alpha f(a), |\alpha| \leq k; \ \ D^{i+j} f(a) \cdot (e_s^i, e_{s+1}^j),\ i, j = 1, \ldots, k,\ s = 2, \ldots, n-2; \\ &\quad D^{i+j} f(a) \cdot (e_1^i, e_2^j), \qquad i = 1, \ldots, 2k,\ j = 1, \ldots, k; \quad i + j \leq 2k \\ &\quad D^{i+j} f(a) \cdot (e_{n-1}^i, e_n^j),\ i = 1, \ldots, k,\ j = 1, \ldots, 2k, \quad i + j \leq 2k. \end{aligned}$$

(and the corresponding changes for a defective vertex).

The notation $D^{i+j} f(a_s) \cdot (a_s a_{s-1}^i, a_s a_{s+1}^j)$ is for directional derivatives at a_s evaluated along $a_s a_{s-1}$ and $a_s a_{s+1}$. More precisely, following [11],

$$D^2 f(x) \cdot (\mu, \nu) = \partial^{2,0} f(x) \mu_x\ \nu_x + \partial^{1,1} f(x) [\mu_x\ \nu_y + \mu_y\ \nu_x] + \partial^{0,2} f(x) \mu_y\ \nu_y,$$

and more generally, for any $\mathbf{T} = (t_1, t_2, \ldots, t_p)$ with $t_i = (\xi_{i1}, \xi_{i2}, \ldots, \xi_{in}) \in \mathbb{R}^n$,

$$\begin{aligned} &D^p f(x) \cdot (\mathbf{T}) = D^p f(x) \cdot (t_1, t_2, \ldots, t_p) = \\ &\qquad \sum_{\mathbf{J}} D_{j_1} D_{j_2} \ldots D_{j_p} f(x)\, \xi_{1,j_1} \xi_{2,j_2} \cdots \xi_{p,j_p} \end{aligned} \tag{3.4}$$

where the sum $\sum_{\mathbf{J}}$ is to be understood for all the n^p different sequences $\mathbf{J} = (j_1, j_2, \ldots, j_p)$ of integers in $\{1, 2, 3, \ldots, n\}$, and $D^p f(x) \cdot (\mu^q, \nu^{p-q})$ denotes the particular case where q of the t_i's are equal to the same vector μ, and $p - q$ of them are equal to ν.

§3. Nested Families of Polynomial Finite Element Spaces

3.1 The h–version

Now we are in situation to consider a first family of nested finite element spaces. Using the data defined in the previous section, the resulting space is $\mathcal{S}_d^k(\Delta)$. At this point, any triangle can then be split in 4 isometric subtriangles, using as new vertices the midpoints of the edges and it is easy to see that we can iterate this process. If we denote by Δ_1 the original triangulation and Δ_2 the new one, and so on, we have

$$\mathcal{S}_d^k(\Delta_1) \subset \mathcal{S}_d^k(\Delta_2) \subset \mathcal{S}_d^k(\Delta_3) \subset \cdots.$$

For example, and for C^1 continuity, we will start with $\mathcal{S}_5^1(\Delta_1)$ as in [10], then built $\mathcal{S}_5^1(\Delta_2), \mathcal{S}_5^1(\Delta_3)$ and so forth. The hierarchical basis of piecewise linear functions defined on triangles is also a special case of this h–version [24].

3.2. The p-Version

Another but possibly not so familiar approach for the finite element approximation is to use what Babuska [4] called the *p-version* instead of the usual *h-version*, which roughly means that instead of having the size of the elements going to 0 when refining the triangulation (for the h-version), it is the degree of the local polynomials which goes to ∞, keeping the same triangulation.

Now we need to go back to the construction of the local interpolant for polynomial triangular finite element. It is shown in [13,17] that another way of generalizing Argyris triangle is to define triangular finite elements of type $(k, p, 4k+p+1)$, for C^k continuity, where p is the number of time the function itself is given on each edge of a triangle, and $4k + p + 1$ is the degree of the polynomial. The general result is proved in [13]:

Theorem 3.1. *Let p , k , κ be 3 integers such that $p \geq 0, \kappa \geq k \geq 0$, and f a function in $C^{k+\kappa}(\bar{\Omega})$. Let $N = 2(k+\kappa)+p+1$ and T be a triangle of the triangulation Δ of Ω with a_1, a_2, a_3 its vertices , $\sigma_1, \sigma_2, \sigma_3$ the edges of T, and ν_s the normal to σ_s, for s=1,2,3. Let Π_Σ^N be a polynomial of degree N such that*

$$\partial^\alpha \Pi_\Sigma^N(a_s) = \partial^\alpha f(a_s) \quad , \qquad |\alpha| \leq k+\kappa, \quad s = 1,2,3$$

$$\text{(3.1)} \qquad \frac{\partial^i}{\partial \nu_s^i} \Pi_\Sigma^N(Q_{rs}^i) = \frac{\partial^i}{\partial \nu_s^i} f(Q_{rs}^i), \quad \begin{cases} s = 1,2,3, \\ i = 0,1,\ldots,k \\ r = 1,2,\ldots,p+i \end{cases}$$

where Q_{rs}^i are points on σ_s such that $Q_{rs}^i \neq a_j$, j=1,2,3, and $Q_{rs}^i \neq Q_{ts}^i$ whenever $t \neq r$. Let g be the function defined on $\bar{\Omega}$ by $g_{|\Sigma} = \Pi_\Sigma^N$, for each $\Sigma \in \Delta$. Then $g \in C^k(\bar{\Omega})$.

Remark: This theorem defines only the first type data which are necessary to obtain C^k triangular elements. One has to add the data which are necessary to ensure the N–unisolvency of the interpolation problem. They can be defined as derivatives in one interior point or from any equivalent set of linear forms.

It is possible to use this result for the construction of a nested family of triangular elements, keeping the same vertices, the same order of continuity k, and $\kappa = k$ and going from a step to another step by changing p in $p+1$ and the relevant data at the appropriate points Q^i_{rs}. Of course, we start with a convenient data set concerning the derivatives at the vertices of Δ (as defined in the previous section for a vertex which can be interior or on the boundary, defective or not) and evaluate the data in (3.1) if necessary.

Given p with $N = 4k + p + 1$, the nested family will be

$$\mathcal{S}^k_N(\Delta) \subset \mathcal{S}^k_{N+1}(\Delta) \subset \mathcal{S}^k_{N+2}(\Delta) \subset \cdots \subset \mathcal{S}^k_{N+i}(\Delta) \subset \cdots.$$

For example, for C^1 continuity, we will start again with $\mathcal{S}^1_5(\Delta)$ as in [10], then built $\mathcal{S}^1_6(\Delta), \mathcal{S}^1_7(\Delta)$ and so forth.

§4. Nested Families of Rational Finite Element of Type (k, N, χ, p)

A rational finite element of type (k, N, χ, p) is obtained by adding a rational ("bubble") function to the polynomial partial finite element of type (k, N, χ, p), *i.e.*, the polynomial part which does not involve second type data.

Thus, if $\mathcal{Q}$ is the rational interpolant associated to the rational finite element, we write $\mathcal{Q} = \mathcal{P} + \mathcal{R}$ where $\mathcal{P}$ is a polynomial and $\mathcal{R}$ is an irreducible rational function. By construction, $\mathcal{P}$ takes care of the data at the vertices and eventually along the edges at the Q^i_{rs}. The rational $\mathcal{R}$ is such that $\mathcal{PR} = 0$, $\mathcal{QP} = \mathcal{P}$ and $\mathcal{Q}$ corresponds to a finite element of class C^{k+1}.

Using the Bézout Theorem [13,1] it can be proved that given two polynomials p and q of any degree n in two variables such that $\frac{p}{q}$ is irreducible and $(\frac{\partial^m}{\partial \nu_i^m}[\frac{p}{q}]) \equiv 0$ along the edges of a triangle T, for all $m \leq k$ then p can be divided by Λ^{k+1}, where Λ is the product of the barycentric coordinates related to T.

It can be proved [1,18] that one can write $\mathcal{R}$ as

$$\mathcal{R} = \frac{1}{q}\Lambda^{k+1} \sum_{i=1}^{3} \lambda_{i+1}\lambda_{i-1} \sum_{s=1}^{k+p+1} p_{is},$$

where the p_{is} are linearly independent polynomials of degree $\leq k + p$ in the variable $(1 - \lambda_i) = \lambda_{i+1} + \lambda_{i-1}$, and q is a polynomial such that, if we define the functions r_{is} by

$$r_{is}(M) = \frac{1}{q}\lambda_i^{k+1}\lambda_{i+1}^{k+2}\lambda_{i-1}^{k+2}p_{is}, \qquad i = 1, 2, 3,$$

then r_{is} must satisfy

$$1)\quad \left(\frac{\partial^{k+1} r_{is}}{\partial \nu_i^{k+1}}\right)_{\sigma_i} \text{ is polynomial of degree } N-k-1,$$

$$2)\quad \frac{\partial^{k+1} r_{is}}{\partial \nu_i^{k+1}} = B \times \frac{1}{q}\lambda_{i+1}^{k+2}\lambda_{i-1}^{k+2} p_{is} + \mathcal{O}(\lambda_i).$$

There are various choices for q, but it seems that the simplest ones are

1) for $\chi = k$, $\quad q(M) = (\lambda_1\lambda_2 + \lambda_2\lambda_3 + \lambda_3\lambda_1)^2$

2) for $\chi = k+1$, $\quad q(M) = \lambda_1\lambda_2 + \lambda_2\lambda_3 + \lambda_3\lambda_1$

We conclude with two examples. For a more complete study, see [1].

1) *For* $k = 0, \chi = 1$ *and* $p = 0$, we can built a rational finite element which is C^1 and uses the following 12 degrees of freedom

$$\Sigma_K = \{\partial^\alpha u(a_i), \quad (\partial_{\nu_i} u)(b_i), \quad i = 1,2,3, \quad |\alpha| \leq 1\}.$$

This set is **P**-unisolvent, where $\mathbf{P} = \mathbb{P}_3 \oplus \mathrm{Span}\{r_{i,1}, i = 1,2,3\}$ with

$$r_{i,1} = \frac{\lambda_i \lambda_{i+1}^2 \lambda_{i+2}^2}{\lambda_1\lambda_2 + \lambda_2\lambda_3 + \lambda_3\lambda_1} \qquad i = 1,2,3.$$

2) *For* $k = 1, \chi = 1$ *and* $p = 0$, we obtain a rational element of class C^2 which involves only 27 degrees of freedom, and no derivatives of order higher than 2 (compare to the Argyris family, cf. [13], where a polynomial C^2 element has to be of degree 9 and involve 55 degrees of freedom, including derivatives of order 4 at the vertices). In the rational case, we use

$$\Sigma_K = \{\partial^\alpha u(a_i), |\alpha| \leq 2; \quad (\partial_{\nu_i}^m u)(Q_{i,s}), \quad i = 1,2,3, \; s = 1, \cdots, 2 \;\},$$

where Q_{rs}^i are points on σ_s such that $Q_{rs}^i \neq a_j$, j=1,2,3, and $Q_{rs}^i \neq Q_{ts}^i$ whenever $t \neq r$.

This set is **P**-unisolvent, where $\mathbf{P} = \mathbb{P}_5 \oplus \mathrm{Span}\{r_{i,s}, i = 1,2,3; s = 1,2\}$ where, for $i = 1,2,3$,

$$r_{i,1} = \frac{\lambda_i^2 \lambda_{i+1}^3 \lambda_{i+2}^3 (2\lambda_{i+1} - \lambda_{i+2})}{(\lambda_1\lambda_2 + \lambda_2\lambda_3 + \lambda_3\lambda_1)^2},$$

$$r_{i,2} = \frac{\lambda_i^2 \lambda_{i+1}^3 \lambda_{i+2}^3 (2\lambda_{i+2} - \lambda_{i+1})}{(\lambda_1\lambda_2 + \lambda_2\lambda_3 + \lambda_3\lambda_1)^2}.$$

It is obvious that those rational elements can be used in order to built a nested sequence of spaces. In the case of C^1 or C^2, the two examples provide a basis for the spaces. In the case of $C^k, k \geq 2$, one has to replace the data at the vertices, namely $\partial^\alpha u(a_i), |\alpha| \leq k + \chi$, by those described in Section 2 for the polynomial case, namely

$$\partial^\alpha f(a), |\alpha| \leq k, \qquad D^{i+j} f(a) \cdot (e_s^i, e_{s+1}^j), \; i,j = 1, \ldots, k, \; s = 1, \ldots, n$$

for an interior vertex a, and

$$\begin{aligned} &\partial^\alpha f(a), |\alpha| \le k; \;\; D^{i+j} f(a) \cdot (e_s^i, e_{s+1}^j), \; i,j = 1,\dots,k, \; s = 2,\dots,n-2, \\ &D^{i+j} f(a) \cdot (e_1^i, e_2^j), \quad i = 1,\dots,2k; j = 1,\dots,k, \quad i+j \le 2k, \\ &D^{i+j} f(a) \cdot (e_{n-1}^i, e_n^j), \; i = 1,\dots,k, \; j = 1,\dots,2k; \quad i+j \le 2k. \end{aligned}$$

for a boundary vertex (and the corresponding changes for a defective vertex). In fact those data defined at the vertices do appear in evaluating $\mathcal{P}$, but not in $\mathcal{R}$, and $\mathcal{P}$ is the same in both the polynomial and the rational case (see [14,16] and [1] for practical formulas). Due to a lack of space, we cannot develop these results here.

References

1. Agbeve, N., *Elements finis triangulaires rationnels de classe C^k*, Ph.D Thesis, University of Nantes, Oct.1993.
2. Alfeld, P. and L. L. Schumaker, The dimension of bivariate splines spaces of smoothness r for degree $d \ge 4r+1$, Constr. Approx. **3** (1987), 189–197.
3. Argyris, J. H., I. Fried, and D. W. Scharpf, The TUBA family of plate elements for the matrix displacement method, Aero. J. Royal Aeronautic Society **72** (1968), 701–709.
4. Babuska, I., B. A. Szabo and I. N. Katz, The p version of the finite element method, SIAM J. Numer. Anal., vol **18** (1981), 516–545.
5. Bell, K., A refined triangular plate bending element, Internat. J. Numer. Methods Engrg. **1** (1969), 101–122.
6. Bernadou, M. and J. M. Boisserie, *The Finite Element Method in Thin Shell Theory*, Birkhäuser, Boston, Basel, 1982.
7. Chui, C. K. and M. J. Lai, On Bivariate Super Vertex Splines, Constr. Approx. **6** (1990), 399–419.
8. Ciarlet, P. G., *The Finite Element Method for Elliptic Problems*, North–Holland, Amsterdam,1978.
9. Ciavaldini, J. F. and J. C. Nedellec., Sur l'élément fini de Fraeijs de Veubeke & Sanders, RAIRO Analyse Numérique R2, vol**4**,(1974).
10. Dahmen,W., P. Oswald and X. Q. Shi, C^1-hierarchical bases, J.of Comp. and Appl. Math.**51** (1994),37–56.
11. Dieudonné, J., *Eléments d'Analyse, Tome 1, Fondements de l'analyse moderne*, Gauthier–Villars, Paris, 1968 (English translation: Foundation of Modern Analysis, Academic Press, New York, 1960).
12. Le Méhauté, A., Taylor interpolation of order n at the vertices of a triangle, in *Approximation Theory and Applications*, Z. Ziegler (ed.), Academic Press, New York, 1981, 171–185.

13. Le Méhauté, A., *Interpolation et Approximation par des fonctions polynomiales par morceaux dans* $\mathbb{R}^n$, Thèse d'Etat, Université de Rennes, 1984.
14. Le Méhauté, A., An efficient algorithm for C^k-simplicial finite element interpolation in $\mathbb{R}^n$, CAT Report #111, Texas A&M University, March 1986.
15. Le Méhauté, A., Unisolvent Interpolation in $\mathbb{R}^n$ and the Simplicial Polynomial Finite Element Method, in *Topics in Multivariate Approximation*, C. K. Chui, L. L. Schumaker, and F. Utreras (eds.), Academic Press, New York, 1987, 141–152.
16. Le Méhauté, A., An efficient algorithm for C^k-simplicial finite element interpolation in $\mathbb{R}^d$, in *Multivariate Approximation and Interpolation*, W. Haussmann and K. Jetter (eds.), ISNM **94**, Birkhäuser, 1990, 179–192.
17. Le Méhauté, A., A finite element approach to surfaces reconstruction, in *Computation of Curves and Surfaces*, W. Dahmen, M. Gasca and C. Micchelli (eds.), Kluwer Academic Pub., 1990, 237–274.
18. Le Méhauté, A., and K. N. Agbeve, Rational C^{k+1} Finite elements in $\mathbb{R}^2$, in *Wavelets, Images and Surface Fitting*, Laurent, Le Méhauté, Schumaker, eds, A.K. Peters, Wellesley, 1994, 335–342.
19. Le Méhauté, A., Nested Sequences of Triangular Finite Elements Spaces, in *Multivariate Approximation, Recent Trends and Results*, Haußmann, Jetter, Reimer, eds, Akademie Verlag, 1997, 132–142.
20. Morgan, J., and R. Scott, A nodal basis for C^1 piecewise polynomials of degree ≥ 5, Math.Comp. **29** (1975), 736–740.
21. Oswald, P., Hierarchical conforming finite element methods for the biharmonic equation, SIAM J. Numer. Anal. **29** (1992), 1610–1625.
22. Sablonnière, P., Composite finite elements of class C^2, in *Topics in Multivariate Approximation*, Chui, Schumaker, Utreras, eds., Academic Press, 1987, 207–217.
23. Schumaker, L. L., Numerical aspects of spaces of piecewise polynomials on triangulation, CAT Report #91,Texas A&M University, 1985.
24. Yserentant, H., On the multilevel splitting of finite element spaces, Numer. Math. **49** (1986), 379–412.
25. Ženíšek, A., A general theorem on triangular C^n elements, RAIRO Anal. Numer. R2 **4** (1974), 119–127.

Alain Le Méhauté
Département de Mathématiques
Université de Nantes
2 rue de la Houssinière, BP 92208,
44322 Nantes Cedex 3, FRANCE
`alm@math.univ-nantes.fr`

Computing Natural Division Lines on Free-Form Surfaces Based on Measured Data

Gábor Lukács and László Andor

Abstract. This paper suggests a technique for reverse engineering applications by means of which certain extremal points on a surface can be identified. Based on triangulated point data with normal vector and curvature estimations at these points, a descriptive method is presented which makes it possible to recognize points of ridges or sub-parabolic lines on a surface.

§1. Introduction

An important problem in reverse engineering is segmentation [11]. Assume that we have a large point cloud which has been sampled from a complex free-form shape. We would like to partition or segment the data points into subsets, to each of which localized surface fitting can be applied. A good segmentation should reflect the underlying structure of the object in a natural manner. Ideally, an algorithm should be designed which produces a natural patchwork automatically for a given surface just on the basis of the geometric data. However, it has turned out that this is an unrealistic aim. The reason for this is, that no matter how accurate our measurement is, a priori knowledge is required which is usually based on the *functionality* of the surface elements. Thus, after some analysis, one usually ends up at an interactive system by which one can define curves and fit surface patches locally approximating a subset of points in the point cloud. One method of curve definition is to select points on the surface which approximate a network of special dividing lines, called ridges. A ridge line consists of those points where one of the principal curvatures takes an extremal value on the corresponding line of principal curvature.

A simple method will be shown for establishing regions which contain ridge lines. This method is based on a triangulated data set, and estimates

Mathematical Methods for Curves and Surfaces II
Morten Dæhlen, Tom Lyche, and Larry L. Schumaker (eds.), pp. 319–326.

ISBN 0-8265-1315-8.

of normal vectors and principal curvature values. In addition, these regions containing ridges are tightly associated with canal surfaces. Based on this observation, it will be shown that our method can be utilised to find automatically those regions which make up special variable radius blends or—in particular—general translational or rotational surfaces.

The outline of the paper is the following. After having defined the ridges and sub-parabolic lines, we show their usage in segmentation of traditional geometric surfaces. Subsequently, an algorithm is shown for establishing those points on a triangulation which belong to a ridge or to a sub-parabolic line. Finally, an approach is proposed which helps to overcome problems which arise if principal curvatures are near to zero in a region.

§2. Ridges in Vision

In computer vision and cartography there has been a long history of defining so-called ridges. Taking the viewing direction into account as "vertical", these approaches have tried to employ the notion of ridge and course from topographic landscape description, see [2,6]. One of the simplest forms of these definitions is where a point on a topographic relief map is considered a point of a ridge or a course if it has the maximum or minimum slope along its isohypse (the line of equal height, terms of [6]).

The question arises whether these topographic ridges can be used for general surface reconstruction purposes. The answer is negative. Koenderinck and van Doorn showed in [6] that the above condition is only necessary and not sufficient, that is, this definition strictly does *not* correspond to the notion of "ridge and course" in the idealized physics of drainage in the terrestrial landscape. What is even worse it has been shown in [6] that no correct "local" ridge and course definitions exist in this sense.

More significantly, this description depends on a particular "vertical" or "viewing" direction. Note that in surface fitting for CAGD, neither measurement techniques nor object geometry depend on any single direction. Put differently, in a general surface reconstruction context, instead of an interpretation of a given picture, we would like to obtain an interpretation of a spatial point-cloud which has possibly been set up by several measurements. Thus, here another concept of special surface lines is required which depends only on the internal geometry of the surfaces.

§3. Ridges in Differential Geometry

The notion of ridges from classical differential geometry is a natural candidate for the role of the above mentioned special line. Its usage in surface design and analysis has been proposed several years ago (see [5,8,10]), but there were few attempts to compute its approximation in a scattered data context (see e.g. [4,7]).

The ridge is a natural generalization of the vertex of curves for surfaces. The vertices of a (planar) curve are the points of *extrema* of curvature (see

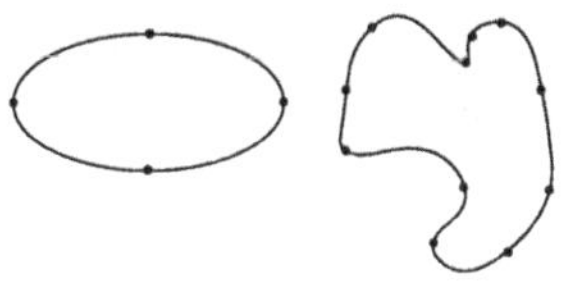

Fig. 1. Vertices of a curve: extrema of curvature.

Figure 1). This is a useful and descriptive notion which helps the partitioning of curves into curvature monotonous pieces.

Let us recall the definition of ridges. Let the surface $\mathbf{p}(\mathbf{u},\mathbf{v})$ be parameterized along its *principal curvature lines.* We shall assume that the first principal curvature line goes along u, the second along v. Let

- $\mathbf{n}$ be the normal vector of the surface,
- κ_1, κ_2 be the u and v principal curvatures (signed with respect to $\mathbf{n}$),
- r_1, r_2 be the u and v principal curvature radii ($r_i = 1/\kappa_i$),
- f^u, f^v denote partial differentiation by u and v.

Definition 1. *The set of points on the surface $\mathbf{p}(\mathbf{u},\mathbf{v})$ where κ_1 is stationary along u, i.e. $\kappa_1^u = 0$, is called the u-ridge of the surface. Similarly, the set of points where $\kappa_2^v = 0$ is called the v-ridge of the surface.*

Definition 2. *The set of points on the surface $\mathbf{p}(\mathbf{u},\mathbf{v})$ where κ_2 is stationary along u, i.e. $\kappa_2^u = 0$, is called the u-sub-parabolic line of the surface. Similarly, the set of points where $\kappa_1^v = 0$ is called the v-sub-parabolic line of the surface.*

Ridges are tightly connected to the evolutes of the surface.

Definition 3. *We shall call the u-evolute (focal surface) of $\mathbf{p}(\mathbf{u},\mathbf{v})$ the surface formed by the principal u-curvature centres. These are the points $\mathbf{p} + \mathbf{r_1}\mathbf{n}$. Similarly, the v-evolute is the collection of the points $\mathbf{p} + \mathbf{r_2}\mathbf{n}$. We shall say that the point $\mathbf{p}(\mathbf{u},\mathbf{v}) + \mathbf{r_1}(\mathbf{u},\mathbf{v})\mathbf{n}(\mathbf{u},\mathbf{v})$ is above $\mathbf{p}(\mathbf{u},\mathbf{v})$.*

Generically, the evolute has a fold-type singularity above a ridge, which can be clearly seen in Figure 2. This motivates the following definition.

Definition 4. *The line above a ridge on the evolute is called the* rib *of the evolute.*

In practice a surface may have several umbilic points or even areas where the two principal curvatures values are the same. For this reason the distinction between the u and v principal curvatures and consequently the consistent labeling of principal curvature lines is a demanding task. Therefore, let us consider the points on a surface as being *either u or v* ridges. This "symmetric" ridge network is particularly useful for partitioning the surface for the following reasons:

- Ridges are natural dividing lines between regions of monotonous principal curvature. (The topological structure of this ridge network can be well understood from [10].)

- Ridges are more robust than geodesics and lines of principal curvature, if we slightly *deform* the surface they will *deform* in contrast with geodesics and lines of curvature which *reform* (Morris, [7]).
- They can be considered as zero width blends. In fact if

$$\kappa_1^u \equiv 0 \quad \text{along a } u \text{ line} \quad \Rightarrow \quad \text{curvature line is a circle,}$$

that is like a canal surface (the envelope of a family of variable radius balls). (For the proof see e.g. [1]). Note that if $\kappa_1^u(u,v) = 0$ for an open (u,v) rectangle, then the evolute is deformed to a curve in this rectangle. In fact the following theorem is true.

Theorem 1. *If one of the two evolutes deforms to a curve then the surface is the envelope of varying radius balls centered on that curve. If both evolutes deform to curves, then the surface is a Dupin cyclide (a surface of which both principal curvature lines are circles).*

The proof can be found in [1] or [12]. One should note that these remarkable facts help us to reduce the complexity of certain segmentation problems provided that reliable curvature estimations can be made. In fact all the rotational and translational surfaces are special canal surfaces. On the other hand, if a surface part lies entirely on *both* ridges, then the surface is a Dupin-cyclide, that is the envelope of spheres touching three fixed spheres in space. In particular all circular cylinders, cones, spheres, planes and tori are Dupin-cyclides. It is well known that the two ribs (spine-curves) are two *conics* (an ellipse and a hyperbola) lying in orthogonal planes (see e.g. [8]).

Thus, if we were able to classify surface points which belong to both (u and v) ridges or just to one (u or v) ridge or none of the ridges, then we could distinguish among three major surface types, respectively:

1) surfaces of traditional engineering type (circular cylinders, circular cones, spheres, tori (being special Dupin-cyclides) and eventually "real" Dupin-cyclides),
2) variable and constant radius blend surfaces (more exactly surface areas of extended, spherical rolling ball blends),
3) surfaces with no inherent rotational and translational invariance.

In the first two types of surfaces the segmentation problem of the *surfaces* can be transformed to segmenting their evolutes or *ribs*. In 1) one should search for conics (in particular straight lines) among the rib points of the surfaces. In this way one may end up at a method similar to that described by Yokoya and Levin [13]. See also [3] on this subject.

The main problem remaining is how to determine if a surface point belongs to a ridge or not. This is the subject of the following section.

§4. Determination of Ridges

Computing of ridge points will be based upon the "focusing" effect of evolutes near the ribs, namely that the surface element of the evolute degenerates to zero at the ribs.

Let E, F, G be the coefficients of the first fundamental form of the surface $\mathbf{p(u,v)}$. Let $\widehat{E}, \widehat{F}, \widehat{G}$ be the coefficients of the first fundamental form of the u-evolute.

Theorem 2. *With the above notations,*

$$\widehat{E} = (r_1^u)^2, \qquad \widehat{F} = r_1^u r_1^v, \qquad \widehat{G} = (r_1^v)^2 + G\left(1 - r_1/r_2\right)^2.$$

The proof can be found e.g. in Weatherburn [12]. In this manner the squared surface element of the evolute is

$$\widehat{E}\widehat{G} - \widehat{F}^2 = (r_1^u)^2 G\left(1 - r_1/r_2\right)^2.$$

Now assume that we have a triangulation of the surface to be reverse engineered. This triangular approximation induces a triangulation on each sheet of evolute. (Note that in order to get reasonable approximations, here one should usually take a *coarser* triangulation for this purpose than the one used for the *curvature* estimations.)

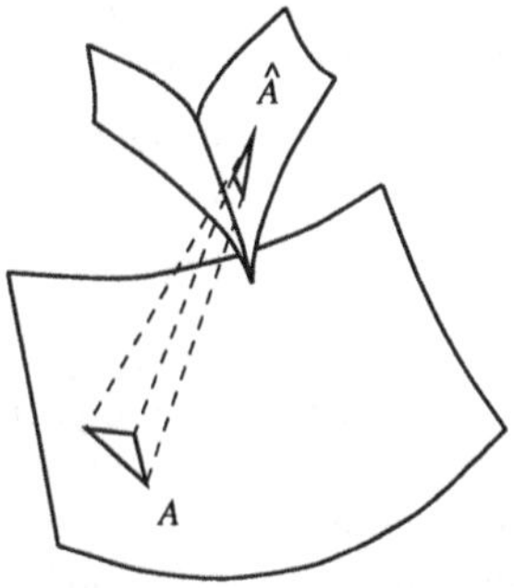

Fig. 2. Ribs and areas.

Suppose that A is the area of a small triangle on $\mathbf{p(u,v)}$ and $\widehat{A}$ is the area of the "lifted" triangle on the u evolute (see Figure 2). Consider the ratio

$$\Phi_1 = \frac{\widehat{A}}{A} \approx \frac{\sqrt{\widehat{E}\widehat{G} - \widehat{F}^2}}{\sqrt{EG - F^2}} = r_1^u \left(1 - \frac{r_1}{r_2}\right) / \sqrt{E}. \tag{1}$$

Assume that none of the principal curvatures is zero. Now Φ_1 is small if and only if

1) the derivative of the curvature by the *arc-length* of the respective curvature line is small ($r_1^u = -\kappa_1^u/\kappa_1^2$) or
2) we are near to an umbilic, i.e. $r_1 \approx r_2$.

In this way by setting some threshold, the Φ_1 ratio can be used for ridge detection. In fact if one uses Φ_1 instead of r_1^u, one includes umbilic points

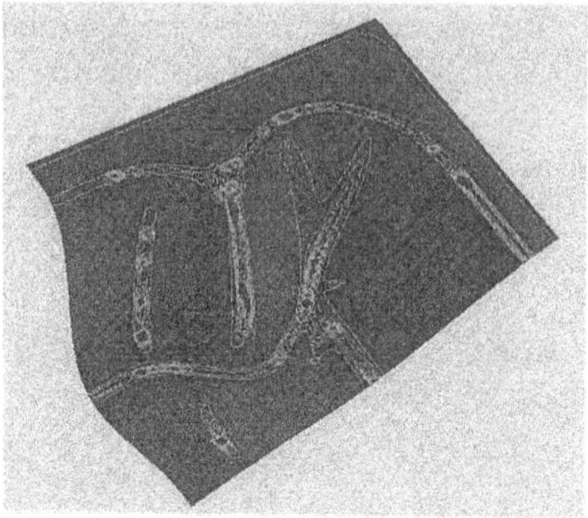

Fig. 3. Ridge levels.

as well, which does not make a difference since ridge lines start from certain umbilics in any case (see [10]). One can see isolines of this function in Figure 3.

Observe that the usage of the approximation (1) does not require the principal directions; the curvature values suffice. The determination of sub-parabolic points is based upon similar ideas.

Theorem 3. *The Gaussian curvature of the u-evolute is*

$$\widehat{K} = -\frac{1}{(r_1 - r_2)^2}\frac{r_2^u}{r_1^u}. \tag{2}$$

The proof can be found in Weatherburn [12] as well. (2) explains the origin of the sub-parabolic term. If we are in a parabolic point on the evolute, where $\widehat{K} = 0$ then $r_2^u = 0$, that is, it corresponds to a sub-parabolic point on the surface. (It is the geodesic inflection of the v curvature line.)

The Gaussian curvature can be approximated by the area of the Gauss map of the evolute divided by the evolute surface area, that is, $\widehat{K} \approx \Psi_1 = \widehat{\Gamma}/\widehat{A}$ where $\widehat{\Gamma}$ is the solid angle subtended by the normals of the evolute. By setting some threshold for Ψ_1, this formula can be used for searching sub-parabolic lines. Since the normals of the u-evolute are the corresponding u principal directions (see e.g. [10,12]) the approximation requires the estimation of the *principal directions* as well.

§5. Parabolic Points

One should notice that by means of the above mentioned methods, one cannot directly handle surfaces containing parabolic points. In this case since evolute points lie "at infinity", quantities like $\widehat{A}$ in (1) cannot be calculated at all. However, this difficulty can be overcome.

Let us compute the inversion of the measured $\mathbf{p}$ points with respect to some fixed sphere of radius ρ and centered in $\mathbf{c}$

$$\mathbf{p} \to \mathbf{p}^* = \mathbf{c} + \frac{\rho^2}{|\mathbf{p} - \mathbf{c}|^2}(\mathbf{p} - \mathbf{c}).$$

$\mathbf{c}$ is called the pole of the inversion. This transformation maps planes or spheres into planes or spheres, straight lines or circles into straight lines or circles, Dupin cyclides into Dupin cyclides. It preserves angles, maps principal curvature lines into principal curvature lines, osculating spheres or planes into osculating spheres or planes, umbilics into umbilics, ridges into ridges. The transformation formulae for normals and curvatures are simple, e.g:

$$\kappa_1^* = -\frac{|\mathbf{p}-\mathbf{c}|^2}{\rho^2}\kappa_1 - \frac{2\langle \mathbf{n}, \mathbf{p}-\mathbf{c}\rangle}{\rho^2}, \tag{3}$$

where $\mathbf{n}$ is the normal in $\mathbf{p}$ (for the proof see e.g. [12]). Thus, ridge lines are invariant with respect to inversion while parabolic (and sub-parabolic) points are not, which makes it possible to transform the problem in a world where evolutes have no infinite points.

In fact (3) shows that in general the principal curvatures obtained after inversion will not be zero. It is easy to see that the following theorem holds.

Theorem 4. *If one inverts a surface with respect to a pole which does not lie on any principal osculating sphere (or plane), then no principal curvature will be zero on the inverted surface.*

This means that in general, using some a priori knowledge on the point-cloud, one can find a suitable pole. One can invert all the points and all the estimated curvatures, compute Φ_1^* and Φ_2^* for the inverted surface, and find the ridges.

§4. Summary and Further Research

A signaling function has been proposed for areas where ridge or sub-parabolic lines may lie based on a triangular approximation of the surface and normal vector and principal curvature value estimates. This function can be used for separating traditional geometry from blends, rotational and translational surfaces and "real" sculptured surface parts. Further research can be initiated for constructing these special curves where the candidate points lie. New segmentation methods can be developed which are based on the classification of degenerate evolutes.

Acknowledgments. The authors are particularly grateful to Tamás Várady, Computer and Automation Research Institute, Budapest, who took the initiative in the research in this field and suggested many fruitful ideas. Special thanks to Ralph R. Martin at the University of Wales, College of Cardiff and to the whole team at the Cardiff Dept. of Computer Science who promoted this work while the first author was a visiting researcher in Cardiff for some months. The research was partially supported by a Copernicus grant No. 1068 of European Union, and the Hungarian National Science Research Foundation (OTKA, No. 16420).

References

1. Blaschke, W., *Vorlesungen über Differentialgeometrie I*, Springer-Verlag, Berlin, 1930.
2. Haralick, R. M., Ridges and valleys on digital images, Computer Vision, Graphics and Image Processing, **22** (1983), 28–38.
3. Jelencsics, M., Segmentation of measured point-clouds based on the distribution of centres of principal curvature, Computer and Automation Research Institute, Hung. Acad. of Sci., Budapest, GML Studies 1997/6.
4. Kent, J. T., K. V. Mardia, and J. M. West, Ridge curves and shape analysis, in *7th British Machine Vision Conference* 1996, R. B. Fisher and E. Trucco (eds.), 43–52.
5. Koenderinck, J. J., *Solid Shape*, The MIT Press, Cambridge, 1990.
6. Koenderinck, J. J. and A. J. van Doorn, Local features of smooth shapes: Ridges and courses, SPIE, **2031** (1993), Geometric Methods in Computer Vision II, 2–13.
7. Morris, R., The sub-parabolic lines of a surface, in *The Mathematics of Surfaces VI*, G. Mullineux, (ed.), Clarendon Press, Oxford, 1996, 79–102.
8. Nutbourne, A. N. and R. R. Martin, *Differential Geometry Applied to Curve and Surface Design Vol. 1*, Ellis Horwood Ltd, Chichester, 1988.
9. Porteous, I. R., Ridges and umbilics of surfaces, in *The Mathematics of Surfaces II*, R. R. Martin, (ed.), Clarendon Press, Oxford, 1987, 447–458.
10. Porteous, I. R., *Geometric Differentiation for the Intelligence of Curves and Surfaces*, Cambridge University Press, 1994.
11. Várady, T., R. R. Martin, and J. Cox. Reverse engineering of geometric models—an introduction, Computer-Aided Design **29** (1997), 255–268.
12. Weatherburn, C. E., *Differential Geometry of Three Dimensions*, volume I, Cambridge University Press, Cambridge, 1955, First Edition, 1927.
13. Yokoya, N. and M. D. Levine, Volumetric description of solids of revolution in a range image, in *10th International Conference on Pattern Recognition*, vol. 1, IEEE Computer Society Press, 1990, 303–308.

Gábor Lukács
Computer and Automation Research Institute
Hungarian Academy of Sciences, Budapest
POB 63, Budapest, H-1518, HUNGARY
`lukacs@sztaki.hu`

László Andor
CADMUS Ltd., Budapest
Kende utca 13-17, Budapest, H-1111, HUNGARY
`andor@geo.sztaki.hu`

Evaluation and Subdivision Algorithms for Totally Positive Systems

E. Mainar and J. M. Peña

Abstract. It is well known that normalized totally positive (NTP) bases provide shape preserving representations for designing curves using control polygons. The Bernstein basis is the NTP basis with optimal shape preserving properties. For a general space with a NTP basis, there exists also a basis with optimal shape preserving properties which has been called a B-basis. Here we use the concept of B-basis in order to generalize the de Casteljau algorithm. These generalized algorithms are evaluation algorithms, satisfy also a subdivision property, and inherit many good properties of the de Casteljau algorithm.

§1. Introduction and Motivation

Given a system $\boldsymbol{u_n} = (u_0, \dots, u_n)$ of functions defined on $[a, b]$ and $\boldsymbol{P_n} = (P_0, \dots, P_n)^T \in (\mathbb{R}^k)^{n+1}$, a curve $\gamma(t)$ may be defined by $\gamma(t) = \boldsymbol{u_n}(t)\boldsymbol{P_n} = \sum_{i=0}^n u_i(t) P_i$. The points $P_0, \dots, P_n$ are called *control points* and the polygon $P_0 \cdots P_n$ with vertices $P_0, \dots, P_n$ is called the *control polygon* of γ. In Computer Aided Geometric Design the functions $u_0, \dots, u_n$ are usually nonnegative and $\sum_{i=0}^n u_i(t) = 1\ \forall t \in [a, b]$, and in this case we say that $(u_0, \dots, u_n)$ is a *blending system*. Geometric properties of the systems of functions can be derived from their collocation matrices.

Definitions 1.1. *Given a system $(u_0, \dots, u_n)$ of functions defined on $[a, b]$ and $t_0 < t_1 < \cdots < t_m$ in $[a, b]$, the* collocation matrix *of $u_0, \dots, u_n$ at $t_0 < \cdots < t_m$ is the matrix*

$$M\begin{pmatrix} u_0, \dots, u_n \\ t_0, \dots, t_m \end{pmatrix} := (u_j(t_i))_{i=0,\dots,m; j=0,\dots,n}, \quad a \le t_0 < t_1 < \cdots < t_m \le b.$$

If all the minors of a matrix A are nonnegative, then A is called totally positive (TP). *A system of functions $(u_0, \dots, u_n)$ is called TP if all its collocation*

Mathematical Methods for Curves and Surfaces II
Morten Dæhlen, Tom Lyche, and Larry L. Schumaker (eds.), pp. 327–334.

ISBN 0-8265-1315-8.

matrices are TP. A TP and blending system of functions is called normalized totally positive *(NTP).*

Let us observe that the collocation matrix of an NTP system is TP and stochastic (each row sums up to 1). It is well-known (cf. [5,2]) that NTP systems lead to shape preserving representations in the sense that many shape properties of the control polygon are inherited by the corresponding curve.

TP bases with optimal shape preserving properties among all TP bases of the space have been called B-bases: the curve γ better imitates the shape of the control polygon with respect to a B-basis than the shape of the control polygon with respect to any other TP basis (see [2,3,7]).

Definition 1.3. *A totally positive basis $\boldsymbol{b}$ of a vector space of functions $\mathcal{U}$ is said to be a* B-basis *if for any totally positive basis $\boldsymbol{u}$ of $\mathcal{U}$ there exists a totally positive matrix A such that $\boldsymbol{u} = \boldsymbol{b}A$.*

In [3] the concept of B-basis was defined in terms of two sequences of vector subspaces of $\mathcal{U}(I)$:

$$\mathcal{U} = \mathcal{L}^0(\mathcal{U}) \supset \mathcal{L}(\mathcal{U}) \supset \cdots \supset \mathcal{L}^n(\mathcal{U}), \quad \mathcal{U} = \mathcal{R}^0(\mathcal{U}) \supset \mathcal{R}(\mathcal{U}) \supset \cdots \supset \mathcal{R}^n(\mathcal{U}) \tag{1.1}$$

(see pp. 641-642 for the definition and construction of these subspaces). Then a TP basis $\boldsymbol{b_n} = (b_0, \ldots, b_n)$ is a B-basis if and only if $b_i \in L^i(\mathcal{U}) \cap R^{n-i}(\mathcal{U})$. By Theorem 3.6 of [3], a space of functions with a TP basis always has a B-basis and by Theorem 4.2 of [3], a space of functions with an NTP basis always has a unique normalized B-basis. Examples of B-bases are the Bernstein basis in the case of the space $\mathbf{P}^k([a,b])$ of polynomials of degree less than or equal to n on an interval $[a,b]$ and the B-spline basis in the case of the corresponding polynomial spline space.

We are going to define algorithms in order to evaluate curves $\gamma(t) = \sum_{i=0}^{n} u_i(t)P_i$ at a point t_0, where $\boldsymbol{u_n} = (u_0, \ldots, u_n)$ is an NTP system defined on I and $t_0 \in I$. Resembling the classical de Casteljau algorithm for the evaluation of Bézier curves (i.e., using the Bernstein basis of polynomials) we define the following triangular schemes:

Algorithm 1.1

for $j := 0$ **to** n
$P_j^n := P_j$
for $i := n-1$ **to** 0 **step** -1
  for $j := 0$ **to** i
  $P_j^i := (1-\alpha_j^i)P_j^{i+1} + \alpha_j^i P_{j+1}^{i+1}$
  for $j := i+1$ **to** n
  $P_j^i := P_j^{i+1}$

Algorithm 1.1′

for $j := 0$ **to** n
$Q_j^n := P_j$
for $i := n-1$ **to** 0 **step** -1
  for $j := 0$ **to** $n-i-1$
  $Q_j^i := Q_j^{i+1}$
  for $j := n-i$ **to** n
  $Q_j^i := (1-\beta_j^i)Q_{j-1}^{i+1} + \beta_j^i Q_j^{i+1}$

where $0 \le \alpha_j^i < 1$ and $0 < \beta_j^i \le 1$, for all $j = 0,\ldots,n$ and $i = 0,\ldots,n$.

Definition 1.4. *We say that an algorithm of the form 1.1 (resp., 1.1′) is a* generalized de Casteljau algorithm *at t_0 if it satisfies*

$$P_0^0 = \gamma(t_0) \quad (\text{resp. } Q_n^n = \gamma(t_0)). \tag{1.2}$$

Let us observe that the algorithms of the form 1.1 and 1.1′ are particular cases of corner cutting algorithms (cf. [5-7,2]). They are formed by convex linear combinations of two consecutive points. Each step of an algorithm of the form 1.1 calculates a new polygon: $\boldsymbol{P_n^i} = (P_0^i \dots P_n^i)$. This can be represented by an upper bidiagonal matrix U_i:

$$U_i = \begin{pmatrix} 1-\alpha_0^i & \alpha_0^i & & & & & \\ & \ddots & \ddots & & & & \\ & & 1-\alpha_i^i & \alpha_i^i & & & \\ & & & 1 & 0 & & \\ & & & & \ddots & \ddots & \\ & & & & & 1 & 0 \\ & & & & & & 1 \end{pmatrix} \tag{1.3}$$

which is nonsingular, stochastic and TP. Analogously, each step of an algorithm of the form 1.1′ calculates a new polygon: $\boldsymbol{Q_n^i} = (Q_0^i \cdots Q_n^i)$ and the corresponding lower bidiagonal matrix L_i is also nonsingular, stochastic and TP. We can represent the algorithms in matrix form:

Algorithm 1.1	**Algorithm 1.1'**
$\boldsymbol{P_n^n} := \boldsymbol{P_n}$	$\boldsymbol{Q_n^n} := \boldsymbol{P_n}$
for $i := n-1$ **to** 0 **step** -1	**for** $i := n-1$ **to** 0 **step** -1
$\quad \boldsymbol{P_n^i} := U_i \boldsymbol{P_n^{i+1}}$	$\quad \boldsymbol{Q_n^i} := L_i \boldsymbol{Q_n^{i+1}}$

The matrix $U := U_0 U_1 \dots U_{n-1}$ (resp., $L := L_0 L_1 \dots L_{n-1}$) is nonsingular, stochastic, TP and upper (resp., lower) triangular. An algorithm of the form 1.1 or 1.1′ calculates a polygon defined by $\boldsymbol{P_n^0} := U\boldsymbol{P_n}$ or $\boldsymbol{Q_n^0} := L\boldsymbol{P_n}$, respectively.

In the case of the de Casteljau algorithm, the polygon $\boldsymbol{P_n^0}$ is the control polygon of the curve with respect to the Bernstein basis on the subinterval $[t_0, 1] \subset [0, 1]$ and the polygon $\boldsymbol{Q_n^0}$ is the control polygon of the curve with respect to the Bernstein basis on the subinterval $[0, t_0] \subset [0, 1]$.

Therefore, the de Casteljau algorithm calculates the control polygon of the curve with respect to the normalized B-basis of the space of polynomials on the subinterval $[0, 1] \cap [t_0, +\infty)$ or with respect the normalized B-basis of the space of polynomials on the subinterval $[0, 1] \cap (-\infty, t_0]$. This property is known as the subdivision property of the de Casteljau algorithm. This motivates the following definition.

Definition 1.5. *We say that an algorithm which manipulates the control polygon of a curve* $\gamma(t) = \sum_{i=0}^n u_i(t)P_i$ *and obtains the control polygon of the curve with respect the normalized B-basis of the space generated by the restrictions of the functions* $\boldsymbol{u_n} = (u_0, \ldots, u_n)$ *to the subinterval* $I \cap [t_0, +\infty)$ *or* $I \cap (-\infty, t_0]$ *has the* subdivision property to the right *or the* subdivision property to the left *respectively.*

A normalized B-basis of a space satisfies by Proposition 3.12 of [3] the *end-point interpolation property*, that is the generated curve interpolates its own control polygon at the end points: $\sum_{i=0}^n P_i u_i(a) = P_0$, $\sum_{i=0}^n P_i u_i(b) = P_n$. Due to this property, an algorithm with the subdivision property evaluates the generated curve γ at the point t_0. The subdivision property of the generalized de Casteljau algorithms will be studied in this paper.

§2. Main Results

The existence of generalized de Casteljau algorithms will be based on the following result.

Proposition 2.1. *Let* $\mathcal{U}(I)$ *be a vector space of functions defined on an interval* I *with an NTP basis, let* $t_0 \in I$, $I_0 := I \cap [t_0, +\infty)$, *and let* $\mathcal{U}(I_0)$ *be the vector space generated by the restrictions of the functions of* $\mathcal{U}(I)$ *to* I_0. *If* $\boldsymbol{u_n}$ *is the normalized B-basis of* $\mathcal{U}(I)$ *and the restrictions of the functions of the system are linearly independent on the subinterval* $I_0 := I \cap [t_0, +\infty)$, *then the unique normalized B-basis* $\boldsymbol{w_n} = (w_0, \cdots, w_n)$ *of the space* $\mathcal{U}(I_0)$ *has the property that the matrix of the change of basis* U *such that* $\boldsymbol{u_n} = \boldsymbol{w_n} U$ *is nonsingular, upper triangular, stochastic and TP.*

Proof: The restrictions of the functions of $\boldsymbol{u_n}$ to I_0 form an NTP system of $\mathcal{U}(I_0)$. Then by the first part of the proof of Theorem 3.6 of [3] there exists a TP basis $\tilde{\boldsymbol{w}}_{\boldsymbol{n}} = (\tilde{w}_0, \cdots, \tilde{w}_n)$ of $\mathcal{U}(I_0)$ given by $(u_0, \cdots, u_n) = (\tilde{w}_0, \cdots, \tilde{w}_n)\tilde{U}$ where $\tilde{U} = (\tilde{u}_{ij})_{0 \le i,j \le n}$ is a nonsingular, TP, upper triangular matrix with unit diagonal and $\tilde{w}_i \in L^i(\mathcal{U}(I_0))$ for all $i \in \{0, \ldots, n\}$. $\tilde{U}$ can be factorized as $\tilde{U} = DU$ with D diagonal matrix with the row sums of $\tilde{U}$ as diagonal entries and $U = (u_{ij})_{0 \le i,j \le n}$ a stochastic and TP matrix. The system defined by $\boldsymbol{w_n} := \tilde{\boldsymbol{w}}_{\boldsymbol{n}} D$ is TP on I_0 and normalized:

$$\sum_{i=0}^n w_i(t) = \sum_{i=0}^n d_i \tilde{w}_i(t) = \sum_{i=0}^n \sum_{j=0}^n \tilde{u}_{ij} \tilde{w}_i(t) = \sum_{j=0}^n (\sum_{i=0}^n \tilde{u}_{ij} \tilde{w}_i(t)) = \sum_{j=0}^n u_j(t) = 1.$$

Clearly $w_i \in L^i(\mathcal{U}(I_0))$ and $\boldsymbol{u_n} = \boldsymbol{w_n} U$. Since $\boldsymbol{u_n}$ is the normalized B-basis of $\mathcal{U}(I)$, $u_i \in R^{n-i}(\mathcal{U}(I))$. By the construction of the subspaces of (1.1) in [3], one can easily deduce that $R^{n-i}(\mathcal{U}(I)) \subset R^{n-i}(\mathcal{U}(I_0))$ and so $u_i \in R^{n-i}(\mathcal{U}(I_0))$. Since U is upper triangular, we can deduce that, for each $i \in \{0, \ldots, n\}$, w_i is a linear combination of the functions u_0, $u_1, \ldots$, u_i. Since each of them belongs to the space $R^{n-i}(\mathcal{U}(I_0))$, we conclude that $w_i \in R^{n-i}(\mathcal{U}(I_0))$. Therefore, $w_i \in L^i(\mathcal{U}(I_0)) \cap R^{n-i}(\mathcal{U}(I_0))$ for all $i \in \{0, \ldots, n\}$ and the system $\boldsymbol{w_n}$ is the normalized B-basis of $\mathcal{U}(I_0)$. □

Remark 2.2. In order to obtain generalized de Casteljau algorithms, it will be necessary to factorize the upper (resp., lower) triangular, stochastic and TP matrix U of the previous proposition as a product of n bidiagonal and stochastic matrices U_i (resp., L_i). This can be done by means of an elimination process as in the proof of Theorem 1 of [6]. This elimination process has been called Neville elimination and studied systematically in a series of recent papers. In [4] Neville elimination was used to prove the uniqueness of these factorizations in terms of bidiagonal matrices and in [1] was studied the stability of this elimination process when applied to TP matrices.

Corollary 2.3. *Let $\mathcal{U}(I)$ be an $(n+1)$-dimensional vector space of functions defined on a real interval I with normalized B-basis $\boldsymbol{u_n}$. Given $t_0 \in I$ such that the restrictions of the basis functions to the interval $I_0 := [t_0, +\infty) \cap I$ are linearly independent, there exists a generalized de Casteljau algorithm at t_0 for the evaluation of any curve $\gamma(t) := \boldsymbol{u_n}(t)\boldsymbol{P_n}$, which has the subdivision property to the right.*

Proof: By Proposition 2.1 the normalized B-basis $\boldsymbol{v_n}$ of $\mathcal{U}(I_0)$ satisfies $\boldsymbol{u_n} = \boldsymbol{v_n}U$, where U is stochastic, TP and upper triangular matrix. By Remark 2.2 we can factorize $U = U_0U_1 \dots U_{n-1}$ with U_i of the form (1.3) for all $i \in \{0, \dots, n-1\}$. For any curve $\gamma(t) := \boldsymbol{u_n}(t)\boldsymbol{P_n}$, we have

$$\gamma(t) = \boldsymbol{u_n}(t)\boldsymbol{P_n} = \boldsymbol{v_n}(t)U\boldsymbol{P_n} = \boldsymbol{v_n}(t)U_0U_1 \dots U_{n-1}\boldsymbol{P_n}.$$

Particularizing at t_0, we derive $\gamma(t_0) = \boldsymbol{v_n}(t_0)U_0U_1 \dots U_{n-1}\boldsymbol{P_n}$. Due to the endpoint interpolation property of $\boldsymbol{v_n}$, which is the B-basis of $\mathcal{U}(I_0)$, we have $\boldsymbol{v_n}(t_0) = (1, 0, \dots, 0) = \mathbf{e}_0$, and so $\gamma(t_0) = \mathbf{e}_0U_0U_1 \dots U_{n-1}\boldsymbol{P_n}$. Now we can define an Algorithm 1.1 for the evaluation of γ at t_0. Furthermore, it has the subdivision property to the right since the final polygon obtained $\boldsymbol{P_n^0} = U_0U_1 \dots U_{n-1}\boldsymbol{P_n} = U\boldsymbol{P_n}$ is the control polygon of γ with respect to the normalized B-basis of $\mathcal{U}(I_0)$. □

A similar result to Proposition 2.1 using the interval $I_0' := (-\infty, t_0] \cap I$ would lead to a lower triangular matrix L, whose factorization in terms of bidiagonal matrices would lead to the following result:

Corollary 2.3′. *If $\mathcal{U}(I)$ and $\boldsymbol{u_n}$ satisfy the hypotheses of Corollary 2.3, for any $t_0 \in I$ such that the restrictions of the basis functions to the interval $I_0' := (-\infty, t_0] \cap I$ are linearly independent there exists a generalized de Casteljau algorithm at t_0 for the evaluation of any curve $\gamma(t) := \boldsymbol{u_n}(t)\boldsymbol{P_n}$, which has the subdivision property to the left.*

The previous results provided generalized de Casteljau algorithms for B-bases. The corresponding algorithms for NTP bases are considered in the following corollary.

Corollary 2.4. *Let $\mathcal{U}(I)$ be an $(n+1)$-dimensional vector space of functions defined on a real interval I with an NTP basis $\boldsymbol{u_n}$. Given $t_0 \in I$ such that the restrictions of the basis functions to the interval $I_0 := [t_0, \infty) \cap I$ are linearly independent, there exists a generalized de Casteljau algorithm at t_0 of the form 1.1 for the evaluation of any curve $\gamma(t) := \boldsymbol{u_n}(t)\boldsymbol{P_n}$. Furthermore, the* *previous algorithm can be followed by an algorithm of the form 1.1′ to give an algorithm with the subdivision property to the right.*

Proof: If $\boldsymbol{u_n}$ is an NTP system of $\mathcal{U}(I_0)$ then $\boldsymbol{u_n} = \boldsymbol{v_n}K$, where K is stochastic and TP matrix and $\boldsymbol{v_n}$ is the normalized B-basis of the space.

We can factorize $K = LU$ with L and U stochastic, TP, lower and upper triangular matrices respectively(cf. Theorem 1 of [6] or [4]). Under these conditions the factorization is unique. Since L is stochastic and lower triangular, it satisfies $\mathbf{e}_0 L = \mathbf{e}_0$ where $\mathbf{e}_0 = (1, \ldots, 0)$ and then $\mathbf{e}_0 LU = \mathbf{e}_0 U$. Given a curve $\gamma(t) := \boldsymbol{u_n}(t)\boldsymbol{P_n}$, we have that $\gamma(t) = \boldsymbol{v_n}(t)K\boldsymbol{P_n} = \boldsymbol{v_n}(t)LU\boldsymbol{P_n}$. By Remark 2.2, we can factorize $U = U_0U_1 \ldots U_{n-1}$ with U_i of the form (1.3) for all $i \in \{0, \ldots, n-1\}$ and particularizing at t_0 we derive

$$\gamma(t_0) = \boldsymbol{v_n}(t_0)LU\boldsymbol{P_n} = \mathbf{e}_0 LU\boldsymbol{P_n} = \mathbf{e}_0 U\boldsymbol{P_n} = \mathbf{e}_0 U_0 U_1 \ldots U_{n-1}\boldsymbol{P_n}.$$

The matrices U_i allow us to construct an algorithm for the evaluation of γ at t_0. We can also factorize $L = L_0L_1 \ldots L_{n-1}$ and define the algorithm

$$\begin{array}{ll} \boldsymbol{P_n^n} := \boldsymbol{P_n} & \boldsymbol{Q_n^n} := \boldsymbol{P_n^0} \\ \textbf{for } i := n-1 \textbf{ to } 0 \textbf{ step } -1 & \textbf{for } i := n-1 \textbf{ to } 0 \textbf{ step } -1 \\ \quad \boldsymbol{P_n^i} := U_i \boldsymbol{P_n^{i+1}} & \quad \boldsymbol{Q_n^i} := L_i \boldsymbol{Q_n^{i+1}} \end{array}$$

which has the subdivision property to the right since the polygon obtained $\boldsymbol{Q_n^0}$ satisfies $\boldsymbol{Q_n^0} = LU\boldsymbol{P_n}$ and we can express the curve as $\gamma(t) = \boldsymbol{v_n}\boldsymbol{Q_n^0}$ for all $t \in I_0$. □

Analogously to Corollary 2.3′ we can also derive

Corollary 2.4′. *If $\mathcal{U}(I)$ and $\boldsymbol{u_n}$ satisfy the hypotheses of Corollary 2.4, for any $t_0 \in I$ such that the restrictions of the basis functions to the interval $I_0' := (-\infty, t_0] \cap I$ are linearly independent there exists a generalized de Casteljau algorithm at t_0 of the form 1.1′ for the evaluation of any curve $\gamma(t) := \boldsymbol{u_n}(t)\boldsymbol{P_n}$. Furthermore, the previous algorithm can be followed by an algorithm of the form 1.1 to give an algorithm with the subdivision property to the left.*

Example 2.5. *Given $k = 2m+1$, let $\boldsymbol{p_k} = (p_0, \ldots, p_k)$ be the system of polynomial functions defined on $[0,1]$ by*

$$p_i(t) = \binom{m+1}{i} t^i (1-t)^{m+1}, \quad p_{i+m+1}(t) = p_{m-i}(1-t), \quad i = 0, \ldots, m.$$

The system is known as the generalized Ball basis *of $\mathbf{P}^k([0,1])$ and it is an NTP basis (cf. [7]). Let us consider $m = 1$ and $t_0 \in I$. Since the restrictions to the subinterval $[t_0, 1]$ are linearly independent, $\boldsymbol{p_3} = \boldsymbol{b_3}K$ where $\boldsymbol{b_3}$ is the Bernstein basis of the polynomials of degree less than or equal to 3 on the interval $[t_0, 1]$ and the nonsingular, stochastic, TP matrix K is*

$$K = \begin{pmatrix} (1-t_0)^2 & 2t_0(1-t_0)^2 & 2t_0^2(1-t_0) & t_0^2 \\ \frac{1}{3}(1-t_0)^2 & \frac{2}{3}(1-t_0)^2 & \frac{4}{3}t_0(1-t_0) & \frac{1}{3}t_0(2+t_0) \\ 0 & 0 & \frac{2}{3}(1-t_0) & \frac{1}{3}(1+2t_0) \\ 0 & 0 & 0 & 1 \end{pmatrix}.$$

We can factorize $K = LU$ as

$$K = \begin{pmatrix} 1 & 0 & 0 & 0 \\ \frac{1}{3} & \frac{2}{3} & 0 & 0 \\ 0 & 0 & 1 & 0 \\ 0 & 0 & 0 & 1 \end{pmatrix} \begin{pmatrix} (1-t_0)^2 & 2t_0(1-t_0)^2 & 2t_0^2(1-t_0) & t_0^2 \\ 0 & (1-t_0)^3 & t_0(1-t_0)(2-t_0) & t_0 \\ 0 & 0 & \frac{2(1-t_0)}{3} & \frac{1+2t_0}{3} \\ 0 & 0 & 0 & 1 \end{pmatrix}.$$

Furthermore, $U = U_0U_1U_2$ with

$$U_0 = \begin{pmatrix} 1-t_0 & t_0 & 0 & 0 \\ 0 & 1 & 0 & 0 \\ 0 & 0 & 1 & 0 \\ 0 & 0 & 0 & 1 \end{pmatrix}, \quad U_1 = \begin{pmatrix} \frac{3(1-t_0)}{3-2t_0} & \frac{t_0}{3-2t_0} & 0 & 0 \\ 0 & \frac{1-t_0}{1+2t_0} & \frac{3t_0}{1+2t_0} & 0 \\ 0 & 0 & 1 & 0 \\ 0 & 0 & 0 & 1 \end{pmatrix},$$

$$U_2 = \begin{pmatrix} \frac{3-2t_0}{3} & \frac{2}{3}t_0 & 0 & 0 \\ 0 & (1+2t_0)(1-t_0)^2 & t_0^2(3-2t_0) & 0 \\ 0 & 0 & \frac{2}{3}(1-t_0) & \frac{1}{3}(1+2t_0) \\ 0 & 0 & 0 & 1 \end{pmatrix}.$$

The algorithm defined by

$$\begin{aligned} &\boldsymbol{P_3^3} := \boldsymbol{P_3} \\ &\textbf{for } i := 2 \textbf{ to } 0 \textbf{ step } -1 \\ &\qquad \boldsymbol{P_n^i} := U_i \boldsymbol{P_n^{i+1}} \end{aligned}$$

is a generalized de Casteljau algorithm at t_0. The algorithm proposed in [8] does not satisfy our definition of generalized de Casteljau algorithm because it is not given by a triangular scheme. If we complement the previous triangular scheme with $\boldsymbol{Q_n} := L\boldsymbol{P_n^0}$, we get an algorithm with the subdivision property to the right, i.e. the polygon obtained $\boldsymbol{Q_n}$ is the control polygon of the curve $\gamma(t) := \sum_{i=0}^n P_i p_i(t)$ with respect to the system $\boldsymbol{b_3}$ and then $\gamma(t) = \sum_{i=0}^n Q_i b_i(t)$ for all $t \in [t_0, 1]$.

Remark 2.6. Let $\mathcal{U}(I)$ be a vector space of functions defined on a real interval I with a TP basis and $\boldsymbol{u_n}$ be the normalized B-basis of $\mathcal{U}(I)$. Let us suppose that $t_0 \in I$ and the system $\boldsymbol{u_n}$ is linearly independent on the interval $I_0 := [t_0, +\infty) \cap I$. By Proposition 2.1, $\boldsymbol{u_n} = \boldsymbol{v_n} U$, where $\boldsymbol{v_n}$ is the normalized B-basis of $\mathcal{U}(I_0)$ and U is a stochastic, TP and upper triangular matrix. By Corollary 2.3 we can define a Generalized de Casteljau algorithm for the evaluation of any curve $\gamma(t) := \sum_{i=0}^n P_i u_i(t)$ at t_0 by means of the factorization $U = U_0U_1 \dots U_{n-1}$. Let us assume now that $[a, b]$ is such that $t_0 \in [a, b] \subset I$, and the system $\boldsymbol{u_n}$ is also linearly independent on $[t_0, b]$. Now we can also use Corollary 2.4 in order to derive a generalized de Casteljau algorithm for the evaluation of the curve. In fact we have $\boldsymbol{u_n} = \boldsymbol{w_n} \tilde{K}$, where $\boldsymbol{w_n}$ is the normalized B-basis of $\mathcal{U}([t_0, b])$ and $\tilde{K}$ a stochastic, TP matrix. $\tilde{K} = \tilde{L}\tilde{U}$ with $\tilde{L}$ ($\tilde{U}$ resp.) a stochastic, TP, lower triangular (upper triangular resp.) matrix. We can also factorize $\tilde{U}$ in terms of upper bidiagonal matrices

$\tilde{U} = \tilde{U}_0\tilde{U}_1 \dots \tilde{U}_{n-1}$ and define a generalized de Casteljau algorithm for the evaluation of the curve γ at t_0. Let us see that the algorithms obtained by applying Corollary 2.3 and Corollary 2.4 coincide. Since $\boldsymbol{u_n}$ is linearly independent on $[t_0, b]$, the same property holds for $\boldsymbol{v_n}$. Similarly to Proposition 2.1 we could prove that $\boldsymbol{v_n} = \boldsymbol{w_n} L$ where L is a stochastic, lower triangular, TP matrix. We have that $\boldsymbol{u_n} = \boldsymbol{v_n} U = \boldsymbol{w_n} LU$. Then $K = LU = \tilde{L}\tilde{U}$ and by the uniqueness of the factorization we conclude that $U = \tilde{U}$ and $L = \tilde{L}$. Taking into account that the factorizations of U and $\tilde{U}$ in terms of bidiagonal matrices are unique (see [4]), we conclude that the resulting algorithms coincide.

Acknowledgements. *Both authors were partially supported by DGICYT PB96-0730 and by the EU project CHRX-CT94-0522.*

References

1. Alonso, P., Gasca, M. and Peña, J. M., Backward error analysis of Neville elimination, Appl. Numer. Math. **23** (1997), 193–204.
2. Carnicer, J. M. and Peña, J. M., Shape preserving representations and optimality of the Bernstein basis, Advances in Computational Mathematics **1** (1993), 173–196.
3. Carnicer, J. M. and Peña, J. M., Totally positive bases for shape preserving curve design and optimality of B-splines, Comput. Aided Geom. Design **11** (1994), 635–656.
4. Gasca, M. and Peña, J. M., A matricial description of Neville elimination, with applications to total positivity, Linear Algebra Appl. **202** (1994), 33–45.
5. Goodman, T. N. T., Shape preserving representations, in *Mathematical methods in Computer Aided Geometric Design*, T. Lyche and L. L. Schumaker (eds.), Academic Press, Boston, 1989, 333–351.
6. Goodman, T. N. T., and Micchelli, C.A., Corner cutting algorithms for the Bezier representation of free form curves, Linear Alg. Appl. **99** (1988), 225–252.
7. Goodman, T. N. T., and Said, H. B., Shape preserving properties of the generalized Ball basis, Comput. Aided Geom. Design **8** (1991), 115–121.
8. Said, H.B., A generalized Ball curve and its recursive algorithm, ACM Transactions on Graphics **8** (1989), 360–371.

E. Mainar and J. M. Peña
Departamento de Matemática Aplicada
Universidad de Zaragoza, 50009 Zaragoza, SPAIN
`esme@posta.unizar.es`, `jmpena@posta.unizar.es`

A Parametric Hybrid Triangular Bézier Patch

Stephen Mann and Matthew Davidchuk

Abstract. We describe a parametric hybrid Bézier patch that, in addition to blending interior control points, blends boundary control points. This boundary blend is necessary to generalize a functional cross-boundary construction that relies on the natural parameterization of the functional setting. When interpolating irregularly scattered data and when increasing the tessellation of the data mesh, the new scheme shows improvement over representative parametric data fitting schemes.

§1. Introduction

A large number of local parametric triangular surface schemes have been developed over the past fifteen years (see [10] for a survey of such schemes). These schemes are local in that changes to part of the data only affect portions of the surface near the changed data. Surprisingly, all of these schemes exhibit similar shape defects. On closer inspection, it is seen that these schemes all have a large number of free parameters that are set using simple heuristics. By manually adjusting these parameters, one can improve the shape of the surfaces [7].

One way to improve automatically the shape of the constructed surfaces is to use variational methods. Several authors have used such schemes to improve the shape of the constructed surfaces, but usually at a high computational cost due to the global nature of the solution (e.g., [11]). Similarly, we can use *local optimization* methods to set the free parameters and improve the shape, although the results are not as good as the global methods [8].

In this paper, we will investigate a local, non-optimization method for improving the construction of the cross-boundary derivatives in a triangular parametric scheme. This is a generalization of the hybrid scattered data fitting scheme of Foley and Opitz [2] (their method is similar to a result independently developed by Goodman and Said [4]). The resulting surfaces show large improvement in shape over other local, parametric, triangular surface fitting techniques. More precisely, given a triangle of data (a set of three vertices with normals), our scheme constructs a hybrid, parametric patch that

Mathematical Methods for Curves and Surfaces II
Morten Dæhlen, Tom Lyche, and Larry L. Schumaker (eds.), pp. 335–342.

ISBN 0-8265-1315-8.

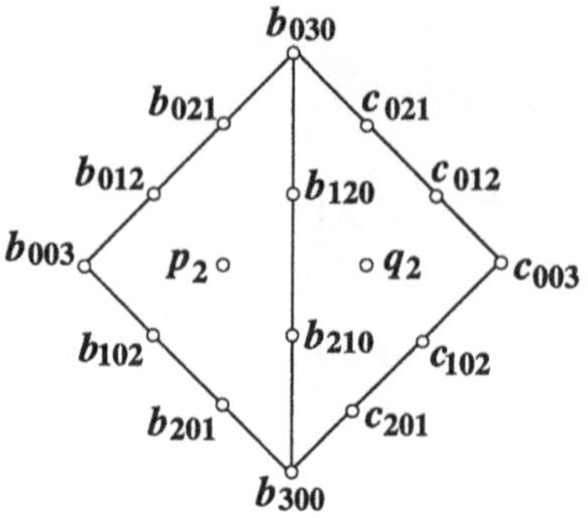

Fig. 1. Domain control net for the Foley-Opitz hybrid Bézier patch.

interpolates the positions and normals at the corners. When used to fill a triangular polyhedron, the resulting surface patches will meet with tangent plane continuity.

§2. The Foley-Opitz Scheme

Foley and Opitz [2] present a method for interpolation of scattered data above the plane using a "hybrid" cubic Bézier patch based on Nielson's scheme [12]. A hybrid cubic patch is similar to a cubic Bézier patch, except the interior control point is a rational blend of three points. With the Foley-Opitz method, the cubic patch boundaries are completely determined by the triangle vertices and normals. The three inner control points are constructed using a C^1 cross boundary construction that gives the hybrid patch cubic precision.

Figure 1 shows the domain control net for two neighboring triangles. $\boldsymbol{p_2}$ is one of the three interior control points associated with the left triangle and $\boldsymbol{q_2}$ is one of the three interior control points associated with the right triangle.

Foley and Opitz compute $\boldsymbol{p_2}$ as follows. Let r, s, and t be the barycentric coordinates of $\boldsymbol{c_{003}}$ with respect to $\boldsymbol{b_{003}}$, $\boldsymbol{b_{030}}$, and $\boldsymbol{b_{003}}$. If both patches if Figure 1 form a single cubic, then from subdividing Bézier cubics it can be shown that

$$\boldsymbol{p_2} = \big(\boldsymbol{c_{102}} + \boldsymbol{c_{012}} - r^2(\boldsymbol{b_{300}} + \boldsymbol{b_{210}}) - 2rs(\boldsymbol{b_{210}} + \boldsymbol{b_{120}}) - 2rt\boldsymbol{b_{201}} - 2st\boldsymbol{b_{021}} \\ - s^2(\boldsymbol{b_{120}} + \boldsymbol{b_{030}}) - t^2(\boldsymbol{b_{102}} + \boldsymbol{b_{012}})\big)/(2(r+s)t).$$

The point $\boldsymbol{q_2}$ is forced by continuity conditions to be

$$\boldsymbol{q_2} = r\boldsymbol{p_2} + s\boldsymbol{b_{120}} + t\boldsymbol{b_{210}}.$$

When applied to data that does not come from a cubic, the Foley-Opitz construction of $\boldsymbol{p_2}$ and $\boldsymbol{q_2}$ ensures that the two triangles have a C^1 join along their common border. Identical calculations would be made to ensure C^1 continuity across the remaining two edges giving three settings for the interior control points of each of the two patches.

Fig. 2. Isophotes of interpolants to the Franke data set.

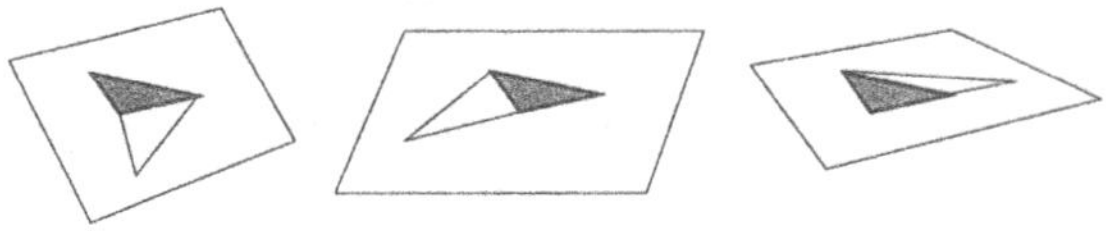

Fig. 3. One plane per patch pair.

The three interior points ($\boldsymbol{p}_0$, $\boldsymbol{p}_1$, $\boldsymbol{p}_2$) are blended with Nielson's rational blend functions,

$$a_i(t_0, t_1, t_2) = \frac{t_j t_k}{t_i t_j + t_i t_k + t_j t_k}, \quad i \neq j \neq k. \tag{1}$$

giving

$$\boldsymbol{b}_{111}(t_0, t_1, t_2) = a_0(t_0, t_1, t_2)\boldsymbol{p}_0 + a_1(t_0, t_1, t_2)\boldsymbol{p}_1 + a_2(t_0, t_1, t_2)\boldsymbol{p}_2.$$

After blending, we are left with a 10 point cubic Bézier patch, which is evaluated at (t_0, t_1, t_2) in the standard way.

Figure 2 shows isophote plots [5] of the Franke data set [3] (on the left), and isophote plots of this data interpolated with Clough-Tocher (middle) and Foley-Opitz (right) patches. The C^1 discontinuities in the isophote lines occur along patch boundaries, and are visible as shape artifacts in shaded images [9]. The Foley-Opitz interpolant is generally smoother than the Clough-Tocher interpolant.

§3. A Hybrid, Parametric, Cubic Scheme

Our goal is to create a parametric version of the Foley-Opitz scheme to bring its good surface quality to the parametric setting. There are two problems we must solve to create this parametric version of the Foley-Opitz method: Finding local parameterizations for the Foley-Opitz cross-boundary method, and finding blend functions for the resulting points that have the appropriate properties.

The Foley-Opitz cross-boundary construction relies on a natural parameterization between patch pairs since the barycentric coordinates of neighboring patches with respect to each other are key in determining the tangent plane fields. In the parametric setting, there is no predefined association between patch domains. Therefore, some association between neighboring patch domains must be made in order to use Foley's tangent plane field construction.

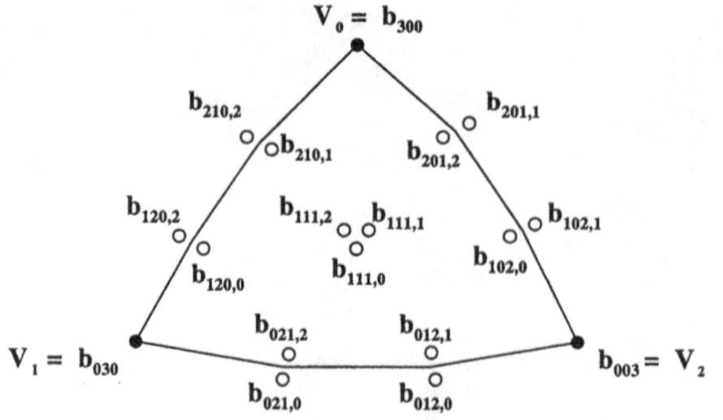

Fig. 4. Domain control net for parametric hybrid patch.

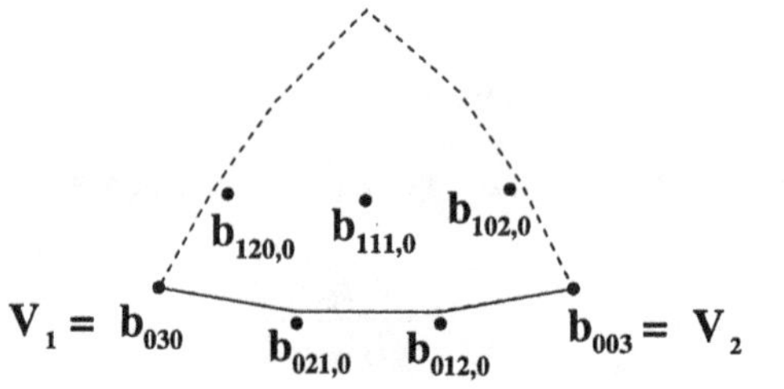

Fig. 5. Tangent plane field control points along the $b_{030}b_{003}$ edge.

Our approach is to choose a plane for each patch pair (Figure 3), project the corner points of both patches onto the plane, and then perform Foley's C^1 construction. Three sets of control points are calculated for each hybrid patch – each set representing a C^1 construction along one triangle edge. The three sets of control points will share the same triangle corner vertices but in general differ in the rest of the boundary and interior control points.

We must blend both boundary and interior control points to produce the final interpolant. Figure 4 shows the domain control net for a parametric version of Foley's scheme. The structure is similar to Foley's – the control points are organized like the control points of a cubic triangular Bézier patch except a group of control points correspond to a single, regular cubic control point.

When constructing the tangent plane field along a particular boundary, only two parametric hybrid patches are involved and consequently only two Bézier patches are needed for the construction. Each Bézier patch contributes seven control points to a parametric hybrid patch, as in Figure 5. These control points determine the tangent plane field along that boundary. To construct the seven points, a plane is chosen as a parameterization for the two Bézier patches allowing the construction as in Foley's functional scheme. Once the plane is chosen, the Bézier patch control points are completely determined by the triangle vertices and the associated normals – Hermite interpolation over a plane completely determines the cubic boundary curves and Foley's

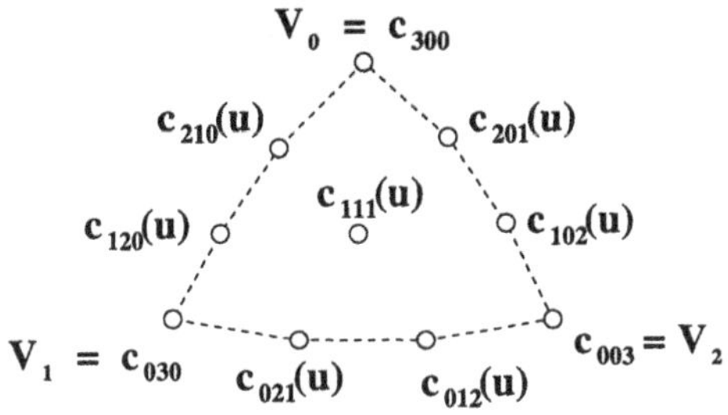

Fig. 6. Control net after blending parametric hybrid control points.

cross boundary construction determines the interior control points.

After the control points from three Bézier patches are calculated, they can be used to formulate the parametric hybrid patch. However, blending is not as straightforward as in the functional case – boundary points are included in the blend. Figure 6 illustrates the control net for the parametric patch. Each point $\boldsymbol{c}_{ijk}$ is a rational blend of the associated control points from the three Bézier patches shown in Figure 4. The points $\boldsymbol{c}_{300}, \boldsymbol{c}_{030}$, and $\boldsymbol{c}_{003}$ are constants, the remaining boundary control points are rational blends of two Bézier patch boundary control points, and the interior point $\boldsymbol{c}_{111}$ is a blend of three points.

The blend formulation must preserve the important properties of the three underlying Bézier patches, boundary curves and cross boundary derivatives, when evaluating along the edges. The corner control points are constants so no blending function is needed. The blending function for the interior control points is the same as used in Foley's functional construction, giving us

$$\boldsymbol{c}_{111}(\boldsymbol{u}) = a_0(\boldsymbol{u})\boldsymbol{b}_{111,0} + a_1(\boldsymbol{u})\boldsymbol{b}_{111,1} + a_2(\boldsymbol{u})\boldsymbol{b}_{111,2},$$

where the a_i are defined in (1). The boundary control points are blended to give two properties:

1) When evaluated along a boundary the parametric Foley control points become the control points of one of the three Bézier patches
2) The tangent plane field of the parametric hybrid patch along a boundary matches the tangent plane field along the same boundary of one of the three Bézier patches.

An asymmetric blend of the following form has both of these properties

$$h_{ij}(u_0, u_1, u_2) = \frac{(1-u_i)u_j{}^2}{(1-u_i)u_j{}^2 + (1-u_j)u_i{}^2}. \tag{2}$$

This blending function is used to weight all the non-vertex boundary Bézier control points:

$$c_{ijk}(u_0, u_1, u_2) = h_{ij}(u_0, u_1, u_2)\boldsymbol{b}_{\boldsymbol{ijk},\boldsymbol{i}} + h_{ji} b_{ijk,k} \boldsymbol{b}_{\boldsymbol{ijk},\boldsymbol{j}},$$

for ijk being any permutation of 012. For example, the V_1V_2 boundary control points would be

$$c_{012}(u_0,u_1,u_2) = h_{01}(u_0,u_1,u_2)b_{012,0} + h_{10}(u_0,u_1,u_2)b_{012,1},$$

$$c_{021}(u_0,u_1,u_2) = h_{02}(u_0,u_1,u_2)b_{021,0} + h_{20}(u_0,u_1,u_2)b_{021,2}.$$

When evaluated along the V_1V_2 edge $u_0 = 0$, we get

$$c_{012}(0,u_1,u_2) = b_{012,0}, \qquad c_{021}(0,u_1,u_2) = b_{021,0},$$

since (2) gives $h_{0j}(0,u_1,u_2) = 1$ and $h_{i0}(0,u_1,u_2) = 0$. The V_1V_2 boundary curve is the cubic Bézier curve given by the control points b_{030}, $b_{021,0}$, $b_{012,0}$, and b_{003}, which gives us C^0 continuity. This blend also gives us C^1 continuity, as described in Davidchuk's thesis [1].

The control points $\{b_{300}, b_{030}, b_{003}, b_{\vec{i},j}\}$ are blended to form the ten control points of a standard cubic Bézier patch, labeled as $c_{\vec{i}}(u)$. The Bézier patch defined by the $c_{\vec{i}}$ is then evaluated at u. The concise definition is

$$F(u) = \sum_{|\vec{i}|=3} B^3_{\vec{i}}(u)c_{\vec{i}}(u).$$

§4. Choice of Plane

There is freedom in the choice of the projection plane, and there are some restrictions. The orientation, not the position, of the plane determines the positions of the control points thus giving two rotational degrees of freedom. The plane should be constructed geometrically from the given information – triangle vertices and normals. The two Bézier patches must not have overlapping domains on the plane so the orientation is restricted to being "underneath" both patches.

One failsafe method of choosing a plane is to take the plane perpendicular to the bisecting plane of the two neighboring triangles and that contains their common edge. With this choice of plane, the projection of the triangles will lie on opposite sides of the projection of their common edge, so the patch domains will never overlap.

However, a better choice is the plane perpendicular to the average of the normals at the two data points, and either date point. Although this construction is not guaranteed to give us a valid plane (i.e., the two projection of the two triangles along the edge may overlap), it does in general give us better shaped surfaces [1].

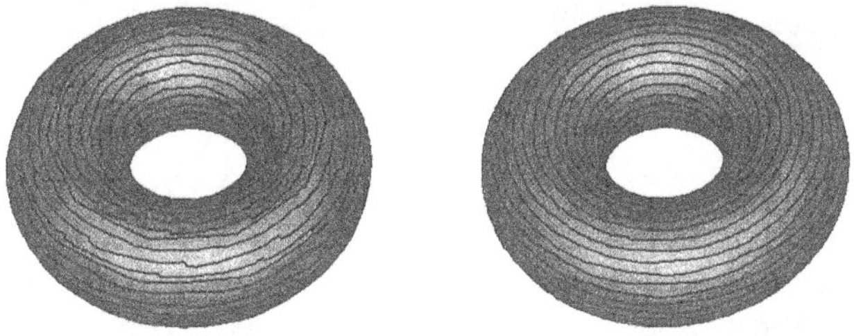

Fig. 7. Triangular Gregory patches and our scheme fit to a torus.

Fig. 8. Triangular Gregory patches and our scheme fit to a cat data set.

§5. Results

To test our surface construction method, we used it to fit patches to a cat data set (where normals are estimated) and to samplings of a torus (where normals come from the torus). We compared our method to Triangular Gregory patches [6]. In Figure 7, we see isophote plots of both methods fit to a 10x10 sampling of the torus. The isophotes for our method are noticeably smoother.

In Figure 8, we see shaded images of both methods fit to a cat data set. This data set has 366 vertices and 698 faces. Normals to the vertices were estimated by a simple averaging of face normals. Many of the shape artifacts that appear on triangular Gregory patches our not present on the surface constructed by our scheme.

One drawback to our scheme is that as the tangent planes on either side of an edge become perpendicular to the edge, the interior points of the boundary curve move towards infinity. Thus, normal estimation becomes a critical step.

Acknowledgments. This research was funded by the Natural Sciences and Engineering Research Council of Canada

References

1. Davidchuk, M., A Parametric Hybrid Triangular Bézier Patch, Master's thesis, University of Waterloo (Waterloo ON Canada), 1997.
2. Foley, T. A. and K. Opitz, Hybrid Cubic Bezier Triangle Patches, in *Mathematical Methods in Computer Aided Geometric Design II,* T. Lyche and L. Schumaker (eds), Academic Press, New York, 1992, 275–286.
3. Franke, R., A critical comparison of some methods for interpolation of scattered data, Report NPS-53-79-003, Naval Postgraduate School, 1979.
4. Goodman, T. N. T. and H. B. Said, A C^1 triangular interpolant suitable for scattered data interpolation, Commun. Appl. Numer. Methods **7** (1991), 479–485.
5. Hagen, H., S. Hahmann, T. Schreiber, Y. Nakajima, B. Wördenweber, and P. Hollemann-Grundetedt, Surface interrogation algorithms, Comp. Graphics and Applics. **12** (1992), 53–60.
6. Longhi, L., Interpolating patches between cubic boundaries, Technical Report T.R. UCB/CSD 87/313, University of California, Berkeley, 1986.
7. Mann, S., Surface Approximation Using Geometric Hermite Patches, dissertation, University of Washington, 1992.
8. Mann, S., Using local optimization in surface fitting, in *Mathematical Methods for Curves and Surfaces,* Morten Dæhlen, Tom Lyche, and Larry L. Schumaker (eds), Vanderbilt University Press, Nashville & London, 1995, 323–332.
9. Mann, S., Cubic precision Clough-Tocher interpolation, submitted for publication.
10. Mann, S., C. Loop, M. Lounsbery, D. Meyers, J. Painter, T. DeRose, and K. Sloan, A survey of parametric scattered data fitting using triangular interpolants, in *Curve and Surface Design,* H. Hagen (ed), SIAM Publications, SIAM, Philadelphia PA, 1992, 145–172.
11. Moreton, H. P. and C. H. Séquin, Functional Optimization for Fair Surface Design, Computer Graphics (ACM SIGGRAPH) **26** (1992), 167–176.
12. Nielson, G. M., A transfinite, visually continuous, triangular interpolant, in *Geometric Modeling: Algorithms and New Trends,* G. E. Farin (ed), SIAM Publications, Philadelphia, 1987, 235–245.

Stephen Mann and Matthew Davidchuk
Computer Science Department
University of Waterloo,
Waterloo, ON, N2L 2R7, CANADA
`smann@cgl.uwaterloo.ca, mdavidch@cgl.uwaterloo.ca`

Comonotone Parametric Hermite Interpolation

C. Manni and M. L. Sampoli

Abstract. We present a general approach to produce local schemes for C^1 Hermite interpolation. The constructed interpolant is a parametric curve which is comonotone with the data. We shall study in detail the case where one component is piecewise quadratic while the other one is cubic. The approximation order is also investigated.

§1. Introduction

In this paper we describe a method for the construction of a C^1 Hermite interpolant which is comonotone with the data.

Let (x_i, f_i, f_i'), $i = 0, \ldots, n$, be given with $x_0 < x_1 < \cdots < x_n$, $f_i = f(x_i)$, and $f_i' = f'(x_i)$. We put

$$h_i = x_{i+1} - x_i, \quad \Delta_i = f_{i+1} - f_i, \quad i = 0, \ldots, n-1.$$

The data are monotone increasing (decreasing) in $[x_i, x_{i+1}]$ if

$$\Delta_i > 0 \ (< 0), \quad f_i', \ f_{i+1}' \geq 0, \ (\leq 0), \quad \text{or } \Delta_i = 0, \quad f_i' = f_{i+1}' = 0. \tag{1}$$

Our goal is to construct a C^1 function, $s(x)$, such that $s(x_i) = f_i$, $s'(x_i) = f_i'$, which is comonotone with the data, i.e. $s'(x) \geq 0$, (≤ 0) in $[x_i, x_{i+1}]$ if the data are monotone increasing (decreasing) in the interval.

Comonotone interpolation, and in general shape preserving interpolation, has been well studied (see for example [1,5,7,10] and references quoted therein). Regarding Hermite interpolation, it is well-known that comonotonicity cannot be achieved either assuming $s(x)$ to be a polynomial with a preassigned degree in each subinterval $[x_i, x_{i+1}]$, or considering $s(x)$ a piecewise polynomial with one additional knot in each subinterval, [3].

To overcome these difficulties, in [7] and [8] the problem is situated in the parametric setting, considering the graph of the function $x \to s(x)$ as the support of a parametric curve $(x(t), y(t))$, where $x(t)$ is strictly increasing. With this approach the required interpolant $s(x(t))$, $x \in [x_i, x_{i+1}]$, is constructed

Mathematical Methods for Curves and Surfaces II
Morten Dæhlen, Tom Lyche, and Larry L. Schumaker (eds.), pp. 343–350.

ISBN 0-8265-1315-8.

as a single-valued parametric curve with cubic [7], or piecewise quadratic [8], components. Even if such schemes produce *visually pleasing* solutions to the given problem, each of them has some drawbacks. The first one requires the solution of a third degree equation in order to compute the interpolant value at a given fixed point $x = \bar{x}$. In addition, a fourth approximation order is guaranteed only in those intervals $[x_i, x_{i+1}]$ that do not contain or approach to a zero of the first derivative of f. On the other hand, in the second scheme only a second order equation has to be solved to compute $s(\bar{x})$, where $\bar{x}$ is fixed. However, this computational simplification is paid for by a reduction in the approximation order that, in this case, is only three.

In this paper, after a general presentation of the basic ideas behind the parametric techniques to solve comonotone interpolation, we shall propose a method constructed with the aim of exploiting the structural simplicity of piecewise quadratics along with the good accuracy of cubics.

More precisely, the scheme presented here is based on piecewise quadratics with one additional knot in each subinterval $[x_i, x_{i+1}]$ for the x–component and on cubics for the y–component. The comonotonicity of the curve is controlled by the magnitude of the tangent vector at the extremes of each subinterval: since the region of monotonicity for each component is known, we propose a simple local algorithm to determine such magnitude by using necessary and sufficient conditions. The method produces visually pleasing interpolants whose shape can be easily controlled, and provides at least a fourth order approximation to a C^4 function, under much more less restrictive hypotheses on the first derivative than in [7].

It is worthwhile to observe that, even if parametric curves are used, as only quadratic expressions are considered for the x-component, the computational cost of our scheme is comparable with that of existent nonparametric methods, especially those based on rationals [6], or on polynomial splines with at least two interior points in each subinterval, [5]. We should notice also that the last two approaches are comparable with the one we are going to describe even regarding their performances. Moreover, while such methods can be hardly generalized to shape preserving interpolation in two dimensions, the parametric setting seems to be a powerful tool also for the bivariate case (see, for instance [2] for an application to comonotone bivariate problems), so that the analysis of parametric schemes in one dimension can be indeed interesting in order to extend the results obtained to the bidimensional case.

The remainder of the paper is divided into three sections. In Section 2 we describe the general parametric interpolation scheme. In Section 3 we detail the quadratic–cubic scheme choosing the free parameters which determine the comonotonicity of the interpolant and investigating its approximation order. Eventually, Section 4 outlines some numerical examples.

§2. The Parametric Setting

For each subinterval $[x_i, x_{i+1}]$, we construct the interpolant by considering the graph of the function $x \to s(x)$ as the support of a planar parametric curve whose components belong to given linear spaces, say V_x and V_y, respectively.

Let $\mathbb{P}_r$ be the linear space of polynomials of degree less than or equal to r, and let V_x, V_y be real linear spaces of functions such that

$$\mathbb{P}_0 \subset V_k \subset C^r[0,1], \; r \geq 1, \; \dim V_k = 4, \quad k = x, y.$$

To ensure the interpolation problem has only one solution, for each pair (d_0, d_1) we assume there exists a unique $v_k(t; d_0, d_1) \in V_k$, $k = x, y$, such that

$$v_k(0; d_0, d_1) = 0, \; v_k(1; d_0, d_1) = 1, \; v_k'(0; d_0, d_1) = d_0, \; v_k'(1; d_0, d_1) = d_1.$$

For each pair $(h_i^{(0)}, h_i^{(1)})$ it is then possible to construct a parametric curve

$$\begin{cases} x = X_i(t; h_i^{(0)}, h_i^{(1)}) \\ y = Y_i(t; h_i^{(0)}, h_i^{(1)}) \end{cases}, \quad i = 0, \ldots, n-1, \quad 0 \leq t \leq 1, \tag{2}$$

such that

$$\begin{gathered} X_i(t; h_i^{(0)}, h_i^{(1)}) \in V_x, \qquad Y_i(t; h_i^{(0)}, h_i^{(1)}) \in V_y, \\ X_i(r; h_i^{(0)}, h_i^{(1)}) = x_{i+r}, \;\; Y_i(r; h_i^{(0)}, h_i^{(1)}) = f_{i+r}, \\ \frac{d}{dt} X_i(t; h_i^{(0)}, h_i^{(1)})|_{t=r} = h_i^{(r)}, \;\; \frac{d}{dt} Y_i(t; h_i^{(0)}, h_i^{(1)})|_{t=r} = h_i^{(r)} f'_{i+r}, \end{gathered} \quad r = 0, 1. \tag{3}$$

The values $h_i^{(0)}, h_i^{(1)}$ determine the magnitude of the tangent vectors at the extremes and they can be chosen to control the monotonicity of each component.

We notice now that if it is possible to find pairs $(h_i^{(0)}, h_i^{(1)})$ such that

$$\frac{d}{dt} X_i(t; h_i^{(0)}, h_i^{(1)}) > 0, \; \Delta_i \frac{d}{dt} Y_i(t; h_i^{(0)}, h_i^{(1)}) \geq 0, \;\; i = 0, \ldots, n-1, \tag{4}$$

then the first component in (2) implicitly defines a function $t = t(x)$, and thus $Y_i(t; h_i^{(0)}, h_i^{(1)})$ can be expressed as a function of x. In this case, putting

$$s(x) = Y_i(t(x); h_i^{(0)}, h_i^{(1)}), \;\; x_i \leq x \leq x_{i+1}, \; i = 0, \ldots, n-1, \tag{5}$$

from (2)–(4) we have that (5) defines a function of class $C^1[x_0, x_n]$ which satisfies the interpolation conditions. Concerning the comonotonicity, since

$$\frac{d}{dx} s(x) = \frac{dY_i}{dt} \frac{dt}{dx}, \quad x_i \leq x \leq x_{i+1},$$

then the monotonicity of $s(x)$ in $[x_i, x_{i+1}]$ is determined by the monotonicity of $Y_i(t; h_i^{(0)}, h_i^{(1)})$ in $[0,1]$, and so it follows from the second inequality of (4).

We can see (cf. [9]) that if $\Delta_i \neq 0$, conditions (4) are fulfilled iff

$$\left(\frac{h_i^{(0)}}{h_i}, \frac{h_i^{(1)}}{h_i} \right) \in \hat{M}(V_x), \quad \left(\frac{f_i' h_i^{(0)}}{\Delta_i}, \frac{f_{i+1}' h_i^{(1)}}{\Delta_i} \right) \in M(V_y), \tag{6}$$

where $M(V_k)$ ($\hat{M}(V_k)$) is the region of (strict) monotonicity for V_k (cf. [3]), i.e.,

$$\begin{aligned} M(V_k) &= \{(d_0,d_1) \in \mathbb{R}^2 \ : \ v_k'(t;d_0,d_1) \geq 0, \ 0 \leq t \leq 1\} \\ \hat{M}(V_k) &= \{(d_0,d_1) \in \mathbb{R}^2 \ : \ v_k'(t;d_0,d_1) > 0, \ 0 \leq t \leq 1\} \end{aligned}, \quad k = x, y.$$

Specifying suitable spaces for V_x and V_y, we can obtain several schemes with different performances both for the approximation order and for the graphical results.

§3. Parametric Quadratic–Cubic Interpolation

In this section we study in detail a *mixed* parametric scheme which couples the simplicity of a quadratic expression for the x–component, with the good approximation order of the cubics for the y–component (see [7,8]).

According to the previous section we consider for the x-component the linear space of C^1 quadratic splines with one additional knot at θ, i.e.

$$V_x = \{q(t) \in C^1 : q(t)|_{[0,\theta]} \in \mathbb{P}_2, q(t)|_{[\theta,1]} \in \mathbb{P}_2, \ \theta = 0.5, t \in [0,1]\},$$

and, regarding the y-component, we consider $V_y = \mathbb{P}_3$. Then (cf. [3]),

$$\hat{M}(V_x) = \{(d_0,d_1) \in \mathbb{R}^2 : 0 < d_0, d_1, \ d_0 + d_1 < 4\}, \tag{7}$$

while $M(\mathbb{P}_3)$ is given, [1, 3], by the convex hull of $\{(0,0)\} \bigcup \mathcal{E}$ where

$$\mathcal{E} = \{(x,y) \in \mathbb{R}^2 : \ x^2 + y^2 + xy - 6x - 6y + 9 \leq 0\}.$$

Combining the two spaces, the functions in (2) become

$$\begin{aligned} X_i(t;h_i^{(0)},h_i^{(1)}) &= x_i + h_i Q_1^{(0)}(t) + h_i^{(0)} Q_0^{(1)}(t) + h_i^{(1)} Q_1^{(1)}(t), \\ Y_i(t;h_i^{(0)},h_i^{(1)}) &= f_i + \Delta_i H_1^{(0)}(t) + h_i^{(0)} f_i' H_0^{(1)}(t) + h_i^{(1)} f_{i+1}' H_1^{(1)}(t), \end{aligned}$$

where $Q_r^{(j)}(t)$ and $H_r^{(j)}(t)$ are, respectively, the cardinal basis for Hermite interpolation in the chosen spaces. More precisely, $Q_r^{(j)} \in V_x$, $H_r^{(j)} \in V_y$ and

$$\begin{aligned} Q_r^{(j)}(s) &= \delta_{rs}\delta_{0j}, \ \frac{d}{dt} Q_r^{(j)}(t)|_{t=s} = \delta_{rs}\delta_{1j}, \\ H_r^{(j)}(s) &= \delta_{rs}\delta_{0j}, \ \frac{d}{dt} H_r^{(j)}(t)|_{t=s} = \delta_{rs}\delta_{1j}, \end{aligned} \quad j,r,s = 0,1.$$

We may observe that if $h_i^{(0)} = h_i^{(1)} = h_i$ we obtain $X_i(t;h_i,h_i) = x_i + h_i t$, thus $Y_i(t;h_i,h_i) = Y_i(t(x);h_i,h_i) = s(x)$, $x_i \leq x \leq x_{i+1}$, is the cubic Hermite polynomial that interpolates the data (x_i, f_i, f_i'), $(x_{i+1}, f_{i+1}, f_{i+1}')$.

According to the previous section we want to determine $h_i^{(0)}, h_i^{(1)} > 0$, $i = 0, \ldots, n-1$, such that (4) are satisfied. From (6) and (7), we have that the first inequality of (4) is fulfilled if and only if

$$0 < h_i^{(0)}, \ h_i^{(1)}, \ \ 0 < h_i^{(0)} + h_i^{(1)} < 4h_i. \tag{8}$$

Concerning the second inequality in (4), in view of (6) we have to look at $M(\mathbb{P}_3)$. Clearly, it is possible to choose an infinite number of pairs $h_i^{(0)}$, $h_i^{(1)} > 0$, such that

$$\left(\frac{h_i^{(0)} f_i'}{f_{i+1} - f_i}, \frac{h_i^{(1)} f_{i+1}'}{f_{i+1} - f_i} \right) \in M(\mathbb{P}_3), \tag{9}$$

if the data satisfy (1) with $\Delta_i \neq 0$. Choosing different parameters determines not only the shape of the interpolant but also its approximation properties. To obtain a satisfactory approximation order we try to choose $h_i^{(j)}, j = 0, 1$, as close as possible to h_i so that the corresponding $s(x)$ is *close* to the piecewise cubic Hermite polynomial interpolating the data.

We now provide a local algorithm to choose $h_i^{(0)}$, $h_i^{(1)}$ in such a way that (8) and (9) hold. Distances and projections are computed according to the uniform norm.

Algorithm 1.

0) *Given* $(f_i,\ f_i',\ i = 0, \ldots, n,)$, $(h_i,\ \tau_i \geq 0,\ i = 0, \ldots, n-1)$.

1) *Put* $h_i^{(0)} = h_i^{(1)} = h_i,\ \tau = \tau_i$.

2) *If* $f_i \neq f_{i+1}$:

   1) *put* $(a, b) = \left(\frac{h_i f_i'}{f_{i+1} - f_i}, \frac{h_i f_{i+1}'}{f_{i+1} - f_i} \right)$;

   2) *if* $(a, b) \notin M(\mathbb{P}_3)$:

      1) *If* $a \geq 1, b \geq 1$ *compute the projection* (α, β) *of* (a, b) *onto* $M(\mathbb{P}_3)$.

      2) *If* $b < 1$ *compute the intersection* (α, β) *between the straight line* $y = b$ *and* $\partial M(\mathbb{P}_3)$.

      3) *If* $a < 1$ *compute the intersection* (α, β) *between the straight line* $x = a$ *and* $\partial M(\mathbb{P}_3)$.

      4) *Compute* $(\bar{\alpha}, \bar{\beta})$ *such that* $\alpha\bar{\beta} - \beta\bar{\alpha} = 0$ *and* $\bar{\alpha} + \bar{\beta} = 3$.

      5) *If* $\|(\bar{\alpha}, \bar{\beta}) - (\alpha, \beta)\| > \tau \|(\alpha, \beta) - (a, b)\|$ *compute* $(\bar{\alpha}, \bar{\beta}) \in M(\mathbb{P}_3)$ *such that* $\alpha\bar{\beta} - \beta\bar{\alpha} = 0$ *and* $\|(\bar{\alpha}, \bar{\beta}) - (\alpha, \beta)\| = \tau \|(\alpha, \beta) - (a, b)\|$.

      6) *If* $f_{i+j}' \neq 0$, *determine* $h_i^{(j)}$ *such that* $(\bar{\alpha}, \bar{\beta}) = \left(\frac{h_i^{(0)} f_i'}{f_{i+1} - f_i}, \frac{h_i^{(1)} f_{i+1}'}{f_{i+1} - f_i} \right)$.

We remark that a simple projection of (a, b) onto $M(\mathbb{P}_3)$ does not respond to our goals. Indeed if (a, b) is too close to the axes, this could give $h_i^{(j)} > h_i$, see Fig. 1, which can cause (8) not being satisfied. On the contrary, by a straightforward application of elementary geometry, applying Algorithm 1, $h_i^{(j)} \leq h_i$, $j = 0, 1$, so that (8) are always satisfied.

Algorithm 1 reduces to that proposed in [7] if $\tau_i = 0$ and gives, in this case, parameters $h_i^{(j)}$ as close as possible to h_i (see also [4]). However, this means that the y-component of the interpolant can have first derivatives at the extremes of each subinterval laying on the boundary of the region of monotonicity, so that the plot can show points with a horizontal tangent. In order to avoid this unpleasant graphical performance we can put $\tau_i > 0$. This forces the first derivatives of Y_i at $t = 0, 1$ to the interior of $M(\mathbb{P}_3)$ (step 2.5) without any reduction of the approximation order, if τ_i is bounded as $h_i \to 0$ (Theorem 2). Then the values τ_i act as tension parameters [8], (Section 4).

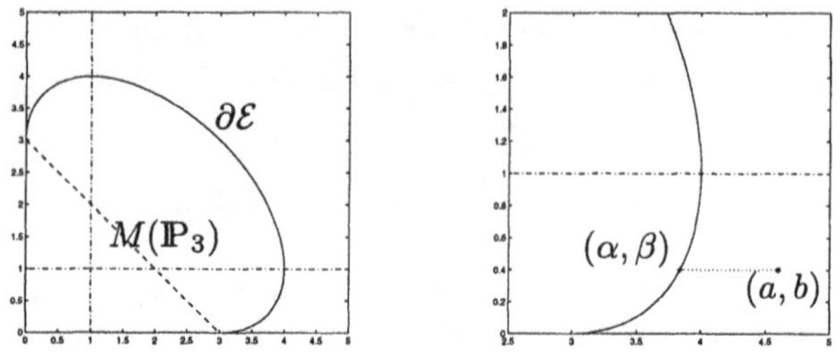

Fig. 1. Sketch of Alg. 1: the region outside $M(\mathbb{P}_3)$ is divided into several parts.

The lower bound $\bar{\alpha} + \bar{\beta} \geq 3$, avoiding too small values for $h_i^{(j)}$, ensures pleasant plots and a good approximation order for C^l functions, $l = 2, 3$, [4]. We note that the method is local, indeed a change on the data at x_i is reflected only in the interval $[x_{i-1}, x_{i+1}]$.

Let us now investigate the approximation order of the constructed interpolant. For the sake of brevity we consider only the increasing case.

Theorem 2. *Let $f \in C^l[x_i, x_{i+1}]$, $l = 2, 3$, be monotone increasing and $h_i^{(0)}$, $h_i^{(1)}$, be obtained from Algorithm 1. Then $s(x)$ given by (5) is comonotone with the data and there exist constants K_l depending on f such that*

$$|s(x) - f(x)| \leq K_l h_i^l, \quad x \in [x_i, x_{i+1}],\ 0 \leq i \leq n-1,\ l = 2, 3.$$

Furthermore, let $f \in C^4[x_0, x_n]$ be monotone increasing in $[x_i, x_{i+1}]$, such that whenever $f'(\xi) = f^{(2)}(\xi) = f^{(3)}(\xi) = f^{(4)}(\xi) = 0$, $\xi \in [x_0, x_n]$, there exist positive constants $\delta_\xi < 1$, C^+, C^-, and $s, r \geq 3$, such that

$$\begin{aligned} C^-(x-\xi)^s \leq f'(x) \leq C^+(x-\xi)^s, &\quad x \in [\xi, \xi + \delta_\xi), \\ C^-(\xi - x)^r \leq f'(x) \leq C^+(\xi - x)^r, &\quad x \in (\xi - \delta_\xi, \xi]. \end{aligned} \tag{10}$$

Then, as $h_i \to 0$, there exists a constant K depending on f and τ_i, such that

$$|s(x) - f(x)| \leq K h_i^4, \quad x \in [x_i, x_{i+1}],\ 0 \leq i \leq n-1.$$

In addition, if $f'(x) > m_i > 0$, $x \in [x_i, x_{i+1}]$, we obtain

$$|s'(x) - f'(x)| \leq K h_i^3, \quad x \in [x_i, x_{i+1}],\ 0 \leq i \leq n-1.$$

The proof of the above theorem can be obtained using arguments similar to those considered in [7, Theorem 6], and [8, Lemma 2] provided that, for the second part, the following crucial lemma has been proved.

Lemma 3. *Let $f \in C^4[x_0, x_n]$, be monotone increasing in $[x_i, x_{i+1}]$ fulfilling the hypotheses of Theorem 2 and $h_i^{(0)}$, $h_i^{(1)}$ obtained from Algorithm 1. Then as $h_i \to 0$ there exists a constant K_1, depending on f and τ_i, such that:*

$$|h_i^{(j)} - h_i| f_i' \leq K_1 h_i^4,\ |h_i^{(j)} - h_i| f_{i+1}' \leq K_1 h_i^4,\ |h_i^{(j)} - h_i| \frac{\Delta_i}{h_i} \leq K_1 h_i^4, \quad j = 0, 1.$$

Proof: For the sake of brevity, we just provide a sketch of the proof which is based on an careful investigation of the position of (a,b) with respect to the region of monotonicity $M(\mathbb{P}_3)$ as $h_i \to 0$. For full details see [9].

If $f'(x) \neq 0$ in $[x_0, x_n]$, as $h_i \to 0$, $(a,b) \in M(\mathbb{P}_3)$. On the other hand, from the hypotheses, f' has at most a finite number of zeroes in $[x_0, x_n]$. Let ξ the zero of f' closest to $[x_i, x_{i+1}]$. If $d(\xi, [x_i, x_{i+1}]) > h_i$, then there exist constants C_1, C_2 such that $C_1 < a/b < C_2$ and the thesis follows from elementary geometry and from [4, Lemma 4.2] and [7, Lemma 7].

If $d(\xi, [x_i, x_{i+1}]) < h_i$, we have to distinguish different cases according to the multiplicity l of ξ as a zero of f'. If $l = 1$, as $h_i \to 0$, $(a,b) \in M(\mathbb{P}_3)$. If $l = 2$, let be $x_i = \xi + \eta_i$, $x_{i+1} = \xi + \nu_i$ and $u = \nu_i/\eta_i$. Let us assume for instance $|u| < 1$. From Taylor expansion, as $h_i \to 0$, (a,b) approaches

$$\left(\frac{3}{u^2+u+1}, \frac{3u^2}{u^2+u+1}\right) \in \partial\mathcal{E}.$$

Then, if $|u| \geq 0.5$, $(a,b) \in M(\mathbb{P}_3)$ or $C_1 \leq a/b \leq C_2$. On the other hand, if $|u| < 0.5$, (a,b) can be arbitrary close to the abscissas outside $M(\mathbb{P}_3)$. In this case, we have $R = a^2 + b^2 + ab - 6a - 6b + 9 > 0$. From the Taylor expansion we can obtain, after some manipulation, that $(\Delta_i/h_i)^2 R = O(\eta_i^4 \nu_i)$. Then, if $\tau_i = 0$, from steps (2.2) and (2.5) of Algorithm 1, $h_i = h_i^{(1)}$ while

$$(h_i - h_i^{(0)}) f_i' = \frac{2\Delta_i R}{\sqrt{12b - 3b^2 + 4R} + \sqrt{12b - 3b^2}} = O(h_i \eta_i^3),$$

and the thesis follows. Analogously we can conclude the result for $\tau_i > 0$. If $l \geq 3$ the claim follows immediately from (10). □

§4. Numerical Examples

In this section we present the performance of the proposed scheme in two classical tests. Other tests can be found in [9]. In the first example we consider the data from the following function, [5],

$$f(x) = \frac{1 + (x+\alpha)^{\frac{1}{3}}}{2 - (x+\alpha)^{\frac{1}{3}}}, \quad \alpha = 10^{-4},$$

at the abscissas

$$x_i: \ -1 \ -0.8 \ -0.6 \ -0.4 \ -0.2 \ -0.04 \ 0.04 \ 0.2 \ 0.4 \ 0.6 \ 0.8 \ 1.$$

Figure 2, left, shows the classical Hermite piecewise cubic interpolant (dotted line) along with the resulting interpolant for $\tau_i = 0$, $i = 0, \ldots, 11$, (dashed line), and the interpolant obtained with $\tau_6 = 0.1$ and $\tau_7 = 0.15$ (solid line). On the right is shown a close up. We may observe clearly how the values τ_i act as tension parameters. Other fittings for this test can be found in [5].

In the second example we consider the following data [10]:

x_i :	0.5	1.5	2	2.5	3	4	6	8	9	10
f_i :	5	10	25	35	30	21	20	20	10	5
f_i' :	5	9.069	23.684	10.0	−57.786	−2.9125	0	0	−6.428	−5

Also in this example we may observe how, considering values for τ_i greater than zero, we obtain better results as the interpolant is visually more pleasing.

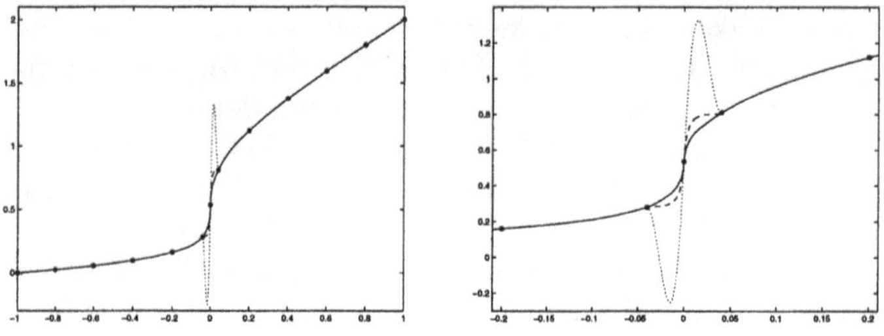

Fig. 2. Ex. 1: '... ' piecewise cubic; '- -' $\tau_i = 0$; '–' $\tau_6 = 0.1,\ \tau_7 = 0.15$.

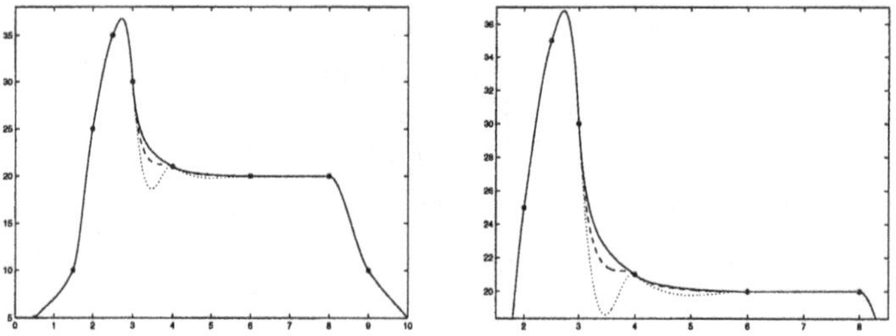

Fig. 3. Ex. 2: '... ' piecewise cubic; '- -' $\tau_i = 0$; '–' $\tau_5 = 0.4$, $\tau_6 = 0.2$.

References

1. Carlson, R. E., and F. N. Fritsch, Monotone piecewise cubic interpolation, SIAM J. Numer. Anal. **17** (1980), 238–246.
2. Costantini, P., and C. Manni, C^1 comonotone parametric interpolation of non–gridded data, J. Comput. Appl. Math. **75** (1996), 147–169.
3. Edelman, A., and C. A. Micchelli, Admissible slopes for monotone and convex interpolation, Numer. Math. **51** (1987), 441–458.
4. Eisenstat, S. C., K. R. Jackson, and J. W. Lewis, The order of monotone piecewise cubic interpolation, SIAM J. Numer. Anal. **22** (1985), 1220–1237.
5. Gasparo, M. G., and R. Morandi, Piecewise cubic monotone interpolation with assigned slopes, Computing **46** (1991), 355–365.
6. Gregory, J. A., and R. Delbourgo, Piecewise rational quadratic interpolation to monotonic data, IMA J. Numer. Anal. **2** (1982), 123–130.
7. Manni, C., C^1 comonotone Hermite interpolation via parametric cubics, J. Comput. Appl. Math. **69** (1996), 143–157.
8. Manni, C., Parametric shape preserving Hermite interpolation by piecewise quadratics, in *Advanced Topics in Multivariate Approximation*, F. Fontanella, K. Jetter and P. J. Laurent (eds.), Singapore, 1996, 211–226.
9. Manni, C., and M. L. Sampoli, Accurate comonotone parametric Hermite interpolation, Technical Report **2/97**, University of Florence (1997).
10. Pruess, S., Shape preserving C^2 cubic spline interpolation, IMA J. Numer. Anal. **13** (1993), 493-507.

"Discrete" G^1 Assembly of Patches Over Irregular Meshes

T. Matskewich, O. Volpin and M. Bercovier

Abstract. A new method for building smooth free form surfaces is presented. It is based on quadrilateral patches and a plate energy type. Smoothness is achieved by approximation of the G^1 continuity conditions. Deviation from the actual G^1 conditions is analyzed. The results are demonstrated on several large unstructured meshes using Bézier patches.

§1. Introduction

The present work deals with G^1 smoothness problems arising in construction of free form surfaces over arbitrary quadrilateral meshes (by approximation or interpolation of the vertices). In order to achieve global smoothness, G^1 conditions must be satisfied at the vertices and common edges. Two problems arise: how to linearize the G^1 condition at vertices, and how to get a G^1 condition along a common edge. A new solution is given for the first question based on the regularity of mappings. It is compared to a solution based on a priori computation of a normal direction at the vertex. The second problem leads either to higher degree patches or subdivision. A discrete approach is introduced by imposing C^1 conditions at the edges control points. The resulting surface is not G^1 in general, hence the terminology "discrete G^1 construction". However, an analytical analysis shows that the deviation from G^1 will be small. Examples are given to illustrate this.

§2. Problem Definition

Given an arbitrary mesh (see Figure 1a for a valid input mesh example) consisting of convex quadrilateral elements, the goal is to build a smooth surface interpolating or approximating this mesh. Consider a parametric space or more exactly a "logical" space $\mathbf{Q}$ made of a 3D mesh of quadrilateral convex patches defined by

$$\mathbf{Q} = \cup_{k=1}^{N} Q_k \,. \qquad (1)$$

Mathematical Methods for Curves and Surfaces II
Morten Dæhlen, Tom Lyche, and Larry L. Schumaker (eds.), pp. 351–358.

ISBN 0-8265-1315-8.

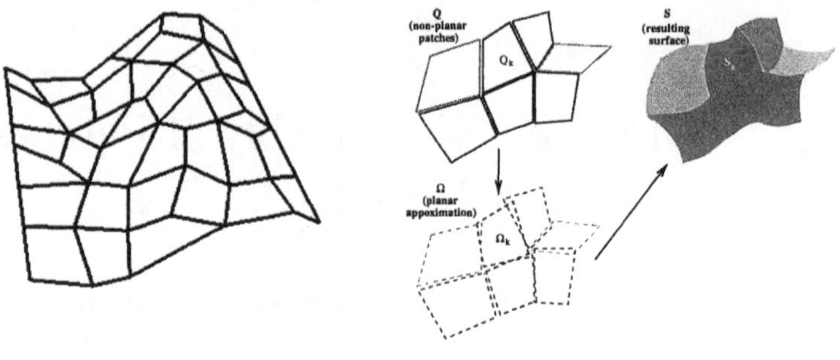

Fig. 1a. A valid input mesh example, **1b.** Construction of the parametric space.

Figure 1b suggests a construction of the parametric space. Composite free form surfaces which interpolate or approximate a given mesh are constructed by assembling tensor product type patches (such as non-uniform B-splines or Bézier patches). Details of such a construction are given in [1].

The present approach is based on the concept of Thin Plate Energy, i.e. functionals that are not directly derived from the patch standard parameterization, and depend on the definition of a local reference frame for each element. Hence each mesh element Q_k is initially approximated by a planar quadrilateral element Ω_k. The reference parametric space

$$\boldsymbol{\Omega} = \cup_{k=1}^{N} \Omega_k \tag{2}$$

for the mesh is then defined as the union of the planar quadrilaterals, each of which is an image of the reference domain $\Omega' = [0,1] \times [0,1]$ under a bilinear transformation. For each Ω_k, a local coordinate system is introduced. The displacement in the normal direction is defined as in the classical FEM ([4]. This displacement defines the energy functional at the element level. The resulting surface minimizes the global energy functional, which is built as the sum of the local functionals. In the global coordinate system, patches of the resulting surface can be written in the form

$$S_k = \sum_{i=0}^{n} \sum_{j=0}^{m} P_{i,j}^k \varphi_{i,j}(u,v), \quad (u,v) \in [0,1] \times [0,1], \tag{3}$$

where $P_{i,j}^k$ are nodal displacements in the global coordinate system, corresponding to the patch k and $\varphi_{i,j}$ - shape functions.

Additional geometrical conditions must be imposed in order to obtain a smooth surface, as will be discussed in section 4.

§3. The Local Coordinate System and the Energy Functional

Now consider one element Q_k of the mesh. The first step is to approximate this non-planar element by a planar quadrilateral Ω_k as close as possible to the original one, and such that the normal to the constructed quadrilateral is roughly a "normal" to the initial element. The plate energy will be defined relative to this new plane. A local coordinate system is defined so that the Z-direction coincides with the normal direction of the planar element Ω_k. By analogy with the thin plate approximation of shells for surface construction, one first defines a local functional at the element level (E_k) and the global energy functional for the whole mesh is taken as the sum of the local functionals over all mesh elements.

The local patch energy is constructed in two steps. The functional formulation is given on the plane quadrilateral related to the underlying mesh. The unknown displacement field is decomposed into a bending displacement $\bar{z}(\hat{x},\hat{y})$ (normal to the local frame defined by Ω_k) and an in-plane displacement $\hat{x},\hat{y}$. (Construction of the local energy functional is described in detail in [2].) For example, the following functional (corresponding to the thin plate energy) can be considered:

$$E_k = \int\int_{\Omega_k} \left(\frac{\partial^2 \bar{z}}{\partial \hat{x}^2} + \frac{\partial^2 \bar{z}}{\partial \hat{y}^2}\right)^2 - 2(1-\nu)\left(\frac{\partial^2 \bar{z}}{\partial \hat{x}^2}\frac{\partial^2 \bar{z}}{\partial \hat{y}^2} - \left(\frac{\partial^2 \bar{z}}{\partial \hat{x}\partial \hat{y}}\right)^2\right) d\hat{x}d\hat{y}, \tag{4}$$

where ν is the Poisson's coefficient, $0 < \nu < 0.5$. It is important to note that the quadratic functional is independent of any underlying parameterization.

At this point $\bar{z}(\hat{x},\hat{y})$ over each patch is defined in the normal direction of its local plane element only. Since normal directions differ for each element, the final surface does not satisfy C^0 continuity along the edge curves. In order to achieve C^0 continuity a local reparametrization in the Ω_k plane is introduced. The initial energy functional (equation (4)) is modified so as to minimize the deviation from this bilinear description. (For details see [2].)

§4. "Discrete" G^1 Continuity Conditions

4.1 Smoothness Condition at n-nodes Vertex

A necessary and sufficient conditions to obtain G^1 smoothness at a common n-curve vertex is that the tangents to the boundaries of all n patches sharing that vertex lie in the same plane.

An implementation of this condition for the interpolation problem proposed in [3] is to fix the normal to the tangent plane in order to linearize the G^1 smoothness condition. Hence, the normal N_{mean} can be chosen as a mean value of the normals to the triangles, formed by the edges sharing the vertex, or as a weighted sum of such normals, where the weight is proportional to the area of the respective triangle.

Then linearized smoothness constraints of the form

$$(\xi_k - r, N_{mean}) = 0 \tag{5}$$

for $k = 1, \ldots, n$ are imposed. Here r is a common vertex of n mesh elements, and $r_1, \ldots, r_n$ are the end vertices of the edges, starting at vertex r, ξ_k is the

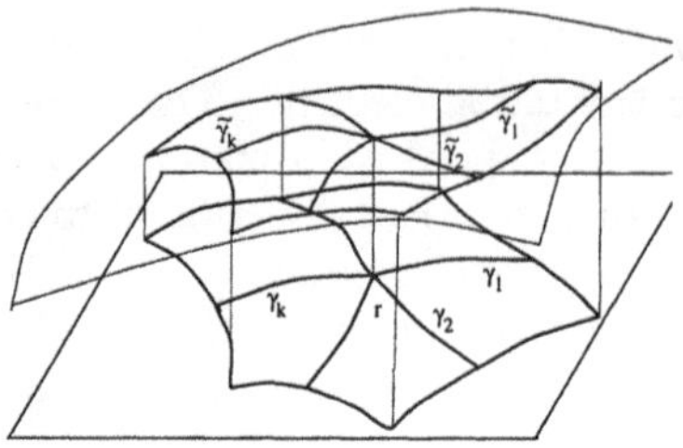

Fig. 2. Condition of C^1 concatenation at the common vertex in the case of a planar initial element.

Bézier control point on the edge k nearest to the vertex r, and (,) denotes scalar product of vectors. This approach gives reasonably good results for interpolation, but may not be suitable for other problems. For example, when a load is applied, fixing the normals of the initial surface (instead of the yet unknown final one) will lead to an incorrect solution. Such problems can be avoided by introducing smoothness conditions at the vertex without imposing the normal direction. This can be done directly as explained below.

Given two smooth manifolds S and $\tilde{S}$ and a C^1 diffeomorphism $\Phi : S \longrightarrow \tilde{S}$ between them, let $r \in S$ be a vertex, let $\gamma_1, \dots, \gamma_n$ be smooth curves passing through that vertex r on the manifold S, and let $\tau_1, \dots, \tau_n$ be the tangents to these curves at the vertex r. If $\tilde{\gamma}_1, \dots, \tilde{\gamma}_n$ are images of the curves $\gamma_1, \dots, \gamma_n$ under the diffeomorphism Φ and $\tilde{\tau}_1, \dots, \tilde{\tau}_n$ are the respective tangents to these curves, then the following result is true.

Lemma. *Under a C^1 diffeomorphism, if $\tau_k = \alpha_k\tau_1 + \beta_k\tau_2$ for some coefficients α_k and β_k ($k = 3, \dots, n$), then $\tilde{\tau}_k = \alpha_k\tilde{\tau}_1 + \beta_k\tilde{\tau}_2$ for the same coefficients α_k and β_k.*

In other words, a C^1 diffeomorphism preserves the relations between tangents to the curves passing through a common vertex. Moreover, if $\gamma_1, \dots, \gamma_n$ initially lie in a common plane, then C^1 continuity at the common vertex implies that (see Figure 2)

$$\tilde{\gamma}_k' = \frac{S_{1k}}{S_{12}}\tilde{\gamma}_2' - \frac{S_{2k}}{S_{12}}\tilde{\gamma}_1', \tag{6}$$

for $k = 3, \dots, n$, where γ_k' or $\tilde{\gamma}_k'$ denote tangent vectors to the correspondent curves at the common vertex and S_{ij} is the area of the triangle with the sides γ_i' and γ_j'. Exactly the same relations hold for the vectors γ_k, γ_1 and γ_2 in the plane. This means that for an initial planar mesh, equation (6) is not only a necessary, but also a sufficient condition for C^1 continuity of the "deformed" surface at the vertex. By analogy, in the present approach one fixes the relations between tangents at the vertex instead of fixing the direction of the normal, which is then determined according to the minimization of the energy functional.

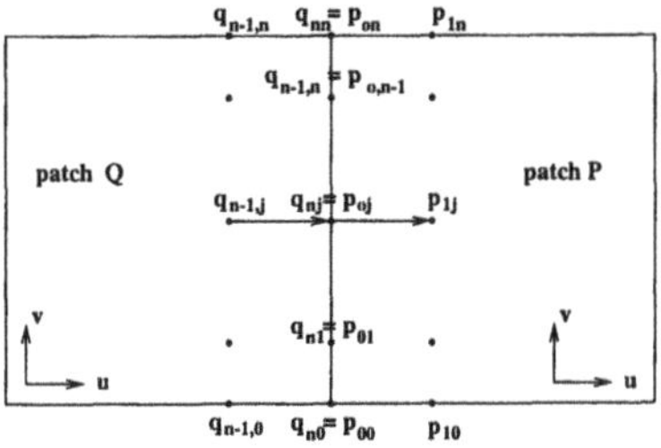

Fig. 3. Smoothness condition along common edge.

In the case when only the initial mesh (and not the whole surface) is given and it is necessary to solve an interpolation or approximation problem, the following algorithm is applied:

• The mean normal N_{mean} at the vertex is computed and is taken as the initial approximation to the normal of the tangent plane at the vertex.

• Edges sharing the vertex are projected onto the plane passing through the given vertex and perpendicular to N_{mean}. Let $p_1, \ldots, p_n$ be the projections of the end vertices $r_1, \ldots, r_n$ of the edges, starting at the common vertex. Then $p_k \quad (k = 1, \ldots, n)$ is given by

$$p_k = r_k + (r - r_k, N_{mean}) \frac{N_{mean}}{||N_{mean}||^2}. \tag{7}$$

Now it is possible to compute coefficients c_{1k} and $c_{2k} \quad (k = 3, \ldots, n)$ of the relations between projections such that

$$p_k - r = c_{1k}(p_1 - r) + c_{2k}(p_2 - r). \tag{8}$$

These coefficients are expressed by

$$c_{1k} = -\frac{S_{2k}}{S_{12}}, \qquad c_{2k} = \frac{S_{1k}}{S_{12}}, \tag{9}$$

where S_{jk} is the area of the triangle with vertices p_j, r and p_k.)

• We now assume the relations (9) are given. When all elements are described by Bézier patches, the geometric conditions at the vertex are equivalent to the linear equations

$$\xi_k - r = c_{1k}(\xi_1 - r) + c_{2k}(\xi_2 - r), \quad (k = 3, \ldots, n), \tag{10}$$

which involve the common control point at the vertex and the control points nearest to it and lying on the edges. Such equations will be treated as constraints. Thus, the direction of the normal is free and is established by the energy functional minimization.

Fig. 4. Two views of mesh structure.

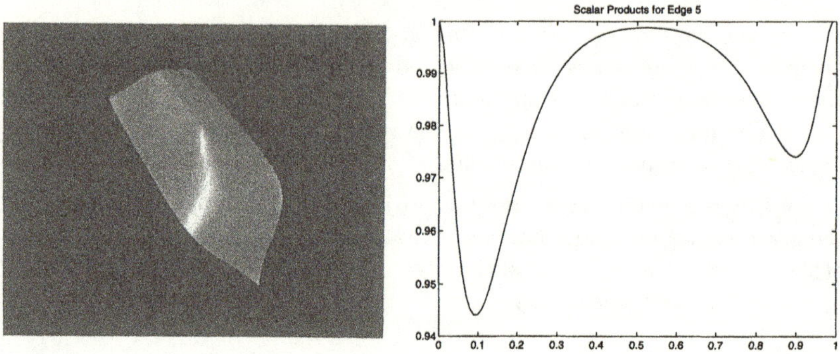

Fig. 5a. Resulting surface and **b.**deviation graph.

4.2 Smoothness Conditions Along Common Edges

Let P and Q be two neighboring elements represented in the Bézier form by control points p_{ij} and q_{ij}, and suppose the parameterization is chosen as shown in Figure 3. In the present approach along the common edge the following (pseudo C^1) linearized smoothness conditions are imposed:

$$q_{nj} - q_{n-1,j} = p_{1j} - p_{0j} \quad j = 1, \dots, n-1. \tag{11}$$

This doesn't result in a G^1 surface in general, but for such a choice of smoothness conditions, along the common edge we have

$$\frac{\partial Q}{\partial v}(1, v) \equiv \frac{\partial P}{\partial v}(0, v). \tag{12}$$

In the second direction we obtain

$$\begin{array}{c} \frac{\partial Q}{\partial u}(1,v) - \frac{\partial P}{\partial u}(0,v) = \\ n\left[v^n(2q_{n0} - q_{n-1,0} - p_{10}) + (1-v)^n(2q_{nn} - q_{n-1,n} - p_{1n})\right]. \end{array} \tag{13}$$

This means that the deviation in the second direction is bounded by its maximal "inconsistency" at vertices, and this deviation is relatively small when v is near the middle of the segment of parameterization. Since at the ends of the segments the G^1 condition is a consequence of (12), it will compensate relatively large deviations in the second direction at the vertices.

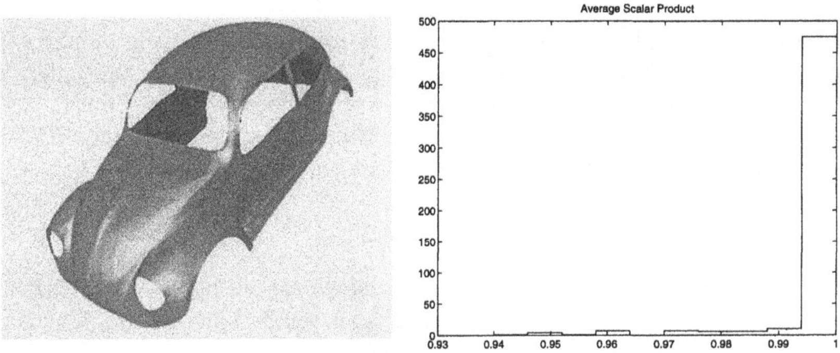

Fig. 6. Volkswagen.

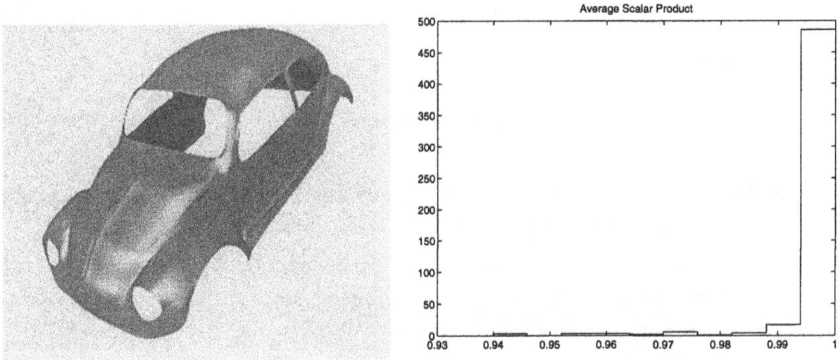

Fig. 7. Volkswagen.

§5. Examples

In all the examples below the elementary patches are biquartic Bézier patches.

5.1 Study of Deviation from G^1

The simple example shown in Figures 4 and 5 illustrates the behavior of the deviation from G^1 in relation to the mesh structure. Taking the common edge at the center of the mesh, the graph in Figure 5b gives the variation of the cosine of the angle between the normals for the two adjacent patches. This graph clearly shows that the maximal deviations are given by the maxima of the polynomials (see (13)). It can be seen that the deviation between normals is small in the middle of the edge (and normals coincide at the ends of the edge). This deviation increases when the adjacent edges form a relatively sharp angle, and non-collinearity at the vertex influences the deviation locally only.

5.2 A Volkswagen

This example shows the behavior of the discrete G^1 method and the influence of the linearization method at the vertices. The free form surface over the given quadrilateral mesh is constructed by minimizing the plate energy and approximating the mesh vertices. In Figure 6a the normals are fixed. The corresponding statistics on the deviation from G^1 are given in Figure 6b. The new method where the normals are free to move is given in Figure 7. There is a notable improvement in the overall smoothness.

§6. Conclusion

For very large assemblies of patches, continuous smoothness conditions can be replaced by "discrete" ones, and G^1 conditions can be linearized in a "natural" way that does not impose arbitrary constraints on the surface. The present method is currently being extended to surface reconstruction starting from a given triangulation such as obtained in reverse engineering.

Acknowledgments. We would like to thank Dr. L. Kobbelt for providing us the Volkswagen mesh example.

References

1. M. Bercovier and O. Volpin, G^1 Hierarchical Bézier surface over irregular mesh, Leibnitz Inst. Tech. Rep., CS97-3, Jerusalem, 1996.
2. O. Volpin and M. Bercovier and T. Matskewich, Approximation/ Interpolation of smooth free from surfaces based on plate type energy, Leibnitz Inst. Tech. Rep., TR97-12, Jerusalem, 1997.
3. M. Bercovier and O. Volpin and T. Matskewich, Globally G^1 free form surfaces using 'real' plate energy invariant energy methods, in *Curves and Surfaces with Applications in CAGD*, A. Le Méhauté, C. Rabut, and L. L. Schumaker (eds.), Vanderbilt Univ. Press, Nashville, 1997, 25–34.
4. O. C. Zienkiewicz, *The Finite Element Method in Engineering Science*, McGraw Hill Publishing, London, 1971.

Tanya Matskewich and Oleg Volpin
Institute of Computer Science
Hebrew University, Jerusalem, 91904, ISRAEL
`fisa@cs.huji.ac.il`
`oleg@cs.huji.ac.il`

Michel Bercovier
Pole Universitaire Leonard de Vinci
92916 Paris la Defense cedex, FRANCE
`Michel.Bercovier@devinci.fr`

Joining Cyclide Patches Along Quartic Boundary Curves

Christoph Mäurer and Rimvydas Krasauskas

Abstract. We show how to join cyclide patches G^1-continuous along boundary curves of degree 4. The method is based on a Laguerre geometric approach. In comparison to the well-known cyclide blends along boundary circles, we obtain more degrees of freedom, which can be used for surface design.

§1. Introduction

Dupin cyclides have been recognized in CAGD as blending surfaces, and are used in applications such as motion planning and geometric modeling. They have a lot of nice geometric properties (see e.g.[9,10]), but on the other hand there are only a few degrees of freedom to manipulate the surface. That makes more difficult to apply cyclide patches to approximation schemes. The standard way to construct a patchwork of cyclides, is to use rational biquadratic representations, and to join them G^1-continuous along the boundary circles. In this paper we present a method to join cyclide patches along curves of degree 4. The appropriate cyclide representation for our method are triangular quartic patches and tensor product patches of degree $(2,4)$ and $(4,4)$. In section 4 we construct a patchwork of two triangular quartic patches, which is bounded by four circles. It has more degrees of freedom in comparison to the standard principal cyclide patch.

A Dupin cyclide in standard orientation can be described with the parameter representation

$$\begin{aligned} y_1 &= (d[c - a\cos\phi\cos\psi] + b^2\cos\phi)/(a - c\cos\phi\cos\psi) \\ y_2 &= (b\sin\phi\ [a - d\cos\psi])/(a - c\cos\phi\cos\psi) \qquad \psi,\phi \in [0,2\pi] \qquad (1) \\ y_3 &= (b\sin\psi\ [c\cos\phi - d])/(a - c\cos\phi\cos\psi) \end{aligned}$$

The constants a, b, c are not independent, since $c^2 = a^2 - b^2$ holds. The isoparametric lines of (1) are the curvature lines of the cyclide. They are circles.

Mathematical Methods for Curves and Surfaces II
Morten Dæhlen, Tom Lyche, and Larry L. Schumaker (eds.), pp. 359–366.

ISBN 0-8265-1315-8.

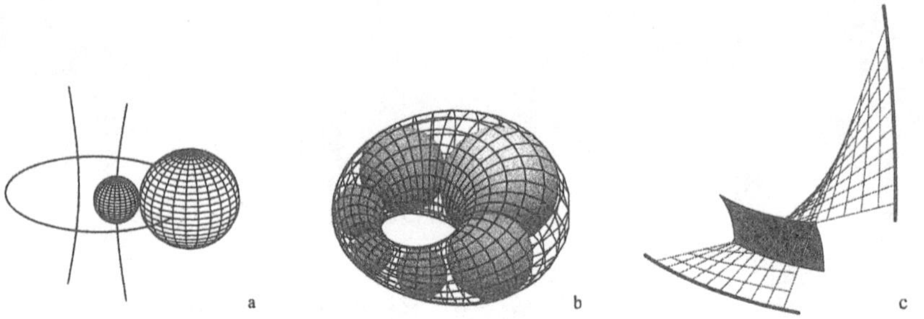

Fig. 1. Geometry of the cyclide.

There are 2 families of enveloping spheres S_ϕ and S_ψ, which are centered at the conics

$$\begin{aligned} \boldsymbol{e}(\phi) &= (a\cos\phi, b\sin\phi, 0)^T, \qquad \phi \in [0, 2\pi), \\ \boldsymbol{h}(\psi) &= (c\sec\psi, 0, b\tan\psi)^T, \qquad \psi \in [0, 2\pi). \end{aligned} \tag{2}$$

The radii of the spheres are given by $r_\phi = d - c\cos\phi$ and $r_\psi = d - a\sec\psi$. The conics (2) are called focal conics, because the focal points of one conic are the vertices of the other. The spheres S_ϕ and the spheres S_ψ are touching each other in the surface point $\boldsymbol{y}(\phi, \psi)$. Thus, the cyclide is a canal surface with the spine curve $\boldsymbol{e}(\phi)$ or $\boldsymbol{h}(\psi)$. Figure 1a shows the focal conics and one sphere of each family. The interpretation of a cyclide as canal surface with spine curve $\boldsymbol{e}(\phi)$ is visualized in Figure 1b.

§2. Cyclographic Model of Laguerre Geometry

Since spheres play an important role in describing cyclides, we need a "good" model to describe spheres. Here we use the so called cyclographic model of Laguerre geometry. It was recognized in the recent paper of Pottmann and Peternell [8], that Laguerre geometry can be a very helpful tool for applications in geometric design. In the cyclographic model an oriented sphere with center $\boldsymbol{m} = (m_1, m_2, m_3)^T$ and signed radius r is represented as a point $(y_1, y_2, y_3, y_4)^T = (m_1, m_2, m_3, r)^T \in \mathbb{R}^4$. The space of points $\mathbb{R}^3$ is embedded in $\mathbb{R}^4$ as hyperplane $y_4 = 0$. The inner product

$$\langle \boldsymbol{a}, \boldsymbol{b} \rangle_{pe} := a_1 b_1 + a_2 b_2 + a_3 b_3 - a_4 b_4 \tag{3}$$

defines a pseudo euclidian metric of $\mathbb{R}^4$. The pseudo euclidian distance $d = \sqrt{\langle \boldsymbol{a}-\boldsymbol{b}, \boldsymbol{a}-\boldsymbol{b} \rangle_{pe}}$ of two points $\boldsymbol{a}, \boldsymbol{b}$ is equal to the tangential distance of the corresponding spheres. In our context, lines whose direction vector $\boldsymbol{a}$ fulfills the equation $\langle \boldsymbol{a}, \boldsymbol{a} \rangle_{pe} = 0$ are important. They are called isotropic lines and they represent a pencil of spheres in oriented contact. The common point of these spheres corresponds to the intersection of the isotropic line with the hyperplane $y_4 = 0$. All isotropic lines through a fixed point $\boldsymbol{a}$ form an isotropic cone $\Gamma(\boldsymbol{a})$. It is described with the implicit equation $\langle \boldsymbol{y} - \boldsymbol{a}, \boldsymbol{y} - \boldsymbol{a} \rangle_{pe} = 0$. Consider a real continuous curve $\boldsymbol{p}(t) \in \mathbb{R}^4$ in the cyclographic model. The curve describes a one-parametric family of spheres. The canal surface, defined as the envelope of these spheres, is called the cyclographic image of $\boldsymbol{p}(t)$. The spine curve of the canal surface is given as the orthogonal projection $\widehat{\boldsymbol{p}}(t)$ of $\boldsymbol{p}(t)$ onto $\mathbb{R}^3$. Now we want to describe a Dupin cyclide in the cyclographic model. Since the cyclide has two families of enveloping spheres, it can be represented by two curves in $\mathbb{R}^4$. With the standard transformations $\tan(\phi/2) = v_1/v_0$ and $\tan(\psi/2) = u_1/u_0$ we can derive from (2) a rational parametrization of the curves $\boldsymbol{f}_e, \boldsymbol{f}_h$, whose cyclographic image is the Dupin cyclide of the normal form (1). Using the projective extension $\mathbb{P}^4$ of $\mathbb{R}^4$ with homogeneous coordinates $x_i/x_0 = y_i$ we get:

$$\begin{aligned} \boldsymbol{f}_e(v_0\!:\!v_1) &= (v_0^2+v_1^2, a(v_0^2-v_1^2), 2bv_0v_1, 0, d(v_0^2+v_1^2) - c(v_0^2-v_1^2))^T, \\ \boldsymbol{f}_h(u_0\!:\!u_1) &= (u_0^2-u_1^2, c(u_0^2+u_1^2), 0, 2bu_0u_1, d(u_0^2-u_1^2) - a(u_0^2+u_1^2))^T. \end{aligned} \tag{4}$$

Note that the orthogonal projections $\widehat{\boldsymbol{f}}_e$ and $\widehat{\boldsymbol{f}}_h$ give the focal conics of the cyclide. The "infinite points" of the conics (4) are lying in the absolute quadric $\Omega : x_0 = 0, x_1^2 + x_2^2 + x_3^2 - x_4^2 = 0$. Thus, $\boldsymbol{f}_e$ and $\boldsymbol{f}_h$ are circles (pe-circles) in the pseudo euclidian metric, which is defined by (3). It is well known from classical geometry (see [6,8] or [3]), that this property characterizes cyclides in the cyclographic model:

Lemma 1. *The set of Dupin cyclides in* $\mathbb{R}^3$ *is corresponding to the set of pe-circles in the cyclographic model of Laguerre geometry.*

The pe-circles are invariant under Laguerre transformations. Thus, in the cyclographic model it is possible to map cyclides on cyclides in a linear way. This property is used in [3] to derive more simple normal forms and implicit equations of cyclides. Also, the mentioned paper presents an algorithm, how to compute the second pe-circle of a cyclide, if the first pe-circle is given.

Consider an arbitrary Dupin cyclide with pe-circles $\boldsymbol{f}_e(v_0\!:\!v_1)$ and $\boldsymbol{f}_h(u_0\!:\!u_1)$. We define a hypersurface $\boldsymbol{H}$ as set of all lines joining both pe-circles:

$$\boldsymbol{H}(s_0\!:\!s_1, u_0\!:\!u_1, v_0\!:\!v_1) := s_0\boldsymbol{f}_e(v_0\!:\!v_1) + s_1\boldsymbol{f}_h(u_0\!:\!u_1) \tag{5}$$

Any sphere, defined as a point of $\boldsymbol{f}_e$, is in oriented contact with any sphere, corresponding to a point on $\boldsymbol{f}_h$. Thus, all lines of the two-parametric family $\boldsymbol{H}$ are isotropic lines. The points on $\boldsymbol{H}$ correspond to all spheres, which are in oriented contact with the cyclide. Any line of $\boldsymbol{H}$ corresponds to the pencil of spheres touching one fixed point on the cyclide. The sphere with radius zero in this pencil is a surface point on the cyclide. Thus, the intersection $\boldsymbol{H}(s_0 : s_1, u_0 : u_1, v_0 : v_1) \cap \{x_4 = 0\}$ gives a parameter representation $\boldsymbol{z}(u_0 : u_1, v_0 : v_1)$ of the Dupin cyclide. Note that $s_0 : s_1$ can be easily eliminated, since it is linear in $\boldsymbol{H}$. The orthogonal projection $\widehat{\boldsymbol{H}}$ in $\mathbb{R}^3$ gives the normal congruence of the cyclide, which is formed by all lines intersecting the cyclide perpendicular.

§3. Bézier Representation of Cyclide Patches

We can also use the hypersurface $\boldsymbol{H}$ to derive Bézier representations of cyclide patches. Suppose that the pe-circles are given in homogeneous coordinates as Bézier curves

$$\boldsymbol{f}_e(v) = \sum_{i=0}^{2} \boldsymbol{c}_i B_i^2(v), \qquad \boldsymbol{f}_h(u) = \sum_{j=0}^{2} \boldsymbol{d}_j B_j^2(u), \qquad u, v \in [0,1] \tag{6}$$

To get a Bézier representation of the corresponding cyclide patch, we have to intersect $s_0 \boldsymbol{f}_e + s_1 \boldsymbol{f}_h$ with the hyperplane $x_4 = 0$. This yields the rational biquadratic Bézier representation:

$$\boldsymbol{z}(u,v) = \sum_{i=0}^{2}\sum_{j=0}^{2} (d_j^{(4)} \boldsymbol{c}_i - c_i^{(4)} \boldsymbol{d}_j) B_i^2(u) B_j^2(v), \qquad u, v \in [0,1], \tag{7}$$

where $d_j^{(4)}$ denotes the fourth component of $\boldsymbol{d}_j$. The geometric interpretation of (7) can be summarized:

Theorem 2. *The homogeneous Bézier points $\boldsymbol{b}_{ij}$ of a rational biquadratic cyclide patch are obtained as the intersection of the nine lines, connecting the homogeneous Bézier points $\boldsymbol{c}_i$ and $\boldsymbol{d}_j$, $i, j \in \{0, 1, 2\}$ of the two corresponding pe-circles, with the hyperplane $x_4 = 0$.*

All parameter lines of the rational biquadratic cyclide patch are circles. For our purpose we need cyclide patches with quartic boundary curves. Here we focus on those triangular patches, which are obtained by subdividing a biquadratic patch along a quartic "diagonal" curve $\boldsymbol{q}(t)$ into two pieces. With the substitution

$$u = t, \quad v = \frac{\lambda t}{1 + (\lambda - 1)t}, \quad \lambda \in \mathbb{R}_+ \tag{8}$$

we obtain a family $\mathcal{Q}_1$ of quartic curve $\boldsymbol{q}(t)$ joining the opposite vertices $\boldsymbol{b}_{00}$ and $\boldsymbol{b}_{22}$ of the rectangular patch. For $\lambda = 1$ the homogeneous Bézier points of $\boldsymbol{q}(t)$ read as follows:

$$\boldsymbol{b}_0 = \boldsymbol{b}_{00}, \boldsymbol{b}_1 = \frac{1}{2}(\boldsymbol{b}_{10} + \boldsymbol{b}_{01}), \boldsymbol{b}_2 = \frac{1}{6}(\boldsymbol{b}_{20} + 4\boldsymbol{b}_{11} + \boldsymbol{b}_{02}), \boldsymbol{b}_3 = \frac{1}{2}(\boldsymbol{b}_{12} + \boldsymbol{b}_{21}), \boldsymbol{b}_4 = \boldsymbol{b}_{22}$$

The other curves of the family $\mathcal{Q}_1$ may be obtained by a reparametrization $\boldsymbol{d}_j \mapsto \lambda^j \boldsymbol{d}_j$ of the pe-circle $\boldsymbol{f}_e$ (or equivalent by the reparametrization $\boldsymbol{b}_{ij} \mapsto \lambda^j \boldsymbol{b}_{ij}$ of the cyclide patch). Now we subdivide the rectangular cyclide patch along the curve $\boldsymbol{q}(t)$. The Bézier representation of both triangular quartic patches can be computed with a standard method [1] or with a construction scheme for triangular cyclide patches [2,5].

§4. G^1-continuous Cyclide Patchwork

In order to describe curves of the family $\mathcal{Q}_1$ in terms of the cyclographic model, we define the two dimensional surface

$$\boldsymbol{R}(s_0\!:\!s_1,t) := s_0\boldsymbol{f}_e(v_0(t)\!:\!v_1(t)) + s_1\boldsymbol{f}_h(u_0(t)\!:\!u_1(t)), \tag{9}$$

by substituting linear polynomials $u_0(t), u_1(t), v_0(t), v_1(t)$ into $\boldsymbol{H}$. It has the following properties: The intersection $\boldsymbol{R}(s_0 : s_1, t) \cap \{x_4 = 0\}$ gives a quartic curve $\boldsymbol{q}(t)$ on the cyclide. These quartic curves form a three-parametric family $\mathcal{Q}_3$, which contains the family $\mathcal{Q}_1$. The orthogonal projection $\widehat{\boldsymbol{R}}$ describes the normal surface of the cyclide along $\boldsymbol{q}(t)$. Figure 1c shows a cyclide patch, a quartic curve $\boldsymbol{q}(t)$ and the corresponding normal surface $\widehat{\boldsymbol{R}}$, which joins both focal conics of the cyclide. It is a ruled surface of parametric degree $(1,2)$ and carries exactly one family of conics.

Lemma 3. *The isoparametric curves $s_0\!:\!s_1 = \text{const}$ on $\boldsymbol{R}(s_0\!:\!s_1,t)$ are exactly the pe-circles on $\boldsymbol{R}$.*

Proof: $\boldsymbol{R}(s_0 : s_1, t)$ is a ruled surface in $\mathbb{P}^4$. The straight lines $t = \text{const}$ are isotropic lines, because they correspond to pencil of touching spheres. Thus, the intersection points of $\boldsymbol{R}$ with $x_4 = 0$ lie in Ω and the conics $s_0 : s_1 = \text{const}$ are pe-circles. But there are no other conics on $\boldsymbol{R}$, because under the orthogonal projection the conics of $\boldsymbol{R}$ are mapped onto the conics of $\widehat{\boldsymbol{R}}$. Since there is just a one-parametric family of conics (see [3] for details), we have no further candidates for pe-circles on $\boldsymbol{R}$. □

All points on $\boldsymbol{R}$ correspond to all spheres touching the cyclide along the curve $\boldsymbol{q}(t)$. The pe-circles on $\boldsymbol{R}$ correspond exactly to the cyclides touching the given cyclide $\boldsymbol{H} \cap \{x_4 = 0\}$ along $\boldsymbol{q}(t)$. With the previous lemma we get:

Theorem 4. *Fix a quartic curve $\boldsymbol{q}(t)$ of type $\mathcal{Q}_3$ on a Dupin cyclide $\mathcal{C}$. Then all cyclides touching $\mathcal{C}$ along $\boldsymbol{q}(t)$ constitute a one-parametric family, corresponding to the family of pe-circles on $\boldsymbol{R}$.*

Note, that for any pe-circle $\boldsymbol{f} := \boldsymbol{R}(s_0\!:\!s_1 = \text{const}, t)$ there exists a second one $\boldsymbol{f}^* := \boldsymbol{R}(s_0^* : s_1^* = \text{const}, t)$, which describe the same cyclide. It is useful to know the relation between $s_0 : s_1$ and $s_0^* : s_1^*$, because for computing a Bézier representation of the cyclide we need both pe-circles. Any line joining $\boldsymbol{f}$ and $\boldsymbol{f}^*$ is an isotropic line. Thus, necessarily $\boldsymbol{f}^*$ lies in the intersection $\Gamma(\boldsymbol{a}_1) \cap \Gamma(\boldsymbol{a}_2)$, where $\boldsymbol{a}_1$ and $\boldsymbol{a}_2$ are arbitrary points on $\boldsymbol{f}$. On the other hand $\Gamma(\boldsymbol{a}_1) \cap \Gamma(\boldsymbol{a}_2)$ is lying in a hyperplane

$$P(\boldsymbol{a}_1, \boldsymbol{a}_2) :\ 2\langle \boldsymbol{y}, \boldsymbol{a}_1 - \boldsymbol{a}_2\rangle_{pe} + \langle \boldsymbol{a}_2, \boldsymbol{a}_2\rangle_{pe} - \langle \boldsymbol{a}_1, \boldsymbol{a}_1\rangle_{pe} = 0. \tag{10}$$

The parameter $s_0^*\!:\!s_1^*$ can be computed by intersecting an arbitrary isotropic line $\boldsymbol{R}(s_0^* : s_1^*, t = \text{const})$ with the hyperplane $P(\boldsymbol{a}_1, \boldsymbol{a}_2)$. This gives a linear equation for $s_0^*\!:\!s_1^*$.

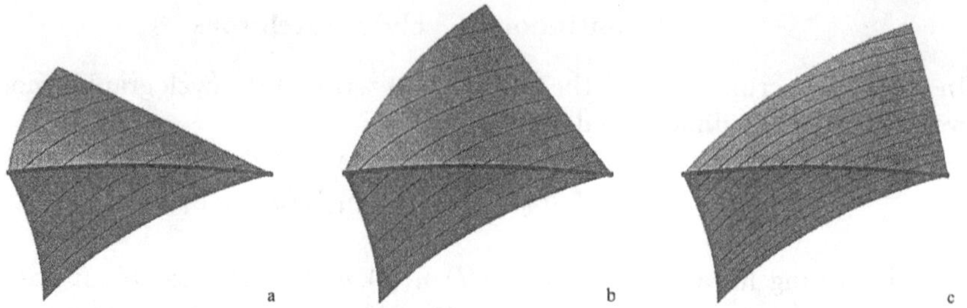

Fig. 2. Cyclide patchwork I.

The result can be used to construct a new kind of cyclide patchwork:

1) Input: pe-circles $\boldsymbol{f}_e$, $\boldsymbol{f}_h$ of a cyclide Bézier patch $\mathcal{C}_1$ and a quartic diagonal curve $\boldsymbol{q}(t)$, defined with equation (8) in the parameter domain of $\mathcal{C}_1$.
2) Determine the first pe-circle $\boldsymbol{f}$ of the new cyclide patch $\mathcal{C}_2$ as isoparametric line of $\boldsymbol{R}$.
3) Compute the second pe-circle $\boldsymbol{f}^*$ of $\mathcal{C}_2$ as described before.
4) Use (7) to compute the Bézier representation of $\mathcal{C}_2$.
5) Subdivide each patch $\mathcal{C}_1$, $\mathcal{C}_2$ into two triangular quartic patches.
6) Output: Patchwork of two triangular cyclide patches bounded by four circles.

This construction has two degrees of freedom: one in choosing the diagonal curve $\boldsymbol{q}(t)$ (which is determined with the parameter λ in (8)) and one in choosing the isoparametric curve on the surface $\boldsymbol{R}$. Figures 2a-c show such a cyclide patchwork. One triangular quartic patch (dark) is fixed for all three cases. The second part (light) of the patchwork is computed with different isoparametric lines of $\boldsymbol{R}$ in step 2 of the algorithm. In Figures 3a and 3b different parameter λ are used to construct different cyclide patchwork. It offers more flexibility as the standard biquadratic cyclide patch. Srinivas, Kumar and Dutta [10] have presented an approximation method with piecewise cyclide patches. An interesting application could be to use the new cyclide patchwork for this design scheme. The new degrees of freedom can be used to improve the shape of the surface.

In order to apply this new patchwork to a design scheme, we need more information about the vertices of the patchwork. It is well known, that the 4 vertices of a standard biquadratic cyclide patch are lying on a circle. The same property has the new patchwork:

Theorem 5. *Let $\tilde{\mathcal{C}}$ be the family of cyclide patches with vertices $\boldsymbol{b}_{00}$, $\tilde{\boldsymbol{b}}_{20}$, $\tilde{\boldsymbol{b}}_{02}$, $\boldsymbol{b}_{22}$, which are touching a given cyclide patch $\mathcal{C}$ with vertices $\boldsymbol{b}_{00}$, $\boldsymbol{b}_{20}$, $\boldsymbol{b}_{02}$, $\boldsymbol{b}_{22}$ along a quartic curve through $\boldsymbol{b}_{00}$ and $\boldsymbol{b}_{22}$, Then the family of vertices $\tilde{\boldsymbol{b}}_{02}$ (resp. $\tilde{\boldsymbol{b}}_{20}$) moves on a circle containing the vertices of $\mathcal{C}$.*

Proof: The families of pe-circles of $\tilde{\mathcal{C}}$ may have the homogeneous Bézier

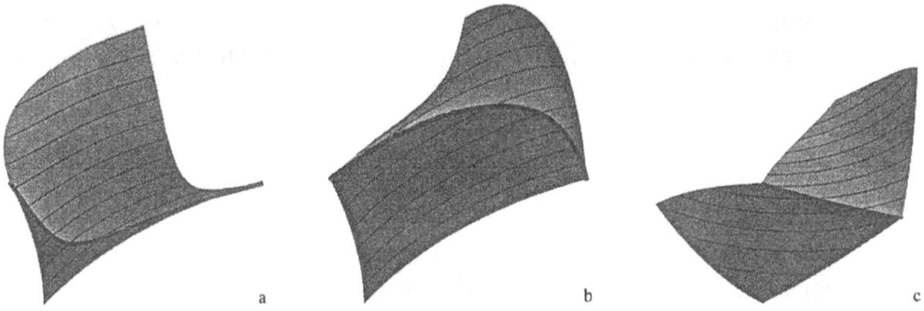

Fig. 3. Cyclide patchwork II.

representations

$$f(s_0\!:\!s_1,t) = \sum_{i=0}^{2} \tilde{c}_i(s_0\!:\!s_1)B_i^2(t), \quad f^*(s_0^*\!:\!s_1^*,t) = \sum_{i=0}^{2} \tilde{d}_i(s_0^*\!:\!s_1^*)B_i^2(t).$$

Consider the family of lines $l := \lambda_0\tilde{c}_0(s_0\!:\!s_1) + \lambda_1\tilde{d}_2(s_0^*\!:\!s_1^*)$ and note that l is linear in $\lambda_0\!:\!\lambda_1$, $s_0\!:\!s_1$, $s_0^*\!:\!s_1^*$. Since the relation between $s_0\!:\!s_1$ and $s_0^*\!:\!s_1^*$ is linear to, we can represent l as bilinear surface in $\lambda_0\!:\!\lambda_1$ and $s_0\!:\!s_1$. If $s_0\!:\!s_1$ is moving, the Bézier point $\tilde{b}_{02}$ is moving on the intersection curve of l with the hyperplane $x_4 = 0$, which is a conic. In fact, this conic is a circle because its infinite points are lying on the absolute circle $x_0 = x_4 = x_1^2 + x_2^2 + x_3^2 = 0$. Since b_{00}, b_{22} and b_{02} are contained in l (they lie on the generators $\lambda_1/\lambda_0 = 0$, $\lambda_0/\lambda_1 = 0$ and $s_1/s_0 = 0$ resp.) and in the hyperplane $x_4 = 0$, the circle contains all vertices of the patch $\mathcal{C}$. □

On the one hand the property of Theorem 5 can be mentioned as a restriction, on the other hand it allows to apply a design scheme – like that of Srinivas, Kumar, Dutta [10] – to the patchwork. One possibility to avoid the property, that the vertices of the patchwork are lying on a circle, is to use patches of degree $(2,4)$ or $(4,4)$. This kind of cyclide patches have been analyzed in [2,4,5]. Figure 3c shows two biquartic cyclide patches, which are joined G^1-continuous along their common boundary curve.

Remark 6. *The normal surface $\widehat{R}(s_0\!:\!s_1,t)$ defines a one-parametric family of conics. By moving $s_0\!:\!s_1$ the conic moves from the focal ellipse $\widehat{f}_e$ to the focal hyperbola $\widehat{f}_h$ of the cyclide. In the limit case the conic is a parabola. It is the focal parabola of a* parabolic *cyclide. Therefore we have found one parabolic cyclide in the family of touching cyclides. Vice versa one may ask, how to construct to a given parabolic cyclide a second cyclide, which touches the first along a curve of degree higher than 2. This question can be answered with similar Laguerre geometric methods. The main difference is that parabolic cyclides carry rational quartic and rational cubic curves. More details can be found in* [3, 7].

Acknowledgments. This work has been supported in part of the DFG (Deutsche Forschungs Gemeinschaft). The second author wish to thank Nordic Information Office and Royal Norwegian Embassy in Vilnius for supporting the participation in the Lillehammer Conference.

References

1. Goldman, R. N., and D. J. Filip, Conversion from Bézier rectangles to Bézier triangles, Computer-Aided Design **19** (1987), 25–27.
2. Krasauskas, R., Universal parametrizations of some rational surfaces, in *Curves and Surfaces with Applications in CAGD*, A. Le Méhauté, C. Rabut, and L. L. Schumaker (eds.), Vanderbilt Univ. Press, Nashville, 1997, 231–238.
3. Krasauskas, R., and C. Mäurer, A Laguerre geometric approach to cyclides, preprint.
4. Mäurer, C., Generalized parameter representations of tori, Dupin cyclides and supercyclides, in *Curves and Surfaces with Applications in CAGD*, A. Le Méhauté, C. Rabut, and L. L. Schumaker (eds.), Vanderbilt Univ. Press, Nashville, 1997, 295–302.
5. Mäurer, C., Rationale Bézier-Kurven und Bézier-Flächenstücke auf Dupinschen Zykliden, PhD-thesis, Darmstadt University of Technology, 1997.
6. Müller, E., and L. Krames, *Vorlesungen über Darstellende Geometrie II*, Deuticke, Leipzig und Wien, 1929.
7. Peternell, M., and H. Pottmann, Designing rational surfaces with rational offsets, in *Advanced topics in Multivariate Approximation*, F. Fontanella, K. Jetter, and P.-J. Laurent (eds.), World Scientific, Singapore, 1996, 1–12.
8. Pottmann, H., and M. Peternell, Applications of Laguerre geometry in CAGD *I*, technical report no. 30, Institut für Geometrie, TU Wien, 1996.
9. Pratt, M.J., Cyclides in computer aided design *II*, Comput. Aided Geom. Design **12** (1995), 221–242.
10. Srinivas, Y. L., K. P. Kumas, and D. Dutta, Surface design using cyclide patches, Computer-Aided Design **28** (1996), 263–276.

Christoph Mäurer
Ship Design Laboratory
National Technical University of Athens, GREECE
`maeurer@deslab.naval.ntua.gr`

Rimvydas Krasauskas
Department of Mathematics
Vilnius University
Vilnius, LITHUANIA
`Rimvydas.Krasauskas@maf.vu.lt`

Canal Surfaces and Inversive Geometry

Marco Paluszny and Katja Bühler

Abstract. A canal surface is the envelope of a 1-parameter family of spheres. The latter are presented in terms of curves in projective 4-space. Properties of the curve are translated into geometric properties of the corresponding envelope. Special attention is given to curves of degree three and the circumstances under which the implicit degree of the envelope drops.

§1. Introduction

Canal surfaces are an important tool in geometric modelling, specially in blending and joining of surface primitives along circular profiles. These include planes, cones, cylinders, Dupin cyclides and arbitrary surfaces of revolution, and have been studied by [1,5,6,7,8] among others. We look at the presentation of canal surfaces as curve segments in four dimensional space, using the pole-polar relationship between points outside a sphere and the usual spheres and planes in Euclidean 3-space. This has been developed recently for use in CAGD in [2], although as a fact of classical geometry it goes back to the European geometers: Darboux, Möbius, Lie and Klein.

§2. Notation and Preliminary Facts

The following considerations take place in $\overline{\mathbb{E}}^4$, the projective closure of 4-dimensional Euclidean space $\mathbb{E}^4$. Its points are described by homogeneous coordinates $\mathfrak{x} = (x_0, x_1, x_2, x_3, x_4)^T \in \mathbb{R}^5 \backslash \{(0,0,0,0,0)^T\}$, where x_0 is the homogenizing coordinate. The 3-dimensional projective Euclidean space $\overline{\mathbb{E}}^3$ will be identified with the hyperplane $x_4 = 0$ of the $\overline{\mathbb{E}}^4$, its points are described by homogeneous coordinates $\mathfrak{x} = (x_0, x_1, x_2, x_3)^T \in \mathbb{R}^4 \backslash \{(0,0,0,0)^T\}$. Points of the Euclidean space $\mathbb{E}^3$ are described by inhomogeneous coordinates $\boldsymbol{x} = (x, y, z)^T = (\frac{x_1}{x_0}, \frac{x_2}{x_0}, \frac{x_3}{x_0})^T \in \mathbb{R}^3$.

Mathematical Methods for Curves and Surfaces II
Morten Dæhlen, Tom Lyche, and Larry L. Schumaker (eds.), pp. 367–374.

ISBN 0-8265-1315-8.

The Möbius Hypersphere and the Stereographic Projection

The unit sphere of $\overline{\mathbb{E}}^4$, here called Möbius hypersphere and denoted by Ψ, is given by the equation

$$\Psi: \qquad \langle \mathbb{x}, \mathbb{x} \rangle = -x_0^2 + x_1^2 + x_2^2 + x_3^2 + x_4^2 = 0,$$

where $\langle \mathbb{x}, \mathbb{y} \rangle := -x_0 y_0 + x_1 y_1 + x_2 y_2 + x_3 y_3 + x_4 y_4$ denotes the bilinear form associated to Ψ. The point $\mathbb{n} = (1,0,0,0,1)^T \in \Psi$ is called the north pole and the hyperplane $x_0 = x_4$ the north hyperplane of the Möbius hypersphere.

For any point $\mathbb{p} \in \overline{\mathbb{E}}^4$, its polar hyperplane with respect to Ψ is given by the equation $\langle \mathbb{p}, \mathbb{x} \rangle = 0$. It lies outside the Möbius hypersphere iff $\langle \mathbb{p}, \mathbb{p} \rangle < 0$, on Ψ iff $\langle \mathbb{p}, \mathbb{p} \rangle = 0$ and inside of Ψ iff $\langle \mathbb{p}, \mathbb{p} \rangle < 0$.

The map

$$\begin{array}{cccc} \phi: & \mathbb{E}^3 & \longrightarrow & \Psi - \{\mathbb{n}\} \\ & \boldsymbol{x} = \frac{1}{w}\begin{pmatrix} x \\ y \\ z \end{pmatrix}; w \neq 0 & \longmapsto & \begin{pmatrix} x^2+y^2+z^2+w^2 \\ 2xw \\ 2yw \\ 2zw \\ x^2+y^2+z^2-w^2 \end{pmatrix} \end{array}$$

is called the stereographic projection with respect to the Möbius hypersphere Ψ and the north pole $\mathbb{n}$. ϕ maps a point $\boldsymbol{x}$ in $\mathbb{E}^3 \subset \overline{\mathbb{E}}^3$ to the second intersection of the line through $\boldsymbol{x}$ and the north pole with Ψ.

The stereographic projection is bijective and maps 2-spheres and planes into plane sections with Ψ. Its inverse is given by

$$\begin{array}{cccc} \phi^{-1}: & \Psi - \{\mathbb{n}\} & \longrightarrow & \mathbb{E}^3 \\ & \mathbb{x} = \begin{pmatrix} x_0 \\ x_1 \\ x_2 \\ x_3 \\ x_4 \end{pmatrix}; x_0 \neq x_4 & \longmapsto & \frac{1}{x_0 - x_4}\begin{pmatrix} x_1 \\ x_2 \\ x_3 \end{pmatrix}. \end{array}$$

The generalized stereographic projection ψ is introduced to give a relation between the spheres and planes of $\overline{\mathbb{E}}^3$ (which consists of the ordinary spheres and planes of $\mathbb{E}^3$ and the plane at infinity) and the points of $\overline{\mathbb{E}}^4$ as follows:

> An arbitrary sphere or a plane of $\overline{\mathbb{E}}^3$ denoted by S is mapped by the stereographic projection into the intersection of the Möbius hypersphere with a hyperplane H_S. The generalized stereographic projection maps S to the pole of the hyperplane H_S.

More precisely, a sphere of $\mathbb{E}^3$ of center $\boldsymbol{m} = (a,b,c)^T$ and finite radius r, corresponds under ψ to the point $\mathbb{x} \in \overline{\mathbb{E}}^4$ given by

$$\mathbb{x} = \begin{pmatrix} a^2+b^2+c^2-r^2+1 \\ 2a \\ 2b \\ 2c \\ a^2+b^2+c^2-r^2-1 \end{pmatrix}$$

A plane of $\mathbb{E}^3$, of equation $ax + by + cz = d$ corresponds to the point $\mathbb{x} = (d, a, b, c, d)^T$ and the plane at infinity corresponds to the north pole $\mathbb{n}$.

And conversely any point of $\overline{\mathbb{E}}^4$ corresponds to a sphere or a plane in $\overline{\mathbb{E}}^3$: given $\mathbb{x} = (x_0, x_1, x_2, x_3, x_4)^T \in \overline{\mathbb{E}}^4$ such that $x_0 \neq x_4$, the corresponding sphere of the $\mathbb{E}^3$ has

$$\text{center } \boldsymbol{m} = \frac{1}{x_0 - x_1}\begin{pmatrix} x_1 \\ x_2 \\ x_3 \end{pmatrix} \text{ and radius } r = \frac{\langle \mathbb{x}, \mathbb{x} \rangle^{\frac{1}{2}}}{|\langle \mathbb{n}, \mathbb{x} \rangle|}.$$

If $x_0 = x_4$ the corresponding sphere in $\mathbb{E}^3$ degenerates to a plane with equation $x_1 x + x_2 y + x_3 z = x_0$. And, if $\mathbb{x} = \mathbb{n}$ the corresponding plane is the plane at infinity.

We conclude this section with a summary of the correspondence ψ. Let $\mathbb{p} \in \overline{\mathbb{E}}^4$.

If $\mathbb{p} \notin$ north hyperplane and lies...	...$\psi^{-1}(\mathbb{p})$ is a ...
on Ψ	point sphere of $\mathbb{E}^3$
outside Ψ	sphere of real radius of $\mathbb{E}^3$
inside Ψ	sphere of imaginary radius of $\mathbb{E}^3$

If $\mathbb{p} \in$ north hyperplane and...	...$\psi^{-1}(\mathbb{p})$ is ...
$\mathbb{p} = \mathbb{n}$	the plane at infinity
$\mathbb{p} \neq \mathbb{n}$	a plane of $\mathbb{E}^3$

Envelopes

Given a 1-parameter family of spheres or planes $f(\boldsymbol{x}, t) = 0$ in $\mathbb{E}^3$, its envelope is defined as the union of all the circles (and lines) of intersection of infinitesimally neighboring pairs of spheres of the series (it follows that every circle of the envelope touches it and a sphere of the series tangentially). The envelope can be computed calculating the bigger set of characteristic points. These points are gotten by solving for each t the equations

$$f(\boldsymbol{x}, t) = 0; \frac{\partial}{\partial t} f(\boldsymbol{x}, t) = 0.$$

Eliminating the parameter t in both equations directly or by using Sylvester resultants gives the implicit equation satisfied by the characteristic points. Supposing F_t is a family of spheres or planes, the envelope is part of the set of characteristic points, but not all characteristic points are points of the envelope and they have to be separated by using the above definition of the envelope.

The intersection of two infinitesimal neighboring spheres or planes of F_t is called a composing circle or composing line and is a subset of the set of characteristic points.

§3. Rational Curves in $\overline{\mathbb{E}}^4$ and Series of Spheres/Planes in $\overline{\mathbb{E}}^3$

The simplest curves in $\overline{\mathbb{E}}^4$ are lines. These correspond to linear families of spheres or planes which are classically known as coaxial pencils (see [4]). A pencil of spheres is called elliptic (parabolic, hyperbolic) if all spheres have one circle (one point, one complex circle) in common. This circle is called carrying circle. It is real iff the line does not intersect the Möbius hypersphere.

A rational curve of degree n in $\overline{\mathbb{E}}^4$ may be written in homogeneous Bézier form

$$\mathbb{y}(t) = \sum_{i=0}^{n} \mathbb{b}_i B_i^n(t), \qquad t \in I \subseteq \mathbb{R},$$

where $B_i^n(t)$ are the Bernstein polynomials of degree n and $\mathbb{b}_i \in \mathbb{R}^5$ are the homogeneous Bézier points.

Applying the correspondence ψ^{-1} to $\mathbb{y}(t)$ determines a 1-parameter family of spheres (planes) of degree n in $\overline{\mathbb{E}}^3$ which could be manipulated through the $\mathbb{b}_i$.

In the next sections we assume that $\mathbb{y}$ is regular and of degree greater than one. We study the geometric properties of the envelope.

§4. Algebraic Degree

Theorem 1. *Assuming that $n > 1$ the envelope of $\psi^{-1}(\mathbb{y}(t))$ is an algebraic surface of degree $4n-4$ or less and contains the circle at infinity $2n-2$ or less times.*

Proof: $\langle \mathbb{p}, \mathbb{x} \rangle = 0$ is the equation of the polar plane of an arbitrary point of $\overline{\mathbb{E}}^4$ with respect to Ψ. Thus, the resultant equation

$$\operatorname{Res}(\langle \mathbb{y}(t), \mathbb{x} \rangle, \langle \dot{\mathbb{y}}(t), \mathbb{x} \rangle) = 0 \tag{1}$$

is satisfied by the characteristic points of the envelope of the polar planes of the points of $\mathbb{y}(t)$.

Computing the Sylvester resultant (see [3]) gives the following result

$$\begin{aligned}&\operatorname{Res}(\langle \mathbb{y}(t), \mathbb{x} \rangle, \langle \dot{\mathbb{y}}(t), \mathbb{x} \rangle = \\ &(\sum_{i=0}^{n} (-1)^{n-i} \binom{n}{i} \langle \mathbb{b}_i, \mathbb{x} \rangle)(\text{homogeneous expression in } \langle \mathbb{b}_i, \mathbb{x} \rangle, \\ &\qquad\qquad i = 0, \ldots, n \text{ of degree } 2n-2).\end{aligned} \tag{2}$$

Substituting $\mathbb{x}$ by $\phi^{-1}(\mathbb{x})$ is equivalent to map the intersection of the envelope of the polar planes by inverse stereographic projection to $\mathbb{E}^3$. This substitution changes equation (1) to the equation of the characteristic points of the series of spheres or planes corresponding to $\mathbb{y}(t)$.

The first factor of the resultant represents now a sphere (or plane), which corresponds to $\mathbb{y}(t_0); t_0 = \infty$. Thus, the sphere is part of the family $\psi^{-1}(\mathbb{y}(t))$.

Considering our definition of envelope it follows that this sphere is not part of the envelope and its degree is $4n-4$.

The substitution above replaces the $\langle \mathbb{b}_i, \mathbb{x}\rangle$ in (2) by $(b_4^i - b_0^i)(x^2+y^2+z^2)^2 + 2(b_1^i x + b_2^i y + b_3^i z) - (b_4^i + b_0^i)$, where $\mathbb{b}_i = (b_0^i, b_1^i, b_2^i, b_3^i, b_4^i)^T$. The second factor of the resultant is homogeneous of degree $2n-2$ in the $\langle \mathbb{b}_i, \mathbb{x}\rangle$, $i = 0,\ldots,n$. Thus, the summand of highest order in the equation of the envelope is $\lambda(x^2+y^2+z^2)^{2n-2}$ where λ depends on the Bézier points. So the surface contains the circle at infinity $2n-2$ times. □

§5. Composing Circles

A tangent of a regular curve in $\overline{\mathbb{E}}^4$ touches it in two infinitesimally neighboring points, which correspond to two infinitesimally neighboring spheres (or planes) in $\overline{\mathbb{E}}^3$. The circle (or line) of intersection of these two spheres (planes) is the carrying circle (line) of the pencil corresponding to the tangent and it is one of the composing circles (lines) of the envelope. So we have the following

Theorem 2. *Let $\mathbb{y}(t) \in \overline{\mathbb{E}}^4$ be a regular curve of degree $n\rangle 1$. The composing circles/lines of the envelope of the family of spheres/planes corresponding to the curve are the carrying circles/lines of the pencils of spheres/planes corresponding to the tangents of $\mathbb{y}$.*

It follows that

- if two tangents to the curve intersect Ψ (non tangentially) then the envelope is not donought like and might even be disconnected, because it contains complex composing circles.
- if one tangent touches the Möbius hypersphere, the envelope has a singular point, because it contains a composing circle degenerated to a point.
- if a tangent lies in the north hyperplane, which means the curve touches the north hyperplane, the envelope is an unbounded surface because it contains a composing line.
- if the curve lies in the north hyperplane the envelope is a ruled surface.

The composing circles might be computed directly from the curve $\mathbb{y}(t)$. In fact, the tangent line at $t = t_0$ is determined by two points $\mathbb{a}$ and $\mathbb{b}$ provided by the de Casteljau recursion. If this line does not lie on the north hyperplane then the radius ρ of the carrying circle of the corresponding pencil is given by

$$\rho^2 = \frac{\mathbb{a}^2\mathbb{b}^2 - \langle \mathbb{a}, \mathbb{b}\rangle^2}{(\langle \mathbb{a}, \mathbb{n}\rangle \mathbb{b} - \langle \mathbb{b}, \mathbb{n}\rangle \mathbb{a})^2}, \text{ where } \mathbb{x}^2 \text{ stands for } \langle \mathbb{x}, \mathbb{x}\rangle.$$

The circle lies in the radical plane of the pencil, which is the plane corresponding to the intersecting point $\langle \mathbb{a}, \mathbb{n}\rangle \mathbb{b} - \langle \mathbb{b}, \mathbb{n}\rangle \mathbb{a}$ of $\mathbb{a} \wedge \mathbb{b}$ with the north hyperplane. In a similar fashion one can characterize the carrying line of the pencil of planes corresponding to a projective line lying on the north hyperplane. Further we can make the following remarks about the composing circles or lines of an envelope corresponding to a curve $\mathbb{y}(t)$.

- If $\mathbb{y}(t)$ spans a 2-plane which does not live in the north hyperplane, all spheres of the corresponding family are orthogonal to a pencil of spheres. So, all the composing circles of the envelope are orthogonal to the radical plane of this pencil, hence it is a plane of symmetry of the envelope.
- If the 2-plane contains the north pole, then its polar line lies in the north hyperplane and goes through $\mathbb{n}$. Hence it represents a pencil of parallel planes, and it therefore follows that the circular envelope is a surface of revolution.
- If $\mathbb{y}(t)$ spans a 3-plane, then all the composing circles are orthogonal to a fixed sphere.

Here the polar line or conjugated line of a 2-plane E in $\overline{\mathbb{E}}^4$ with respect to a quadric is the line of intersection of all polar hyperplanes of points of E.

§6. Constructing Low Algebraic Degree Canal Surfaces Using Bézier Curves in $\overline{\mathbb{E}}^4$

For the purpose of this paper we define canal surface to be the envelope of a series of spheres (possibly of infinite radius). In particular our definition of canal surface includes surfaces that contain lines.

We look at curves in $\overline{\mathbb{E}}^4$, whose corresponding series of spheres have envelopes of low algebraic degree. The first interesting case is that of conics which have been treated in [2]. We consider next the case of curves of degree three, which produce surfaces of algebraic degree less or equal to eight.

For $n = 3$, Theorem 1 reads:

Theorem 3. *Rational curves of degree three in $\overline{\mathbb{E}}^4$ generate canal surfaces of degree less or equal than eight, containig the circle at infinity not more than four times.*

An arbitrary curve of degree three in $\overline{\mathbb{E}}^4$ might be written in homogeneous Bézier form

$$\mathbb{y}(t) = \sum_{i=0}^{3} \mathbb{b}_i B_i^3(t); \qquad t \in I \subseteq \mathbb{R} \tag{3}$$

with the homogeneous Bézier points $\mathbb{b}_i = (b_0^i, b_1^i, b_2^i, b_3^i, b_4^i)^T$, $i = 0, \ldots, 3$.

The envelope of the corresponding family of spheres computed as in the proof of Theorem 1 is determined by

$$(-6a_0a_1a_2a_3 + a_0^2a_3^2 - 3a_1^2a_2^2 + 4a_0a_2^3 + 4a_1^3a_3) = 0, \tag{4}$$

where $a_i = (b_4^i - b_0^i)(x^2 + y^2 + z^2)^2 + 2(b_1^i x + b_2^i y + b_3^i z) - (b_4^i + b_0^i)$; $i = 0, \ldots, 3$ if no $\mathbb{b}_i$ lies in the north hyperplane. This equation does not contain the component that represents the sphere $(-a_0 + 3a_1 - 3a_2 + a_3) = 0$, which by definition is not part of the envelope.

If one of the control points $\mathbb{b}_i$ lies in the north hyperplane, $b_4^i - b_0^i = 0$ and the corresponding polynomial is reduced to a polynomial of degree one

$a_i = 2(b_1^i x + b_2^i y + b_3^i z) - (b_4^i + b_0^i)$. If $\mathbb{b}_i = \mathbb{n}$ its correspoding polynomial is reduced to the constant $a_i = -(b_4^i + b_0^i)$.

This suggests to search for conditions under which the degree of the envelope drops.

Degree dropping

Let $\mathbb{y}(t)$ be a curve, that contains a point $\mathbb{y}(t_0)$ lying in the north hyperplane. Without loosing generality the Bézier representation (3) can be choosen in a way that $\mathbb{b}_0 \equiv \mathbb{y}(t_0)$ and by looking at the a_i's in (3) we observe,

- if $\mathbb{b}_0$ and $\mathbb{b}_1$ lie in the north hyperplane, (4) is of degree less or equal than seven.
- if $\mathbb{b}_0 = \mathbb{n}$ and $\mathbb{b}_1$ lies in the north hyperplane or if $\mathbb{b}_0, \mathbb{b}_1$ and $\mathbb{b}_2$ are lying in the north hyperplane, (4) has degree at most six.
- if $\mathbb{b}_0 = \mathbb{n}$ and $\mathbb{b}_1, \mathbb{b}_2$ are lying in the north hyperplane, (4) is of degree less or equal than five.
- if $\mathbb{b}_0, \mathbb{b}_1, \mathbb{b}_2$ and $\mathbb{b}_3$ are lying in the north hyperplane, the degree of (4) is at most four.

This can be rephrased in a parameter independent fashion.

Theorem 4. *A regular curve of degree three of* $\overline{\mathbb{E}}^4$ *generates in* $\mathbb{E}^3$ *a surface of degree less or equal than*

- *seven, if it touches the north hyperplane at a point which is not the north pole.*
- *six, if it touches the north hyperplane at the north pole or it has an osculating 2-plane that lies in the north hyperplane at a point distinct from the north pole.*
- *five, if it touches the north hyperplane at the north pole and has an osculating 2-plane that lies in the north hyperplane.*
- *four, if it lies in the north hyperplane.*

§7. Constructing Tubes

A circular contact in $\mathbb{E}^3$ is a pair of a sphere and a circle lying on it. Its corresponding figure in $\overline{\mathbb{E}}^4$ is a point and a line through it.

The problem of constructing piecewise canal surfaces consists of interpolating circular contacts given in pairs, i.e. constructing a surface that is tangent to a series of spheres along prescribed circles. In $\overline{\mathbb{E}}^4$ this problem is reduced to the construction of a piecewise rational curve that is tangent to pairs of lines at given points. Higher order continuity between the curves will be translated into higher order smoothness of the tube at the joins.

Piecewise canal surfaces that are G^1 at the joins can be constructed in the following way. Consider two G^1-connected Bézier curves $k_1 : \mathbb{y}(t) = \sum_{i=0}^{n} \mathbb{b}_i B_i^n(t)$, $t \in [0,1]$ and $k_2 : \mathbb{z}(t) = \sum_{i=0}^{m} \mathbb{c}_i B_i^m(t)$, $t \in [0,1]$ with $\mathbb{b}_n = \mathbb{c}_0$; and $\mathbb{b}_n \wedge \mathbb{b}_{n-1} = \mathbb{c}_0 \wedge \mathbb{c}_1$ that don't touch the north hyperplane. k_1

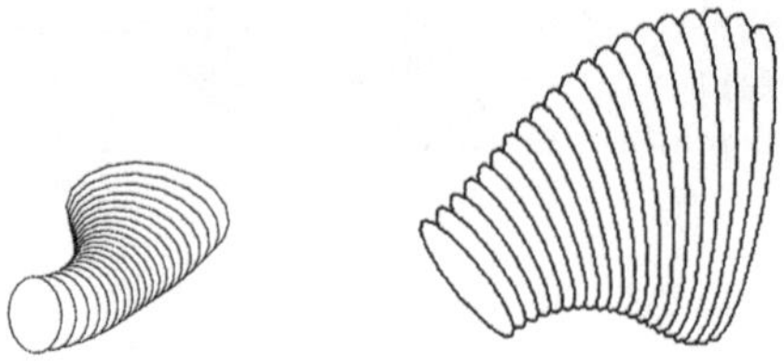

Fig. 1. Segments of canal surfaces corresponding to segments of conics.

determines a canal surface C_1 that interpolates the circular contacts given by the spheres corresponding to $\mathbb{b}_0$ and $\mathbb{b}_n$, along the circles corresponding to the lines $\mathbb{b}_0 \wedge \mathbb{b}_1$ and $\mathbb{b}_n \wedge \mathbb{b}_{n-1}$, respectively. In the same way k_2 determines a canal surface C_2. The sphere $\psi^{-1}(\mathbb{b}_n) = \psi^{-1}(\mathbb{c}_0)$ touches both surfaces at the circle corresponding to $\mathbb{b}_n \wedge \mathbb{b}_{n-1}$. Thus, the surfaces C_1 and C_2 have G^1 contact along the circle $\psi^{-1}(\mathbb{b}_n \wedge \mathbb{b}_{n-1})$.

References

1. Boehm, W., On cyclides in geometric modeling, Comput. Aided Geom. Design **7** (1990), 243–255.
2. Boehm, W. and M. Paluszny, General cyclides as joining pipes, Preprint. 1997. (http://euler.ciens.ucv.ve/~marco/articles/pipe.html)
3. Chionh, E. W., and R. N. Goldmann, Elimination and Resultants. Part 1: Elimination and Bivariant Resultants, IEEE Comp. Graph. Appl. **15** (1995), 69–77.
4. Coxeter, H. S. and S. L. Greitzer, *Geometry Revisited*, Random House, 1967.
5. Lü, W., and H. Pottmann, Pipe surfaces with rational spine curve are rational, Comput. Aided Geom. Design **13** (1996), 621–628.
6. Peternell, M., and H. Pottmann, Computing rational parametrizations of canal surfaces, Journal of Symbolic Computation **23** (1997), 255–266.
7. Pottmann, H., and M. Peternell, Applications of Laguerre geometry in CAGD I, Tech. Report Nr.30, 1996. Institut für Geometrie. Technische Universität Wien.
8. Pratt, M. J., Cyclides in computer aided geometric design, Comput. Aided Geom. Design **7** (1990), 221–242.

Marco Paluszny and Katja Bühler
Centro de Computación Gráfica y Geometría Aplicada
Apartado 47809, Los Chaguaramos, Caracas 1041, VENEZUELA
`marco@ciens.ucv.ve`
`katja@ciens.ucv.ve`
`katja@ira.uka.de`

From Wavelets to Multiwavelets

Gerlind Plonka and Vasily Strela

Abstract. This paper gives an overview of recent achievements of the multiwavelet theory. The construction of multiwavelets is based on a multiresolution analysis with higher multiplicity generated by a scaling vector. The basic properties of scaling vectors such as L^2-stability, approximation order and regularity are studied. Most of the proofs are sketched.

§1. Introduction

Wavelet theory is based on the idea of *multiresolution analysis* (MRA). Usually it is assumed that an MRA is generated by one scaling function, and dilates and translates of only one wavelet $\psi \in L^2(\mathbb{R})$ form a stable basis of $L^2(\mathbb{R})$. This paper considers a recent generalization allowing several wavelet functions $\psi_1, \ldots, \psi_r$. The vector $\mathbf{\Psi} = (\psi_1, \ldots, \psi_r)^{\mathrm{T}}$ is then called a multiwavelet.

Multiwavelets have more freedom in their construction and thus can combine more useful properties than the scalar wavelets. Symmetric scaling functions constructed by Geronimo, Hardin, and Massopust [16] (see Figure 1) have short support, generate an orthogonal MRA, and provide approximation order 2. These properties are very desirable in many applications but cannot be achieved by one scaling function. Thus, multiwavelets can be useful for various practical problems [11,53].

Our purpose is to give a survey of the basic ideas of multiwavelet theory and to show how naturally multiwavelets generalize the scalar ones. We start with a simple example of piecewise linear multiwavelets and the definition of an MRA. Then we discuss necessary properties of a function vector $\mathbf{\Phi} = (\phi_1, \ldots, \phi_r)^{\mathrm{T}}$ in order to generate an MRA. In particular, $\mathbf{\Phi}$ has to be *refinable* and L^2-*stable*, i.e., $\mathbf{\Phi}$ can be seen as a stable solution vector of a *matrix refinement equation.* Such a vector is then called a scaling vector or multiscaling function. The *mask symbol* of the matrix refinement equation is closely associated with the scaling vector. Similarly to the scalar case, the mask symbol and two linear operators, the *transition operator* and the *subdivision operator*, are the main tools in multiwavelet theory.

Mathematical Methods for Curves and Surfaces II
Morten Dæhlen, Tom Lyche, and Larry L. Schumaker (eds.), pp. 375–399.

ISBN 0-8265-1315-8.

In Section 2, we discuss matrix refinement equations with compactly supported, L^2-stable solutions. Section 3 is devoted to the approximation properties of scaling vectors. We show how approximation comes from the Strang-Fix conditions, and that it implies certain factorization of the mask symbol. This factorization generalizes zeros at $\omega = \pi$ of the scalar symbols.

In Section 4 we give an overview of basic methods of estimation of the scaling vectors' smoothness. Aiming to explain the main ideas behind the theory, we will sketch the proofs of most of the assertions. Unfortunately, space does not allow us to consider more applied aspects of multiwavelets, such as decomposition and reconstruction algorithms, construction of biorthogonal bases, lifting, and preprocessing. We only mention that the application of multiwavelets is a fast developing field, and the literature on this subject is rapidly growing.

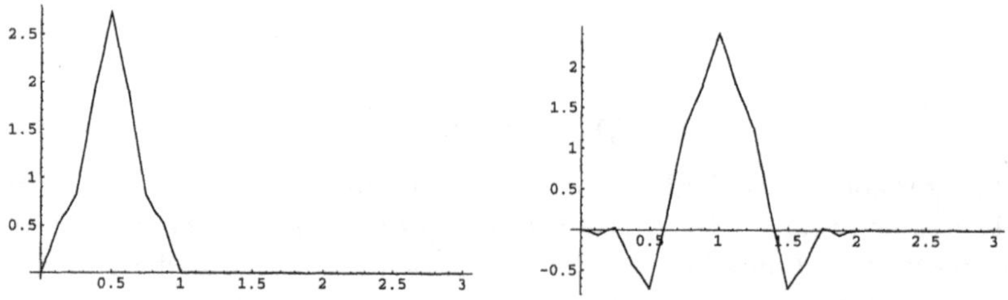

Fig. 1. GHM symmetric orthogonal multi–scaling function with approximation order 2.

1.1. Example of Linear Multiwavelets

Let us start with a simple example taken from [1]. Consider two piecewise linear functions

$$\phi_1(t) = \begin{cases} 1, & 0 \le t < 1, \\ 0, & \text{otherwise}, \end{cases} \qquad \phi_2(t) = \begin{cases} 2, \sqrt{3}(t - \frac{1}{2}), & 0 \le t < 1, \\ 0, & \text{otherwise}, \end{cases}$$

(see Figure 2). Their integer translates $\phi_1(\cdot - l), \phi_2(\cdot - l)$, $l \in \mathbb{Z}$ form an orthonormal basis of the closed subspace $V_0 \subset L^2(\mathbb{R})$ containing functions piecewise linear on integer intervals. Furthermore, let V_j be the closed subspace of $L^2(\mathbb{R})$ spanned by $2^{j/2}\,\phi_1(2^j \cdot - l)$, $2^{j/2}\,\phi_2(2^j \cdot - l)$ ($l \in \mathbb{Z}$), and containing all functions which are piecewise linear on the intervals $[2^{-j}l,\ 2^{-j}(l+1)]$ ($l \in \mathbb{Z}$). It is easy to see that

$$\begin{bmatrix} \phi_1 \\ \phi_2 \end{bmatrix} = \begin{bmatrix} 1 & 0 \\ -\frac{\sqrt{3}}{2} & \frac{1}{2} \end{bmatrix} \begin{bmatrix} \phi_1(2\cdot) \\ \phi_2(2\cdot) \end{bmatrix} + \begin{bmatrix} 1 & 0 \\ \frac{\sqrt{3}}{2} & \frac{1}{2} \end{bmatrix} \begin{bmatrix} \phi_1(2\cdot -1) \\ \phi_2(2\cdot -1) \end{bmatrix}.$$

This relation represents the fact that $\phi_1, \phi_2 \in V_0 \subset V_1$. Analogously,

$$V_j \subset V_{j+1} \quad (j \in \mathbb{Z}). \tag{1}$$

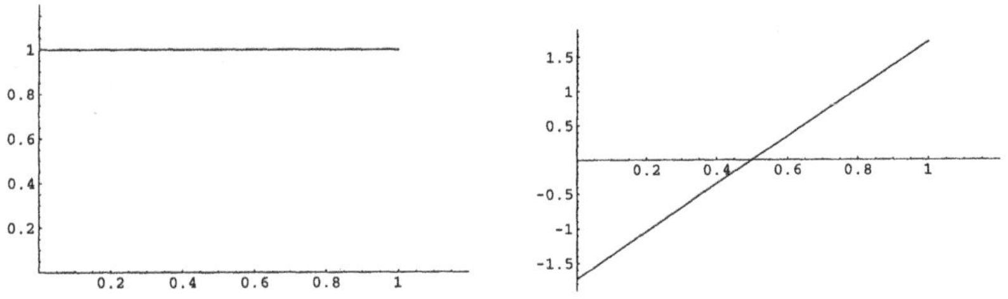

Fig. 2. Piecewise linear orthogonal scaling functions ϕ_1, ϕ_2 with approximation order 2.

Orthogonal projections $f_j(t)$ of any function $f \in L^2(\mathbb{R})$ on subspaces V_j are successive piecewise linear approximations converging to $f(t)$ as j goes to ∞. Thus,

$$\overline{\bigcup_{j=-\infty}^{\infty} V_j} = L^2(\mathbf{R}), \qquad \bigcap_{j=-\infty}^{\infty} V_j = \{0\}. \tag{2}$$

This nested structure of subspaces $\{V_j\}_{j\in\mathbb{Z}}$ is usually referred to as a multiresolution analysis (MRA). In our case it is generated by two functions ϕ_1, ϕ_2 and thus is of multiplicity 2. Observe that the spaces V_j considered here cannot be spanned by integer translates of only one function ϕ.

Consider two more piecewise linear functions

$$\psi_1(t) = \begin{cases} 6t-1, & 0 \le t < 1/2, \\ 6t-5, & 1/2 \le t < 1, \\ 0, & \text{otherwise}, \end{cases} \qquad \psi_2(t) = \begin{cases} 2\sqrt{3}(2t-\frac{1}{2}), & 0 \le t < 1/2, \\ -2\sqrt{3}(2t-\frac{3}{2}), & 1/2 \le t < 1, \\ 0, & \text{otherwise}, \end{cases}$$

(see Figure 3). Let W_j ($j \in \mathbb{Z}$) be the closed subspaces of $L^2(\mathbb{R})$ spanned by $2^{j/2}\psi_1(2^j\cdot -l)$ and $2^{j/2}\psi_2(2^j\cdot -l)$ ($l \in \mathbb{Z}$).

The integer translates $\psi_1(\cdot - l), \psi_2(\cdot - n), l, n \in \mathbb{Z}$ are orthogonal to each other and to the integer translates of ϕ_1, ϕ_2 which makes W_0 orthogonal to V_0. ψ_1 and ψ_2 are piecewise linear on half integer intervals, thus $W_0 \subset V_1$. In particular,

$$\begin{bmatrix} \psi_1 \\ \psi_2 \end{bmatrix} = \begin{bmatrix} \frac{1}{2} & \frac{\sqrt{3}}{2} \\ 0 & 1 \end{bmatrix} \begin{bmatrix} \phi_1(2\cdot) \\ \phi_2(2\cdot) \end{bmatrix} + \begin{bmatrix} -\frac{1}{2} & \frac{\sqrt{3}}{2} \\ 0 & -1 \end{bmatrix} \begin{bmatrix} \phi_1(2\cdot -1) \\ \phi_2(2\cdot -1) \end{bmatrix}.$$

Finally, generators of V_1 are linear combinations of translates of $\phi_1, \phi_2, \psi_1, \psi_2$,

$$\phi_1(2\cdot) = \frac{1}{2}\phi_1 - \frac{\sqrt{3}}{4}\phi_2 + \frac{1}{4}\psi_1, \quad \phi_1(2\cdot -1) = \frac{1}{2}\phi_1 + \frac{\sqrt{3}}{4}\phi_2 - \frac{1}{4}\psi_1,$$

$$\phi_2(2\cdot) = \frac{1}{4}\phi_2 + \frac{\sqrt{3}}{4}\psi_1 + \frac{1}{2}\psi_2, \quad \phi_2(2\cdot -1) = \frac{1}{4}\phi_2 + \frac{\sqrt{3}}{4}\psi_1 - \frac{1}{2}\psi_2.$$

Hence, V_1 is an orthogonal sum of V_0 and W_0. Analogously, it follows that $V_j \oplus W_j = W_{j+1}$ for all $j \in \mathbb{Z}$, and

$$L^2(\mathbb{R}) = \bigoplus_{j\in\mathbb{Z}} W_j = \operatorname{clos}_{L^2} \operatorname{span}\{2^{j/2}\psi_1(2^j\cdot -l),\ 2^{j/2}\psi_2(2^j\cdot -l) : l, j \in \mathbb{Z}\}.$$

The above implies that the dilates and translates of ψ_1, ψ_2 form an orthogonal basis of $L^2(\mathbb{R})$.

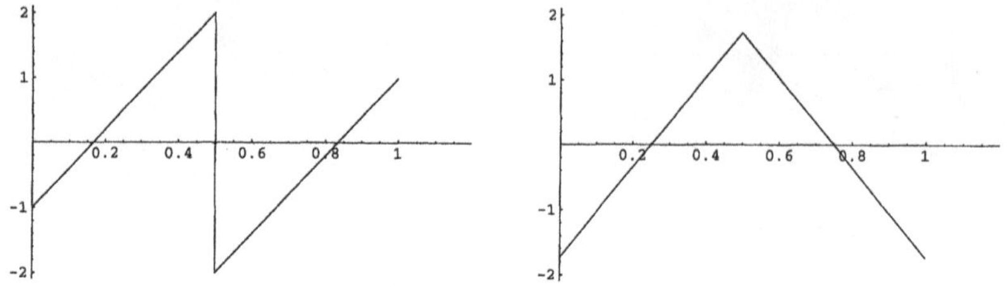

Fig. 3. Piecewise linear orthogonal wavelets ψ_1, ψ_2.

1.2 Multiresolution Analysis with Multiplicity r

Generally, a sequence of closed subspaces $\{V_j\}_{j\in\mathbb{Z}}$ of $L^2(\mathbb{R})$ is called a multiresolution analysis (MRA) of multiplicity r if (1) and (2) are satisfied, and if there exists a vector $\mathbf{\Phi} = (\phi_1, \ldots, \phi_r)^{\mathrm{T}}$ of L^2-functions such that $2^{j/2}\,\phi_\nu(2^j\cdot -l)$ ($\nu = 1, \ldots r$, $l \in \mathbb{Z}$) form an L^2-stable basis of V_j. The scaling spaces V_j are *finitely generated* $2^{-j}\,\mathbb{Z}$*-translation-invariant subspaces of* $L^2(\mathbb{R})$. Function vectors $\mathbf{\Phi}$ that generate an MRA with multiplicity $r \geq 1$, are called scaling vectors or multi-scaling functions.

Once an MRA $\{V_j\}_{j\in\mathbb{Z}}$ is given, we define the *wavelet spaces* W_j as orthogonal complements of V_j in V_{j+1}. The wavelet spaces W_j are also finitely generated $2^{-j-1}\mathbb{Z}$-translation-invariant subspaces of $L^2(\mathbb{R})$. Moreover, the structure of an MRA implies that W_j is the closure of the span of $2^{j/2}\psi_\nu(2^j\cdot -l)$ ($\nu = 1, \ldots, r$, $l \in \mathbb{Z}$) if only

$$W_0 = \mathrm{clos}_{L^2}\,\mathrm{span}\{\psi_\nu(\cdot - l) : \nu = 1, \ldots, r-1,\, l \in \mathbb{Z}\}$$

can be shown. If we can find a function vector $\mathbf{\Psi} = (\psi_1, \ldots, \psi_r)^{\mathrm{T}}$ such that $\{\psi_\nu(\cdot - l) : \nu = 1, \ldots, , r,\, l \in \mathbb{Z}\}$ forms an L^2-stable basis of W_0, then $\{2^{j/2}\,\psi_\nu(2^j\cdot -l) : \nu = 1, \ldots, r,\, l, j \in \mathbb{Z}\}$ forms an L^2-stable basis of $L^2(\mathbb{R})$.

In the Fourier domain, the problem of finding the multiwavelet $\mathbf{\Psi}$ can be reduced to an algebraic problem of matrix completion (see e.g. [36,51]). Moreover, if $\mathbf{\Phi}$ is compactly supported, then a corresponding compactly supported wavelet vector can be found.

Since $\psi_\nu \in W_0 \subset V_1$ is a linear combination of the dilated components of the scaling vector $\mathbf{\Phi}$, most properties of the multiwavelet $\mathbf{\Psi}$ are determined by the properties of $\mathbf{\Phi}$.

A scaling vector $\mathbf{\Phi}$ has to satisfy some special properties which are induced by the conditions of the MRA. Another set of properties such as compact support and smoothness of the components ϕ_ν of $\mathbf{\Phi}$ and polynomial reproduction in V_j is required by applications.

Condition (1) of the MRA implies that $\mathbf{\Phi}$ needs to satisfy a *matrix refinement equation* of the form

$$\mathbf{\Phi}(t) = \sum_{l\in\mathbb{Z}} P_l\,\mathbf{\Phi}(2t - l). \qquad (3)$$

Here $\boldsymbol{P}_l$ are $r \times r$ *mask* coefficient matrices. A function vector $\boldsymbol{\Phi}(t)$ satisfying (3) is called refinable. Application of the Fourier transform to (3) leads to

$$\widehat{\boldsymbol{\Phi}}(\omega) = \boldsymbol{P}\left(\frac{\omega}{2}\right) \widehat{\boldsymbol{\Phi}}\left(\frac{\omega}{2}\right), \tag{4}$$

where $\boldsymbol{P}$ denotes the symbol of the mask $\{\boldsymbol{P}_l\}_{l \in \mathbb{Z}}$,

$$\boldsymbol{P}(\omega) := \frac{1}{2} \sum_{l \in \mathbb{Z}} \boldsymbol{P}_l \, e^{-i\omega l}. \tag{5}$$

Here, the Fourier transform is taken componentwise, i.e., $\hat{\boldsymbol{\Phi}} = (\hat{\phi}_1, \ldots, \hat{\phi}_r)^{\mathrm{T}}$ with $\hat{\phi}_\nu(\omega) := \int_{-\infty}^{\infty} \phi_\nu(t) \, e^{-i\omega t} \, \mathrm{d}t$.

For simplicity we assume that the sum in the refinement equation (3) is finite, or equivalently, the symbol $\boldsymbol{P}$ is a matrix of trigonometric polynomials $\boldsymbol{P}(\omega) = 2^{-1} \sum_{l=0}^{N} \boldsymbol{P}_l \, e^{-i\omega l}$. Being interested in the refinable function vectors, we will view the refinement equation as a functional equation. The finite mask already implies that solutions of (3) are compactly supported in the sense that each component ϕ_ν $(\nu = 1, \ldots, r)$ is compactly supported.

The second condition on $\boldsymbol{\Phi}$ induced by the MRA is the L^2-stability. We say that a function vector $\boldsymbol{\Phi} \in (L^2(\mathbb{R}))^r$ is L^2-stable if there are constants $0 < A \leq B < \infty$ such that

$$A \sum_{l=-\infty}^{\infty} \boldsymbol{c}_l^\star \boldsymbol{c}_l \leq \| \sum_{l=-\infty}^{\infty} \boldsymbol{c}_l^\star \boldsymbol{\Phi}(\cdot - l) \|_{L^2}^2 \leq B \sum_{l=-\infty}^{\infty} \boldsymbol{c}_l^\star \boldsymbol{c}_l \tag{6}$$

holds for any vector sequence $\{\boldsymbol{c}_l\}_{l \in \mathbb{Z}} \in l^2(\mathbb{Z})^r$. Here $l^2(\mathbb{Z})^r$ denotes the set of sequences of vectors $\boldsymbol{c}_l \in \mathbb{R}^r$ with $\sum_{l=-\infty}^{\infty} \boldsymbol{c}_l^\star \boldsymbol{c}_l < \infty$ and $\boldsymbol{c}_l^\star$ stands for $\overline{\boldsymbol{c}_l}^{\mathrm{T}}$.

Let us introduce the autocorrelation symbol

$$\boldsymbol{\Omega}(\omega) := \sum_{l=-\infty}^{\infty} \left(\langle \phi_\mu, \phi_\nu(\cdot - l) \rangle_{L^2} \right)_{\mu,\nu=1}^{r} e^{-i\omega l}.$$

The Poisson summation formula gives

$$\boldsymbol{\Omega}(\omega) = \sum_{n=-\infty}^{\infty} \hat{\boldsymbol{\Phi}}(\omega + 2\pi n) \, \hat{\boldsymbol{\Phi}}(\omega + 2\pi n)^\star \quad a.e.$$

Observe that the autocorrelation symbol is positive semidefinite, and for compactly supported $\boldsymbol{\Phi}$, it is a matrix of trigonometric polynomials. L^2-stability of $\boldsymbol{\Phi}$ is equivalent to the assertion that $\boldsymbol{\Omega}(\omega)$ is strictly positive definite for all real ω (see [18]). In the Fourier domain, L^2-stability of $\boldsymbol{\Phi}$ is ensured if and only if the sequences $\{\hat{\phi}_\nu(\omega + 2\pi l)\}_{l \in \mathbb{Z}}$ $(\nu = 1, \ldots, r)$ are linearly independent for each $\omega \in \mathbb{R}$ (see [29]).

Finally, let us consider the union and intersection properties (2) of the MRA. As shown in [17] the equality $\mathrm{clos}_{L^2} \cup_{j\in\mathbb{Z}} V_j = L^2(\mathbb{R})$ is satisfied if and only if $\cup_{\nu=1}^r \mathrm{supp}\, \hat{\phi}_\nu = \mathbb{R}$ (modulo a null set). This condition obviously holds for compactly supported $\boldsymbol{\Phi}$. The intersection condition $\cap_{j\in\mathbb{Z}} V_j = \{0\}$ follows for nested sequences V_j if $\sum_{\nu=1}^r |\hat{\phi}_\nu(\omega)| > 0$ *a.e.* in some neighborhood of the origin (see [17], Theorem 3). Indeed, this condition is already satisfied if $\boldsymbol{\Phi}$ is refinable and L^2-stable (see Section 2).

Let us summarize: In order to obtain a scaling vector $\boldsymbol{\Phi}$ generating an MRA, it is enough to find a compactly supported, L^2-stable function vector which is a solution of a matrix refinement equation (3). In other words, we are looking for a mask $\{\boldsymbol{P}_l\}$ (a symbol $\boldsymbol{P}(\omega)$), such that a compactly supported L^2-solution vector of (3) is L^2-stable. This problem will be considered in Section 2.

Applications usually require several other features such as exact polynomial reproduction in V_j (vanishing moments of the wavelets) and smoothness of the elements of V_j. The corresponding properties of the scaling vector $\boldsymbol{\Phi}$ are considered in Sections 3 and 4.

1.3. Transition Operator and Subdivision Operator

Besides the mask, two other important tools for studying the properties of a scaling vector are the transition operator T and the subdivision operator S. For a given mask symbol $\boldsymbol{P}(\omega) = 2^{-1}\sum_{l=0}^N \boldsymbol{P}_l\, e^{-i\omega l}$, the transition operator $T : (L^2_{2\pi})^{r\times r} \to (L^2_{2\pi})^{r\times r}$ acting on $r\times r$ matrices $\boldsymbol{H}(\omega)$ with 2π-periodic quadratic integrable entries is defined by

$$T\boldsymbol{H}(2\omega) := \boldsymbol{P}(\omega)\boldsymbol{H}(\omega)\boldsymbol{P}^\star(\omega) + \boldsymbol{P}(\omega+\pi)\boldsymbol{H}(\omega+\pi)\boldsymbol{P}^\star(\omega+\pi). \qquad (7)$$

Observe that the autocorrelation symbol $\boldsymbol{\Omega}(\omega)$ is an eigenmatrix of T corresponding to the eigenvalue 1. For $\boldsymbol{H} \in (L^2_{2\pi})^{r\times r}$, we find

$$(T^n\boldsymbol{H})(\omega) = \sum_{l=0}^{2^n-1} \boldsymbol{\Pi}_n(\omega+2\pi l)\, \boldsymbol{H}(\frac{\omega+2\pi l}{2^n})\, \boldsymbol{\Pi}_n(\omega+2\pi l)^\star, \qquad (8)$$

where $\boldsymbol{\Pi}_n(\omega) := \prod_{j=1}^n \boldsymbol{P}(2^{-j}\omega)$. If $\mathbb{H}_M$ is the space of $r\times r$ matrices of trigonometric polynomials of degree at most M, then for $M \geq N$, $\mathbb{H}_M$ is invariant under the action of T.

The transition operator T is a linear operator, and $T : \mathbb{H}_M \to \mathbb{H}_M$ can be represented by an $r^2(2M+1)\times r^2(2M+1)$ matrix $\boldsymbol{T}$

$$\boldsymbol{T} := (\frac{1}{2}\boldsymbol{A}_{2k-j})_{j,k=-M}^M$$

with

$$\boldsymbol{A}_j := \sum_{l=0}^M \boldsymbol{P}_{l-j} \otimes \boldsymbol{P}_l,$$

where $\boldsymbol{A} \otimes \boldsymbol{B}$ stands for the Kronecker product of matrices $\boldsymbol{A} = (a_{jk})_{j,k=1}^{r} \in \mathbb{C}^{r\times r}$ and $\boldsymbol{B} \in \mathbb{C}^{r\times r}$,

$$\boldsymbol{A} \otimes \boldsymbol{B} := \begin{bmatrix} a_{11}\boldsymbol{B} & \dots & a_{1r}\boldsymbol{B} \\ \vdots & \ddots & \vdots \\ a_{r1}\boldsymbol{B} & \dots & a_{rr}\boldsymbol{B} \end{bmatrix}$$

(see [33,41]). As in the scalar case, the spectral properties of the transition operator govern stability and regularity of the scaling vectors.

The subdivision operator $S = S_{\boldsymbol{P}}$ associated with the refinement mask $\{\boldsymbol{P}_l\}_{l\in\mathbb{Z}}$ of $\boldsymbol{\Phi}$ is a linear operator on $l(\mathbb{Z})^r$ defined by

$$(S_{\boldsymbol{P}}\boldsymbol{c})_\alpha = \sum_{\beta\in\mathbb{Z}} \boldsymbol{P}^\star_{\alpha-2\beta}\, \boldsymbol{c}_\beta, \tag{9}$$

where $\boldsymbol{c} \in l(\mathbb{Z})^r$. Here $l(\mathbb{Z})^r$ denotes the set of sequences with arbitrary vectors of $\mathbb{R}^r$ as entries. With the help of the double infinite matrix

$$\boldsymbol{L} := (\boldsymbol{P}_{\alpha-2\beta})_{\alpha,\beta=-\infty}^{\infty}$$

we find $S_{\boldsymbol{P}}\boldsymbol{c} = \boldsymbol{L}^\star\, \boldsymbol{c}$. Note that with the infinite vector $\boldsymbol{F} := (\dots, \boldsymbol{\Phi}(\cdot + 1)^{\mathrm{T}}, \boldsymbol{\Phi}(\cdot)^{\mathrm{T}}, \boldsymbol{\Phi}(\cdot-1)^{\mathrm{T}}, \dots)^{\mathrm{T}}$ the refinement equation (3) can formally be written in the vector form $\boldsymbol{L}\,\boldsymbol{F}(2\cdot) = \boldsymbol{F}$.

Assuming that $\boldsymbol{c} \in l^2(\mathbb{Z})^r$, we can consider the Fourier series $\hat{\boldsymbol{c}}(\omega) := \sum_l \boldsymbol{c}_l\, e^{-i\omega l}$. Then (9) leads to

$$(\widehat{S_{\boldsymbol{P}}\boldsymbol{c}})(\omega) = 2\, \boldsymbol{P}(\omega)^\star\, \hat{\boldsymbol{c}}(2\omega).$$

In the scalar case, there is a simple connection between transition and subdivision operators, namely

$$(\widehat{S_{|P|^2}c})(\omega) = (T^\star \hat{c})(\omega) = 2\, |P(\omega)|^2\, \hat{c}(2\omega)$$

for $c \in l^2(\mathbb{Z})$. Here $T^\star$ denotes the adjoint operator of T. In the vector case, we can find a similar close relation (see e.g. [47]): Let for the $r \times r$ matrix $\boldsymbol{H}$ with columns $\boldsymbol{h}_1, \dots, \boldsymbol{h}_r$, let the vec-operator be defined as

$$\operatorname{vec} \boldsymbol{H} = (\boldsymbol{h}_1^{\mathrm{T}}, \dots, \boldsymbol{h}_r^{\mathrm{T}})^{\mathrm{T}} \in \mathbb{C}^{r^2}.$$

Introduce the scalar product of two matrices $\boldsymbol{H}, \boldsymbol{G} \in L^2(\mathbb{R})^{r\times r}$ by

$$\begin{aligned} \langle \boldsymbol{H}, \boldsymbol{G}\rangle_2 := \langle \operatorname{vec} \boldsymbol{H}, \operatorname{vec} \boldsymbol{G}\rangle &:= \sum_{j=1}^{r^2} \langle (\operatorname{vec} \boldsymbol{H})_j, (\operatorname{vec} \boldsymbol{G})_j\rangle_{L^2} \\ &= \sum_{j=1}^{r^2} \frac{1}{2\pi} \int_{-\pi}^{\pi} (\operatorname{vec} \boldsymbol{H})_j(\omega)\, \overline{(\operatorname{vec} \boldsymbol{G})_j(\omega)}\, \mathrm{d}\omega, \end{aligned}$$

and recall the well-known property of Kronecker products $\operatorname{vec}(\boldsymbol{A}\boldsymbol{X}\boldsymbol{B}) = (\boldsymbol{B}^{\mathrm{T}} \otimes \boldsymbol{A}) \operatorname{vec} \boldsymbol{X}$ for $\boldsymbol{A}, \boldsymbol{B}, \boldsymbol{X} \in \mathbb{C}^{r\times r}$. Then we find

$$\begin{aligned}
\langle T\boldsymbol{H}, \boldsymbol{G}\rangle_2 &= \Big\langle \operatorname{vec}\Big(\boldsymbol{P}(\tfrac{\cdot}{2})\,\boldsymbol{H}(\tfrac{\cdot}{2})\,\boldsymbol{P}(\tfrac{\cdot}{2})^\star \\
&\quad + \boldsymbol{P}(\tfrac{\cdot}{2}+\pi)\,\boldsymbol{H}(\tfrac{\cdot}{2}+\pi)\,\boldsymbol{P}(\tfrac{\cdot}{2}+\pi)^\star\Big), \operatorname{vec}\boldsymbol{G}\Big\rangle \\
&= \Big\langle \Big(\overline{\boldsymbol{P}(\tfrac{\cdot}{2})} \otimes \boldsymbol{P}(\tfrac{\cdot}{2})\Big) \operatorname{vec}\boldsymbol{H}(\tfrac{\cdot}{2}) \\
&\quad + \Big(\overline{\boldsymbol{P}(\tfrac{\cdot}{2}+\pi)} \otimes \boldsymbol{P}(\tfrac{\cdot}{2}+\pi)\Big) \operatorname{vec}\boldsymbol{H}(\tfrac{\cdot}{2}+\pi), \operatorname{vec}\boldsymbol{G}\Big\rangle \\
&= \sum_{j=1}^{r^2} \frac{1}{2\pi} \int_{-2\pi}^{2\pi} \Big[\Big(\overline{\boldsymbol{P}(\tfrac{\omega}{2})} \otimes \boldsymbol{P}(\tfrac{\omega}{2})\Big) \operatorname{vec}\boldsymbol{H}(\tfrac{\omega}{2})\Big]_j \,\overline{(\operatorname{vec}\boldsymbol{G})_j(\omega)}\, \mathrm{d}\omega \cdot \\
&= \frac{2}{2\pi} \sum_{k=1}^{r^2} \int_{-\pi}^{\pi} [\operatorname{vec}\boldsymbol{H}(\omega)]_k \Big[\Big(\overline{\boldsymbol{P}(\omega)} \otimes \boldsymbol{P}(\omega)\Big)^\star \,\overline{\operatorname{vec}\boldsymbol{G}(2\omega)}\Big]_k \mathrm{d}\omega \\
&= 2\langle \operatorname{vec}\boldsymbol{H}, \big(\overline{\boldsymbol{P}} \otimes \boldsymbol{P}\big)^{\mathrm{T}} \operatorname{vec}\boldsymbol{G}(2\cdot)\rangle.
\end{aligned}$$

On the other hand, for $\boldsymbol{a}, \boldsymbol{b} \in l^2(\mathbb{Z})^r$,

$$\begin{aligned}
\langle \boldsymbol{H}, (\widehat{S\boldsymbol{a}})\,(\widehat{S\boldsymbol{b}})^\star\rangle_2 &= 4\,\langle \operatorname{vec}\boldsymbol{H}, \operatorname{vec}(\boldsymbol{P}^\star\,\hat{\boldsymbol{a}}(2\cdot)\,\hat{\boldsymbol{b}}(2\cdot)^\star\,\boldsymbol{P})\rangle \\
&= 4\,\langle \operatorname{vec}\boldsymbol{H}, \big(\overline{\boldsymbol{P}} \otimes \boldsymbol{P}\big)^{\mathrm{T}} \operatorname{vec}(\hat{\boldsymbol{a}}\hat{\boldsymbol{b}}^\star)(2\cdot)\rangle.
\end{aligned}$$

Together, these observations imply the relation

$$2\,T^\star(\hat{\boldsymbol{a}}\,\hat{\boldsymbol{b}}^\star) = (\widehat{S\boldsymbol{a}})\,(\widehat{S\boldsymbol{b}})^\star.$$

Both operators play a crucial role in characterization of scaling vectors and lead to deep insight in the structure of the solutions of matrix refinement equations. As in the scalar case, the subdivision operator implies an efficient algorithm for the iterative computation of $\boldsymbol{\Phi}$ in the time domain. The corresponding subdivision algorithm is closely related to the cascade algorithm (see e.g. [38], Theorem 2.1, [39]). Actually, there is no reason to restrict the subdivision operator to the L^2-case, general solution vectors with components in L^p can also be handled (see e.g. [12,30,31]). However, in $L^2(\mathbb{R})$, the transition operator often provides simpler results. Since both, T and S are linear operators, their spectral properties are computable by considering their representing matrices.

§2. Existence, Uniqueness, and Stability of Scaling Vectors

In this section we summarize some basic results on existence and uniqueness of solutions $\boldsymbol{\Phi}$ of (3) in terms of the symbol $\boldsymbol{P}(\omega)$. We also are going to characterize the L^2-stability of $\boldsymbol{\Phi}$.

Let us assume that the mask symbol of $\mathbf{\Phi}$ is given in the form

$$\boldsymbol{P}(\omega) := \frac{1}{2}\sum_{k=0}^{N} \boldsymbol{P}_k e^{-i\omega k}. \tag{10}$$

The following theorem states a necessary and sufficient condition for the existence of a solution vector of (3).

Theorem 1. *The matrix refinement equation* (3) *has a compactly supported distributional solution vector* $\mathbf{\Phi}$ *if and only if* $\boldsymbol{P}(0)$ *has an eigenvalue of the form* 2^n, $n \in \mathbb{Z}, n \geq 0$.

Proof: The necessary part of Theorem 1 was first obtained in [21]. Since $\mathbf{\Phi}$ is compactly supported, $\hat{\mathbf{\Phi}}$ is analytic. Hence, $\hat{\mathbf{\Phi}} \neq \mathbf{0}$ implies that there is an integer $\alpha \geq 0$ with $\mathrm{D}^\alpha \hat{\mathbf{\Phi}}(0) \neq \mathbf{0}$ and $\mathrm{D}^n \hat{\mathbf{\Phi}}(0) = \mathbf{0}$ for $n = 0, \ldots, \alpha - 1$. Thus,

$$\begin{aligned}\mathrm{D}^\alpha \hat{\mathbf{\Phi}}(0) &= \mathrm{D}^\alpha \left[\boldsymbol{P}(\tfrac{\omega}{2})\,\hat{\mathbf{\Phi}}(\tfrac{\omega}{2})\right]\Big|_{\omega=0} = \frac{1}{2^\alpha}\sum_{n=0}^{\alpha}\binom{\alpha}{n}\mathrm{D}^{n-\alpha}\boldsymbol{P}(0)\,\mathrm{D}^n\hat{\mathbf{\Phi}}(0) \\ &= \frac{1}{2^\alpha}\boldsymbol{P}(0)\,\mathrm{D}^\alpha\hat{\mathbf{\Phi}}(0).\end{aligned}$$

The sufficient part was proved in [34]. Consider

$$\hat{\mathbf{\Phi}}_n(\omega) = \prod_{j=1}^{n} \boldsymbol{P}(\frac{\omega}{2^j})\,\hat{\mathbf{\Phi}}_0(\frac{\omega}{2^n})$$

with suitable compactly supported $\mathbf{\Phi}_0$. It can be shown that $\hat{\mathbf{\Phi}}_n$ converges for $n \to \infty$ to an entire function $\hat{\mathbf{\Phi}}$ with polynomial growth. Moreover, for $n > 0$, the solution $\mathbf{\Phi}$ of the refinement equation (3) with symbol $\boldsymbol{P}(\omega)$ is n-th distributional derivative of the solution of the dilation equation with symbol $\frac{1}{2^n}\boldsymbol{P}(\omega)$. □

Theorem 1 is analogous to the condition of existence in the scalar case $r = 1$. However, unlike the scalar case when the uniqueness (up to multiplication by a constant) of a distributional solution is also guaranteed, in the vector case ($r > 1$) the number of linearly independent solutions of the matrix refinable equation (3) is determined by the multiplicity of the eigenvalue 2^n of $\boldsymbol{P}(0)$.

Without loss of generality, we restrict ourselves to the case $n = 0$ and assume that 1 is the only eigenvalue of $\boldsymbol{P}(0)$ of the form 2^n, $n \in \mathbb{Z}$, $n \geq 0$. (As we will see later, L^2-stability implies that the spectral radius of $\boldsymbol{P}(0)$ is equal to 1). Then uniqueness can be ensured as follows (see [21,34]).

Theorem 2. *Let* 1 *be the only eigenvalue of* $\boldsymbol{P}(0)$ *of the form* 2^n, $n \in \mathbb{Z}$, $n \geq 0$. *The matrix refinement equation (3) has a unique (up to a constant) solution* $\mathbf{\Phi}$ *with* $\widehat{\mathbf{\Phi}}(0) = \boldsymbol{r}$ *if and only if 1 is a simple eigenvalue of* $\boldsymbol{P}(0)$ *and* $\boldsymbol{r} = \boldsymbol{P}(0)\boldsymbol{r}$. *In particular,*

$$\hat{\mathbf{\Phi}}(\omega) = \lim_{L\to\infty}\prod_{j=1}^{L}\boldsymbol{P}\left(\frac{\omega}{2^j}\right)\boldsymbol{r}. \tag{11}$$

The convergence of the infinite product (11) is also studied in [9,23,56].

We are especially interested in L^2-stable solutions. Let us introduce the following definition. A matrix (or a linear operator) $\boldsymbol{A}$ is said to satisfy Condition E if it has a simple eigenvalue 1 and the moduli of all its other eigenvalues are less then 1. First we observe some necessary conditions (see e.g. [12,25,34]).

Theorem 3. *Let $\boldsymbol{\Phi}$ be a compactly supported, L^2-stable solution vector of (3). Then for the corresponding symbol $\boldsymbol{P}(\omega)$ we have*

(a) *$\boldsymbol{P}(0)$ satisfies Condition E.*

(b) *There exists a nonzero vector $\boldsymbol{y} \in \mathbb{R}^r$ such that $\boldsymbol{y}^\star \boldsymbol{P}(0) = \boldsymbol{y}^\star$ and $\boldsymbol{y}^\star \boldsymbol{P}(\pi) = \boldsymbol{0}^{\mathrm{T}}$.*

Proof: Recall that the autocorrelation symbol $\boldsymbol{\Omega}$ is an eigenmatrix of the transition operator T to the eigenvalue 1 and, by L^2-stability, $\boldsymbol{\Omega}(\omega)$ is invertible for all $\omega \in \mathbb{R}$. For a given eigenvalue λ of $\boldsymbol{P}(0)$ with left eigenvector $\boldsymbol{y}^\star$ we find

$$\begin{aligned}\boldsymbol{y}^\star\boldsymbol{\Omega}(0)\,\boldsymbol{y} &= \boldsymbol{y}^\star(T\boldsymbol{\Omega})(0)\,\boldsymbol{y}\\ &= \boldsymbol{y}^\star\,\boldsymbol{P}(0)\,\boldsymbol{\Omega}(0)\,\boldsymbol{P}(0)^\star\,\boldsymbol{y} + \boldsymbol{y}^\star\,\boldsymbol{P}(\pi)\,\boldsymbol{\Omega}(\pi)\,\boldsymbol{P}(\pi)^\star\,\boldsymbol{y}\\ &\geq |\lambda|^2\,\boldsymbol{y}^\star\boldsymbol{\Omega}(0)\,\boldsymbol{y}\end{aligned}$$

which means that $|\lambda| \leq 1$. For $\lambda = 1$, $\boldsymbol{y}^\star\,\boldsymbol{P}(\pi)\,\boldsymbol{\Omega}(\pi)\,\boldsymbol{P}(\pi)^\star\,\boldsymbol{y} = 0$, and since $\boldsymbol{\Omega}(\pi)$ is invertible, $\boldsymbol{y}^\star\,\boldsymbol{P}(\pi) = 0$. The hypothesis that there are other eigenvalues of $\boldsymbol{P}(0)$ on the unit circle can be shown to contradict the L^2-stability. □

Necessary conditions for existence of L^2-stable (or more general L^p-stable) solution vectors of (3) can also be given in terms of the subdivision operator (see e.g. [31], Theorem 2.1).

Obviously, a necessary condition for L^2-stability of $\boldsymbol{\Phi}$ is that the entries of $\boldsymbol{\Phi}$ are in $L^2(\mathbb{R})$. The following theorem generalizes the results of Villemoes [54].

Theorem 4. *Let $\boldsymbol{P}(\omega)$ be a matrix of trigonometric polynomials satisfying the assertions (a), (b) of Theorem 3 and $\boldsymbol{P}(0)\boldsymbol{r} = \boldsymbol{r}$. Let $\boldsymbol{U}$ be an invertible matrix with first column $\boldsymbol{r}$ such that $\boldsymbol{U}^{-1}\,\boldsymbol{P}(0)\,\boldsymbol{U}$ is the Jordan matrix of $\boldsymbol{P}(0)$ with leading entry 1. Further, let $\boldsymbol{\Phi}$ be the corresponding solution vector of (3) with $\hat{\boldsymbol{\Phi}}(0) = \boldsymbol{r}$. Then $\boldsymbol{\Phi} \in L^2(\mathbb{R})^r$ if and only if there exists an $r \times r$ matrix $\boldsymbol{H} \in \mathbb{H}_N$ satisfying $T\boldsymbol{H} = \boldsymbol{H}$, and the leading entry of $\boldsymbol{U}^{-1}\,\boldsymbol{H}(0)\,(\boldsymbol{U}^{-1})^\star$ is positive.*

Proof: We refer to [34], Proposition 3.14. If $\boldsymbol{y}^\star$ is a left eigenvector of $\boldsymbol{P}(0)$ corresponding to the eigenvalue 1 with $\boldsymbol{y}^\star\,\boldsymbol{r} = 1$, then $\boldsymbol{U}^{-1}$ has $\boldsymbol{y}^\star$ as its first row. Theorem 3 implies that $\boldsymbol{y}^\star\,\boldsymbol{P}(\pi) = 0$. Repeated application of (4) easily leads to $\boldsymbol{y}^\star\,\hat{\boldsymbol{\Phi}}(2\pi l) = \delta_{0,l}$. If $\boldsymbol{\Phi} \in L^2(\mathbb{R})^r$ then its autocorrelation symbol $\boldsymbol{\Omega}$ is a matrix of trigonometric polynomials in $\mathbb{H}_N$ satisfying $T\boldsymbol{\Omega} = \boldsymbol{\Omega}$ and

$$\boldsymbol{y}^\star\,\boldsymbol{\Omega}(0)\boldsymbol{y} = \boldsymbol{y}^\star \sum_{l=-\infty}^{\infty} \hat{\boldsymbol{\Phi}}(2\pi l)\,\hat{\boldsymbol{\Phi}}(2\pi l)^\star\,\boldsymbol{y} = 1.$$

Conversely, assume that there is a matrix $\boldsymbol{H} \in \mathbb{H}_N$ with $T\boldsymbol{H} = \boldsymbol{H}$, and the first entry of $\boldsymbol{U}^{-1}\boldsymbol{H}(0)(\boldsymbol{U}^{-1})^\star$ is $c > 0$. Set $\boldsymbol{\Pi}_n(\omega) := \prod_{j=1}^n \boldsymbol{P}(2^{-j}\omega)$, $\boldsymbol{\Pi}(\omega) := \lim_{n\to\infty} \boldsymbol{\Pi}_n(\omega)$ and observe that, by Theorem 3(a), $\boldsymbol{\Pi}(\omega)\,\boldsymbol{U} = (\hat{\boldsymbol{\Phi}}(\omega), \boldsymbol{0}, \ldots, \boldsymbol{0})$. Hence,

$$\boldsymbol{\Pi}(\omega)\,\boldsymbol{H}(0)\,\boldsymbol{\Pi}(\omega)^\star = \boldsymbol{\Pi}(\omega)\boldsymbol{U}\,\boldsymbol{U}^{-1}\,\boldsymbol{H}(0)\,(\boldsymbol{U}^{-1})^\star\,\boldsymbol{U}^\star\,\boldsymbol{\Pi}(\omega) = c\,\hat{\boldsymbol{\Phi}}(\omega)\hat{\boldsymbol{\Phi}}(\omega)^\star.$$

Using (8) and $T\boldsymbol{H} = \boldsymbol{H}$ we get

$$\begin{aligned} c\int_{-\infty}^{\infty} \hat{\boldsymbol{\Phi}}(\omega)\,\hat{\boldsymbol{\Phi}}(\omega)^\star\,\mathrm{d}\omega &= \lim_{n\to\infty}\int_{-2^n\pi}^{2^n\pi} \boldsymbol{\Pi}_n(\omega)\,\boldsymbol{H}(\frac{\omega}{2^n})\,\boldsymbol{\Pi}_n(\omega)^\star\,\mathrm{d}\omega \\ &= \lim_{n\to\infty}\int_{-\pi}^{\pi} (T^n\boldsymbol{H})(\omega)\mathrm{d}\omega = \int_{-\pi}^{\pi} \boldsymbol{H}(\omega)\mathrm{d}\omega, \end{aligned}$$

and the assertion follows. □

As in the scalar case, there are three different methods for characterization of necessary and sufficient conditions of L^2-stability of $\boldsymbol{\Phi}$. The first is based on spectral properties of the transition operator T associated with $\boldsymbol{P}$ (see e.g. [48]). Using the representing matrix of T (see e.g. [41]), the resulting stability condition can be seen as a generalization of Lawton's criteria [35] for scaling functions. Another way is to try to find explicit conditions on the mask symbol $\boldsymbol{P}$. This idea generalizes the criteria of Cohen [8] and that of Jia and Wang [32]. However, these conditions are rather complicated. One has not only to struggle with noncommuting matrix products, but also needs to ensure the algebraic linear independence of the components ϕ_ν of $\boldsymbol{\Phi}$ and their translates in terms of $\boldsymbol{P}(\omega)$ [24,43,55]. Neither of these problems occurs in the scalar case.

Finally, there is a close relation between stability and convergence of subdivision schemes (see e.g. [10,12,38,39]).

We only present the stability conditions in terms of the transition operator, proved in [48].

Theorem 5. *The refinable function vector $\boldsymbol{\Phi}$ is L^2-stable if and only if its symbol $\boldsymbol{P}(0)$ satisfies Condition E, and the corresponding transition operator T restricted to $\mathbb{H}_N$ satisfies Condition E, where the eigenmatrix corresponding to the eigenvalue 1 is positive definite for all $\omega \in \mathbb{R}$.*

Proof: The sufficiency of the assertions is obvious: Since $\boldsymbol{P}(0)$ satisfies Condition E, Theorem 2 implies that the matrix refinement equation (3) has a unique, compactly supported solution vector $\boldsymbol{\Phi}$. Let $\boldsymbol{H} \in \mathbb{H}_N$ be a positive definite eigenmatrix of T with eigenvalue 1, then by Theorem 4, $\boldsymbol{\Phi} \in L^2(\mathbb{R})^r$. Thus, the autocorrelation symbol $\boldsymbol{\Omega}$ of $\boldsymbol{\Phi}$ exists, and it equals $\boldsymbol{H}$ up to multiplication by a constant.

The necessity is more complicated. The assertion on $\boldsymbol{P}(0)$ follows from Theorem 3(a). Theorem 4 implies that T possesses an eigenvalue 1. If there were an eigenvalue λ of T with $|\lambda| > 1$, $T\tilde{\boldsymbol{H}} = \lambda\tilde{\boldsymbol{H}}$, then since $\boldsymbol{\Pi}(\omega)\,\tilde{\boldsymbol{H}}(0)\,\boldsymbol{\Pi}(\omega)^\star = \tilde{c}\,\hat{\boldsymbol{\Phi}}(\omega)\,\hat{\boldsymbol{\Phi}}(\omega)^\star$ with some finite constant $\tilde{c}$, the divergence of

$$\| \int_{-\infty}^{\infty} \boldsymbol{\Pi}(\omega)\,\tilde{\boldsymbol{H}}(0)\,\boldsymbol{\Pi}(\omega)^\star\,\mathrm{d}\omega\| = \| \lim_{n\to\infty} \int_{-\pi}^{\pi} T^n \tilde{\boldsymbol{H}}(\omega)\,\mathrm{d}\omega\|$$

would contradict the assertion $\boldsymbol{\Phi} \in L^2(\mathbb{R})^r$ (see the proof of Theorem 4). Similar arguments apply for showing that 1 is the only eigenvalue of T on the unit circle. Finally, the assertion follows by recalling that the autocorrelation $\boldsymbol{\Omega}$ of $\boldsymbol{\Phi}$ is positive definite and satisfies $T\boldsymbol{\Omega} = \boldsymbol{\Omega}$. □

§3. Approximation Order and Factorization of the Symbol

In this section we give an overview of the polynomial reproduction properties of a scaling vector $\boldsymbol{\Phi}$. We start with definitions and notation. Let $S_0(\boldsymbol{\Phi})$ denote the linear space of all functions of the form $\sum_{\alpha\in\mathbb{Z}} \boldsymbol{b}_\alpha^\star\,\boldsymbol{\Phi}(\cdot - \alpha)$, where $\boldsymbol{b} \in l(\mathbb{Z})^r$ is an arbitrary vector sequence on $\mathbb{Z}$.

Assume Π_k is the space of algebraic polynomials of degree at most k. A vector sequence $\boldsymbol{y} \in l(\mathbb{Z})^r$ is called a polynomial vector sequence of degree k, if there exists a vector of algebraic polynomials $\boldsymbol{Y} \in (\Pi_k)^r$ with $\boldsymbol{y}_l = \boldsymbol{Y}(l)$ $(l \in \mathbb{Z})$.

We say that a function vector $\boldsymbol{\Phi}$ has accuracy k if $\Pi_{k-1} \subseteq S_0(\boldsymbol{\Phi})$. A function vector $\boldsymbol{\Phi} \in (L^2(\mathbb{R}))^r$ is said to provide approximation order k if for any sufficiently smooth function $g \in L^2(\mathbb{R})$,

$$\operatorname{dist}(g,\, S_h(\boldsymbol{\Phi})) = O(h^k).$$

Here $S(\boldsymbol{\Phi}) = L^2(\mathbb{R}) \cap S_0(\boldsymbol{\Phi})$ and

$$S_h(\boldsymbol{\Phi}) := \{f(\cdot/h) : f \in S(\boldsymbol{\Phi})\}.$$

As shown in [27], $\boldsymbol{\Phi}$ provides approximation order k if and only if it has accuracy k.

There are two equivalent methods for characterization of accuracy of a scaling vector $\boldsymbol{\Phi}$; the first method uses the subdivision operator S, the second gives conditions on the symbol $\boldsymbol{P}(\omega)$ in the Fourier domain.

Theorem 6. *Let $\boldsymbol{\Phi} = (\phi_1, \ldots, \phi_r)^{\mathrm{T}}$ be a compactly supported, L^2-stable scaling vector with a finite refinement mask $\{\boldsymbol{P}_l\}_{l\in\mathbb{Z}}$. Then following statements are equivalent:*

(i) *$\boldsymbol{\Phi}$ provides approximation order k.*

(ii) *There exists vector sequences $\boldsymbol{y}_n = \{(\boldsymbol{y}_n)_\alpha\}_{\alpha\in\mathbb{Z}} \in l(\mathbb{Z})^r$ $(n = 0, \ldots k-1)$ such that*

$$\sum_{\alpha\in\mathbb{Z}} (\boldsymbol{y}_n)_\alpha^\star\,\boldsymbol{\Phi}(t-\alpha) = t^n \quad (t \in \mathbb{R}, n = 0, \ldots, k-1). \tag{12}$$

These series converge absolutely and uniformly on any compact interval of $\mathbb{R}$. *In particular,* $\boldsymbol{y}_n$ *are polynomial vector sequences of degree* n *with*

$$(\boldsymbol{y}_n)_\alpha = \sum_{l=0}^{n} \binom{n}{l} \alpha^{n-l} (\boldsymbol{y}_l)_0. \tag{13}$$

(iii) *There is a unique* superfunction $f \in S(\boldsymbol{\Phi})$ *which is a finite linear combination of shifts* $\boldsymbol{\Phi}(\cdot - l)$ $(l = 0, \ldots, k-1)$, *and satisfies*

$$\mathrm{D}^\mu f(2\pi l) = \delta_{\mu,0}\, \delta_{l,0} \quad (l \in \mathbb{Z}, \mu = 0, \ldots, k-1). \tag{14}$$

(iv) *There exists a nontrivial vector of trigonometric polynomials* $\boldsymbol{A}(\omega) := \sum_{l=0}^{k-1} \boldsymbol{a}_l\, e^{-i\omega l}$ *with coefficient vectors* $\boldsymbol{a}_l \in \mathbb{R}^r$, *such that the symbol* $\boldsymbol{P}$ *of* $\boldsymbol{\Phi}$ *satisfies*

$$\begin{aligned} \mathrm{D}^n [\boldsymbol{A}(2\omega)^\star\, \boldsymbol{P}(\omega)]|_{\omega=0} &= \mathrm{D}^n\, \boldsymbol{A}(0)^\star, \\ \mathrm{D}^n [\boldsymbol{A}(2\omega)^\star\, \boldsymbol{P}(\omega)]|_{\omega=\pi} &= \boldsymbol{0}^{\mathrm{T}}. \end{aligned} \tag{15}$$

for $n = 0, \ldots, k-1$.

(v) *There exists a polynomial vector sequence* $\boldsymbol{b}$ *satisfying*

$$S_{\boldsymbol{P}}\, \boldsymbol{b} = \left(\frac{1}{2}\right)^{k-1} \boldsymbol{b}$$

and $\boldsymbol{b}^{\mathrm{T}}\, \hat{\boldsymbol{\Phi}}(0)$ *has degree* $k-1$.

Proof: We give a sketch of the proof.
(i) $\Rightarrow$ (ii): The existence of sequences $\boldsymbol{y}_n \in l(\mathbb{Z})^r$ satisfying (12) follows directly from the definition of accuracy. Only their polynomial structure (13) has to be shown. For $n = 0$ we find (by replacing of t by $t+1$)

$$1 = \sum_{\alpha \in \mathbb{Z}} (\boldsymbol{y}_0)_\alpha^\star\, \boldsymbol{\Phi}(t - 1 - \alpha) = \sum_{\alpha \in \mathbb{Z}} (\boldsymbol{y}_0)_{\alpha+1}^\star\, \boldsymbol{\Phi}(t - \alpha).$$

Hence $(\boldsymbol{y}_0)_\alpha = (\boldsymbol{y}_0)_{\alpha+1}$ for all $\alpha \in \mathbb{Z}$. For $n > 0$ the assertion easily follows by induction (see [42], Lemma 2.1). Here we need to use the assumed L^2-stability of $\boldsymbol{\Phi}$ which also applies to polynomial sequences.
(ii)$\Rightarrow$(iii): Take the trigonometric polynomial vector

$$\boldsymbol{A}(\omega) = \sum_{l=0}^{k-1} \boldsymbol{a}_l\, e^{-i\omega l} \tag{16}$$

with $\boldsymbol{a}_l \in \mathbb{R}^r$ determined by

$$\mathrm{D}^n \boldsymbol{A}(0) = i^n\, (\boldsymbol{y}_n)_0 \quad (n = 0, \ldots, k-1). \tag{17}$$

Consider the function $f \in S(\boldsymbol{\Phi})$ with $\hat{f}(\omega) := \boldsymbol{A}(\omega)^\star\, \hat{\boldsymbol{\Phi}}(\omega)$. Then f is the desired superfunction satisfying (14).

For $n = 0$, (12) and Poisson summation formula imply

$$1 = \sum_{\alpha \in \mathbb{Z}} \boldsymbol{A}(0)^\star \, \boldsymbol{\Phi}(t-\alpha) = \boldsymbol{A}(0)^\star \sum_{j \in \mathbb{Z}} e^{2\pi i j x} \, \hat{\boldsymbol{\Phi}}(2\pi j).$$

Hence, $\hat{f}(0) = \boldsymbol{A}(0)^\star \, \hat{\boldsymbol{\Phi}}(0) = 1$ and $\hat{f}(2\pi j) = \boldsymbol{A}(0)^\star \, \hat{\boldsymbol{\Phi}}(2\pi j) = 0$ for $j \in \mathbb{Z} \setminus \{0\}$. For $n > 0$, an induction proof using (12) and differentiated versions of Poisson summation formula can be applied (see [42], Theorem 2.2). The uniqueness of f is shown in [6], Theorem 4.2.

(iii) $\Rightarrow$ (i): This statement is true since f satisfies the Strang-Fix conditions of order k.

(iii) $\Leftrightarrow$ (iv): We show that the trigonometric polynomial vector $\boldsymbol{A}(\omega)$ determined by (16)–(17) also satisfies the conditions (15). First, note that $\hat{f}(2\omega) = \boldsymbol{A}(2\omega)^\star \, \hat{\boldsymbol{\Phi}}(2\omega) = \boldsymbol{A}(2\omega)^\star \, \boldsymbol{P}(\omega) \, \hat{\boldsymbol{\Phi}}(\omega)$. Using (14), we find for $l \in \mathbb{Z}$

$$0 = \hat{f}(4\pi l + 2\pi) = \boldsymbol{A}(0)^\star \, \boldsymbol{P}(\pi) \, \hat{\boldsymbol{\Phi}}(2\pi l + \pi).$$

The linear independence of the sequences $\{\hat{\phi}_\nu(2\pi l + \pi)\}_{l \in \mathbb{Z}}$ $(\nu = 1, \ldots, r)$ implies $\boldsymbol{A}(0)^\star \, \boldsymbol{P}(\pi) = \boldsymbol{0}^{\mathrm{T}}$. To get the second equality in (15) for $n > 0$, we proceed by induction, using $\mathrm{D}^n \hat{f}(4\pi l + 2\pi) = 0$.

The first equality in (15) for $n = 0$ follows by comparison of $\hat{f}(4\pi l) = \boldsymbol{A}(0)^\star \, \boldsymbol{P}(0) \, \hat{\boldsymbol{\Phi}}(2\pi l) = \delta_{0,l}$ and $\hat{f}(2\pi l) = \boldsymbol{A}(0)^\star \, \hat{\boldsymbol{\Phi}}(2\pi l) = \delta_{0,l}$. For $n > 0$, the assertion is a consequence of $\mathrm{D}^n \hat{f}(4\pi l) = \mathrm{D}^n \, \hat{f}(2\pi l) = 0$ (see [40], Theorem 2). Conversely, take $\boldsymbol{A}(\omega)$ of the form (16) satisfying (15) for $n = 0, \ldots, k-1$ and with $\boldsymbol{A}(0)^\star \, \hat{\boldsymbol{\Phi}}(0) = 1$. This last equality is only a normalization, since $\boldsymbol{A}(0)^\star$ and $\hat{\boldsymbol{\Phi}}(0)$ are left and right eigenvector of $\boldsymbol{P}(0)$ to the simple eigenvalue 1, respectively. (Indeed, $\boldsymbol{A}$ is already uniquely determined by these conditions.) Then it can be shown that f determined by $\hat{f}(\omega) = \boldsymbol{A}(\omega)^\star \, \hat{\boldsymbol{\Phi}}(\omega)$ is the superfunction.

(ii) $\Rightarrow$ (v): Let $\boldsymbol{b} := \boldsymbol{y}_{k-1}$ with $\boldsymbol{y}_{k-1}$ as in Theorem 6 (ii). On the one hand, with $\gamma = 2l + n$,

$$\begin{aligned} t^{k-1} &= \sum_{l \in \mathbb{Z}} \boldsymbol{b}_l^\star \boldsymbol{\Phi}(t-l) = \sum_{l \in \mathbb{Z}} \boldsymbol{b}_l^\star \sum_{n \in \mathbb{Z}} \boldsymbol{P}_n \, \boldsymbol{\Phi}(2t - 2l - n) \\ &= \sum_{\gamma \in \mathbb{Z}} (S_{\boldsymbol{P}} \boldsymbol{b})_\gamma^\star \, \boldsymbol{\Phi}(2t - \gamma), \end{aligned}$$

where (3) and (9) are used. On the other hand,

$$t^{k-1} = \frac{1}{2^{k-1}} (2t)^{k-1} = \frac{1}{2^{k-1}} \sum_{l \in \mathbb{Z}} \boldsymbol{b}_l^\star \, \boldsymbol{\Phi}(2t - l).$$

Comparison yields $S_{\boldsymbol{P}} \boldsymbol{b} = \frac{1}{2^{k-1}} \boldsymbol{b}$. By definition, $\boldsymbol{b}$ is a polynomial sequence and

$$\boldsymbol{b}_\alpha^\star \, \hat{\boldsymbol{\Phi}}(0) = \sum_{l=0}^{k-1} \binom{k-1}{l} \alpha^{k-1-l} (\boldsymbol{y}_s)_0^\star \, \hat{\boldsymbol{\Phi}}(0)$$

is of degree $k-1$ if and only if $(\boldsymbol{y}_0)_0^\star \hat{\boldsymbol{\Phi}}(0) \neq 0$. This is a direct consequence of the Poisson summation formula (applied to (12) for $n = 0$).

(v) $\Rightarrow$ (i): Let $\boldsymbol{b}$ satisfy the conditions of Theorem 6(v), and set

$$p := \sum_{\alpha \in \mathbb{Z}} \boldsymbol{b}_\alpha^\star \, \boldsymbol{\Phi}(\cdot - \alpha).$$

Then

$$p = \sum_{\alpha \in \mathbb{Z}} (S_{\boldsymbol{P}} \boldsymbol{b})_\alpha^\star \, \boldsymbol{\Phi}(2 \cdot - \alpha) = \left(\frac{1}{2}\right)^{k-1} \sum_{\alpha \in \mathbb{Z}} \boldsymbol{b}_\alpha^\star \, \boldsymbol{\Phi}(2 \cdot - \alpha) = \left(\frac{1}{2}\right)^{k-1} p(2\cdot).$$

Since $\boldsymbol{b}$ is a polynomial vector sequence of degree $k-1$, it follows that $\nabla^k_{(1/2)^n} p = 0$ for all positive integers n. Here $\nabla_h f := f - f(\cdot - h)$ and $\nabla_h^k f := \nabla_h^{k-1}(\nabla_h f)$. Using these properties of p, one can find that $p(t) = ct^{k-1}$ with some constant $c \neq 0$ (see [30], Theorem 3.1). This means that $S_0(\boldsymbol{\Phi})$ contains the monomial t^{k-1}. Since $S_0(\boldsymbol{\Phi})$ is $\mathbb{Z}$-translation-invariant, it contains $1, \ldots, t^{k-1}$, i.e., $\boldsymbol{\Phi}$ has accuracy k. □

We now give some further explanatory remarks.

1) In [26], Jia already characterized the L^∞-approximation order of FSI spaces $S_0(\boldsymbol{\Phi})$ in terms of Strang–Fix conditions implied on a single element $f \in S_0(\boldsymbol{\Phi})$. In the special case when $\boldsymbol{\Phi} = \phi$ is a single generator, Ron [46] showed that ϕ provides L^∞-approximation order k if and only if $S_0(\phi)$ contains Π_{k-1}. This statement was extended to the FSI spaces and L^p-approximation order in [27]. Observe that this result is not longer true for shift-invariant spaces on $\mathbb{R}^d$ with $d > 1$ (see [2]). There is a rich literature generalizing these results, including extensions to the multivariate case and to noncompactly supported function vectors (see e.g. [3,4,5,6,15,20,28] and references therein).

2) The investigation of the approximation properties of FSI spaces in [4,5,6] was focused on the superfunction approach. In [6], de Boor, DeVore and Ron succeeded to construct a superfunction f of $S(\boldsymbol{\Phi}) \subset L^2(\mathbb{R}^d)$ with special properties. In the univariate case, this superfunction can be constructed directly (see Theorem 6). The symbol $\boldsymbol{A}$ of the superfunction is then used to derive direct conditions for the mask symbol $\boldsymbol{P}$ (see Theorem 6(iv)). Conversely, the conditions (iv) on the mask symbol $\boldsymbol{P}$ can be successively applied to determine $\mathrm{D}^\mu \boldsymbol{A}(0)$ $\mu = 0, \ldots$ and hence to derive the accuracy of $\boldsymbol{\Phi}$ directly from the mask. Conditions of type (iv) were independently found in [22,37,42]. These approximation results can be generalized to the multivariate case with general dilation matrices (see [7]).

3) Already in 1994, Strang and Strela [50] had observed that accuracy k of $\boldsymbol{\Phi}$ implies that the double infinite matrix $\boldsymbol{L} = (\boldsymbol{P}_{\alpha - 2\beta})_{\alpha,\beta \in \mathbb{Z}}$ has eigenvalues $1, \frac{1}{2}, \ldots, \left(\frac{1}{2}\right)^{k-1}$ with corresponding eigenvectors of special structure. Since $\boldsymbol{L}^\star$ is the representing matrix of the subdivision operator $S_{\boldsymbol{P}}$, their results are closely related to the conditions of Theorem 6(v).

4) Jia, Riemenschneider and Zhou [30] showed that it suffices to consider the eigenvalue $\left(\frac{1}{2}\right)^{k-1}$ of $S_{\boldsymbol{P}}$ and its eigenvector sequence $\boldsymbol{b}$ only (see Theorem 6 (v)). Furthermore, [30] generalizes the result to the case of nonstable function vectors; then $S_{\boldsymbol{P}}\boldsymbol{b} - \left(\frac{1}{2}\right)^{k-1} \boldsymbol{b}$ has to be contained in the linear space $\{\boldsymbol{a} \in l(\mathbb{Z})^r : \sum_{\alpha\in\mathbb{Z}} \boldsymbol{a}_\alpha^\star \boldsymbol{\Phi}(\cdot - \alpha) = 0\}$.

In the scalar case $r = 1$, the conditions (iv) of Theorem 6 simplify to

$$\begin{aligned} \mathrm{D}^n \boldsymbol{P}(\pi) &= 0 \quad (n = 0, \ldots, k-1), \\ \boldsymbol{P}(0) &= 1. \end{aligned} \tag{18}$$

They directly imply a factorization of $\boldsymbol{P}(\omega)$ in the form

$$P(\omega) = \left(\frac{1 + e^{-i\omega}}{2}\right)^k Q(\omega). \tag{19}$$

In the vector case, the accuracy of $\boldsymbol{\Phi}$ also implies a factorization, but it is more complicated.

Theorem 7. *Let $\boldsymbol{\Phi}$ be a compactly supported, L^2-stable scaling vector with (finite) mask symbol $\boldsymbol{P}$. If $\boldsymbol{\Phi}$ provides approximation order k, then there exists an $r \times r$ matrix $\boldsymbol{C}(\omega)$ of the form $\boldsymbol{C}(\omega) = \sum_{l=0}^{k} \boldsymbol{C}_l \, e^{-i\omega l}$ satisfying*

$$\det \boldsymbol{C}(\omega) = \text{const}\,(1 - e^{-i\omega})^k$$

such that

$$\boldsymbol{P}(\omega) = \frac{1}{2^k}\, \boldsymbol{C}(2\omega)\, \boldsymbol{Q}(\omega)\, \boldsymbol{C}(\omega)^{-1}, \tag{20}$$

where $\boldsymbol{Q}$ is a matrix of trigonometric polynomials. In particular,

$$\det \boldsymbol{P}(\omega) = \left(\frac{1 + e^{-i\omega}}{2^r}\right)^k \det \boldsymbol{Q}(\omega).$$

For the proof we refer to [42]. As shown in [42], a matrix $\boldsymbol{C}$ can be constructed explicitly, and it can be factored into $\boldsymbol{C} = \boldsymbol{C}_1 \ldots \boldsymbol{C}_{k-1}$, where each $\boldsymbol{C}_l$ (corresponding to a change of approximation order by 1) has an analogous structure. If $\boldsymbol{A}(\omega)$ is a symbol of trigonometric polynomials of the form (16) satisfying (15) for $n = 0, \ldots, k-1$, then

$$\mathrm{D}^n [\boldsymbol{A}(\omega)^\star \, \boldsymbol{C}(\omega)]|_{\omega=0} = \boldsymbol{0}^{\mathrm{T}} \qquad (n = 0, \ldots, r-1) \tag{21}$$

(cf. [45]). The factorization matrix $\boldsymbol{C}$ is not unique. A general characterization of factorization matrices is presented in [44,45,52].

Using a reverse version of the factorization theorem, a procedure for construction of multi-scaling functions with special properties (symmetry, compact support, arbitrary approximation order) is given in [44]. We remark that there is a close connection between the spectrum of the symbol $\boldsymbol{P}(0)$ and that of the inner matrix $\boldsymbol{Q}(0)$. More exactly, $\boldsymbol{P}(0)$ possesses the spectrum $\{1, \mu_1, \ldots, \mu_{r-1}\}$, if $\boldsymbol{Q}(0)$ possesses the spectrum $\{1, 2^k\mu_1, \ldots, 2^k\mu_{r-1}\}$ (see [9], Lemma 4.3, [52], Lemma 2.2).

The factorization can be transferred into the time domain, using the representing matrix of the subdivision operator $\boldsymbol{L}^\star$ (see [38]).

The inner matrix $\boldsymbol{Q}$ can be considered as a mask symbol of a distribution vector $\boldsymbol{\Psi}$ such that

$$\hat{\boldsymbol{\Phi}}(\omega) = \frac{1}{(i\omega)^k}\, \boldsymbol{C}(\omega)\, \hat{\boldsymbol{\Psi}}(\omega).$$

In the scalar case $r = 1$, when $C(\omega) = (1 + e^{-i\omega})^k$, we observe that $\hat{\boldsymbol{\Phi}}(\omega) = \hat{B}_k(\omega)\, \hat{\boldsymbol{\Psi}}(\omega)$, i.e., $\boldsymbol{\Phi} = B_k \star \boldsymbol{\Psi}$, where B_k is the cardinal B-spline of order k with support $[0, k]$. In the vector case, a similar convolution result is not obvious, however, if $\boldsymbol{\Phi}$ has accuracy k, a special linear combination $\hat{\boldsymbol{F}}(\omega) = \boldsymbol{M}(\omega)\hat{\boldsymbol{\Phi}}(\omega)$ (with some invertible matrix $\boldsymbol{M}$ of trigonometric polynomials) can be found, such that $\hat{\boldsymbol{F}}(\omega) = \hat{B}_k(\omega)\, \hat{\boldsymbol{\Psi}}_0(\omega)$, where $\boldsymbol{\Psi}_0$ is a refinable vector of distributions (see [45]).

§4. Regularity of Scaling Vectors

In this section we analyze the regularity of a scaling vector $\boldsymbol{\Phi}$. Again, let us use the symbol $\boldsymbol{P}(\omega)$ for a matrix of trigonometric polynomials, and suppose $\boldsymbol{\Phi}$ is L^2-stable.

Throughout this section we assume that $\boldsymbol{\Phi}$ provides approximation order k. This assumption makes sense since a stable scaling vector $\boldsymbol{\Phi}$ with all components in C^{k-1} always provides approximation order k (see [9], Lemma 2.2). Many papers dealing with the smoothness of scaling functions or scaling vectors rely on the given approximation order and the corresponding factorization of the mask symbol. Unfortunately, in the multivariate setting, no factorization properties are known, and the ideas based on factorization cannot be generalized to that case.

Similarly to the scalar case, we can use the product representation of $\hat{\boldsymbol{\Phi}}$,

$$\hat{\boldsymbol{\Phi}}(\omega) = \lim_{n\to\infty} \prod_{j=1}^{n} \boldsymbol{P}\left(\frac{\omega}{2^j}\right) \boldsymbol{r}, \tag{22}$$

where $\boldsymbol{r}$ is a right eigenvector of $\boldsymbol{P}(0)$ to the eigenvalue 1. By Theorem 2, this representation is unique up to multiplication by a constant.

The simplest method to find regularity estimates of $\boldsymbol{\Phi}$ is to consider the decay of its Fourier transform. Let us briefly recall the situation in the scalar case ($r = 1$). Let $P(\omega)$ be a trigonometric polynomial of the form

$$P(\omega) = \left(\frac{1 + e^{-i\omega}}{2}\right) Q(\omega),$$

where $Q(0) = 1$ and $Q(\pi) \neq 0$. Assume that the corresponding scaling function ϕ is given by $\hat{\phi}(\omega) = \prod_{j=1}^{\infty} P(\omega/2^j)$. Exploiting the factorization, we find

$$|\hat{\phi}(\omega)| \le C\,(1+|\omega|)^k \prod_{j=1}^{\infty} |Q(\omega/2^j)|.$$

Together with estimates of the type $\sup_\omega |Q(2^{k-1}\omega)\dots Q(\omega)| \le B^l$, we obtain

$$|\hat{\phi}(\omega)| \le C\,(1+|\omega|)^{-k+\log B}$$

(see [13,14]). This idea can be generalized to the vector case (see [9]). Suppose that $\boldsymbol{\Phi} \in (L^2(\mathbb{R}))^r$ is compactly supported and

$$\|\hat{\boldsymbol{\Phi}}(\omega)\| := (\sum_{\nu=1}^{r} |\hat{\phi}_\nu(\omega)|^2)^{1/2}.$$

For $\boldsymbol{H} = (h_{\mu,\nu})_{\mu,\nu=1}^r \in (L^2_{2\pi})^{r\times r}$ let $\|\boldsymbol{H}(\omega)\| = (\sum_{\mu,\nu=1}^r |h_{\mu,\nu}(\omega)|^2)^{1/2}$. Then the following theorem holds.

Theorem 8. *Let $\boldsymbol{P}$ be a finite mask symbol of a scaling vector $\boldsymbol{\Phi}$ such that*

$$\boldsymbol{P}(\omega) = \frac{1}{2^k}\,\boldsymbol{C}(2\omega)\,\boldsymbol{Q}(\omega)\,\boldsymbol{C}(\omega)^{-1}.$$

Here $\boldsymbol{C}(\omega)$ and $\boldsymbol{Q}(\omega)$ are matrices of trigonometric polynomials satisfying the conditions of Theorem 7. Suppose that $\rho(\boldsymbol{Q}(0)) < 2$, and for $l \ge 1$, let

$$\gamma_l := \frac{1}{l}\log_2 \sup_\omega \|\boldsymbol{Q}(2^{l-1}\omega)\dots \boldsymbol{Q}(\omega)\|.$$

Then there exists a constant C such that for all $\omega \in \mathbb{R}$

$$\|\hat{\boldsymbol{\Phi}}(\omega)\| \le C\,(1+|\omega|)^{-k-\gamma_l}. \tag{23}$$

Proof: The idea of the proof is based on the following observation. Using the factorization (20) in the infinite product (22), one can see that

$$\hat{\boldsymbol{\Phi}}(\omega) = \lim_{n\to\infty} \frac{1}{2^{kn}}\,\boldsymbol{C}(\omega)\,\boldsymbol{Q}(\frac{\omega}{2})\dots\boldsymbol{Q}(\frac{\omega}{2^n})\,\boldsymbol{C}(\frac{\omega}{2^n})\,\boldsymbol{r},$$

and with $\boldsymbol{E}(\omega) := (1-e^{-i\omega})^k\,\boldsymbol{C}(\omega)^{-1}$,

$$\hat{\boldsymbol{\Phi}}(\omega) = \lim_{n\to\infty}\left\{\left(\frac{1}{2^n(1-e^{-i\omega/2^n})}\right)^k \boldsymbol{C}(\omega)\prod_{j=1}^{n}\boldsymbol{Q}(\frac{\omega}{2^j})\,\boldsymbol{E}(\frac{\omega}{2^n})\,\boldsymbol{r}\right\}.$$

Since $\lim_{n\to\infty}|2^{-n}(1-e^{-i\omega/2^n})^{-1}| = |\omega|^{-1}$ for $\omega \ne 0$, it follows that

$$\|\hat{\boldsymbol{\Phi}}(\omega)\| \le C\,(1+|\omega|)^{-k}\,\|\boldsymbol{C}(\omega)\|\,\lim_{n\to\infty}\|\prod_{j=1}^{n}\boldsymbol{Q}(\frac{\omega}{2^j})\,\boldsymbol{E}(\frac{\omega}{2^n})\,\boldsymbol{r}\|. \tag{24}$$

We show that $\boldsymbol{E}(0)\,\boldsymbol{r}$ is a right eigenvector of $\boldsymbol{Q}(0)$: By assumption, there exists a polynomial vector $\boldsymbol{A}(\omega) = \sum_{l=0}^{k-1} \boldsymbol{a}_l\, e^{-i\omega l}$ with $\mathrm{D}^n[\boldsymbol{A}(\omega)^\star\, \boldsymbol{C}(\omega)]|_{\omega=0} = \mathbf{0}^{\mathrm{T}}$ for $n = 0, \ldots, k-1$. In particular, $\boldsymbol{A}(0)^\star\, \boldsymbol{C}(0) = \mathbf{0}^{\mathrm{T}}$. Since the eigenvalue 0 of $\boldsymbol{C}(0)$ has geometric multiplicity 1 and $\boldsymbol{C}(0)\,\boldsymbol{E}(0) = \boldsymbol{E}(0)\,\boldsymbol{C}(0) = \mathbf{0}$, we find $\boldsymbol{E}(0) = \boldsymbol{a}\,\boldsymbol{A}(0)^\star$ where $\boldsymbol{a}$ is a suitable right eigenvector of $\boldsymbol{C}(0)$ to the eigenvalue 0. Then $\boldsymbol{E}(0)\,\boldsymbol{P}(0) = \boldsymbol{E}(0)$, because $\boldsymbol{A}(0)^\star\, \boldsymbol{P}(0) = \boldsymbol{A}(0)^\star$. On the other hand, the factorization (20) also implies

$$\boldsymbol{E}(0)\,\boldsymbol{P}(0) = \boldsymbol{Q}(0)\,\boldsymbol{E}(0).$$

Thus, $\boldsymbol{Q}(0)\boldsymbol{E}(0)\,\boldsymbol{r} = \boldsymbol{E}(0)\,\boldsymbol{r}$.

As in the scalar case, the problem is now reduced to the estimation of the infinite matrix product of the inner matrix $\boldsymbol{Q}$. For a complete proof, see [9], Theorem 4.1. □

From (23) it follows that the components of $\boldsymbol{\Phi}$ are continuous if $\boldsymbol{P}$ satisfies the conditions of the Theorem with $\gamma_l < k-1$ for some l. Estimates similar to Theorem 8 were independently found in [52].

The brute force method presented above usually does not provide sharp estimates for the smoothness of the components of the scaling vector. It gives only lower bounds.

In the second part of this section, we are going to consider a more refined method based on the transition operator. Let the Sobolev exponent s of $\boldsymbol{\Phi}$ be defined by

$$s = \sup\{\delta : \int_{-\infty}^{\infty} \|\hat{\boldsymbol{\Phi}}(\omega)\|^2\,(1+|\omega|^2)^\delta\,\mathrm{d}\omega < \infty\}.$$

Assume that the symbol $\boldsymbol{P}(\omega)$ satisfies the conditions of polynomial reproduction, i.e., there exists a vector $\boldsymbol{A}(\omega) = \sum_{l=0}^{k-1} \boldsymbol{a}_l\, e^{-i\omega l}$, such that for $n = 0, \ldots, k-1$

$$\mathrm{D}^n[\boldsymbol{A}(2\omega)^\star\, \boldsymbol{P}(\omega)]|_{\omega=0} = \mathrm{D}^n \boldsymbol{A}(0)^\star,$$
$$\mathrm{D}^n[\boldsymbol{A}(2\omega)^\star\, \boldsymbol{P}(\omega)]|_{\omega=\pi} = \mathbf{0}^T.$$

According to Theorem 7, this assumption implies a factorization of $\boldsymbol{P}$:

$$\boldsymbol{P}(\omega) = \frac{1}{2^k}\,\boldsymbol{C}(2\omega)\,\boldsymbol{Q}(\omega)\,\boldsymbol{C}(\omega)^{-1},$$

such that $\mathrm{D}^n[\boldsymbol{A}(\omega)^\star\, \boldsymbol{C}(\omega)]|_{\omega=0} = \mathbf{0}^{\mathrm{T}}$ for $n = 0, \ldots, k-1$.

Consider the transition operator $T = T_{\boldsymbol{P}}$ as given in (7). We want to show the basic idea of how to estimate regularity by spectral properties of T restricted to special subspaces. Recall that for $\boldsymbol{P}(\omega)$ of the form (10) the space $\mathbb{H}_M$ of $r \times r$ matrices of trigonometric polynomials of degree at most M is for $M \geq N$ a finite-dimensional space invariant under the action of T.

We need some definitions. For each matrix of 2π-periodic functions $\boldsymbol{H}(\omega) = (h_{i,j}(\omega))_{i,j=1}^r \in (L^2_{2\pi})^{r\times r}$, let

$$\|\boldsymbol{H}\|_F^2 := \sum_{1\leq i,j\leq r} \|h_{i,j}\|_2^2$$

be the Frobenius norm of $\boldsymbol{H}$, where $\|\cdot\|_2$ denotes the usual norm in $L^2_{2\pi}$. Further, let the norm of a linear operator $T : (L^2_{2\pi})^{r\times r} \to (L^2_{2\pi})^{r\times r}$, restricted to a subspace $\mathbb{H}$ of $(L^2_{2\pi})^{r\times r}$, be defined by

$$\|T|_{\mathbb{H}}\| := \sup_{H\in\mathbb{H}\setminus\{0\}} \frac{\|TH\|_F}{\|H\|_F}.$$

The spectral radius $\rho_{\mathbb{H}}$ of T restricted to $\mathbb{H}$ satisfies

$$\rho_{\mathbb{H}} = \lim_{n\to\infty} \|(T|_{\mathbb{H}})^n\|^{1/n}.$$

Introduce the smallest closed subspace $\mathbb{H}_0 \subseteq \mathbb{H}_N$ invariant under T and containing the matrix $\boldsymbol{C}\boldsymbol{C}^\star$, where $\boldsymbol{C}$ is the factorization matrix in (20) satisfying (21). Obviously, $\mathbb{H}_0$ is finite-dimensional. Further, observe that by (8), for each $n \in \mathbb{N}$,

$$T^n(\boldsymbol{C}\boldsymbol{C}^\star)(\omega) = \sum_{l=0}^{2^n-1} \boldsymbol{\Pi}_n(\omega+2\pi l)\, \boldsymbol{C}(\frac{\omega+2\pi l}{2^n})\, \boldsymbol{C}(\frac{\omega+2\pi l}{2^n})^\star\, \boldsymbol{\Pi}_n(\omega+2\pi l)^\star,$$

where $\boldsymbol{\Pi}_n(\omega) := \prod_{j=1}^n \boldsymbol{P}(2^{-j}\omega)$. Using that $\boldsymbol{Q}(\omega) = 2^k\, \boldsymbol{C}(2\omega)^{-1}\, \boldsymbol{P}(\omega)\, \boldsymbol{C}(\omega)$ we find $\boldsymbol{C}(\omega)^{-1}\, \boldsymbol{\Pi}_n(\omega)\, \boldsymbol{C}(2^{-n}\omega) = 2^{-kn} \prod_{j=1}^n \boldsymbol{Q}(2^{-j}\omega)$, and hence

$$\begin{aligned} T^n(\boldsymbol{C}\boldsymbol{C}^\star)(\omega) &= \frac{1}{4^{kn}} \boldsymbol{C}(\omega) \sum_{l=0}^{2^n-1} (\prod_{j=1}^n \boldsymbol{Q}(\frac{\omega+2\pi l}{2^j}))(\prod_{j=1}^n \boldsymbol{Q}(\frac{\omega+2\pi l}{2^j}))^\star \boldsymbol{C}(\omega)^\star \\ &= 4^{-kn}\, \boldsymbol{C}(\omega)\, T^n_{\boldsymbol{Q}}(\boldsymbol{I})(\omega)\, \boldsymbol{C}(\omega)^\star, \end{aligned}$$

where $T_{\boldsymbol{Q}}$ denotes the transition operator corresponding to $\boldsymbol{Q}(\omega)$, and $\boldsymbol{I}$ is the identity matrix. In particular, since $\mathrm{D}^n[\boldsymbol{A}(\omega)^\star\, \boldsymbol{C}(\omega)]|_{\omega=0} = \boldsymbol{0}^{\mathrm{T}}$, it follows that for all $\boldsymbol{H} \in \mathbb{H}_0$,

$$\mathrm{D}^n[\boldsymbol{A}(\omega)^\star\, \boldsymbol{H}(\omega)]|_{\omega=0} = \boldsymbol{0}^{\mathrm{T}} \quad (n = 0, \ldots, k-1).$$

Consider the smallest subspace $\mathbb{H}_1$ of $\mathbb{H}_N$ which is invariant under $T_{\boldsymbol{Q}}$ and contains the identity matrix.

Lemma 9. *Let $\boldsymbol{H} \in \mathbb{H}_1$ be an eigenmatrix of $T_{\boldsymbol{Q}}$ to the eigenvalue λ. Then $\tilde{\boldsymbol{H}} = \boldsymbol{C}\,\boldsymbol{H}\,\boldsymbol{C}^\star$ is an element of $\mathbb{H}_0$ and it is an eigenmatrix of $T_{\boldsymbol{P}}$ to the eigenvalue $4^{-k}\lambda$.*

Proof: From $\boldsymbol{C}(2\omega)\, \boldsymbol{Q}(\omega) = 2^k\, \boldsymbol{P}(\omega)\, \boldsymbol{C}(\omega)$ we find

$$\begin{aligned} &T_{\boldsymbol{P}}\tilde{\boldsymbol{H}}(2\omega) \\ &= \boldsymbol{P}(\omega)\,\boldsymbol{C}(\omega)\,\boldsymbol{H}(\omega)\,\boldsymbol{C}(\omega)^\star\,\boldsymbol{P}(\omega)^\star + \\ &+ \boldsymbol{P}(\omega+\pi)\,\boldsymbol{C}(\omega+\pi)\,\boldsymbol{H}(\omega+\pi)\,\boldsymbol{C}(\omega+\pi)^\star\,\boldsymbol{P}(\omega+\pi)^\star \\ &= 4^{-k}\,\boldsymbol{C}(2\omega)\;[\boldsymbol{Q}(\omega)\,\boldsymbol{H}(\omega)\,\boldsymbol{Q}(\omega)^\star + \boldsymbol{Q}(\omega+\pi)\,\boldsymbol{H}(\omega+\pi)\,\boldsymbol{Q}(\omega+\pi)^\star]\;\boldsymbol{C}(2\omega)^\star \\ &= 4^{-k}\,\boldsymbol{C}(2\omega)\,(T_{\boldsymbol{Q}}\boldsymbol{H})(\omega)\,\boldsymbol{C}(2\omega)^\star \\ &= 4^{-k}\,\boldsymbol{C}(2\omega)\,\lambda\,\boldsymbol{H}(\omega)\,\boldsymbol{C}(2\omega)^\star = 4^{-k}\lambda\,\tilde{\boldsymbol{H}}(2\omega). \quad \square \end{aligned}$$

Indeed, there is a close connection between $\mathbb{H}_0$ and $\mathbb{H}_1$; $\boldsymbol{H} \in \mathbb{H}_1$ if and only if $\boldsymbol{C}\,\boldsymbol{H}\,\boldsymbol{C}^\star \in \mathbb{H}_0$.

Theorem 10. *Let $\mathbf{\Phi}$ be a compactly supported, L^2-stable scaling vector with mask symbol $\mathbf{P}(\omega)$. Further, let $\mathbf{\Phi}$ provide approximation order k leading to the factorization (20) of $\mathbf{P}$ with $\mathbf{C}$ and $\mathbf{Q}$ as in Theorem 7. Denote by $T_{\mathbf{P}}$ and $T_{\mathbf{Q}}$ the transition operators corresponding to $\mathbf{P}$ and $\mathbf{Q}$, and let the spaces $\mathbb{H}_0$ and $\mathbb{H}_1$ be defined as above. Let ρ_0 be the spectral radius of $T_{\mathbf{P}}$ restricted to $\mathbb{H}_0$ and ρ_1 the spectral radius of $T_{\mathbf{Q}}$ restricted to $\mathbb{H}_1$. Then the Sobolev exponent s is:*

(1) $s = \frac{-\log \rho_0}{2\log 2}$.

(2) $s = k - \frac{\log \rho_1}{2\log 2}$.

Proof: For the proof of (2) we can follow the line of proof of Theorem 5.1 in [9]. As was observed in (24), the factorization (20) implies

$$\|\hat{\mathbf{\Phi}}(\omega)\| \le C\,(1+|\omega|)^{-k}\,\|\mathbf{G}(\omega)\|$$

with $\mathbf{G}(\omega) = \lim_{n\to\infty} \mathbf{G}_n(\omega)$ and $\mathbf{G}_n(\omega) := \mathbf{Q}(2^{-1}\omega)\ldots\mathbf{Q}(2^{-n}\omega)\,\mathbf{E}(0)\,\mathbf{r}$.

Using the definition of s, it follows that $s \le k-\gamma$ if the estimate

$$\int_{-2^n\pi}^{2^n\pi} \|\mathbf{G}_n(\omega)\|^2 \mathrm{d}\omega \le C\,2^{2n\gamma}$$

is true. Indeed,

$$\begin{aligned}\int_{-\pi}^{\pi} (T_{\mathbf{Q}}^n \mathbf{I})(\omega)\,\mathrm{d}\omega &= \int_{-2\pi}^{2\pi} \mathbf{Q}(\tfrac{\omega}{2})\,(T^{n-1}\mathbf{I})(\tfrac{\omega}{2})\,\mathbf{Q}(\tfrac{\omega}{2})^\star\,\mathrm{d}\omega = \ldots \\ &= \int_{-2^n\pi}^{2^n\pi} \mathbf{Q}(\tfrac{\omega}{2})\ldots\mathbf{Q}(\tfrac{\omega}{2^n})\,\mathbf{Q}(\tfrac{\omega}{2^n})^\star\ldots\mathbf{Q}(\tfrac{\omega}{2})^\star\,\mathrm{d}\omega\end{aligned}$$

implies

$$\int_{-2^n\pi}^{2^n\pi} \|\mathbf{G}_n(\omega)\|^2\,\mathrm{d}\omega \le C_\epsilon\,(\rho_1+\epsilon)^n \le \tilde{C}_\epsilon\,2^{2n\gamma}$$

for $\gamma > \frac{\log\rho_1}{2\log 2}$. Hence, it follows that $s \le k - \frac{\log \rho_1}{2\log 2}$. Moreover, analogously as e.g. in [31], the L^2-stability of $\mathbf{\Phi}$ implies (2). The equality (1) is a simple consequence of (2) since $\rho_0 = 4^{-k}\,\rho_1$. □

We conclude this section with several remarks.

1) Accuracy of order k implies eigenvalues 2^{-n} $(n = 0,\ldots,k-1)$ of the transition operator $T = T_{\mathbf{P}}$ (see e.g [33], Theorem 2.2.). By restriction of $\mathbb{H}_N$ to the subspace $\mathbb{H}_0$ we get rid of the eigenvalues of T that are related to polynomial reproduction. Another way to suppress these eigenvalues is to use the factorization of the symbol, and then consider the transition operator $T_{\mathbf{Q}}$ of the inner matrix $\mathbf{Q}$.

2) In [48] and [33], the subspace

$$\mathbb{H}_{\mathbf{A}} := \{\mathbf{H} \in \mathbb{H}_N : \mathbf{H} = \mathbf{H}^\star,\ \mathrm{D}^n[\mathbf{A}(\omega)^\star\,\mathbf{H}(\omega)]|_{\omega=0} = \mathbf{0}^{\mathrm{T}},\ n = 0,\ldots,k-1\},$$

of $\mathbb{H}_N$ was introduced. It can be shown that $\mathbb{H}_{\boldsymbol{A}}$ is invariant under the action of T and $\mathbb{H}_0 \subseteq \mathbb{H}_{\boldsymbol{A}}$. The use of $\mathbb{H}_{\boldsymbol{A}}$ has the advantage that a matrix factorization of $\boldsymbol{P}$ need not be known, only the symbol vector $\boldsymbol{A}(\omega)$ of the superfunction is applied (see [34,48]). Moreover, instead of $\mathbb{H}_{\boldsymbol{A}}$, one can also consider the smallest $T_{\boldsymbol{P}}$ invariant subspace $\mathbb{H}_{\boldsymbol{I}}$ of $\mathbb{H}_M$ ($M = \max(N, k')$) which contains the matrices $(1-\cos\omega)^{k'}\, \boldsymbol{e}_j\, \boldsymbol{e}_j^{\mathrm{T}}$ ($j = 1, \ldots r$), where $\boldsymbol{e}_j$ are the usual unit vectors. This subspace $\mathbb{H}_{\boldsymbol{I}}$ applied in Theorem 3.4 of [31] does not need any information on approximation properties of $\boldsymbol{\Phi}$ (up to an estimate of the approximation order k, since k' should be chosen $\geq k$).

In case of an L^2-stable scaling vector, Jia, Riemenschneider and Zhou [31] showed that $s = -\frac{\log \rho_{\boldsymbol{I}}}{2 \log 2}$, where $\rho_{\boldsymbol{I}}$ is the spectral radius of T restricted to $\mathbb{H}_{\boldsymbol{I}}$. Since the factorization is not involved when using the subspaces $\mathbb{H}_{\boldsymbol{A}}$ or $\mathbb{H}_{\boldsymbol{I}}$, the corresponding method can be transferred to the multivariate setting.

3) The computation of the eigenvalues of the transition operator of magnitude smaller than the spectral radius is numerically unstable. Factorization greatly improves numerical stability!

4) Smoothness estimates can also be obtained in terms of the subdivision operator. This approach is taken in [38] using the factorization technique, and in [31] without using factorization. Results in [31] and [38] are also obtained in the general L^p-space (Besov-exponents) and in the L^∞-case (Hölder-exponents).

5) Unfortunately, all methods above do not allow to estimate the smoothness of each entry of $\boldsymbol{\Phi}$ separately, but only the common smoothness of all functions in $\boldsymbol{\Phi}$. A first attempt to tackle this problem can be found in [47].

References

1. Alpert, B. K., and V. Rokhlin, A fast algorithm for the evaluation of Legendre expansions, SIAM J. Sci. Statist. Comput. **12** (1991), 158–179.
2. de Boor, C., and R. Q. Jia, A sharp upper bound on the approximation order of smooth bivariate pp functions, J. Approx. Theory **72** (1993), 24–33.
3. de Boor, C., and A. Ron, Fourier analysis of approximation orders from principal shift-invariant spaces, Constr. Approx. **8** (1992), 427–462.
4. de Boor, C., R. A. DeVore, and A. Ron, Approximation from shift-invariant subspaces of $L_2(\mathbb{R}^d)$, Trans. Amer. Math. Soc. **341** (1994), 787–806.
5. de Boor, C., R. A. DeVore, and A. Ron, The structure of finitely generated shift-invariant spaces in $L_2(\mathbb{R}^d)$, J. Funct. Anal. **119** (1994), 37–78.
6. de Boor, C., R. A. DeVore, and A. Ron, Approximation orders of FSI spaces in $L_2(\mathbb{R}^d)$, Constr. Approx., to appear.
7. Cabrelli, C., C. Heil, and U. Molter, Accuracy of several multidimensional refinable distributions, preprint.

8. Cohen, A., Ondelettes, analyses multirésolutions et filtres miroir en quadrature, Ann. Inst. H. Poincaré, Anal. non linéaire **7** (1990), 439–459.
9. Cohen, A., I. Daubechies, and G. Plonka, Regularity of refinable function vectors, J. Fourier Anal. Appl. **3** (1997), 295–324.
10. Cohen, A., N. Dyn, and D. Levin, Stability and inter-dependence of matrix subdivision schemes, in *Advanced Topics in Multivariate Approximation* (F. Fontanella, K. Jetter, P. J. Laurent, eds.) World Scientific, Singapore, 1996, 33-45.
11. Cotronei, M., L. B. Montefusco, and L. Puccio, Multiwavelet analysis and signal processing, IEEE Trans. on Circuits and Systems II, to appear.
12. Dahmen, W., and C. A. Micchelli, Biorthogonal wavelet expansions, Constr. Approx. **13** (1997), 293–328.
13. Daubechies, I., Orthornormal bases of wavelets with compact support, Comm. Pure Appl. Math. **41** (1988), 909-996.
14. Daubechies, I., *Ten Lectures on Wavelets*, SIAM, Philadelphia, 1992.
15. Dyn, N., I. R. H. Jackson, D. Levin, and A. Ron, On multivariate approximation by the integer translates of a basis functions, Israel J. Math. **78** (1992), 95–130.
16. Geronimo, J. S., D. P. Hardin, and P. R. Massopust, Fractal functions and wavelet expansions based on several scaling functions, J. Approx. Theory **78** (1994), 373–401.
17. Goodman, T. N. T., Construction of wavelets with multiplicity, Rediconti di Matematica, Serie VII, Vol. 15, Roma (1994), 665–691.
18. Goodman, T. N. T., and S. L. Lee, Wavelets of multiplicity r, Trans. Amer. Math. Soc. **342** (1994), 307–324.
19. Goodman, T. N. T., S. L. Lee, and W. S. Tang, Wavelets in wandering subspaces, Trans. Amer. Math. Soc. **338** (1993), 639–654.
20. Halton, E. J., and W. A. Light, On local and controlled approximation order, J. Approx. Theory **72** (1993), 268–277.
21. Heil, C., and D. Colella, Matrix refinement equations: Existence and uniqueness, J. Fourier Anal. Appl. **2** (1996), 363-377.
22. Heil, C., G. Strang, and V. Strela, Approximation by translates of refinable functions, Numer. Math. **73** (1996), 75–94.
23. Hervé, L., Multi-resolution analysis of multiplicity d: Application to dyadic interpolation, Appl. Comput. Harmonic Anal. **1** (1994), 199–315.
24. Hogan, T. A., Stability and independence of shifts of finitely many refinable functions, J. Fourier Anal. Appl., to appear.
25. Hogan, T. A., A note on matrix refinement equations, SIAM J. Math. Anal., to appear.
26. Jia, R. Q., A characterization of the approximation order of translation invariant spaces of functions, Proc. Amer. Math. Soc. **111** (1991), 61–70.

27. Jia, R. Q., Shift-invariant spaces on the real line, Proc. Amer. Math. Soc. **125** (1997), 785–793.
28. Jia, R. Q., and J. J. Lei, Approximation by multiinteger translates of a function having global support, J. Approx. Theory **72** (1993), 2–23.
29. Jia, R. Q., and C. A. Micchelli, On linear independence of integer translates of a finite number of functions, Proc. Edinburgh Math. Soc. **36** (1992), 69–85.
30. Jia, R. Q., S. D. Riemenschneider, and D. X. Zhou, Approximation by multiple refinable functions, preprint, 1996.
31. Jia, R. Q., S. D. Riemenschneider, and D. X. Zhou, Smoothness of multiple refinable functions and multiple wavelets, preprint, 1997.
32. Jia, R. Q., and J. Z. Wang, Stability and linear independence associated with wavelet decompositions, Proc. Amer. Math. Soc. **117** (1993), 1115–1124.
33. Jiang, Q., On the regularity of matrix refinable functions, SIAM J. Math. Anal., to appear.
34. Jiang, Q., and Z. Shen, On existence and weak stability of matrix refinable functions, Constr. Approx., to appear.
35. Lawton, W., Necessary and sufficient conditions for constructing orthonormal wavelet bases, J. Math. Phys. **32** (1991), 57–61.
36. Lawton, W., S. L. Lee, and Z. Shen, An algorithm for matrix extension and wavelet construction, Math. Comp. **65** (1996), 723–737.
37. Lian, J. A., Characterization of the order of polynomial reproduction for multiscaling functions, in *Wavelets and Multilevel Approximation*, C. K. Chui, L. L. Schumaker, eds., World Scientific, Singapore, 1995, 251–258.
38. Micchelli, C. A., and T. Sauer, Regularity of multiwavelets, Advances in Comp. Math. **7** (1997), 455-545.
39. Micchelli, C. A., and T. Sauer, On vector subdivision, preprint 1997.
40. Plonka, G., Approximation properties of multi-scaling functions: A Fourier approach, Rostocker Math. Kolloq. **49** (1995), 115–126.
41. Plonka, G., On stability of scaling vectors, in *Surface Fitting and Multiresolution Methods*, A. Le Méhauté, C. Rabut, L. L. Schumaker, eds., Vanderbilt University Press, Nashville, 1997, 293–300.
42. Plonka, G., Approximation order provided by refinable function vectors, Constr. Approx. **13** (1997), 221–244.
43. Plonka, G., Necessary and sufficient conditions for orthonormality of scaling vectors, in *Multivariate Approximation and Splines*, G. Nürnberger, J. W. Schmidt, and G. Walz, eds., ISNM Vol. 125, Birkhäuser, Basel, 1997, 205–218.
44. Plonka, G., and V. Strela, Construction of multi-scaling functions with approximation and symmtry, SIAM J. Math. Anal., to appear.

45. Plonka, G., and A. Ron, A new factorization technique for the matrix mask of univariate refinable functions, in preparation.

46. Ron, A., A characterization of the approximation order of multivariate spline spaces, Studia Math. **98** (1991), 73–90.

47. Ron, A., and Z. Shen, The sobolev regularity of refinable functions, preprint 1997.

48. Shen, Z., Refinable function vectors, SIAM J. Math. Anal., to appear.

49. Strang, G., and T. Nguyen, *Wavelets and Filter Banks*, Wellesley, Cambridge, 1996.

50. Strang, G., and V. Strela, Orthogonal multiwavelets with vanishing moments, J. Optical Eng. **33** (1994), 2104–2107.

51. Strang, G., and V. Strela, Short wavelets and matrix dilation equations, IEEE Trans. Acoustic, Speech, and Signal Processing **43** (1995), 108–115.

52. Strela, V., Multiwavelets: regularity, orthogonality and symmetry via two-scale similarity transform, Stud. Appl. Math., to appear.

53. Strela, V., P. N. Heller, G. Strang, P. Topiwala, and C. Heil, The application of multiwavelet filter banks to image processing, IEEE Trans. Acoustic, Speech, and Signal Processing, to appear.

54. Villemoes, L., Energy moments in time and frequency for two-scale difference equation solutions and wavelets, SIAM J. Math. Anal. **23** (1992), 1519–1543.

55. Wang, J., Stability and linear independence associated with scaling vectors, SIAM J. Math. Anal., to appear.

56. Zhou, D. X., Existence of multiple refinable distributions, Michigan Math. J. **44** (1997), 317–329.

Gerlind Plonka
Fachbereich Mathematik
Universität Duisburg
47048 Duisburg, GERMANY
`plonka@math.uni-duisburg.de`

Vasily Strela
Department of Mathematics
Imperial College
180 Queen's Gate
London SW7 2BZ, U.K.
`strela@ic.ac.uk`

An Analysis of Biarc Algorithms for 2-D Curves

Janet F. Poliakoff, Yew Kee Wong, and Peter D. Thomas

Abstract. The approximation of a smooth curve using circular arcs can be achieved by breaking the curve into segments and replacing each segment with two arcs in such a way that all adjacent arcs join smoothly. A new method is reported for the general solution to the problem of finding suitable arcs. The method uses a geometrical construction and allows the range of values for an acceptable solution to be identified. From this general solution we derive three particular algorithms. The algorithm which minimises the change in curvature is shown to put an upper bound on the curvature and to be more widely applicable. The use of this curvature minimising algorithm to approximate a parametric cubic spline can be shown theoretically to give good results for curves of small local deflection.

§1. Introduction

A smooth curve can be approximated by breaking it down into short segments to be approximated by circular arcs. In order to produce a smooth curve, adjacent arcs must meet a common tangent. In general it is not possible to fit a single arc between two points with a given tangent direction at each end. However, in all but the most extreme cases, it is possible to fit a pair of arcs which meet at an intermediate point with a common tangent, as illustrated in Figure 1. The above constraints are not sufficient to determine the two arcs uniquely. In general, it is possible to choose the direction of the common tangent to the two arcs within certain limits. Various criteria have been proposed for selecting the most appropriate solution, some of which suffer from limitations as we describe in Sections 2 and 5.

Although there is no precise definition of the fairness (or aesthetic quality) of a curve, there is a consensus [2,9,10] that fairness is closely related to the way the curvature varies along the curve; ideally the curvature is a continuous piecewise monotone function with as few monotone pieces as possible. In particular, sudden large changes in curvature are perceived as imperfections. Such a view of fairness leads naturally to a criterion involving curvature. We

Mathematical Methods for Curves and Surfaces II
Morten Dæhlen, Tom Lyche, and Larry L. Schumaker (eds.), pp. 401–408.

ISBN 0-8265-1315-8.

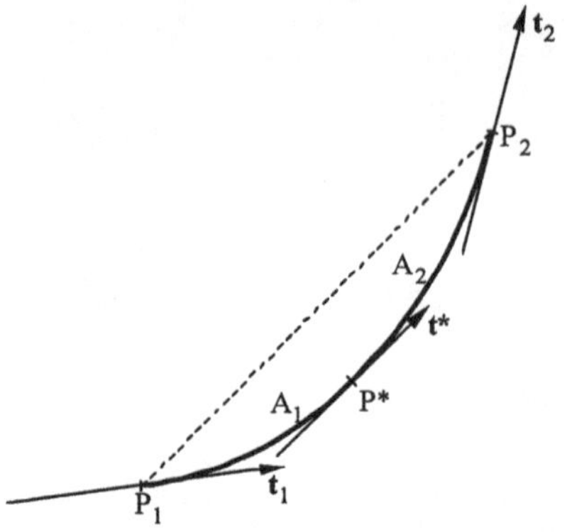

Fig. 1. Suppose that $\mathbf{t}_i$ is the tangent to the curve at P_i. We require two arcs, A_1 and A_2, such that the tangent to A_i at P_i matches $\mathbf{t}_i$, for $i = 1, 2$. If A_1 meets A_2 at P^*, then the two arcs must have common tangent, $\mathbf{t}^*$ say, at P^* .

show below that the algorithm derived for the "curvature" criterion satisfies this natural criterion and avoids the limitations described above.

A variety of strategies have been proposed for achieving a satisfactory curve from a set of data points using the biarc method. In order to fit biarcs, the direction of the tangent at each data point is needed. Sabin placed an extra requirement that the curvature should not change at the data points, thus producing fewer arcs [8], but the jumps in curvature between arcs tend to be larger. Schönherr used global minimisation of curvature discontinuity [11] and then simplified the method to obtain a linearized solution.

When a smooth curve is to be fitted through data points, the parametric cubic spline with chord length parametrisation has been found to be highly effective [2]. However, small inaccuracies in the coordinates of the points can cause undulations in the curve. We have developed a fairing algorithm for parametric cubic splines to remove such undulations [5]. After fairing, we obtained good results using the faired data points and the tangent directions at those points for biarc fitting using the curvature criterion. As described in Section 7, we have extended the monotone curvature method of Su et al. [13] to the curvature criterion to show that a fair curve will be obtained using this criterion for splines of small local deflection [7].

§2. Biarcs

In the case of biarcs, three different criteria have been proposed for selecting the most appropriate solution:

(i) the common tangent to the two arcs should be parallel to the chord joining the two ends of the segment [3],

(ii) the difference between the radii of the two arcs should be as small as possible (i.e. $|r_2 - r_1|$ is a minimum) [1,4],

(iii) the difference between the curvatures of the two arcs should be as small as possible (i.e. $|\frac{1}{r_2} - \frac{1}{r_1}|$ is a minimum) [1].

A number of other criteria have been proposed which have been shown to be equivalent to one of these three [12,13].

Although algorithms based on all three criteria have been implemented, there are limitations to the range of cases to which criteria (i) and (ii) can be applied. They cannot be used when the curve is inflected, and a different ad hoc method had to be used in such cases [1,4]. Furthermore, method (i) is not applicable in some non-inflected cases.

There is no lower bound on the size of the radii of arcs obtained by methods (i) and (ii). In real applications, however, radii below a certain size may not be acceptable because, for example, offsetting for CNC cutting machines requires arcs above a certain radius. Karow [3] avoids this problem by replacing an arc by a straight line when its radius is too small.

This paper presents a new but simple geometrical construction which can be used to derive an algorithm for a general solution, following Su et al. [13] by taking the direction of the common tangent as an additional parameter. We determine the range of directions of the common tangent which give acceptable solutions *in every case*, and derive a simple algorithm for criterion (iii) which is more widely applicable.

In order to demonstrate the use of this geometrical method, we first apply it in a simpler case which is equivalent to criterion (i). Then, we extend this method to the general case, and establish the conditions needed for an acceptable solution. By contrast, Karow [3] merely assumes a solution exists.

§3. The Geometrical Solution for Criterion (i)

We follow Karow [3] and solve for the two radii in terms of $|P_1P_2|$ and angles α_i, as shown in Figure 2. Let the angle between the tangent $\mathbf{t}_i$ at P_i and P_1P_2 be α_i (anticlockwise). We allow the α_i to be in the range $(-\pi, \pi)$, whereas Karow restricted to $(-\frac{\pi}{2}, \frac{\pi}{2})$. If $\alpha_1 = \alpha_2$, we obtain a single arc (or line when $\alpha_1 = \alpha_2 = 0$). We now assume that $\alpha_1 \neq \alpha_2$.

If the arcs have centres M_i and meet at P^* with common tangent $\mathbf{t}^*$ which is parallel to P_1P_2, then P^* lies at the intersection of the bisectors of the two angles. If $\alpha_1 = 0$ and $\alpha_2 \neq 0$, then $P^* = P_2$ and the first arc is a straight line while the second has radius zero, giving a "corner" at P_2 and the approximation is not smooth. Therefore, we assume that both $\alpha_i \neq 0$.

In order to avoid unacceptable solutions, we require that P^* lie on the positive part of each bisector, i.e. angle $P^*\hat{P}_1P_2 = \frac{1}{2}\alpha_1$, rather than $\frac{1}{2}\alpha_1 \pm \pi$, and similarly for $P_1\hat{P}_2P^*$. Figure 3 shows an example of what can happen if this restriction is lifted. We deduce that if α_1 and α_2 lie either both in $(-\pi, 0)$ or both in $(0, \pi)$ then the two positive bisectors will be on the same side of P_1P_2, so an acceptable solution for P^* will exist. We obtain for $i = 1, 2$:

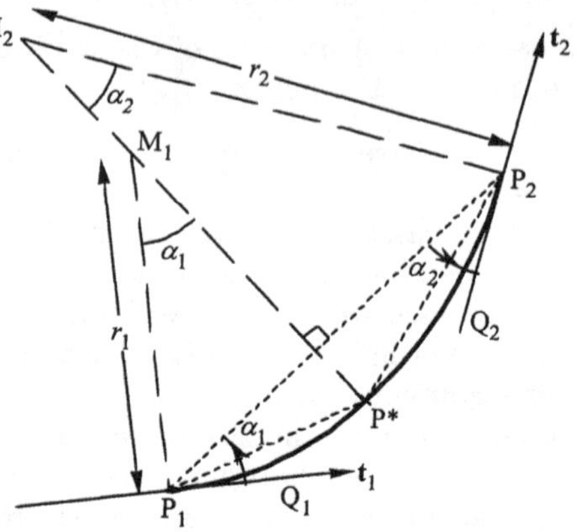

Fig. 2. For criterion (i) the tangent $\mathbf{t}^*$ at P^* is assumed to be parallel to P_1P_2. The two arcs, A_1 and A_2, have centres M_1 and M_2 and the tangents $\mathbf{t}_1$ and $\mathbf{t}_2$, are represented by P_1Q_1 and Q_2P_2, respectively .

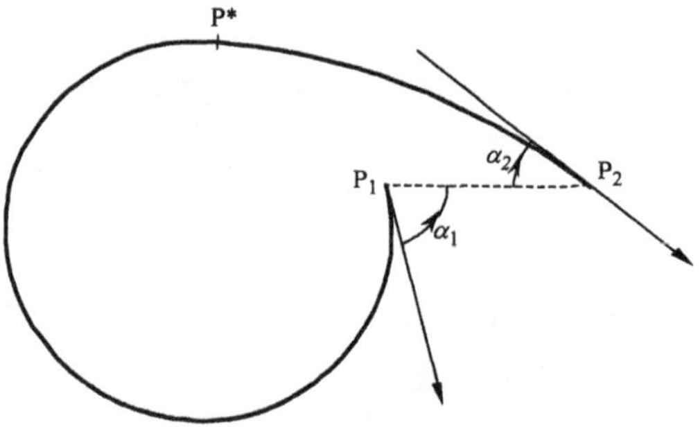

Fig. 3. An example of an unacceptable solution, obtained when P^* is allowed to be on the negative part of the bisector of angle $Q_1\hat{P}_1P_2$. The curve loops "behind" P_1 before reaching P_2 .

$$r_i = \frac{|P_1P_2| \sin \frac{1}{2}\alpha_{1-i}}{2 \sin \frac{1}{2}(\alpha_1 + \alpha_2) \sin \frac{1}{2}\alpha_i}. \tag{1}$$

When $\alpha_1 = \alpha_2$, equations (1) reduce to

$$r_1 = r_2 = \frac{|P_1P_2|}{2 \sin \alpha_1} \tag{2}$$

Thus, for criterion (i) the solution is given by equations (1) for all α_1 and α_2

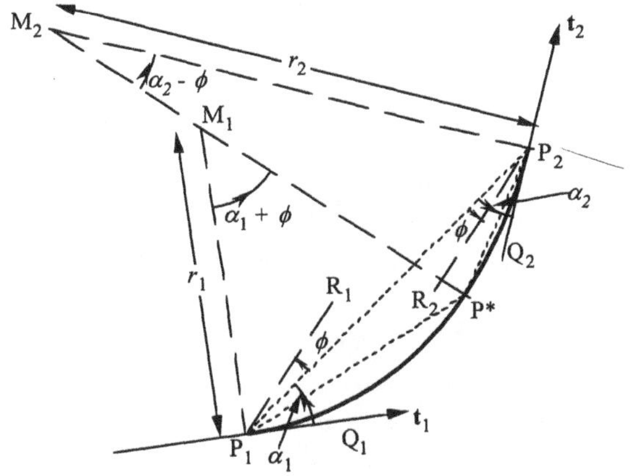

Fig. 4. The situation is similar to that in Figure 2, except that we now assume that $\mathbf{t}^*$ is of angle ϕ to P_1P_2. We introduce points $R_1$and R_2 so that P_1R_1 and R_2P_2 are both parallel to $\mathbf{t}^*$, so both are at angle ϕ to P_1P_2 .

in the range $(-\pi, \pi)$, provided that α_1 and α_2 have the same sign or both are zero. In addition, if the angles α_1 and α_2 are both less than $\frac{\pi}{2}$ in magnitude, then $P^*\hat{P}_1P_2$ and $P_1P_2P^*$ are both less than $\frac{\pi}{4}$, so the distance of P^* from P_1P_2 is at most $\frac{1}{4}|P_1P_2|$. However, if α_1 is fixed and α_2 tends to 0, then r_1 tends to 0, i.e. there is no lower limit to the magnitude of the radii.

§4. The General Geometrical Solution

We now assume that the tangent $\mathbf{t}^*$ makes an angle ϕ with P_1P_2, measured anticlockwise, as shown in Figure 4. We assume that the α_i are in the range $(-\pi, \pi)$. As in the simple case, the solution when $\alpha_1 = \alpha_2$ is a single arc (or line), so we assume that $\alpha_1 \neq \alpha_2$. The situation is similar to that shown in Figure 2 but with additional points R_i so that P_1R_1 and R_2P_2 are both at an angle ϕ to P_1P_2. In this more general case the construction consists of bisecting the two angles $Q_1\hat{P}_1R_1$ and $R_2\hat{P}_2Q_2$. We have investigated the range of values of ϕ which give a sensible solution for given α_1 and α_2 (i.e. the two positive bisectors will meet). We obtain

$$r_i = \frac{|P_1P_2| \sin \frac{1}{2}(\alpha_{1-i} \pm \phi)}{2 \sin \frac{1}{2}(\alpha_1 + \alpha_2) \sin \frac{1}{2}(\alpha_i \pm \phi)}. \tag{3}$$

Equations (3) reduce to equation (2) when $\alpha_1 = \alpha_2$. We now look at the case when both angles $P^*\hat{P}_1P_2$ and $P_1\hat{P}_2P^*$ are zero. Then $\alpha_1 = -\alpha_2$ and

$\phi = \alpha_1 = -\alpha_2$, so P^* lies between P_1 and P_2 and

$$r_1 - r_2 = \frac{|P_1P_2|}{2\sin\alpha_1}. \tag{4}$$

Thus, equations (3) and (4) represent the general solution to the problem for α_1 and α_2 in $(-\pi, \pi)$. If $\alpha_1 = \alpha_2$, we have a single arc (or line). If $\alpha_1 = -\alpha_2 \neq 0$, then $\phi = \alpha_1 = -\alpha_2$ and (4) gives the solution with P^* anywhere on P_1P_2. In all other cases ϕ must lie strictly between α_1 and $-\alpha_2$ and the solution is given by (3), with an infinite radius for a straight line.

§5. Particular Algorithms

Criterion (i) means $\phi = 0$ and equations (3) reduce to equations (1). The requirement that ϕ lie strictly between α_1 and $-\alpha_2$ means that α_1 and α_2 have the same sign; the only other possibility is $\alpha_1 = \alpha_2 = 0$.

For criterion (ii) we seek stationary values for $r_2 - r_1$ from equations (3), giving $\phi = \frac{1}{2}(\alpha_1 - \alpha_2)$ or $\frac{1}{2}(\alpha_1 - \alpha_2) \pm \pi$ and, restricting α_1 and α_2 to have magnitude less than $\frac{\pi}{2}$, we obtain:

$$r_i = \frac{|P_1P_2| \sin\frac{1}{4}(3\alpha_{1-i} - \alpha_i)}{2\sin\frac{1}{2}(\alpha_1 + \alpha_2)\sin\frac{1}{4}(\alpha_1 + \alpha_2)}. \tag{5}$$

It can then be shown [5] that α_1 and α_2 must have the same sign and satisfy $\frac{1}{3}|\alpha_2| < |\alpha_1| < 3|\alpha_2|$. Also, if α_1 is kept fixed and α_2 tends to $3\alpha_1$, then r_2 tends to 0, i.e. there is no lower limit to the magnitude of the radii.

§6. Algorithm for Criterion (iii)

For criterion (iii) we seek stationary values for the expression for the change in curvature, $\frac{1}{r_2} - \frac{1}{r_1}$, from equations (3) giving $\phi = \frac{1}{2}(\alpha_1 - \alpha_2)$ and

$$r_i = \frac{|P_1P_2|}{4\cos\frac{1}{4}(\alpha_1 + \alpha_2)\sin\frac{1}{4}(3\alpha_i - \alpha_{1-i})}. \tag{6}$$

If $\alpha_1 \neq -\alpha_2$, ϕ is always in the required range and a solution will always exist. When $\alpha_1 = -\alpha_2$ P^* lies on the line P_1P_2 and the change in curvature is minimised when P^* is at the midpoint of P_1P_2 [5] with

$$r_1 = -r_2 = \frac{|P_1P_2|}{4\sin\alpha_1}. \tag{7}$$

Equations (6) apply when $\alpha_1 = \alpha_2$ and equation (7) when $\alpha_1 = -\alpha_2$. The cases $\alpha_1 = \alpha_2 = 0$, $\alpha_1 = 3\alpha_2$ and $\alpha_2 = 3\alpha_1$ are also catered for, with an infinite radius meaning a straight line.

Algorithm (iii) also allows us to make some deductions about the position of P^* and the magnitudes of the radii. We obtain $P^*\hat{P}_1P_2 = P_1\hat{P}_2P^* = \frac{1}{4}(\alpha_1 + \alpha_2)$, i.e. P^* is equidistant from P_1 and P_2 and $|r_i| \geq \frac{1}{4}|P_1P_2|$.

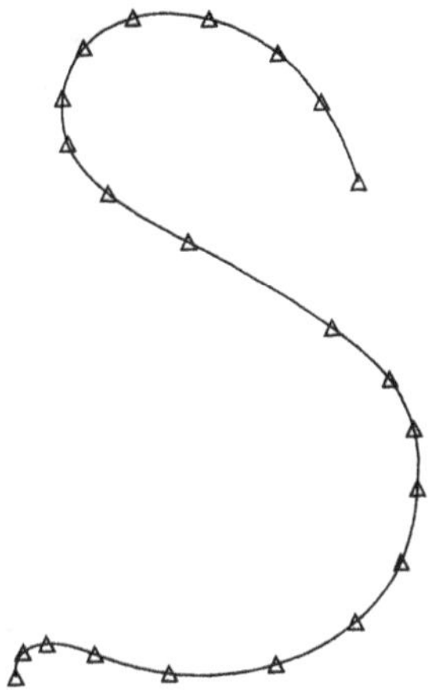

Fig. 5. An example of the use of criterion (iii) for a faired parametric cubic spline with small local deflection .

§7. The Monotone Curvature Property

Su et al. [13] have shown that a parametric cubic spline with chord length parametrisation through a sequence of data points can be approximated very closely by the local cubic spline through the same points provided that it is of small local deflection. They also showed that for criterion (i) biarc approximation of a local cubic spline has the monotone curvature property. This property can be extended to biarcs using criterion (iii) [7]. Again we can deduce, for such a cubic spline, not only that an inflected biarc segment cannot arise from a non-inflected segment but also that the biarc approximation cannot have a greater number of monotone pieces than the spline [7]. Figure 5 shows an example of a parametric cubic spline which has been faired and then approximated using criterion (iii).

§8. Conclusions

Our geometrical method produces a general algorithm with the above three algorithms as special cases. The algorithm minimising the change in curvature between the two arcs gives a solution which is not only more satisfactory aesthetically but also does not suffer from the limitations of the others. It applies to both inflected and non-inflected cases and a numerical lower bound can be given for the radii. In addition, the use of this curvature-minimising algorithm has the minimum curvature property for a parametric cubic spline of small local deflection. This means that a fair biarc approximation can be achieved for such a spline which is itself fair, provided that it is of small local deflection.

Acknowledgments. We thank Axiomatic Technology and Pacer Systems Ltd. for supporting this work, Mr. C. M. Voisey, for helpful mathematical discussions, and Mr. D. J. Moore and Mr. S. Abrahams for technical support.

References

1. Bolton, K. M., Biarc curves, Computer Aided Design **7** (1975), 89–92.
2. Farin, G., *Curves and Surfaces for Computer Aided Geometric Design* (second edition), Academic Press, London, 1990.
3. Karow, P., *Digital Formats for Typefaces*, URW Verlag, Hamburg, 1987.
4. Parkinson, D. B. and D. N. Moreton, Optimal biarc-curve fitting, Computer-Aided Design **23** (1991), 411–419.
5. Poliakoff, J. F., *The Digital Representation of Two-dimensional Cutter Paths*, PhD Thesis, The Nottingham Trent University, 1992.
6. Poliakoff, J. F., An improved algorithm for automatic fairing of non-uniform parametric cubic splines, Computer-Aided Design **28** (1996), 59–66.
7. Poliakoff, J. F., to be published.
8. Sabin, M., *The Use of Piecewise Forms for Numerical Representation of Shape*, PhD Thesis, Hungarian Academy of Sciences, Budapest, 1977.
9. Sapidis, N., and G. Farin, Automatic fairing algorithm for B-spline curves, Computer-Aided Design **22** (1990), 121–129.
10. Sapidis, N., Towards automatic shape improvement of curves and surfaces for computer graphics and CAD/CAM applications, in *Progress in Computer Graphics* (Volume 1) Zobrist, G. W. and Sabharwal, C. (eds) Alex Publishing Corporation, Norwood, New Jersey, 1992, 216–253.
11. Schönherr, J., Smooth Biarc Curves, Computer Aided Design **25** (1993) 365–370.
12. Shippey, G. A., *Technical Reports IEG/67/3 & 15*, Ferranti Ltd., 1967.
13. Su B-Q., and Liu D-Y., *Computational Geometry - Curve and Surface Modeling*, transl. Chang G-z., Academic Press, 1989.

Janet F. Poliakoff, Yew Kee Wong, and Peter D. Thomas
Department of Computing
The Nottingham Trent University
Burton Street
Nottingham, NG1 4BU, UK
tel: +44 (0)115 948 6538
fax: +44 (0)115 948 65184
`jfp@doc.ntu.ac.uk`

Two-sided Patches Suitable for Inclusion in a B-Spline Surface

Malcolm Sabin

Abstract. The transfinite 2-sided patches introduced in a previous paper [3] are used as a basis for patches with given Bézier parametric boundaries and cross-slope behaviour. It is shown how the patches may be used to fill 2-sided holes in a network which uses B-spline-like control to achieve C^1 continuity everywhere. The patches constructed have trimmed rational Bézier representations and have no singularity at the corners.

§1. Introduction: Domain and Notation

We define a transfinite patch as a map from part of the uv-plane (the domain) into xyz-space, such that the specified boundary of the domain maps into a specified boundary in xyz-space. The domain we use for a two-sided patch is the region shown in Fig. 1:

$$-u(1-u) \le v \le u(1-u).$$

The boundaries of the domain can be expressed in terms of a parameter t by the equations

$$\begin{aligned} u &= t; \qquad v = t(1-t), \\ u &= t; \qquad v = -t(1-t). \end{aligned} \tag{1}$$

The idea that the boundary can be curved in parameter space is essentially that of Wachspress [4], who defined finite elements over a domain with curved boundaries. His approach became neglected in favour of Irons' iso-parametric elements[1]. The boundaries that we shall take in xyz-space are expressed as Bézier curves, and in the first instance we take quadratics. Because we are starting from a transfinite definition, the same procedure works for any order.

The boundaries of the domain map into the curves

$$\begin{aligned} P_1(t) &= A(1-t)^2 + 2B(1-t)t + Ct^2, \\ P_2(t) &= A(1-t)^2 + 2D(1-t)t + Ct^2. \end{aligned} \tag{2}$$

Mathematical Methods for Curves and Surfaces II
Morten Dæhlen, Tom Lyche, and Larry L. Schumaker (eds.), pp. 409–416.

ISBN 0-8265-1315-8.

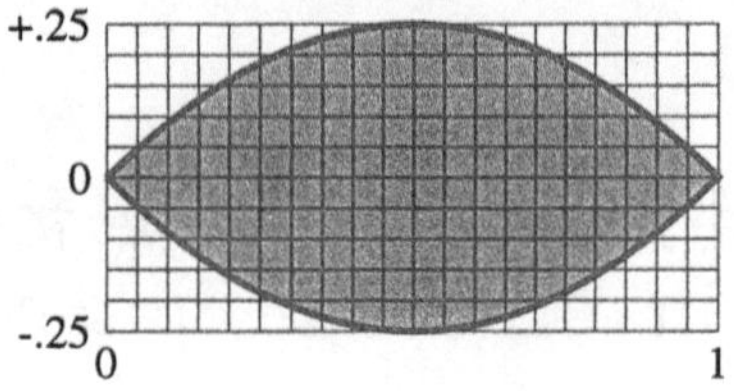

Fig. 1. Domain in uv-space.

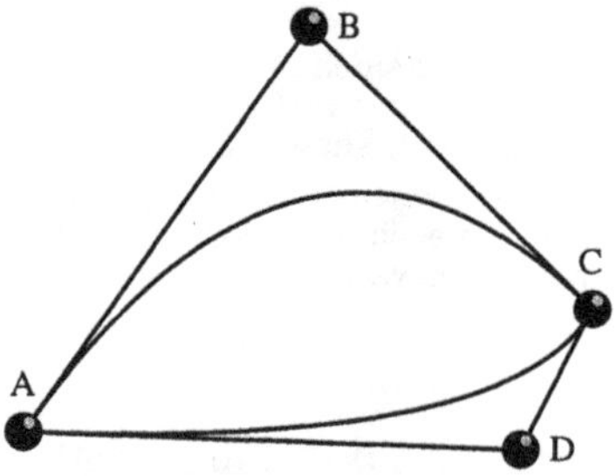

Fig. 2. Boundaries in xyz-space.

Because the points A, B, C and D as shown in Figure 2 control the boundaries and the boundaries control the patch, we may take A, B, C and D as control points of the patch itself.

A final notational point is that from this point on, the notation $\bar{.}$ is used to denote the complement $(1-.)$ This simplifies the algebra a great deal, and lets the natural symmetry of the formulation shine through. Thus, the domain boundaries become

$$\begin{aligned} v &= u\bar{u}, \\ v &= -u\bar{u}, \end{aligned} \tag{3}$$

and the patch boundaries

$$\begin{aligned} P_1(t) &= A\bar{t}^2 + 2B\bar{t}t + Ct^2, \\ P_2(t) &= A\bar{t}^2 + 2D\bar{t}t + Ct^2. \end{aligned} \tag{4}$$

§2. Elaboration of the Ruled Surface

The first transfinite two-sided patch in [3] is just the ruled surface between points of corresponding parameter on the two boundary curves. This gives the surface equation

$$P(u,v) = \frac{(u\bar{u}+v)P_1(u) + (u\bar{u}-v)P_2(u)}{2u\bar{u}}, \tag{5}$$

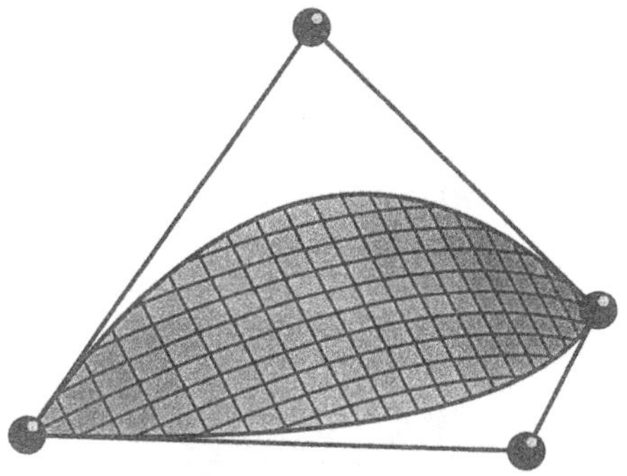

Fig. 3. Ruled patch in xyz-space.

which by substitution of the equations for the boundaries expands into

$$P(u,v) = A\bar{u}^2 + Cu^2 + 2\left[\frac{B+D}{2}\right]u\bar{u} + [B-D]v. \tag{6}$$

This can be expressed as a Bézier patch linear in v and quadratic in u. This uses the range -0.25 to +0.25 in v. It is more convenient to use the range 0 to 1, and so we change variable by making the substitution $v = (w - \bar{w})/4$. The untrimmed patch then becomes

$$P(u,v) = [\bar{u}^2 \quad 2\bar{u}u \quad u^2] \begin{bmatrix} [4A-B+D]/4 & [4A+B-D]/4 \\ [3D+B]/4 & [3B+D]/4 \\ [4C-B+D]/4 & [4C+B-D]/4 \end{bmatrix} \begin{bmatrix} \bar{w} \\ \bar{w} \end{bmatrix}. \tag{7}$$

Note how there are no singularities at A and C, as there would be if a four-sided patch equation were used, with coincident control points.

§3. Cross Derivatives

Suppose that we require a C^1 relationship with a biquadratic patch adjacent across P_1. One of the parameters of this patch will just be t, as it contains the common boundary as an iso-parameter line. Call the other s.

We need to define a direction in the domain of our 2-sided patch which will map into real space to the direction of increasing s. The choice here is arbitrary (subject to avoiding singular or degenerate configurations), but we choose to make this orthogonal in the domain to the boundary curve P_1.

Since $\partial u/\partial t = 1$ and $\partial v/\partial t = \bar{t} - t$, we can set the derivatives with respect to s to be

$$\begin{aligned} \partial u/\partial s &= -(\bar{t} - t) = u - \bar{u}, \\ \partial v/\partial s &= 1. \end{aligned} \tag{8}$$

This definition gives s increasing as we leave the domain, crossing into the adjacent patch.

(After developing the theory based on this choice I noticed that it can probably be related to the ideas of Jorg Hahn, who used Riemann surfaces as a way of achieving the symmetry of n-sided patches when he was working with John Gregory. Our domain is a double covering of the image of a unit square under the conformal transformation $w = z^2$.)

Substituting these expressions into the derivative of $P(u, v)$ obtained from (6), we get

$$\begin{aligned} dP(u,v)/ds &= \frac{dP(u,v)}{du}\frac{\partial u}{\partial s} + \frac{dP(u,v)}{dv}\frac{\partial v}{\partial s} \\ &= 2\left[[A-D]\bar{u}^2 + 2\left[B - \frac{A+C}{2}\right]u\bar{u} + [C-D]u^2\right]. \end{aligned} \tag{9}$$

This has the quadratic form that we seek, and we can therefore achieve slope continuity with an adjacent patch by choosing the control points of the latter appropriately.

We want the coefficient of the $2u\bar{u}$ term to be envisaged as a vector to B from a twist control point T, and

$$T = [A + C]/2$$

fulfils this role although we cannot yet control it independently of A, B, C and D. Note that because the expression for T is algebraically symmetric, it plays the same role for both boundaries. Our continuity arguments therefore apply to both boundaries.

The twist vector $\partial^2 P/\partial s \partial t$ at A is proportional to $[T - B] - [D - A]$, which can also be expressed in the symmetric form

$$[T + A] - [B + D]$$

or as $[T-D]-[B-A]$, which is the twist vector as seen from the opposite side of the 2-sided patch. We are thus avoiding all twist incompatibility problems at vertex A (and by symmetry also at C).

We now wish to embed our patch more actively into its surroundings to get a B-spline rather than a Bézier effect.

In particular, we can determine A, B, C, D and T from the twist control points P, Q, R and S at the centres of adjacent biquadratic patches:

$$\begin{aligned} T &:= (2S + 2Q + P + R)/6 \\ A &:= (P + Q + S + T)/4 \\ B &:= (Q + T)/2 \\ C &:= (R + Q + S + T)/4 \\ D &:= (S + T)/2. \end{aligned} \tag{10}$$

This generation of the Bézier points from the twist control points in the style of [2] ensures that the surface as a whole is C^1.

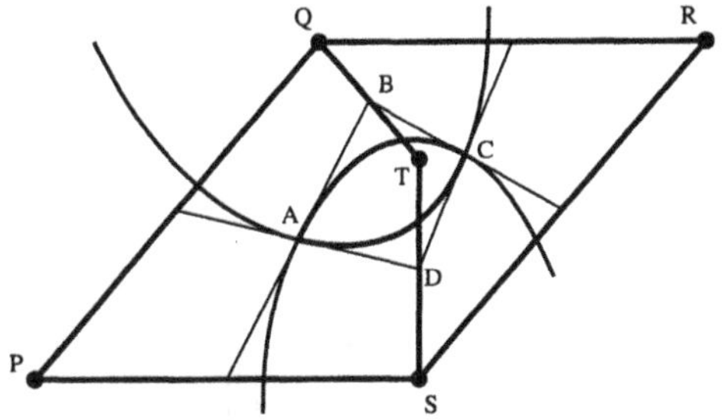

Fig. 4. Using twist points as controls.

§4. Bubble Functions

A bubble function is one which is zero on the boundary of the domain. We can therefore add any multiple of it to our surface equation without spoiling the interpolation of the boundary. The simplest bubble function is

$$\phi(u,v) = (\bar{u}u - v)(\bar{u}u + v) = \bar{u}^2u^2 - v^2,$$

but unfortunately this has a quartic cross-derivative, and so we cannot use it to modify a quadratic cross-derivative. However, the rational bubble function

$$\phi = (u^2\bar{u}^2 - v^2)/2(u^2 + \bar{u}^2) \tag{11}$$

is much more cooperative. On the boundary P_1, where $\phi = 0$, we have

$$d\phi/ds = -2u\bar{u}\,,$$

which is exactly what we need for adjusting quadratic cross-slope.

We want to convert this to a tensor product form. The range in v that we want to include in our patch is -.25 to .25, and so again we change variables by setting $v = (w - \bar{w})/4$,

$$\begin{aligned} \phi &= (u^2\bar{u}^2 - ((w - \bar{w})/4)^2)/2(u^2 + \bar{u}^2) \\ &= \left(16u^2\bar{u}^2 - (\bar{w}^2 + w^2 - 2\bar{w}w)\right)/32(u^2 + \bar{u}^2). \end{aligned} \tag{12}$$

The numerator is converted to tensor product form by multiplying the first term by $(\bar{w} + w)^2$ and the second term by $(\bar{u} + u)^4$, while the denominator is converted by multiplying by $(\bar{u} + u)^2(\bar{w} + w)^2$. This is just order raising. This gives the function, expressed in terms of its Bernstein coefficients:

$$\phi = \frac{[\bar{u}^4 \quad 4\bar{u}^3u \quad 6\bar{u}^2u^2 \quad 4\bar{u}u^3 \quad u^4] \begin{bmatrix} -1 & 1 & -1 \\ -2\left(\frac{1}{2}\right) & 2\left(\frac{1}{2}\right) & -2\left(\frac{1}{2}\right) \\ +5\left(\frac{1}{3}\right) & 11\left(\frac{1}{3}\right) & +5\left(\frac{1}{3}\right) \\ -2\left(\frac{1}{2}\right) & 2\left(\frac{1}{2}\right) & -2\left(\frac{1}{2}\right) \\ -1 & 1 & -1 \end{bmatrix} \begin{bmatrix} \bar{w}^2 \\ 2\bar{w}w \\ w^2 \end{bmatrix}}{32[\bar{u}^4 \quad 4\bar{u}^3u \quad 6\bar{u}^2u^2 \quad 4\bar{u}u^3 \quad u^4] \begin{bmatrix} 1 & 1 & 1 \\ 1/2 & 1/2 & 1/2 \\ 1/3 & 1/3 & 1/3 \\ 1/2 & 1/2 & 1/2 \\ 1 & 1 & 1 \end{bmatrix} \begin{bmatrix} \bar{w}^2 \\ 2\bar{w}w \\ w^2 \end{bmatrix}} \tag{13}$$

§5. Complete Quadratic Formulation

We really want to be able to control T independently of the other twist control points. We can do this by taking the ruled surface and adding some vector multiple of the bubble function. The amount of bubble function to be added is just the vector to the desired twist control point from the point $[A + C]/2$, the twist control point of the ruled surface.

In order to add the bubble function, which is rational quartic/quadratic, we convert the ruled surface (7) to the same form, with the same denominator, $(\bar{u}^2 + u^2)$, which can then be brought up to quartic order by order-raising, as shown in (13). We convert the numerator from (7) by just multiplying it by $(\bar{u}^2 + u^2)(\bar{w} + w)$.

Because the ruled surface and the bubble function now share the same denominator, we can add them by just adding their numerators, giving the control point matrices:

$$\begin{bmatrix} A + [D - B]/4 & A & A + [B - D]]/4 \\ \left(\frac{1}{2}\right)[3D + B]/4 & \left(\frac{1}{2}\right)[B + D]/2 & \left(\frac{1}{2}\right)[3B + D]/4 \\ \left(\frac{1}{3}\right)[2A + 2C + D - B]/4 & \left(\frac{1}{3}\right)[A + C]/2 & \left(\frac{1}{3}\right)[2A + 2C + B - D]/4 \\ \left(\frac{1}{2}\right)[3D + B]/4 & \left(\frac{1}{2}\right)[B + D]/2 & \left(\frac{1}{2}\right)[3B + D]/4 \\ C + [D - B]/4 & C & C + [B - D]/4 \end{bmatrix}$$

$$+ (1/32)\,[T - [A + C]/2] \begin{bmatrix} -1 & +1 & -1 \\ -\left(\frac{1}{2}\right)2 & +\left(\frac{1}{2}\right)2 & -\left(\frac{1}{2}\right)2 \\ +\left(\frac{1}{3}\right)5 & +\left(\frac{1}{3}\right)11 & +\left(\frac{1}{3}\right)5 \\ -\left(\frac{1}{2}\right)2 & +\left(\frac{1}{2}\right)2 & -\left(\frac{1}{2}\right)2 \\ -1 & +1 & -1 \end{bmatrix}. \tag{14}$$

Thus, the patch can be converted from the original five control points A, B, C, D and T to the Bézier points of an untrimmed patch. If the original five control points are derived from the central twist points of an array of patches by the normal quadratic B-spline to Bézier conversion, the trimmed patch will meet all of its neighbours with C^1 continuity, as exemplified in Figure 5.

As mentioned previously, the symmetry of the formulation avoids all problems with incompatibility of twists.

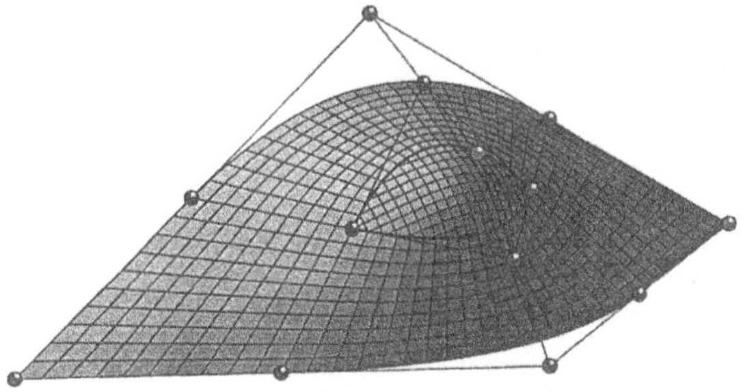

Fig. 5. Final patch with neighbours.

§6. Summary

The derivation here has been carried out for quadratic boundaries and cross-slopes. The starting point was a transfinite patch, and so generalisation to higher order boundaries follows exactly the same process. In order to match higher order cross-derivatives, it will be necessary to create multiple bubble functions, with cross slopes matching the required quadratic form. For a cubic, for example, two bubble functions are needed, whose coefficients depend on the twist control points in the 2-sided patch corresponding to the two vertices. This has not been done here, but is not expected to produce any great difficulty. What will be much harder is to create a C^2 match across the boundaries of the 2-sided patch.

The generalisation which will work trivially is the unequal interval version, provided that the same rule as for tensor product surfaces applies, that the interval ratio across a patch boundary curve cannot vary from one part of the curve to another.

Acknowledgments. This work was supported by the Royal Society of London under their Visiting Industrial Fellowships scheme.

References

1. Ergatoudis, I., B. M. Irons, and O. C. Zienckiewicz, Curved isoparametric 'quadrilateral' elements for finite element analysis, Int. J. Solids and Structures **4** (1968), 31–42.
2. Sabin, M. A., n-sided patches suitable for inclusion in a B-spline surface, *Proc Eurographics 1983*, 57–70.
3. Sabin, M. A., Transfinite Surface Interpolation, in *Mathematics of Surfaces VI*, Mullineux (ed.), Clarendon Press, London, 1995, 517–534.
4. Wachspress, E. L., *A Rational Finite Element Basis*, Academic Press, New York, 1975.

Malcolm Sabin
Numerical Geometry Ltd.
26 Abbey Lane, Lode,
Cambridge CB5 9EP, ENGLAND
malcolm@geometry.demon.co.uk

Malcolm Sabin
Department of Applied Mathematics and Theoretical Physics
Silver Street
Cambridge, ENGLAND
mas33@damtp.cam.ac.uk

Optimal Geometric Hermite Interpolation of Curves

Robert Schaback

Abstract. Bernstein–Bézier two–point Hermite G^2 interpolants to plane and space curves can be of degree up to 5, depending on the situation. We give a complete characterization for the cases of degree 3 to 5 and prove that rational representations are only required for degree 3.

§1. Introduction and Overview

We consider recovery of curves from irregularly sampled data. If the curves are to be represented by NURBS, we want to generate representations which are minimal in the following sense:

1) they should have a minimal number of knots,
2) their degree should be as small as possible,
3) polynomial pieces are preferred over rational ones, and
4) the sampling and reconstruction process should be independent of the parametrization.

Then evaluation algorithms are fast, and additional knot elimination will not be necessary. Furthermore, using minimal degrees usually helps to preserve shape properties. Within the above setting, this paper continues the presentation [14] given at the Biri conference, and we solve some problems posed there. In particular, the examples of [14] showed that the degree of G^2 piecewise polynomial or rational curve interpolants must necessarily be at least five in general, while there are cases that work with degree three (de Boor, Höllig, and Sabin [1], Höllig [9]) and degree four (Peters [11]). Here, we focus on the problem of determining the minimal degree that works in each specific situation, and we give a complete classification. However, we omit Lagrange interpolation and confine ourselves to two–point Hermite interpolation in $\mathbb{R}^2$ or $\mathbb{R}^3$. The presentation will mainly be in geometric terms; it started from a complicated algebraic analysis with 19 different cases as provided by C. Schütt [16].

Mathematical Methods for Curves and Surfaces II
Morten Dæhlen, Tom Lyche, and Larry L. Schumaker (eds.), pp. 417–428.

ISBN 0-8265-1315-8.

Our results will show that all degrees up to five actually occur in specific cases. If degrees four or five are necessary, one can get away with polynomial pieces. Rational pieces are sometimes required in cases that work with degree three (e.g. Höllig [9]). We provide a geometric approach to the full classification and illustrate it by numerous examples. However, the different cases depend crucially (and nonlinearly) on the sampled data, and the transition from one case to another may be discontinuous or may involve singularities. We exhibit some examples for this behavior, and provide a general technique for regularization of singular situations. However, shape–preserving properties and approximation orders still are open research problems.

§2. Hermite Data for Two–point G^2 Interpolation

We want to recover a regular and smooth curve f from data at two different positions $y_0,\ y_1 \in \mathbb{R}^3$. If the curve f is locally reparametrized over $[0, 1]$, we require positional interpolation in the form

$$f(i) = y_i, \qquad i = 0, 1. \tag{1}$$

To make interpolated pieces G^1 continuous, we assume normalized tangent directions $r_0,\ r_1 \in S^2 := \{r \in \mathbb{R}^3\ :\ \|r\|_2 = 1\}$ to be given such that

$$\frac{f'(i)}{\|f'(i)\|_2} = r_i, \qquad i = 0, 1. \tag{2}$$

Note that this is intended to generate interpolants that are regular at the data positions. Furthermore, the fixed orientation of the tangents helps to avoid cusps. The tangent directions r_i generate tangent rays R_i and tangent lines L_i defined as

$$R_i := \{y_i + (-1)^i \alpha_i r_i\ :\ \alpha_i \in \mathbb{R}_{>0}\}$$
$$L_i := \{y_i + \beta_i r_i\ :\ \beta_i \in \mathbb{R}\}$$

at y_i for $i = 0, 1$. Note that we let R_1 point "backwards" to simplify subsequent arguments.

There are several possibilities for prescribing second order data for G^2 continuity. To be compatible with the purely planar situation, we require normalized binormal vectors $\eta_0,\ \eta_1 \in S^2$ to be given together with nonnegative real curvature values κ_0, κ_1. The given vectors η_i should make sure that the orthogonality conditions

$$\eta_i \perp f'(i),\ \eta_i \perp f''(i), \qquad i = 0, 1 \tag{3}$$

hold. Thus, η_i is orthogonal to the two vectors which span the linear part of the osculating hyperplane P_i at $y_i = f(i)$, if they are linearly independent. Note, however, that we do not assume this extra condition to be satisfied. The additional scalar G^2 interpolation conditions then are

$$\kappa_f(i) = \frac{\|f'(i) \times f''(i)\|_2}{\|f'(i)\|_2^3} = \kappa_i, \qquad i = 0, 1,$$

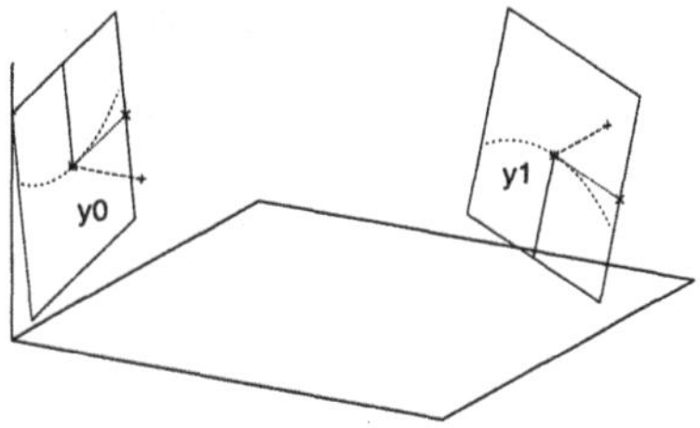

Fig. 1. Data for G^2 interpolation.

but it is more convenient to use the vectorial conditions

$$\frac{f'(i) \times f''(i)}{\|f'(i)\|_2^3} = \kappa_i \eta_i, \qquad i = 0, 1 \tag{4}$$

which also account for a proper orientation of the osculating hyperplanes with respect to the curve data. Altogether, we thus consider $\kappa_i \eta_i$ for $i = 0, 1$ to be the composite second–order data for G^2 continuity in the sense of (4). The vectors η_i should always be normalized to length one, and $\kappa_i = 0$ must hold in the degenerated case where $f'(i)$ and $f''(i)$ are linearly dependent. This may look complicated at first sight, but it has the advantage that there is always a (generalized) osculating hyperplane P_i at $y_i = f(i)$ well–defined by its normal η_i. We shall use the term osculating hyperplane in the above sense. Note that this setting also works for purely planar curves, and it models G^2 continuity of adjacent interpolating curve pieces by ensuring continuity of osculating hyperplanes.

The linear part of the osculating hyperplane P_i at y_i is spanned by tangent directions r_i and normal vectors $n_i := -r_i \times \eta_i \in S^2$ such that r_i, n_i, and $\eta_i = r_i \times n_i$ form a Frénet frame at y_i. Note that the n_i are defined indirectly here via the vectors r_i and η_i. Figure 1 shows the arrangement of full G^2 data with the osculating hyperplanes, but without the tangent rays. The prescribed curvature values are indicated by small circles in the osculating hyperplanes. These circles lie in the open halfspaces

$$H_i := \{y_i + \beta_i r_i + \gamma_i n_i \; : \; \beta_i \in \mathbb{R}, \; \gamma_i \in \mathbb{R}_{>0}\}$$

within the osculating hyperplanes P_i, containing the normal vectors n_i, and we call these halfspaces admissible for reasons that will soon be apparent. Note that the boundary of the halfspace H_i is the tangent line L_i. The admissible halfspaces in Figure 1 are indicated by the normals n_0 and n_1 pointing up and down in the osculating planes at y_0 and y_1, respectively.

§3. Necessary Conditions

Let us first look at conditions under which there is a polynomial or rational solution of degree $n \geq 2$ in Bernstein–Bézier form. It must have control points $b_0, \ldots, b_n \in \mathbb{R}^3$ with the interpolation conditions (1) for positions satisfied by

$$b_0 = y_0, \ b_n = y_1. \tag{5}$$

For G^1 data, the conditions (2) imply that b_1 and b_{n-1} must lie on the tangent rays R_0 and R_1 at $y_0 = b_0$ and $y_1 = b_n$ in the directions r_0 and $-r_1$, respectively:

$$b_1 = y_0 + \alpha_0 r_0, \ b_{n-1} = y_1 - \alpha_1 r_1, \tag{6}$$

where the real numbers α_0, α_1 must be positive to ensure regularity of the Bernstein–Bézier interpolant at the data positions.

Now let us look at second–order conditions. The control points b_0, b_1, b_2 must lie in the osculating hyperplane P_0 at y_0, while b_{n-2}, b_{n-1}, b_n lie in the osculating hyperplane P_1 at y_1, respectively. This implies

$$\begin{aligned} b_2 &= y_0 + \beta_0 r_0 + \gamma_0 n_0, \\ b_{n-2} &= y_1 - \beta_1 r_1 + \gamma_1 n_1, \end{aligned} \tag{7}$$

with real numbers β_i, γ_i, $i = 0, 1$, and we shall see next that the γ_i must be nonnegative in order to account for (4). Using standard rational Bernstein–Bézier notation with positive weights $w_0, \ldots w_n$ from *e.g.* [2,3,10], conditions (4) are equivalent to

$$\begin{aligned} \kappa_0 &= \frac{n-1}{n} \frac{w_0 w_2}{w_1^2} \frac{\gamma_0}{\alpha_0^2}, \\ \kappa_1 &= \frac{n-1}{n} \frac{w_n w_{n-2}}{w_{n-1}^2} \frac{\gamma_1}{\alpha_1^2}, \end{aligned} \tag{8}$$

since the vectorial part is already accounted for by (7). This implies nonnegativity of the γ_i. But we must be careful in case $\kappa_i = 0$. This implies $\gamma_i = 0$, and forces $b_{2+i(n-4)}$ to lie on the tangent line L_i at y_i. Since we want to make sure that the control points b_2 and b_{n-2} always lie in certain sets H_0 and H_1 at y_0 and y_1, respectively, we have to redefine H_i by

$$H_i := \{y_i + \beta_i r_i + \operatorname{sgn}(\kappa_i)\gamma_i n_i \ : \ \beta_i \in \mathbb{R}, \ \gamma_i \in \mathbb{R}_{>0}\}.$$

Thus, H_i degenerates into the tangent line L_i at y_i in case $\kappa_i = 0$, but is an open halfspace if $\kappa_i > 0$. As a warm–up exercise let us first look at the case $n = 5$.

Theorem 1. *There always is a polynomial solution of degree 5.*

Proof: For $n = 5$, the placement of the points b_0, b_1, b_2 to satisfy the interpolation conditions at y_0 is independent of the data at y_1, since these only affect

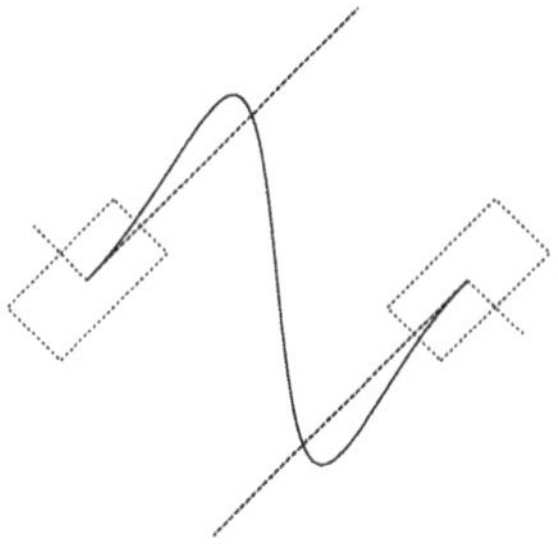

Fig. 2. Planar data for degree 5.

b_3, b_4, b_5. One can pick any real value of β_0 and any two values of $\alpha_0 > 0$ and $\gamma_0 \geq 0$ that are related by (8) in the special form

$$5\kappa_0\alpha_0^2 = 4\gamma_0 \tag{9}$$

for a degree 5 polynomial. The same thing is done on the other side. Thus, there always is a solution, and the degrees of freedom can be visualized by simply picking any two points b_2 and $b_3 = b_{n-2}$ in the admissible sets H_0 and H_1 at y_0 and y_1, respectively. Equations (9) will then place the points b_1 and $b_4 = b_{n-1}$ on the tangent lines to satisfy the curvature requirements. However, they allow additional degrees of freedom for these control points if zero curvature data are prescribed. □

Note that a practical solution to this interpolation problem will require some strategy to handle the excessive degrees of freedom. We shall comment on this in Section 8. Furthermore, the above discussion shows that there always is a variety of polynomial solutions of degree 5 that can be used instead of solutions with smaller degrees that we construct later. Note finally that the degree 5 case can even occur in purely planar situations, namely if tangents are parallel and inflection points are required (see Figure 2). Nonplanar cases with minimal degree 5 will follow in the next section.

§4. Cases of Degree 4 or Less

Let us now derive necessary conditions for cases that can possibly be handled by polynomial or rational pieces of degree $n \leq 4$. By degree elevation, it suffices to look at $n = 4$. Then we see from the preceding discussion that the control point $b_2 = b_{n-2}$ must lie in the intersection of the admissible sets H_0 and H_1 at y_0 and y_1. This yields

Theorem 2. *The minimal degree solution is of degree 5 (and a polynomial) iff the admissible sets H_0 and H_1 do not intersect.*

Proof: It remains to show that we can construct an interpolant of degree at most 4 if there is a nonempty intersection. But this is clear from the degree

5 technique and our definition of the sets H_i. Instead of (9) we now use

$$4\kappa_0\alpha_0^2 = 3\gamma_0. \tag{10}$$

Any choice of $b_2 = b_{n-2}$ from the intersection will work. This technique is partially covered by the work [11] of Peters. □

§5. Cases of Degree 3 or Less

If there is a solution of degree 3, there is a solution of degree 4 by degree elevation. Therefore the intersection $S = H_0 \cap H_1$ is not empty. But for degree $n = 3$ we have

$$b_1 \in R_0 \cap \overline{H_1},\ b_2 \in R_1 \cap \overline{H_0},$$

and this implies

$$b_1 \in R_0 \cap \overline{S},\ b_2 \in R_1 \cap \overline{S}. \tag{11}$$

Theorem 3. *There is a solution of degree 3 or less iff the sets $R_0 \cap \overline{S}$ and $R_1 \cap \overline{S}$ are not empty. In general, this solution is rational and uniquely dependent on the choice of points from (11).*

Proof: We have to carry out the construction of the interpolant if both sets in (11) are not empty. All control points $b_0, \ldots, b_3$ are now assumed to satisfy either (1) or (11). This defines positive values of the α_i and nonnegative values of the γ_i with $\operatorname{sgn}(\gamma_i) = \operatorname{sgn}(\kappa_i)$ for $i = 0, 1$ via (6) and (7). Thus, one cannot assume (8) to be satisfied with equal weights. But when going over to a rational solution, one can fix $w_1 = w_2 = 1$ and pick positive values of w_0 and w_3 to satisfy (8). This coincides with the technique introduced by Höllig [9]. □

§6. Geometric Characterizations

To arrive at a more geometric description of the above cases, we have to look at the possible forms of nonempty intersections S of the two admissible sets H_0 and H_1. To this end, we first check the nonempty intersection of the two osculating hyperplanes P_0 and P_1. It can be a single line or a full hyperplane. The latter case implies full planarity of the problem and will be discussed later in Section 7. Let the intersection of the osculating hyperplanes be a line that we call the pivot line from now on. On the pivot line, the nonempty intersection S of the two admissible sets H_0 and H_1 must be a simply connected set S. Either $\overline{S}$ consists of a single point or $\overline{S}$ contains an open subset of the pivot line. The latter case cannot occur if both curvature values are prescribed to be zero. It provides multiple solutions of degree 4.

Solvability with degree at most 3 will require both tangent rays R_0 and R_1 to hit the pivot line within the set $\overline{S}$. This will usually fix the two control points b_1 and b_2 uniquely, and the central piece of the control net will lie

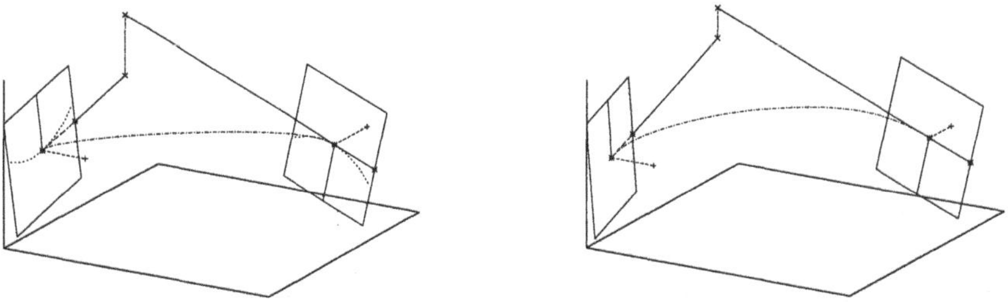

Fig. 3. Rational Solutions of Degree 3.

completely in $\overline{S}$ on the pivot line, since S is simply connected. This is why the degree 3 case usually has a unique rational solution in the nonplanar case.

For illustration, look at the left case of Figure 3 which has the same data as Figure 1. The tangent rays R_0 and R_1 are now emanating out of the osculating hyperplanes, and they hit the vertical pivot line in the background. The normals at the data locations indicate that the admissible halfspace on the left is pointing upward, while the one on the right is pointing downward. Thus, the intersection S of the two halfspaces is an open bounded interval on the pivot line. The two boundary points of S lie on the tangent rays, and they precisely form the sets of (11). Consequently, there is a unique rational solution of degree 3, as drawn into the figure. The two tangent rays and the set S on the pivot line form the control polygon. This situation is typical for cases with degree 3, and note that the solution looks unexpectedly "straight". This is due to the fact that it cares for the boundary curvature by weight adjustment, introducing a tension–like effect. The picture on the right simply has smaller curvature requirements, and the tension effect disppears.

§7. Planar Data

For purely planar cases, the above theorems apply without modification. But the preceding paragraph needs another argument, since now the osculating hyperplanes always coincide and there is no pivot line. For nonzero curvature data, the set S is an intersection of two halfspaces in the plane, and this intersection will in general be a cone. Since the tangents are the boundary lines of the halfspaces, the vertex of the cone will in general be at the intersection of the tangents. The two tangents split the plane into four cones, and one of those is S, depending on the direction of the normals at the data positions. Each point of S, when used as b_2, will each lead to a degree 4 solution.

Reduction to degree 3 is based on the sets in (11), and these are parts of the bounding rays of the cone, because the tangent rays are parts of the boundary lines of the admissible halfspaces. The admissible sections of the boundary of the cone are precisely the intersections of the tangent rays with

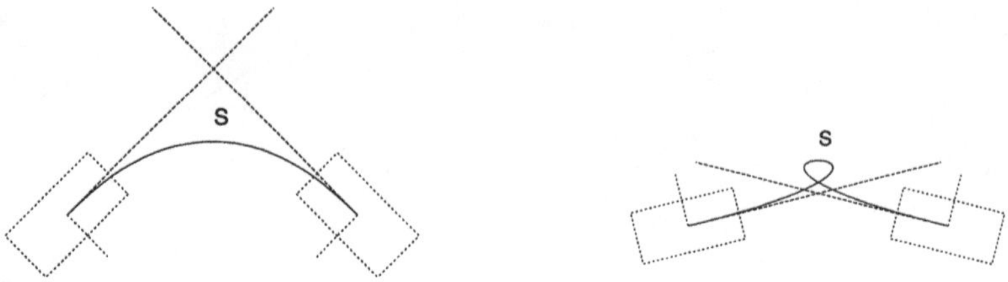

Fig. 4. Planar polynomial solutions of degree 3.

the cone's boundary.

In contrast to the space curve case, the degree 3 solution in the planar case will thus be nonunique. One can try to use the freedom to get away with a polynomial solution. But this is a hazardous task, since it leads to two quadratic equations with two unknowns that have to be solved on a restricted domain. The paper [1] of de Boor, Höllig, and Sabin treats the special case of fully convex data, while the short unpublished note [13] works for a simple inflection point. The approach of this paper can be easily applied to generate various other situations in which a degree 3 polynomial solution may exist.

Figure 4 shows two planar cases where the set S forms a nondegenerate two–dimensional cone. On the left we have the situation treated by de Boor, Höllig, and Sabin in [1]. The cone S is pointing downwards, and the control points b_1, b_2 can vary along the tangent rays up to their intersection. The picture on the right has the cone directed upwards because both normals are pointing upwards, and the control points can vary on the tangent rays from the intersection up to infinity, leading to a cusp if they are sufficiently far up. This actually happens if they are moved to generate the shown polynomial solution.

However, since the sets of (11) are nonempty in general, one can get away with a variety of degree 3 rational solutions, and good ones will keep the control points at reasonable distances from the data positions. The papers [4,5,6,7,8] by T. N. T. Goodman and his collaborators deal with methods to pick a suitable solution with shape preserving properties. For instance, the right–hand case of Figure 4 could better be handled by a rational solution with control points near the vertex of the cone S. The cusp would be avoided, but bringing the control points too near to the vertex will result in a solution looking like a straight line, because the weights at the positions will become large in order to cope with curvature data prescribed there. See Figure 6 for a similar situation in $\mathbb{R}^3$.

§8. Regularization of Degree 5 Solutions

For picking a useful solution in case of ambiguities we adopt a very simple scheme that first distinguishes between "good" and "bad" data. Imagine that at the position y_0 a large curvature $\kappa_0 > 0$ is required. The curvature radius $R_0 = \kappa_0^{-1}$ then is very small, and the curve will have second order contact to

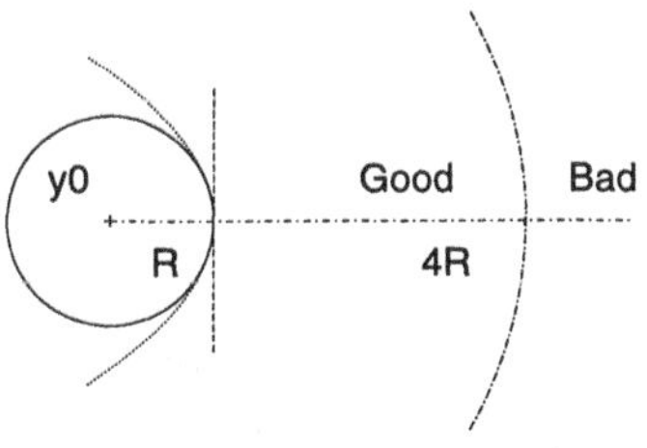

Fig. 5. Regions for good and bad y_1.

the circle of radius R_0 with center $c_0 := y_0 + R_0 n_0$ in the osculating hyperplane. To allow some leeway, one could say that it should locally stay within a "trust region" of, say, radius $4R_0$ around the position y_0. But if the second position y_1 is further away, we should decide that this is a "bad" case. Look at Figure 5 for illustration.

More precisely, we fix a constant $K > 1$ and say that the data are bad at y_0 if $\kappa_0 \|y_1 - y_0\|_2 > K$. The same can be done at y_1. For regularizing cases with degree 4 or 3, we cannot split the problem into two separate parts, and then we call a data set bad if

$$\sqrt{\kappa_0 \kappa_1} \|y_1 - y_0\|_2 > K.$$

The main implication of good and bad situations is that we use different strategies to handle them. The good cases are regularized towards good approximation properties for dense samples. Note that for a dense sample of data from a smooth curve we have only good pieces, since the distances get small while the curvature values are bounded.

The bad pieces are regularized for good asymptotic behavior when curvature tends to infinity. This sounds strange, but it can be done by moving control points near to the data locations, and the resulting curves simply interpolate bad data pieces by nearly straight curves with "high tension". This avoids unreasonable large loops, cusps, and wiggles, keeping the control points near the data locations.

Let's study the details for degree 5, and we can do this around y_0 locally, fixing the control points b_0, b_1, b_2 properly. For a dense sample from a smooth curve, the distances $\|b_{i+1} - b_i\|_2$ are each about $A/5$, where A is the arclength of the curve between y_0 and y_1. This arclength can be estimated by

$$a = \delta \left(1 + \frac{\kappa_0^2 \delta^2}{24}\right)$$

for small values of $\delta = \|y_0 - y_1\|_2$ up to terms of order five in δ (see [12] for details). Using (6) and (8), we take $\alpha_0 = a/5$, $\beta_0 = 2a/5$, and pick $\gamma_0 = a^2 \kappa_0 / 20$ from (9). Since $a\kappa_0$ stays bounded by our criterion for "good" pieces, this works nicely for dense data samples. A proper convergence analysis would be an interesting project.

For "bad" cases we want to make α_0 proportional to κ_0^{-1} in order to let the control point b_1 move towards b_0 for extremely large curvature values. Using this proportionality and making the transition to the "good" case continuous, a short calculation yields the recipe

$$\begin{aligned} c &= \frac{K}{4}\left(1+\frac{K^2}{24}\right) \\ \alpha_0 &= \frac{4c}{5\kappa_0} \\ \beta_0 &= 2\alpha_0 \\ \gamma_0 &= \frac{4c^2}{5\kappa_0} \end{aligned}$$

satisfying (9). Note that α_0, β_0, and γ_0 are proportional to κ_0^{-1}.

For true space curves, the cases of degree 3 usually are unique. For degree 4, the regularization technique needs modification, because the control point b_2 is restricted to S on the pivot line. After α_0 and γ_0 are estimated as above, one can in general find a point b_2^0 on the pivot line satisfying (8). The same can be done from the other side, resulting in some point b_2^1. But these points need not lie in S, though they often are good guesses for the actual point b_2. If just one is feasible, we take it, and if both are feasible, we take the one that is nearer to its corresponding tangent ray. If both are unfeasible (this is a rare exception), then $\overline{S}$ must be a bounded interval on the pivot line, and we take the mean of its bounds, ignoring our previous estimates. This complicated strategy works well in most cases, but improvements seem to be possible. A comparison with the strategy of Peters [11] is still missing. Details of the regularization technique will be found in [15].

§9. Computational Remarks

If data from a smooth space curve are sampled, one can observe that most pieces can be handled by rational interpolation of degree 3, *i.e.*, with Höllig's technique from [9]. The exceptions usually are solvable by polynomials of degree 4, and it needs special tricks to produce cases where degree 5 actually is needed. The interpolation is stable, especially if the above regularization technique is used. The whole process seems to be largely shape–preserving due to the technique of picking the smallest possible degree.

§10. Transitions between Cases

It would be desirable to have an interpolation process that depends smoothly on the data. But this is not possible if we insist on minimal degrees and if we always prefer polynomial pieces over rational ones. The main reason for this is that the rational solution of degree 3 often is unique, but cannot in general

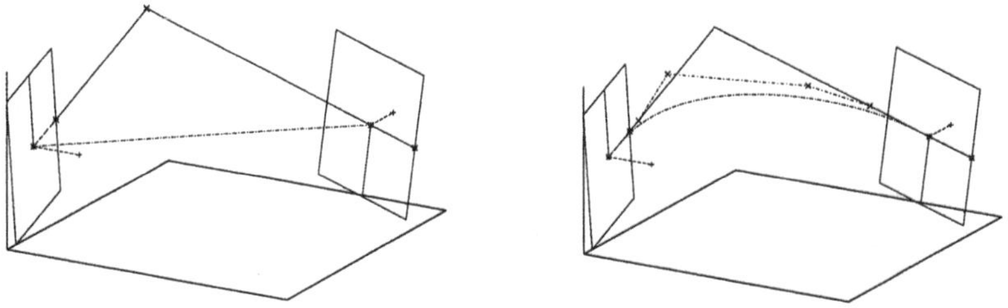

Fig. 6. Neighboring solutions of degree 3 and 5.

go continuously over into a polynomial of degree 4 or 5. The left picture of Figure 6 shows a rational degree 3 case with an extremely small interval S on the pivot line. A small rotation of one of the osculating planes lets the interval disappear, and then one gets the degree 5 solution on the right–hand side. But this solution cannot in general be near to a degree three truly rational curve with nearly coalescing inner control points. The latter gets so much tension that it finally looks very much like a straight line. This special transition situation forces the control points b_1 and b_2 to lie nearly on the tangent rays from the opposite position. This in turn yields very small values of the α_i and γ_i, and the curvature requirements are satisfied by extremely large weights at $y_0 = b_0$ and $y_1 = b_3$, making the solution look like a straight line.

Note how the degree 5 solution is smoothed out by our regularization. The tangent rays make up most of the control polygon for degree 3 due to the strong tension effect. However, for degree 5 the control polygon nicely uses the leeway in the admissible halfspaces.

Acknowledgments. Help in proofreading was provided by Christoph Schröder.

References

1. deBoor, C., K. Höllig, and M. Sabin, High accuracy geometric Hermite interpolation, Comput. Aided Geom. Design **4** (1987), 269–278.
2. Farin, G. *Curves and Surfaces for Computer-Aided Geometric Design*, Academic Press, San Diego, 1988.
3. Farin, G. Rational curves and surfaces, in *Mathematical Methods in Computer Aided Geometric Design*, T. Lyche and L. L. Schumaker (eds.), Academic Press, New York, 1989, 215–238.
4. Goodman, T. N. T., Shape preserving interpolation by parametric rational cubic splines, *Numerical Mathematics Singapore 1988*, R. P. Agarwal, Y. M. Chow, and S. J. Wilson (eds.), International Series of Numerical Mathematics, Vol. 86, Birkhäuser, Basel, 1988, 149–158.

5. Goodman, T. N. T., B. H. Ong, and K. Unsworth, Constrained interpolation using rational cubic splines, Univ. of Dundee, Computer Sciences Report CS90/01, 1990.
6. Goodman, T. N. T., and K. Unsworth, Shape preserving interpolation by parametrically defined curves, SIAM J. Numer. Anal. **25** (1988), 1453–1465.
7. Goodman, T. N. T., and K. Unsworth, Shape preserving interpolation by curvature continuous parametric curves, Comput. Aided Geom. Design **5** (1988), 323–340.
8. Goodman, T. N. T., and K. Unsworth, An algorithms for generating shape preserving parametric interpolating curves using rational cubic splines, Univ. of Dundee, Computer Science Report CS 89/01, 1989.
9. Höllig, K., Algorithms for rational spline curves, ARO-Report 88–1, 1988, 287–300.
10. Hoschek, K., and D. Lasser, *Fundamentals of Computer Aided Geometric Design*, A K Peters, Wellesley, 1993.
11. Peters, J., Local generalized Hermite interpolation by quartic C^2 space curves, ACM Trans. on Graphics **8** (1989), 235–242.
12. Schaback, R. Interpolation in $\mathbb{R}^2$ by piecewise quadratic visually C^2 Bezier polynomials, Comput. Aided Geom. Design **6** (1989), 219–233.
13. Schaback, R. Remarks on high accuracy geometric Hermite interpolation, preprint, Göttingen, 1989.
14. Schaback, R. Rational curve interpolation, in *Mathematical Methods in Computer Aided Geometric Design II*, T. Lyche and L. L. Schumaker (eds.), Academic Press, New York, 1992, 517–535.
15. Schröder, C., Regularisierung von G^2–Hermite–Interpolanten für Raumkurven, Diplomarbeit, Univ. Göttingen, 1997.
16. Schütt, C., GC^2–Hermite–Interpolation im $\mathbb{R}^3$ mit gradreduzierten Kurven, Diplomarbeit, Univ. Göttingen, 1995.

R. Schaback
Georg–August–Universität Göttingen
Institut für Numerische und Angewandte Mathematik
Lotzestraße 16–18
D–37083 Göttingen, GERMANY
`schaback@math.uni-goettingen.de`
`http://www.num.math.uni-goettingen.de/schaback`

An Interesting Class of Polynomial Vector Fields

Gerik Scheuermann, Hans Hagen, and Heinz Krüger

Abstract. The visualization of vector fields is one of the most important topics in visualization. Of special interest over the last years have been topology-based methods. We present a method to construct polynomial vector fields after determine the critical points and their index to generate test data and help analysing vector field topology. The idea is based on Clifford algebra and analysis. We start with a formulation of a plane real vector field in Clifford algebra giving more topological information directly from the formulas than the usual description in Cartesian coordinates. The theorem presented here is an essential improvement to our previous result in [9], and has implications to general polynomial vector fields.

§1. Introduction

Scientific Visualization deals often with a large amount of data. Especially in the visualization of vector fields one usually has the problem that it is nearly impossible to show the whole data. One has to reduce the information in a useful way. One very successful approach is to look for the topology of the field and visualize its essential parts [4].

The testing of new algorithms in this direction was done up to now by comparing with handdrawings based on experiments. The problem is that in the literature [1,3,7] one only finds an analysis of linear fields and some special equations, where the position, number and type of critical points is not arbitrary. The reason is probably that it is difficult to analyse non-linear fields in the classical way by computing the integral curves. We have found and published a theorem in [9] which allows a description of the topology of some polynomial vector fields. This was used in [10] to derive a new algorithm for vector field visualization which can detect and visualize higher order singularities. In this paper we give an essential generalization of this theorem with a similar result for a larger class of polynomial vector fields. We also can show how the analysis of a arbitrary polynomial vector field can be simplified.

This result is obtained by using the Clifford algebra and analysis. They are described in Sections 2 and 3. The result is derived and proved in Section 4. The last section gives some examples showing how one can actually construct non-trivial polynomial vector fields by this result which should be useful for testing algorithms.

Mathematical Methods for Curves and Surfaces II

Morten Dæhlen, Tom Lyche, and Larry L. Schumaker (eds.), pp. 429–436.

ISBN 0-8265-1315-8.

§2. The Clifford Algebra

We need the Clifford algebra only in two dimensions, so for simplicity we should stay there. Let $\{e_1, e_2\}$ be an orthogonal basis of $\mathbb{R}^2$ with standard scalar product. The Clifford algebra $\mathcal{G}_2$ is then the $\mathbb{R}$-algebra of maximal dimension containing $\mathbb{R}$ and $\mathbb{R}^2$ such that for each vector $x \in \mathbb{R}^2$, $x^2 = \|x\|^2$.

This implies the following rules

$$e_j^2 = 1 \quad j = 1,2$$
$$e_1e_2 + e_2e_1 = 0.$$

We get a 4-dimensional algebra with the vector basis $\{1, e_1, e_2, e_1e_2\}$. We now have the real vectors

$$xe_1 + ye_2 \in \mathbb{R}^2 \subset \mathcal{G}_2$$

and the real numbers

$$x1 \in \mathbb{R} \subset \mathcal{G}_2,$$

both in the algebra.

We may also define the projections $\langle \cdot \rangle_k$, $k = 0, 1, 2$ by

$$\begin{aligned} \langle \cdot \rangle_0 : \mathcal{G}_2 &\to \mathbb{R} \subset \mathcal{G}_2 \\ a1 + be_1 + ce_2 + de_1e_2 &\mapsto a1 \\ \langle \cdot \rangle_1 : \mathcal{G}_2 &\to \mathbb{R}^2 \subset \mathcal{G}_2 \\ a1 + be_1 + ce_2 + de_1e_2 &\mapsto be_1 + ce_2 \\ \langle \cdot \rangle_2 : \mathcal{G}_2 &\to \mathbb{R}e_1e_2 \subset \mathcal{G}_2 \\ a1 + be_1 + ce_2 + de_1e_2 &\mapsto de_1e_2. \end{aligned}$$

For two vectors one can then describe the new product by already known products. Let $v = v_1e_1 + v_2e_2$, $w = w_1e_1 + w_2e_2$ be two vectors, $v_1, v_2, w_1, w_2 \in \mathbb{R}$. Then

$$\begin{aligned} vw &= (v_1w_1 + v_2w_2)1 + (v_1w_2 - w_1v_2)e_1e_2 \\ &= \langle vw \rangle_0 + \langle vw \rangle_2 \\ &= v \cdot w + v \wedge w, \end{aligned}$$

where $\cdot$ denotes the scalar (inner) product and $\wedge$ denotes the outer product of Grassmann. This unification of inner and outer product is the starting point of the geometric interpretation. We will not need it here, see [5] for a good introduction.

More important for us is that the complex numbers can also be canonically embedded by recognizing $(e_1e_2)^2 = -1$, so we set

$$i := e_1e_2.$$

Then

$$a1 + bi \in \mathbb{C} \subset G_2$$

is a subalgebra. The next section introduces a bit of Clifford analysis.

§3. Clifford Analysis

Here we again need only the two-dimensional case, so we limit our definitions to that case to avoid technical overload. Our basic maps will be multivector fields

$$A : \mathbb{R}^2 \to \mathcal{G}_2$$
$$r \mapsto A(r)$$

A Clifford vector field is just a multivector field with values in $\mathbb{R}^2 \subset \mathcal{G}_2$

$$v : \mathbb{R}^2 \to \mathbb{R}^2 \subset \mathcal{G}_2$$
$$r = xe_1 + ye_2 \mapsto v(r) = v_1(x,y)e_1 + v_2(x,y)e_2.$$

The directional derivative of A in direction $b \in \mathbb{R}^2$ is defined by

$$A_b(r) = \lim_{\epsilon \to 0} \frac{1}{\epsilon}[A(r + \epsilon b) - A(r)],$$

for $\epsilon \in \mathbb{R}$. This allows the definition of the vector derivative of A at $r \in \mathbb{R}^2$ by

$$\partial A(r) : \mathbb{R}^2 \to \mathcal{G}_2$$
$$r \mapsto \partial A(r) = \sum_{k=1}^{2} g^k A_{g_k}(r).$$

This is independent of the basis $\{g_1, g_2\}$ of $\mathbb{R}^2$.

The integral in Clifford analysis is defined as follows : Let $M \subset \mathbb{R}^2$ be an oriented r-manifold and $A, B : M \to \mathcal{G}^2$ be two piecewise continous multivector fields. Then one defines

$$\int_M AdXB = \lim_{n \to \infty} \sum_{i=0}^{n} A(x_i)\Delta X(x_i)B(x_i),$$

where $\Delta X(x_i)$ is a r-volume in the usual Riemannian sense. This allows the definition of the Poincaré-index of a vector field v at $a \in \mathbb{R}^2$ as

$$ind_a v = \lim_{\epsilon \to 0} \frac{1}{2\pi i} \int_{S^1_\epsilon} \frac{v \wedge dv}{v^2},$$

where S^1_ϵ is a circle of radius ϵ around a.

§4. Analysis of Polynomial Vector Fields

For our result it is necessary to look at $v : \mathbb{R}^2 \to \mathbb{R}^2 \subset \mathcal{G}_2$ in a different way. Let $z = x + iy$, $\bar{z} = x - iy$. This means

$$x = \frac{1}{2}(z + \bar{z}), \qquad y = \frac{1}{2i}(z - \bar{z}).$$

We get

$$\begin{aligned} v(r) &= v_1(x,y)e_1 + v_2(x,y)e_2 \\ &= [v_1(\frac{1}{2}(z+\bar{z}), \frac{1}{2i}(z-\bar{z})) - iv_2(\frac{1}{2}(z+\bar{z}), \frac{1}{2i}(z-\bar{z}))]e_1 \\ &= E(z,\bar{z})e_1, \end{aligned}$$

where

$$\begin{aligned} E : \mathbb{C}^2 &\to \mathbb{C} \subset G_2 \\ (z,\bar{z}) &\mapsto v_1(\frac{1}{2}(z+\bar{z}), \frac{1}{2i}(z-\bar{z})) - iv_2(\frac{1}{2}(z+\bar{z}), \frac{1}{2i}(z-\bar{z})) \end{aligned}$$

is a complex-valued function of two complex variables. The idea is now to analyse E instead of v and get topological results directly from the formulas in some interesting cases.

Let us first assume that E and also v is linear.

Theorem 1. *Let*

$$v(r) = (az + b\bar{z} + c)e_1$$

be a linear vector field. For $|a| \neq |b|$ it has a unique zero at $z_0 e_1 \in \mathbb{R}^2$. For $|a| > |b|$ has v one saddle point with index -1. For $|a| < |b|$ it has one critical point with index 1. The special types in this case can be got from the following list:

1) *$Re(b) = 0 \Leftrightarrow$ circle at z_0.*
2) *$Re(b) \neq 0$, $|a| > |Im(b)| \Leftrightarrow$ node at z_0.*
3) *$Re(b) \neq 0$, $|a| < |Im(b)| \Leftrightarrow$ spiral at z_0.*
4) *$Re(b) \neq 0$, $|a| = |Im(b)| \Leftrightarrow$ focus at z_0.*

In cases 2) – 4) one has a sink for $Re(b) < 0$ and a source for $Re(b) > 0$.

For $|a| = |b|$ one gets a whole line of zeros.

Proof: A computation of the derivatives of the components v_1, v_2 and a comparison with the classic classification gives this result. □

We included this easy theorem to show that this description gives topological information more directly. Let us look now at the general polynomial case

Theorem 2. *Let $v : \mathbb{R}^2 \to \mathbb{R}^2 \subset \mathcal{G}_2$ be an arbitrary polynomial vector field with isolated critical points. Let $E : \mathbb{C}^2 \to \mathbb{C}$ be the polynomial so that $v(r) = E(z,\bar{z})e_1$. Let $F_k : \mathbb{C}^2 \to \mathbb{C}$, $k = 1,\ldots,n$ be the irreducible components of E, so that $E(z,\bar{z}) = \prod_{k=1}^{n} F_k$. Then the vector fields $w_k : \mathbb{R}^2 \to \mathbb{R}^2$, $w_k(r) = F_k(r)e_1$ also have only isolated zeros $z_1,\ldots,z_m$. These are then the zeros of v, and for the Poincaré-indices we have*

$$ind_{z_j} v = \sum_{k=1}^{n} ind_{z_j} w_k.$$

Proof: The w_k have only isolated zeros because otherwise v would have also not isolated zeros. It is also obvious that a zero of a w_k is a zero of v and a zero of v must be a zero of one of the w_k.

For the derivatives we get

$$\frac{\partial E}{\partial z} = a \sum_{k=1}^{n} \frac{\partial F_k}{\partial z} \prod_{l=1, l \neq k}^{n} F_l$$

$$\frac{\partial E}{\partial \bar{z}} = a \sum_{k=1}^{n} \frac{\partial F_k}{\partial \bar{z}} \prod_{l=1, l \neq k}^{n} F_l.$$

For the computation of the Poincaré-index we assume $z_j = 0$ after a change of the coordinate system and that ϵ is so small that there are no other zeros inside S^1_ϵ. We get

$$\begin{aligned} ind_{z_j} v =& \frac{1}{2\pi i} \int_{S^1_\epsilon} \frac{v \wedge dv}{v^2} \\ =& \frac{1}{2\pi i} \int_{S^1_\epsilon} \frac{1}{v^2} < a \prod_{k=1}^{n} F_k e_1 [dza \sum_{k=1}^{n} \frac{\partial F_k}{\partial z} \prod_{l=1, l\neq k}^{n} F_l + \\ & d\bar{z} a \sum_{k=1}^{n} \frac{\partial F_k}{\partial \bar{z}} \prod_{l=1, l \neq k}^{n} F_l] e_1 >_2 \\ =& \frac{1}{2\pi i} \int_{S^1_\epsilon} \frac{1}{v^2} < a\bar{a} \sum_{k=1}^{n} F_k \overline{(dz \frac{\partial F_k}{\partial z} + d\bar{z} \frac{\partial F_k}{\partial \bar{z}})} \prod_{l=1, l\neq k}^{n} F_l \bar{F}_l >_2 \\ =& \sum_{k=1}^{n} \frac{1}{2\pi i} \int_{S^1_\epsilon} \frac{1}{F_k \bar{F}_k} < F_k e_1 (dz \frac{\partial F_k}{\partial z} + d\bar{z} \frac{\partial F_k}{\partial \bar{z}}) e_1 >_2 \\ =& \sum_{k=1}^{n} ind_{z_j} F_k e_1 \\ =& \sum_{k=1}^{n} ind_{z_j} w_k. \quad \square \end{aligned}$$

For experiments it is nice to use linear factors because one has good insights into their behavior from Theorem 1.

Theorem 3. *Let $v : \mathbb{R}^2 \to \mathbb{R}^2 \subset \mathcal{G}_2$ be the vector field $v(r) = E(z, \bar{z}) e_1$ with*

$$E(z, \bar{z}) = \prod_{k=1}^{n} (a_k z + b_k \bar{z} + c_k), \qquad |a_k| \neq |b_k|,$$

and let z_k be the unique zero of $a_k z + b_k \bar{z} + c_k$. Then v has zeros at z_j, $j = 1, \ldots, n$, and the Poincaré-index of v at z_j is the sum of the indices of the $(a_k z + b_k \bar{z} + c_k) e_1$ at z_j.

Proof: Special case of Theorem 2. $\square$

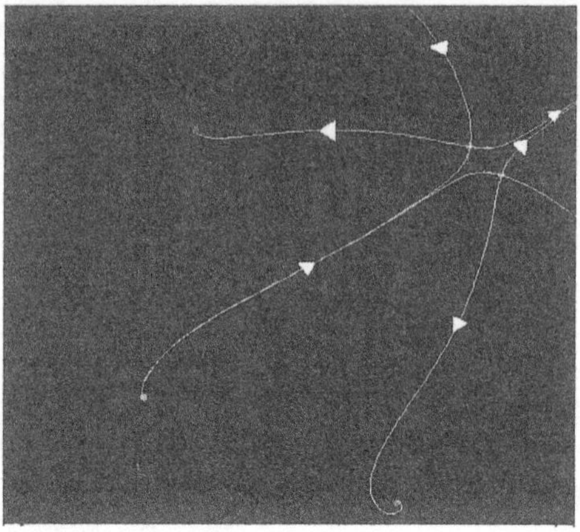

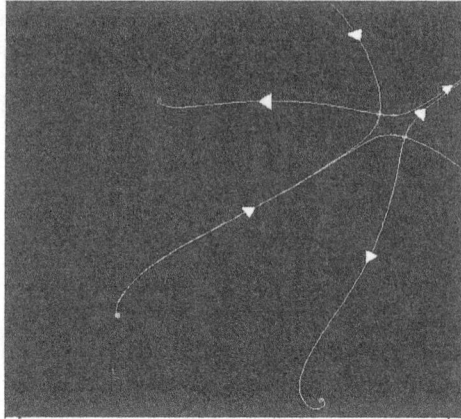

Fig. 1. Two saddles and three index 1 singularities.

§5. Examples

The following examples show the topological structure of vector fields with the algebraic structure of Theorem 3. They illustrate its usefulness for the testing of topological vector field visualization algorithms by showing that one has full control over position and index of the critical points.

Our first example is the vector field

$$\begin{aligned} v(r) =& (\bar{z} - \overline{(-0.51 - 0.51i)})(\bar{z} - \overline{(0.38 - 0.92i)})(z - (0.63 + 0.46i)) \\ & (z - (0.74 + 0.35i))(\bar{z} - \overline{(-0.33 + 0.52i)})e_1 \end{aligned}$$

in the square $[-1, 1] \times [-1, 1]$ in Cartesian coordinates. Figure 1 shows that the field has saddles at $(0.63, 0.46)$ and $(0.74, 0.35)$ produced by the factors with z. There are further singularities of index 1 at $(-0.51, -0.51)$, $(0.38, -0.92)$ and $(-0.33, 0.52)$ produced by the factors with $\bar{z}$. The sampled arrows indicate the direction of the unit vector field at a 20×20 quadratic grid.

The second example is

$$\begin{aligned} v(r) =& (\bar{z} - \overline{(-0.58 - 0.64i)})(\bar{z} - \overline{(0.51 + 0.27i)})(z - (0.68 - 0.59i)) \\ & (\bar{z} - \overline{(-0.12 - 0.84i)})^2(z - (-0.11 - 0.72i))(z - (0.74 + 0.35i))e_1 \end{aligned}$$

again in the square $[-1, 1] \times [-1, 1]$ in Cartesian coordinates. Figure 2 shows that the field has saddles at $(0.68, -0.59)$,$(-0.11, -0.72)$ and $(0.74, 0.35 - 0.18, 0.46)$ produced by the factors with z. There are further singularities of index 1 at $(-0.58, -0.64)$ and $(0.51, 0.27)$ produced by the factors with $\bar{z}$. There is also an index 2 singularity at $(-0.12, -0.84)$ stemming from the squared factor. Again the sampled arrows indicate the direction of the unit vector field.

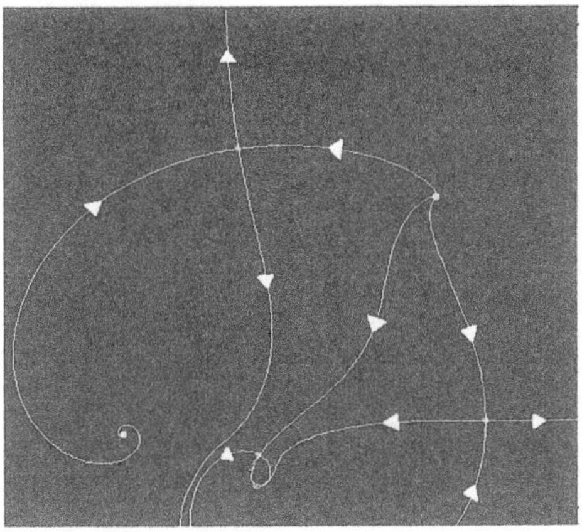

Fig. 2. Three saddles, two index 1 and one index 2 singularity.

Acknowledgments. This work was partly made possible by financial support by the Deutscher Akademischer Auslandsdienst (DAAD). The first author got a "DAAD-Doktorandenstipendium aus Mitteln des zweiten Hochschulsonderprogramms" for his stay at the Arizona State University from Oct. 96 to Jan. 97. We also want to thank Alyn Rockwood, Greg Nielson and David Hestenes from Arizona State University for many comments, suggestions and inspiration. Special thanks go to Shoeb Bhinderwala who wrote the excellent user interface to create the pictures in the examples [2].

References

1. Arnold, V. I., *Gewöhnliche Differentialgleichungen*, Deutscher Verlag der Wissenschaften, Berlin, 1991.
2. Bhinderwala, S. A., Design and visualization of vector fields, master thesis, Arizona State University, 1997.
3. Guckenheimer, J., and P. Holmes, *Nonlinear Oszillations, Dynamical Systems and Linear Algebra*, Springer, New York, 1983.
4. Helman, J. L., and L. Hesselink, Visualizing vector field topology in fluid flows, IEEE Computer Graphics and Applications **11** (1991), 36–46.
5. Hestenes, D., *New Foundations for Classical Mechanics*, Kluwer Academic Publishers, Dordrecht, 1986.
6. Krüger, H., and M. Menzel, Clifford-analytic vector fields as models for plane electric currents, in *Analytical and Numerical Methods in Quaternionic and Clifford Analysis*, W. Sprössig, K. Gürlebeck (eds.), Seiffen, 1996.
7. Hirsch, M. W., and S. Smale, *Differential Equations, Dynamical Systems and Linear Algebra*, Academic Press, New York, 1974.

8. Milnor, J. W., *Topology from the Differentiable Viewpoint*, The University Press of Virginia, Charlottesville, 1965.
9. Scheuermann, G., H. Hagen, and H. Krüger, Clifford algebra in vector field visualization, preprint.
10. Scheuermann, G., H. Hagen, H. Krüger, M. Menzel, and A. P. Rockwood, Visualization of higher order singularities in vector fields, IEEE Visualization 1997, to appear.

Gerik Scheuermann and Hans Hagen
Dept. of Computer Science
University of Kaiserslautern
Postfach 3049
D-67653 Kaiserslautern, GERMANY
`scheuer@informatik.uni-kl.de`
`hagen@informatik.uni-kl.de`

Heinz Krüger
Dept. of Physics
University of Kaiserslautern
Postfach 3049
D-67653 Kaiserslautern, GERMANY
`krueger@physik.uni-kl.de`

Interpolation with Developable Strip–Surfaces Consisting of Cylinders and Cones

Mike Schneider

Abstract. Developable surfaces (ruled surfaces of Gaussian curvature $k \equiv 0$) can be constructed by combinations of cylinders, cones and tangent surfaces of space curves. In this paper we will evolve a strategy for interpolating a given curve and auxiliary generators with developable spline surfaces consisting of cylinders and cones of arbitrary degree. We will also work out a criterion for the existence of surfaces free from singular points in a prescribed strip given by two parallel planes.

§1. Introduction

As developable surfaces can be formed by isometric bending of a flat sheet of material, they occur in many industrial applications [4], especially in automotive ones as binder (blankholder) surfaces for sheet metal forming processes [4,11,14]. Developable surfaces are a special subclass of ruled surfaces and are either cylinders, cones, tangent surfaces, i.e. surfaces generated by the tangent lines of a space curve, or combinations of these three surface types. Recently, several geometric methods for constructing developable surfaces in CAD environments have evolved, whereas two basically diverse approaches have gained acceptance: one of them uses the dual representation of developable surfaces, introduced by Hoschek [5]. Based on this representation (using plane coordinates), projective design approaches [2,3], interpolation methods [7,8,9,13] and approximation techniques [7,8] have been developed. The other approach, resting upon point coordinates, e.g. has been applied by Lang and Röschel to characterize all developable rational $(1, n)$–Bézier surfaces [10], by Aumann to interpolate a given planar cubic Bézier curve and two auxiliary lines with a developable Bézier patch [1], and by Leopoldseder and Pottmann to construct spline surfaces composed of right circular cones [12].

As done by the last two authors, we also will use developable surfaces consisting of cones (and as their limit of cylinders) only, whereas the cones utilized in this paper will be of arbitrary degree (also denoted as generalized

Mathematical Methods for Curves and Surfaces II
Morten Dæhlen, Tom Lyche, and Larry L. Schumaker (eds.), pp. 437–444.

ISBN 0-8265-1315-8.

cones), i.e. cones arising from the connection of any curves with fixed points. Since spline surfaces consisting of cones (and cylinders) only have a finite number of singular points, the singularities of such surfaces can be controlled easily. As most industrial applications require the developable surfaces to be free from singularities in prescribed areas, this is a main advantage of cone spline surfaces in comparison with developable surfaces, which contain tangent surfaces as well. By using splines consisting of cones only, we will present a geometric approach which is a generalization of Aumann's method [1]. Given an arbitrary space curve and some (suitable) lines, we will construct an at least G^1–continuous developable surface interpolating the given elements and free from singularities in a regarded strip. We will also specify a criterion which the given lines have to fulfill so that the strip–surface contains no singularities.

§2. Interpolation with Developable Strip–Surfaces

Using a purely geometric approach which does not depend on any kind of basis functions for the given curve and the required developable surface, we will now investigate the following problem:

Problem. *An arbitrary space curve* $\mathbf{X}$ *and some lines* $\mathbf{g}_j$ *(at least two: one at the beginning and one at the end of the curve) belonging to suitable parameter values are given and should be interpolated by an at least* G^1*–continuous developable surface* $\mathbf{Y}$ *consisting of generalized cones and cylinders. Additionally, two given parallel planes* $\mathbf{B}_1$ *and* $\mathbf{B}_2$ *should embed the curve and bound the area of the surface, which is postulated to be free from singular points (see Fig. 1).*

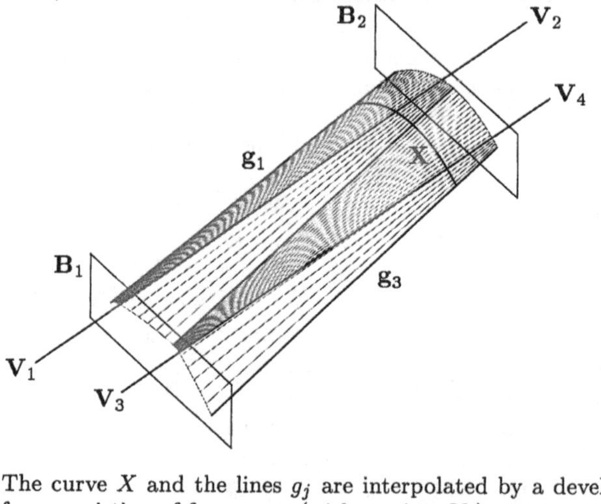

Fig. 1. The curve X and the lines g_j are interpolated by a developable strip–surface consisting of four cones (with vertices V_k).

In the following, we are only interested in the part of the developable surface which lies within the strip terminated by the planes $\mathbf{B}_k$. This part of the surface is called a (developable) strip–surface.

Our aim now is to construct two cones (as limit cylinders) between each pair of consecutive given lines. As this approach is a local one, we only have to analyze the particular case of two given generators $\mathbf{g}_1$ and $\mathbf{g}_2$ at the curve's endpoints $\mathbf{P}_1$ and $\mathbf{P}_2$. Besides, we can assume that the two generators do not cross each other (even at infinity), as in this case the resulting surface (generally) consists of only one cone (resp. one cylinder). Hence, we demand two cones which each interpolate one of the given generators as well as coherent parts of the curve. These parts should contain precisely one common point $\mathbf{P}$. To get an at least G^1–continuous surface, we have to connect the two cones along one common generator with coinciding tangent planes. Since developable surfaces have stationary tangent planes along their generators and, additionally, the tangent planes of both cones coincide at the common point $\mathbf{P}$, it is sufficient to determine a line which contains the vertices of both cones as well as the point $\mathbf{P}$ (shown in Fig. 1). Changing the point of view, we ascertain that to each point $\mathbf{P}$ of the curve (except $\mathbf{P}_1$ and $\mathbf{P}_2$) there is exactly one line which fulfills this condition. It is the line of intersection of the two planes $\mathbf{E}_1 = \mathbf{g}_1 \cup \mathbf{P}$ and $\mathbf{E}_2 = \mathbf{g}_2 \cup \mathbf{P}$. As the line $\mathbf{E}_1 \cap \mathbf{E}_2$ lies in both planes $\mathbf{E}_k$, it crosses the generator $\mathbf{g}_1$ as well as $\mathbf{g}_2$ (maybe at infinity) and under construction even the point $\mathbf{P}$. The resulting intersection points of $\mathbf{E}_1 \cap \mathbf{E}_2$ and $\mathbf{g}_1$ resp. $\mathbf{g}_2$ can then be used as the vertices of the requested cones.

Now we want to detect those points $\mathbf{P}$ of the curve which lead to vertices lying outside the regarded strip, as only these vertices result in strip–surfaces which are free from singularities. For this, we introduce two pencils of planes

$$\Sigma_j : \quad \mathbf{E}_j(\lambda_j) = (1-\lambda_j)\mathbf{E}_{j,\min} + \lambda_j \mathbf{E}_{j,\max} \quad (j = 1, 2),$$

whereas the planes $\mathbf{E}_{j,\min}$ and $\mathbf{E}_{j,\max}$ (represented in homogeneous coordinates) contain the generator $\mathbf{g}_j$ and, additionally, form the smallest part of Σ_j, which encloses the complete curve (as outlined in Fig. 2a). Since it is necessary for the following argumentation that the curve $\mathbf{X}$ is only touched but not crossed by the planes $\mathbf{E}_{j,\min}$ and $\mathbf{E}_{j,\max}$, we have to restrict the selectable curves as follows: a curve is acceptable if and only if there are two planes which contain $\mathbf{g}_1$ resp. $\mathbf{g}_2$ in such a way that the curve is not crossed by any of the two planes. As all curves can be cut into segments fulfilling this condition, the above requirement is no intrinsic limitation. With the above pencil's definition the following statements now hold:

- $\forall\ \lambda_j \in \mathbb{R}$: $\mathbf{E}_j(\lambda_j)$ crosses the generator $\mathbf{g}_i$ $(i \in \{1,2\} - j)$
- if $\lambda_j \in [0,1]$: $\mathbf{E}_j(\lambda_j)$ crosses the curve $\mathbf{X}$.

Furthermore, each pencil can be divided into two regions: one whose planes cross the opposite generator within the strip and one whose planes cross outside. These two parts of pencil Σ_j are separated by the planes

$$\mathbf{E}_j(\lambda_{j,k}) = \mathbf{g}_j \cup \mathbf{S}_{i,k}$$

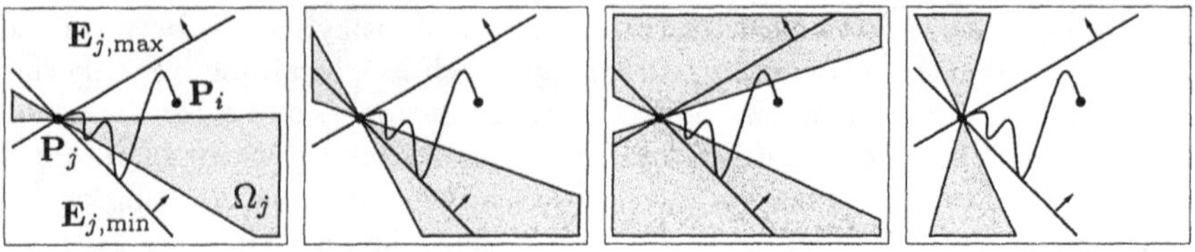

(a) (b) (c) (d)

Fig. 2 a – d. Basic situations of the pencil's Σ_j admissible region Ω_j.

which, additionally to the given line $\mathbf{g}_j$, contain the intersection points

$$\mathbf{S}_{i,k} = \mathbf{g}_i \cap \mathbf{B}_k.$$

Since the point $\mathbf{P}_i$ lies inside the regarded strip, the part of the pencil which contains $\mathbf{P}_i$ also encloses all planes crossing $\mathbf{g}_i$ within the strip. The other part of the pencil, which should be denoted as the *admissible region* Ω_j of pencil Σ_j, includes all planes crossing outside and belonging to suitable vertices. Accordingly, all points of the curve which lie in the admissible region Ω_j lead to cones with vertices lying outside the strip.

Figures 2a – 2d show outlines of all basic situations which can occur for admissible regions with respect to the planes enclosing the curve. As the admissible region in the last figure contains no point of the curve (except $\mathbf{P}_j$), there is no cone whose vertex lies outside the strip. In the remaining three situations all points of the curve which lie within the admissible regions Ω_j define useful cones interpolating the generator $\mathbf{g}_j$. But, the associated cones at $\mathbf{g}_i$ do not necessarily have to be permissible as well. After all, we can establish the following lemma:

Lemma 1. *The above problem has acceptable solutions if and only if the area of intersection of the admissible regions Ω_1 and Ω_2 (belonging to the two pencils of planes Σ_1 resp. Σ_2) contains at least one point of the curve.*

If more than one point of the curve lies within the intersection of the two admissible regions Ω_1 and Ω_2, more than one solution exists, wherefore additional criteria can be introduced to choose the best (referring to the desired application) of the possible regular strip–surfaces.

For curves given as rational Bézier curves or as rational B–spline curves, this strategy has the main advantage of obtaining the degree. If the curve has degree k, then the constructed strip–surface will be of degrees $(1, k)$. This is based on the well-known property of rational B–spline curves to be *projectively invariant* [6]. To use this property, the curve has to be subdivided into those segments being interpolated by a single cone. Each segment has to be projected into the planes $\mathbf{B}_1$ and $\mathbf{B}_2$. The center of the used central projection thus is the vertex of the cone belonging to the curve segment. Finally, the projected curve segments lying in the plane $\mathbf{B}_j$ have to be combined to a new spline curve. The two obtained curves can be used as boundary curves of a (developable) ruled surface represented as $(1, k)$–B–spline surface.

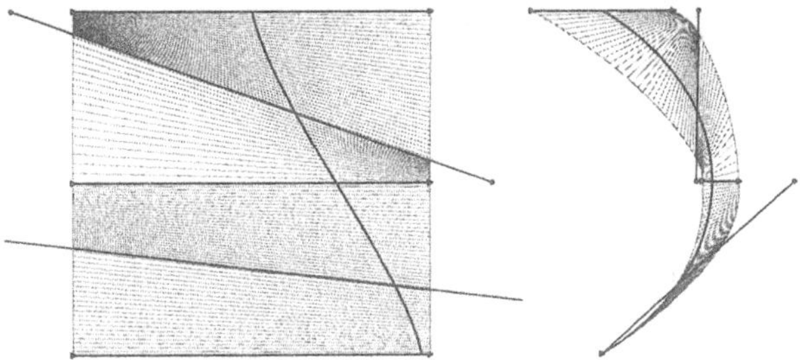

Fig. 3. A cubic Bézier curve, which is similar to a helix, and three arbitrarily chosen lines interpolated by a developable strip–surface consisting of four cones (represented as $(1,3)$–B–spline surface).

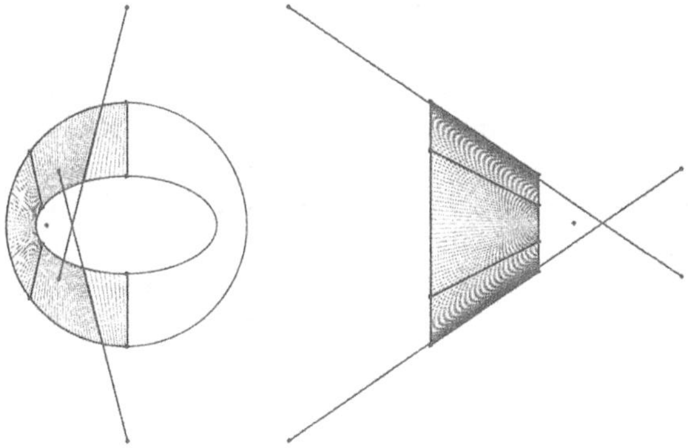

Fig. 4. The curve X (being half a circle) and four given generators interpolated by a spline surface combined of five cones. The cone spline surface approximates the developable surface defined by the circle and the ellipse.

Figures 3 and 4 show two examples basing on (rational) Bézier curves. The given lines $\mathbf{g}_j$ (drawn within the strip only) and the computed vertices of the cones are also pictured, whereas each pair of associated vertices is connected by a line segment.

§3. The Existence of Regular Strip–Surfaces

In this section we will characterize the generators which lead to acceptable strip–surfaces in the sense of Section 2. By doing this we change the given elements of the previous section: instead of fixing both lines $\mathbf{g}_1$ and $\mathbf{g}_2$, we start now only with the generator $\mathbf{g}_1$ at the beginning of the curve. Our aim is to develop a strategy for detecting all lines $\mathbf{g}_2$ which guarantee the existence of strip–surfaces free from singularities. According to Lemma 1, such solutions exist if and only if the two admissible regions Ω_1 and Ω_2 contain at least one common point of the curve. Therefore, the idea of the following strategy is to choose admissible regions fulfilling this condition, and to construct suitable generators $\mathbf{g}_2$ for the chosen regions.

First, we select out of the pencil of planes Σ_1 an arbitrary plane $\mathbf{E}_1(\lambda_{1,j})$ intersecting the curve and not containing the point $\mathbf{P}_2$. This plane then shall bound the admissible region Ω_1, i.e. it is desired to enclose the point of intersection $\mathbf{S}_{2,j}$ of the requested generator $\mathbf{g}_2$ and the plane $\mathbf{B}_j$. Another plane containing the point $\mathbf{S}_{2,j}$ is the plane

$$\mathbf{E}_2(\lambda_{2,i}) = \mathbf{g}_2 \cup \mathbf{S}_{1,i} \quad (i \in \{1,2\} - \{j\})$$

which is part of the pencil Σ^* given by the line connecting $\mathbf{P}_2$ and $\mathbf{S}_{1,i}$. By choosing $\mathbf{E}_2(\lambda_{2,i})$ out of Σ^* in such a way that it contains a point of the curve lying in Ω_1, the existence of (at least) one regular solution is assured (accordingly to Lemma 1). Since by this choice of $\mathbf{E}_1(\lambda_{1,j})$ and $\mathbf{E}_2(\lambda_{2,i})$ also the points

$$\mathbf{S}_{2,j} = \mathbf{E}_1(\lambda_{1,j}) \cap \mathbf{E}_2(\lambda_{2,i}) \cap \mathbf{B}_j$$

and $\mathbf{P}_2$ are fixed, the generator $\mathbf{g}_2$ is already determined.

In order to get all allowable generators $\mathbf{g}_2$ (for fixed $\mathbf{E}_1(\lambda_{1,j})$), we have to generalize this strategy in the following way: instead of choosing the plane $\mathbf{E}_2(\lambda_{2,i})$ from Σ^* (almost) arbitrarily, we determine its extremal positions, i.e. we determine those two planes of the pencil Σ^* which enclose the smallest part of Σ^* containing all points of the curve which lie within Ω_1. One of these planes contains the point $\mathbf{P}_1$ as well. Its intersection point $\mathbf{S}_{2,j}^{(1)}$ with the planes $\mathbf{E}_1(\lambda_{1,j})$ and $\mathbf{B}_j$ is always determined by

$$\mathbf{S}_{2,j}^{(1)} \equiv \mathbf{S}_{1,j}.$$

The other plane can be calculated iteratively. By computing the point of intersection $\mathbf{S}_{2,j}^{(2)}$ which belongs to the second plane, the line segment

$$\mathbf{s}_j = \overline{\mathbf{S}_{1,j}\mathbf{S}_{2,j}^{(2)}}$$

determines the region at which the requested generator $\mathbf{g}_2$ is allowed to cross the plane $\mathbf{B}_j$.

If the admissible regions of Σ_1 and Σ^* (adopting the role of Σ_2) are both of the principal form shown in Fig. 2c this line segment could be extended, eventually. First, we can easily check whether this situation is given or not. By choosing the plane $\mathbf{E}_1(\lambda_{1,j})$, the plane

$$\mathbf{E}_2(\lambda_{2,j}) = \mathbf{g}_2 \cup \mathbf{S}_{1,j}$$

is also fixed, as it contains the same line of intersection with $\mathbf{B}_j$ as $\mathbf{E}_1(\lambda_{1,j})$ and, additionally, contains the point $\mathbf{P}_2$. If the plane $\mathbf{E}_2(\lambda_{2,j})$ does not cross resp. touch the curve, we get no auxiliary possibilities for the choice of generator $\mathbf{g}_2$. Otherwise, we temporary exchange the roles of the curve's ends, transfer the part of $\mathbf{E}_1(\lambda_{1,j})$ to $\mathbf{E}_2(\lambda_{2,j})$ and calculate the two extremal positions for the plane $\mathbf{E}_1(\lambda_{1,i})$ (analogously to the determination of $\mathbf{E}_2(\lambda_{2,i})$).

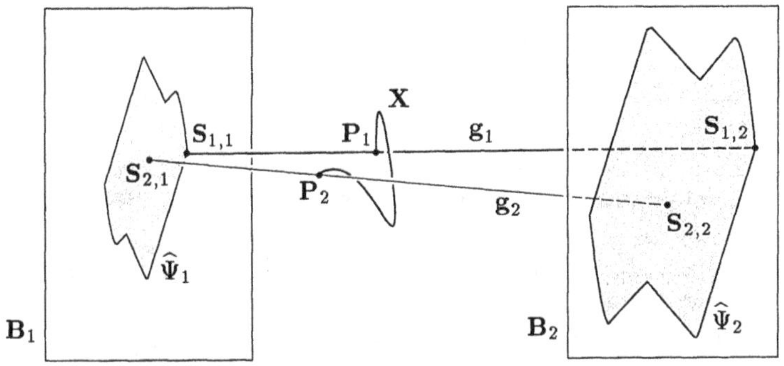

Fig. 5. The maximum regions $\widehat{\Psi}_j$, the generator g_2 is allowed to cross the planes B_j, are determined by the space curve X and the line g_1.

From the two detected planes $\mathbf{E}_1(\lambda_{1,i})$, only the one which does not contain $\mathbf{P}_2$ can lead to new solutions. Since it contains the point $\mathbf{S}_{1,i}$ as well as the point $\mathbf{S}_{2,i}$, it fixes another plane

$$\widehat{\mathbf{E}}_2(\lambda_{2,i}) = \mathbf{P}_2 \cup \mathbf{S}_{2,i} \cup \mathbf{S}_{1,i}$$

which crosses the planes $\mathbf{E}_1(\lambda_{1,j})$ and $\mathbf{B}_j$ at the point $\widehat{\mathbf{S}}_{2,j}$. This intersection point either lies at the line segment $\mathbf{s}_j$, in this case no more choices for $\mathbf{g}_2$ do exist, or it does not, whereby the admissible possibilities for the choice of $\mathbf{g}_2$ will be extended to the line segment

$$\mathbf{s}_j = \overline{\mathbf{S}_{1,j}\widehat{\mathbf{S}}_{2,j}}.$$

By rotating the plane $\mathbf{E}_1(\lambda_{1,j})$ around the generator $\mathbf{g}_1$ and connecting all line segments $\mathbf{s}_j$, we obtain the region Ψ_j of plane $\mathbf{B}_j$, in which the point $\mathbf{S}_{2,j}$ can be chosen, whereby the generator $\mathbf{g}_2$ will be determined.

Now, only the so far excluded case in which the plane $\mathbf{E}_1(\lambda_{1,j})$ contains the point $\mathbf{P}_2$ remains. Since in this case the points $\mathbf{P}_2$ as well as $\mathbf{S}_{2,j}$ are enclosed in plane $\mathbf{E}_1(\lambda_{1,j})$, even the generator $\mathbf{g}_2$ completely lies in this plane and hence crosses the generator $\mathbf{g}_1$ (maybe at infinity). In this case we get all solutions consisting of one cone (resp. one cylinder) only. The admissible intersection points of generator $\mathbf{g}_2$ and plane $\mathbf{B}_j$ lie at the line segment between $\mathbf{S}_{1,j}$ and the point which arises by crossing the plane $\mathbf{B}_j$ with the line given by the points $\mathbf{S}_{1,i}$ and $\mathbf{P}_2$.

Employing this strategy in the end for both $j = 1$ and $j = 2$, projecting the permissible region Ψ_2 into the plane $\mathbf{B}_1$ (via central projection with center $\mathbf{P}_2$), and combining this projected region with the region Ψ_1 leads to the maximum admissible region $\widehat{\Psi}_1$, at which the generator $\mathbf{g}_2$ is allowed to cross the plane $\mathbf{B}_1$. By choosing the point $\mathbf{S}_{2,1}$ out of $\widehat{\Psi}_1$ the existence of at least one regular developable surface consisting of two cones is guaranteed.

Fig. 5 shows an example containing a space curve $\mathbf{X}$, the given generator $\mathbf{g}_1$, and the calculated regions of intersection $\widehat{\Psi}_j$ between all allowable second generators $\mathbf{g}_2$ and the planes $\mathbf{B}_j$, which bound the strip.

References

1. Aumann, G., Interpolation with developable Bézier patches, Comput. Aided Geom. Design **8** (1991), 409–420.
2. Bodduluri, R. M. C., and B. Ravani, Geometric design and fabrication of developable surfaces, ASME Adv. Design Autom. **2** (1992), 243–250.
3. Bodduluri, R. M. C., and B. Ravani, Design of developable surfaces using duality between plane and point geometries, Computer-Aided Design **10** (1993), 621–632.
4. Frey, W. H., and D. Bindschadler, Computer aided design of a class of developable Bézier surfaces, General Motors R&D Publication 8057, 1993.
5. Hoschek, J., Dual Bézier curves and surfaces, in *Surfaces in Computer Aided Geometric Design*, R. E. Barnhill and W. Boehm (eds.), North Holland (Amsterdam), 1983, 147–156.
6. Hoschek, J., and D. Lasser, *Fundamentals of Computer Aided Geometric Design*, A. K. Peters, Wellesley MA, 1993.
7. Hoschek, J., and H. Pottmann, Interpolation and approximation with developable B–spline surfaces. in *Mathematical Methods for Curves and Surfaces*, M. Dæhlen, T. Lyche, and L. L. Schumaker (eds.), Vanderbilt University Press, Nashville, 1995, 255–264.
8. Hoschek, J., and M. Schneider, Interpolation and approximation with developable surfaces, in *Curves and Surfaces with Applications in CAGD*, A. Le Méhauté, C. Rabut, and L. L. Schumaker (eds.), Vanderbilt Univ. Press, Nashville, 1997, 185–202.
9. Kim, T., Design of rational curves and developable surfaces, Ph D Thesis, Arizona State University, Tempe, 1996.
10. Lang, J., and O. Röschel, Developable $(1, n)$–Bézier surfaces, Comput. Aided Geom. Design **9** (1992), 291–298.
11. Lange, K., *Umformtechnik Band 3*, Springer, 1990, 413–425.
12. Leopoldseder, S., and H. Pottmann, Cone spline surfaces, Technical Report No. 44, TU Wien, 1997.
13. Pottmann, H., and G. Farin, Developable rational Bézier and B–spline surfaces, Comput. Aided Geom. Design **12** (1995), 513–531.
14. Vogt, C. D., Analyse und Werkzeuge zur Unterstützung des Auslegungsprozesses von Ziehstufen großer freiformflächiger Blechbauteile, Ph D Thesis, ETH Zürich, 1994.

Mike Schneider
Dept. of Mathematics
Darmstadt University of Technology
Schloßgartenstraße 7
D–64289 Darmstadt, GERMANY
`mschneider@mathematik.tu-darmstadt.de`

Global Illumination Computations with Hierarchical Finite Element Methods

Hans-Peter Seidel, Philipp Slusallek, and Marc Stamminger

Abstract. Since the beginning of computer graphics, one of the primary goals has been to create convincingly realistic images of three-dimensional environments that are impossible to distinguish from photographs of the real scene. The goal to create photo-realistic images has lead to the development of completely new algorithms and software techniques for dealing with the inherent geometric and optical complexity of real world scenes. This paper gives an overview of advanced algorithms for photo-realistic rendering and in particular discusses hierarchical techniques for global illumination computations.

§1. Introduction

Photo-realistic rendering has a strong influence on many areas of computer graphics and this influence seems to become even larger with 3D techniques entering the consumer market. Although the advances in photo-realistic rendering are no longer as spectacular and visible to the public as with early ray-tracing or radiosity images, it is a very active field of computer graphics research with substantial new developments. This article tries to highlight some of the new developments and points out areas of future research. Some of the important application areas of photo-realistic rendering include

- **Lighting and Luminaire Design:** The design of new luminaires is a complex and costly process that is often tightly coupled with the lighting design of buildings. In particular for the latter case, it is usually not feasible to test design decisions at a model in order to avoid costly mistakes. The simulation of the distribution of light in luminaires, rooms, and buildings allows to increase the quality as well as reduce the cost and time to market.
- **Virtual Reality:** The main goal of virtual reality is the immersion of a user in a virtual environment. This immersion depends to a large extent on the generation of adequate optical stimuli. The real-time requirements

Mathematical Methods for Curves and Surfaces II

Morten Dæhlen, Tom Lyche, and Larry L. Schumaker (eds.), pp. 445–460.

ISBN 0-8265-1315-8.

of virtual reality still impose severe limitations on the realism that can be achieved, but precomputed illumination and visual detail are an important part of convincing environments.

- **Computer Augmented Reality:** This extended form of virtual reality combines virtual environments with real world (video) data. This poses the additional problem of simulating optical and illumination effects from real objects on objects in the virtual environment and vice-versa.
- **3D Prototypes:** Virtual mockups are used more and more to reduce the cost of manufacturing. While standard hardware-generated images are sufficient for a simple geometric analysis, increased realism is required to analyze features such as the overall design, the finish of surfaces, and the visual impressions of the final product for marketing.
- **Games:** Today, convincing realism is an important factor for the success of a computer game title. More and more games use techniques from realistic image synthesis such as precomputed global illumination, shadow maps, as well as advanced mirror and refraction effects to increase the illusion of reality.
- **3D on the Web:** With techniques like VRML that allow for distributed 3D environments on the web, the demand for realistic renderings of environments is steadily increasing. Casual users are much more critical of the quality of computer presentations. Online 3D catalogues and shopping malls require much more realism than what is offered today.
- **Ubiquitous 3D:** 3D technology is just entering the consumer market now and many new applications will certainly be developed in this area. Many of them will benefit greatly from realistic image generation techniques.

Software for photo-realistic rendering can be thought of as consisting of three main components: scene description, lighting computation, and image generation. The first component deals with the problem of describing 3D environments with the necessary level of detail and physical accuracy that allows for generating realistic images. This component is also concerned with the interface that a rendering system offers for importing scene description from external modeling programs.

A key component of convincingly realistic presentations is the simulation of the rich details that are due to lighting effects, such as soft shadows and correct interreflections between objects. The proper simulation of these lighting effects can avoid the "computer" or "artificial" touch of computer generated images. Once the distribution of light in a scene has been computed, we can take a snapshot of the light field with a virtual camera and generate an image from it. This last component of photo-realistic rendering deals with problems like hidden surface removal, shading, motion-blur, and anti-aliasing.

Each of these components has its own challenges, some of which will be covered here. Special emphasis will be put on physically correct lighting computations. Many of the subtle lighting effects can and have been simulated manually by graphic designers and animation experts. Particularly in the film and animation industry, such experts have created remarkably realistic images and film sequences often without even knowing the physics behind it. However, as 3D graphics spreads into everyday life, those experts are no longer available. This increases the demand for automatic and correct simulation of optical and lighting effects, which stimulated much of the research presented here.

The key issue in this trend is the ability to correctly simulate the physics of light, the interaction of light with matter, and the transport of light in an environment. A full and accurate simulation is only required by a few application areas, such as lighting and interior design, as well as illumination engineering. But the same techniques can also be used to compute a coarse approximation to the correct illumination. Such an automatically computed approximation will be sufficiently realistic for many applications. Adaptively controlling the accuracy of computations allows for trading quality for speed in applications. This adaptive control also requires that lighting systems can switch between rendering algorithms that offer different quality/speed trade-offs.

In recent years, photo-realistic rendering has seen a dramatic development of faster, more accurate, and robust solution techniques. These techniques include Finite Element and Monte-Carlo solvers for integral equations, and the adoption of wavelets, multi-grid, and multi-resolution techniques. New algorithms exploit a number of advanced geometric algorithms and data structures for efficiently solving the problem of transfer and representation of light in complex environments.

The speed and the robustness of the available algorithms are now good enough for use in industrial-strength applications. This is supported by the fact that most companies offering animation systems are currently integrating advanced lighting simulation into their products (e.g. Alias|Wavefront, Soft-Image). However, there is still a large gap between recent research results and the techniques available in todays commercial systems. One reason for this gap is that system integration and other practical aspects of new algorithms often receive little attention. These aspects will very likely become much more important in the future as new techniques will be used in a broader scope.

The methods and techniques that have been developed for realistic lighting computations are not restricted to generating realistic images only. These methods can easily be modified to work with other forms of radiation (like heat, radar, radio, etc.). Furthermore, many of the advanced techniques used in realistic lighting computations like efficient hierarchical methods, clustering, and variance reduction techniques can equally be applied to other engineering applications where they are less widely known.

Lets start with a brief review of the underlying physics and mathematics, before we discuss some of the new techniques for computing realistic images.

§2. Fundamentals

The fundamental quantity in physically-based rendering is radiance. The quantity radiance $L(y, \omega, \nu)$ is the power radiated from a point y in a certain direction ω per unit projected area perpendicular to that direction and per unit solid angle. The units of radiance are $[W\ m^{-2}\ sr^{-1}]$. The dependency on the frequency of the radiation is usually ignored in lighting computations, and the equations are solved for each frequency band separately.

The distribution of light in a scene is completely described by the radiance distribution in space. All other quantities can be derived from it. Radiance is invariant along its direction of propagation, provided there are no absorption, emission, or scattering effects by participating media (like smoke or fog). Additionally, the amount of light received at the retina of the human eye is proportional to the radiance emitted by the viewed surface. These properties make radiance the appropriate quantity for illumination computations in a rendering system.

Light is emitted at light sources and interacts with other surfaces in the environment, being absorbed, reflected, or transmitted. The propagation of light in the absence of participating media is fully described by the radiance equation [5]

$$L(y,\omega) = L_e(y,\omega) + \int_\Omega f_{rt}(V\omega_i, y, \omega) L_i(y, V\omega_i) |\cos\theta_y\ d\omega_i|. \qquad (1)$$

It states that the radiance reflected at a point y in direction ω is the sum of the self emitted radiance L_e and the reflected radiance. The latter is computed by integrating over the fraction of incident light from all directions that is scattered into the outgoing direction ω. The scattering characteristics of a surface is described by the bidirectional reflection distribution function f_{rt} (BRDF). The incident light L_i itself contains contributions from light reflected off other surfaces, thus coupling it to outgoing radiance L at all other visible points in the environment. It is this global nature of the radiance equation and the embedded problem of visibility together with complex scene arrangements that makes it so hard to compute true photo-realistic images.

We are often not interested in the full *directional* distribution of light at a point. In this case, we can eliminate the directional parameter and switch to radiosity $B(y) = \int_{\Omega^+} L(y,\omega)\cos\theta_y d\omega$ for describing the light distribution. This results in the radiosity equation

$$B(y) = B_e(y) + \frac{\rho(y)}{\pi} \int_{x\in S} B(x) G(x,y) V(x,y) dx, \qquad (2)$$

where we changed the integration domain to all surfaces in the environment. This explicitly adds the visibility function V and the geometric form factor $G(x,y) = \cos\theta_x \cos\theta_y / |x-y|^2$ to the integrand [5].

2.1. Overview of Lighting Techniques

Computing the illumination in some environment requires the solution of the radiance or radiosity equations (1) and (2). As Fredholm equations of the second kind [1], they cannot be solved analytically, except for trivial cases.

Instead, numerical algorithms for computing a solution must be used, all of which are derived from two basic numerical approaches: *finite element* and *Monte-Carlo* (or point sampling) techniques. In the following we concentrate on the finite elelement approach and discuss it in more detail.

In the finite element approach, the illumination is described as a function representing outgoing radiance or radiosity over the surfaces in the scene. This technique takes advantage of the fact that reflected radiance and radiosity are usually piecewise smooth. Finite element techniques are based on a subdivision of the environment into small surface patches and compute a functional representation of the illumination on each of them.

With this approach finite element algorithms try to exploit coherence in the light distribution in the environment. Finite element techniques are in a sense at the opposite side of the solution spectrum compared to Monte-Carlo algorithms: computation at points versus surface patches, deterministic versus stochastic computations, and explicit versus implicit use of coherence. Unfortunately, approximating the light distribution by smooth functions results in biased solutions — a problem that true Monte-Carlo methods can avoid completely. However, this bias is generally not a problem for most applications.

After representing the distribution of light in the finite-dimensional function space, we can use the standard Galerkin technique to reduce the radiance or radiosity equation to a linear system describing the exchange of light between surface patches. Although many standard techniques are available for solving linear systems, only very few of them can actually be used. Compared to other disciplines, the linear system used for lighting computations have some unique features: The computation of the coefficients is very costly and requires knowledge about the scene as a whole. Depending on the complexity of the scene, the full matrix would become extremely large. It is thus mandatory to avoid computing or even storing coefficients if at all possible. This requirement has lead to a number of unique approaches for representing and solving the linear system that will be described below.

§3. Hierarchical Techniques

This section concentrates on the use of hierarchical finite element techniques. The use of hierarchies has made finite element techniques practical and applicable to a large variety of virtual environments. It allows for scaling the algorithms with scene complexity, available computation time, and required accuracy. The main issues are the time and memory complexity of the algorithm.

In contrast to pure Monte-Carlo techniques, which require little to no memory for storing intermediate results, the finite element algorithms must store the representation of illumination that is computed for all surfaces. For complex scenes this can become a considerable problem. Furthermore, the matrix of the linear system must be represented in some form during the solution process. Given a number of patches n in the environment an early finite element algorithm computed the light transport between any pair of them, thus resulting in quadratic complexity $O(n^2)$ in the number of patches.

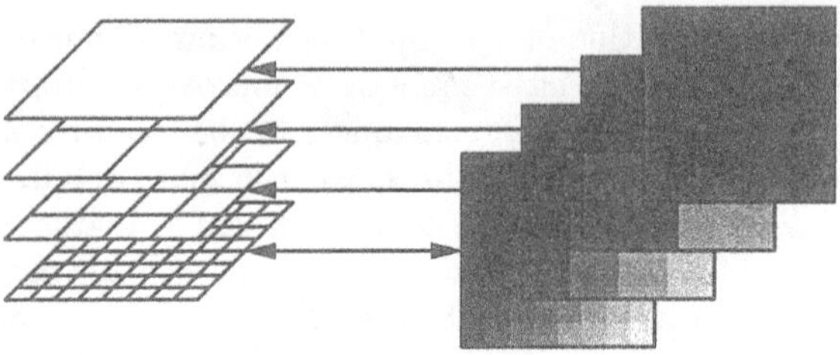

Fig. 1. Representation of illumination at different levels of a hierarchy. Patches in higher levels represent the average illumination of their children.

The introduction of hierarchical algorithms changed the situation for finite element algorithms dramatically. Hierarchical techniques allow for automatically selecting the level of detail at which light should be transported between patches. Together with clustering techniques, described in the next section, this allows for algorithms that have only *linear* complexity $O(n)$ in the number of input surfaces. Lets take a closer look on how this can be achieved.

3.1. The Basic Hierarchical Algorithm

Finite element algorithms require the discretization of the environment into patches. For a given functional representation of illumination on a patch, the accuracy of the computation is determined by the discretization of the environment. Many small patches allow for a more detailed and accurate computation. What is required is an adaptive discretisation into smaller patches where it significantly improves accuracy, for example along shadow boundaries or in other areas where illumination changes rapidly.

The hierarchical approach allows for exactly this kind of adaptive computation. Instead of discretizing the environment into a fine mesh of patches before starting the computation, the algorithms refines the initial coarse set of surfaces during the course of the computation wherever this becomes necessary. For recursive refinement, patches are usually represented in the form of quadtrees on triangles, quadrilaterals, or other parametric surfaces (see Figure 1).

When considering the transport of light between surfaces, the algorithm chooses the appropriate level of discretization that is required for meeting a user selected error criterion. For two nearby patches that exchange a large amount of radiation, which may result in large illumination gradients on the receiver, a detailed subdivision of either the sender or the receiver or both is required. On the other hand, if the two surfaces are far apart or exchange only very little energy, it might be sufficient to compute the exchange between large patches in upper levels of the quadtrees [11] (see Figure 2).

During the computation the hierarchical algorithms automatically selects the coarsest level at which the accuracy requirements can be met. The same surface may receive and send illumination at various levels of the discretization hierarchy. It is therefore necessary to keep the different levels consistent.

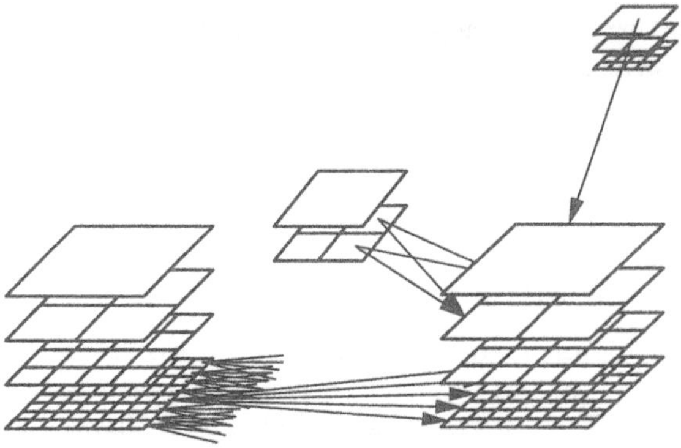

Fig. 2. Light is transported via links between various levels in two hierarchies depending on the configuration of the sending and receiving patch. A refinement procedure selects the coarsest level at which the transport can be computed within a user selected error threshold.

If energy is received at some level, this energy must also be distributed or *pushed* down to the children of the receiving patch. On the other hand, the energy must also be propagated or *pulled* to upper levels in the hierarchy. Starting at the leaf nodes in the hierarchy, parent patches then compute their illumination by summing up the energy of their children. The whole process can be described as a single *push/pull* procedure, that keeps the different representations of illumination consistent [11].

3.2. Representing and Solving the Linear System

A matrix is certainly not an appropriate way to represent the exchange coefficients of the hierarchical system. The complex interactions between different levels in the hierarchy can best be represented by storing the exchange coefficients explicitly as *links* between the two patches. The flexible structure of this representation allows for simple additions and deletion of the links. These dynamic changes would be difficult to implement in a matrix structure.

The solution of the linear system now proceeds by using an iterative solver for linear systems. The most commonly used iterative solver is the Gauss-Seidel algorithm, which considers each surface in turn and *gathers* light via all links that connect to the hierarchy on the receiver. Similarly, the Southwell algorithm can be used, which *shoots* energy via these links to other surfaces. However, due to the fact that the adaptively created links all transport similar amounts of energy, the original advantage of this solver is lost [4,9,11].

Before energy is actually transported via a link, the iterative solver must check the accuracy of the link. Due to the changes in illumination, links may become inaccurate and may need to be refined during the course of the iterations. The refinement of links is guided by a refiner which is discussed in more detail in the following subsection and in Section 5.

Many more iterative solvers are known for linear systems. Most of these algorithms are difficult to adapt to the link representation of the exchange coefficients. Furthermore, *the adaptive creation of links together with the computation of the correct exchange coefficients by far dominates the solution time* compared to the actual transport of energy via the links. Thus, somewhat faster iterative solvers would not have much effect and would complicate the hierarchical algorithm compared to the simple Gauss-Seidel and Southwell solvers.

3.3. Refinement Criteria

The crucial part of any hierarchical algorithm is the refiner (often also called oracle). It determines if an exchange coefficient between two patches can adequately be approximated within a user selected error threshold. Otherwise the patches need to be subdivided. The particular design of the refiner determines to a very large extent the accuracy, memory requirements, and running time of a hierarchical algorithm. As a consequence, it is not meaningful to compare hierarchical algorithms without specifying the particular design of the refiner and the chosen parameters.

For hierarchical radiosity computations several different refiners have been proposed. The most important refiners use the exchange coefficient itself (F-refinement) [11], resulting radiance on the sender (BF-refinement), or reflected flux (BFA-refinement) for estimating the error that is committed by transporting light across a link. The BFA-refiner seems to work best for all kinds of environments and is used in most implementations.

These refiners are based on the observation that the error of a link can be bounded above by a constant factor times the exchange coefficient of the link. Weighting this error with the transported energy and the area of the receiver results in the last two refiners. The BFA-refiner is well behaved and has the added benefit of automatically maintaining a minimum size for patches, which usually is an extra parameter for the other two refiners. Several modified versions of these refiners have also been proposed (e.g. [12]).

3.4. Wavelets and Higher Order Function Spaces

Most of the hierarchical algorithms use simple constant basis functions for representing illumination on a patch, requiring only a single coefficient for this representation. The Galerkin construction of finite element algorithms also works for higher order basis functions [5]. The formal mathematical basis for a hierarchical construction is the wavelet theory that describes hierarchical function spaces [7]. Hierarchical wavelet algorithms have been used in various ways in rendering, e.g. for radiosity computations [10,16], or for full radiance computations [3,15,14].

Higher order functional representations allow for using less but larger patches, because these functions have a higher approximation power. This can result in reduced subdivision and consequently in less patches and links. However, higher order functions are also much more costly in terms of computing and storing exchange coefficients. For instance bi-quadratic basis functions require storing 81 exchange coefficients per link (between each pair of the nine

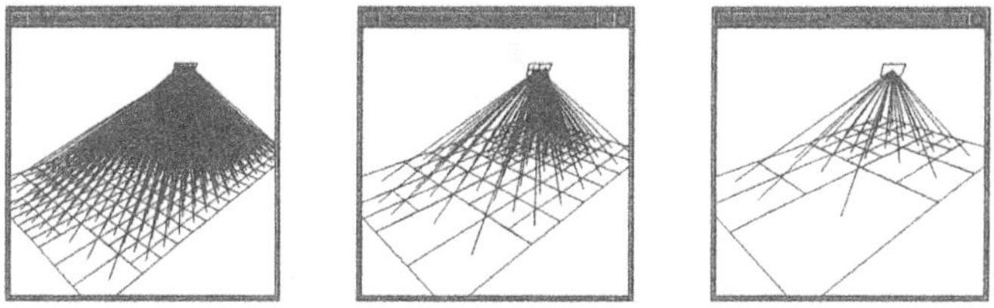

Fig. 3. Links between two patches generated by using Multi-Wavelets of various orders. Constant: 1 coefficient per link (left), Bi-linear: 16 coefficients (center), bi-quadratic: 81 coefficients (right).

coefficients on each patch), instead of storing only a single coefficient per link as with constant basis functions (see Figure 3).

This drastic increase in storage requirements is the main disadvantage of higher order wavelet algorithms, which becomes most apparent at shadow boundaries. Even the larger approximation power of bi-quadratic or bi-cubic functions cannot adequately represent these discontinuity features and still requires subdivision resulting in additional links that are much more expensive than in the constant case. In these situations the larger approximation power cannot compensate the increased memory requirements for storing links. One possibility to avoid subdivision in these situations is by approximately representing visibility separately using a shadow mask [23,32].

§4. Clustering

The main problem with hierarchical algorithms as described in the previous section is that the algorithms still have quadratic complexity in the number of input surfaces. The algorithms must still consider each pair of input surfaces even if the energy transfer between them is negligible. For scenes with many small surfaces instead of a few large surfaces, this *initial linking* can easily dominate the total computation time. The problem is that the benefits of linear complexity of the hierarchical algorithms are only exploited during adaptive refinement. A simple solution would be to avoid initial linking wherever possible [12].

Another approach extends the hierarchical approach. In order to use the benefits of adaptive computation of hierarchical algorithms, we must extend the hierarchy that exists below the level of a surface to a full hierarchy covering the whole scene. This can be done by grouping related surfaces into *clusters* (see Figure 4). These clusters can in turn be grouped hierarchically into larger clusters until the whole scene is contained in a single root cluster [18,24].

Although "hand made" cluster hierarchies often perform slightly better [28], automatic clustering algorithms are preferred, because they are applicable to general and complex scenes. Here, hierarchical bounding volume algorithms seem to give best results [19,28] compared to octrees or k-d trees. The latter two often construct clusters that consist of unrelated patches and can result in large lighting error.

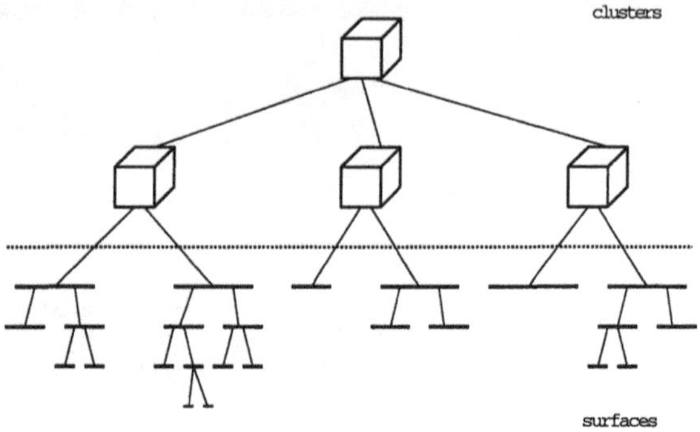

Fig. 4. Extending the hierarchical structure above the level of input patches by clustering. This creates a single hierarchy for the complete scene.

Fig. 5. The model of a station consisting of approximately 32000 patches and 24 light sources. The lighting simulation based on clusters was computed in less than 10 minutes on a MIPS R8K CPU. Only 0.02 percent of the potential links where actually created. The image on the right shows the clustering hierarchy that was created.

The important feature of a cluster is that it can receive and emit energy for all of its contents. This allows to exchange light between a large group of surfaces in a single operation. An important difference to the use of planar surfaces is that a cluster may exchange energy with itself. Consequently, the processing starts by refining the first link in the scene, which is a self link from the root cluster to itself. Since the exchange of light at this level would generally introduce too much error, the refiner would replace this link, open up the cluster, and instead link its child clusters or surfaces before recursing.

This clustering strategy allows for quickly computing coarse approximations in a short amount of time. By choosing a large error threshold only a few large cluster would ever be considered. Although this would result in a coarse approximation, it may already allow a first estimate of the lighting effect in a scene (see Figure 5). By adjusting the threshold the user can continuously change the level of accuracy and speed of the computation. This is a valuable feature in particular for interactive applications.

Even if all surfaces in a cluster are purely diffuse, the cluster will receive and reflect light with a directionally dependent characteristic. However, in the simplest case, the light exchange of a cluster and its environment is modeled as that of an isotropic volume that emits and scatters light in all directions evenly. Although this is an extremely crude approximation, it already works quite well in many cases. In order to avoid some of the resulting artifacts, light energy received by a cluster can immediately be propagated to all surfaces in its hierarchy. This allows to use the direction of the incident light while it is still known and computing its contribution to each surface separately [24]. Note, that this is different and still much more efficient than computing the form factor to each surface separately.

The cluster algorithms can also be extended for using non-isotropic interactions with its environment by storing the directional distribution of emitted energy with each cluster [20].

4.1. Visibility

Clusters cannot only be used to receive and emit energy, they are also a valuable help for visibility computations. The tree of clusters can be used for computing visibility hierarchically. If a visibility ray traverses a cluster, a special visibility refiner decides whether the cluster can be approximated by a semi-transparent volume or whether it must be opened and the ray intersected with its children. The properties of a cluster that are used for this decision can directly be computed from its contents [19,18].

Clusters are a natural extension of the previous hierarchical algorithms in rendering. By extending the hierarchy to the whole scene, the benefits of the hierarchical approach can optimally be exploited. Unfortunately, correct handling of clusters is more difficult due to their non-isotropic reflection characteristics. This area offers many opportunities for further research.

§5. Refiners Based on Bounded Transport Computations

As already noted, the refiner is the crucial part of any hierarchical algorithm. Early refiners for lighting simulations used simple criteria and point sampling to estimate light transport. By accurately bounding the input parameters of the computation, it is possible to obtain tight but conservative bounds on the amount of light transport over a receiving patch. These bounds offer a new opportunity for guiding the refinement procedure for normal as well as for hierarchical algorithms [27,26].

One problem with the refiners usually is that they only estimate an upper bound of some transport quantity (i.e. the form factor or radiosity). The lower bound is always assumed to be zero. This approach ignores much of the information that is available or can easily be computed for a link [24]. Such information includes geometric quantities like the extent of sender and receiver, their distance, and the variation of normal vectors across each surface. The variation of normal vectors is of particular importance for the ability to treat curved surfaces efficiently.

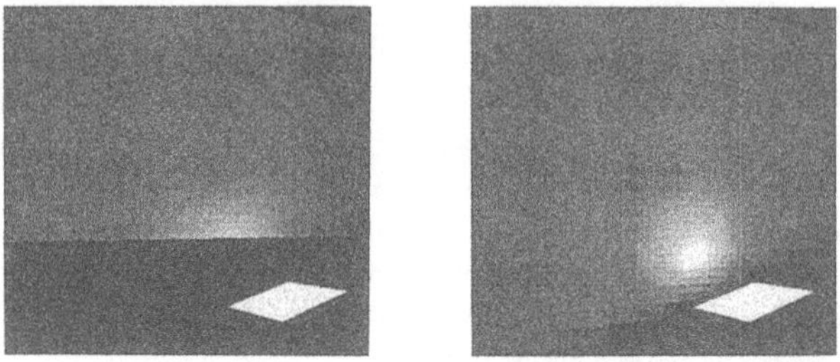

Fig. 6. Sampling problem for a small light source located directly in front of a large patch causing a small but bright spot: Any point sampling strategy is likely to miss the spot and omit important light contributions during subdivision (left). Correctly bounding the transport quantity allows for locating hot spots on a patch (right).

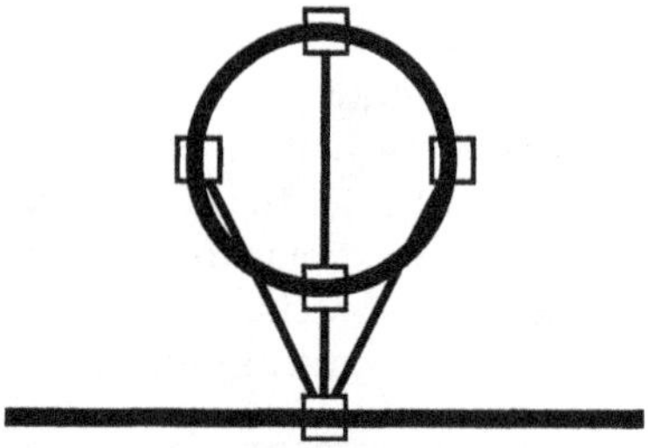

Fig. 7. Sampling problems for light transport between a planar patch and a curved surfaces (2D slice): The estimation of light transport is very unreliable, depending on the actual sampling location on the curved surface and the planar patch.

A second problem of these refiners is the way the transport quantities are computed. These refiners generally use some form of point sampling for estimating them. They often use only a single sample for computing the exchange coefficient at the "center" of a patch, e.g. using the disc to point form factor approximation. The transport quantity may then be further modulated by estimating visibility of the link with sample rays that are sent between the two patches in question.

The sampling problem already becomes apparent for a small light source located directly in front of a large patch (see Figure 6). Although there might be a considerable transport of light to and from a small spot on the large patch, any point sampling strategy across the large patch is likely to miss this hot-spot. Another example for the sampling problem becomes apparent if we consider curved surfaces (see Figure 7). In this case, the geometric parameters (locations, distances, and normals), on which the computation of the exchange coefficients is based, vary strongly over the surface, again making any point sampling strategy highly unreliable.

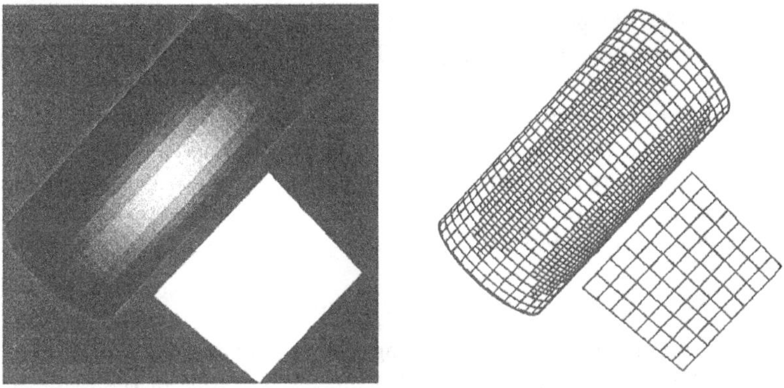

Fig. 8. A square lighting a cylinder. Note that the subdivision on the cylinder at the center coarser the in the neighborhood. This is due to the slower varying cosine term in the center.

A solution is to compute upper and lower bounds for any of the parameters that are used as input to the computation of the exchange coefficients. If tight conservative bounds are computed for the input parameters, the resulting exchange coefficients can also be bounded by applying interval or affine arithmetic [13,6,25]. The location of points and their normals can efficiently be bound with the cone of normals technique [17] and bounding boxes. Based on this data, bounds on the set of connecting rays, cosines factors, distances, and solid angles can be computed. Bounds can also be computed for visibility between patches. Finally, these terms are combined for obtaining a conservatively bounded estimate of the exchange coefficient [27].

During the computation, triple values <min, estimate, max> are maintained that denote an upper and lower bound as well as an estimate of the correct value. These values can then be used by the new refinement strategy for deciding where to refine links. Links are refined, when the difference between upper and lower bound becomes to large. The new refiner also has enough information for deciding whether to refine the sender or the receiver [27]. The decision is based on their relative influence on the error. This is an important benefit over previous refiners, where no such information is available.

Figure 8 shows the effect of this new refinement strategy for the light exchange between a planar patch and a cylinder. The cylinder is refined less in the center, where most of energy is transfered, but the illumination is mostly constant. This is where the older refiners would concentrate their effort. Instead the new refiner shifts the focus to areas with higher illumination gradients as indicated by larger differences between upper and lower bounds. Thus, it refines near the silhouette edges of the cylinder where there is still a large amount of received light and where both the cosine factor at the cylinder and the distance varies considerably.

The interface to geometric objects for obtaining the required geometric information can be kept very simple. This allows a wide variety of objects to implement it, including free form surfaces [27]. As shown in [26] the same strategy can be extended to work equally well for environment with clusters. Although it is more difficult to compute bounds for clusters, the algorithms and refiners still benefit considerably for having accurate bounds available. The bounds can also be used to estimate the self illumination of non-convex curved surfaces and clusters [27].

§6. Conclusion

Photo-realistic rendering has come a long way in reaching the goal of generating images that are hard to distinguish from photographs of the same scene. In the process of this research activity many interesting new techniques have been developed that may also be applied in other engineering applications.

In particular the use of modern mathematical approaches like hierarchical representations and algorithms combined with the clever use of efficient data structures and programming techniques has led to new robust, accurate, and efficient FE algorithms that allow to simulate the physical process on which photo-realistic image generation is based.

Together with approaches that increase the practicality of these techniques this brings photo-realistic rendering and global illumination to a new level of applicability and usability. This will hopefully result in new and better tools for the user of systems that can make use of photo-realistic rendering.

References

1. Atkinson, K. E., *A Survey of Numerical Methods for the Solution of Fredholm Integral Equations of the Second Kind*, Society for Industrial and Applied Mathematics, Philadelphia, 1976.
2. Chen, Shenchang Eric, Holly E. Rushmeier, Gavin Miller, and Douglas Turner, A progressive multi-pass method for global illumination, Computer Graphics (SIGGRAPH '91 Proceedings) **25** (1991), 165–174.
3. Christensen, Per H., Eric J. Stollnitz, David Salesin, and Tony D. DeRose, Wavelet radiance, in *Fifth Eurographics Workshop on Rendering*, Darmstadt, June 1994, 287–301.
4. Cohen, Michael, Shenchang E. Chen, John R. Wallace, and Donald P. Greenberg, A progressive refinement approach to fast radiosity image generation, Computer Graphics (SIGGRAPH '88 Proceedings) **22** (1988), 75–84.
5. Cohen, Michael F., and John R Wallace, *Radiosity and Realistic Image Synthesis*, Academic Press, 1993.
6. Comba, João L. D., and Jorge Stolfi, Affine arithmetic and its applications to computer graphics, in *Anais do VII Sibgrapi*, 1993, 9–18. Available from http://www.dcc.unicamp.br/~stolfi/EXPORT/papers/affine-arith.

7. Daubechies, Ingrid, *Ten Lectures on Wavelets*, SIAM Philadelphia, Pennsylvania, 1992.
8. Gibson, S., and R. J. Hubbold, Efficient hierarchical refinement and clustering for radiosity in complex environements, Computer Graphics Forum **15** (1996), 297–310.
9. Gortler, Steven, Michael F. Cohen, and Philipp Slusallek, Radiosity and relaxation methods, IEEE Computer Graphics & Applications **14** (1994), 48–58.
10. Gortler, Steven, Peter Schröder, Michael Cohen, and Pat M. Hanrahan, Wavelet radiosity, Computer Graphics (SIGGRAPH '93 Proceedings) **27** (1993), 221–230.
11. Hanrahan, Pat, David Salzmann, and Larry Aupperle, A rapid hierarchical radiosity algorithm, Computer Graphics (SIGGRAPH '91 Proceedings) **25** (1991), 197–206.
12. Holzschuch, Nicolas, François Sillion, and George Drettakis, An efficient progressive refinement strategy for hierarchical radiosity, in *Photorealistic Rendering Techniques (Proceedings Fifth Eurographics Workshop on Rendering)*, Darmstadt, Germany, June 1994. Springer, 343–357.
13. Moore, Ramon E., *Interval Analysis*, Prentice-Hall, 1966.
14. Pattanaik, Sumant, and Kadi Bouatouch, Haar wavelet: A solution to global illumination with general surface properties, in *Photorealistic Rendering Techniques (Proceedings of Fourth Eurographics Workshop on Rendering)*, Eurographics, Springer, June 1994, 281–294.
15. Schröder, Peter, *Wavelet Algorithms for Illumination Computations*, PhD thesis, Princeton University, November 1994.
16. Schröder, Peter, Steven Gortler, Michael F. Cohen, and Pat Hanrahan, Wavelet projections for radiosity, *Proceedings on the Fourth Eurographics Workshop on Rendering*, June 1993, 95–104.
17. Shirman, Leon A., and Salim S. Abi-Ezzi, The cone of normals technique for fast processing of curved patches, Computer Graphics Forum (EUROGRAPHICS '93 Proceedings) **12** (1993), 261–272.
18. Sillion, François, A unified hierarchical algorithm for global illumination with scattering volumes and object clusters, IEEE Transactions on Visualization and Computer Graphics **1** (1995).
19. Sillion, François, and George Drettakis, Feature-based control of visibility error: A multi-resolution clustering algorithm for global illumination, Computer Graphics (SIGGRAPH '95 Proceedings) **29** (1995), 145–152.
20. Sillion, François, George Drettakis, and Cyril Soler, A clustering algorithm for radiance calculation in general environments, in *Rendering Techniques '95 (Proceedings of Sixth Eurographics Workshop on Rendering)*, P. M. Hanrahan and W. Purgathofer (eds.), Springer, August 1995, 196–205.

21. Sillion, François, and Claude Puech, A general two-pass method integrating specular and diffuse reflection, Computer Graphics (SIGGRAPH '89 Proceedings) **23** (1989), 335–344.

22. Sillion, François, and Claude Puech, *Radiosity & Global Illumination*, Morgan Kaufmann, 1994.

23. Slusallek, Philipp, Michael Schröder, Marc Stamminger, and Hans-Peter Seidel, Smart links and efficient reconstruction for wavelet radiosity, in *Rendering Techniques '95 (Proceedings Sixth Eurographics Workshop on Rendering)*, Dublin, June 1995. Springer, 240–251.

24. Smits, Brian, James Arvo, and Donald Greenberg, A clustering algorithm for radiosity in complex environments, Computer Graphics (SIGGRAPH '94 Proceedings) **28** (1994), 435–442.

25. Snyder, John M., Interval analysis for computer graphics, in Computer Graphics (SIGGRAPH '92 Proceedings) **26** (1992), 121–130.

26. Stamminger, Marc, Philipp Slusallek, and Hans-Peter Seidel, Bounded clustering – finding good bounds on clustered light transport, Technical Report TR-97-1, Universität Erlangen, IMMD 9, 1997.

27. Stamminger, Marc, Philipp Slusallek, and Hans-Peter Seidel, Bounded radiosity – illumination on general surfaces and clusters, Computer Graphics Forum (EUROGRAPHICS '97 Proceedings) **16** (1997).

28. Stamminger, Marc, Philipp Slusallek, and Hans-Peter Seidel, Isotropic clustering for hierarchical radiosity — implementation and experiences, in *Proceedings Fifth International Conference in Central Europe on Computer Graphics and Visualization — WSCG '97*, 1997.

29. Wallace, John R., Michael F. Cohen, and Donald P. Greenberg, A two-pass solution to the rendering equation: A synthesis of ray tracing and radiosity methods, Computer Graphics (SIGGRAPH '87 Proceedings) **21** (1987), 311–320.

30. Ward, Gregory J., The RADIANCE lighting simulation and rendering system, Computer Graphics (SIGGRAPH '94 Proceedings) **28** (1994), 459–472.

31. Willmott, Andrew J., and Paul S. Heckbert, An empirical comparison of progressive and wavelet radiosity, in *Rendering Techniques '97*, J. Dorsey and Ph. Slusallek (eds.), Saint-Etienne, June 1997. Springer, 175–186.

32. Zatz, Harold R., Galerkin radiosity: A higher order solution method for global illumination, Computer Graphics (SIGGRAPH '93 Proceedings) **27** (1993), 213–220.

Philipp Slusallek, Marc Stamminger, and Hans-Peter Seidel
Universität Erlangen
Am Weichselgarten 9
91058 Erlangen, GERMANY
`slusallek,stamminger,seidel@informatik.uni-erlangen.de`

Boundary Curves in Reverse Engineering

Vibeke Skytt

Abstract. Reverse engineering is the process of transforming measured points into a CAD-model. Creating the edges of a CAD-model from the available information is a great challenge. Two approaches are presented: the first is based on point clouds, the second on intersecting extended surfaces. The problem of numerically intersecting tangential surfaces is also analysed.

§1. Introduction

A boundary representation CAD-model consists of curves and surfaces as geometric entities, and topological entities representing the relationship between the curves and the surfaces. Two approaches to generate the curves are addressed in this paper: generate the curves as intersection curves between already defined surfaces; define the curves prior to surface generation and use curve information to generate the surfaces. B-spline curves and surfaces are used in this context.

In Section 2 the concept of reverse engineering is presented. Section 3 covers curve generation from scattered data. Section 4 is devoted to the generation of curves from surfaces, while Section 5 looks at tangential intersections. Conclusions will be presented in Section 6.

§2. Reverse Engineering

Conceptually, reverse engineering is the process of transforming a set of measurement points into a CAD-model. The quality of the final model is highly dependent on measurement noise in addition to density and distribution of the points. Noise puts a restriction on how tightly the surfaces of the final CAD-model can approximate the data points.

Two major aspects in reverse engineering are segmenting the point set into subsets each suitable for being approximated by one surface (see [1,8]),

Mathematical Methods for Curves and Surfaces II
Morten Dæhlen, Tom Lyche, and Larry L. Schumaker (eds.), pp. 461–468.

ISBN 0-8265-1315-8.

Fig. 1. The original point set and the feature points.

and generating the surfaces maintaining some continuity constraints between adjacent surfaces. In this paper we concentrate on C^0 continuity.

Figure 1 (left) shows an example data set. It consists of 10139 unevenly distributed points. The first step is data preprocessing. Outliers are removed, and in this case the boundary is cleaned up to achieve better conditions for the creation of boundary curves. Dense data sets may be filtered.

The next step is to order the points by triangulation. The data set is projectable, and Delaunay triangulation is used. The first version of the triangulation will be convex. It is necessary to remove long boundary edges from the triangulation in order to ensure that the triangulation boundary corresponds to the boundary of the model.

Now, feature points are computed using information from the triangulation. The feature points are: boundary points which are original data points lying at inner or outer boundaries of the data set; points representing sharp edges in the model; points representing discontinuities in curvature; points representing inflection curves in the model. Figure 1 (right) shows feature points.

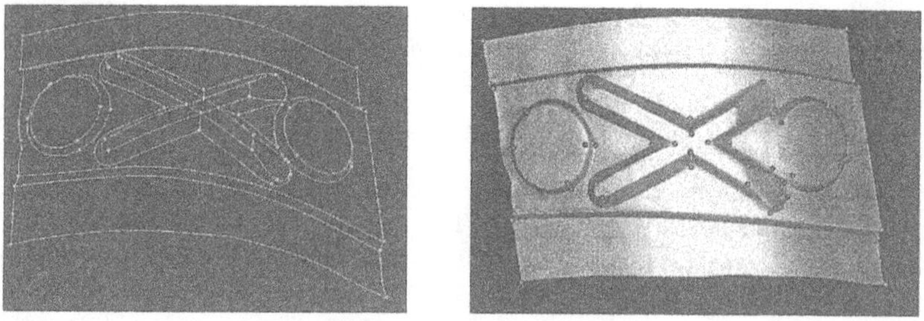

Fig. 2. The edges and the surfaces in the model.

The feature points are only approximate. Nevertheless, they give valuable information when segmenting the point set, i.e. positioning the edges in the CAD-model. Feature curves are positioned by user interaction, and information from the feature points and the triangulation is available, see Figure 2 (left) for feature curves.

The original data set is split into subsets, where each subset is surrounded by a loop of feature curves. Each subset of data points is approximated by a surface. Figure 2 (right) illustrates the model after surface generation. The final step is quality control of the model, and iteration of the surface generation in order to improve accuracy and smoothness.

§3. Curve Creation from the Point Cloud

In this section we will describe the strategy for curve creation employed in the example, in more detail. First we look at qualities we want the curves to satisfy. The curves must be smooth because the quality of the final surfaces is very dependent on the curve quality as the surfaces interpolate the curves. For the same reason, the curves must be represented with as little data as possible. The curves must also be consistent with the point set. If the information represented as curves does not match the point information, the surfaces will not be able to approximate the point set within the given tolerance.

The starting point for the curve creation is a given guiding curve indicating the position of the curve. The guiding curve can be modified by approximating nearby feature points of a specified type. Then the guiding curve is projected onto the point cloud in order to produce a curve. The curve generation procedure is divided into two steps: the actual curve generation; and the creation of information prior to the curve generation.

The curves are generated by a least squares approximation with a smoothness term. Given a sequence of points $\{a_r\}_{r=1}^R$, find a spline curve $f(t) = \sum_{i=1}^n c_i B_{i,k,\tau}$ such that f satisfies

$$\min[\omega_1 \int_{t_0}^{t_n} [f''(t)]^2 dt + \omega_2 \int_{t_0}^{t_n} [f'''(t)]^2 dt + \omega_3 \sum_{r=1}^{R} (f(t_r) - a_r)^2].$$

Here $\{c_i\}_{i=1}^n$ are the unknown coefficients of the curve, τ denotes the knot vector and the curve has order k. The curve is parametrized on the interval t_0, t_n, while ω_1, ω_2 and ω_3 are weights. Initially the data points are parametrized by chord length parametrization to define the parameter values t_r of the data points a_r. An initial knot vector is chosen. New knots may be added iteratively in order to approximate the points within a given tolerance.

The critical point in this step is the proper choice of input points for the curve approximation. How can we produce these points? They must be derived from points belonging to the data set, but cannot in general belong to the point set themselves. We define a surface from the points of the data set, and project points evaluated on the guiding curve onto the surface. Thus, a sequence of input points for the curve generation step, is generated.

Several surface definitions might be used for this purpose, for instance the already defined Delaunay triangulation. Evaluating a linear triangulation is stable. On the other hand, points evaluated on a piecewise linear surface will not necessarily be well suited for being approximated by a smooth function. This is in particular the case if the linear segments are large. The method is

Fig. 3. The Delaunay triangulation of the point set.

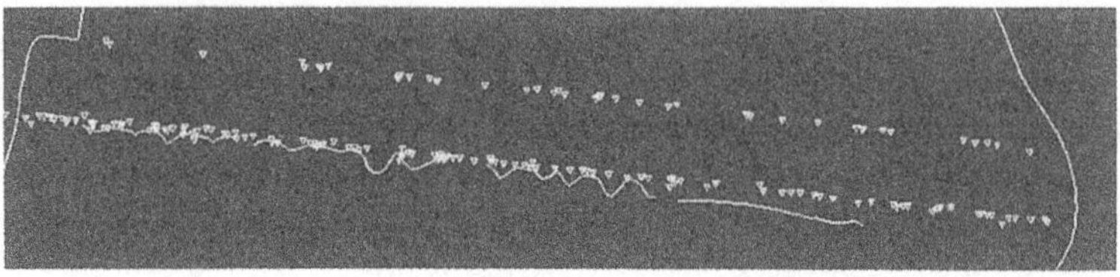

Fig. 4. Curves from Delaunay triangulation and radial basis function surfaces.

also dependent on the behavior of the triangulation. Figure 3 shows the triangulation of a steep area in an otherwise well projectable data set. The curve made from evaluating this triangulation, is the leftmost curve in Figure 4.

To avoid a linear behaviour of the data points, we can define a smooth function interpolating the nodes of the triangulation and evaluate this function instead of the triangulation itself. This will in general improve the quality of the evaluated points, but when performing interpolation, there is a risk of an overshoot, and thereby an oscillating point sequence, if the triangulation is not completely consistent. This method is also dependent on the quality of the triangulation, see the second curve in Figure 4.

By approximating the nodes of the triangulation instead of interpolating them, the risk of overshoot is reduced. We will, on the other hand, not interpolate points belonging to the original point set.

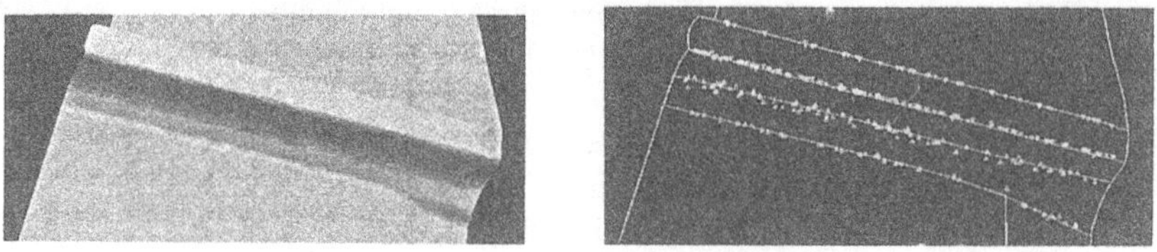

Fig. 5. Data-dependent triangulation.

In Figure 5, a data-dependent triangulation defined from the same point set, is shown. Simulated annealing is used to produce the triangulation, see [7]. Figure 5 also shows a set of curves produced by linear evaluation of this triangulation. A data-dependent triangulation represents features in the model much better than the Delaunay triangulation. However, this triangulation is dependent on sufficient data points in critical areas. Often the improvement of the triangulation is less satisfactory at the boundary, see [4].

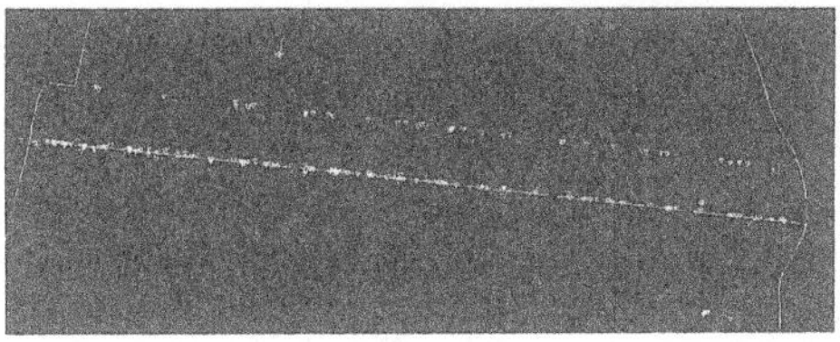

Fig. 6. Curve generated from evaluating a spline surface.

Interpolation of scattered data can be performed by other surfaces than triangulations. In Figure 4 curves made from evaluation of two radial basis function surfaces, i.e. cubic and multiquadric basis functions, are shown. A subset of the original data set lying close to the guiding curve is extracted, and radial basis functions with global support are used on this subset. The surface is given as $f(X) = \sum_{i=1}^{N} c_i\phi(||X - X_i||_2)$, where the c_i's are coefficients, $X_i = (x_i, y_i)$ and $f(X_i) = z_i$. The reason for the bad result is that the radial basis surface is defined from distances in the parameter plane, which in this case was defined as the projection plane of the entire data set, and not from distances in $\mathbb{R}^3$.

Finally, in Figure 6, we see a curve created from points evaluated on a spline surface approximating a set of data points close to the guiding curve. The resulting curve is very smooth, but using an approximative method, we have lost some accuracy control.

In the introductory example, evaluation of the point cloud was performed by evaluating either the Delaunay triangulation or a data dependent triangulation. Experience shows that these choices work well in many cases.

The edges of the CAD-model will in general be boundary curves of the surface. There is the possibility for an exact match between surfaces. C^0 continuity is ensured by matching spline spaces and coefficients of adjacent surfaces. On the other hand, the surface boundaries are fixed. The requirements on the continuity between adjacent surfaces, which is defined by properties of the CAD-system receiving the model, in general implies that having defined the common boundary of two surfaces, one surface may not be changed at the boundary without violating the continuity regarding the other surface.

§4. Generating Curves as Intersection Curves between Surfaces

Generating curves directly from scattered data points involves estimation of data points for curve approximation. This estimation involves a risk. Scattered data points are better suited for surface generation. Then the data set may be approximated directly without the estimation of new points.

Regardless of the method for curve generation, the original data set has to be segmented by splitting it into subsets each being suitable for being approximated by one surface. The triangulation and the feature points are valuable information also in this case. In order to perform segmentation feature curves generated as in Section 3 might be used. However, in this context the quality requirements are different. The curves must be well placed, and they must be smooth, but the distance between a curve and the point cloud is of less interest. The data size of the curves is also not as critical as before. The data points lying inside a loop of feature curves define one subset of points.

Surfaces are generated as before, but in this case there is no need to interpolate boundary curves. Thus, given a well-behaved data set, there are no inconsistencies in the input values for the surface generation, and the chance of getting a well-behaved surface is improved.

The two subsets of points to be approximated by neighboring surfaces will be disjoint, and least squares approximation might shrink the surface compared to the point set. Thus, neighboring surfaces will in general not touch within the tolerance defining continuity of the model across patch boundaries. In order to get a continuous model, the surfaces are extended across edges shared by two surfaces. Extensions are unstable operations, and to keep maximum control of the result, the extension should be done linearly. Finally, the curves are produced as intersections between adjacent surfaces. These intersection curves will trim the surface.

It is necessary to perform intersections between two sculptured surfaces, and these intersections may be tangential. If the point set was split along this curve because there is a tangent plane discontinuity along the curve, then the intersection will be transversal. However, a point set might also be split along discontinuities in curvature or inflection curves, or even to avoid making surfaces of a very complex shape. In these cases, the intersection will be tangential. Moreover, the continuity across a boundary may change from C^0 to G^1 along the boundary curve. This will produce an intersection problem which is partly transversal and partly tangential.

§5. Tangential Intersections

Given two spline surfaces $f(s,t)$ and $g(u,v)$, we want to find all intersections between these two surfaces. We want to find s, t, u and v such that $f(s,t) = g(u,v)$. We might also want to find the positions where the two surfaces lie closer to each other than a given tolerance without really intersecting, i.e. we want to find s, t, u and v such that $|f(s,t) - g(u,v)| < \epsilon$.

An intersection problem between two surfaces is often solved by a two step method. First all branches of the intersection curves are found by a recursive divide and conquer algorithm, see [2]. Then the true intersection curves are approximated by a marching procedure. Typical for tangential intersections, and almost tangential intersections, is that the two surfaces lie closer to each other than the given tolerance in large areas along an intersection curve. In [6], tangential intersections are addressed.

Assume that the recursive part of the intersection algorithm has been able to find a point p on the intersection curve between the two surfaces. The next step is to march along the intersection curve passing through p. The tangent direction of this curve at p is given by the cross product between the surface normals in the point. Along a tangential intersection curve, the value of this cross product will be zero. Along a well-defined tangential intersection

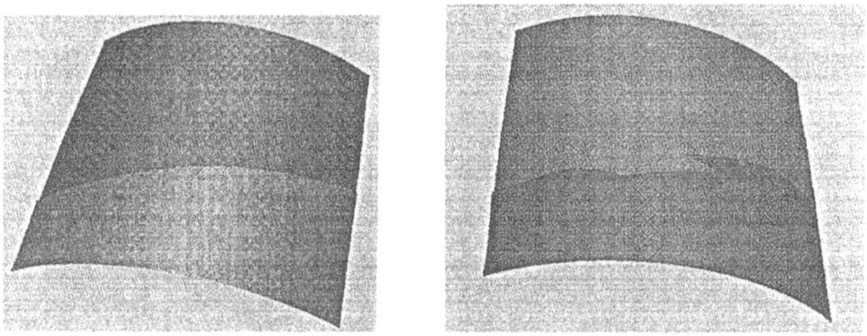

Fig. 7. Two surfaces approximating a point set, before and after extension.

curve, higher order derivatives can give some information about the behavior of the intersection curve, but is a tangential intersection curve well behaved in this context?

We have isolated two subsets of points from the original data points. Two neighboring surfaces, f and g, are constructed by approximating the two point sets within a tolerance α. Both surfaces are linearly extended across their common boundary. The surfaces are expected to intersect in a flat area. Now we want to compute all intersections between f and g using the tolerance ϵ where $\epsilon << \alpha$. Making one cross-section through the extended surfaces, we might find that the distance between the surfaces at the expected intersection curve is close to $2 * \alpha$. Making a different cross-section, we might find a similar distance between the surfaces, but the ordering of the surfaces might have changed. Obviously, there must be an intersection curve between these cross-sections, where the direction of the intersection curve is close to orthogonal to the expected intersection curve. This is the case for Figure 7. A point set is split and approximated by two surfaces. Before extension, the surfaces did not intersect. After extension, the surfaces intersect in 5 curves. The intersections are found by recursively subdividing the surfaces until each branch of the intersection curves intersects the boundaries of the subsurfaces. Then marching may be performed.

In a similar situation we might have created a well-behaved tangential intersection. We might have created a set of intersection curves lying mainly across the expected intersection curve. We might have created a partial coincidence. In that case a smooth curve lying in the area of partial concidence, defined for instance by user interaction, may serve as the intersection curve. There is also a possibility that we did not create any intersection at all. The situation is very unstable, and small perturbations in the surfaces may lead to drastic changes in the intersections between the surfaces.

§6. Conclusion

Where two surfaces are expected to meet in a sharp edge, the intersection approach is likely to lead to surfaces of better quality, but creation of tangential intersections in the situation described above should be avoided. The risk of creating a different topology than intended is very high, and even if our intersection algorithm were able to find the intersections between the surfaces, the real intersections would probably not correspond to the intended one. In this situation more control is required when defining the geometry.

On the other hand, fixing the boundaries of the surfaces to an a priori created curve net, puts too much emphasis on the curves. If one surface may be considered as more important than its neighbors, this surface may define the boundary curves, and the neighboring surfaces can adapt to these curves. Otherwise, it is necessary to decrease the importance of the boundary curves. Define the surfaces given the curve net as boundary curves. Then let the boundaries of the surfaces be adjusted to achieve better consistency with the data points, using C^0-continuity between adjacent surfaces as a side condition for the surface modification.

References

1. Besl, P. J., Jain, R., Segmentation through variable-order surface fitting, IEEE PAMI, **14** (1988), 239–256.
2. Dokken, T., V. Skytt and A-M. Ytrehus, Recursive Subdivision and Iteration in Intersections and Related Problems, in *Mathematical Methods in Computer Aided Geometric Design*, T. Lyche and L. L. Schumaker (eds.), Academic Press, New York, 1989, 207–214.
3. Dokken, T., Aspects of Intersection Algorithms and Approximation, Thesis for Doctor of Philosophy, University of Oslo, 1997.
4. Dyn, N., Levin, D., Rippa, S., Boundary correction for piecewise linear interpolation defined over data-dependent triangulations, Journal of Computational and Applied Mathematics **39** (1992), 179–192.
5. Hjelle, Ø. , Explicit Surfaces in Siscat, Report STF42 A96008, SINTEF Applied Mathematics, 1996.
6. Markot R. P., Magedson R. L., Solutions of tangential surface and curve intersections, Computer-Aided Design **21** (1989), 421–429.
7. Schumaker, L. L., Computing optimal triangulations using simulated annealing, Comput. Aided Geom. Design **10** (1993), 329–345.
8. Varady, T., Martin R. R., Cox J., Reverse Engineering of geometric models, an introduction, Computer-Aided Design **29** (1997), 225–268.
9. Zudel, A.K, Surface–Surface Intersection: Loop Destruction Using Bézier Clipping and Pyramidal Bounds. Thesis for Doctor of Philosophy, Brigham Young University, 1994.

Vibeke Skytt
SINTEF Applied Mathematics
P. O. Box 124, Blindern
N-0314 Oslo, NORWAY
Vibeke.Skytt@math.sintef.no

Representation of Images by Surfaces and Higher Dimensional Manifolds in Non-Euclidean Space

Nir Sochen and Yehoshoua Y. Zeevi

Abstract. In image analysis, processing and understanding, it is highly desirable to be able to process the image and feature domains by methods that are specific to these domains. We show how the geometrical framework for scale-space flows is most convenient for this, and we demonstrate, as an example, how one can switch continuously between the L_1 and L_2 norms for processing of different image and grey level domains. The parameter that interpolates between the norms is the contrast normalization, taken here as a local function of the image embedding space. The resulting flow is spatial and/or grey level preserving and can be used for conditional denoising and segmentation. This example demonstrates the ability of the proposed framework to incorporate context or task data, furnished by either the human user or an active vision subsystem, in a coherent and convenient way.

§1. Introduction

In a variety of applications of denoising, smoothing and enhancement of images, it is advantageous to have simple and automatic "buttons" that enable control of the amount of local smoothing in the image or feature spaces according to some a priori knowledge of the task at hand. We present and implement a method that employs the recently proposed geometrical framework for non-linear scale-space methods [1]. In this framework an image is treated as an embedding of a manifold in a higher dimensional manifold. A grey level image is considered in this framework as a two-dimensional surface (i.e. the graph of $I(x, y)$) **embedded** in the three-dimensional space whose coordinates are (x, y, I).

It was suggested [1] that the non-linear scale space can be treated as a gradient descent with respect to a functional integral that depends on the geometry (i.e. the metric) of the image surface as well as on the embedding and the geometry of the embedding space. In the examples treated in the past,

Mathematical Methods for Curves and Surfaces II
Morten Dæhlen, Tom Lyche, and Larry L. Schumaker (eds.), pp. 469–476.

ISBN 0-8265-1315-8.

it was assumed that the embedding space is Euclidean and that the system of coordinates that describes it is Cartesian [1,2]. In fact, the geometry of the embedding space is flexible and can be determined according to the high level task that one has in mind. We view the geometry of the embedding space as the interface between the high-level task and the low-level process to be implemented.

This paper provides the simplest and most intuitive example where a high level constraint is introduced via a change of the embedding space geometry. We treat grey level images as an embedding of a 2D surface in a 3D space whose coordinates are (x, y, I). We then control locally in x,y, and I the contrast normalization ratio that parameterizes the way we choose to normalize the feature coordinate I with respect to the spatial coordinates x and y. Making the contrast normalization ratio a local function of (x, y, I) enables us to control the rate of diffusion in the spatial (i.e. x and y) domain as well as along the feature coordinate I. A similar idea, of controlling locally the gain in the image formation process, was used by Zeevi et al [5][6].

The paper is organized as follows: Section 2 reviews several concepts from Riemannian geometry needed in the sequel. Section 3 presents the geometrical framework for scale space, following the approach presented in [1]. A detailed example for a grey level image is given in Section 4. Section 5 deals with the numerical implementation of the method and present results. In Section 6 we discuss the potential for further research.

§2. Polyakov Action and Harmonic Maps

The basic concept of Riemannian differential geometry is distance. The natural question in this context is how do we measure distances? We will first take the important example $\mathbf{X} : \Sigma \to \mathbb{R}^3$. Denote the local coordinates on the two-dimensional manifold Σ by (σ^1, σ^2), these are analogous to arc length for the one-dimensional manifold, i.e. a curve, see Fig. 1. The map $\mathbf{X}$ is explicitly given by $(X^1(\sigma^1, \sigma^2), X^2(\sigma^1, \sigma^2), X^3(\sigma^1, \sigma^2))$.

Since the local coordinates σ^i are curvilinear, the squared distance is given by a positive definite symmetric bilinear form called the *metric* whose components are denoted by $g_{\mu\nu}(\sigma^1, \sigma^2)$:

$$ds^2 = g_{\mu\nu} d\sigma^\mu d\sigma^\nu \equiv g_{11}(d\sigma^1)^2 + 2g_{12} d\sigma^1 d\sigma^2 + g_{22}(d\sigma^2)^2, \tag{1}$$

where we used Einstein summation convention in the second equality. We will denote the inverse of the metric by $g^{\mu\nu}$, so that $g^{\mu\nu} g_{\nu\gamma} = \delta^\mu_\gamma$, where δ^μ_γ is the Kronecker delta. We choose to work with the induced metric

$$g_{\mu\nu}(\sigma^1, \sigma^2) = h_{ij}(\mathbf{X}) \partial_\mu X^i \partial_\nu X^j, \tag{2}$$

where $i, j = 1, ..., \dim M$ are being summed over, and $\partial_\mu X^i \equiv \frac{\partial X^i(\sigma^1, \sigma^2)}{\partial \sigma^\mu}$.

Take for example a grey level image which is, from our viewpoint, the embedding of a surface described as a graph in $\mathbb{R}^3$:

$$\mathbf{X} : (\sigma^1, \sigma^2) \to (x = \sigma^1, y = \sigma^2, z = I(\sigma^1, \sigma^2)), \tag{3}$$

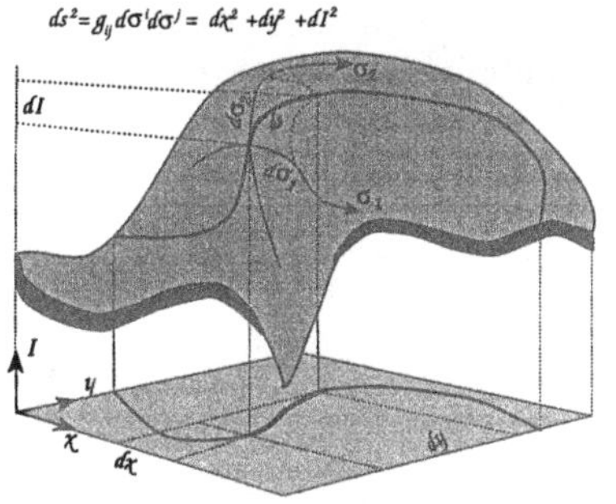

Fig. 1. Length element of a surface curve ds, may be defined either as a function of a local metric defined on the surface (σ_1, σ_2), or as a function of the coordinates of the space in which the surface is embedded (x, y, I) .

where (x, y, z) are Cartesian coordinates. Using (2), we get

$$(g_{\mu\nu}) = \begin{pmatrix} 1+I_x^2 & I_x I_y \\ I_x I_y & 1+I_y^2 \end{pmatrix}, \tag{4}$$

where we used the identification $x \equiv \sigma^1$ and $y \equiv \sigma^2$ in the map $\mathbf{X}$. Next we provide a measure on the space of these maps.

The Measure On Maps

The diffusion equation we are going to use is derived as a gradient descent of an action functional. The functional in question depends on *both* the image manifold and the embedding space. Denote by $(\Sigma, (g_{\mu\nu}))$ the image manifold and its metric and by $(M, (h_{ij}))$ the space-feature manifold and its metric, then the map $\mathbf{X} : \Sigma \to M$ has the following weight [3]:

$$S[X^i, g_{\mu\nu}, h_{ij}] = \int d^m\sigma \sqrt{g} g^{\mu\nu} \partial_\mu X^i \partial_\nu X^j h_{ij}(\mathbf{X}), \tag{5}$$

where m is the dimension of Σ, g is the determinant of the image metric and $g^{\mu\nu}$ is the inverse of the image metric. The range of indices is $\mu, \nu = 1, \ldots, \dim\Sigma$, and $i, j = 1, \ldots, \dim M$. The metric of the embedding space is h_{ij}.

The Polyakov action is the generalization of the L_2 norm to curved spaces. Here, $d^m\sigma\sqrt{g}$ is the volume element (area element for d=2) of Σ – the image manifold , and $g^{\mu\nu}\partial_\mu X^i \partial_\nu X^j h_{ij}(\mathbf{X})$ is the generalization of $|\nabla I|^2$ to maps between non-Euclidean manifolds. Note that the volume element as well as the rest of the expression is invariant under reparameterization, that is, $\sigma^\mu \to \tilde{\sigma}^\mu(\sigma^1, \sigma^2)$. The Polyakov action depends, really, on the geometrical objects and not on the way we describe them via our parameterization of the coordinates.

Given the above functional, we have to choose the minimization. We may choose, for example, to minimize only with respect to the embedding. In this case the metric $g_{\mu\nu}$ is treated as a parameter and can be chosen at will. Another choice is to minimize only with respect to the feature coordinates of the embedding space, or we may choose to minimize the image metric as well. These different choices yield different flows. Some flows are recognized as existing methods like the heat flow, a generalized Perona-Malik flow, or the mean-curvature flow. Other choices are new and will be described below in detail.

Another important point is the choice of the embedding space and its geometry. In general, we need information about the task at hand in order to fix the right geometry. Consider for example the grey level images. It is clear that the intensity I has to be considered differently from x and y. In fact, the relative scale of I with respect to the spatial coordinates (x, y) is to be specified. This can be interpreted as taking the metric of the embedding space as

$$(h_{ij}) = \begin{pmatrix} 1 & 0 & 0 \\ 0 & 1 & 0 \\ 0 & 0 & \beta^2 \end{pmatrix}. \tag{6}$$

We will see below that different limits of this ratio β interpolate between the flows that originate from the Euclidean L_1 and L_2 norms.

Using standard methods in variational calculus, the Euler-Lagrange (EL) equations, with respect to the embedding, are (see [1] for derivation):

$$-\frac{1}{2\sqrt{g}}h^{il}\frac{\delta S}{\delta X^l} = \frac{1}{\sqrt{g}}\partial_\mu(\sqrt{g}g^{\mu\nu}\partial_\nu X^i) + \Gamma^i_{jk}\partial_\mu X^j \partial_\nu X^k g^{\mu\nu}, \tag{7}$$

where Γ^i_{jk} are the Levi-Civita connection's coefficients with respect to the metric h_{ij} (defined in Eq. (17)), that describes the geometry of the embedding space.

We view scale-space as the gradient descent:

$$X^i_t \equiv \frac{\partial X^i}{\partial t} = -\frac{1}{2\sqrt{g}}h^{il}\frac{\delta S}{\delta X^l}. \tag{8}$$

A few remarks are in order. First, note that we used our freedom to multiply the Euler-Lagrange equations by a strictly positive function and a positive definite matrix.

This factor is the simplest one that does not change the minimization solution, while giving a reparameterization invariant expression. This choice guarantees that the flow is geometric and does not depend on the parameterization. The operator that is acting on X^i in the first term of (7) is the natural generalization of the Laplacian from flat spaces to manifolds and is called the Laplace-Beltrami operator, or in short Beltrami operator, denoted by Δ_g. When the embedding is in a Euclidean space with Cartesian coordinate system, the connection elements are zero. If the embedding space is not Euclidean, we have to include the Levi-Civita connection term since it is not identically zero any more.

§3. The Beltrami Flow

A grey level image is regarded as a map $f : \Sigma \to \mathbb{R}^3$, where Σ is a two-dimensional manifold, and the flow is natural in the sense that it minimizes the action functional with respect to I and (g_{ij}), while being reparameterization invariant. The coordinates X and Y are parameters from this viewpoint, and are identified as above with σ^1 and σ^2, respectively. The result of the minimization is the Beltrami operator acting on I:

$$I_t = \Delta_g I \equiv \frac{1}{\sqrt{g}} \partial_\mu (\sqrt{g} g^{\mu\nu} \partial_\nu I) = H N_{\hat{I}},$$

where the metric is the induced one given in (2), H is the mean curvature and $N_{\hat{I}}$ is the component, in the I direction, of the vector normal to the surface.

The geometrical meaning is obvious; each point on the image surface moves with a velocity that depends on the mean curvature and the I component of the normal to the surface at that point. Since along the edges the normal to the surface lies almost entirely in the X-Y plane, I changes very little on the edges while the flow drives other regions of the image towards a minimal surface in a more rapid rate.

§4. Local Contrast Normalization

There is an implicit parameter of the embedding in the above analysis. We described the embedding as

$$\{X^1(\sigma^1, \sigma^2) = \sigma^1,\ X^2(\sigma^1, \sigma^2) = \sigma^2,\ X^3(\sigma^1, \sigma^2) = I(\sigma^1, \sigma^2)\}, \tag{9}$$

but, actually the contrast normalization is arbitrary and we could as well choose

$$\tilde{X}^3(\sigma^1, \sigma^2) = \beta I(\sigma^1, \sigma^2). \tag{10}$$

The line element in this case is

$$d\tilde{s}^2 = dx^2 + dy^2 + d(\tilde{X}^3)^2 = dx^2 + dy^2 + \beta^2 dI^2. \tag{11}$$

We interpret the scaling of the intensity as a change in the metric of the embedding space:

$$(h_{ij}) = \begin{pmatrix} 1 & 0 & 0 \\ 0 & 1 & 0 \\ 0 & 0 & \beta^2 \end{pmatrix}. \tag{12}$$

The limits of very large and very small β correspond to the L_1 and L_2 norms, respectively, as can be easily seen from the Euler-Lagrange equations (16)

$$\begin{aligned} \lim_{\beta \to 0} I_t(\beta) &= I_{xx} + I_{yy}, \\ \lim_{\beta \to \infty} \beta^2 I_t(\beta) &= \frac{I_y^2 I_{xx} - 2 I_x I_y I_{xy} + I_x^2 I_{yy}}{(I_x^2 + I_y^2)^2}. \end{aligned} \tag{13}$$

The main idea is to let now β be a **local** function of x, y and I such that different regions will be treated differently. As an example we choose a contrast normalization function $\beta(x, y, I)$ such that for dark regions, β is large and the flow is L_1-like, while for more luminous regions, β is small and the flow is L_2-like. There are many reasonable choices for the function β. We will use

$$\beta(x, y, I) = Ae^{-b_1(x-x_0)^2 - b_2(y-y_0)^2 - b_3(I-I_0)^2}, \tag{14}$$

with A and b's constants to be defined.

The effect of the function β effect is to impose slow diffusion in an ellipsoid around (x_0, y_0, I_0), with axes b_1, b_2 and b_3. More distant areas diffuse at the normal Beltrami rate while in very distant areas where $\beta < 1$, the diffusion is faster then the Beltrami rate and approaches the linear scale-space rate. From a geometrical viewpoint, it amounts to choosing an appropriate embedding space. The metric of the embedding space is, under these circumstances, not constant:

$$(h_{ij}) = \begin{pmatrix} 1 & 0 & 0 \\ 0 & 1 & 0 \\ 0 & 0 & \beta(x, y, I)^2 \end{pmatrix}. \tag{15}$$

This equation means that the distance along the Intensity axis depends on the point (x, y, I) where it is measured. In other words our embedding space is not Euclidean any more.

As a consequence of the fact that we are working now on a curved manifold as our embedding manifold, we have to take into account that the Levi-Civita connection of our three-dimensional manifold does not vanish, and need to add this contribution to the Euler Lagrange equation. Note also that the two-dimensional image induced metric is different from the one we had in the Euclidean case:

$$\begin{aligned} \Delta_{g(\beta)} I = \frac{1}{g^2}((1 + \beta^2 I_x^2) I_{yy} - 2\beta^2 I_x I_y I_{xy} + (1 + \beta^2 I_y^2) I_{xx} \\ - \beta(\beta_x I_x + \beta_y I_y)(I_x^2 + I_y^2)) \end{aligned}, \tag{16}$$

where $g = \det(g_{\mu\nu}) = 1 + \beta^2(I_x^2 + I_y^2)$. The Levi-Civita connection is defined as follows:

$$\Gamma^i_{jk} = \frac{1}{2} h^{il} (\partial_j h_{lk} + \partial_k h_{jl} - \partial_l h_{jk}), \tag{17}$$

where there is an implicit sum over l. In our case h_{ij} is given in (15), and using the above definition we get

$$\Gamma^3_{ij} \partial_\mu X^i \partial_\nu X^j g^{\mu\nu} = \frac{1}{\beta g}(2\beta_x I_x + 2\beta_y I_y + \beta_I (I_x^2 + I_y^2)). \tag{18}$$

Collecting all the terms, we get the flow

$$I_t = \Delta_{g(\beta)} I + \frac{1}{\beta g} \left(2\beta_x I_x + 2\beta_y I_y + \beta_I (I_x^2 + I_y^2)\right). \tag{19}$$

This can be written in an explicit way by performing the derivatives on the function β given by (14).

§5. Experimental Results

We show in this section the results obtained with our algorithm and compare it to the Beltrami flow. There is an inherent problem in comparing any two different flows. This is because of the lake of widely acceptable objective criterion(s) with which we can measure the performance of a given algorithm. The evaluation is also task dependent which complicate things further. Nevertheless it is very easy, by simply looking at the sequences $I(x, y, t)$, from the two algorithms to realize the differences between the flows.

We take the "Lena" image with "side illumination" effect achieved by adding a tilted plane to the image. (See Fig. 2).

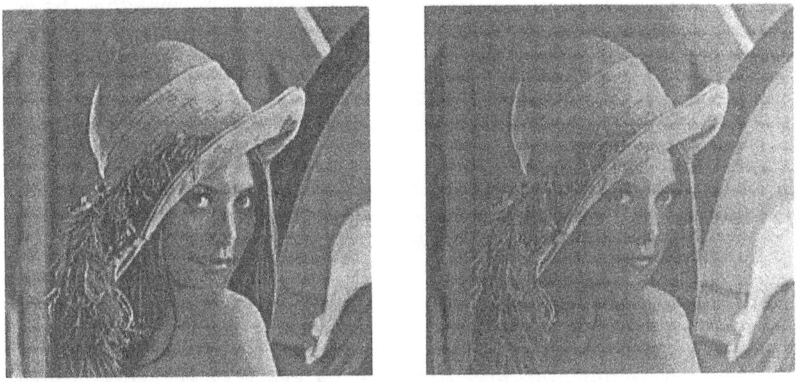

Fig. 2. Left image: original 256x256 image, Right image: Lena + a tilted plane.

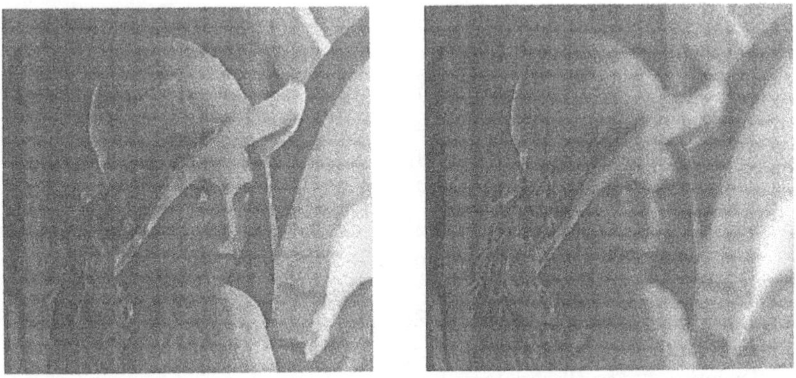

Fig. 3. Left image: Lena + plane + 20 iterations of the Beltrami flow, Right image: Lena + a plane + 20 iterations of the non Euclidean flow.

The "tilted Lena" was processed with β in the form (14) with parameters chosen as $b_1 = b_2 = 0$, $b_3 = 0.0025$ and $A = 3$. The derivatives of β can be done analytically while for those of the intensity we used the central differences scheme. We use an explicit Euler scheme for solving the partial differential equation. The result of the non-Euclidean flow after 20 iterations is compared to the usual Beltrami flow in Fig. 3.

As expected and clearly seen in Fig. 3, our flow diffuses the brightest area of the image (particularly in the right side of the image) faster than the Beltrami flow does. It is comparable in the center of the image, and it diffuses slower in the darker regions, notably in the left side of the image.

§6. Directions for Further Research

The idea of controlling image processing and analysis by changing the geometry of the embedding space is only in its beginning. We present in this paper only preliminary results to demonstrate the principles in as simple setting as possible. More realistic applications need much more work in order to understand (by a hybrid of analytical and experimental methods) the right geometry of the embedding space appropriate for a given task. A study of color space that incorporates psychophysical research will be reported elsewhere. Other directions are texture and various invariance properties. These questions are under current study.

References

1. Sochen N., Kimmel R. and Malladi R., A general framework for low level vision, IEEE Trans. IP, 1997, in press.
2. Kimmel R., Malladi R. and Sochen N., Images as embedded maps and minimal surfaces: Movies, color, texture, and volumetric medical images, IEEE Trans. PAMI, submitted.
3. Polyakov A. M., Quantum geometry of bosonic strings, Physics Letters **103B** (1981), 207–210.
4. Kreyszing E., *Differential Geometry*, Dover Publications, Inc., New York, 1991.
5. Zeevi Y. Y., Ginosar R. and Hilsenrath O., "Wide dynamic range camera", U.S.A. patent No. 5,144,442, 1992.
6. Zeevi Y. Y., Ginosar R. and Hilsenrath O., "Dynamic image representation system", U.S.A. patent No. 5,420,637, 1995.

Nir Sochen and Yehoshua. Y. Zeevi
Faculty of Electrical Engineering
Technion – The Technology Institute of Israel
Technion City, Haifa, 32000, ISRAEL
`sochen,zeevi@ee.technion.ac.il`

Discrete Smoothing D^m-splines: Applications to Surface Fitting

J. J. Torrens

Abstract. In this paper we review the basic theory of discrete smoothing D^m-splines for Lagrange problems, including an improved convergence result and error estimates. Likewise, we consider several applications to surface fitting problems.

§1. Introduction

Discrete smoothing D^m-splines were introduced in connection with Lagrange aproximation by smoothing D^m-splines over a bounded domain in $\mathbb{R}^n$. These latter splines, due to M. Atteia (cf. [7,8]), are the solutions of certain minimization problems of quadratic functionals in a Sobolev space. Except in the univariate case, in general such splines cannot be described in terms of known functions. However, using the finite element method, it is possible to discretize the minimization problems and to find suitable approximations, the so-called discrete smoothing D^m-splines (cf. [4]).

This paper focuses on discrete smoothing D^m-splines and their application to several surface fitting problems (for discrete interpolating D^m-splines, see [4] and [6]). Section 2 is devoted to the main topics of the theory: definition, variational characterization, computation, convergence and error estimates. Section 3 provides several examples of application to the construction of explicit surfaces of class C^k. In the remainder of this section, we introduce some notation and preliminary results.

Let $n \in \mathbb{N}^*$. For any $m \in \mathbb{N}$ and any nonempty open subset ω of $\mathbb{R}^n$, we denote by $H^m(\omega)$ the usual Sobolev space of order m over ω, endowed with the norm

$$\|v\|_{m,\omega} = \left(\sum_{|\alpha| \leq m} \int_\omega |\partial^\alpha v(x)|^2 \, dx \right)^{1/2},$$

Mathematical Methods for Curves and Surfaces II

Morten Dæhlen, Tom Lyche, and Larry L. Schumaker (eds.), pp. 477–484.

ISBN 0-8265-1315-8.

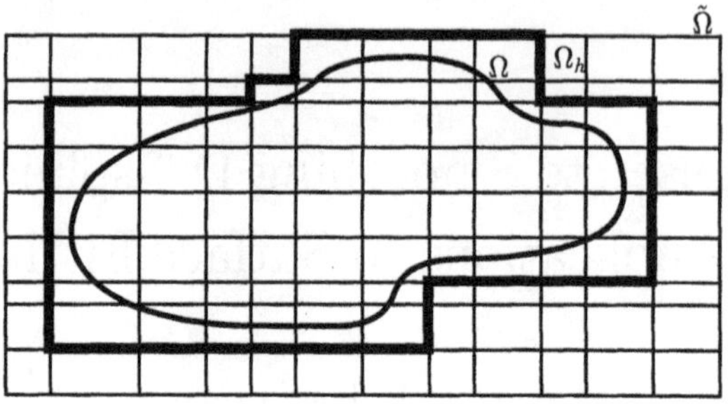

Fig. 1. The open sets Ω, Ω_h and $\tilde{\Omega}$.

with $\alpha = (\alpha_1, \dots, \alpha_n) \in \mathbb{N}^n$, $|\alpha| = \alpha_1 + \cdots + \alpha_n$, and $\partial^\alpha v$ being the α-th derivative of v in the distribution sense. We shall also use the semi–norms

$$|v|_{j,\omega} = \left(\sum_{|\alpha|=j} \int_\omega |\partial^\alpha v(x)|^2 \, dx \right)^{1/2}, \quad 0 \le j \le m,$$

and the inner semi–product associated with the semi–norm of order m:

$$(u, v)_{m,\omega} = \sum_{|\alpha|=m} \int_\omega \partial^\alpha u(x) \, \partial^\alpha v(x) \, dx.$$

If ω has a Lipschitz–continuous boundary (in the J. Nečas [14] sense), it is well known that for any $\mu \in \mathbb{N}$ such that $m > \frac{n}{2} + \mu$, $H^m(\omega)$ is a subset of $C^\mu(\overline{\omega})$ with continuous injection, where $C^\mu(\overline{\omega})$ stands for the Banach space of continuous functions of class C^μ over $\overline{\omega}$, endowed with the norm

$$\|v\|_{C^\mu(\overline{\omega})} = \max_{|\alpha| \le \mu} \sup_{x \in \omega} |\partial^\alpha v(x)| \, .$$

Finally, for any $l \in \mathbb{N}$, P_l denotes the space of all real polynomials of degree $\le l$.

§2. Basic Theory on Discrete Smoothing D^m-splines

2.1. Definition and Characterization

Let m and n be two positive integers such that $m > \frac{n}{2}$, and let Ω be an open, bounded, connected, nonempty subset of $\mathbb{R}^n$ with Lipschitz–continuous boundary. Suppose we are given

- a bounded, polygonal, open set $\tilde{\Omega} \subset \mathbb{R}^n$ such that $\Omega \subset \tilde{\Omega}$;
- a bounded set H of real positive numbers such that $0 \in \overline{H}$;
- for any $h \in H$, a tesselation $\tilde{\mathcal{T}}_h$ of $\overline{\tilde{\Omega}}$ with elements of diameter $\le h$, and a finite element space $\tilde{V}_h$, constructed on $\tilde{\mathcal{T}}_h$, such that $\tilde{V}_h \subset H^m(\tilde{\Omega})$ (for the definition of finite element spaces, see [9].

For all $h \in H$, let Ω_h be the interior of the union of elements $K \in \tilde{\mathcal{T}}_h$ such that $K \cap \Omega \ne \emptyset$ (see Fig. 1), and, finally, let V_h be the space of restrictions to Ω_h of functions in $\tilde{V}_h$, which is a finite dimensional subspace of $H^m(\Omega_h)$.

Remark 1. Observe that for any $h \in H$, $\Omega \subset \Omega_h \subset \tilde{\Omega}$, and that as $h \to 0$, the measure of $\Omega_h \setminus \Omega$ tends to 0. Likewise, in practice every tesselation $\tilde{\mathcal{T}}_h$ is made with n-simplices or n-rectangles of diameter $\leq h$. The space $\tilde{V}_h$ is then constructed on $\tilde{\mathcal{T}}_h$ from a generic Hermite finite element (K, P_K, Σ_K) of class $C^{k'}$, with $k' \geq 1$, such that $P_K \subset H^m(K)$ and $m \leq k' + 1$. Hence, $V_h \subset H^m(\Omega_h) \cap C^{k'}(\overline{\Omega}_h)$.

Suppose we are given a set $A = \{a_1, \dots, a_N\} \subset \overline{\Omega}$ and a vector $\beta \in \mathbb{R}^N$. For all $\varepsilon > 0$ and $h \in H$, let $J_{\varepsilon h}$ be the functional defined in $H^m(\Omega_h)$ by

$$J_{\varepsilon h}(v) = \langle \rho v - \beta \rangle^2 + \varepsilon |v|^2_{m,\Omega_h},$$

where $\langle \cdot \rangle$ stands for the Euclidean norm in $\mathbb{R}^N$ and $\rho v = (v(a_i))_{1 \leq i \leq N}$. We remark that the vector ρv is well defined, since $H^m(\Omega_h) \subset C^0(\overline{\Omega}_h)$ and $A \subset \overline{\Omega} \subset \overline{\Omega}_h$.

Then for any $h \in H$ and $\varepsilon > 0$, a V_h-discrete smoothing D^m-spline relative to A, β and ε is any solution of the problem: find $\sigma_{\varepsilon h}$ such that

$$\sigma_{\varepsilon h} \in V_h \text{ and } J_{\varepsilon h}(\sigma_{\varepsilon h}) = \inf_{v \in V_h} J_{\varepsilon h}(v). \tag{1}$$

Theorem 2. *If A contains a P_{m-1}-unisolvent subset, then for any $h \in H$ and $\varepsilon > 0$, problem (1) admits a unique solution $\sigma_{\varepsilon h}$ which is also the unique solution of the variational problem: find $\sigma_{\varepsilon h}$ such that*

$$\left| \begin{array}{l} \sigma_{\varepsilon h} \in V_h, \\ \forall v \in V_h, \langle \rho \sigma_{\varepsilon h}, \rho v \rangle + \varepsilon (\sigma_{\varepsilon h}, v)_{m,\Omega_h} = \langle \beta, \rho v \rangle, \end{array} \right. \tag{2}$$

where $\langle \cdot , \cdot \rangle$ denotes the Euclidean inner product in $\mathbb{R}^N$.

Proof: It suffices to endow V_h with the norm $[\![v]\!] = \left(\langle \rho v \rangle^2 + |v|^2_{m,\Omega_h} \right)^{1/2}$ and then apply the Lax–Milgram Lemma. □

Remark 3. For any $h \in H$ and $\varepsilon > 0$, problem (1) is obviously a discretization of the problem: find σ_ε such that

$$\sigma_\varepsilon \in H^m(\Omega) \text{ and } J_\varepsilon(\sigma_\varepsilon) = \inf_{v \in H^m(\Omega)} J_\varepsilon(v), \tag{3}$$

with $J_\varepsilon(v) = \langle \rho v - \beta \rangle^2 + \varepsilon |v|^2_{m,\Omega}$, whose solution is called *smoothing D^m-spline over Ω*. The variational problem associated to (3) is discretized by (2). Observe that this spline, due to M. Atteia, is different from the *smoothing D^m-spline over $\mathbb{R}^n$* (for $m = 2$, *thin plate spline*) introduced by J. Duchon, which is obtained by minimizing J_ε with

$$|v|^2_m = \sum_{|\alpha| = m} \frac{m!}{\alpha!} \int_{\mathbb{R}^n} |\partial^\alpha v(x)|^2 \, dx$$

instead of $|v|^2_{m,\Omega}$, in the Beppo Levi space $D^{-m}L^2(\mathbb{R}^n)$ (cf. [11]).

2.2. Computation

Now let $h \in H$, and let $w_1, \ldots, w_M$ be a basis of the space V_h. Thus $\sigma_{\varepsilon h} = \sum_{j=1}^{M} \alpha_j w_j$ for some $\alpha_1, \ldots, \alpha_M \in \mathbb{R}$. From (2), it follows that $\alpha = (\alpha_j)_{1 \le j \le M}$ is the solution of the linear system $(\mathcal{A}^T \mathcal{A} + \varepsilon \mathcal{R})\alpha = \mathcal{A}^T \beta$, where $\mathcal{A} = \big(w_j(a_i)\big)_{\substack{1 \le i \le N \\ 1 \le j \le M}}$, and $\mathcal{R} = \big((w_j, w_i)_{m,\Omega_h}\big)_{1 \le i,j \le M}$. The coefficient matrix $\mathcal{A}^T \mathcal{A} + \varepsilon \mathcal{R}$ is banded, symmetric and positive definite. We remark that the order of the linear system is $M = \dim V_h$ (independent of the number N of data points).

2.3. Convergence and Error Estimates

Suppose we are given a set D of real positive numbers for which 0 is an accumulation point, and for every $d \in D$, an ordered set A^d of $N = N(d)$ points in $\overline{\Omega}$ such that

$$\sup_{x \in \Omega} \delta(x, A^d) = d, \tag{4}$$

δ being the Euclidean distance in $\mathbb{R}^n$.

Remark 4. Hypothesis (4) implies that D is bounded and that $\cup_{d \in D} A^d$ is a dense subset of $\overline{\Omega}$. Likewise, for d sufficiently small, A^d contains a P_{m-1}-unisolvent subset. In fact, without loss of generality, we can suppose that this is true for all $d \in D$.

Assume that the family of finite element spaces $(\tilde{V}_h)_{h \in H}$ is constructed in such a way that the following result, due to P. Clément (cf. [10]), holds:

There exist an integer $m' \ge m$ and, for all $h \in H$, a linear operator $\Pi_h : L^2(\tilde{\Omega}) \to \tilde{V}_h$ such that:
(i) $\exists C > 0$, $\forall h \in H$, $\forall l = 0, \ldots, m' - 1$, $\forall v \in H^{m'}(\tilde{\Omega})$,

$$\left(\sum_{K \in \tilde{\mathcal{T}}_h} |v - \Pi_h v|_{l,K}^2 \right)^{1/2} \le C\, h^{m'-l} |v|_{m',\tilde{\Omega}}; \tag{5}$$

(ii) $\forall v \in H^{m'}(\tilde{\Omega})$, $\displaystyle\lim_{h \to 0} \left(\sum_{K \in \tilde{\mathcal{T}}_h} |v - \Pi_h v|_{m',K}^2 \right)^{1/2} = 0.$

Remark 5. The result of P. Clément assumes that the family of tesselations $(\tilde{\mathcal{T}}_h)_{h \in \mathcal{H}}$ is regular (cf. [9]), and that the generic finite element (K, P_K, Σ_K) of the family $(\tilde{V}_h)_{h \in \mathcal{H}}$ satisfies the condition $P_{m'}(K) \subset P_K \subset H^{m'}(K)$ and a "uniformity" property of the basis functions (cf. [10,15]) fulfilled by the usual finite elements (for example, in the bivariate case, by the elements of Bogner–Fox–Schmit, Argyris or Bell).

Assume, in addition, that the families $(A^d)_{d \in D}$ and $(\tilde{\mathcal{T}}_h)_{h \in \mathcal{H}}$ satisfy the relation

$$\exists C > 0,\ \forall (d,h) \in D \times H,\ \forall K \in \tilde{\mathcal{T}}_h : \frac{\operatorname{card} A^d \cap K}{\operatorname{meas} K} \le C\, d^{-n},$$

which bounds above the density of data points in every element of every tesselation $\tilde{\mathcal{T}}_h$. Finally, considering h and ε as functions of d, suppose that

$$\varepsilon = o(d^{-n}),\ d \to 0, \tag{6}$$

and that

$$\frac{h^{2m'}}{d^n \varepsilon} = o(1),\ d \to 0. \tag{7}$$

In these conditions we have a convergence result that improves those previously presented in [4,5]. The proof, due to R. Arcangéli, still remains unpublished, but it is quite similar to that of Theorem 6.3 in [6].

Theorem 6. *Let f be a function in $H^{m'}(\Omega)$, with m' given in (5). For all $d \in D$, let $\sigma_{\varepsilon h}^d$ be the V_h-discrete smoothing D^m-spline relative to A^d, $(f(a))_{a\in A^d}$ and ε. Then*

$$\lim_{d\to 0} \|\sigma_{\varepsilon h}^d - f\|_{m,\Omega} = 0.$$

Finally, we state an error estimate.

Theorem 7. *Suppose in addition that*

$$\exists C > 0,\ \varepsilon \geq C\, d^{2m-n},\ d \to 0. \tag{8}$$

Then, as $d \to 0$, we have

(i) $\forall l = 0, \ldots, m-1,\ |\sigma_{\varepsilon h}^d - f|_{l,\Omega} = O\left(\left(\frac{\varepsilon}{N}\right)^{\frac{m-l}{2m}}\right)$,

(ii) $|\sigma_{\varepsilon h}^d - f|_{m,\Omega} = o(1)$.

Proof: Proceed as in [13] for the proof of Theorem 4.2, but using Theorem 6 and the error estimates established in Theorem 2.1 of [12]. □

Remark 8. By (6) and (8), there exists positive constants C and C^* such that, for d sufficiently small, $Cd^{2m-n} \leq \varepsilon \leq C^* d^{-n}$. Hence, as $d \to 0$, ε can tend to 0 or even to $+\infty$, but not too fast.

Remark 9. Suppose that (8) holds and that $(\tilde{\mathcal{T}}_h)_{h\in H}$ satisfies the "inverse assumption" (see [9]). Then (7) is satisfied if, as $d \to 0$, $h^{2m'}/d^{2m} \to 0$ or, equivalently, $N^m/M^{m'} \to 0$, with $N = \operatorname{card} A^d$ and $M = \dim V_h$. In other words, M should be a bit greater than $N^{m/m'}$. When $m' > m$ (that is the practical case), this allows to consider finite element spaces V_h with dimension M much less than N. For example, suppose for $n = 2$ that V_h is constructed from the Argyris' triangle of class C^1. Hence, $m' = 5$ and $m = 2$. If $N = 10^5$, then $N^{m/m'} = 100$, and thus we can take M much less than N. In conclusion, discrete smoothing D^m-splines are well suited for fitting large sets of data.

§3. Applications to Surface Fitting

The basic theory presented in Section 2 can be applied, directly or in a modified form, to several problems of surface reconstruction that frequently appear in oil prospecting, geology or geophysics. For these problems, interpolation methods are not convenient, due to the availability of large amounts of data. In the sequel, we shall retain the notation of Sections 2.1 and 2.2, with $n = 2$, and we shall denote by f a function sufficiently regular defined on $\overline{\Omega}$.

Let us begin with a classical problem: Given a finite set $A \subset \overline{\Omega}$ and the set of values $\{f(a) \mid a \in A\}$, find an approximant $\Phi \in C^k(\overline{\Omega})$ of the function f, with $k = 1$ or 2 (e.g. find a smooth approximating surface of the explicit surface $z = f(x, y)$). It is clear that we can merely take Φ as a V_h-discrete smoothing D^m-spline relative to A, $\beta = (f(a))_{a \in A}$ and ε for suitable values of m, h and ε. Of course, the finite element spaces V_h should be constructed from a generic finite element of class $C^{k'}$ with $k' \geq k$ (see Remark 1). In practice, one has $k = k'$ and $m = k' + 1$. See [2] for more details and numerical examples.

This last problem can be extended in two directions: introducing 1st order Hermite data (values of the first derivatives of f), and allowing discontinuities on a subset F of $\overline{\Omega}$. In geology, the Hermite data are usually deduced for every point from two angles, dip and strike, that indicate the position of the tangent plane to the surface. Likewise, discontinuities on F model geological faults: vertical faults correspond to discontinuities of f, while direct oblique faults are associated with discontinuities of the first derivatives of f. In this setting, the theory given in Section 2 is no longer valid: modifications are needed in order to model the set F, to construct finite element spaces of functions with possible discontinuities on F, and to properly define the functional $J_{\varepsilon h}$. All these aspects are considered in [6]. The reader is referred to this paper for additional explanations, a convergence result and numerical examples.

Another kind of problem arises when the data is continuously distributed on curves or subdomains of $\overline{\Omega}$. The case of curves is discussed in [1]. A possible application is the fit of f from a set of level curves. In the remainder of this section we detail the case of subdomains, first studied in [3]. To be concrete, we consider the following problem:

Let ω be an open subset of Ω. Given the restriction of the function f to ω, find an approximant $\Phi \in C^k(\overline{\Omega})$ of f, with $k = 1$ or 2. (9)

In most practical cases, ω is a union of a finite collection $\omega_1, \ldots, \omega_N$ of connected open subsets of Ω. In some sense, problem (9) can be seen as a problem of blending the surface patches $S_1, \ldots, S_N$, where, for $j = 1, \ldots, N$, $S_j = \{(x, y, f(x, y)) \mid (x, y) \in \omega_j\}$ (see Fig. 2).

Assume that $f \in H^m(\Omega)$ for some integer $m \geq 2$. Since f is completely known on ω, the key idea to solve problem (9) is to consider a discrete version of problem (3), where, in this case, $J_\varepsilon(v) = \|v - f\|_{0,\omega}^2 + \varepsilon |v|_{m,\Omega}^2$. For this,

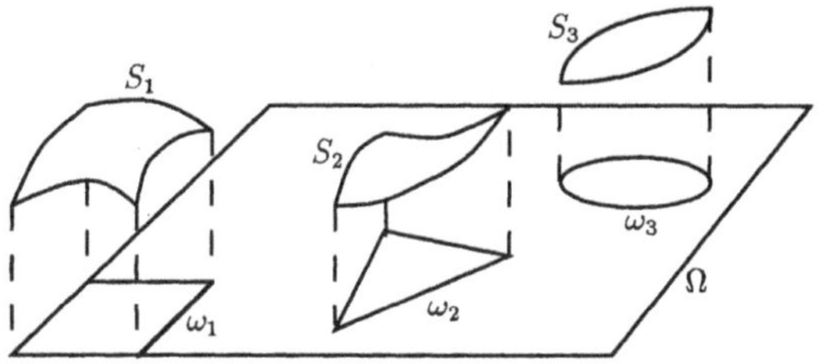

Fig. 2. Surface patches.

assume that we have a quadrature formula

$$\iint_\omega v(x,y)\,dx\,dy \sim \ell(v) = \sum_{i=1}^{L} \lambda_i v(\xi_i), \tag{10}$$

with nodes $\xi_1, \ldots, \xi_L$ in $\overline{\omega}$ and strictly positive weights $\lambda_1, \ldots, \lambda_L$. Since quadrature formulae for triangles are well–established (see [9]), if ω is polygonal, (10) can be easily obtained: first, make a triangulation τ of $\overline{\omega}$, then apply a quadrature formula on every element of τ, and finally add all such formulae.

Let $(V_h)_{h\in H}$ be a family of finite element spaces constructed as in Section 2.1. Then, for any $h \in H$ and $\varepsilon > 0$, we consider problem (1) with $J_{\varepsilon h}(v) = \ell((v-f)^2) + \varepsilon |v|^2_{m,\Omega_h}$. If $\{\xi_1, \ldots, \xi_L\}$ contains a P_{m-1}-unisolvent subset in every connected component of $\overline{\omega}$, it is readily seen that problem (1) has now a unique solution, also characterized by the relations

$$\left|\begin{array}{l}\sigma_{\varepsilon h} \in V_h,\\ \forall v \in V_h,\ \ell(\sigma_{\varepsilon h}\, v) + \varepsilon(\sigma_{\varepsilon h}, v)_{m,\Omega_h} = \ell(f\, v).\end{array}\right.$$

Expressing again $\sigma_{\varepsilon h}$ as a linear combination of a basis $\{w_1, \ldots, w_M\}$ of V_h, we deduce that the coefficient vector $\alpha = (\alpha_j)_{1\le j\le M}$ is the solution of the linear system $(\mathcal{A}^T\Lambda\mathcal{A} + \varepsilon\mathcal{R})\alpha = \mathcal{A}^T\Lambda\beta$, where $\mathcal{A} = \big(w_j(\xi_i)\big)_{\substack{1\le i\le L\\ 1\le j\le M}}$, $\Lambda = \operatorname{diag}(\lambda_1, \ldots, \lambda_L)$, $\mathcal{R} = \big((w_j, w_i)_{m,\Omega_h}\big)_{1\le i,j\le M}$, and $\beta = (f(\xi_j))_{1\le j\le L}$.

Once $\sigma_{\varepsilon h}$ has been computed, problem (9) is solved by setting $\Phi = \sigma_{\varepsilon h}$. For an indepth analysis of this problem, a convergence result (which strongly depends on the exactness of the quadrature formula (10)), and numerical examples, we refer the reader to [3].

Acknowledgments. This work has been supported in part by the Gobierno de Navarra (Research Project BON n° 33–18/3/94).

References

1. Apprato, D., and R. Arcangéli, Ajustement spline le long d'un ensemble de courbes, Math. Model. Numer. Anal. **25** (1991), 193–212.

2. Apprato, D., R. Arcangéli, and R. Manzanilla, Sur la construction de surfaces de classe C^k à partir d'un grand nombre de données de Lagrange, Math. Model. Numer. Anal. **21** (1987), 529–555.

3. Apprato, D., and C. Gout, Spline fitting on surface patches, Publication URA CNRS 1204 n° 95/2, Pau, 1995.

4. Arcangéli, R., D^m-splines sur un domaine borné de $\mathbb{R}^n$, Publication URA CNRS 1204 n° 86/2, Pau, 1986.

5. Arcangéli, R., Some applications of discrete D^m-splines, in *Mathematical Methods in Computer Aided Geometric Design*, T. Lyche and L. L. Schumaker (eds.), Academic Press, New York, 1989, 35–44.

6. Arcangéli, R., R. Manzanilla, and J. J. Torrens, Approximation spline de surfaces de type explicite comportant des failles, Math. Model. Numer. Anal. **31** (1997), 643–676.

7. Atteia, M., Fonctions–splines généralisées, C. R. Acad. Sci. Paris **261** (1966), 2149–2142.

8. Atteia, M., Existence et détermination des fonctions–splines à plusieurs variables, C. R. Acad. Sci. Paris **262** (1966), 575–578.

9. Ciarlet, P. G., *The Finite Element Method for Elliptic Problems*, North–Holland, 1978.

10. Clément, P., Approximation by finite element functions using local regularization, Rev. Française Automat. Informat. Rech. Opér. R–2 (1975), 77–84.

11. Duchon, J., Interpolation des fonctions de deux variables suivant le principe de la flexion des plaques minces, Rev. Française Automat. Informat. Rech. Opér., Anal. Numer. **10** (1976), 5–12.

12. López de Silanes, M. C., and D. Apprato, Estimations de l'erreur d'approximation sur un domain borné de $\mathbb{R}^n$ par D^m-splines d'interpolation et d'ajustement discrètes, Numer. Math. **53** (1988), 367–376.

13. López de Silanes, M. C., and R. Arcangéli, Sur la convergence des D^m-splines d'ajustement pour des données exactes ou bruités, Rev. Mat. Univ. Complut. Madrid **4** (1991), 279–294.

14. Nečas, J., *Les Méthodes Directes en Théorie des Équations Elliptiques*, Masson, 1967.

15. Strang, G., Approximation in the finite element method, Numer. Math. **19** (1972), 81–98.

J. J. Torrens
Departamento de Matemática e Informática
Universidad Pública de Navarra
Campus de Arrosadía s/n
31006 Pamplona, SPAIN
jjtorrens@upna.es

Spherical Pythagorean-Hodograph Curves

Kenji Ueda

Abstract. Geometric inversion preserves Pythagorean hodographs. Stereographic projection, that is, a composition of the three mappings: translations, inversion and dilatation, maps Pythagorean hodograph (PH) plane curves onto the unit sphere. These stereographic image curves are spherical PH curves. The rotated curves and the geodesic offset curves of a spherical curve are also spherical PH curves. A great number of PH plane curves can be obtained as the preimage curves of such derived spherical PH curves.

§1. Introduction

Curves on the unit sphere are not only the intersections between the unit sphere and other objects and the unit tangent vectors to space curves, but also the representations of spatial motions or rotations. Thus, spherical curves are very important in computer aided geometric design, computer graphics and kinematics. They are often represented by quarternion curves [7], due to their compactness and their convenience in application to rotation.

Rational spherical curves can be constructed as images of rational curves under stereographic projection [4]. As mentioned in [6], stereographic projection preserves Pythagorean-hodograph (PH) curves [1,3] that have rational speed. So we are able to obtain different PH curves as the preimage of the rotated stereographic image of a PH curve. Furthermore, we can consider the rational offset curves of the spherical PH curves as well as PH plane curves.

This paper presents the construction of spherical PH curves and their applications with the help of quarternion calculus. First, it is shown that inversion and stereographic projection preserve PH curves in $\mathbb{R}^n$. Then the rotation of spherical PH curves, which are the stereographic image of PH plane curves, and the geodesic offset curves of spherical curves and their properties are presented.

Mathematical Methods for Curves and Surfaces II
Morten Dæhlen, Tom Lyche, and Larry L. Schumaker (eds.), pp. 485–492.

ISBN 0-8265-1315-8.

§2. Inversion Preserves Pythagorean Hodographs

The unit sphere in $\mathbb{R}^{n+1}$ is usually denoted by S^n and is called the *unit n-sphere.* A parametric curve $\mathbf{p}(t)$ in $\mathbb{R}^n$ is transformed into $\mathbf{q}(t) = \iota(\mathbf{p}(t))$ by *inversion* $\iota : \mathbb{R}^n \to \mathbb{R}^n$ in the unit $(n-1)$-sphere, namely,

$$\mathbf{p}(t) = (p_1(t), \cdots, p_i(t), \cdots, p_n(t)) \mapsto \mathbf{q}(t) = \frac{\mathbf{p}(t)}{\sum_{j=1}^n p_j(t)^2}. \tag{1}$$

Each component of the derivative $\mathbf{q}'(t)$ of the image curve $\mathbf{q}(t)$ becomes

$$q_i'(t) = \left(\frac{p_i(t)}{\sum_{j=1}^n p_j(t)^2}\right)' = \frac{p_i'(t)\left(\sum_{j=1}^n p_j(t)^2\right) - p_i(t)\left(\sum_{j=1}^n 2p_j(t)p_j'(t)\right)}{\left(\sum_{j=1}^n p_j(t)^2\right)^2}. \tag{2}$$

The parametric speed $|\mathbf{q}'(t)|$, which is the absolute value of the derivative $\mathbf{q}'(t)$, is expressed using the original curve $\mathbf{p}(t)$ and its speed $|\mathbf{p}'(t)|$ as follows,

$$|\mathbf{q}'(t)| = \sqrt{\sum_{i=1}^n \left[\left(\frac{p_i(t)}{\sum_{j=1}^n p_j(t)^2}\right)'\right]^2} = \frac{\sqrt{\sum_{i=1}^n p_i'(t)^2}}{\sum_{j=1}^n p_j(t)^2} = \frac{|\mathbf{p}'(t)|}{|\mathbf{p}(t)|^2}. \tag{3}$$

Hence, the image curve $\mathbf{q}(t)$ has rational speed, assuming that the original curve $\mathbf{p}(t)$ has rational speed. Inversion maps one polynomial or rational PH curve of degree d into another rational PH curve of degree $2d$ in $\mathbb{R}^n$.

The stereographic projection σ_+ (or σ_-) of $\{0\} \times \mathbb{R}^n$ with pole $\mathbf{N} = (0, \cdots, 0, 1)$ (or $\mathbf{S} = (0, \cdots, 0, -1)$) onto the unit n-sphere S^n is the map $\sigma_\pm$: $\mathbb{R}^n \to S^n$ defined as

$$\mathbf{r}(t) = (r_1(t), \cdots, r_i(t), \cdots, r_n(t)) \mapsto \mathbf{s}(t) = \frac{(2\mathbf{r}(t), |\mathbf{r}(t)|^2 - 1)}{|\mathbf{r}(t)|^2 \pm 1}. \tag{4}$$

The inverse map $\sigma_+^{-1} : S^n \backslash \{\mathbf{N}\} \to \mathbb{R}^n$ (or $\sigma_-^{-1} : S^n \backslash \{\mathbf{S}\} \to \mathbb{R}^n$) is given by

$$\mathbf{s}(t) = (s_1(t), \cdots, s_i(t), \cdots, s_{n+1}(t)) \mapsto \mathbf{r}(t) = \frac{(r_1(t), \cdots, r_i(t), \cdots, r_n(t))}{1 \mp r_{n+1}(t)}. \tag{5}$$

Since the stereographic projection is expressed using the inversion ι as

$$(s_1(t), \cdots, s_i(t), \cdots, s_{n+1}(t) - 1) = 2\iota((\mathbf{r}(t), -1)), \tag{6}$$

the mappings $\sigma_\pm$ and $\sigma_\pm^{-1}$ preserve Pythagorean hodographs of PH curves.

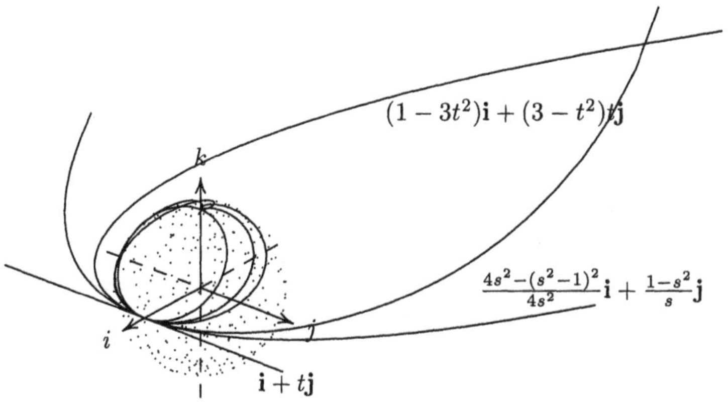

Fig. 1. Spherical PH curves obtained as the stereographic images of PH plane curves.

§3. PH Curves as Quarternion-Valued Functions

A quarternion is an extension of the complex number [5] and can be expressed as the elements of the 4-space $\mathbb{R}^4$ of the span $\{1, \mathbf{i}, \mathbf{j}, \mathbf{k}\}$, which follows the rules of production $\mathbf{i}\,\mathbf{j} = -\mathbf{j}\,\mathbf{i} = \mathbf{k}$, $\mathbf{j}\,\mathbf{k} = -\mathbf{k}\,\mathbf{j} = \mathbf{i}$, $\mathbf{k}\,\mathbf{i} = -\mathbf{i}\,\mathbf{k} = \mathbf{j}$ and $\mathbf{i}^2 = \mathbf{j}^2 = \mathbf{k}^2 = -1$. The conjugate $\overline{\mathbf{q}}$, norm $|\mathbf{q}|$ and inverse $\mathbf{q}^{-1}$ of a quarternion $\mathbf{q} = q + q_x\,\mathbf{i} + q_y\,\mathbf{j} + q_z\,\mathbf{k}$ are defined as

$$\overline{\mathbf{q}} = q - q_x\,\mathbf{i} - q_y\,\mathbf{j} - q_z\,\mathbf{k}, \qquad |\mathbf{q}|^2 = \mathbf{q}\overline{\mathbf{q}} = q^2 + q_x^2 + q_y^2 + q_z^2, \qquad \mathbf{q}^{-1} = \frac{\overline{\mathbf{q}}}{|\mathbf{q}|^2}. \tag{7}$$

A curve $\mathbf{p}(t)$ in $\mathbb{R}^4$ is expressed as a quarternion-valued function with a real parameter t as follows:

$$\mathbf{p}(t) = p(t) + p_x(t)\,\mathbf{i} + p_y(t)\,\mathbf{j} + p_z(t)\,\mathbf{k}. \tag{8}$$

The inverted curve $\mathbf{q}(t)$ is given by

$$\mathbf{q}(t) = \iota(\mathbf{p}(t)) = \frac{\mathbf{p}(t)}{|\mathbf{p}(t)|^2} = \overline{\mathbf{p}}^{-1}(t). \tag{9}$$

A curve $\mathbf{r}(t) = u(t)\,\mathbf{i} + v(t)\,\mathbf{j}$ in $\mathbb{R}^2$ is transformed to a spherical curve $\mathbf{s}(t)$ in S^2 under the stereographic projection σ_+, that is,

$$\mathbf{s}(t) = \sigma_+(\mathbf{r}(t)) = \frac{2u(t)\,\mathbf{i} + 2v(t)\,\mathbf{j} + (u(t)^2 + v(t)^2 - 1)\,\mathbf{k}}{u(t)^2 + v(t)^2 + 1}. \tag{10}$$

The speed of the stereographic image curve $\mathbf{s}(t)$ of the plane curve $\mathbf{r}(t)$ becomes

$$|\mathbf{s}'(t)| = \sqrt{\frac{4\,(u'(t)^2 + v'(t)^2)}{(u^2(t) + v^2(t) + 1)^2}} = 2\frac{|\mathbf{r}'(t)|}{|\mathbf{r}(t)|^2 + 1}. \tag{11}$$

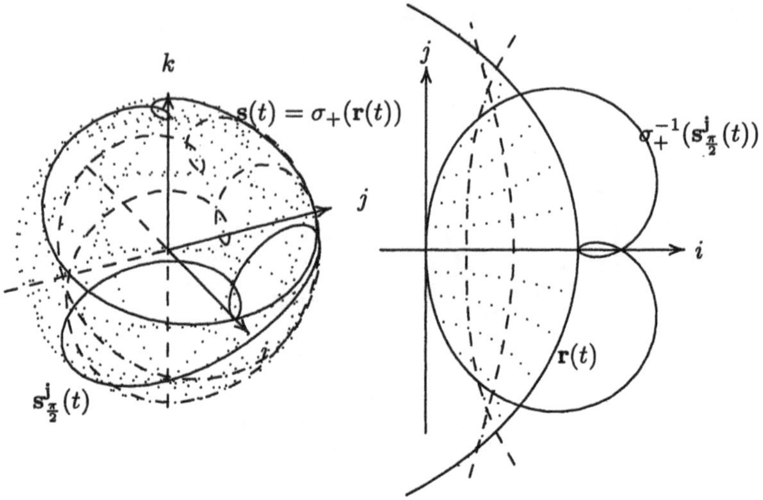

Fig. 2. Rotated stereographic images of a PH cubic and their preimage curves.

Figure 1 illustrates the spherical PH curves of three PH plane curves: a straight line, a parabola, which is parameterized rationally, and a Tschirnhausen cubic.

Conversely, the preimage curve $\mathbf{r}(t)$ of a spherical curve

$$\mathbf{s}(t) = x(t)\,\mathbf{i} + y(t)\,\mathbf{j} + z(t)\,\mathbf{k}, \qquad (x(t)^2 + y(t)^2 + z(t)^2 = 1) \tag{12}$$

in $\mathbb{R}^3$ is expressed as

$$\mathbf{r}(t) = \sigma_+^{-1}(\mathbf{s}(t)) = \frac{x(t)}{1 - z(t)}\,\mathbf{i} + \frac{y(t)}{1 - z(t)}\,\mathbf{j}, \tag{13}$$

and the speed $|\mathbf{r}'(t)|$ is given by

$$|\mathbf{r}'(t)| = \sqrt{\frac{x'(t)^2 + y'(t)^2 + z'(t)^2}{(1 - z(t))^2}} = \frac{|\mathbf{s}'(t)|}{|1 - z(t)|}. \tag{14}$$

The relationship between $\mathbf{r}'(t)$ and $\mathbf{s}'(t)$ enables the Hermite interpolation on the 2-sphere with stereographic image curves of Hermite interpolation with a PH plane curve [2].

§4. Rotation of Spherical PH Curves

Usually, pure vector quarternions $\mathbf{q} = q_x\,\mathbf{i} + q_y\,\mathbf{j} + q_z\,\mathbf{k}$ are used to express 3D space $\mathbb{R}^3$. The inner product and outer products in vector calculus can be

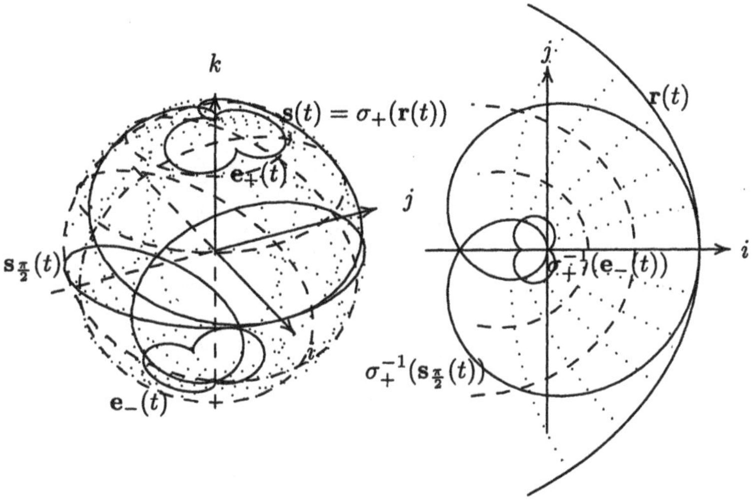

Fig. 3. Geodesic offsets of the stereographic image of a PH cubic and their preimages.

expressed using pure vector quarternions. The inner product $\mathbf{p} \cdot \mathbf{q}$ and outer product $\mathbf{p} \times \mathbf{q}$ of two pure vector quarternion $\mathbf{p}$ and $\mathbf{q}$ can be defined as

$$\mathbf{p} \cdot \mathbf{q} \equiv -\frac{\mathbf{pq} + \mathbf{qp}}{2}, \qquad \mathbf{p} \times \mathbf{q} \equiv \frac{\mathbf{pq} - \mathbf{qp}}{2}. \tag{15}$$

Hence, the product $\mathbf{pq}$ of two pure vector quarternions is equal to $-\mathbf{p}\cdot\mathbf{q}+\mathbf{p}\times\mathbf{q}$.

It is well known that spatial rotation is expressed compactly by using quarternion calculus. Let $\mathbf{u}$ be a unit pure vector quarternion,

$$\mathbf{u} = \alpha\,\mathbf{i} + \beta\,\mathbf{j} + \gamma\,\mathbf{k}, \qquad (\alpha^2 + \beta^2 + \gamma^2 = 1). \tag{16}$$

The expression $(\cos\varphi + \sin\varphi\,\mathbf{u})\,\mathbf{q}\,(\cos\varphi + \sin\varphi\,\mathbf{u})^{-1}$ is the result of the rotation of the pure vector quarternion $\mathbf{q}$ about the pure vector quarternion $\mathbf{u}$ through 2φ [5].

Therefore the rotated curve $\mathbf{s}_\theta^{\mathbf{u}}(t)$ of the stereographic image curve $\mathbf{s}(t) = \sigma_+(\mathbf{r}(t))$ about the axis $\mathbf{u}$ through θ becomes

$$\mathbf{s}_\theta^{\mathbf{u}}(t) = \left(\cos\frac{\theta}{2} + \sin\frac{\theta}{2}\mathbf{u}\right) \mathbf{s}(t) \left(\cos\frac{\theta}{2} - \sin\frac{\theta}{2}\mathbf{u}\right). \tag{17}$$

Varying angle θ, we can derive various PH plane curves $\sigma_+^{-1}(\mathbf{s}_\theta^{\mathbf{u}}(t))$ from the original PH curve $\mathbf{r}(t)$. As a rotation of the unit sphere stereographically

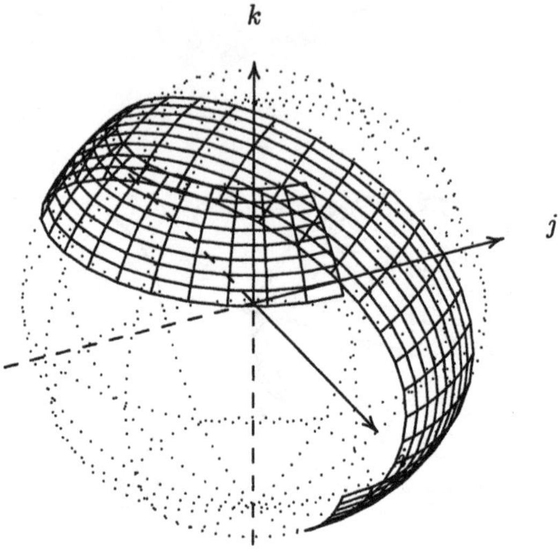

Fig. 4. A spherical zone surface.

induces an elliptic Möbius transformation of the plane, the derived curves are also obtained by Möbius transformations [8]. Figure 2 illustrates the derivation of PH plane curves using the rotation.

The spherical PH curves $\mathbf{s}(t)$ in S^2 is also a spherical curve in S^3, because $\mathbf{S}(t) = \sigma_+(\mathbf{s}(t)) = \sigma_-(\mathbf{s}(t)) = \mathbf{s}(t)$. Let $\mathbf{U}$ be a unit quarternion,

$$\mathbf{U} = \sqrt{1-\alpha^2-\beta^2-\gamma^2} + \alpha\,\mathbf{i} + \beta\,\mathbf{j} + \gamma\,\mathbf{k}, \qquad (\alpha^2+\beta^2+\gamma^2 \le 1). \tag{18}$$

Since both the quarternion products $\mathbf{U}\,\mathbf{S}(t)$ and $\mathbf{S}(t)\,\mathbf{U}$ express rotations in $\mathbb{R}^4$ [9], $\mathbf{U}\,\mathbf{S}(t)$ and $\mathbf{S}(t)\,\mathbf{U}$ are spherical PH curves in S^3 and the speeds $|(\mathbf{U}\,\mathbf{S}(t))'|$ and $|(\mathbf{S}(t)\,\mathbf{U})'|$ are equal to $|\mathbf{S}'(t)|$. Hence, the preimage curves $\sigma_\pm^{-1}(\mathbf{U}\,\mathbf{s}(t))$ and $\sigma_\pm^{-1}(\mathbf{s}(t)\,\mathbf{U})$ are also PH curves in $\mathbb{R}^3$. An example of the rotation of spherical curves in S^3 will be shown in the next section.

§5. Offset Curves of Spherical PH Curves

The offset curve on the unit 2-sphere is the geodesic offset curve. As a geodesic offset curve is the set of points at a fixed distance d along the geodesic perpendicular to a curve, the offset point is a rotation of a curve point about the great circle perpendicular to the tangent to the spherical curve. Therefore, the offset curve at a geodesic distance d of the spherical PH curve $\mathbf{s}(t)$ can be obtained by replacing the rotation axis $\mathbf{u}$ in (17) by the unit tangent $\mathbf{s}'(t)/|\mathbf{s}'(t)|$. Then the geodesic offset curve $\mathbf{s}_d(t)$ becomes

$$\mathbf{s}_d(t) = \left(\cos\frac{d}{2} + \sin\frac{d}{2}\frac{\mathbf{s}'(t)}{|\mathbf{s}'(t)|}\right)\mathbf{s}(t)\left(\cos\frac{d}{2} - \sin\frac{d}{2}\frac{\mathbf{s}'(t)}{|\mathbf{s}'(t)|}\right), \tag{19}$$

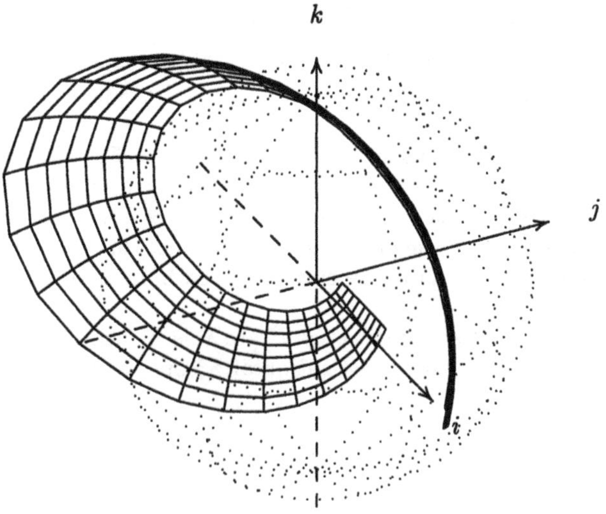

Fig. 5. Preimage surface of the rotated stereographic image of a spherical zone surface.

and is reduced to

$$\mathbf{s}_d(t) = \cos d\, \mathbf{s}(t) + \sin d \left(\frac{\mathbf{s}'(t)}{|\mathbf{s}'(t)|} \times \mathbf{s}(t) \right). \tag{20}$$

This shows that a spherical PH curve also has rational geodesic offset curves.

An evolute is the locus of the centers of the osculating circles. The stereographic image of the osculating circles of a plane curve is also the osculating circles of the stereographic image curve, because stereographic projection is conformal. There are two antipodal evolute curves $\mathbf{e}_+(t)$ and $\mathbf{e}_-(t)$,

$$\mathbf{e}_\pm(t) = \pm \frac{\mathbf{s}'(t) \times \mathbf{s}''(t)}{|\mathbf{s}'(t) \times \mathbf{s}''(t)|} \tag{21}$$

for a spherical curve $\mathbf{s}(t)$ as the locus of the centers of the osculating circles. While the evolutes of PH plane curves are also PH curves, the evolutes of spherical PH curves are not rational in general.

Figure 3 illustrates the geodesic offset curve of the stereographic image of a PH plane curve and their preimage curves with the evolute curves.

We can construct spherical zone surfaces $\mathbf{z}_d(s,t)$ by parametrizing the trigonometric functions in the geodesic offset curve of (19) as

$$\mathbf{z}_{\frac{\pi}{2}}(s,t) = \frac{1-s^2}{1+s^2}\, \mathbf{s}(t) + \frac{2s}{1+s^2} \left(\frac{\mathbf{s}'(t)}{|\mathbf{s}'(t)|} \times \mathbf{s}(t) \right). \tag{22}$$

Figure 4 illustrates a subdivided surface patch of the spherical zone surface $\mathbf{z}_{\frac{\pi}{2}}(s,t)$, and Figure 5 illustrates the preimage surface of the surface patch rotated in $\mathbb{R}^4$, i.e., $\sigma_+^{-1}(\mathbf{U}\, \mathbf{z}_{\frac{\pi}{2}}(s,t))$. While the preimage surface is also a spherical surface, the isoparametric curves are no longer geodesic offset curves.

§6. Conclusion

In the space $\mathbb{R}^n$, a PH curve derives many PH curves by applying transformations which preserve the Pythagorean hodographs. These transformations include such mappings as dilatation, rotation, offset, and inversion.

Stereographic projections transform PH curves in $\mathbb{R}^n$ into spherical PH curves in $\mathbb{R}^{n+1}$, and vice versa. By using the compositions of the transformations and projections, we can obtain many PH curves from one PH curve. As proof of this, almost all of the curves in the figures of this paper were derived from a PH cubic, which is an image curve mapped from an obvious PH curve, the straight line.

It was also shown that rotations in S^3 are able to deform not only curves but also surfaces in $\mathbb{R}^3$. The investigation of S^3 operations is the subject of further research.

References

1. Farouki, R. T. and T. Sakkalis, Pythagorean hodographs, IBM J. Res. Develop. **34** (1990), 736–752.
2. Farouki R. T., and C. A. Neff, Hermite interpolation by Pythagorean-hodograph quintics, Math. Comp. **64** (1995), 1589–1609.
3. Farouki, R. T. and H. Pottmann, Polynomial and rational Pythagorean-hodograph curves reconciled, in *The Mathematics of Surfaces VI*, G. Mullineux (ed.) 1996, 355–378.
4. Hoschek, J., Bézier curves and surface patches on quadrics, in *Mathematical Methods in Computer Aided Geometric Design II*, T. Lyche and L. L. Schumaker (eds.), Academic Press 1991, 331–342.
5. Kantor, I. L. and A. S. Solodovnikov, *Hypercomplex Numbers: An Elementary Introduction to Algebras*, Springer, Heidelberg, 1989.
6. Peternell, M. and H. Pottmann, Applications of Laguerre geometry in CAGD II, Technical Report Nr. 29, Institut für Geometrie, TU Wien 1996.
7. Shoemake, K., Animating rotation with quarternion curves, Computer Graphics (SIGGRAPH '85 Proceedings) **19** (1985), 245–254.
8. Ueda, K., Deformation of plane curves preserving Pythagorean hodographs, in *Proc. Information Visualization IV'97*, IEEE Computer Society Press, 1997, 71–76.
9. Ward, J. P., *Quarternions and Cayley Numbers: Algebra and Applications*, Kluwer Academic Publishers, 1997.

Kenji Ueda
Ricoh Company, Ltd.
1-1-17, Koishikawa, Bunkyo-ku, Tokyo 112-0002, JAPAN
ueda@src.ricoh.co.jp

On Interactive Design Using The PDE Method

Hassan Ugail, Malcolm I. G. Bloor, and Michael J. Wilson

Abstract. Interactive design of practical surfaces using a technique for computer-aided design known as the partial differential equation(PDE) method is considered. Due to the boundary-value nature of the method, the boundary conditions imposed around the edges of the surface control the internal shape of the surface. Closed form analytic solutions of the PDE are obtained which facilitate interactive manipulation of the surfaces in real time.

§1. Introduction

In CAD it is usual to generate surfaces using mathematical representations based upon polynomial functions of two parameters. These include Bezier surfaces [2,11,13], B-splines [11,13,17] and rational B-splines [14]. Such surfaces can be defined using an array of points [15], a set of plane cross-sectional curves [16], or a network of curves [1]. In any case, due to the large number of control points often required to define such surfaces, it can be a difficult task to manipulate such surfaces in order to create a desired shape. This especially is the case if the surface need to be defined and manipulated in an interactive environment. In this paper we consider an alternative approach to surface design, based upon the solutions of partial differential equations known as the PDE Method.

This paper discusses the techniques that allow the scope of the PDE method [3,4] to be extended for interactive design [6]. Since the method treats surface design as a boundary-value problem, this ensures surfaces can be defined using a small set of design parameters. Moreover, the surfaces created in this manner tend to be smooth and fair.

We show how a user, who may not necessarily be familiar with the mathematics of partial differential equations and the influence of the boundary conditions upon their solutions, is able to construct complex geometries using the usual computer devices. It is worth mentioning that the examples presented in various parts of the paper are created interactively in real time.

Mathematical Methods for Curves and Surfaces II
Morten Dæhlen, Tom Lyche, and Larry L. Schumaker (eds.), pp. 493–500.

ISBN 0-8265-1315-8.

§2. The PDE Method

The PDE method regards the generation of the parametric surface patch

$$\underline{X}(u,v) = (x(u,v), y(u,v), z(u,v)) \tag{1}$$

as the solution to a set of partial differential equations for each of the Cartesian coordinates x, y and z. The coordinates are regarded as a mapping from a region of R^2 in a domain Ω in (u,v) parameter space, onto the surface $\underline{X}(u,v)$ in a Cartesian coordinate system (E^3). The parameters u and v define a curvilinear coordinate system on the surface $\underline{X}(u,v)$. The surface patch $\underline{X}(u,v)$ is obtained by posing suitable boundary conditions along the boundary $\partial\Omega$ of the domain Ω. We assume $\underline{X}(u,v)$ to be the solution of a partial differential equation of the general form

$$D^m_{u,v}(\underline{X}) = \underline{f}(u,v), \tag{2}$$

where $D^m_{u,v}$ is a partial differential operator of order m in independent variables u and v. The choice of $D^m_{u,v}$ is taken to be elliptic. The PDE chosen for the work described in this paper is

$$\left(\frac{\partial^2}{\partial u^2} + a^2\frac{\partial^2}{\partial v^2}\right)^2 \underline{X}(u,v) = 0. \tag{3}$$

Eq.(3) is solved over the finite region Ω of the (u,v) parameter space, subject to the boundary conditions on the solution which relate how $\underline{X}(u,v)$ and its normal derivatives $\frac{\partial \underline{X}}{\partial \underline{n}}$ vary along $\partial\Omega$. With periodic boundary conditions, v being the periodic parameter, the analytic solution of (3) can be written as

$$\underline{X}(u,v) = \underline{A}_0(u) + \sum_{n=1}^{\infty}[\underline{A}_n(u)\cos(nv) + \underline{B}_n(u)\sin(nv)] \tag{4}$$

with

$$\underline{A}_0 = \underline{a}_{00} + \underline{a}_{01}u + \underline{a}_{02}u^2 + \underline{a}_{03}u^3 \tag{5}$$

$$\underline{A}_n = \underline{a}_{n1}e^{anu} + \underline{a}_{n2}e^{anu} + \underline{a}_{n3}e^{-anu} + \underline{a}_{n4}e^{-anu} \tag{6}$$

$$\underline{B}_n = \underline{b}_{n1}e^{anu} + \underline{b}_{n2}e^{anu} + \underline{b}_{n3}e^{-anu} + \underline{b}_{n4}e^{-anu}, \tag{7}$$

where $\underline{a}_{n1}$, $\underline{a}_{n2}$, $\underline{a}_{n3}$, $\underline{a}_{n4}$, $\underline{b}_{n1}$ $\underline{b}_{n2}$, $\underline{a}_{n3}$, $\underline{a}_{n4}$ are vector-valued constants, determined by the imposed boundary conditions at $u = 0$ and $u = 1$.

The boundary conditions imposed on the solution are of the form

$$\underline{X}(0,v) = \underline{P}_0(v) \tag{8}$$

$$\underline{X}(1,v) = \underline{P}_1(v) \tag{9}$$

$$\underline{X}_u(0,v) = \underline{d}_0(v) \tag{10}$$

$$\underline{X}_u(1,v) = \underline{d}_1(v). \tag{11}$$

§3. Interactive Design of a Simple PDE Surface

Previous work using PDE method has shown that surfaces with complex geometries such as ship hulls [12], marine propellors [10] and aircraft [7] can usually be generated from a small number of freeform surfaces. The aim of this paper is to show how a user is able to use the method to create such geometries in an interactive environment. Here we show how a user can impose the necessary boundary conditions, in the form of space curves, in order to construct a PDE surface.

For many surfaces of practical significance the boundary conditions can be expressed by a finite number analytic functions. However, in the case of interactive design, it is often more convenient to define the boundary curves in terms of periodic cubic B-splines by input from the mouse. A discrete Fourier analysis of the curves are then performed in order to obtain coefficients for the appropriate number of Fourier modes. The basic idea is then to approximate the surface solution given by (4) by an expression of the form

$$X(u,v) = \underline{A}_0(u) + \sum_{n=1}^{N}[\underline{A}_n(u)\cos(nv) + \underline{B}_n(u)\sin(nv)] + \underline{R}(u,v) \qquad (12)$$

with a *remainder* function defined by

$$\underline{R}(u,v) = \underline{r}_1 e^{wu} + \underline{r}_2 e^{wu} + \underline{r}_3 e^{-wu} + \underline{r}_4 e^{-wu}, \qquad (13)$$

where $\underline{r}_1$, $\underline{r}_2$, $\underline{r}_3$, $\underline{r}_4$ are vector-valued constants. Here N is finite and in a typical problem it is not large at all, e.g. considering the first five modes are often more than adequate with regard to small approximation errors. More details are given in reference [9].

Taking the parameter space to be the region $(0 \le u \le 1, 0 \le v \le 2\pi)$, Fig. 1(a) shows a typical set of space curves corresponding to $\underline{P}_0(v)$ and $\underline{P}_1(v)$. Note these curves are the results of geometric transformations applied interactively to an initially defined plane curve. The initial curves are defined in terms of B-splines by input from the mouse or curves chosen from a standard library of curve shapes. At this stage of the design session, the user is free to choose the parameterisation in terms of v (which is periodic along the curve), i.e. the point on each curve at which $v = 0$.

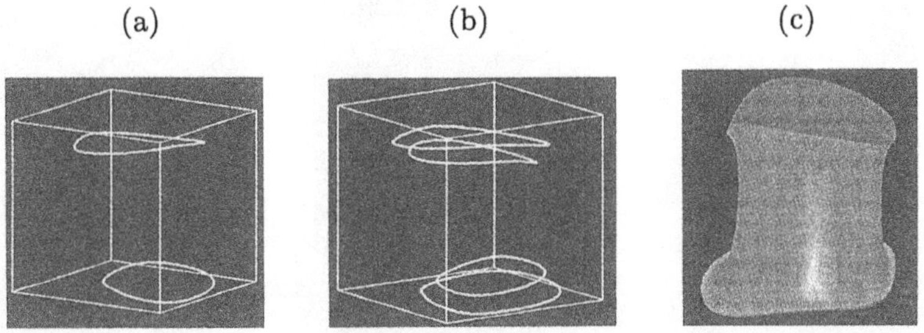

Fig. 1. Boundary curves with the coresponding PDE surface.

We now show how to define the boundary conditions $\underline{d}_0(v)$ and $\underline{d}_1(v)$ on the function $\frac{\partial X}{\partial \underline{n}}$. There are several ways this can be done. The idea here is to define a vector field, which we will refer to as the *derivative vector field*, along each of the positional boundary curves at $u = 0, 1$. One way of doing this is to define a curve in E^3 near each boundary $u = 0, 1$. This curve can be obtained by perturbating the postional boundary curve. The difference between each point on this newly defined curve and an associated point on the positional boundary curve will determine both the magnitude and direction of the derivative vector field. Another way of defining the derivative vector field is to define a vector field at each boundary $u = 0, 1$ which is normal to the positional boundary curve.

The solution to the PDE effectively determines the surface points in E^3 in terms of u, v. It is an easy procedure to calculate normals to these surface points. Fig. 1(c) shows the resulting surface produced on a Silicon Graphics workstation for the boundary conditions shown in Fig. 1(b).

§4. Interactive Manipulation of the Surface

It is often unlikely that a designer will be able to create exactly the required surface shape with the initial definition of boundary conditions. This would especially be true if the design process is carried out from *scratch*. Thus, it is desirable for the user to have control over the shape of the surface, once it has been defined.

The surface solution obtained from (3) or (12) has a closed form, making its calculation and recalculation computationally very efficient and is in response to the user manipulation. Here we describe how a surface can be manipulated in order to construct the final shape of the surface.

A design parameter that influences the shape of the surface is the smoothing parameter a. Fig. 2 shows the sequence of surfaces resulting upon the interactive change of the size of this parameter. As discussed in earlier work [3], this parameter controls the length-scale over which the boundary conditions influence the interior of the surface. The higher the value of a the more *waist* the surface aquires.

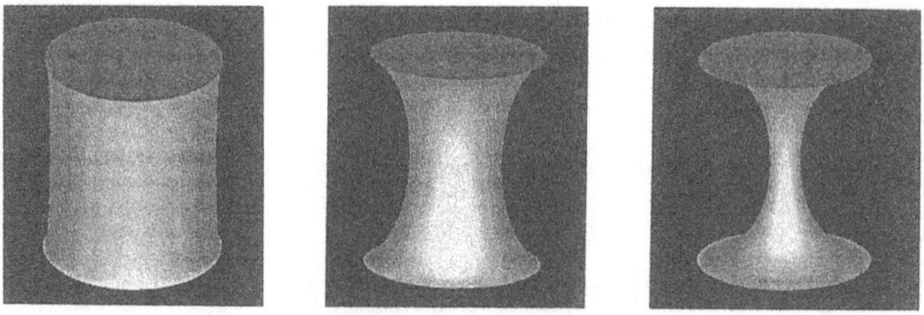

Fig. 2. The effect of smoothing parameter a.

The points on the boundary curves where $v = 0$ is rather important in defining the overall shape of the surface. Fig. 3(b) shows the result of changing

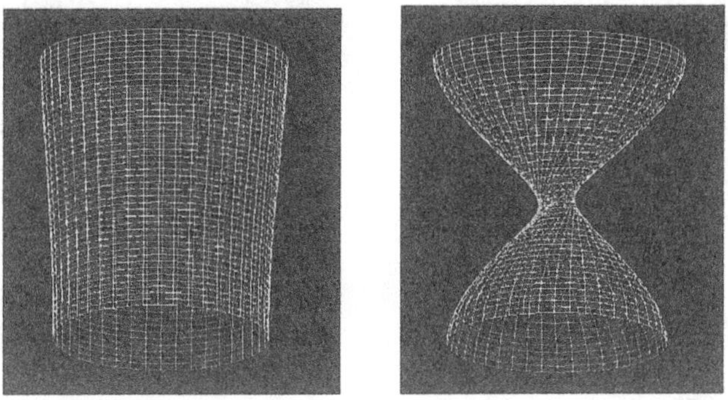

Fig. 3. The effect of changing v parameterisation.

the position of this point at $u = 1$ on the boundary curves corresponding to the surface shown in Fig. 3(a).

The size and the direction of the derivative vectors can also be changed interactively to vary the shape of the surface. Since we are solving a fourth order PDE, we only specify the first order derivatives. However, if more control over the shape surface is required a higher order PDE can be solved, thus providing more control over the surface shape via the derivative conditions.

(a) (b)

§5. Removing Sections of PDE Surfaces

For complex surface constructions it may be required to join multiple patches together with various degrees of continuty. Thus, a given portion of a surface may need to be removed and a new surface blended into the hole. Here we describe a technique by which we show how sections of the surface of desired shapes can be removed.

Since we have points in E^3 of the surface parameterised by u and v, it is a simple procedure to determine the exact position of the surface where the portion needs to be removed. This is done by moving the cursor on the shaded image of the surface. Considering the (u, v) parameter space, a plane curve defined in this space will be guaranteed to lie on the surface. The curve is then manipulated to obtain the desired shape in E^3. Fig. 4 shows some examples where parts have been removed from original PDE surfaces.

Once again the calculation and recalculation of the solution to (3) or (12) is in response to user manipulation, so that the desired shape of the section can be obtained by varying the shape of the curve in the parameter space interactively.

§6. Interactive Design of Practical Surfaces

In this section we show how the techniques described above can be applied to objects such as a tea cup. Such an object can be considered as a collection

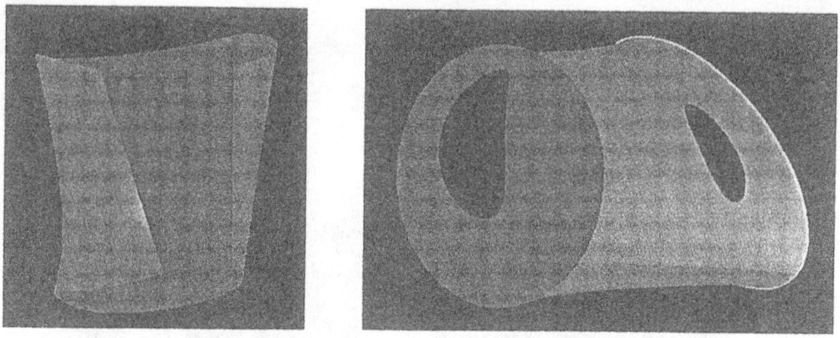

Fig. 4. Examples of sections removed from PDE surfaces.

of several individual surface patches. In the case of a tea cup, the bowl and the handle can be considered as two separate surfaces patches. The positional boundary conditions for the bowl of the tea cup can be defined by two plane curves separated by a distance s. Choosing the appropriate derivative vector fields (which merely depend on the desired shape of the cup), the bowl of the tea cup can be created as a freeform surface patch, say S_1. Two holes in S_1 can then be cut at the positions where the handle meets the bowl of the tea cup. Since the edges of the holes cut are effectively a set of curves in E^3, these, together with the appropriately chosen derivative vector fields will provide the boundary conditions for a new surface patch which will serve as the handle of the tea cup. Fig. 5 shows the final shape of the tea cup designed interactively in a single design session.

Fig. 5. A tea cup designed interactively.

For many practical purposes where the physical properties of an object are crucial from a functional point of view, the geometry of the surface is of paramount importance. For instance, the geometry of an inlet port [8] of an diesel internal combustion engine will influence its engine performance. Hence it may be required to have a technique enabling this geometry to be changed interactively. Fig. 6 shows one such geometry of an inlet port created interactively using two PDE surface patches.

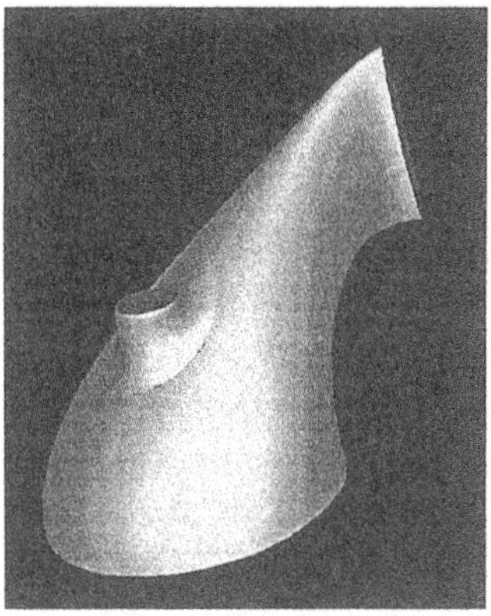

Fig. 6. PDE inlet port geometry created interactively.

Since the PDE surfaces described in this paper are represented using a Fourier based system, data transfer to other CAD systems may be neccesary. More details regarding this issue can be found in [5].

§7. Conclusions

We have described techniques whereby the interactive design using the PDE method is a practical proposition. The solution of the PDE chosen is expressed in terms of a finite number of analytic functions by Fourier analysis of the boundary conditions. Using this technique, we obtain solutions for the PDE surfaces rapidly enough to be able to design interactively in real time. In fact, on Silicon Graphics workstations the surfaces change at the same rate as any alterations in the design parameters effected via the user interface, i.e at moving-picture speed.

The existence of low level parameterisation which arise from the very nature of the PDE method is an advantage so that the user is able to define and manipulate surfaces intuitively. However, more work needs to be done in this area in order to develop techniques for the efficient parameterisation of the surfaces. Consequently the freedom of local modification of surfaces may be introduced so that the user will have even more intuitive control over the shape of the surface in an interactive environment.

Acknowledgments. The authors would like to acknowledge financial assistance for H. Ugail from the UK government's Overseas Research Students (ORS) awards scheme and the award of a University of Leeds Research Scholarship.

References

1. Barnhill, R. E., Surfaces in computer aided geometric design: a survey with new results, Comput. Aided Geom. Design **2** (1985), 1–17.

2. Bezier, P., *The Mathematical Basis of UNISURF CAD System*, Butterworths, London, 1986.
3. Bloor, M. I. G., and M. J. Wilson, Generating blend surfaces using partial differential equations, Computer-Aided Design **21** (1989), 165–171.
4. Bloor, M. I. G., and M. J. Wilson, Using partial differential equations to generate freeform surfaces, Computer-Aided Design **22** (1990), 202–212.
5. Bloor, M. I. G., and M. J. Wilson, Representing PDE surfaces in terms of B-splines, Computer-Aided Design **22** (1990), 324–331.
6. Bloor, M. I. G., and M. J. Wilson, Interactive design using partial diffrential equations, in *Designing Fair Curves and Surfaces*, SIAM, Philadelphia, 1994, 231–251.
7. Bloor, M. I. G., and M. J. Wilson, Efficient parameterization of aircraft geometry, Journal of Aircraft **32** (1995), 1269–1275.
8. Bloor, M. I. G., and M. J. Wilson, Complex PDE surface generation for analysis and manufacture, Computing **10** (1995), 61–77.
9. Bloor, M. I. G., and M. J. Wilson, Spectral approximations to PDE surfaces,Computer-Aided Design **28** (1996), 145–152.
10. Dekanski, C. W., M. I. G. Bloor, and M. J. Wilson, The generation of propellor blades using the PDE method, Journal of ship research **39** (1995), 108–116.
11. Faux, I. D., and M. J. Pratt, *Computational Geometry for Design and Manufacture*, Ellis Horwood, Chichester, UK, 1979.
12. Lowe, T. W., M. I. G. Bloor, and M. J. Wilson, The automatic design of hull surface geometries, Journal of ship research **38** (1994), 319–328.
13. Mortenson, M. E., *Geometric Modelling*, Wiley-Interscience, New York, 1985.
14. Pigel, L., and W. Tiller, Curve and surface constructions using rational B-splines, Computer-Aided Design **19** (1987), 485–498.
15. Rooney, J., and P. Steadman, *Principles of Computer-Aided Design*, Pitman Publishing, London, 1987.
16. Tiller, W., Rational B-splines for Curve and Surface Representation, IEEE Comput. Graphics Applications **3** (1983), 61–69.
17. Woodward, C. D, Skinning Tecniques for Interactive B-spline Surface Interpolation, Computer-Aided Design **20** (1988), 441–451.

Hassan Ugail, Malcolm I. G. Bloor, and Michael J. Wilson
Department of Applied Mathematical Studies,
University of Leeds,
Leeds LS2 9JT, UK
`hassan@amsta.leeds.ac.uk`
`http://www.amsta.leeds.ac.uk/~hassan`

Vertex Blending: Problems and Solutions

Tamás Várady and Chris M. Hoffmann

Abstract. Blending operations significantly restructure the topology of B-rep solid models. The blending process can be classified, by complexity, into three categories: regular, overlap and global blending. We illustrate why already regular blending, the least difficult category, poses significant problems for solid modeling. The difficulties include blend terminations, shape imperfections, connecting trimlines, sequential problems and incomplete blends. The analysis of these examples suggests a blending approach based on a set of simple and intuitive blending rules. We describe an algorithm based on those rules that resolves the known difficult cases. The key aspect of the approach is a vertex blending concept that emphasizes the application of setbacks and so achieves uniformity formerly not possible. We illustrate how the approach overcomes the previous problem cases. Finally, representational issues are revisited and two methods are described that implement setback vertex blends.

§1. Introduction

Blends are smooth transition surfaces between faces of a geometric model which are generated to replace sharp edges and vertices. The significance of blending does not need to be elaborated on; see, e.g., the review papers by Woodwark [18] and Vida et al. [17].

Despite significant and sustained research efforts and despite numerous research and commercial implementations, automated blending remains a research issue. Practicing users of commercial CAD/CAM systems often report failures when performing "irregular" or "too complex" blending operations — either the shape is not pleasing or the system simply rejects the operation for the given blending parameters. Mysterious error messages may be received, and it is difficult to find out why the operation cannot be accomplished — whether it is due to wrong data or whether the implementation is incomplete.

In this paper, we address the problem of blending boundary representation (B-rep) solid models algorithmically, based on a few, intuitive shape parameters such as ranges or radii. We would like to set these parameters and attach them to edges to be blended (and to vertices in exceptional cases). We

Mathematical Methods for Curves and Surfaces II

Morten Dæhlen, Tom Lyche, and Larry L. Schumaker (eds.), pp. 501–527.

ISBN 0-8265-1315-8.

would like to compute a new, evaluated B-rep model that incorporates the specified blends. Blending CSG models in this sense seems almost impossible, since typically an evaluated model is required for attaching blending parameters to edges. Blending E-rep models [6] should be possible once a complete set of blending parameters has been formalized, but we do not address it here.

The difficulties realizing blending operations algorithmically originate from three strongly interrelated issues:

(i) *User interface:* The rules of the blending operations are often not clearly defined. Without them design may become a cumbersome trial and error process. While it may seem straightforward to generate a blending surface between two simple surfaces, the situation is far from trivial if we have a collection of bounded faces around a common vertex and there are various blending parameters attached to the incident edges. Such complex cases will be analyzed later in detail, from a topological point of view. A clear understanding of the blending rules and an easy specification of the entire blending process are very important for users and system developers alike.

(ii) *Topological restructuring:* It must be clearly understood how the original topological structure will be converted into a new structure by the blending. Clearly, this strongly depends on the above rules. There are many special cases that ideally should be described by a uniform blending paradigm. To start an analysis from a topological point of view, we distinguish three basic schemata of deriving a new topological structure in an increasing order of complexity; see Section 2. The first is regular blending where edge blends are local and interfere with each other only at the vertices. The second is overlap blending, for instance when enlarging edge blends, but it is possible to resolve the interference of conflicting blends within the faces involved. Finally, global blending can create transition surfaces between faces that are not necessarily adjacent and distant elements can be smoothly connected as well.

(iii) *Geometric representation of blending surfaces:* In spite of the many publications on blending (see reviews in [18, 17]), there remain open problems concerning their geometric representation. The majority of publications deal with various blends generated between two surfaces. There are publications on various n-sided patches; e.g., [13, 5, 7], and there are polyhedral smoothing approaches in which all surfaces including the vertex blends are automatically created; e.g., [9, 8, 10]. These approaches provide partially the tools, but do not address the more global aspects of vertex blending in solid modeling. Only a few papers investigate the tight relationship between the topology and geometry of vertex blends (see the early paper by Várady et al. [14] and a recent publication by Braid [3]). Almost no paper considers the problems of global blending.

In general, edge and vertex blends must smoothly join the faces of the object with at least G^1 continuity. The shape of the blends must be predictable and aesthetically pleasing; smooth transitions with even internal curvature distribution are required.

In this paper, the term *blending* includes rounding and filleting; that is, material would be added along a convex edge, and would be subtracted along

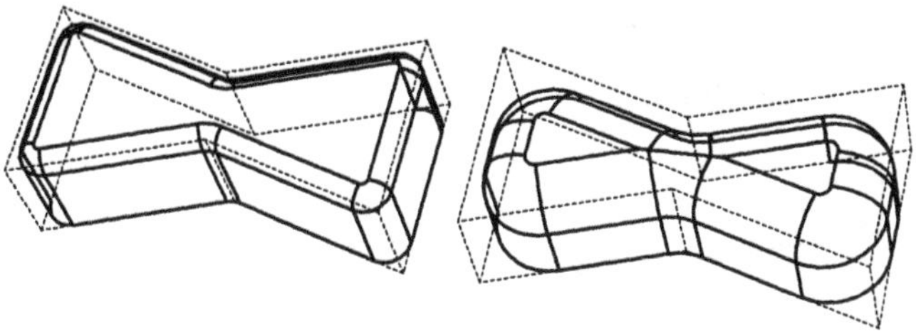

Fig. 1. a - Regular blending, b - selfintersecting edge loop.

a concave edge. Edge blends are bounded by *trimlines* and *profile curves.* Trimlines are curve segments to which the original faces to be blended must be trimmed and profile curves close up the ends of the trimlines. We do not care how the trimlines were generated, i.e., whether from constant or variable radius rolling ball blends, or from range parameters that define the width of the blend. Instead, we assume that a trimline pair is somehow generated for each face pair, and that there is a bijective correspondence between the points on the trimlines and those on the edge.

In our work, setback type vertex blends play a fundamental role. The importance of setbacks has been explained before in [3, 16] as well. Setback vertex blends are bounded by the profile curves of the edge blends and, typically, by *spring curves* that smoothly connect adjacent trimlines in the same face. The position of the profile curves, detailing where the edge blend needs to broadened or narrowed, is determined by the *setback* parameter that measures distance from a vertex of a blended edge to where the edge blend would end and the vertex blend begins.

§2. Classification of Blending Operations

In B-rep solid modelers, blends are normally attached to edges or edge sequences. The blended edges are then replaced by smooth transition surfaces. Depending on the complexity of these operations, the parameters of the blends, and the interference between them, we define three categories of blending.

In *regular blending* there is always a one-to-one map between the topological structures of the original and the blended object (disregarding degeneracies). Each edge to be blended will be replaced by an edge blend face, each vertex by a vertex blend face, as described in detail in Section 5). We assume that the trimlines of the edge blends run in the "vicinity" of the original edges without interfering with other trimlines or internal face boundaries. The edge blends can only join at the vertices. Moreover, the vertex blends are also small and do not interfere with other entities. Accordingly, the original faces involved in blending are shrunk back, but their edge loop structures are preserved - see Figures 1(a), 2(a), and 3(a). More precisely, the edge loop is replaced by a loop of alternating trimlines and spring curves, however, spring curves may degenerate to points.

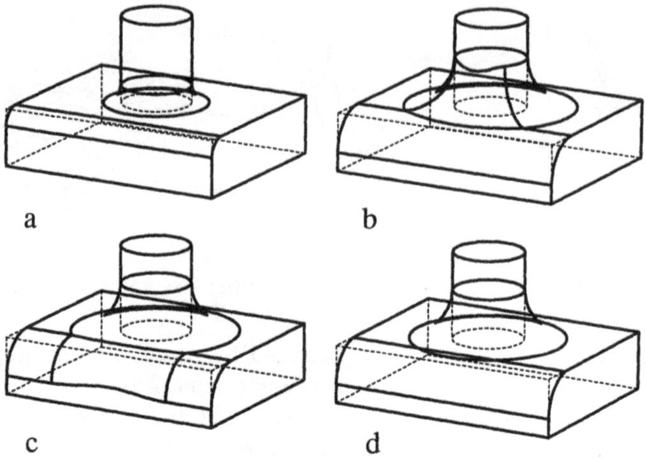

Fig. 2. a - Regular blending, b and c - interfering edge loops, d - overlap blending, original trimlines preserved.

Overlap blending preserves at least portions of each original face of the object, but in a more complex manner than before. The trimlines of the blends must remain within the face but may intersect each other. Therefore, this operation may alter the topology of the face: new loops may be created, other loops may vanish. In the example of Figure 1(b), the blend radius has been enlarged. The new blending boundary loop intersects itself in the middle of the top face splitting the original face into two components. The blends intersect in a sharp edge, clearly not desirable. To smooth the new intersection requires another blend. In the example of Figure 2(d), the blending surfaces for the straight edge and for the circular edge intersect each other only where the two trimlines intersect. In between, the toroidal blend is everywhere above the cylindrical blend, creating a gap in the solid boundary and thus to an invalid solid model.

In overlap blending there exists an inverse mapping that associates with each edge-blend face an edge of the original boundary, and associates with each new vertex-blend face a vertex of the original boundary. More complex blending regions can also also arise that are associated with more than one topological entity in the original boundary. Moreover, new loops are created changing the topology of the faces. Overlap blend surfaces require special geometric constructions, and there remain open problems constructing them algorithmically.

In certain cases it is possible two overcome these difficulties by performing "blends on blends," see for example [12, 3]. In that approach, blends must be generated one after the other sequentially. Unfortunately, new difficulties may arise: how should one choose the right ordering, will symmetry be preserved, are the original trimlines destroyed? For instance, in 2(b) there is an unwanted alteration of the trimline on the cylinder, and in 2(c), there is an unwanted alteration of the trimline on the front face. They result from blending the straight edge first and then the circular edge, in case (b), and blending the circular edge first and then the straight edge, in case (c).

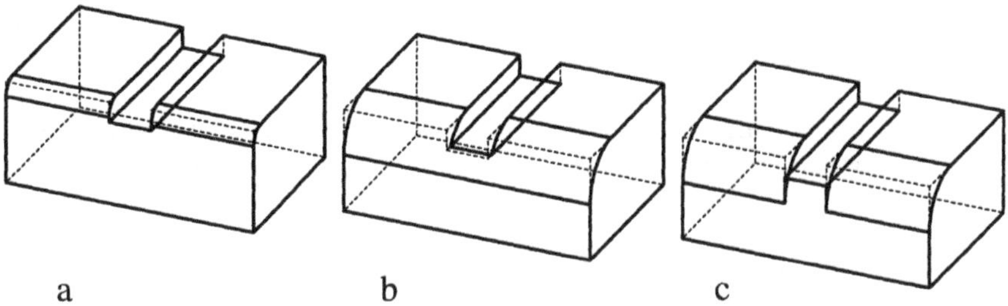

Fig. 3. Global blending and a slot: a - regular case, b and c - two possible solutions.

Global blending operations are most general and may significantly alter the original topology; vertices, edges, edge loops and entire faces may disappear. There is no complete mapping between the old and new topological entities. Semantical problems also arise; an example is shown in Figure 3. If the blending radius is small we obtain regular blends (a). However, if we have a large radius, there are two options to restructure the object. If the blending surface has higher priority than the slot, we obtain object (b), if the slot dominates, we obtain (c). This requires consideration of the sequence in which the blends are constructed, and how to mix them with other local or Boolean operations.

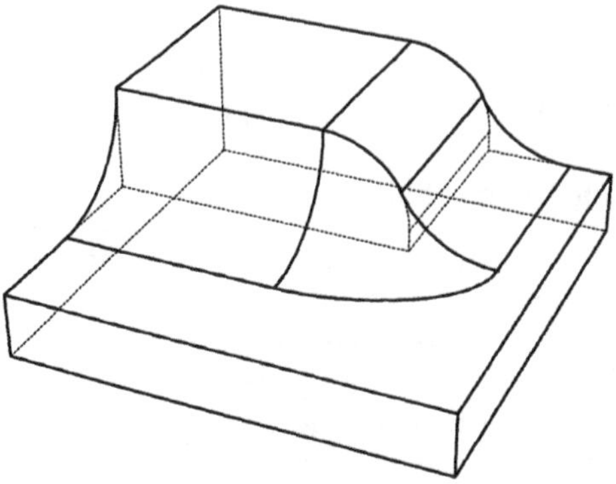

Fig. 4. Global blending - rolling a large ball.

Another example of global blending is a generalized rolling ball operation where rolling a "large" ball is allowed to create a new structure, see Figure 4. Conventional rolling ball blends always touch two adjacent faces during their motion. In our example, the ball is touching first an edge and the bottom face, then another edge and the bottom face, finally the motion ends with a regular face-face rolling. It can be seen that a smooth blend is created, but two relatively small vertical faces of the original object are deleted. Such blends raise another interesting question: while in some cases reasonable objects are obtained, in other cases these operations cannot be applied or only nonsense objects are generated. For a particular global blending solution see the Advanced Blending Husk of the ACIS modeler [1,2].

In the following sections, we restrict our investigations to regular blending. Despite an apparent conceptual simplicity, even in this case substantial difficulties must be overcome.

§3. Problems with Regular Blending

We present an anthology of problem cases for regular blending. Many of these examples cannot be properly handled by current commercial CAD/CAM systems; others can be produced with some difficulty or by applying certain tricks. Although not a complete set, our examples are paradigmatic of most situations in which blending algorithms fails. Our objective is to devise a conceptual method for regular blending that makes no restrictions on the blending configurations. In particular, it should not matter

(i) how many edges to be blended meet at a vertex,
(ii) whether we are dealing with convex, concave or smooth edges,
(iii) what equations or algorithms describe the underlying geometry of the curves and surfaces involved,
(iv) what the angles between the adjacent edges are, e.g. orthogonal, sharp, widely concave, tangential or cuspate,
(v) what blending parameters (radii, ranges) are attached to the individual edges.

We isolate five groups of difficult cases and develop a uniform concept how to handle them in the second part of the paper.

3.1. Terminating Edge Blends

Edge blends need to be terminated at a vertex at which a single blended edge ends. This problem also occurs at incomplete vertex blends when the sharpness of certain edges has to be preserved or when the magnitude of the blends is so small that they are negligible in the geometric model.

A simple idea to terminate edge blends is to intersect them, or their extensions, with the faces that are situated between the unblended edges at the given vertex. These faces are then locally shrunk or extended, and small connecting arcs inserted. The operation is illustrated in Figure 5(a), viz. the top convex edge and the right vertical face; or in (b), viz. the concave edge and the left vertical face, respectively. Unfortunately, this naive approach may often fail. Consider the other ends of these blends. In Figure 5(a), the two adjacent edges at the top are almost parallel and, as a consequence, the edge blend is terminated far away from the common vertex. Similarly, in

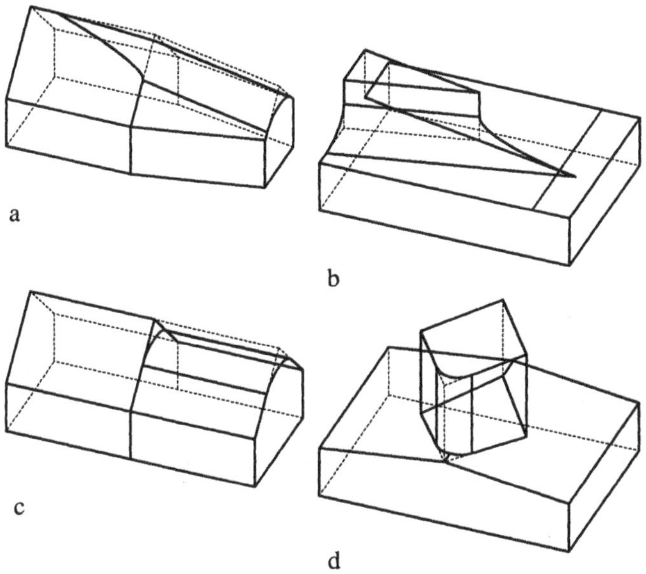

Fig. 5. Edge terminations: a - almost parallel edges, b - large concave angle, c - new facet inserted, d - new edge inserted.

Figure 5(b), the edge being blended is next to another edge having a large concave angle. The result is unacceptable in both cases, particularly when the intersection with the blending surface falls outside the object, as in Figure 5(b) at the marked edge.

Furthermore, there need not be an intersection with the adjacent faces, see Figure 4(c). Here, the blending surface of the top right edge never intersects the top left edge. A small, three-sided face must be inserted to obtain a valid solid boundary. Note that this is not a trivial issue, particularly, if the incoming edges are free-form.

It is also possible that the topology of the object is altered due to blending: For example, in Figure 5(d), we want to blend the convex, vertical edge. The blending operation now requires that two faces intersect in a new edge that did not exist previously. Creating such new edges by blending another edge is intuitively undesirable.

3.2. Shape Imperfections

Assume that all edges meeting at a vertex should be blended. We need to connect adjacent trimlines lying on the same face. A simple way is to intersect the trimlines as shown in Figure 6(a) and Figure 7(a). In both cases, the radii of the blends are more or less of the same magnitude, and smooth three- or four-sided vertex blends are generated. Unfortunately, this naive approach is inappropriate when the blend radii differ by much.

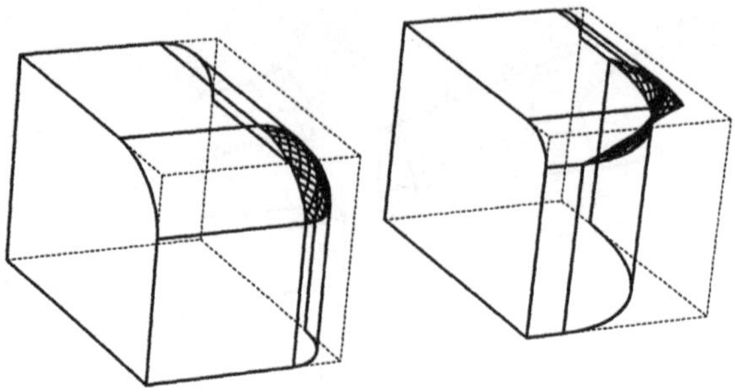

Fig. 6. Connecting trimlines - suitcase corner: a - even, b - uneven radii.

In Figure 6(b) the top two edges are blended with a small radius and the vertical edge by a large one. Notice that the intersection of the trimlines are too close to the vertex, extending well beyond the vertical cylindrical blend. As a consequence, a nasty, tongue-like shape is obtained, and is clearly not acceptable. In Figure 7(b), a similar situation leads to a self-intersection.

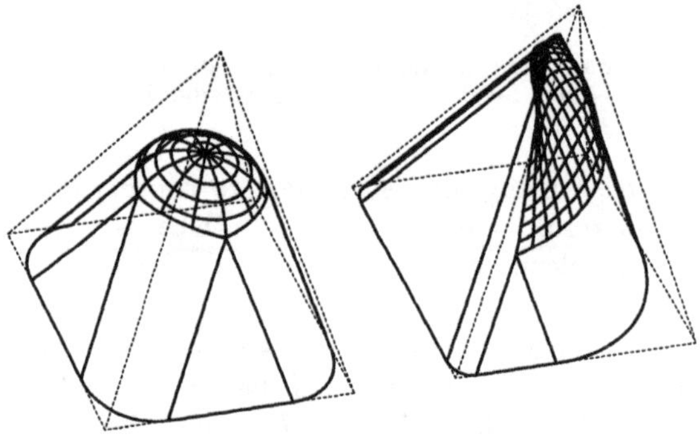

Fig. 7. Connecting trimlines - pyramid: a - even, b - uneven radii.

3.3. Connecting Trimlines

Connecting trimlines can cause problems at certain angles. Consider for example Figure 8(a). The intersection of the trimlines is too distant from the vertex and interferes with the trimline along the opposite edge. As before, the distant trimline intersection may cause not only an undesirable shape but an invalid solid boundary. The situation is exacerbated when the depth of the block is diminished.

In the example of Figure 8(b), a cylindrical and a tangential planar face meet along a smooth edge. Accordingly, we have two tangential edge sequences to blend — one consisting of two convex edges, the other of two concave edges. Assume that the assigned radii are unequal. As shown, the trimlines will not intersect in the concave case and are parallel. Here, the solid boundary must

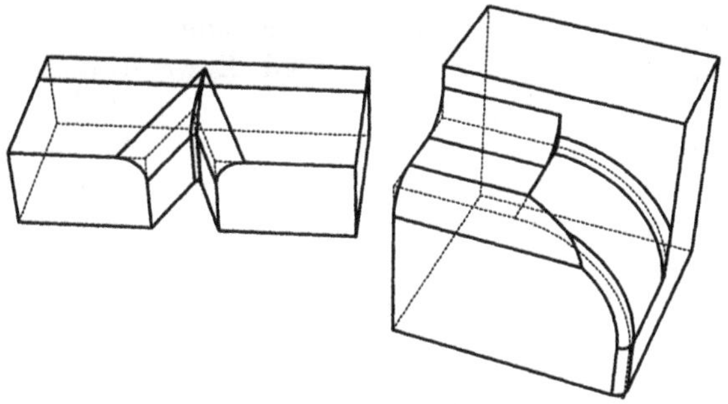

Fig. 8. Connecting trimlines: a - too far from the vertex, b - smooth edges .

be closed with a small facet, clearly an undesirable solution. In the convex case, the blends may intersect each other. This results in a valid boundary but the solution is questionable because the blend extends beyond the trim curve on the cylindrical face.

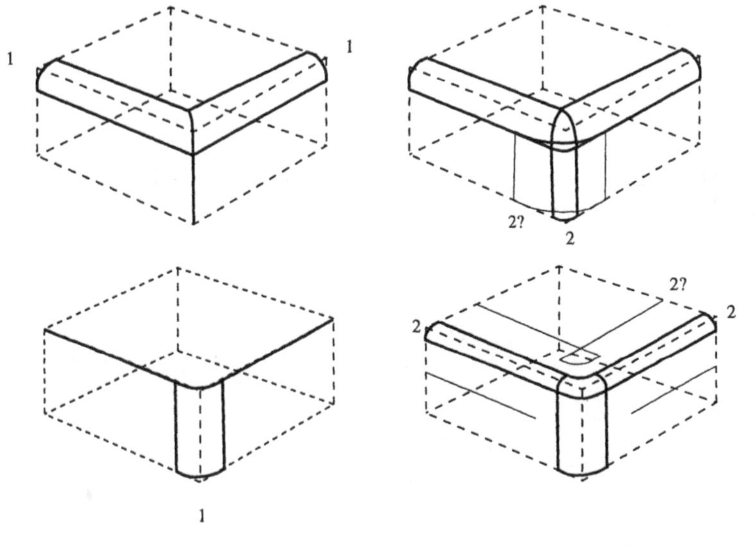

Fig. 9. Sequential problems: a - no large radius for the vertical edge, b - no large radius for the top edges .

3.4. Sequentiality Problems

A problem frequently encountered in blending is to devise a valid sequence of the blending operations. A simple example concerns blending the corner of a cube. Assume the incident edges will be blended with different radii. Assume we blend first the edges of the top face, as shown in Figure 9(a). The blends intersect and we obtain an edge sequence consisting of a vertical edge and an elliptical arc. If the blend radius for this sequence is not larger than the other blending radii, we obtain a good blend. However, if the radius is larger, no rolling ball blend can be generated at the vertex. This problem might be avoided if we blend first the vertical edge and then the top edges. However, in many situations we cannot define a good sequence from the magnitudes of the radii. This is particularly true when there are several edge sequences to be blended that cross each other at the vertex. Note that in the case of Figure 9(b), there may be a self-intersection if the top radius is too large.

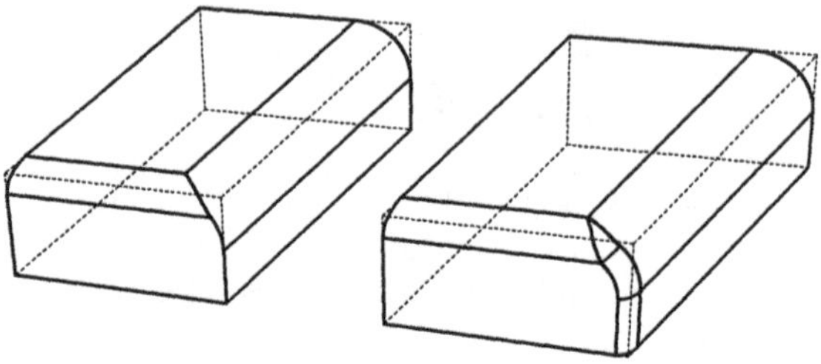

Fig. 10. Sequential problems: a - intersection, b - reblending.

Other problems arise when the two edges to be blended on the top face have different radii, as in Figure 10. Obviously, a prescribed sequence of the blending operations constrains the radii of the consecutive blends. If the vertical blend has the same radius as the left top edge, the operation is possible. Unfortunately, the result will be asymmetric.

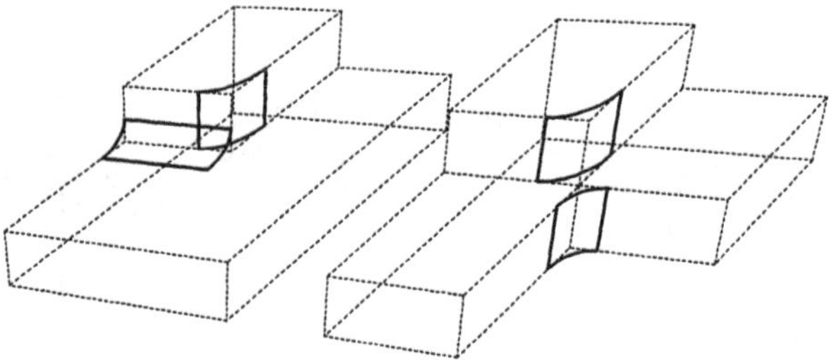

Fig. 11. Incomplete vertex blends - no intersections.

3.5. Incomplete Blend Configurations

An incomplete vertex blends is one in which some, but not all, incident edges are to be blended. The difficulties are similar to the cases investigated before. Intersecting blend faces meeting at the vertex works in some cases, as shown in Figure 9(a) and Figure 10(a), but fails in others. Two failure cases are shown in Figure 11(a) and (b). In the first example, the blends intersect in a point only. In the second example, the two blends do not touch each other.

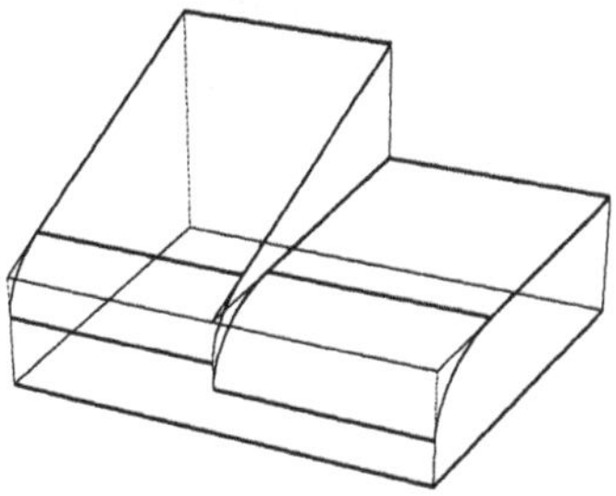

Fig. 12. Incomplete vertex blend - parallel trimlines.

New edge loops and faces, or face extensions, must be created to obtain valid results.

Further difficulties arise at tangential or cuspate edges. In Figure 12 we want to blend two horizontal edges with the same radius. For a rolling ball blend, the trimlines on the front face do not intersect nor will the two blend surfaces. Another example is shown in Figure 13 and is taken from [3]. Two adjacent edges — one convex, the other concave — form a cusp. Here, the problem is caused by the vertical edge between the cylinder and the front face that is not blended. Two possible interpretations are shown.

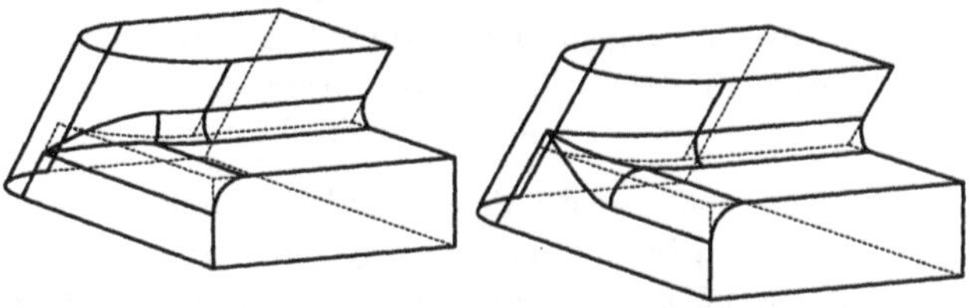

Fig. 13. Incomplete vertex blend - cuspate edges: blend and reblend.

To conclude, a good solid modeling system should allow blending any combination of edges, independent of their type, properties and the sequence of the blending operations. Consider the test object in Figure 26, for instance. It contains convex and concave edges with a cusp and a tangential edge pair. Select any number of edges out of the four, incident to the center vertex, and attach various blend radii. Then all combinations should work in any order. In Section 5 we develop an algorithm for such situations.

§4. Basic Blending Rules

We present a set of blending rules that provide a simple and uniform framework in which to develop complete blending algorithms. The rules allow implementing regular blending in a general and concise manner. In particular, all vertex blends are handled uniformly.

Rule 1. The primary entities blended are *sharp edges* of an object.

Note 1.1. Edges at which two faces meet tangentially are not sharp and are never blended.

Note 1.2. Blendable edges are either convex or concave. If necessary, an edge has been subdivided to establish this property.

Rule 2. Vertex blends are created where edge blends need to be smoothly connected, at a vertex *or* when a single edge is terminated.

Note 2.1. A vertex blend is generally represented by a face. However, it may degenerate to a single point or a single curve segment.

Rule 3. For each edge to be blended appropriate *blending parameters* have been specified that uniquely determine a blending surface. The blending parameters also define two *initial trimline segments.* The segments are bounded portions of the trimlines within the faces adjacent to the edge to be blended.

Note 3.1. Blending parameters may include the radius of a rolling ball blend, or ranges to define intersection curves of an offset surface with the current face, or any quantity by means of which the requested blending surface can be computed (see [17]).

Note 3.2. In regular blending, trimline segments generally run from one winged edge to another one without intersecting any other edge or edge loop; see, e.g., Figure 14(a). Two invalid cases are shown in Figure 14(b) and (c). For concave angles, it is assumed that the initial trimline segments run until they reach the bisector curve (d). The role of the initial and final trimline segments will be explained later in Section 5.

Rule 4. While edge blends are usually defined by the user, vertex blends are *automatically determined* by the modeling system based on the edge blends and certain global shape parameters. *All* vertex blends should be *setback*; i.e. adjacent trimlines on a given face should always be connected by spring curves. For a justification see the next section and the examples that illustrate how formerly difficult cases and poor shapes can be handled simply.

Note 4.1. An option to individually redefine the various shape parameters of a vertex blend may be offered. However, this may lead to an undesirable design practice and should be allowed in exceptional cases only.

Note 4.2. Global parameters of vertex blending may include an overall fullness factor, or default setback enlargement ratios that determine the actual setback values.

Rule 5. Blending a solid object is realized through a sequence of *mark_edge* and *evaluate_blend* operations. The mark_edge operation labels an edge (or group of edges) and attaches blending parameters to them. Often the term *implicit blending* is used for this operation; see [3]. In the evaluate_blend operation all previously marked edges are processed and a new, blended solid model is computed.

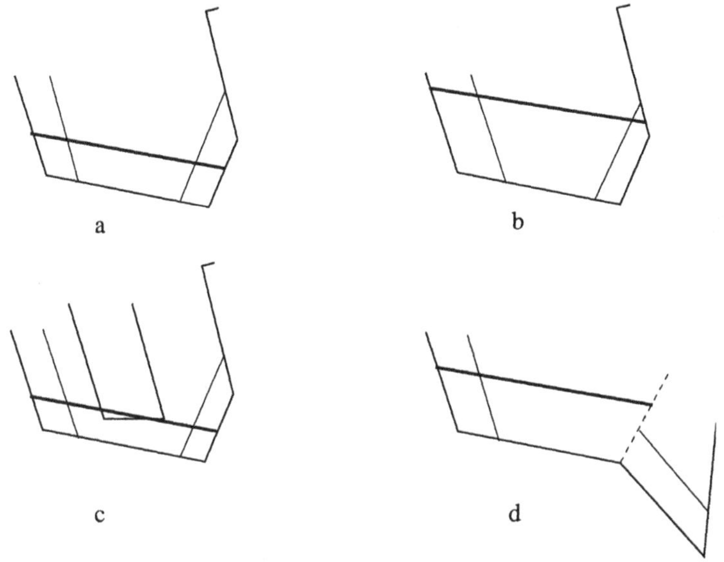

Fig. 14. Initial trimline segments, a - regular case, b and c - invalid cases, d - concave angle.

Note 5.1. There is no need to mark vertices, as this is a consequence of edge blending.

Note 5.2. Prior to evaluate_blend, the blending parameters may be modified or deleted.

Note 5.3. Typically there are several mark_edge operations between the successive evaluate_blends. Evaluating blends without previous attachments represents a void operation.

Note 5.4. The marked edges need not constitute a connected graph.

Note 5.5. In many cases groups of edges are selected for attaching a blending parameter. Possible selections include:

- connected edges that share common vertices with tangent continuity (smooth edge sequence),
- all edges of a loop,
- all edges of a face,
- edges between vertex A and B on a given face,
- and so on.

Rule 6. If all edges belonging to a certain vertex are blended, the shape of the final blend should not depend on the sequence in which the edges were selected, nor on how many evaluate_blend operations were performed. For example, the sequence "select 1, select 3, select 4, select 2, evaluate" should define the same vertex blend as "select 2, select 4, evaluate, select 3, select 1, evaluate."

Note 6.1. This rule cannot be applied if a different operation is performed between two consecutive evaluate_blend operations that modify the current vertex blend or its vicinity, for example a Boolean operation.

Rule 7. Blending should not create new sharp edges or sharp facets. If two edges meet at a common vertex, the transition between them must be *smooth.* A simple example is shown in Figure 20.

Note 7.1. The rule does not apply to profile curve edges in incomplete blend configurations. For an example, see the blend terminations at the three vertices in Figure 8(a) or in Figure 18.

§5. The Algorithm for Regular Blending

Now we develop an algorithm for regular blending. In order to achieve conceptual unity, vertex blends always use setbacks. Prior work on vertex blending has identified the utility of applying setbacks. However, setbacks were always applied as special cases of ordinary vertex blending. Our main point is that the problems previously characterized disappear when setbacks at vertices are used. The description of our algorithm concentrates on the concepts and leaves out some of the details. The reader should have no difficulty working out the missing details after working through the examples in this section.

5.1. Topological Transformations

If n edges meet at a vertex V, the setback vertex blend of V is a $2n$-sided face; see Figure 15. The vertex blend is bounded by n profile curves; each denoted by p. They are curves obtained by reducing the incident edge blends (e). The position of each profile curve is defined by the setback values (sb) associated with each end of the edge blended. The other n boundary curves of the vertex blend are spring curves (s). Each spring curve connects the two adjacent trimlines (t) lying on the same face. By definition, the spring curves always join the trimlines with tangent continuity, and there is at least G^1 continuity between the vertex blend and the adjacent faces of the object. In Figure 15, a corner of a cube is blended. Vertex V is converted into vertex blend VBL_V, edge e_1 is converted into edge blend EBL_1 and face F_C is converted into a reduced F_C.

In the following, the first subscript will always refer to the "parent" entity, i.e. vertex V, edge e_1 or face F_C, and the second subscript, if any, is used to distinguish several entities belonging to the same parent. For example, p_{1V} identifies the profile curve of e_1 at vertex V, while p_{1W} identifies the profile curve of e_1 at vertex W. The profile curves and the spring curves may be degenerate. Degeneracies occur at non-blended edges (zero blend radius) or when trimlines intersect without inserting a spring curve (zero length spring curves). We will denote such a degeneracy by $|p_{1V}| = 0$.

Figure 15(b) shows the vertex blend VBL_V where all incident edges have a nonzero blend radius. There is no degeneracy. The edges are always blended by a four-sided face, bounded by the remaining segments of the trimlines t_{1C}

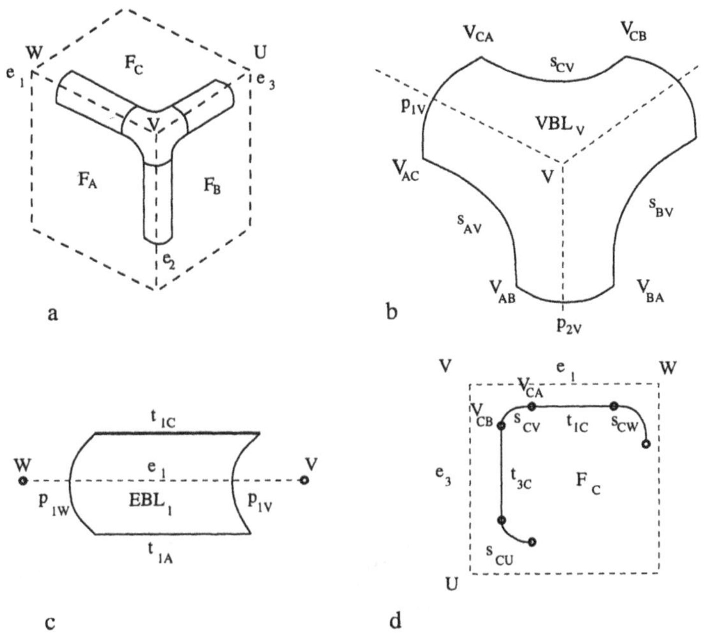

Fig. 15. General case of setback vertex blending.

and t_{1A}, and two profile curves p_{1V} and p_{1W}; see also Figure 15(c). The topological transformation of the edge loop is also straightforward; Figure 15(d). For example, on F_C, the face boundary "vertex U ... edge e_3 ... vertex V ... edge e_1 ... vertex W ..." is changed into "spring curve s_{CU} ... trimline t_{3C} ... s_{CV} ... t_{1C} ... s_{CW} ...," evidently a one-to-one transformation.

5.2. Setbacks

Given an edge e incident to a vertex v, the *minimum setback* for e at v is the minimum distance below which regular blending cannot be performed. In Figure 16, a corner with two small and one large blending radius is shown. From the trimline intersection in face F_C, we conclude that the setback for edge e_1, the upper left edge, cannot be smaller than sb_{1C}. But the trimline intersection in face F_A requires the larger setback sb_{1A}, so that we do not interfere with the blend on the vertical edge e_3. So, the minimum setback would have to be sb_{1A} in this case. If a spring curve is inserted in face F_A, as shown, then the setback would be enlarged to sb_1.

The minimum setback is determined from the distances of the intersections of the trim curves, as $\max(sb_{1C}, sb_{1A}, 0)$. As seen in Figure 17(a) and (b), trimcurve intersections may determine negative distances, i.e. the intersection may take place behind the vertex or there may be no intersection. This would result in dysfunctional blending surfaces, and is avoided by requiring nonnegative minimum setbacks.

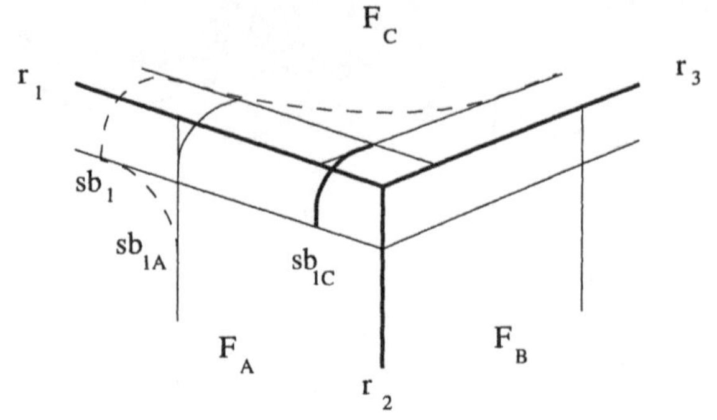

Fig. 16. Setting setbacks.

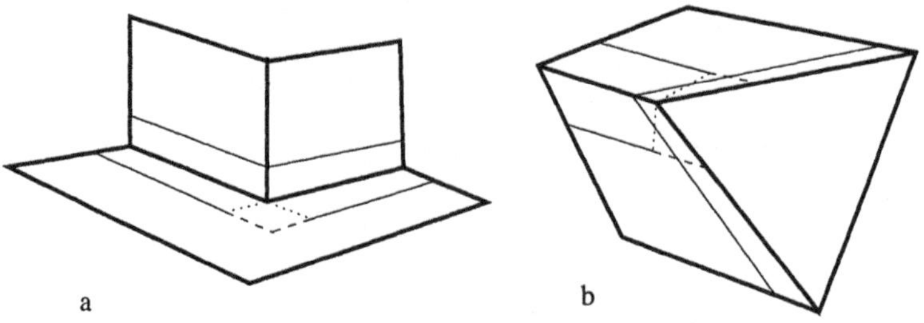

Fig. 17. Negative setbacks.

Often, we want larger setbacks than the minimum value, because they may lead to a more pleasing vertex blend. One possible strategy is to use a setback

$$sb = sb_{min} \times sb\_enl\_ratio + sb\_shift$$

In most of our experiments, the setback shift value is zero so that we effectively multiply the minimum setback by a setback enlargement ratio, a value typically between 1 and 2, or else the setback enlargement ratio is equal to 1 and there is a constant enlargement by the setback shift value. On rare occasions, we have specified individual setback values for edges at certain vertices.

5.3. Blending a Single Edge

Consider blending a single edge with a blend radius r_1, as shown in Figure 18. Here, the blend termination in face F_B corresponds to a single spring curve insertion, and fits perfectly into the general scheme. Moreover, two of the profile curves are zero length for the unblended edges. Therefore, when we blend edge e_1, a general vertex blend data structure is created, assigned to vertex V in our example. The structure provides a complete record of the vertex blend transformations. In our example, we take $sb\_enl\_ratio = 1$, $sb_{1min} = sb_1 = 0, sb_{2min} = sb_2 = r_1, sb_{3min} = sb_3 = r_1$. We obtain $V_{CA} =$

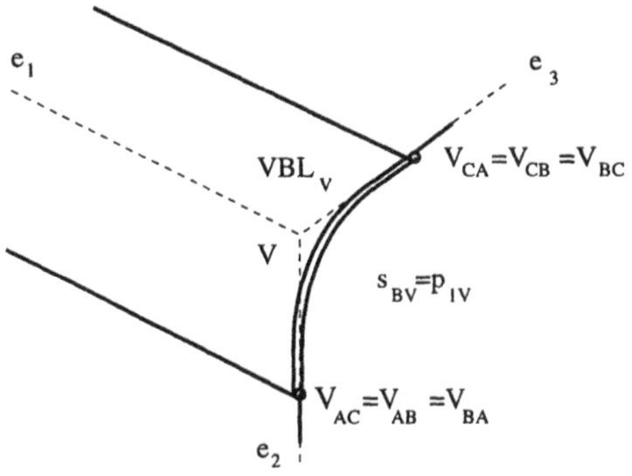

Fig. 18. Blending a single edge.

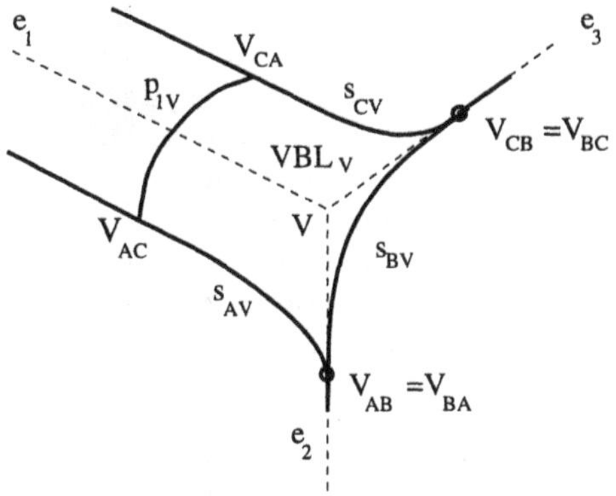

Fig. 19. Blending a single edge with special termination.

$V_{CB} = V_{BC}$, thus $|s_{CV}| = 0$ and $|p_{3V}| = 0$, similarly $V_{AC} = V_{AB} = V_{BA}$, thus $|s_{AV}| = 0$ and $|p_{2V}| = 0$. Note, that in this case the profile curve p_{1V} is identical to the spring curve s_{BV}.

The general scheme of vertex blending makes it possible to terminate the edge blend differently by widening the end section. For example, apply a positive setback for e_1 and use $sb\_enl\_ratio = 1.2$ for e_2 and e_3. In this case we obtain $sb_1 = const$, $sb_2 = sb_3 = r_1 * 1.2$. The result is shown in Figure 19, where now there are three spring curves s_{AV}, s_{BV} and s_{CV}, and two degenerate profile curves p_{2V} and p_{3V} for the unblended edges. In this case s_{BV} is a smooth edge curve.

5.4. Blending Two Edges

Now we blend two edges using r_1 and r_3, as shown in Figure 20. The topological transitions for the faces F_C and F_B are simple. The setback calculation for e_2 requires special care here: since r_1 is larger than r_3, it determines the

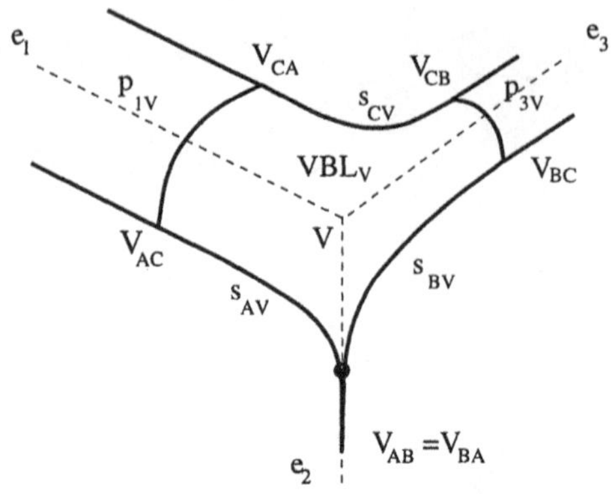

Fig. 20. Blending two edges.

minimum setback sb_2. This value, multiplied by the *sb_enl_ratio*, is used for sb_2. There is only one degeneracy in this case: since e_2 has not been blended, $V_{AB} = V_{BA}$, and therefore $|p_{2V}| = 0$. If e_2 is later marked for blending, the structure of the vertex blend can be modified accordingly (see also Rule 6).

The terminating point of the vertex blend on e_2 is peculiar from a geometric point of view. At V_{AB} three faces come together, namely F_A, F_B and VBL_V. There must be smooth transitions between F_A and VBL_V, and F_B and VBL_V, respectively. This is the case everywhere along the spring curves s_{AV} and s_{BV}, except at the endpoint itself, where there is no unique surface normal on VBL_V. In this case the two spring curves run into the common tangent line of e_2. Such a singularity is well-known in geometric modeling, take for example the apex of a cone. From a representational point of view, such a degeneracy does not cause any problem. However, less advanced geometric modeling systems may find it difficult to compute surface-surface intersections that pass through the singularity. As we will see in the next section, control frame patches are not sensitive to this sort of degeneracies. The example is also shown in Figure 21.

In a successive blending phase, edge e_2 can also be blended; this will lead to the general case discussed before. If, in that case, the blending radius is smaller than r_3, then the minimum setbacks do not have to be adjusted. However, if the blend radius for e_2 is larger than r_3, both setbacks sb_1 and sb_3 must be recalculated. Again, note that the vertex blend is treated uniformly and consistently in every blending phase.

5.5. Resolving Problem Cases

We examine next the difficult cases presented in Section 3. As shown in Figure 22, problems of far intersections disappear since they are now controlled by the setback values; see (a) and (b). Instead of the small strange face of Figure 5(c), now a natural transition is constructed in Figure 22(c).

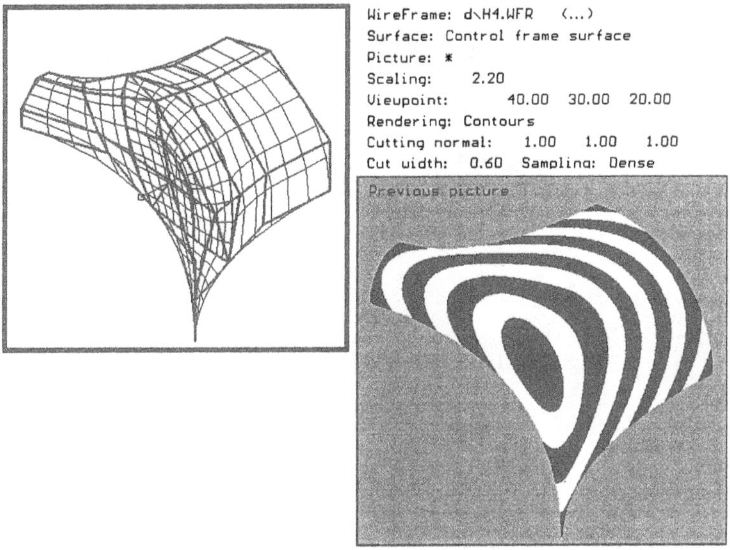

Fig. 21. Incomplete vertex blend with uneven radii.

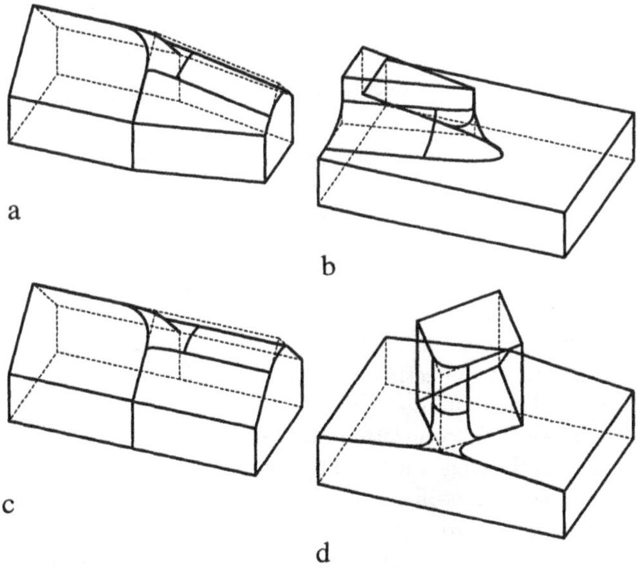

Fig. 22. Edge termination with setback blending.

In contrast to Figure 5(d), our solution of Figure 22(d) avoids the insertion of a new edge, and a more natural new vertex blend provides transition between the blended edge and the rest of the object. Our approach allows

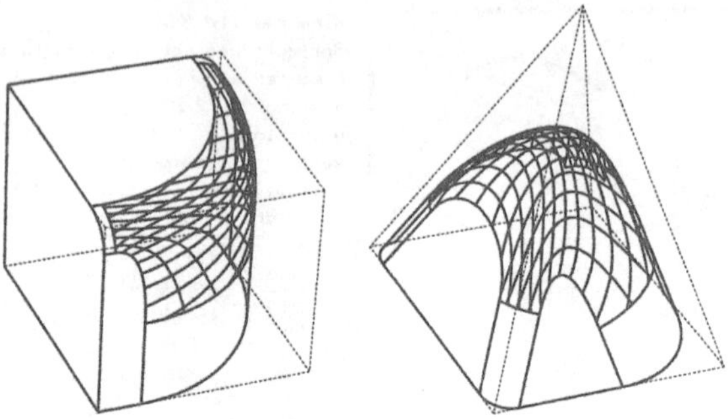

Fig. 23. Improved vertex blends with setbacks.

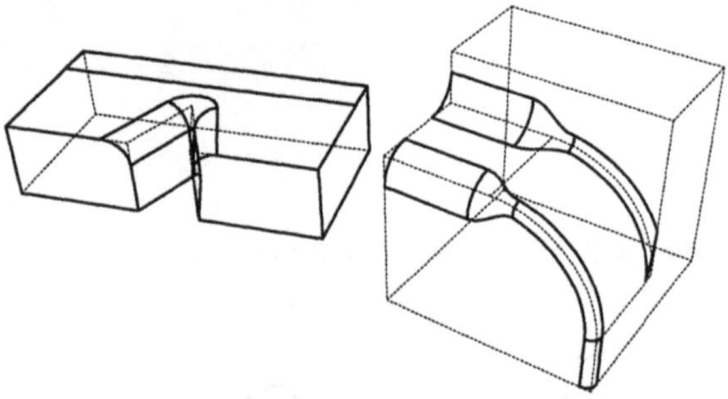

Fig. 24. Connecting trimlines - improved.

us to continue blending the original edges. Otherwise, we would have had to deal with the artificial edge inserted in Figure 5(d).

Next, we examine how to avoid the shape imperfections of Figure 6 and Figure 7. Because of the minimum setback mechanism, the ugly vertex blends of the cube and the pyramid have disappeared; Figure 23(a) and (b). Instead, a pleasing shape is obtained even in the case of widely different radii.

The trimline connection problems of Figure 24 are also avoided by the setback mechanism. Connecting the trimlines by setback curves guarantees that the vertex blend remains relatively close to the vertex, as seen in Figure 24(a). The smooth connections between the two tangential edges that were blended with different radii are presented in (b). Here we have two-edge vertex blends with two profile curves and two spring curves. By fixing the setback values, appropriate transitions are created.

The sequential problems (Figure 9) are clearly solved in our approach. As an illustration see Figure 21. Finally, a few examples of incomplete vertex blends (Figure 11 and Figure 12) are shown together with fully blended versions. In Figure 25, the non-intersecting trimlines were properly connected and the rest of the vertex blend was joined with the unblended edges, see (a). Figure 25(b) shows when all edges are blended. Figure 25(c) shows a previous complex vertex blend. Here, we have a 12-sided vertex blend with four

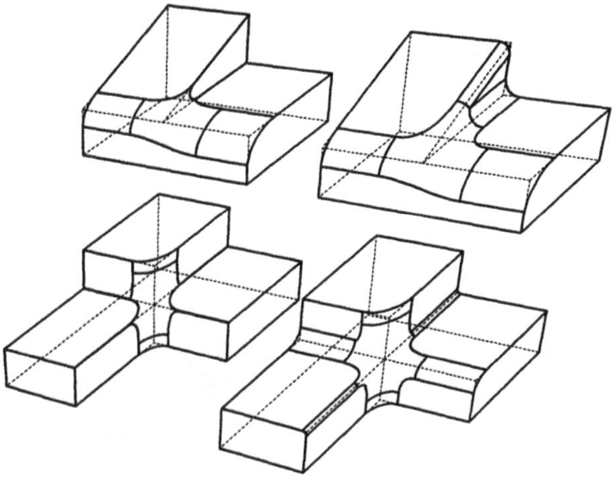

Fig. 25. Connecting trimlines - incomplete and complete vertex blends.

degenerate profile curves. In Figure 25(d), the full 12-sided patch is shown.

To illustrate the versatility of this sort of blending, all possible two-edge blends are presented in Figure 26(b–g). Among the four edges, there is a tangential edge pair and a cuspate edge. As the figure shows, neither the sequence of operations nor the type of edges matters. Figure 26(h) shows an object where all four edges have been blended.

§6. Representational Issues

Different equations can be used to describe a vertex blend, depending on a trade-off between mathematical precision and computational efficiency. We briefly review two solutions suggested by Plowman-Charrot [11] and Várady and Rockwood [15, 16].

6.1. Plowman-Charrot Vertex Blends

This surface representation is an extension of the pentagonal patch of Charrot and Gregory[4], and it has been implemented in the ACIS solid modeler [1]. The domain is a regular n-sided polygon and we would like to interpolate n boundary curves $\mathbf{C}_i(s_i)$ and normal vector functions $\mathbf{N}_i(s_i)$. Each domain point (u, v) is mapped into local parameters (s_i, t_i), where $i = 1, \ldots, n$ and $t_i = 1 - s_i$. The projectors used are shown in Figure 27. The surface is a weighted average of corner interpolants, given as

$$\mathbf{P}(u,v) = \sum_{i=0}^{n-1} \alpha_i(u,v)\mathbf{Q}_i(s_i,t_i),$$

where the weighting functions α_i are rational quadratic functions that are chosen such that they vanish on the polygon sides other than the current

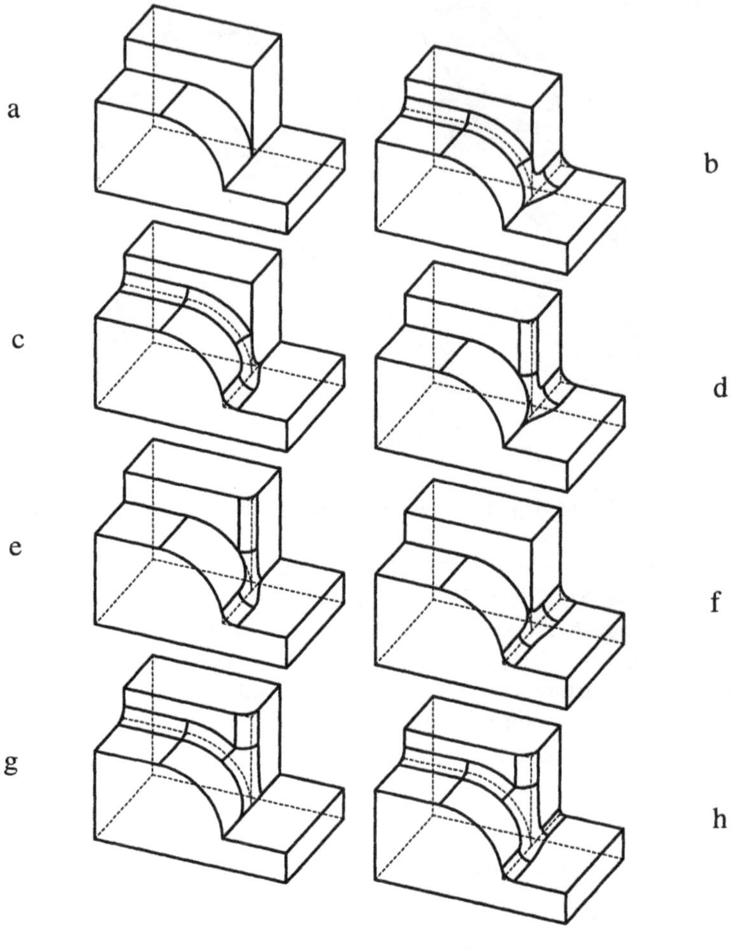

Fig. 26. a - Test object, b – g - Incomplete vertex blends, h - All four edges blended.

corner interpolant. The i-th corner interpolant is given as follows:

$$\mathbf{Q}_i(s_i, t_i) = \mathbf{C}_i(s_i) + \mathbf{C}_{i-1}(s_{i-1}) - \mathbf{C}_i(0) + t_i\mathbf{D}_i^+(s_i) \\ + s_i\mathbf{D}_{i_1}^-(s_{i-1}) - t_i\mathbf{D}_i^+(0) - s_i\mathbf{D}_{i-1}^-(1) - s_it_i\mathbf{T}_i.$$

In the formula above, cross derivative functions are denoted by $\mathbf{D}_i^+(s_i)$ and $\mathbf{D}_{i-1}^-(s_{i-1})$. These are defined using the given normal vector functions; $\mathbf{T}_i$ denotes a twist vector assigned to the i-th corner.

6.2. Control Frame Patches

The control frame patch is a subdivision formulation based on a special construction called the setback split. Usually, but not always, an even number of

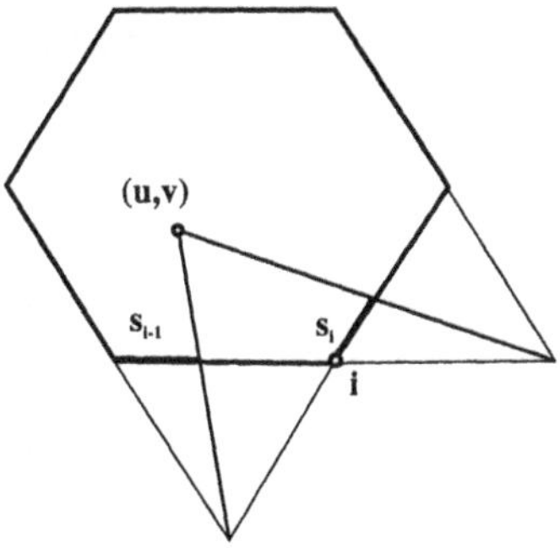

Fig. 27. Parametric domain of the Plowman-Charrot patch.

boundaries is assumed. The polygonal domain is subdivided into rectangular areas, but not according to the usual central split. Instead, this split reflects the philosophy behind creating setback vertex blends. The midpoints of the spring curves are called *base points.* The neighboring base points are connected by *interior boundary curves*, thus an *interior region* is created; see Figure 28. The interior and the edge blends are connected by n *setback patches*, while the interior is split in the traditional way into n interior patches. Therefore, we have here $2n$ rectangular patches as well, but in a different arrangement: the broadening of the edge blends has been separated from the common interior region. The setback patches always connect two half-spring curves, while the interior region is detached from the surrounding primary surfaces except at the base points.

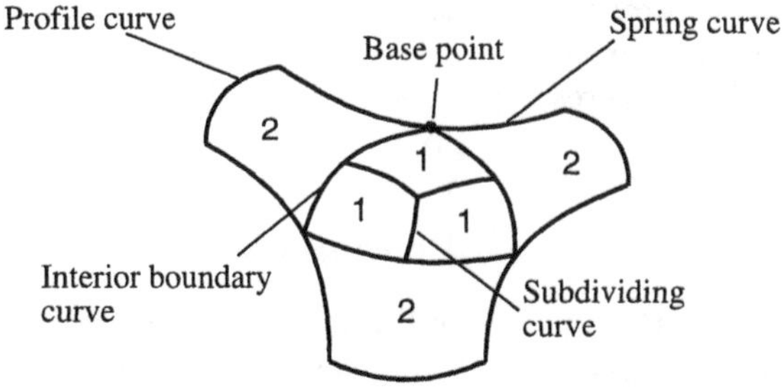

Fig. 28. Setback split (1 - interior, 2 - setback patches).

Another important feature of this scheme is the control frame, which is based algorithmically on repeated chamfering of tangent planes at the end points of the profile curves and at the base points; for details, see [16]. The control frame is similar to a Bézier polygon in that it approximates shape and, in fact, indirectly determines the surfaces quantities of the constituting patches.

By construction, $G1$ continuity between the setback and the interior patches is assured, and each of them is represented by standard Bézier or

B-spline patches. At the base point, three rectangular patches meet, namely two setback and an interior patch. Their twist compatibility is guaranteed by forcing them to match a common surface curvature. This surface curvature is computed based on the control frame and its associated fullness parameters. The mathematical details can be found in the papers cited before. Here, we emphasize that only simple, noniterative numerical methods need to be applied, such as solving linear systems of equations or finding the roots of cubic or quartic polynomials.

As was stated before, the control frame transforms in a straightforward manner according to various parameter setting of the radii and the setback values. In this way, it varies the related surface quantities. In Figure 21, a degenerate 6-sided patch with a zero radius edge is shown. Another example is the "evolution" of a 10-sided vertex blend (5 edges meet with mixed convexity): The related control frames are shown on the left side, vertex blends with slices on the right of Figure 29.

6.3. Discussion

A strength of the Plowman-Charrot scheme is that it assures precise continuity to the surrounding surfaces — independent of the surface types. On the other hand, it is a non-standard formulation. Therefore, surface approximations have to be computed when transferring data between CAD systems, and the computational load becomes demanding if we have many boundaries.

A strength of the control frame patch scheme is that it usually uses low-degree, standard patches. Therefore, it is easy to transfer this surface data between CAD systems. The scheme is computationally very efficient. The weakness of the control frame patch scheme is that, in the current version, it is accurate for planar and quadric surfaces, but it only approximates the spring curves within a prescribed tolerance, in the case of general free-form surfaces.

§7. Conclusions

After analyzing how the topology of B-rep solid models has to be restructured for regular blending, a simple algorithm has been presented to generate vertex blends. These blends are the critical component in regular blending, and are the reason we can achieve a conceptually comprehensible and uniform blending behavior. Thus, a complete algorithmic implementation is possible.

The underlying concept of our approach is based on a few blending rules that clearly define the effect of the blending operations and minimize the dependency on arcane parameter settings and on the sequence in which the blends have been specified. Setback vertex blends are considered here as the most general entities: All special cases arising in regular blending can be derived from them in a way that is comprehensible to any user.

Concerning representational issues, both the Plowman-Charrot and the Varady-Rockwood patches can be used to represent setback vertex blends of

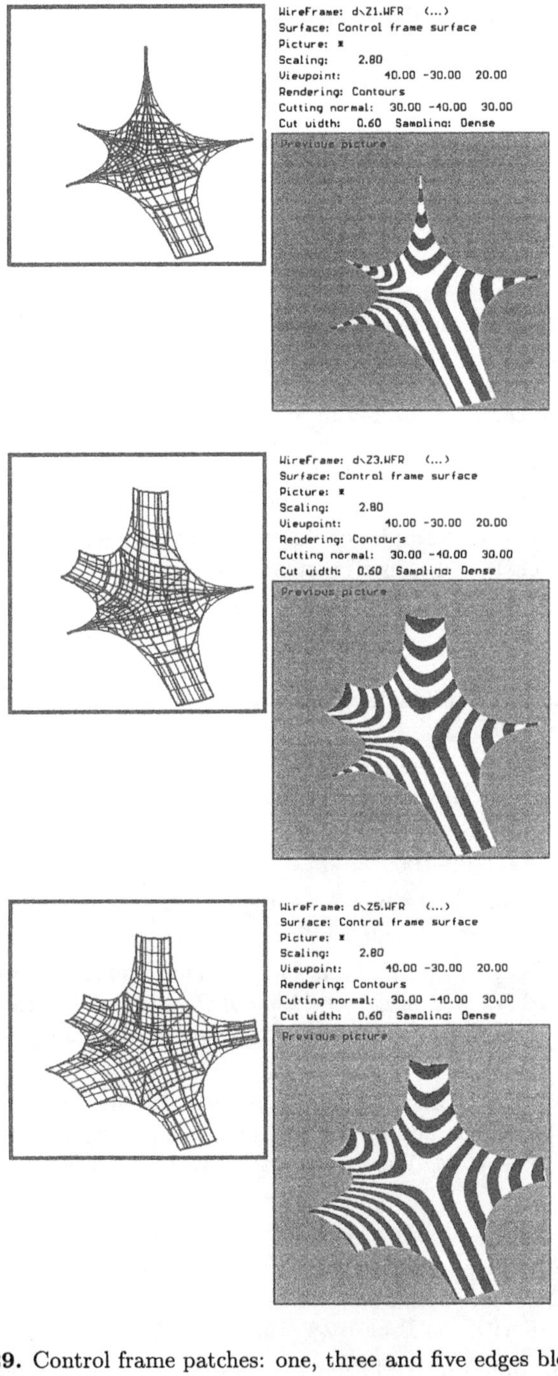
WireFrame: d\Z1.WFR (...)
Surface: Control frame surface
Picture: *
Scaling: 2.80
Vieupoint: 40.00 -30.00 20.00
Rendering: Contours
Cutting normal: 30.00 -40.00 30.00
Cut uidth: 0.60 Sampling: Dense
Previous picture
WireFrame: d\Z3.WFR (...)
Surface: Control frame surface
Picture: *
Scaling: 2.80
Vieupoint: 40.00 -30.00 20.00
Rendering: Contours
Cutting normal: 30.00 -40.00 30.00
Cut uidth: 0.60 Sampling: Dense
Previous picture
WireFrame: d\Z5.WFR (...)
Surface: Control frame surface
Picture: *
Scaling: 2.80
Vieupoint: 40.00 -30.00 20.00
Rendering: Contours
Cutting normal: 30.00 -40.00 30.00
Cut uidth: 0.60 Sampling: Dense
Previous picture

Fig. 29. Control frame patches: one, three and five edges blended.

any complexity. They offer trade-offs between accuracy and computational efficiency.

A number of interesting problems remain for topologically more complicated blending problems. They have been categorized into overlap blends and global blends. A simple, uniform semantics for them, based on a few intuitive parameters, remains an unsolved and interesting problem.

Acknowledgments. Many thanks for inspiring discussions with Ian Braid, Charles Lang (Three-Space Ltd., Cambridge, UK) and Alyn Rockwood (Arizona State University, Tempe). This project was begun at the Geometric Modeling Laboratory, Computer and Automation Research Institute, Budapest, and was completed in a joint effort with the second author at the Computer Science Department, Purdue University in 1996, while the first author was a visiting scholar there, sponsored by a Fulbright Scholarship. This research was supported by the US-Hungarian Joint Science and Technology Fund (No. 396), a COPERNICUS grant (No. 1068) and in part by ONR Contract N00014-96-1-0635 and NSF Grants CDA 92-23502 and CCR 95-05745.

Most of the test examples of this paper were drawn using the ACIS solid modeling kernel [1] and a small vertex blending package.

References

1. ACIS, Application Guide, Version 2.1, Spatial Technology Inc., Boulder, Colorado, 1996.
2. ACIS, Advanced Blending Husk, Development Manual, Version 2.1, Spatial Technology Inc. Boulder, 1996.
3. Braid, I., Non-local blending of boundary models, Computer-Aided Design **29** (1997), 89–100.
4. Charrot, P., J. A. Gregory, A pentagonal surface patch for CAGD, Comput. Aided Geom. Design **1** (1984), 87–94.
5. Gregory, J. A., V. K. H. Lau, and J. Zhou, Smooth parametric surfaces and n-sided patches, in *Computation of Curves and Surfaces*, W. Dahmen, M. Gasca and C. A. Micchelli (eds.), Kluwer Academic Pub., 1990, 457–498.
6. Hoffmann, C. M., and R. Juan, Erep, an editable, high-level representation for geometric design and analysis, in *Geometric and Product Modeling*, P. Wilson, M. Wozny and M.J. Pratt (eds.), North Holland, 1993, 129–164.
7. Loop, C., and T. DeRose, Generalized B-spline surfaces of arbitrary topology, Computer Graphics (SIGGRAPH '90 Proc.) **24** (1990), 347–356.
8. Loop, C., Smooth spline surfaces over irregular meshes, Computer Graphics (SIGGRAPH '94 Proc.) **28** (1994), 303–310.
9. Peters, J., Smooth free-form surfaces over irregular meshes generalizing quadratic splines, Comput. Aided Geom. Design **10** (1993), 347–361.

10. J. Peters, Constructing C1 surfaces of arbitrary topology using biquadratic and bi-cubic splines, in *Designing Fair Curves and Surfaces*, N. S. Sapidis (ed.), SIAM, Philadelphia, 1994, 277–294.

11. Plowman, D., P. Charrot, A practical implementation of vertex blend surfaces using an n-sided patch, in *The Mathematics of Surfaces VI*, G. Mullineux (ed.), Clarendon Press, Oxford, 1996, 67–78.

12. Rockwood, A. P. Blending surfaces in solid modeling, PhD dissertation, Dept. of Applied Mathematics and Theoretical Physics, University of Cambridge, 1987.

13. Várady, T., Survey and new results in n-sided patch generation, in *The Mathematics of Surfaces II*, R. R. Martin (ed.), Oxford University Press, Oxford, 1987, 203–235.

14. Várady, T., R. R. Martin, J. Vida, Topological considerations in blending boundary representation solid models, in *Theory and Practice of Geometric Modeling*, W. Strasser and H.-P. Seidel (eds.), Springer-Verlag, 1989, 205–220.

15. Várady, T., A. P. Rockwood, Vertex blending based on the setback split, in *Mathematical Methods for Curves and Surfaces*, M. Dæhlen, T. Lyche, and L. L. Schumaker (eds.), Vanderbilt University Press, Nashville, 1995, 527–542.

16. Várady, T., A. P. Rockwood, Geometric construction for setback vertex blending, Computer-Aided Design **29** (1997), 413–425.

17. Vida, J., R. R. Martin and T. Várady, A survey of blending methods that use parametric surfaces, Computer-Aided Design **26** (1994), 341–365.

18. Woodwark, J. R., Blends in geometric modeling, in: *The Mathematics of Surfaces II*, R. R. Martin (ed.), Oxford University Press, 1987, 255–297.

Tamás Várady
Computer and Automation Research Institute
Hungarian Academy of Sciences
1111 Budapest, Kende u. 13-17, HUNGARY
`varady@sztaki.hu`

Chris M. Hoffmann
Computer Science Department
Purdue University, West Lafayette, IN 47906, USA
`cmh@cs.purdue.edu`

Adding Constraints to Gordonesque Surfaces

Marshall Walker

Abstract. Given a rectangular parametric mesh consisting of arbitrary curves in $\mathbb{R}^3$, this paper generalizes and extends Gordon methods for constructing a suitable surface which interpolates each of the curves of the mesh. It is shown in this paper that any univariate interpolation method may be used to generate such a surface in such a way that that derivatives across mesh lines may be assigned according to need. The techniques allow the construction of surfaces based on a mix of univariate interpolation methods should such be necessary. Surfaces constructed possess local control if it is also possessed by the univariate methods. They also have a degree of global continuity limited only by that of constituent univariate methods and that of the original mesh.

§1. Introduction

Given a rectangular mesh of curves in $\mathbb{R}^3$, this paper addresses the problem of constructing a suitable surface which interpolates each of the curves of the mesh and for which it is possible to assign partial derivatives across mesh lines.

The classical way of attacking this problem starts with the work of Coons [2] in which a rectangular array of four arbitrary C^1 curves is spanned by a C^1 surface patch that agrees with given cross-boundary derivatives and mixed-partials at the vertices of the array. Given a rectangular mesh with appropriate cross-boundary and mixed partial information, an interpolating surface is then constructed patch by patch. The task of assigning mixed partials was addressed by Gregory [5]. Chiyokura and Kimura [1] adapted Gregory techniques to the handling of Bézier style patches.

Instead of building surfaces as conglomerates of individual patches Gordon [3,4] introduced a technique whereby a certain method of univariate polynomial spline interpolation is extended in such a way as to allow a rectangular mesh to be interpolated as a single surface. Although the Gordon methods dispense with having to specify large amounts of data for each surface patch, they provide no good methods to specify this same data should it be essential.

Mathematical Methods for Curves and Surfaces II

Morten Dæhlen, Tom Lyche, and Larry L. Schumaker (eds.), pp. 529–536.

ISBN 0-8265-1315-8.

Two results are shown in this paper. The first appears in Section 2 where it is shown that any two univariate interpolation methods may be used to develop a family of bivariate mesh interpolation schemes. Such surfaces possess a degree of global continuity limited only by the continuity of the mesh and that of the univariate interpolation scheme. In the case of univariate methods based on polynomial splines of odd degree, this result has been proved by Gordon [3]. The present formulation extends Gordon's techniques allowing one to consider, for example, a univariate interpolation scheme based on exponential splines.

The second result is in Section 3 and states that, in addition to interpolating the given mesh, the methods of Section 2 can be used to create surfaces which also interpolate given values for derivatives across mesh lines. Although this result applies equally well to the surfaces constructed by Gordon, there are no echoes of it in the literature surrounding his work. The result increases the usefulness of the Gordon methods, and in particular allows the possibility of nicely attaching surfaces in such away that a mesh line of one is connected to a curve in another. This application is discussed in Section 4.

§2. Preliminaries

A parametric mesh of curves is defined to be a function with range $\mathbb{R}^3$ whose domain consists of a rectangular grid of finite line segments. An associated interpolating surface is an extension of such a function to a bounding rectangle.

Definition 1. *Given strictly increasing sequences* $u_0, u_1, \cdots, u_m$ *and* $v_0, v_1, \cdots, v_n$, *let*

$$U = \bigcup_{j=0}^{n} \Big([u_0, u_m] \times \{v_j\} \Big) \qquad \text{and} \qquad V = \bigcup_{i=0}^{m} \Big(\{u_i\} \times [v_0, v_n] \Big).$$

Then the associated parametric mesh *is a function* $f : U \cup V \to \mathbb{R}^3$. *Given a parametric mesh* $f : U \cup V \to \mathbb{R}^3$, *an* interpolating surface *is a function*

$$F : [u_0, u_m] \times [v_0, v_n] \to \mathbb{R}^3$$

such that $F|_{X \cup Y} = f$.

Informally, given a sequence of knots and associated data points a univariate interpolation scheme is simply a method of interpolating the data points at the associated knots.

Definition 2. *Given an increasing sequence* $t_0, t_1, \ldots, t_n$, *an associated* univariate interpolation scheme *is a function*

$$h : [t_0, t_n] \times (\mathbb{R}^3)^{n+1} \to \mathbb{R}^3$$

with the property that

$$h(t_i, P_0, \cdots, P_n) = P_i,$$

for each set of points $P_0, \ldots, P_n$ in $\mathbb{R}^3$.

Given a parametric mesh $f : U \cup V \to \mathbb{R}^3$ and univariate interpolation schemes, $h_1 : [u_0, u_m] \times (\mathbb{R}^3)^{m+1} \to \mathbb{R}^3$ and $h_2 : [v_0, v_n] \times (\mathbb{R}^3)^{n+1} \to \mathbb{R}^3$, define extensions $G_i : [u_0, u_m] \times [v_0, v_n] \to \mathbb{R}^3$, $1 \leq i \leq 4$, by

$$\begin{aligned}
G_1(u,v) &= h_1\big(u,\ f(u_0,v), \cdots, f(u_m,v)\big)\\
G_2(u,v) &= h_2\big(v,\ f(u,v_0), \cdots, f(u,v_n)\big)\\
G_3(u,v) &= h_2\big(v,\ G_1(u,v_0), \cdots, G_1(u,v_n)\big)\\
G_4(u,v) &= h_1\big(u,\ G_2(u_0,v), \cdots, G_2(u_m,v)\big).
\end{aligned}$$

The following states that any two univariate interpolation schemes give rise to a family of interpolating surfaces. The result is known in the special case when h_1 and h_2 are polynomial splines of odd degree, [3], and forms the foundation of Gordon's results.

Theorem 1. *Given a parametric mesh $f : U \cup V \to \mathbb{R}^3$ and a function $\alpha : [u_0, u_m] \times [v_0, v_n] \to \mathbb{R}$, the function $F : [u_0, u_m] \times [v_0, v_n] \to \mathbb{R}^3$ defined by*

$$F = G_1 + G_2 - \big((1-\alpha)G_4 + \alpha G_3\big)$$

interpolates the mesh f.

Proof: For $0 \leq i \leq n$,

$$\begin{aligned}
&(1-\alpha(u_i,v))G_4(u_i,v) + \alpha(u_i,v)G_3(u_i,v)\\
&\quad = (1-\alpha(u_i,v))h_1\big(u_i, G_2(u_0,v)\cdots, G_2(u_m,v)\big)\\
&\qquad\qquad + \alpha(u_i,v)h_2\big(v, G_1(u_i,v_0), \cdots, G_1(u_i,v_n)\big)\\
&\quad = (1-\alpha(u_i,v))G_2(u_i,v) + \alpha(u_i,v)h_2\big(v, f(u_i,v_0), \cdots, f(u_i,v_n)\big)\\
&\quad = (1-\alpha(u_i,v))G_2(u_i,v) + \alpha(u_i,v)G_2(u_i,v) = G_2(u_i,v).
\end{aligned}$$

Thus, $F(u_i,v) = G_1(u_i,v) = f(u_i,v)$. Similarly for $0 \leq j \leq n$, $F(u,v_j) = f(u,v_j)$. Thus F interpolates the mesh f. □

The interpolating surface F of the preceding theorem is said to be of *type 1* or *type 2* according to whether α is identically 0 or identically 1.

It is not difficult to show that if the parametric mesh is C^k continuous in the sense that derivatives exist up to order k whenever they can be defined, then the interpolating surface F is also C^k continuous, provided in addition that the constituent univariate interpolation schemes are themselves C^k continuous. Throughout the paper it will be assumed that the various functions mentioned are such that indicated derivatives do in fact exist.

Finally, a standard notational convention will be used — if g is any function from a region of $\mathbb{R}^n$ to $\mathbb{R}^m$, denote its coordinate functions as $g^1, \ldots g^m$ and define

$$\frac{\partial g}{\partial x_i} = \begin{pmatrix} \frac{\partial g^1}{\partial x_i} \\ \vdots \\ \frac{\partial g^m}{\partial x_i} \end{pmatrix}.$$

§3. Adding Constraints

If $f : U \cup V \to \mathbb{R}^3$ is a parametric mesh, we address the problem of creating surfaces of the type developed in Section 2 which also interpolate given derivative values across mesh lines. In particular we shall show that in the case of a type 1 surface, the kth partial derivative of F with respect to u at a point (u_i, v) is the same as the kth partial derivative of the intermediary surface G_1 at the same point. Since G_1 is constructed from a univariate interpolation scheme, it is only necessary to make sure that the chosen univariate interpolation produces correct derivatives at points (u_i, v).

We begin by calculating derivatives of intermediate surfaces, G_1 and G_2. Defining

$$\phi : [u_0, u_m] \times [v_0, v_n] \to [u_0, u_m] \times (\mathbb{R}^3)^{m+1}$$

by

$$\phi : (u, v) \mapsto \big(u,\ f(u_0, v), \cdots, f(u_m, v)\big),$$

it follows that $G_1 = h_1 \circ \phi$ so that

$$DG_1(u_{i_s}, v) = Dh_1\big(\phi(u_{i_s}, v) \cdot D\phi(u_{i_s}, v)\big),\ 1 \le i_s \le m. \tag{1}$$

For the purpose of calculating partial derivatives, denote variables in the domain of ϕ as $u, w_{0,1}, w_{0,2}, w_{0,3}, \cdots w_{m,1}, w_{m,2}, w_{m,3}$, and define for $1 \le i \le 3$ and $0 \le j \le m$

$$\frac{\partial h_1^i}{\partial w_j} = \Big(\frac{\partial h_1^i}{\partial w_{j,1}}, \frac{\partial h_1^i}{\partial w_{j,2}}, \frac{\partial h_1^i}{\partial w_{j,3}}\Big).$$

Letting $\mathbf{0}$ denote the 3×1 matrix of zeros, it follows that

$$D\phi(u_{i_s}, v) = \begin{pmatrix} 1 & 0 \\ \mathbf{0} & \frac{\partial f}{\partial v}(u_0, v) \\ \vdots & \vdots \\ \mathbf{0} & \frac{\partial f}{\partial v}(u_m, v) \end{pmatrix}.$$

Calculating the product in (1) gives

$$\frac{\partial G_1}{\partial u}(u_{i_s}, v) = \frac{\partial h_1}{\partial u}\big(u_{i_s},\ f(u_0, v), \cdots, f(u_m, v)\big), \tag{2}$$

and

$$\frac{\partial G_1}{\partial v}(u_{i_s}, v) = \begin{pmatrix} \sum_{i=0}^m \frac{\partial h_1^1}{\partial w_i}\big(\phi(u_{i_s}, v)\big) \cdot \frac{\partial f}{\partial v}(u_i, v) \\ \sum_{i=0}^m \frac{\partial h_1^2}{\partial w_i}\big(\phi(u_{i_s}, v)\big) \cdot \frac{\partial f}{\partial v}(u_i, v) \\ \sum_{i=0}^m \frac{\partial h_1^3}{\partial w_i}\big(\psi(u_{i_s}, v)\big) \cdot \frac{\partial f}{\partial v}(u_i, v) \end{pmatrix}.$$

Similarly, examining $G_2(u,v) = h_2\big(v,\ f(u,v_0),\cdots,f(u,v_n)\big)$ and defining

$$\psi : [u_0,u_m] \times [v_0,v_n] \to [v_0,v_n] \times (\mathbb{R}^3)^{n+1}$$

by

$$\psi : (u,v) \mapsto \big(v,\ f(u,v_0),\cdots,f(u,v_n)\big),$$

it follows that

$$D\psi(u_{i_s},v) = \begin{pmatrix} 0 & 1 \\ \frac{\partial f}{\partial u}(u_{i_s},v_0) & \mathbf{0} \\ \vdots & \vdots \\ \frac{\partial f}{\partial u}(u_{i_s},v_n) & \mathbf{0} \end{pmatrix}.$$

As before, denoting variables in the domain of h_2 as

$$v, w_{0,1}, w_{0,2}, w_{0,3} \cdots w_{n,1}, w_{n,2}, w_{n,3},$$

and for $1 \le j \le 3$ and $0 \le i \le n$ setting

$$\Big(\frac{\partial h_2^j}{\partial w_i} = \frac{\partial h_2^j}{\partial w_{i,1}}, \frac{\partial h_2^j}{\partial w_{i,2}}, \frac{\partial h_2^j}{\partial w_{i,3}}\Big),$$

we see that

$$\frac{\partial G_2}{\partial v}(u_{i_s},v) = \frac{\partial h_2}{\partial v}\big(v, f(u_{i_s},v_0)\cdots f(u_{i_s},v_n)\big)$$

and

$$\frac{\partial G_2}{\partial u}(u_{i_s},v) = \begin{pmatrix} \sum_{i=0}^{n} \frac{\partial h_2^1}{\partial w_i}\big(\psi(u_{i_s},v)\big)\cdot\frac{\partial f}{\partial u}(u,v_i) \\ \sum_{i=0}^{n} \frac{\partial h_2^2}{\partial w_i}\big(\psi(u_{i_s},v)\big)\cdot\frac{\partial f}{\partial u}(u,v_i) \\ \sum_{i=0}^{n} \frac{\partial h_2^3}{\partial w_i}\big(\psi(u_{i_s},v)\big)\cdot\frac{\partial f}{\partial u}(u,v_i) \end{pmatrix}. \tag{3}$$

In this context, if $\{i_1,i_2,\cdots,i_r\}$ is a subset of $\{0,1,\cdots,m\}$ and if

$$m_{i_s,1}, m_{i_s,2},\cdots,m_{i_s,q_s},\ 1 \le s \le r$$

are functions $m_{i_s,k} : [v_0,v_n] \to \mathbb{R}^3, 1 \le k \le q_s$ which define desired derivative values across meshlines, there is the following theorem.

Theorem 2. *Given a parametric mesh $f : U \cup V \to \mathbb{R}^3$, suppose*

$$h_1 : [u_0,u_m] \times (\mathbb{R}^3)^{m+1} \to \mathbb{R}^3$$

is an univariate interpolation scheme which for $v_0 \le v \le v_n$ satisfies the constraints

$$\frac{\partial h_1^k}{\partial u^k}\big(u_{i_s},\ f(u_1,v),\cdots,f(u_m,v)\big) = m_{i_s,k}(v),\ 1 \le k \le q_s.$$

Then the type 1 surface $F = G_1 + G_2 - G_3$ interpolates the mesh f and $\frac{\partial^k F}{\partial u^k}(u_{i_s}, v) = m_{i_s,k}(v)$, $0 \le k \le q_s$.

Proof: Since $F = G_1 + G_2 - G_3$, then

$$\frac{\partial F}{\partial u}(u_{i_s}, v) = \frac{\partial G_1}{\partial u}(u_{i_s}, v) + \frac{\partial G_2}{\partial u}(u_{i_s}, v) - \frac{\partial G_3}{\partial u}(u_{i_s}, v).$$

By analogy to the above and defining

$$\tilde{\psi} : [u_0, u_m] \times [v_0, v_n] \to [u_0, u_m] \times (\mathbb{R}^3)^{n+1}$$

by

$$\tilde{\psi} : (u, v) \mapsto \left(v,\ G_1(u, v_0), \cdots, G_1(u, v_n)\right),$$

it follows from (2) and (3) that

$$\begin{aligned}\tfrac{\partial F}{\partial u}(u_{i_s}, v) = \ & \tfrac{\partial h_1}{\partial u}\left(u_{i_s},\ f(u_0, v), \cdots, f(u_m, v)\right) + \\ & \begin{pmatrix} \sum_{i=0}^{n}\left(\frac{\partial h_2^1}{\partial w_i}(\psi(u_{i_s}, v)) - \frac{\partial h_2^1}{\partial w_i}(\tilde{\psi}(u_{i_s}, v))\right) \cdot \frac{\partial f}{\partial u}(u, v_i) \\ \sum_{i=0}^{n}\left(\frac{\partial h_2^2}{\partial w_i}(\psi(u_{i_s}, v)) - \frac{\partial h_2^2}{\partial w_i}(\tilde{\psi}(u_{i_s}, v))\right) \cdot \frac{\partial f}{\partial u}(u, v_i) \\ \sum_{i=0}^{n}\left(\frac{\partial h_2^3}{\partial w_i}(\psi(u_{i_s}, v)) - \frac{\partial h_2^3}{\partial w_i}(\tilde{\psi}(u_{i_s}, v))\right) \cdot \frac{\partial f}{\partial u}(u, v_i) \end{pmatrix}.\end{aligned}$$

However, since $\tilde{\psi}(u_{i_s}, v) = \psi(u_{i_s}, v)$, the latter term vanishes so that

$$\frac{\partial F}{\partial u}(u_{i_s}, v) = \frac{\partial h_1}{\partial u}\left(u_{i_s},\ f(u_0, v), \cdots, f(u_m, v)\right)$$

Iterating, it follows that

$$\frac{\partial^k F}{\partial u^k}(u_{i_s}, v) = \frac{\partial h_1^k}{\partial u^k}\left(u_{i_s},\ f(u_0, v), \cdots, f(u_m, v)\right) = m_{i_s,k}(v),$$

$v_0 \le v \le v_n$ and $1 \le k \le q_s$. □

3.1 Remarks

If in the proof of Theorem 2 the interpolatory surface F had been chosen to be of type 2, there would be no simple way of prescribing tangent values across a mesh line $f(\{u_i\} \times [v_0, v_n])$. However, should one desire to assign tangent values across a mesh line of the form $f[u_0, u_m] \times \{v_j\})$, by symmetry, an analogous type 2 surface together with an appropriate univariate interpolation scheme h_2 could be used.

The task of assigning tangent values across mesh lines that intersect is more complicated, but may be accomplished by means of an appropriate blending of type 1 and type 2 surfaces. In particular, if it is desired to assign tangent values simultaneously across mesh lines $f(\{u_i\} \times [v_0, v_n])$ and $f[u_0, u_m] \times \{v_j\})$, one may set $F = G_1 + G_2 - H$, where for instance

$$H(u,v) = \Big(1 - \frac{(v-v_j)^2}{\|(u-u_i, v-v_j)\|^2}\Big) G_4(u,v) + \frac{(v-v_j)^2}{\|(u-u_i, v-v_j)\|^2} G_3(u,v),$$

for $(u,v) \neq (u_i, v_j)$ and $H(u_i, v_j) = f(u_i, v_j)$. Variations of the above are possible should one should wish to assign tangent values across other combinations of mesh lines.

§4. Application

Given a parametric surface $H : D \to \mathbb{R}^3$, where D a connected open subset of $\mathbb{R}^2$, let $\alpha : [a,b] \to D$ be a curve in the domain of H and let $m : [a,b] \to \mathbb{R}^3$ be such that $m(t)$ is orthogonal to $(H \circ \alpha)'(t) = 0$ and $m(t)$ is in the tangent plane of H at the point $H(\alpha(t))$. As in Definition 1, let $f : X \cup Y \to \mathbb{R}^3$ be a parametric mesh with the properties that for a change of variable map $\gamma : [v_0, v_n] \to [a,b]$,

- $f(u_i, v) = H(\alpha(\gamma(v)))$,
- $\frac{\partial f}{\partial u}(u_i, v_j) = m(\gamma(v_j))$, $0 \leq j \leq n$.

Theorem 2 guarantees a surface F interpolating the mesh f and for which along the mesh line $f(\{u_i \times [v_0, v_n]), \frac{\partial F}{\partial u}(u_i, v) = m(\gamma(v))$. Since variation in the surface H in the region of attachment induces related variation in the surface F, it is desirable to have good control over the extent to which the variation in F radiates through the surface. It is recommended that the univariate interpolation scheme

$$h_1 : [u_0, u_m] \times (\mathbb{R}^3)^{m+1} \to \mathbb{R}^3$$

be such that for $v_0 \leq v \leq v_n$, the two segments

$$h_1|_{[u_{i_s-1}, u_{i_s}] \times \{v\}} \text{ and } h_1|_{[u_{i_s}, u_{i_s+1}] \times \{v\}}$$

are Hermite segments so that variation in F can be confined to the region $[u_{i-1}, u_{i+1}] \times [v_0, v_n]$.

References

1. Chiyokura, H., and F. Kimura, Design of solids with free-form surfaces, Computer Graphics (SIGGRAPH '83 Proceedings) **17** (1983), 289–298.
2. Coons, S., Surfaces for computer aided design of space forms, Technical Report MIT, Project MAC, 1968.
3. Gordon, W., Blending-function methods of bivariate and multivariate interpolation and approximation, Technical Report CM834, General Motors Research Laboratories, Warren Michigan, October, 1968.

4. Gordon, W., Spline-blended surface interpolation through curve networks, J. of Math. and Mechanics **18** (1969), 71–87.
5. Gregory, J. A., Smooth interpolation without twist constraints, in *Computer Aided Geometric Design*, R. E. Barnhill and R. F. Riesenfeld (eds.), Academic Press, New York, 1974, 71–87.

Marshall Walker
Department of Computer Science and Mathematics
Atkinson College
York University
North York, M3J 1P3
Ontario, CANADA
`walker@yorku.ca`

Generalized Multiresolution Analysis for Arc Splines

Johannes Wallner

Abstract. In order to approximate a curve by another curve consisting of circular arcs ('arc spline'), we apply a generalized multiresolution analysis on base of trigonometric spline functions to the support function of this curve.

§1. Homogeneous B-Splines and Trigonometric Splines

In this section we recall some facts about the connection between trigonometric B-spline functions and homogeneous polynomial B-spline functions. First, we follow [10,3], then [9,12].

We assume that the reader is familiar with the definition of the space $\mathcal{T}_k(\phi, M)$ of trigonometric splines of order k, based on a knot sequence $\phi = (\phi_0, \dots, \phi_{n-k})$ with $a = \phi_0 < \phi_1 < \dots < \phi_{n-k-1} < \phi_{n-k} = b$, and a vector $M = (m_1, \dots, m_{n-k-1})$ of integer multiplicities m_i which satisfy $1 \le m_i \le k+1$. It is of dimension n and is a special case of an L-spline space. We are going to demonstrate its connections with homogeneous polynomial B-splines.

The polar angle $\theta(u, v)$ of two nonzero vectors $u, v \in \mathbb{R}^2$ is defined to be the smallest $\theta \in \mathbb{R}$, $\theta \ge 0$, such that $\begin{pmatrix} \cos\theta & -\sin\theta \\ \sin\theta & \cos\theta \end{pmatrix} \cdot u = \lambda v$ $(\lambda > 0)$. For $u, v \in \mathbb{R}^2 \setminus 0$ we define the closed interval $[u, v] := \{x \in \mathbb{R}^2 \setminus 0, \ \ \theta(u, x) + \theta(x, v) = \theta(u, v)\}$, and the half-open interval $[u, v) := \{x \in [u, v], \forall \lambda > 0 : x \ne \lambda v\}$. Let $k \ge 0$ and $t = (t_0, \dots, t_{n+k})$ be a sequence of nonzero vectors $\in \mathbb{R}^2$ (knot sequence), such that for all $i = 0, \dots, n-1$

$$\theta(t_i, t_{i+1}) + \theta(t_{i+1}, t_{i+2}) + \dots + \theta(t_{i+k}, t_{i+k+1}) = \theta(t_i, t_{i+k+1}) < \pi \qquad (1)$$

holds, and we always have $\theta(t_i, t_{i+k+1}) > 0$. Because the sequence may wind itself around the origin more than once, we lift it to the universal cover $\widetilde{\mathbb{R}^2}$ of $\mathbb{R}^2 \setminus 0$; i.e. we tacitly assign to each vector of $\mathbb{R}^2 \setminus 0$ a winding number. This will help to avoid ambiguities.

Mathematical Methods for Curves and Surfaces II
Morten Dæhlen, Tom Lyche, and Larry L. Schumaker (eds.), pp. 537–544.

ISBN 0-8265-1315-8.

Definition. *The homogeneous polynomial B-spline basis function* $N_i^0 : \widetilde{\mathbb{R}^2} \to \mathbb{R}$ *of degree* 0 *is the characteristic function of the interval* $[t_i, t_{i+1})$, *and the homogeneous B-spline basis functions* $N_i^k : \widetilde{\mathbb{R}^2} \to \mathbb{R}$ *of degree* $k > 0$ *are defined recursively by*

$$N_i^k(x) = \frac{\det(x, t_i)}{\det(t_{i+k}, t_i)} N_i^{k-1}(x) + \frac{\det(t_{i+k+1}, x)}{\det(t_{i+k}, t_i)} N_{i+1}^{k-1}(x). \qquad (2)$$

It is well known that the trigonometric B-spline functions of order k can be expressed in terms of homogeneous B-spline functions of order k: For $u \in \mathbb{R}$, we define the unit vector $n(u) = (\cos u, \sin u) \in \widetilde{\mathbb{R}^2}$. The angle u determines the appropriate winding number. To a knot angle sequence ϕ with multiplicities M we assign the knot vector sequence

$$\underbrace{n(\phi_0/k), \ldots,}_{k+1 \text{ times}} \underbrace{n(\phi_1/k), \ldots,}_{m_1 \text{ times}} \underbrace{n(\phi_2/k), \ldots,}_{m_2 \text{ times}} \ldots, \underbrace{n(\phi_{n-k}/k), \ldots}_{k+1 \text{ times}}. \qquad (3)$$

Then the linear span of the restriction to the unit circle of the homogeneous polynomial B-spline basis functions $N_i^m | S^1$ equals the space of all functions $f(k \cdot u)$, where $f \in \mathcal{T}_k(\phi, M)$ if the knot sequence fulfills condition (1).

The functions $T_i^k(u) = N_i^k(n(\frac{u}{k}))$, $i = 1, \ldots, n$ corresponding to the homogeneous B-spline basis functions will be called trigonometric B-spline basis functions. It is well known that they form a basis of $\mathcal{T}_k(\phi, M)$.

Because we will need it later, we recall some facts about polar forms of polynomial functions: The polar form of a homogeneous polynomial function $p : \mathbb{R}^2 \to \mathbb{R}$ of degree n is the unique multilinear symmetric function in n variables $S_p : (\mathbb{R}^2)^n \to \mathbb{R}$ which has the property $S_p(t, \ldots, t) = p(t)$ for all $t \in \mathbb{R}^2$. Further, the following is true: Let $p = \sum c_i(t) N_i^r(t)$. If $[t_j, t_{j+1})$ is nonempty, let S_j be the polar form of the restriction $p|[t_j, t_{j+1})$. Then

$$c_i = S_i(t_{i+1}, \ldots t_{i+r}) = S_{i+1}(t_{i+1}, \ldots, t_{i+r}) = \cdots = S_{i+r}(t_{i+1}, \ldots, t_{i+r}), \quad (4)$$

whenever the polar forms S_j are defined. Thus the polar form can be used to calculate the coefficients c_i.

Lemma. *Let* $\alpha_i = (\phi_{i+1} - \phi_i)/2$. *The quadratic trigonometric spline functions possess the following* approximate convex hull property*:*

$$\sum c_i T_i^2(u) \in \text{c.h.}\left(\frac{c_{i-2}}{\cos \alpha_{i-1}}, \frac{c_{i-1}}{\cos \alpha_i}, \frac{c_i}{\cos \alpha_{i+1}}\right) \quad (\phi_i \le u \le \phi_{i+1}). \qquad (5)$$

Here all undefined terms are tacitly assumed to be absent.

Proof: The bivariate polynomial function $x^2 + y^2$ is constant when restricted to the unit circle. Its polar form is the euclidean scalar product. Thus (4) implies that the constant function 1 can be represented as a quadratic trigonometric B-spline function by $1 = \sum \langle t_{i+1}, t_{i+2} \rangle T_i^2(u) = \sum \cos \alpha_{i+1} T_i^2(u)$, where $t = (t_1, \ldots)$ is the knot vector sequence corresponding to ϕ and M according to (3). If $\phi_{i+k} - \phi_i < 2\pi$, we always have $0 \le T_i^k(u), \cos \alpha_i \le 1$. Therefore, the sum $\sum c_i T_i(u) = \sum \frac{c_i}{\cos \alpha_{i+1}} \cos \alpha_{i+1} T_i^2(u)$ is a convex combination. The assertion of the lemma now follows from the fact that for $u \in [\phi_i, \phi_{i+1}]$, at most three terms in the above sum are nonzero. □

§2. A Generalized Multiresolution Analysis with Trigonometric B-Splines

Definition. *Let $I = [a,b] \subset \mathbb{R}$. A generalized multiresolution analysis of $L^2(I)$ is a nested sequence $V_0 \subset V_1 \subset V_2 \subset \cdots$ of closed linear subspaces of $L^2(I)$ such that their union is dense in $L^2(I)$, together with Riesz bases of the spaces V_i and of complements W_i of V_i in V_{i+1}. The basis functions of W_i will be called (pre-)wavelets.*

We will show how to construct a generalized multiresolution analysis with trigonometric B-spline functions of order two. Let $(\phi^{(0)}, M^{(0)}) \le (\phi^{(1)}, M^{(1)}) \le \cdots$, be nested knot angle sequences. We choose the corresponding trigonometric B-spline functions as bases of $V_i = \mathcal{T}_2(\phi^{(i)}, M^{(i)})$. Let the spaces W_i be the orthogonal complements of V_i in V_{i+1} in the sense of L^2.

Let t and t' be knot vector sequences such that t' is a refinement of t. The homogeneous B-spline basis functions corresponding to t and t' will be denoted by N_i and N_i'. Then there are coefficients c_{ij} such that $N_i(x) = \sum c_{ij} N_j'(x)$.

Lemma. *The coefficients c_{ij} are given by the following algorithm: Choose k such that $j \le k \le j+2$ and $[t_k', t_{k+1}')$ is not empty. This interval is contained in an interval $[t_l, t_{l+1})$. Then we have $t_{j+1}' = \alpha t_l + \beta t_{l+1}$, $t_{j+2}' = \gamma t_l + \delta t_{l+1}$ and*

$$c_{ij} = \alpha\gamma N_i(t_l) + \beta\delta N_i(t_{l+1}) + (\alpha\delta + \beta\gamma)\delta_{p,i-1}. \tag{6}$$

Proof: Equation (4) implies that c_{ij} equals $S_k'(t_{j+1}', t_{j+2}')$, where S_k' is the polar form of $N_i|[t_k, t_{k+1})$. Because of the multilinearity and symmetry of the polar form we have $c_{ij} = \alpha\gamma S_k'(t_l, t_l) + \beta\delta S_k'(t_{l+1}, t_{l+1}) + (\alpha\delta + \beta\gamma) S_k'(t_l, t_{l+1})$. Because of $[t_k', t_{k+1}') \subset [t_l, t_{l+1})$, the latter is not empty and we have $S_l = S_k'$, where S_l is the polar form of $N_i|[t_l, t_{l+1})$. It follows that $S_k'(t_l, t_l) = N_i(t_l)$ and $S_k'(t_{l+1}, t_{l+1}) = N_i(t_{l+1})$. Equation (4) implies that $S_l(t_l, t_{l+1})$ equals 1 if $i = l-1$ and 0 if not. The assertion follows. □

Let $\phi_{\max}^{(i)} = \max_j(\phi_{j+1}^{(i)} - \phi_j^{(i)})$ and $\phi_{\min}^{(i)} = \min_j(\phi_{j+1}^{(i)} - \phi_j^{(i)})$. Assume that $\lim_{i\to\infty} \phi_{\max}^{(i)} = 0$. Then the union of the V_i is dense in $L_2(I)$. This is proved in [3], where it is derived from a more general theorem of [10]. In our special case this also follows directly from the approximate convex hull property.

In [3] a basis of the space W_i is constructed which consists of functions ψ of the minimal possible support.

§4. Support Functions

4.1. (Locally) Convex Curves

For simplicity, we restrict ourselves to the case of piecewise C^2 curves. What we are going to do in this section can be done for continuous curves as well, because the restriction of (local) convexity is strong enough to eliminate possible degeneracies, but we use the curvature of the curve to simplify the discussion.

Definition. *Assume that $I = [a, b]$ and the curve $c : I \to \mathbb{R}^2$ is parametrized by arc length. We call c piecewise C^k if there is a discrete set T of parameter values such that for all intervals $(u, v) \subset I \setminus T$ there is a C^k extension of c to an interval $(u - \epsilon, v + \epsilon)$ with an $\epsilon > 0$.*

Definition. *A piecewise C^2 curve is called* locally convex *and of* nonnegative curvature *if for all $t \in I \setminus T$, the curvature is nonnegative, and for all $t \in T$ the turning angle $\theta(\dot{c}_-, \dot{c}_+)$ of the limit tangent vectors is nonnegative. A locally convex curve of nonpositive curvature is defined in the obvious way.*

A unit vector n is called oriented unit normal vector at $u = u_0 \notin T$ if it is perpendicular to $\dot{c}(u_0)$ and points in the direction of $\ddot{c}(u_0)$. If $u = u_1 \in T$ and the curve is of nonnegative curvature, we consider the left- and right-handed limit normal vectors n_1 and n_2 and define all vectors n with $\theta(n_1, n) + \theta(n, n_2) = \theta(n_1, n_2)$ as oriented unit normal vectors. After re-parametrizing the curve such that for a whole interval the point $c(u)$ rests in a point of tangent discontinuity, we can define the piecewise C^1 oriented normal vector field $n(u)$ and the function

$$d_c : I \to \mathbb{R}, \quad u \mapsto \langle c(u), n(u) \rangle. \tag{7}$$

At last we re-parametrize d_c such that its argument is the polar angle of the oriented normal vectors. We will always assume that d_c is parametrized in this way. The function d_c is thus piecewise C^1.

Definition. *The function d_c, parametrized by the polar angle of the oriented normal vectors, is called the* support function *of the (locally) convex curve c.*

If the support function d_c is given, the curve c can be reconstructed as the envelope of the lines $l(u) : \langle x, n(u) \rangle = d_c(u)$. If the envelope is not defined because d_c is not C^1, this was necessarily caused by a straight line segment contained in the curve c.

Not all piecewise C^1 functions, however, are support functions of piecewise C^2 locally convex curves. The following is well known:

Lemma. *The function d is the support function of a (locally) convex curve, if and only if the sign of $d + d''$ is constant.*

4.2 Filter Bank Decomposition of Piecewise Circular Curves

A circle with center $m = (m_1, m_2)$ can be parametrized by $c : \mathbb{R} \to \mathbb{R}^2$, $c(u) = m - rn(u)$. Then $n(u)$ is the oriented unit normal vector to c in the point $c(u)$. The circle's support function d_c is given by $d_c(u) = \langle m, n(u) \rangle - r = m_1 \cos u + m_2 \sin u - r$. This leads to the following definition and lemma:

Definition. *An* arc spline curve *is a C^1 curve which consists of discrete circular arcs.*

Lemma. *The locally convex curve c is an arc spline curve if and only if its support function d_c is a trigonometric B-spline function of order two whose knot vector has only knots of multiplicity one.*

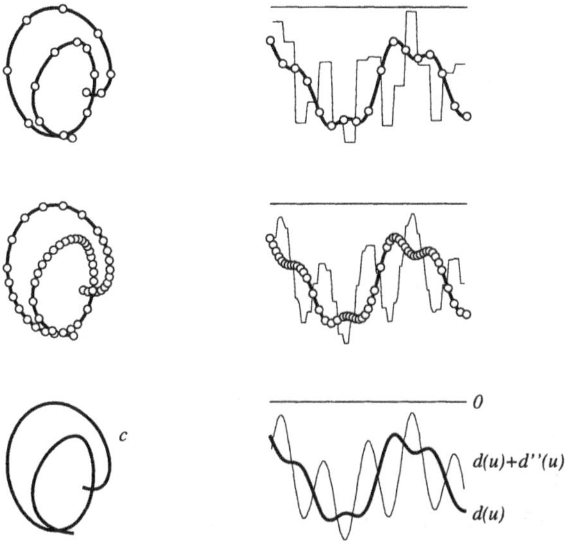

Fig. 1. Approximation of a locally convex curve by arc splines. Left: Curve and arc spline approximants. Right: Support functions d together with the curvature radius functions $d + d''$.

The filter bank algorithm defined above can now be used to define a wavelet transform of the arc spline c. This is defined as the wavelet transform of its support function, using the bases of the spaces W_i as wavelet functions: The orthogonal direct sum $V_{i+k} = V_i \oplus W_i \oplus W_{i+1} \oplus \cdots \oplus W_{i+k-1}$ leads in a natural way to projections $\pi_i : V_{i+k} \to V_i$ and $\rho_i : V_{i+k} \to W_i$. We will call the decomposition $x = \pi_i(x) + \sum_{j=0}^{k} \rho_{i+j}(x)$ $(x \in V_{i+k})$ the filter bank decomposition of x, and the sequence of coefficient vectors of the various projections in the bases selected above, its wavelet transform.

§5. Approximation of Curves by Arc Splines

There are many ways to approximate and interpolate curves and discrete point sets with arc splines. The interested reader is referred to [2,4,5,6,7,8] and the literature cited therein.

5.1. Approximation of Locally Convex Curves

Let a locally convex curve $c : I \to \mathbb{R}^2$ be given. We can approximate its support function $d_c(u)$ by a trigonometric spline function of order two. An obvious choice is the closest approximation in the sense of L^2. Points of tangent discontinuity can be reproduced by choosing appropriate greater multiplicities.

Applying the wavelet transform to $\overline{d}$ and setting all coefficients below some threshold to zero gives an approximation $\widetilde{d}$ of $\overline{d}$ and d_c which is the support function of an arc spline $\widetilde{c}$, if the condition $\widetilde{d}+\widetilde{d}''$ is fulfilled. The points of curvature discontinuity of $\widetilde{c}$ correspond to the points of curvature discontinuity of $\widetilde{d}$ and are contained the union of the points of curvature discontinuity of all basis and wavelet functions which actually contribute to $\widetilde{d}$. The more coefficients of higher index we set to zero, the fewer points of curvature discontinuity the resulting arc spline will have, i.e., the fewer circular arcs it will consist of. Fig. 1 shows an example. The small circles indicate the points of curvature discontinuity.

5.2 Estimates

Definition. *The distance $d(c_1, c_2)$ between locally convex curves $c_i : I \to \mathbb{R}^2$ is the minimal ϵ such that c_1 lies in a closed ϵ-neighborhood of the union of c_2 and its initial and final tangent ray, and vice versa.*

Theorem. *Let $c : I \to \mathbb{R}^2$ be a locally convex piecewise C^2 curve and d_c its support function, and let $\mathcal{T}_2(\phi, M)$ be a trigonometric spline space. Then there exists an arc spline curve $\widetilde{c}$ with support function $d_{\widetilde{c}} \in \mathcal{T}_2(\phi, M)$ such that $d(c,\widetilde{c}) \le C_1\phi^3_{\max}\|d_c\|_\infty + C_2\omega^3_\infty(d_c, \phi_{\max})$. If d_c is in C^3, we have $d(c,\widetilde{c}) \le C_3\phi^3_{\max}(\|d_c\|_\infty + \|D^3 d_c\|_\infty)$. The constants C_i depend on the parameter interval, but not on c, ϕ or M. If $\phi_{\max}/\phi_{\min}$ is bounded, for all knot vectors ϕ with $\phi_{\max}$ small enough, the curve $\widetilde{c}$ does not have points of regression.*

Proof: Let L be the differential operator of degree m which defines the spaces $\mathcal{T}_{m-1}(\phi, M)$. For $m = 3$, we have $L = D(D^2+1)$. Theorem 10.24 of [10] implies that for $j = 0, 1, \ldots, m-1$ there exists a constant C, such that for all admissible (ϕ, M) and all $f \in C^m[a,b]$, the inequality

$$\|D^j(f - Qf)\|_\infty \le C\|Lf\|_\infty \cdot (\phi^m_{\max}/\phi^j_{\min}) \tag{8}$$

holds. Q is a projection onto $\mathcal{T}_{m-1}(\phi, M)$, which is introduced in [10]. By Theorem 10.1 of [10], there is a constant C independent on f, such that $\|Lf\|_\infty \le C(\|f\|_\infty + \|D^m f\|_\infty)$. Letting $j = 0$ in (8), this implies

$$\|f - Qf\|_\infty \le C\phi^m_{\max}(\|f\|_\infty + \|D^m f\|_\infty), \tag{9}$$

with a constant independent of the knot sequence and of f. Because (9) holds for all $\phi_{\max} > 0$, Theorem 2.68 of [10] allows us to conclude that there are constants C_1, C_2 such that for all $f \in C[a,b]$ and all admissible (ϕ, M), there is an $\overline{f} \in \mathcal{T}_{m-1}(\phi, M)$ such that $\|f - \overline{f}\|_\infty \le C_1\phi^m_{\max}\|f\|_\infty + C_2\omega^m_\infty(f, \phi_{\max})$. If $f \in C^m[a,b]$, the modulus of smoothness can be replaced by $\phi^m_{\max}\|D^m f\|_\infty$. Now assume that always $\phi_{\min} \ge k\phi_{\max}$. Letting $j = 2$ and $m = 3$ in (8), for all $f \in C^3([a,b])$ we have

$$\|(1 + D^2)(f - Qf)\|_\infty \le C\phi_{\max}(\|f\|_\infty + \|D^3 f\|_\infty) \tag{10}$$

for all knot sequences with $\phi_{\max}$ small enough, with a constant C independent on f and the knot sequence. □

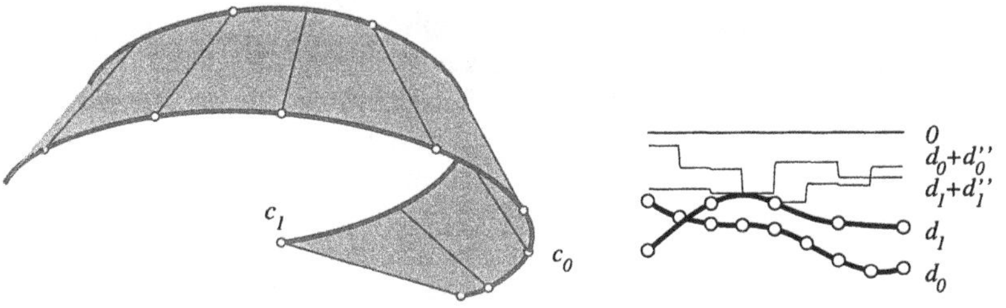

Fig. 2. Left: Approximation of a developable surface by segments of quadratic cones. Right: Support functions d_i of contour curves c_i with their curvature radius functions $d_i + d_i''$.

5.3 Approximation of Curves with Interpolation of Line Elements

We want to approximate the given support function d_c such that both $d_c(a)$ and $d_c'(a)$ are reproduced. This can be done as follows:

Assume that the spaces V_i and W_i are constructed as above on the interval $[a, b]$. Fix an index i and denote the basis functions of V_i by $c_{0i}, c_{1i}, \ldots, c_{n_i i}$. Equation (2) shows that $c_{ki}(a) = 0$ for $j \geq 1$ and $c_{ki}'(a) = 0$ for $k \geq 2$. Let

$$d^* = d_c - (d_c(a)/c_{00}(a))c_{00} \text{ and } d^{**} = d^* - (d^{*\prime}(a)/c_{10}'(a))c_{10} \tag{11}$$

Let $\widetilde{V}_i = \text{span}(c_{2i}, \ldots, c_{n_i i})$ and let $\widetilde{W}_i$ equal the orthogonal complement of $\widetilde{V}_i$ in $\widetilde{V}_{i+1}$. It is easy to find a basis of $\widetilde{W}_i$ The decomposition $\widetilde{V}_n = \widetilde{V}_0 \oplus \widetilde{W}_0 \oplus \cdots \oplus \widetilde{W}_{n-1}$ defines a multiresolution analysis for functions f with $f(a) = f'(a) = 0$.

We therefore approximate d^{**} by a trigonometric spline function in $\widetilde{V}_n$, apply the modified wavelet transform and set all coefficients below some threshold to zero. This gives an approximation $\bar{d}^*$ to d^{**} and

$$\bar{d} = \bar{d}^* + (d_c(a)/c_{00}(a))c_{00} + (d^{*\prime}(a)/c_{10}'(a))c_{10} \tag{12}$$

is an approximation to d_c with the property that $d_c(a) = \bar{d}(a)$ and $d_c'(a) = \bar{d}'(a)$. The algorithm can be modified in an obvious way, if one wants to reproduce the line element at the endpoint b of the interval also.

This can be used to approximate curves with inflection points: Assume that, after some pre-smoothing process, the curve has a discrete set of inflection points which we want to be reproduced after an approximation by arc splines. Now approximate each of the curve's maximal (locally) convex segments separately. In order to fit together, the single arc spline segments have to approximated in such a way that the initial and final line elements are reproduced exactly.

§6. Final Remarks

It should be remarked that the procedure can be applied to dual focal splines as well, because they are defined in terms of trigonometric spline functions.

This makes it possible to define a multiresolution analysis for the special class of rational curves with rational offsets which is studied in [9].

There is also an application to *surfaces*: Two locally convex curves c_1, c_2 in parallel ("horizontal") planes define a developable surface if we join points possessing parallel tangents with straight lines. The multiresolution analysis defined in this paper can be applied to both curves separately and gives a multiresolution analysis for the surface. The approximant will be a developable surface which consists of pieces of quadratic cones all of whose contour lines in horizontal planes are circles. Fig. 2 shows an example.

References

1. Chui C. (Ed.), *Wavelets: A Tutorial in Theory and Applications* (Wavelet Analysis and its Applications, Vol. 2), Academic Press, 1992.
2. Hoschek J., Circular splines, Computer-Aided Design **24** (1992), 611–618.
3. Lyche T., and L. Schumaker, L-Spline wavelets, in *Wavelets: Theory, Algorithms, and Applications*, C. Chui, L. Montefusco, L. Puccio (eds.), Academic Press, 1994, 197–212.
4. Marciniak K., and B. Putz Approximation of spirals by piecewise curves of fewest circular arc segments, Computer-Aided Design **16** (1984), 87–90.
5. Meek D. S., and D. J. Walton, Approximation of discrete data by G^1 arc splines, Computer-Aided Design **24** (1992), 301–306.
6. Meek D. S., and D. J. Walton, Approximating quadratic NURBS curves by arc splines, Computer-Aided Design **25** (1993), 371–376.
7. Meek D. S., and D. J. Walton, Approximating smooth planar curves by arc splines, J. Comput. Appl. Math. **59** (1995), 221–231.
8. Meek D. S., and D. J. Walton, Planar osculating arc splines, Comput. Aided Geom. Design **13** (1996), 653–671.
9. Neamtu M., H. Pottmann, and L. L. Schumaker, Homogeneous splines and rational curves with rational offsets, Technical Report No. 29, Institut für Geometrie, Technische Universität Wien, 1996.
10. Schumaker L. L., *Spline Functions: Basic Theory*, Wiley, New York, 1981.
11. Stollnitz E., T. Derose, and D. Salesin, *Wavelets for Computer Graphics — Theory and Applications*, Morgan Kaufmann Publishers, 1996.
12. Strøm K., B-Splines with homogenized knots, Adv. Comput. Math. **3** (1995), 291–308.

Johannes Wallner
Institut für Geometrie — TU Wien
Wiedner Hauptstraße 8–10/113
A-1040 Vienna, AUSTRIA
`wallner@geometrie.tuwien.ac.at`

Analysis and Visualization of Geometric Curve Properties

Geir Westgaard and Justus Heimann

Abstract. We give an overview of different measures and techniques for investigation of local curve properties. They are based on the curve derivatives, the Frenet Frame, curvature and torsion, and their derivatives. Two promising geometric measures are introduced, namely variation of curvature and variation of torsion. These two measures allow to reveal curve deficiencies that are not detected by curvature and torsion. To judge the quality of a curve, good intuitive visualization techniques are needed. We present visualization techniques showing the geometric properties "close" to the curve.

§1. Introduction

In construction of "smooth" and "good looking" curves (fairing, see e.g [2]), it is important to apply methods for checking the curve quality [4]. We believe that the higher order measures presented here are more sensitive for detecting local shape deficiencies as compared to the corresponding low order measures. Examples are given to show this capability. We will also demonstrate the advantage of displaying local curve properties (measures) "close" to the curve.

§2. Curve Analysis

Let (α, t) be a parameterization of a regular curve, with $\alpha(t) : [a, b] \to \mathbb{R}^3$, and (β, s), the corresponding curve parameterized by arc length s, $\beta(s) : [c, d] \to \mathbb{R}^3$ [1]. We use the notation $\alpha'(t) = d\alpha(t)/dt$, $\alpha''(t) = d\alpha^2(t)/dt^2, \ldots$, to identify derivatives of $\alpha(t)$ up to a certain order.

Before we concentrate on quality measures for curves (curve properties), some well-known [1] basic curve features are worth mentioning. They are the arc length function and accordingly the speed of the curve

$$s(t_0) = \int_a^{t_0} |\alpha'(t)| \; dt, \qquad \frac{ds}{dt} = |\alpha'(t)| \,, \tag{1}$$

Mathematical Methods for Curves and Surfaces II
Morten Dæhlen, Tom Lyche, and Larry L. Schumaker (eds.), pp. 545–552.

ISBN 0-8265-1315-8.

the Frenet Frame $\{\boldsymbol{t}, \boldsymbol{n}, \boldsymbol{b}\}$ (bold face indicates a vector) defining a natural coordinate system (in case the curvature $k \neq 0$) along the curve

$$\boldsymbol{t} = \frac{\alpha'}{|\alpha'|}, \qquad \boldsymbol{b} = \frac{\alpha' \times \alpha''}{|\alpha' \times \alpha''|}, \qquad \boldsymbol{n} = \boldsymbol{b} \times \boldsymbol{t}, \tag{2}$$

and the Frenet Formulas denoting the change of the Frenet Frame along the curve, with e.g. $\boldsymbol{t}' = d\boldsymbol{t}/ds$

$$\boldsymbol{t}' = k \cdot \boldsymbol{n}, \qquad \boldsymbol{n}' = -k \cdot \boldsymbol{t} - \tau \cdot \boldsymbol{b}, \qquad \boldsymbol{b}' = \tau \cdot \boldsymbol{n}\,, \tag{3}$$

where k is the curvature and τ the torsion.

Curve Quality Measures

In the following we focus on parameterization-independent measures we applied to analyze geometric quality of curves.

An often used quality measure is the curvature $k(s)$. For a curve (α, t) the curvature k is given by [1]

$$k = \frac{|\alpha' \times \alpha''|}{|\alpha'|^3}\,. \tag{4}$$

In this paper we would like to move to higher order quality measures. We believe that higher order measures are much more sensitive for detecting local shape deficiencies. In this context, the variation of curvature dk/ds seems to be a promising curve shape property

$$\frac{dk(s)}{ds} = \frac{dk(t)}{dt}\frac{dt}{ds}\,. \tag{5}$$

By focusing on the compact expression (5) in more detail, we end up with

$$\begin{aligned} \frac{dk}{dt} &= \frac{1}{|\alpha'|^3}\left[\frac{(\alpha' \times \alpha''') \cdot (\alpha' \times \alpha'')}{|\alpha' \times \alpha''|} - \frac{3\,|\alpha' \times \alpha''|\,(\alpha'' \cdot \alpha')}{|\alpha'|^2}\right], \\ \frac{dk}{ds} &= k' = \frac{1}{|\alpha'|}\frac{dk}{dt}\,. \end{aligned} \tag{6}$$

In case the curve is no longer planar, another curve property should be taken into account, the torsion $\tau(s)$. The torsion τ for a curve (α, t) is given by $(k \neq 0)$ [1]

$$\tau = \frac{\alpha'''(\alpha' \times \alpha'')}{|\alpha' \times \alpha''|^2}\,. \tag{7}$$

In order to proceed to higher order more sensitive quality measures, we introduce the variation of torsion $d\tau/ds$. For a curve (α, t) the variation of torsion $d\tau/ds$ is given by

$$\frac{d\tau(s)}{ds} = \frac{d\tau(t)}{dt}\frac{dt}{ds}. \tag{8}$$

Computing (8) in more detail, we can see that

$$\frac{d\tau}{dt} = \frac{1}{|\alpha' \times \alpha''|^2}\Bigg[\alpha'''' \cdot (\alpha' \times \alpha'') + \alpha''' \cdot (\alpha' \times \alpha''') - \frac{2\,[(\alpha' \times \alpha''') \cdot (\alpha' \times \alpha'')][\alpha''' \cdot (\alpha' \times \alpha'')]}{|\alpha' \times \alpha''|^2}\Bigg], \tag{9}$$

$$\frac{d\tau}{ds} = \tau' = \frac{1}{|\alpha'|}\frac{d\tau}{dt}\ .$$

We are also dealing with global quality measures giving a real value. Global measures are the result of an integration of the above given local measures along the curve.

§3. Visualization Techniques

In order to estimate the quality of the curve in question using the measures we already defined, good "intuitive" visualization techniques are needed.

Let X_i be one of the following measures along $\alpha(t)$: $|\alpha'|$ = speed, $|\alpha''|$ = 2nd derivative, $|\alpha'''|$ = 3rd derivative, k = curvature, k' = variation of curvature, τ = torsion, τ' = variation of torsion and $|\boldsymbol{t}'|$, $|\boldsymbol{b}'|$, $|\boldsymbol{n}'|$ = variation of the Frenet Frame, that we denote by

$$X = \{|\alpha'|, |\alpha''|, |\alpha'''|, k, k', \tau, \tau', |\boldsymbol{t}'|, |\boldsymbol{b}'|, |\boldsymbol{n}'|\}_\alpha\ . \tag{10}$$

In our examples we visualize geometric curve properties as explicit plots, as spike plots, and as circular tubes locally along the curve.

Explicit plots display the graph of X_i over the parameter space $(t, c\cdot X_i(t))$ or as function of the arc length $(s, c\cdot X_i(s))$, where $c \in \mathbb{R}$ is a scaling factor for visual effects. A drawback of explicit plots is that the measures X are plotted over the parameter, so it is difficult or sometimes impossible to assign a local measure to the equivalent geometric position on the curve, see examples.

However, a more "intuitive" visualization can be achieved by linking the measures X directly to the axes of the Frenet Frame. In this scheme the Frenet Frame is moved stepwise along the curve, changing its orientation and the length of its axes according to the local values of X. The formula is given by

$$\tilde{\boldsymbol{t}}_X = \alpha + c_1\, X_i\, \boldsymbol{t}, \qquad \tilde{\boldsymbol{n}}_X = \alpha + c_2\, X_j\, \boldsymbol{n}, \qquad \tilde{\boldsymbol{b}}_X = \alpha + c_3\, X_k\, \boldsymbol{b}, \tag{11}$$

where $i \neq j \neq k$ and $c_1, c_2, c_3 \in \mathbb{R}$ are any scaling factors. To assign the measures in a "natural" way to the axes of the Frenet Frame, we recommend the following combinations

$$X_i = |\alpha'|, \qquad X_j = k \ \ or \ \ k', \qquad X_k = \tau \ \ or \ \ \tau'\ . \tag{12}$$

Nevertheless, this scheme is promising. We found that in practice, plotting all three axes together along the curve is often confusing to the observer, even when dealing with coloured axes. Therefore we allow to interactively switch off axes. For instance, if only the normal direction linked to the curvature is plotted, this scheme turns into the popular porcupine plot.

For an "intuitive" visualization of X we also consider a circular tube around the curve (examples are also given in [3]). A circle oriented in the normal plane defined by $\{\boldsymbol{n}, \boldsymbol{b}\}$ is moved stepwise along the curve, changing its radius according to the local values of X. The formula is given by

$$Circ\ (\theta, X_i) = \alpha + c \cdot X_i\,(\boldsymbol{n} \cdot \cos\,\theta + \boldsymbol{b} \cdot \sin\,\theta), \tag{13}$$

where $\theta \in [0, 2\pi]$ is an angle positive counterclockwise from the positive normal axis $\boldsymbol{n}$ and $c \in \mathbb{R}$ again any scaling factor. We recommend using different colours for $X_i \geq 0$ and $X_i < 0$.

§4. Examples

Fig. 1 presents plots of the curvature and variation of curvature "close" to the curve. In (a) and (b) both measures are visualized as circular tubes, in (c) and (d) as porcupine plots and in (e) and (f) as explicit plots as function of the parameter t. Black ink indicates positive values and light grey ink indicates negative values of the quality measures. Here we examined a B-spline curve of order five. The curve is planar in the first part while extending into $\mathbb{R}^3$ at the aft part.

Fig. 2 gives the results for torsion and the variation of torsion for the same curve (see Fig. 1). As the pictures (Fig. 2, (a)-(f)) indicate, the torsion and variation of torsion is equal to zero in the planar part of the curve. When focusing on the track of the curvature and the torsion compared to the pattern of variation of curvature and variation of torsion in Fig. 1 and Fig. 2, it can be seen that curvature and torsion are almost "smoothly" distributed along the curve while their derivatives k', τ' are more sensitive to local shape deficiencies. A major advantage of the variational measures is that they are capable of detecting sign changes of the curvature, respectively torsion pattern.

We can derive the same conclusion from Fig. 3, which shows spike plots (left hand side) and explicit plots (right hand side) for a B-spline space curve of order five given by 25 control points. Notice that we are dealing with different viewpoints in (a), (c) and (e), (g), in order to achieve an expressive visualization of the analysis measures.

From a comparison of the curvature plot (b) as a function of t with the corresponding porcupine plot (a), the drawback of explicit plots is obvious. Only by analyzing the explicit plot (b), an observer would get the impression that the maximum curvature appears in the very first part of the curve. But when looking to the corresponding porcupine plot (a), the maximum curvature is located geometrically close to half of the curve length. Such problems occur in case of non "natural" parameterizations, that is, parameterizations far away from an arc length parameterization.

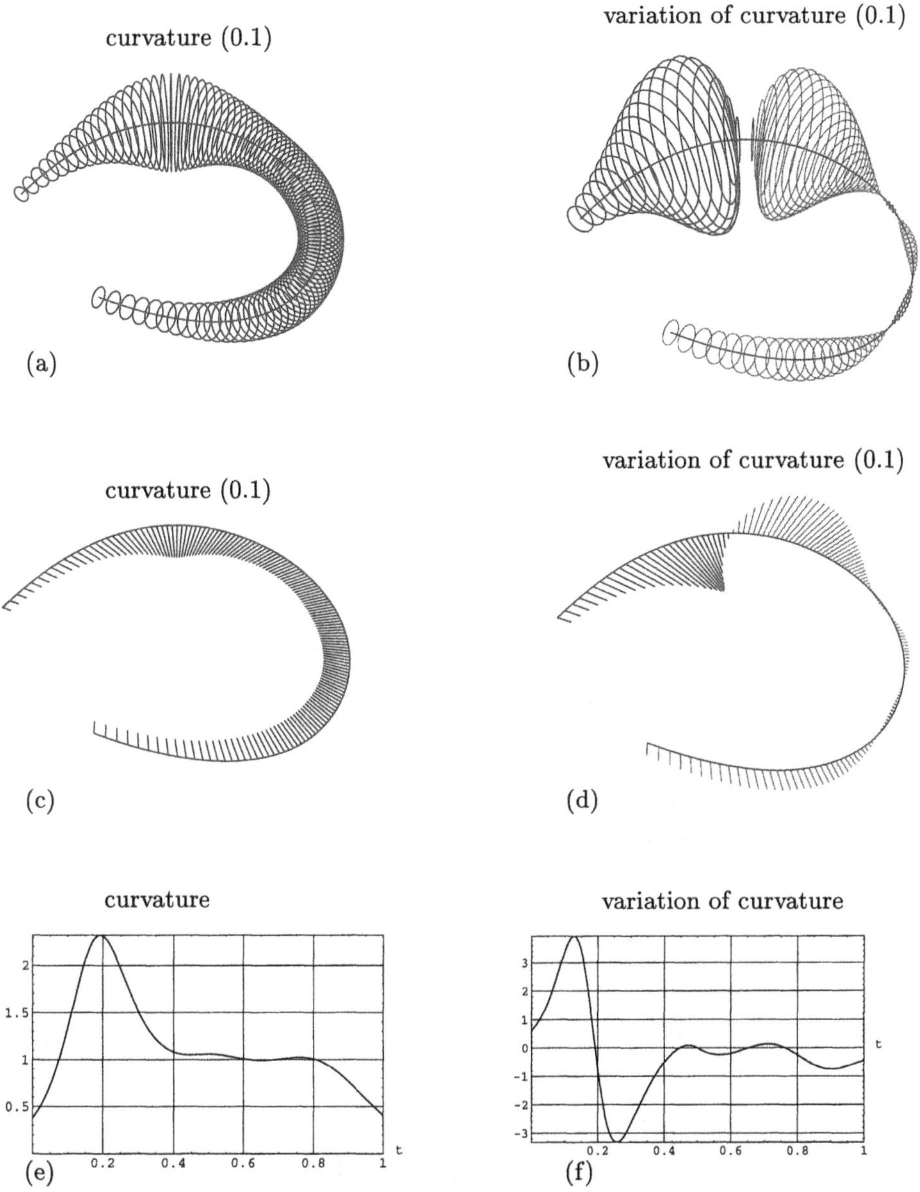

Fig. 1. Example 1, k and k', scaling factors for (a)-(d) are given in brackets.

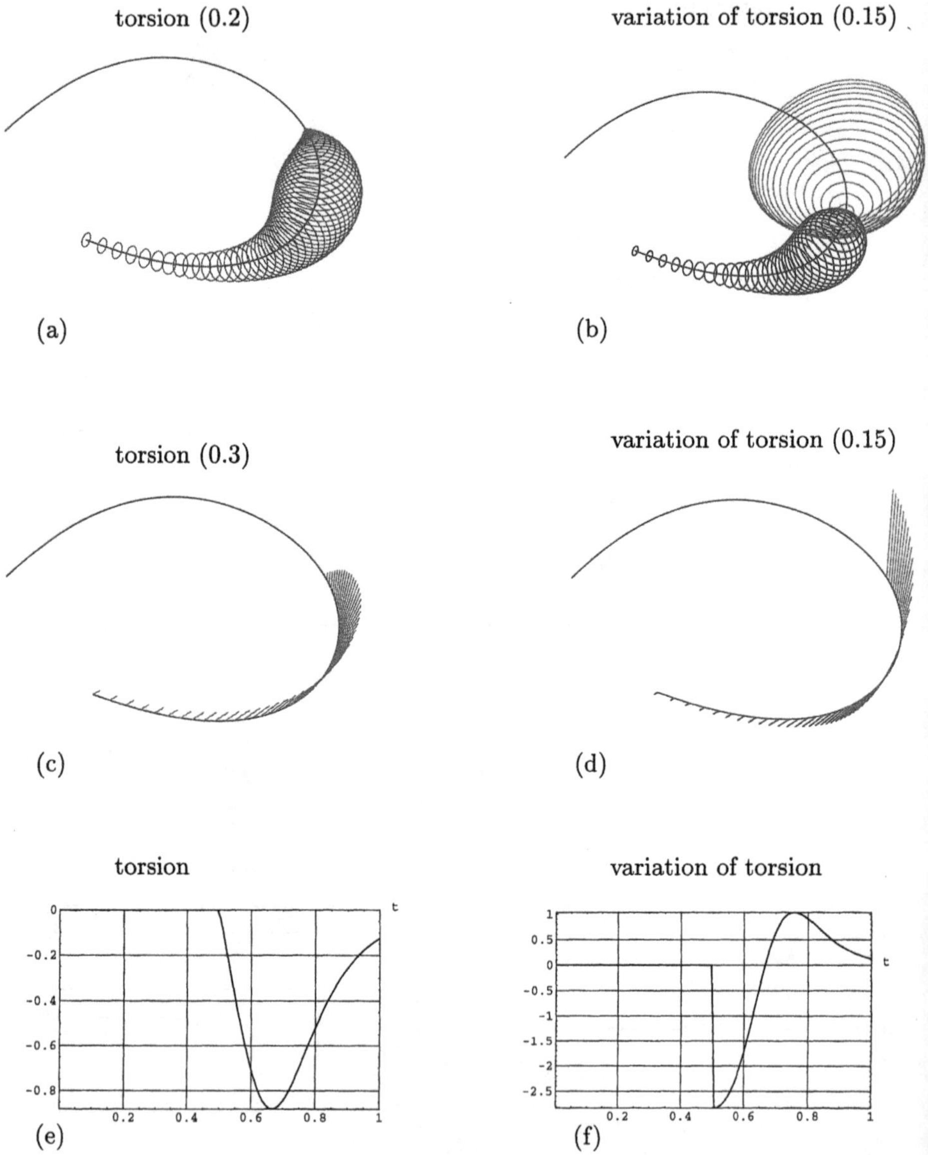

Fig. 2. Example 2, τ and τ', scaling factors for (a)-(d) are given in brackets.

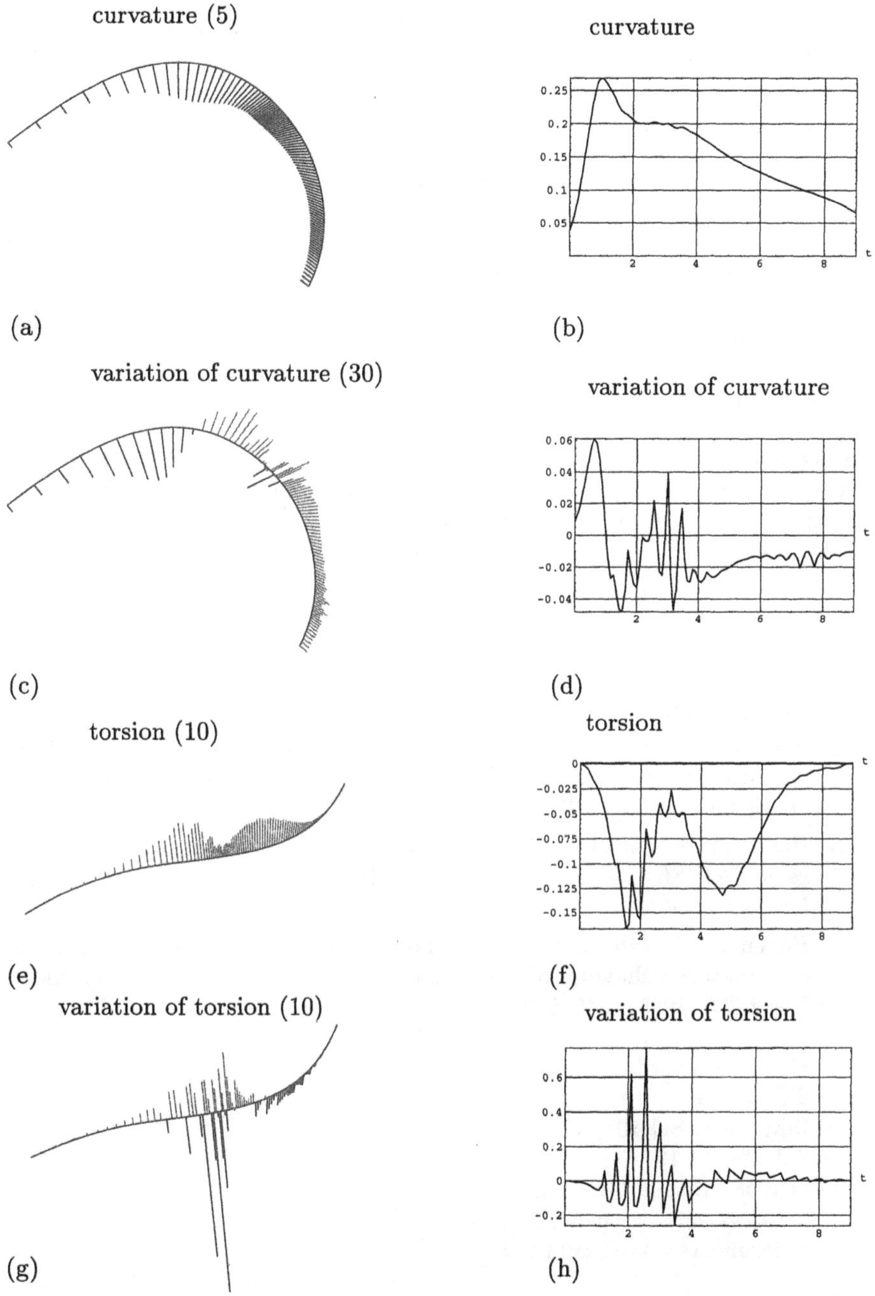

Fig. 3. Example 3, scaling factors are given in brackets.

§5. Summary

We have presented parameter independent measures to detect the quality of curves. We found that especially the higher order measures, variation of curvature and variation of torsion are very sensitive to local shape deficiencies. An advantage of these higher order measures is that they measure sign changes of the curvature, respectively the torsion along the curve. We also illustrated that a visualization of quality measures "close" to the curve is more intuitive to an observer than e.g. an explicit plot as function of the parameter t. A main drawback of explicit plots is that in case of curves parameterized far away from an arc length parameterization, it is difficult or even impossible to assign a local measure to the equivalent geometric position on the curve. We therefore suggest visualizing curve quality measures "close" to the curve as spike plots or as circular tubes, as illustrated in the examples.

Acknowledgments. This work is supported by the HCM - FAIRSHAPE Framework founded by the European Community and Technische Universität Berlin. The first author is also supported by SINTEF Applied Mathematics, Oslo, Norway.

References

1. Do Carmo, M. P., *Differential Geometry of Curves and Surfaces*, Prentice-Hall, Inc., New Jersey, 1976.
2. Eck, M., and J. Hadenfeld, Local Energy Fairing of B-spline Curves, in G. Farin, H. Hagen and H. Noltemeier (eds.), Computing, Supplementum 10, Springer, 1995, 129–147.
3. Hadenfeld J., Fairing of B-spline Curves and Surfaces, in *Advanced Course on FAIRSHAPE*, J. Hoschek and P. Kaklis (eds.), B. G. Teubner, 1996, 59–75.
4. Hagen, H., S. Hahmann, and T. Schreiber, Visualization and computation of curvature behaviour of freeform curves and surfaces, Computer-Aided Design**27** (1995), 545–552.

Geir Westgaard and Justus Heimann
TU Berlin
ISM, Sekr. SG 10
Salzufer 17-19
D-10587 Berlin, GERMANY
`westgaard@ism.tu-berlin.de`
`heimann@ism.tu-berlin.de`

INDEX

E

F

G

H

I

K

L

M

T

V

W

Z

www.ingramcontent.com/pod-product-compliance
Lightning Source LLC
Chambersburg PA
CBHW021427180326
41458CB00001B/164
* 9 7 8 0 8 2 6 5 1 3 1 5 1 *